MATERIALS SURFACE PROCESSING

EUROPEAN MATERIALS RESEARCH SOCIETY SYMPOSIA PROCEEDINGS

Volume 1: Ceramic Materials Research (ed. R.J. Brook)
Volume 2: Photon, Beam and Plasma Assisted Processing (eds. I.W. Boyd and E.F. Krimmel)
Volume 3: Deep Implants (eds. G.G. Bentini, A. Golanski and S. Kalbitzer)
Volume 4: Metastable Alloys: Preparation and Properties (eds. K. Samwer, M. von Allmen, J. Bøttiger and B. Stritzker)
Volume 5: Superconducting and Low-Temperature Particle Detectors (eds. G. Waysand and G. Chardin)
Volumes 6A, 6B: High T_c Superconductors (eds. P.F. Bongers, C. Schlenker and B. Stritzker)
Volume 7: Solid State Ionics (eds. M. Balkanski and C. Julien)
Volume 8: Rare-Earth Permanent Magnets (ed. I.R. Harris)
Volume 9: Defects in Silicon (eds. C.A.J. Ammerlaan, A. Chantre and P. Wagner)
Volumes 10A, 10B: Silicon Molecular Beam Epitaxy (eds. E. Kasper and E.H.C. Parker)
Volume 11: Acoustic, Thermal Wave and Optical Characterization of Materials (eds. G.M. Crean, M. Locatelli and J. McGilp)
Volume 12: Beam Processing and Laser Chemistry (eds. I.W. Boyd and E. Rimini)
Volume 13: Interfaces in Biomaterials Sciences (eds. D. Muster and G. Hastings)
Volumes 14A, 14B: High T_c Superconductor Materials (eds. H.-U. Habermeier, E. Kaldis and J. Schoenes)
Volume 15: Metal Matrix Composites (eds. G. Chadwick and L. Froyen)
Volume 16: Magnetic Thin Films, Multilayers and Superlattices (eds. A. Fert, G. Güntherodt, B. Heinrich, E.E. Marinero and M. Maurer)
Volume 17: Metallurgical Coatings and Materials Surface Modifications (eds. H. Hintermann and J. Spitz)
Volume 18: Surface Processing and Laser Assisted Chemistry (eds. I.W. Boyd, E. Fogarassy and M. Stuke)
Volume 19: Materials for Optoelectronic Devices, OEICs and Photonics (eds. H. Schlötterer, M. Quillec, P.D. Greene and M. Bertolotti)
Volume 20: SiC, Natural and Synthetic Diamond and Related Materials (eds. A.A. Gippius, R. Helbig and J.P.F. Sellschop)
Volume 21: Analytical Techniques for the Characterization of Compound Semiconductors (eds. G. Bastard and H. Oppolzer)
Volume 22: Micronic Integrated Sensors (ed. J.L. Robert)
Volume 23: High Energy and High Dose Ion Implantation (eds. S.U. Campisano, J. Gyulai, P.L.F. Hemment and J.A. Kilner)
Volume 24: Laser Surface Processing and Characterization (ed. I.W. Boyd)
Volume 25: Nuclear Methods in Semiconductor Physics (eds. G. Langouche, J.C. Soares and J.P. Stoquert)
Volume 26: Clays and Hydrosilicate Gels in Nuclear Fields (ed. A. Meunier)
Volume 27: Chemistry of Cements for Nuclear Applications (eds. P. Barret and F.P. Glasser)
Volume 28: Nuclear Materials for Fission Reactors (eds. Hj. Matzke and G. Schumacher)
Volume 29: Modifications Induced by Irradiation in Glasses (ed. P. Mazzoldi)
Volume 30: Polyconjugated Materials (ed. G. Zerbi)
Volume 31: SiGe Based Technologies (eds. E. Kasper, Y. Shiraki and T.P. Pearsall)
Volume 32: Materials Surface Processing (eds. M. Stuke, E.E. Marinero and I. Nishiyama)
Volume 33: Chemistry for Electronic Materials (eds. Y. Pauleau, G. Wahl, K.F. Jensen and T. Hirai)
Volume 34: Semiconductor Materials Analysis and Fabrication Process Control (eds. G. Crean, R. Stuck and J. Woollam)
Volume 35: Synthetic Materials for Non-Linear Optics and Electronics (eds. C. Taliani, Z.V. Vardeny and Y. Maruyama)
Volume 36: CdTe and Related Cd Rich Alloys (ed. R. Triboulet, W.R. Wilcox and O. Oda)
Volume 37: Single Chamber Processing (eds. Y.I. Nissim and A. Katz)

MATERIALS SURFACE PROCESSING

PROCEEDINGS OF SYMPOSIUM B ON
LASER, LAMP AND SYNCHROTRON ASSISTED
MATERIALS SURFACE PROCESSING
OF THE 1992 E-MRS SPRING CONFERENCE

STRASBOURG, FRANCE, JUNE 2–5, 1992

Edited by:

M. STUKE
Max-Planck-Institut für biophysikalische Chemie, Göttingen, Germany

E.E. MARINERO
IBM Almaden Research Labs., San Jose, CA, USA

and

I. NISHIYAMA
NEC Opto-electronics Research Labs., Tsukuba, Japan

1993

NORTH-HOLLAND
AMSTERDAM-LONDON-NEW YORK-TOKYO

This book is printed on acid-free paper.

ISBN: 0 444 89906 5

Published by:

North-Holland
Elsevier Science Publishers B.V.
P.O. Box 211
1000 AE Amsterdam
The Netherlands

Reprinted from:
APPLIED SURFACE SCIENCE 69

The manuscripts for the Proceedings were received by the Publisher: 30 October 1992

Printed in The Netherlands

Preface

This Symposium covered recent trends and progress in the field of Materials Surface Processing by energetic radiation using lasers, lamps and/or synchrotron radiation. One Plenary Speaker (Ehrfeld, Mainz, Germany) and one speaker in the Joint Session (Ehrlich, MIT Lincoln Lab., USA) presented recent results in this field to the participants of all symposia of the EMRS Spring 1992 conference.

Invited speakers treated the fields of laser ablation: Afonso (Spain), Cotell (USA), Fogarassy (France), Habermeier (Germany), Yabe (Japan); laser deposition and related phenomena: Boyd (UK), John (UK), Murata (Japan), van der Bergh (Switzerland); and synchrotron radiation induced processes: Cerrina (USA), Shobatake (Japan). More than 100 participants enjoyed the possibility of using this unique forum in the field of Materials Surface Processing by optical radiation and discussing the key issues ranging from basic phenomena to applications in industry.

The two-day International Equipment Exhibition, held in the Palais Josephine, together with the Poster Session(s) provided the opportunity for the research community to see and discuss the latest innovative machines, devices and services, and for the equipment exhibitors to learn about the needs and trends for laboratory equipment in science and industry.

The stimulating atmosphere in Strasbourg again provided the perfect environment for creative progress. Due to the strong participation by colleagues from Eastern Europe, Strasbourg promises to develop further as an ideal conference site for meeting all international experts. The expert local organization headed by Dr. Siffert, Mademoiselle Cobut and Madame Adloff was of great help.

November 1992 M. Stuke, E.E. Marinero and I. Nishiyama

Conference Organization

Co-Chairmen of the Symposium

M. Stuke (Max-Planck-Institut, Göttingen, Germany)
E.E. Marinero (IBM, San Jose, CA, USA)
I. Nishiyama (NEC, Tsukuba, Japan)

Scientific Committee

S.D. Allen (USA)
Y. Aoyagi (Japan)
F. Bachmann (Germany)
E. Fogarassy (France)
H. Hiraoka (Hong Kong)
M. Hirose (Japan)
H.J. Kahlert (Germany)
C.T. Kao (Taiwan)
U. Kogelschatz (Switzerland)
E.E. Marinero (USA)
T. Miyata (Japan)
I. Nishiyama (Japan)
Y.I. Nissim (France)
R. Osgood (USA)
H.W. Roesky (Germany)
M. Stuke (Germany)

Local Organization

P. Siffert (CRN-PHASE, Strasbourg, France)
M. Brutt (EMRS, Strasbourg, France)

The conference was part of the European Materials Research Society 1992 Spring Meeting

Supporting Organizations and Sponsors

This conference was held under the auspices of:

The Council of Europe
The Commission of the European Communities

It is our pleasure to acknowledge with gratitude the financial assistance provided by:

Banque Populaire (France)
Brasserie de Kronenbourg (France)
Centre de Recherches Nucléaires (France)
Centre National de la Recherche Scientifique (France)
Elsevier Science Publishers B.V. (The Netherlands)
Office du Tourisme de la Ville de Strasbourg (France)
The Commission of the European Communities
The Council of Europe
The European Parliament
Ville de Strasbourg (France)

Contents

Applied Surface Science 69 (1993) 1–6
North-Holland

applied
surface science

Excimer laser polymer ablation: formation of positively charged surfaces and its application into the metallization of polymer films

Hiroyuki Niino and Akira Yabe

National Chemical Laboratory for Industry, Higashi, Tsukuba, Ibaraki 305, Japan

Received 2 June 1992, accepted for publication 30 October 1992

Excimer laser ablation of poly(ethylene terephthalate), polyimide, and polyethersulfone films has been investigated by measuring the surface potential on the films and the velocity distribution in ionic fragments. Upon the irradiation with XeCl or KrF excimer laser in air, the surface potential of the films changed from neutral to positive. By the measurement of velocity distribution in ionic fragments with a Faraday cup assembly, it turned out that the change into a positive potential on the films was mainly caused by the redeposition of cationic fragments onto the surface of the ablated area. The change in surface potential was applied into the area-selective electroless plating (chemical plating) of copper or nickel on the ablated surface of the films.

1. Introduction

Ablative photodecomposition (APD) of organic polymers using pulsed ultraviolet lasers with a high intensity has been widely studied in the last decade [1]. In these studies, it was found that the surface of polymer films was chemically changed by APD upon exposure to excimer lasers [2–4]. APD is of interest in not only photo-etching but also surface modification of various polymers. For example, the ablated surfaces of poly(ethylene terephthalate) (PET) and polyimide (PI) films are able to be modified by imagewise metallization [3] and fluorination [4]. These modifications are related to the finding that the electric conductivity or the chemical reactivity is enhanced by excimer laser radiation. In the present study, we describe ionic behavior on the surfaces and the fragments of PET, PI, and polyethersulfone (PES) films in APD from the viewpoint of surface potential on the films [5] and velocity distribution in ionic fragments measured with an electrostatic voltmeter and a Faraday cup assembly. In addition, we have applied their surface potential behavior into the electroless plating (chemical plating) technique of the ablated films.

2. Experimental

XeCl (308 nm; Lambda Physik; EMG102MSC) and KrF (248 nm; EMG201MSC) excimer lasers were utilized for APD of PET film (semicrystalline phase; 100 μm in thickness; manufactured by Teijin Limited), PI film (amorphous; 60 μm; Du Pont (KAPTON)), and PES film (amorphous; 75 μm; Mitsui Toatsu Chemicals, Inc. (TALPA-1000LC)) [6,7]. It was generally operated with 1–100 shots at 1 Hz to avoid any polymer heating effects. The focused laser beam with a quartz lens (focal length: 10 or 17 cm) was incident perpendicularly on the sample. A desired fluence, determined by measuring the pulse energy with a pyroelectric joulemeter (Gentec Inc.; ED-500), was obtained by adjusting the distance between the lens and the sample so that the energy per pulse was constant, though the irradiated area changed. The irradiation was conducted in the atmosphere of ambient air, vacuum, helium (1 atm), nitrogen (1 atm), or oxygen (1 atm).

Laser-ablated polymer films were analyzed by a surface potential, an X-ray photoelectron spectroscopy (XPS), an electron spin resonance (ESR; JEOL; JES-RE1X (X-band; 9.441 GHz)), and

0169-4332/93/$06.00

Table 1
Surface potentials of PET, PI, and PES films irradiated by XeCl or KrF excimer laser with 100 shots in ambient air and successively treated with deionized water

Surface potential	PET [a)]	PI [b)]	PES [c)]
Before irradiation [d)]	−0.5–0.5 V	0.2–0.4 V	0.0–0.2 V
After irradiation	2.5–3.0 V	1.6–1.8 V	3.0–3.8 V
After irradiation and washing with water	16–25 V	3.5–4.5 V	12–14 V

[a)] Exposure: XeCl laser; fluence: 0.6 J cm^{-2}.
[b)] XeCl laser; 0.1 J cm^{-2}.
[c)] KrF laser; 0.1 J cm^{-2}.
[d)] After washing with ethanol.

with a scanning electron microscope (SEM). Surface potentials were measured with an electrostatic voltmeter (Monroe Electronics, Inc.; model 162 and probe 1017) in a nitrogen atmosphere (1 atm). XPS analysis was performed on a surface analyzer (VG Scientific Ltd.; ESCA LAB Mark2) using Mg Kα radiation.

3.Results and discussion

3.1. Surface potential of polymer films and velocity distribution in ionic fragments

Upon the XeCl laser irradiation with a fluence near the threshold for APD in *air*, the surface potential value of the PET film changed from almost neutral (−0.5 to +0.5 V) to positive (2.5 to 3.0 V). Table 1 shows surface potentials before and after APD. The surface potential usually changed into a higher positive value by APD, although the reproducibility of the potential values was poor. This change into positive potential was also observed on the films of PI and PES, suggesting that this phenomenon by APD in air is common to typical polymers.

Interestingly, it was found that, when the PET film was ablated in a *vacuum*, the difference in the potential before and after APD on the etched area was negligibly small. For the origin of surface potential on the polymer film, there are two mechanistic possibilities: one is a conformational orientation of functional moieties in polymer chain and the other is a formation of ionic species on the polymer surface [8]. In this case, the former is clearly excluded because laser ablation in our system is not carried out in a definite orientation. On the other hand, the critical difference between air and vacuum atmospheres suggests the remarkable participation of ionic fragments in the change of surface potential. The presence of charged fragments has been confirmed by the collection of ejected debris on charged electrodes placed on the polymer surface [9] and by the detection with mass spectroscopy [10], although non-charged products [11] including radical fragments (ref. [1]) are formed in large quantities. When the fragments, which were removed from the PET surface by the APD with a KrF laser, were deposited on a substrate in a vacuum of ca. 10^{-3} Torr, the coated substrate possessed a positive potential (table 2). In the case of APD in air, it has been reported that a part of fragments is redeposited onto the etched surface due to collisions with surrounding gas molecules near the etched region [12,13]. Thus, the origin of surface potential could be explained in terms of the preferential deposition of cationic species compared with anionic fragments during APD.

These assumptions were also supported by the detection of ionic fragments analyzed with a Faraday cup assembly. The schematic drawing of

Table 2
Surface potential of deposited fragments ablated from PET film by KrF laser irradiation in a vacuum

	Surface potential [a)]
Before deposition	0.0–0.5 V
After deposition	1.0–1.5 V

[a)] Ejected fragments were deposited on a fresh PET film.

the Faraday cup is shown in fig. 1. A three-grid optics comprised of G_1 (entrance grid), G_2 (energy barrier grid), and G_3 (repeller grid) was employed for the extraction of the ion beam in fragments during APD. When a cationic species was analyzed with the assembly, the ion collector (C) was set at a negative potential (−50 V) to repel anionic fragments, and the second grid (G_2) was set at a positive potential (0 to 35 V) to explore the kinetic distribution in the cations. For the measurement of negative fragments, the potential of V_1 and V_2 was changed to their opposite polarity. The transient current of the ionic fragments was monitored with an oscilloscope (Iwatsu: TS-8123), and the distance between the target and the Faraday cup was ca. 15 cm.

Fig. 2 shows transient current profiles in the case of a PET film during XeCl laser ablation. Although the signals in fig. 2 were ascribed to both cationic species and metastable neutral fragments, the difference in intensity between the potentials of V_2 (0 to +20 V) corresponded to the pure current of cations because the neutral fragments were decomposed into cation and electron by their bombardment on the collector [14]. The intensity of cations ejected from PET and PES films was larger than that of anions, since

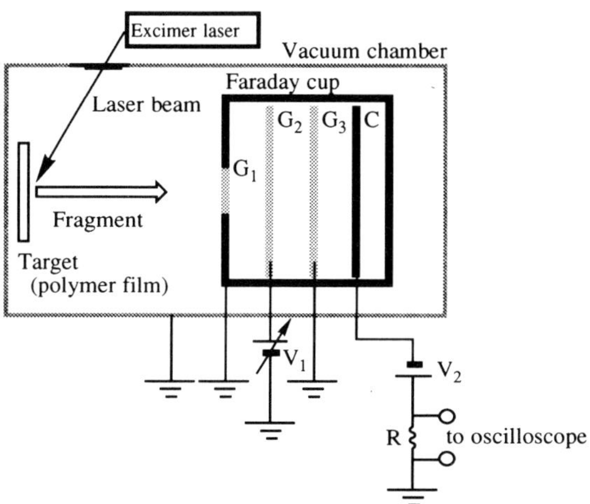

Fig. 1. Schematic. Faraday cup assembly. G_1: entrance grid, G_2: energy barrier grid, G_3: repeller grid, C: ion collector, V_1: energy barrier potential (0–±35 V), V_2: repeller potential (±50 V), R: resistor (500 Ω).

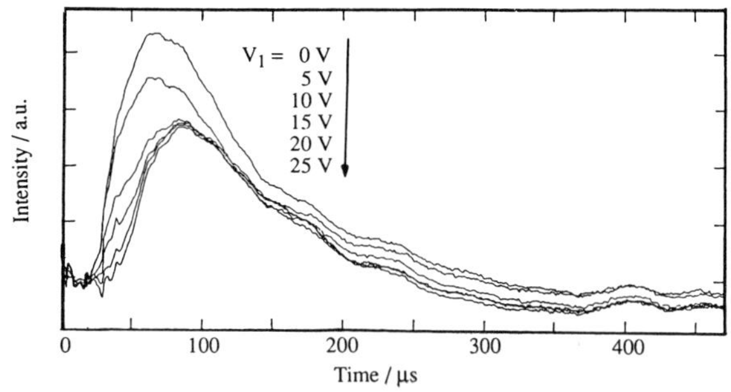

Fig. 2. The transient signals of fragments ejected from PET film during XeCl laser ablation at 0.5 J cm^{-2}, 5×10 Torr (V_1 = 0, 5, 10, 15, 20 and 25 V; V_2 = −50 V; R = 10 kΩ).

the electrons released from the ablation area could not be detected in our system owing to their high velocity (figs. 3a and 3c). Moreover, the velocity of cations decreased with an increase in the pressure of nitrogen in the chamber (fig. 4), indicating that the ablated fragments would suffer many collisions at the high ambient pressure of nitrogen (10 and 20 mTorr). At a higher ambient pressure such as 1 atm, an outgoing shock wave during APD also provides the fragments with backward driving forces in part (ref. [12]). Therefore, cationic fragments were redeposited preferentially onto the exposed area during ablation in air. In the case of a PI film, cations had large intensity in the portion of lower velocity, so that cations could be subjected to redeposition on the film.

3.2. Application to electroless plating

On the basis of the results mentioned above, we intended to apply the surface potential behavior into an electroless plating technique for conventional polymer films containing no catalyst for plating. The procedure of the plating consisted of two steps after APD (scheme 1). First, the ablated film is dipped for 0.5–12 h in an aqueous solution containing a negatively charged colloid composed of palladium (Pd) metal and a surfactant as an activation process [15]. Poly (ethyleneglycol mono-*p*-nonylphenyl) ester or sodium dodecylbenzensulfonate is used as the surfactant of negatively charged colloids. Second, the dipped film is rinsed with deionized water and immersed

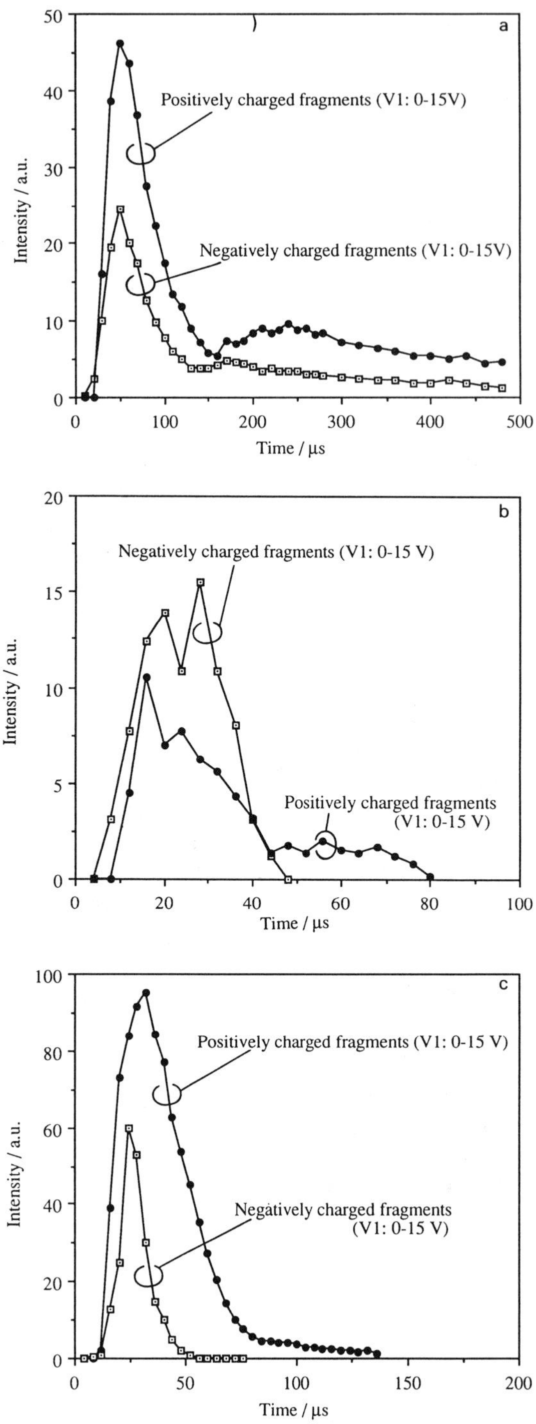

Fig. 3. The velocity distribution in positively and negatively charged fragments of PET, PI, and PES films during APD at the pressure of 5×10^{-5} Torr; (a) PET film (irradiation: XeCl laser, 0.5 J cm^{-2}; $R = 10$ kΩ), (b) PI film (XeCl laser, 0.5 J cm^{-2}), (c) PES film (KrF laser, 0.2 J cm^{-2}).

in a plating bath of copper (Cu) or nickel (Ni) for 3 to 10 min at 25°C [16]. Although it was supposed that the ionic properties of the laser-treated surfaces might vary by dipping in water, the potential value of the irradiated area was kept at a positive value (table 1).

The treatment with the negative colloid solution resulted in the deposition of Cu and Ni metal on the photo-irradiated area. A scanning electron micrograph shows selective-area plating of the Ni metal on the PET film ablated with the XeCl laser in the air (fig. 5). The activity for metallization was dependent on radiative conditions; the exposure of excimer laser with a lower fluence or a longer wavelength was preferable in the case of PET film (e.g. XeCl laser radiation with intensities near the threshold fluence for ablation). The metallization was also observed on the ablated region of PI and PES films.

When an exposed film was treated with a positively charged Pd colloid solution (surfactant: stearyltrimethylammonium chloride), as expected, no metal could be located on the film. It has become apparent by ESR measurements that the adsorption quantity of the negatively charged colloid of iron (Fe) on an ablated PET film was about 15 times larger than that of the positively charged colloid (fig. 6), indicating that the

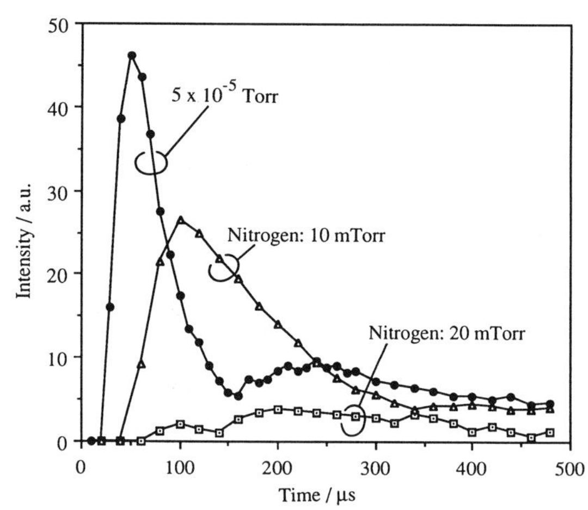

Fig. 4. The dependence of the velocity distribution in cationic fragments of PET film upon the pressure of nitrogen in chamber (irradiation: XeCl laser (0.5 J cm^{-2}); $V_2 = -50$ V; $R = 10$ kΩ).

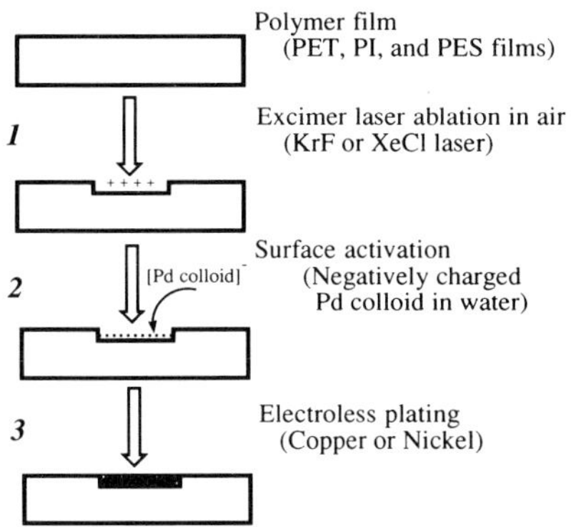

Scheme 1. Schematic diagram on selective-area electroless plating of copper or nickel on polymer films after excimer laser ablation.

Coulombic attraction between ablated films and the charged colloid was essential for our surface activation process.

From the results on the surface potential analysis and plating process, it became apparent that the laser ablation on polymer surfaces led to the change in ionic situation in the exposed area. An attempt was made to clarify the deposition mechanism of fragments on the ablated surface, that is, APD of a PET film was conducted in various surrounding gases at the pressure of 1 atm. After APD in helium, nitrogen, and oxygen atmospheres, the metallization also occurred on the etched surface, whereas no metal was deposited on the film ablated in a vacuum. These results

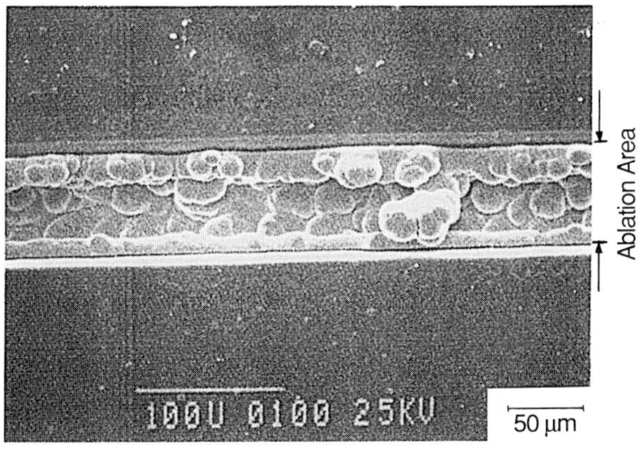

Fig. 5. SEM picture of the biaxially stretched PET film treated by nickel electroless plating after XeCl excimer laser ablation (0.6 J cm^{-2}, 100 shots, in air).

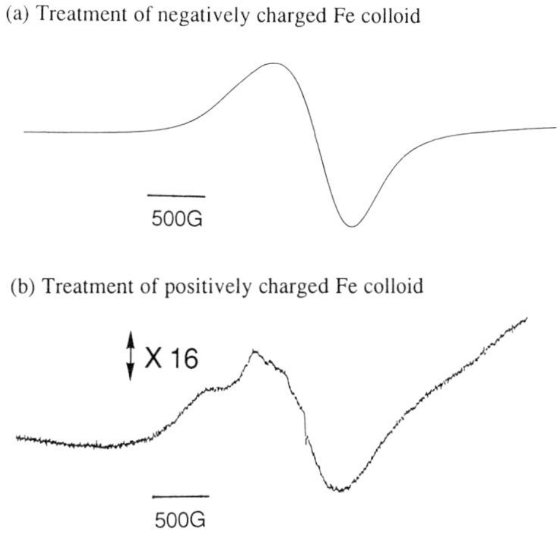

Fig. 6. ESR spectra (X-band) of Fe metal colloid adsorbed on PET film ablated with XeCl laser (0.3 J cm^{-2}, 100 shots, in air); (a) negatively charged Fe colloid, (b) positively charged Fe colloid.

are consistent with the above-mentioned assumptions in the potential analysis. Moreover, the plating process of the films irradiated in nitrogen and air was effected by the fact that they were treated for a shorter time than those irradiated in helium and oxygen environments. This distinction should be attributed to the density of ionic species on the surface of the etched region. This consideration is based on the finding that the nitrogen signal which might be ascribed to amino moieties was detected on the film ablated in nitrogen and air by XPS, suggesting that the deposition of fragments including nitrogen increased the surface density of cationic species during APD. The secondary reactions of fragments with surrounding gases have also been shown by the fast photographic study [17].

Finally, we mention the fact that some morphological micron-ordered structures appear on the axially stretched PET films (semicrystalline phase) after excimer laser ablation [3,18]. Preliminary experiments showed that these microstructures were effective for improvement of adhesion of the plating metal to polymer surface because of anchoring effects.

In conclusion, the selective-area plating on a polymer film by laser ablation could become a fascinating process, because the electroless plating on polymer films is an indispensable process for high-density multichip interconnect substrates of copper conductors in practical use [19].

Acknowledgements

The authors thank Drs. H. Nozoe and M. Matsumoto of NCLI for many helpful discussions on the measurements with the Faraday cup and electrostatic voltmeter.

References

[1] R. Srinivasan and B. Braren, Chem. Rev. 89 (1989) 1303.
[2] S. Küper and M. Stuke, Appl. Phys. B 49 (1989) 211.
[3] H. Hiraoka and S. Lazare, Appl. Surf. Sci. 46 (1990) 264.
[4] R. Srinivasan and S. Lazare, J. Phys. Chem. 90 (1986) 2124.
[5] H. Niino and A. Yabe, Appl. Phys. Lett. 60 (1992) 2697.
[6] H. Niino, A. Yabe, S. Nagano and T. Miki, Appl. Phys. Lett. 54 (1989) 2159.
[7] H. Niino, M. Nakano, S. Nagano, A. Yabe and T. Miki, Appl. Phys. Lett. 55 (1989) 510.
[8] J.V. Sanders and J.A. Spink, Nature 175 (1955) 644.
[9] R.J. von Gutfeld and R. Srinivasan, Appl. Phys. Lett. 51 (1987) 15.
[10] G. Ulmer, B. Hasselberger, H.-G. Busmann and E.E.B. Campbell, Appl. Surf. Sci. 46 (1990) 272.
[11] C.E. Otis, Appl. Phys. B 49 (1989) 455.
[12] G. Koren and U.P. Oppenheim, Appl. Phys. B 42 (1987) 41.
[13] Direct observations of deposition of fragments on ablated area is shown in figs. 3h–3l of: R. Srinivasan, B. Braren, K.G. Casey and M. Yeh, Appl. Phys. Lett. 55 (1989) 2790.
[14] G. Lantschner and A. Niehaus, Chem. Phys. Lett. 23 (1973) 223.
[15] Y. Nakao and K. Kaeriyama, J. Colloid Interface Sci. 110 (1986) 82.
[16] Copper plating bath in 100 ml (pH 11.5): $CuSO_4$, 3 g; Rochelle salt, 14 g; NaOH, 4 g; formaldehyde solution, 16 g. Nickel bath in 100 ml (pH 8–9): $NiSO_4$, 2 g; EDTA, 1.5 g; $(NH_4)_2SO_4$, 4 g; $NaPH_2O_2$, 2 g.
[17] P.E. Dyer and J. Sidhu, J. Appl. Phys. 64 (1988) 4657.
[18] S. Lazare and R. Srinivasan, Polymer 26 (1985) 1297.
[19] T.J. Hirsch, R.F. Miracky and C. Lin, Appl. Phys. Lett. 57 (1990) 1357.

Applied Surface Science 69 (1993) 7–11
North-Holland

applied
surface science

Structure of metallic multilayers studied by X-ray absorption spectroscopy

S. Pizzini [a], F. Baudelet [b], A. Fontaine [a], D. Chandesris [a], H. Magnan [a], A. Fert [c] and C. Marlière [d]

[a] LURE, Bâtiment 209D, Université Paris-Sud, 91405 Orsay, France
[b] Laboratoire de Physique des Solides, Université de Nancy I, Vandoeuvre les Nancy, France
[c] Laboratoire de Physique des Solides, Université de Paris-Sud, 91405 Orsay, France
[d] Institut d'Optique, Université Paris-Sud, 91405 Orsay, France

Received 2 June 1992; accepted for publication 30 October 1992

The use of X-ray absorption spectroscopy (XAS) as a probe of the structure of metallic multilayers is illustrated. The capability of XAS to point out structural anisotropies and strains, essential parameters exhibited by two-dimensional materials, is shown using the examples of Fe/Cu and Co/Cu multilayers.

1. Introduction

Metallic multilayers have recently been the subject of extensive research because of the unusual magnetic properties induced by their two-dimensionality. The discovery of properties such as the giant magnetoresistance of Fe/Cr multilayers, which may radically change the technology of magnetic registration, justifies all the efforts put in the new field of applied physics [1].

In this paper we concentrate on the structural properties of these multilayers and will report some results obtained using X-ray absorption spectroscopy. The potential of this technique applied to the structural characterization of metallic multilayers has been recently demonstrated [2]. It is well known that, by matching the X-ray energy to the energy of a specific absorption edge, XAS allows the local structure around a specific atomic site to be probed selectively. The information obtainable with XAS differs from that obtained with X-ray scattering techniques in that the structural information is site-specific, i.e. it allows the structure around each of the chemical species in a multilayer to be pointed out, and is local, i.e. gives the image of the real rather than the "virtual" crystal [3]. One of the main interests of metallic multilayers is the possibility to stabilize in thin layers crystallographic structures which are unstable in the bulk and exhibit magnetic properties different from those of the bulk material. For example, bcc Cu layers with thicknesses up to ~ 20 Å can be metastable in Fe/Cu multilayers grown on Fe substrates [4]. In the following we give a short review of our latest studies on Fe/Cu and Co/Cu systems [4,5]. For their outstanding magnetic properties, these multilayers have recently been the object of extensive research [6]. The exchange coupling between magnetic layers (Fe) has been shown to exhibit oscillations as a function of the thickness of the nonmagnetic spacer (Cu or Co). These oscillations have been observed by various research groups both for MBE grown and for sputtered multilayers. These multilayers also show a large magnetoresistance in the thickness ranges where the exchange is antiferromagnetic [6].

2. Samples

Fe/Cu multilayers with $t_{Fe} = 21$ Å and $t_{Cu} = 6$, 13, 24, 42 and 60 Å were deposited on a 50 Å Fe buffer on Si(100) substrates at −10°C in a UHV compatible sputtering system. The details of the experimental methods used to characterize these samples and to study their magnetic and transport properties are reported elsewhere [7].

Co/Cu multilayers were prepared by thermal evaporation from tungsten crucibles in an ultra-high vacuum chamber [8]. Co_x/Cu_y multilayers ($Co_{10}Cu_{10}$, Co_5Cu_5, Co_5Cu_3, Co_3Cu_5, Co_3Cu_3, where x (y) is the number of Co (Cu) monolayers) were evaporated on a 250 Å thick Au layer deposited on a float glass substrate. The results of NMR data presented in a previous publication [9] show that in all the samples Co is present in the fcc crystallographic phase, with a large (38%) concentration of stacking faults having a local hcp environment. The quality of the Co/Cu interfaces was controlled by X-ray reflectivity. The presence of a superstructure peak at around 3° is the fingerprint of rather abrupt Co/Cu interfaces (fig. 1a). Calculations show in fact that if the interfacial roughness is larger than 4 Å, the superstructure peak disappears.

The θ–2θ XRD spectrum recorded for the $Co_{10}Cu_{10}$ multilayer (fig. 1b) indicates that the multilayer has a very well defined (111) texture. The observed XRD peaks can be indexed as Au(111) for peak Al, and Cu(111) and Co(111) for peak C1. The peaks labelled S are due to the superstructure of the multilayer. The good agreement between the experiment and the calculation (obtained by modelling the multilayer as a perfectly regular alternance of layers of the two metals) confirm that a "step-like" model is a good approximation of the multilayer structure and that the Co/Cu interfaces are quite abrupt [5].

3. Experiments and data analysis

X-ray absorption spectroscopy (XAS) measurements were carried out at the 1.85 GeV DCI synchrotron radiation source at the Laboratoire pour l'Utilisation de Rayonnement Electromagnetique (LURE) at Orsay. A Si(311) channel cut monochromator was used. The Co, Cu and Fe K edge absorption spectra of the multilayers were recorded in electron yield mode [3]. The spectra

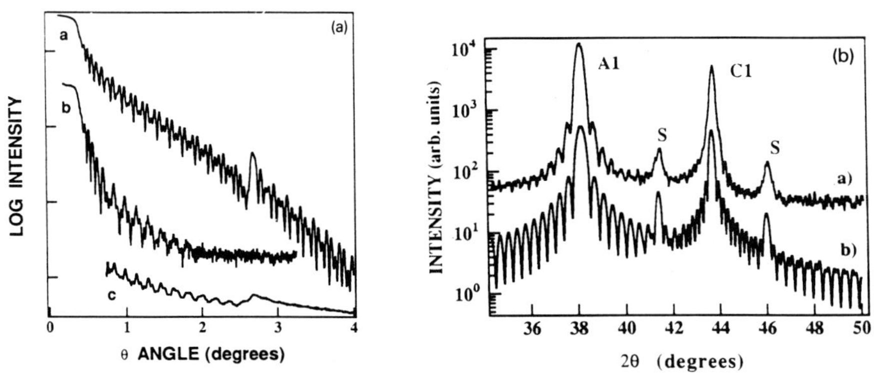

Fig. 1. (a) X-ray reflectivity curves for the (Co_5Cu_3) multilayer: (curve a) simulation using thickness values deduced from quartz balance shifts and rms interface roughnesses of 4 Å; (curve b) experimental data obtained with the high-resolution diffractometer; (curve c) experimental data obtained with a commercial diffractometer. (b) X-ray diffraction θ–2θ curves for the $Co_{10}Cu_{10}$ multilayer; (curve a) experimental data; (curve b) best fit of the experimental data, obtained using the "step model" and the crystal parameters of the bulk metals. A1 is the Au(111) peak; C1 is the convolution of the Cu(111) and the Co(111) peak, S are the diffraction peaks due to the superstructure.

of the model compounds (Co, Cu and Fe foils) were recorded in transmission geometry. The EXAFS spectra of the multilayers were recorded in two geometries, with the sample surface parallel and nearly perpendicular (80°) to the polarization direction of the X-ray beam, respectively. Since the photoelectron scattering amplitude depends (as $\cos^2\theta_i$) on the angle θ_i between the X-ray polarization direction and the position vector of the ith neighbour of the absorbing atom, differences between the crystal structure in the growth plane and in the growth direction can be pointed out by recording the XAS spectra with the sample surface parallel and perpendicular to the polarization direction respectively.

The data were analyzed using standard methods. Fourier transforms (FT) of the EXAFS spectra were performed in the wavevector region of approximately 3–12 Å^{-1}. The spectra were filtered in real space to isolate the first peak of the FT and back-transformed in k-space. The Fourier-filtered spectra were fitted using a plane-wave approximation and experimental phase shifts, which include the curved wave corrections [3]. All details on experiments and analysis can be found in refs. [4,5].

4. Results

4.1. Co / Cu multilayers

Fig. 2 shows the Fourier transforms of the Cu K edge EXAFS spectra obtained for the Co/Cu multilayers, recorded with in-plane X-ray polarization. For each of the multilayers, the positions and the amplitudes of the FT peaks are very close to those of bulk Cu, up to large distances from the central atoms (~ 8 Å). This indicates that the fcc structure of Cu layers is highly ordered. This result is confirmed by the least-squares fits of the EXAFS spectra (table 1) which show that the Debye–Waller factors for the Cu neighbour atoms are identical to those of bulk Cu, within the typical ±0.0005 Å^2 error bar. The results of the fits also show that the nearest Co–Co and Cu–Cu distances are independent of the layer thickness, and are different for in-plane

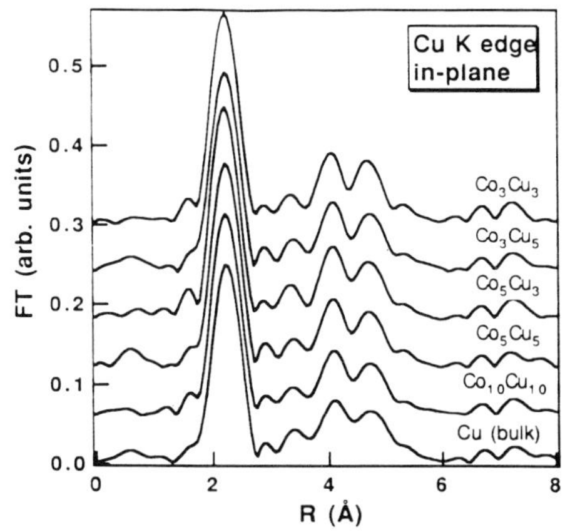

Fig. 2. Fourier transform amplitudes of the Cu K edge EXAFS spectra of the Co_xCu_y multilayers, with the sample surface parallel (in-plane) to the polarization direction of the X-ray beam.

or out-of-plane directions, i.e. the layers' structure is anisotropic. For Co layers, the in-plane distance is 2.52 Å (±0.01 Å) and the out-of-plane distance 2.50 Å (±0.01 Å). For Cu layers, the in-plane distance is 2.54 Å (±0.01 Å) and the

Table 1
Results of the modelling of the Co K and Cu K edge EXAFS spectra recorded for the Co_xCu_y multilayers with the polarization direction parallel (in-plane) and perpendicular (out-of-plane) to the polarization of the X-ray beam

Sample	R_1 (Å) (In-plane)	DW_1 (Å^2) (In-plane)	R_1 (Å) (Out-of-plane)	DW_1 (Å^2) (Out-of-plane)
Cu K edge				
Cu (bulk)	2.556	–	2.556	–
$Co_{10}Cu_{10}$	2.54	0.0001	2.55	0.0004
Co_5Cu_5	2.54	0.0000	2.55	0.0005
Co_3Cu_3	2.54	−0.0005	2.55	0.0000
Co K edge				
Co (bulk)	2.51	–	2.50	–
$Co_{10}Cu_{10}$	2.52	0.0003	2.50	0.0008
Co_5Cu_5	2.52	0.0018	2.51	0.0019
Co_3Cu_3	2.52	0.0023	2.51	0.0023

R_1 are the nearest-neighbour distances. DW_1 are the Debye–Waller factors, relative to those of the model compounds used to obtain experimental phaseshifts: bulk Cu for Cu K edge spectra and bulk Co for the Co K edge spectra. Typical error bars are ±0.0005 Å^2 for DW_1 and ±0.01 Å for R_1. The coordination number for the nearest-neighbour shell has been fixed at 12, as for an fcc or hcp structure.

out-of-plane distance 2.55 Å (±0.01 Å). In the plane of the layers the Co and Cu lattices are strained with respect to the bulk layers (the first neighbour distance is 2.51 Å in Co and 2.556 Å in Cu). This in-plane strain tends to decrease, but not completely eliminate, the mismatch between the two lattices. The deformation tends to expand (by less than 1%) the Co layers, which have a smaller bulk lattice parameter, and to contract by a similar amount the Cu layers. In the direction perpendicular to the Co/Cu interfaces and in both Cu and Co layers, the nearest-neighbour distances are unchanged with respect to the bulk.

The values of the Debye–Waller factors obtained from the fits of the Co K edge spectra (table 1) indicate that the Co layers are significantly more disordered than the Cu layers. While the copper layers tend to form an almost perfect fcc structure, Co grows in the fcc metastable structure imposed by the Cu substrate. As NMR measurements have shown [9] the stacking of the Co layers is not perfect, and the large Debye–Waller factors are probably due to the presence of defects which tend to locally re-establish the hcp structure typical of the bulk Co phase. The Debye–Waller factors tend to decrease as the layer thickness increases and, for the thickest layer, are larger in the direction perpendicular to the Co/Cu interfaces. This could be due to the formation of large ordered hcp domains in the thicker Co layers.

4.2. Fe / Cu multilayers

In fig. 3 the Fe K and Cu K near-edge XAS spectra obtained for the Fe/Cu multilayers (t_{Fe} = 21 Å and t_{Cu} = 6, 13, 24, 42 and 60 Å) are compared with those of Fe and Cu bulk metals. These results show that the structure of the Fe layers is bcc as in the bulk metal and is independent of the thickness of the Cu films in the multilayer. On the other hand, the structure of the Cu layers strongly depends on their thickness. The Cu K near-edge XAS spectra evolve with the thickness, from that typical of a bcc structure for t_{Cu} = 6 Å to that more typical of a fcc structure for t_{Cu} = 60 Å. For t_{Cu} = 60, 42 and 24 Å the near-edge spectra show, even if less marked, the

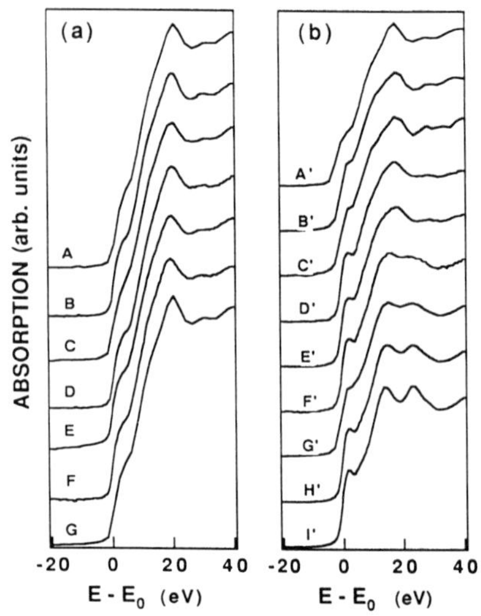

Fig. 3. Fe K edge (a) and Cu K edge (b) near-edge X-ray absorption spectra for the (Fe 15 Å/Cu t_{Cu}) multilayers. (A) t_{Cu} = 13 Å (out-of-plane); (B) t_{Cu} = 13 Å; (C) t_{Cu} = 24 Å; (D) t_{Cu} = 42 Å; (E) t_{Cu} = 60 Å (out-of-plane); (F) t_{Cu} = 60 Å; (G) bulk Fe. (b): (A′) bulk Fe; (B′) t_{Cu} = 6 Å (out-of-plane); (C′) t_{Cu} = 6 Å; (D′) t_{Cu} = 13 Å (out-of-plane); (E′) t_{Cu} = 13 Å; (F′) t_{Cu} = 24 Å; (G′) t_{Cu} = 42 Å; (H′) t_{Cu} = 60 Å; (I′) Cu bulk.

features characteristic of the fcc Cu bulk structure. These observations are confirmed by the FT data (fig. 4). The Fourier transforms of the Cu K spectra for t_{Cu} = 6 Å exhibit the peaks typical of a bcc structure. The large-distance peaks are less marked in the direction perpendicular to the growth plane, probably because of the contribution to the spectra of Fe atoms at the Fe/Cu interface. The Fourier transforms of the Cu K spectra for thicker Cu films (t_{Cu} = 24, 42 and 60 Å) show the large distance peaks characteristics of a fcc structure. The presence of well-defined FT peaks for larger distances indicates the formation of a fairly ordered Cu fcc phase. On the other hand the small amplitude of the FT peaks with respect to those typical of Cu bulk, points to the presence of a static disorder larger than in the bulk Cu phase. The multilayer with t_{Cu} = 13 Å appears to be close to the transition point between the bcc structure for thin Cu films and the bulk-like fcc structure for thicker Cu layers. For this multilayer, while the nearest-neighbour

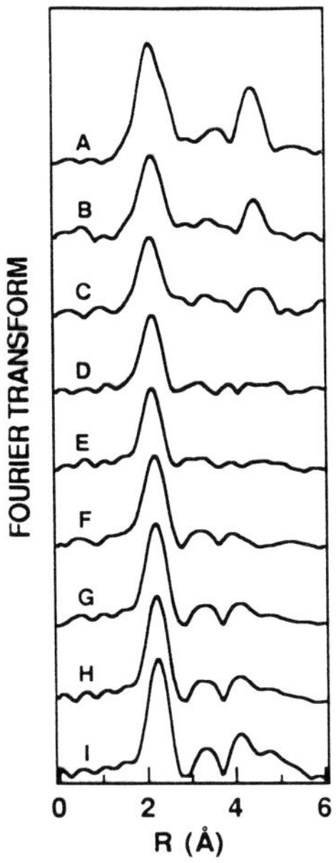

Fig. 4. Fourier transform amplitudes of the EXAFS spectra for (Fe 15 Å/Cu t_{Cu}) multilayers at the CuK edge. Unless specified, the spectra were recorded with the sample surface parallel to the polarization of the X-ray beam: (A) bulk Fe; (B) $t_{Cu}=6$ Å (out-of-plane); (C) $t_{Cu}=6$ Å; (D) $t_{Cu}=13$ Å (out-of-plane); (E) $t_{Cu}=13$ Å; (F) $t_{Cu}=24$ Å; (G) $t_{Cu}=42$ Å; (H) $t_{Cu}=60$ Å; (I) Cu bulk.

distance is similar to that found for $t_{Cu}=6$ Å, no important FT peaks are found for larger distances.

The results of the least-squares fits of the Fourier filtered spectra [4] give more quantitative information about the structural changes within the Cu layers. For $t_{Cu}=6$ and 13 Å the nearest Cu–Cu distance is 2.50–2.52 Å, 1% larger than the Fe–Fe distance in bulk bcc Fe. For $t_{Cu}=13$ Å, the bcc structure is very disordered and beyond this thickness the Cu layers relax to the fcc structure typical of bulk Cu. For $t_{Cu}=24$, 42 and 60 Å the Cu–Cu distance is 2.55 Å, close to the nearest-neighbour distance in fcc Cu (2.556 Å). The process of relaxation is gradual, and a more and more ordered fcc phase is formed as the Cu film thickness increases.

The latter result may explain the "anomalous" behaviour of the Fe/Cu coupling, which has been observed to increase as t_{Cu} increases. Since the electron mean free path is expected to be longer in more ordered structures, the progressive ordering probably accounts for the "anomalous" increase of the magnetoresistance as the Cu thickness goes from 24 to 60 Å.

5. Conclusion

Using the examples of Fe/Cu and Co/Cu multilayers, we have shown the potential XAS to characterize the local structure of metallic multilayers. The result of these studies may constitute an important step in the understanding of the exceptional magnetic properties of these multilayers.

References

[1] L.M. Falicov, D.T. Pierce, S.D. Bader, R. Gronsky, K.B. Hathaway, H.J. Hopster, D.N. Lambeth, S.S.P. Parkin, G. Prinz, M. Salamon, I.K. Schuller and R.H. Victoria, J. Mater. Res. 5 (1990) 1299, and references therein.

[2] F. Baudelet, PhD Thesis, Université Paris-Sud, 1991.

[3] For details about EXAFS see, for example: D.C. Koningsberger and R. Prins, X-ray Absorption. Principles, Applications, Techniques of EXAFS, SEXAFS and XANES, (Wiley, New York, 1988).

[4] S. Pizzini, F. Baudelet, D. Chandesris, A. Fontaine, H. Magnan, J.M. George, F. Petroff, A. Barthélemy, A. Fert, R. Loloee and P.A. Schroeder, Phys. Rev. B 46 (1992) 1253.

[5] S. Pizzini, F. Baudelet, A. Fontaine, M. Galtier, D. Renard and C. Marlière, Phys. Rev. B, to be published.

[6] F. Petroff, A. Barthélemy, D.H. Mosca, D.K. Lottis, A. Fert, P.A. Schroeder, W.P. Pratt, R. Loloee and S. Lequien, Phys. Rev. B 44 (1991) 5355.

[7] J.M. Slaughter, W.P. Pratt and P.A. Schroeder, Rev. Sci. Instrum. 60 (1989) 127.

[8] C. Marlière, D. Renard and J.P. Chauvineau, Thin Solid Films 201 (1991) 317.

[9] J.P. Renard, P. Beauvillain, C. Dupas, K. Le Dang, P. Veillet, E. Velu, C. Marlière and D. Renard, J. Magn. Magn. Mater. 115 (1992) L147.

Applied Surface Science 69 (1993) 12–15
North-Holland

applied
surface science

Designing surface properties of textile fibers by UV-laser irradiation

Thomas Bahners, Wolfgang Kesting and Eckhard Schollmeyer
Deutsches Textilforschungszentrum Nord-West e.V., Frankenring 2, D-4150 Krefeld, Germany

Received 2 June 1992; accepted for publication 30 October 1992

The irradiation of polymeric material by pulsed UV lasers generates characteristic modifications of the surface topography of the polymer beside other effects. The resulting surface properties may have an important impact in textile processing, as they can affect technical properties of synthetic and even natural fibers, e.g. particle and coating adhesion or wetting properties, in a dry process. An overview is given over microscopic and macroscopic properties of the generated surfaces in dependence on laser parameters and material constants. Various understandings of the genesis of the characteristic geometry and structure of irradiated surfaces are compared to experimental observations from polymer films and a number of textile fibers. With respect to specific fiber properties the experimental data give evidence of cooperative transport phenomena in a superheated polymer as the prime driving forces of the effect.

1. Introduction

A new technique for a controlled surface modification of synthetic fibers and films is the irradiation with a pulsed UV-laser. Under certain conditions the irradiation of synthetic fibers with excimer lasers generates a characteristic modification of their surface topography. Many types of synthetic fibers, e.g. poly(ethylene terephthalate) (PET), polyamides (PA) and aramides, expose a characteristic surface modification when irradiated by ArF- and KrF-lasers [1] (fig. 1). The originally smooth surface of these fibers changes its morphology to a rather regular roll-like structure perpendicular to the fiber axis after this treatment.

Results of laboratory experiments dealing with possible textile applications show that the characteristic morphology of UV-laser treated fibers has a potential to design the optical appearance of these fibers [2] and important quantities for textile processing as wetting behaviour and adhesion properties of a fiber or fabric with respect to particulate matter [3], which not only affect processing, but also define a textil's properties in certain applications. Experiments with laser treated PET-fabrics show that the laser induced surface morphology enhances the adhesion of dyestuff pigments in dyeing processes and dust particles on laser irradiated filter fabrics [4]. SEM-investigation revealed that particles are captured in the grooves of the characteristic surface structure, where much higher adhesion forces act as can be expected from a smooth fiber surface.

In order to define industrial realizations of this highly interesting application of excimer lasers it is necessary to understand the basic dependence of resulting surface properties to parameters of the process. A fundamental frame of experimental observations is sketched in this work giving rise to a model of the laser-induced processes which govern this effect. A simple method of simulation by means of cellular automata is presented.

2. Topography of modified surfaces

For the first time Bahners and Schollmeyer [1] studied the dependence of characteristical surface data on laser parameters in the case of

0169-4332/93/$06.00

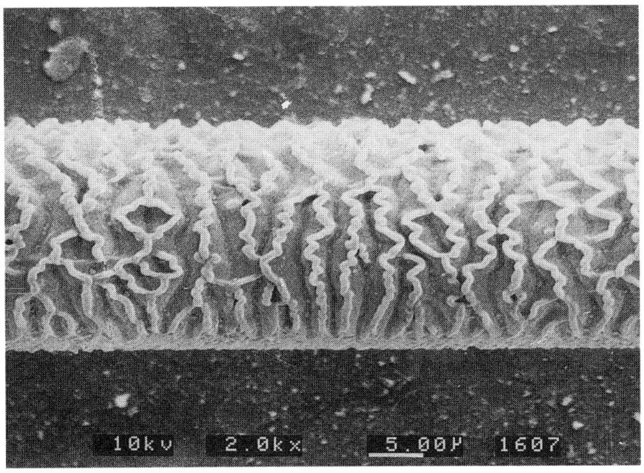

Fig. 1. Laser irradiated PET-fiber as seen in the scanning electron microscope.

commercial poly(ethylene)terephthalate (PET)- and polyamide (PA)-6.6-fibers to great detail. The effect was observed using both 193 and 248 nm excimer lasers in a range of fluences from below the ablation threshold (about 30 mJ/cm^2 at 193 nm for PET) up to more than 150 mJ/cm^2 with no thermal defect occurring.

In all instances the orientation of the rolls was strictly perpendicular to the fibrillar orientation (given by the stress axis) of the fiber. A thermal damage to the polymer occurred only at higher repetition rates, typically at more than 5 Hz.

The geometry of the surface can be described by the mean roll distance $\langle D \rangle$. Characteristicly the mean roll distance exhibits a strong dependence from the number of laser pulses applied up to ten pulses. Further irradiation increases the depth of the structure yielding a coarser structure [5]. Irradiation with several hundreds of pulses eventually leads to the destruction of the fiber, which decays to characteristic elliptical pieces. In all investigated cases the dependence of the mean roll distance $\langle D \rangle$ of the pulse number N_p could be described by a logarithmic expression

$$\langle D \rangle = K_1 \log N_p + K_2,$$

where $\langle D \rangle$ was taken as mean value over 10 to 20 rolls. For given conditions K_1 and K_2 are constant. K_1 is weakly dependent on laser fluence, but sensible for the given absorption depth and specific fiber properties (e.g. draw rate). K_2 gives the morphological dimension after the very first laser shot [5].

It is well known, that the existence of a threshold fluence is a main characteristic of UV-laser induced ablation of polymers surfaces. It has to be noted that a threshold behaviour was also found for the formation of the surface structure [6]. This threshold fluence is not equal to the ablation threshold fluence and not even constant, but coupled to a similar condition concerning the existence of – internal or externally applied – stress fields in the polymeric material. Earlier investigations on elastomeric fibers and PET-films gave evidence for a strong influence of external or internal stress fields [7]. Only when drawn above a certain ratio the UV-laser irradiation effects a morphological modification indicating the necessity of stress fields. Interestingly, the minimum draw ratio changes with fluence, so that in summary non-constant fluence and draw ratio thresholds for the generation of morphology changes exist [6].

3. Laser-induced processes in the macromolecular scale

Several models describing the modification of the surface morphology due to laser irradiation were published, e.g. interpreting the structure generation as synergetic effect [8], as image of underlying crystallinic structures of the polymer [9] or as interference effect between incident beam and reflected light [10].

Based on the experimental data a prime condition for the occurrence of the described effect of laser induced surface roughness is seen by the authors in the very high absorption of PET and other polyaromatic polymers. Typical absorption coefficients are in the order of 10^5 cm^{-1} at wavelengths 193 and 248 nm. Evidence for the importance of the high absorption may be found in the irradiation of polyolefinic materials (poly(propylene) (PP), poly(ethylene) (PE), poly(tetraflourethylene) (PTFE)). The absorption of these materials is up to 2 orders of magnitude smaller than that of PET in the spectral range of 190 to 250 nm. Accordingly, no modification of

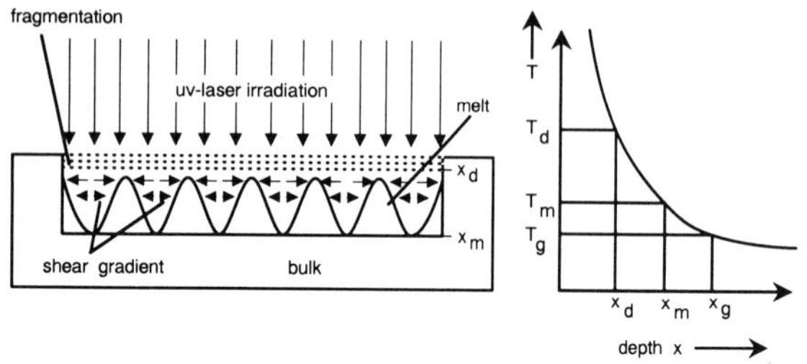

Fig. 2. State of the surface and approximated temperature profile near the polymer surface during UV-laser irradiation. Symbols are: T – local temperature at penetration depth x, T_d – degradation temperature for ablation at etch depth x_d, T_m – melting point at depth x_m, and T_g – glassy point at x_g.

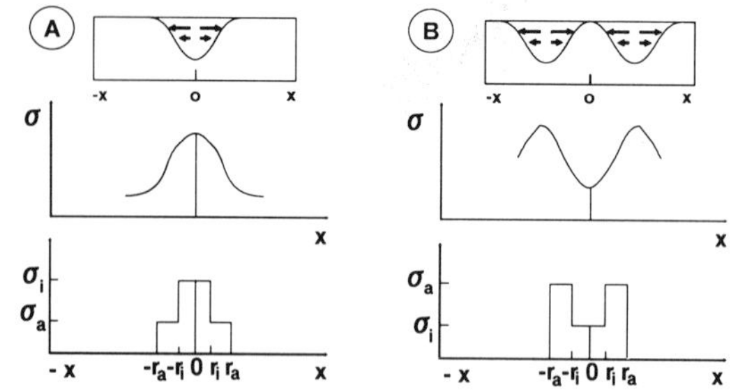

Fig. 3. Schematic illustration of modelling UV-laser induced surface structures by cellular automata. Symbols are: σ – internal stress at position x, σ_i – internal stress at inner circular area with radius r_i, σ_a – internal stress at outer, surrounding area $(r_a - r_i)$. (A) $\sigma_a(r_a) < \sigma_i(r_i)$, surface crack generation at $x = 0$, (B) $\sigma_a(r_a) > \sigma_i(r_i)$, material compression at $x = 0$.

the surface structure was observed after irradiation of PP-fibers with ArF- and KrF-lasers, but heavy thermal damages. Taking advantage of the marked increase of the absorption coefficient of PP below 180 nm only a F_2-laser can be used to induce the same characteristic surface morphology found on PET-fibers [11].

We expect a temperature profile to be induced by thermal contributions to this concentrated deposition of the laser energy, which in a first approximation may be derived from Beer's law (fig. 2). In the time scale involved heat dissipation can be neglected and the temperature field regarded as constant. Absorption coefficients in the regime of 10^5 cm^{-1} result in extreme temperatures at the polymer surface and a steep temperature gradient: estimates for PET at 193 nm give more than 1200 K at the surface and a mean gradient of about 6×10^7 K/cm. Therefore besides ablation – which would affect regions with temperatures in excess of some degradation threshold T_d – a certain layer $(x_m - x_d)$ would form a pool of molten material for which a rough calcalution gives a thickness of about 0.1 μm. The occurrence of melting due to laser irradiation was also indicated by FT-IR-investigation [12].

Corresponding to the temperature gradient according gradients of viscosity and chain mobility in the polymer melt have to be assumed, allowing relaxation of the very high internal stress fields in commercial fibers, again dependent on penetration depth. This results in a complex shear field with huge shear forces at the very surface – these give rise to a rather random distribution of shear induced cracks – and decreasing shear forces with increasing depth. It is assumed that in these far-from-equilibrium conditions the coupling of a thermal and an elastic (shear) field creates cooperative transport phenomena.

In order to give some impression of the potential of this understanding the formalism of cellular automata was chosen, which allows the computer simulation of complex cooperative processes in a rather simple manner [13]. The investigated system is represented by a matrix of discrete elements (unit cell), whose state is described by a small number of specific characteristics, which are related to the problem under investigation. Beginning from random states of all unit cells the history of every single unit cell, e.g.

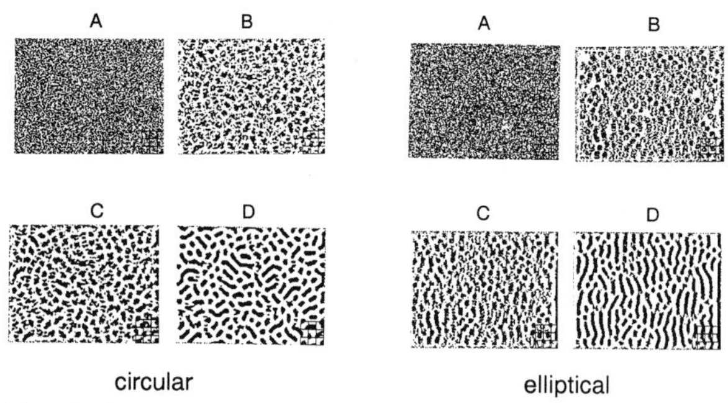

Fig. 4. Pattern formation by cellular automata, left side: circular neighbourhood, right side: elliptical neighbourhood, number of iteration steps: 0 (A), 3 (B), 5 (C), 15 (D).

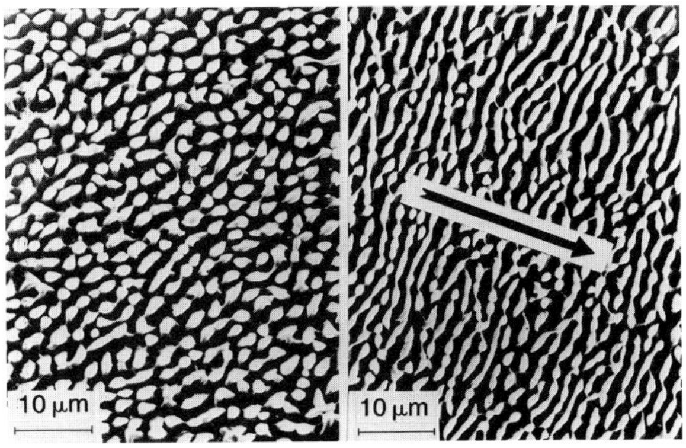

Fig. 5. SEM-pictures of UV-laser irradiated PET-films; irradiation: 248 nm, 20 pulses, 120 mJ/cm^2; left side: original, biaxially stretched film; right side: film monoxially stretched during irradiation, the arrow indicates the stress direction.

a finite volume of a solid, is determined in an iterative procedure by the state of its neighbourhood, which could be some characteristic quantity as mass, temperature, attracting or detracting force, etc. Usually the procedure is continued until the whole system converges into a stable state, which is reached in the case of most physical systems.

With respect to the given problem of material transport and selforganization due to complex shear fields, one has to calculate a possible material transport into or out of a unit cell (position $x = 0$) on the basis the integral shear forces originating from a closely defined ensemble of neighbouring unit cells. As sketched in fig. 3A we assume a material transport out of a cell, if the sum of all masses within the neighbourhood is higher than the mass inside the cell (resulting in an effective positive shear force), and vice versa (fig. 3B). The procedure allows the simulation of one-directional tension fields – as are usually given in fibrous polymers – by definition of an "elliptical" neighbourhood instead of a "circular", which would correspond to isotropic stress fields.

Simulated transport phenomena for isotropic and unaxial stress field are presented in fig. 4 at varying states of the cellular automata (number of iterations). It is interesting to note, that the convergent states of these simulations (denoted "D") show topographies, which are in good agreement to structures found in scanning electron micrographs from unaxially and biaxially ("rather" isotropic) stretched PET-films (fig. 5).

Acknowledgements

We thank the Bundesministerium für Forschung und Technologie (BMFT) for financial support of this work (Project 13 N 5634). The responsibility for the context of this publication lies with the authors.

References

[1] T. Bahners and E. Schollmeyer, Angew. Makromol. Chem. 151 (1987) 19.
[2] D. Knittel, A. Eickmeier, T. Bahners and E. Schollmeyer, Text. Prax. Int. 45 (1990) 236.
[3] T. Bahners, W. Kesting and E. Schollmeyer, in: Excimer Lasers and Applications III, Eds. T. Letardi and L.D. Laude, Proc. SPIE 1503 (1991) 206.
[4] T. Bahners and E. Schollmeyer, Staub-Reinhalt. Luft 47 (1987) 115.
[5] W. Kesting, D. Knittel, T. Bahners and E. Schollmeyer, Appl. Surf. Sci. 54 (1992) 330.
[6] W. Kesting, T. Bahners, D. Knittel and E. Schollmeyer, Angew. Makromol. Chem. 180 (1990) 109.
[7] T. Bahners, A. Bossmann and E. Schollmeyer, Angew. Makromol. Chem. 170 (1989) 203.
[8] T. Bahners and E. Schollmeyer, J. Appl. Phys. 66 (1989) 1884.
[9] Y. Novis, J.J. Pireaux, A. Brezini, E. Petit, R. Caudano, P. Lutgen, G. Feyder and S. Lazare, J. Appl. Phys. 64 (1988) 365.
[10] P.E. Dyer and R.J. Farley, Appl. Phys. Lett. 57 (1990) 765.
[11] W. Kesting, T. Bahners and E. Schollmeyer, J. Polym. Sci., in press.
[12] T. Bahners, D. Knittel, F. Hillenkamp, U. Bahr, C. Benndorf and E. Schollmeyer, J. Appl. Phys. 68 (1990) 1854.
[13] D.A. Young, Math. Biosci. 72 (1984) 51.

Applied Surface Science 69 (1993) 16–19
North-Holland

applied
surface science

Laser-induced surface modification and structure formation of polymers

E. Arenholz, J. Heitz, M. Wagner, D. Bäuerle
Angewandte Physik, Johannes-Kepler-Universität Linz, A-4040 Linz, Austria

H. Hibst and A. Hagemeyer
BASF AG, ZAA / F, D-6700 Ludwigshafen, Germany

Received 2 June 1992; accepted for publication 30 October 1992

Different types of surface modifications and structure formation of PET (poly-ethylene-terephthalate) and PI (polyimide) foils induced by excimer-laser radiation are investigated.

1. Introduction

Laser-induced surface modifications and laser-induced patterning of material surfaces by deposition, etching, and ablation is captivating increasing attention [1]. This paper deals with the excimer-laser processing of poly-ethylene-terephthalate (PET) and polyimide (PI) foils [2]. It is well known that laser radiation with fluences, ϕ, below the ablation threshold, ϕ_{th}, can modify the roughness, crystallinity, resistivity, and chemical composition of polymer surfaces. With fluences $\phi > \phi_{th}$ the ablated material surface often reveals quasiperiodic microstructures. In this paper we present recent results on the laser-enhanced adhesion of metal films on polymer surfaces, on the laser-enhanced surface conductivity of polyimide and on self-developing structures on polymer foils.

2. Laser-enhanced adhesion of metal films on polymer surfaces

The coating of polymer foils with metal films has found increasing importance in connection with information storage on video tapes, floppy disks, etc. The adhesion of the metal film evaporated onto the polymer foil determines the lifetime of the product.

In the present investigations we have proved that the adhesion of metal films on polymer foils can be improved by laser irradiation of the polymer surface prior to metal deposition. As a model system we have chosen metal films of different Co alloys such as CoNi, CoCr, CoNi-oxide etc. on PET foils. Laser irradiation was performed by means of 248 nm KrF and 308 nm XeCl laser light with fluences of, typically, $\phi(248\ \text{nm}) \geq 5$ mJ/cm^2 and $\phi(308\ \text{nm}) \geq 40$ mJ/cm^2, respectively. The number of laser pulses employed was between 1 pulse and 20 pulses. Metal deposition was performed by means of electron-beam evaporation immediately after laser-beam irradiation.

Fig. 1 shows the adhesion force of a 200 nm thick CoNi film on a 50 μm thick PET foil as a function of laser fluence for single-pulse 248 nm KrF-laser irradiation. The adhesion force was measured by a standard peel test (EAA test) [3]. Here, another polymer foil is laminated onto the metal film at a temperature of about 105°C (glass temperature T_G (PET) ~ 67°C) and a pressure of 30 to 100 bar. With this test, the adhesion force is

0169-4332/93/$06.00

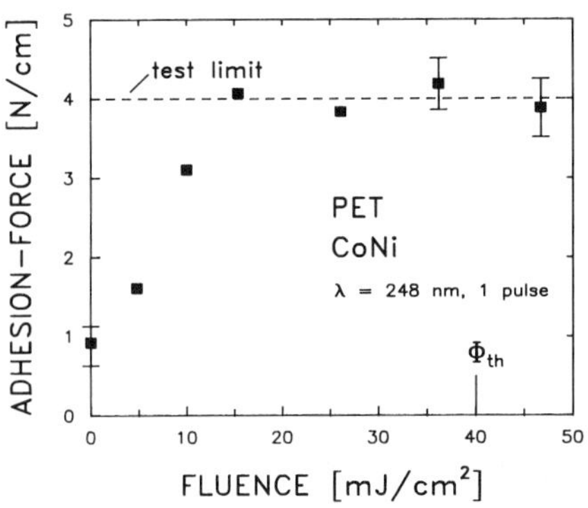

Fig. 1. Adhesion force of CoNi films deposited on PET foils after 248 nm KrF-laser irradiation.

defined by the lift-off of the metal film together with the laminated foil. The figure shows that the adhesion force rapidly increases; it reaches the limit of this test, which is around 4 N/cm, at a fluence of only 15 mJ/cm^2. This clearly demonstrates that the adhesion of the metal film on the polymer surface can significantly be improved by excimer-laser irradiation with fluences that are considerably *below* the threshold fluence for ablation, ϕ_{th}. It is evident that the data point obtained with the non-irradiated foil contains possible changes in adhesion force due to the particular properties of the test employed in these measurements.

3. Laser-enhanced surface conductivity on polyimide

The surface conductivity of some polymers can significantly be increased upon UV-excimer-laser irradiation. In the present experiments we have employed commercial PI foils (Upilex R, Ube Inc.) and 248 nm KrF-laser radiation (FWHM 28 ns, repetition rate 5 pps, spot size 2×20 mm^2). Conductivity measurements have been carried out by using the standard DC four-point probe technique.

Fig. 2 shows the dependence of the surface conductivity on the number of laser pulses. Irradiation has been performed in air with a laser fluence of 55 mJ/cm^2. Because the exact thickness of the laser-modified surface layer is unknown, we have plotted the surface conductivity times the modified layer thickness. The figure shows that after laser irradiation with about 500 pulses the conductivity is increased by 15 orders of magnitude. A similar behavior has recently been found for Kapton (DuPont) [4]. The increase in surface conductivity is ascribed to the formation of a carbon layer on the polymer surface.

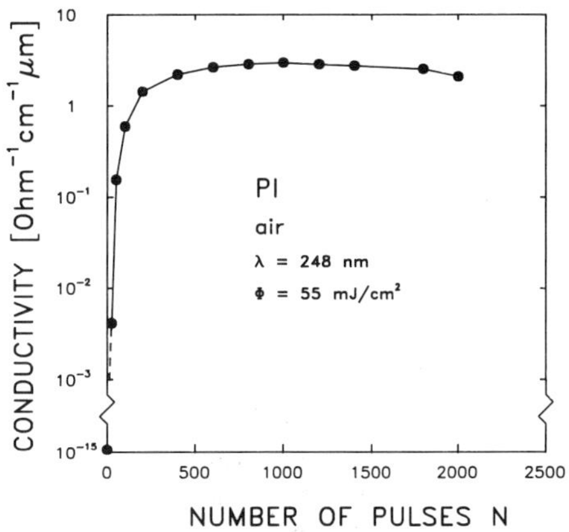

Fig. 2. Dependence of the surface conductivity of PI foils on the number of 248 nm KrF laser pulses. Irradiation has been performed in air with a fluence of 55 mJ/cm^2.

Fig. 3 shows that the maximum in the conductivity appears near the ablation threshold which is $\phi(248\text{ nm}) \approx (44 \pm 3)$ mJ/cm^2 [5]. The effect of different gaseous atmospheres on the conductiv-

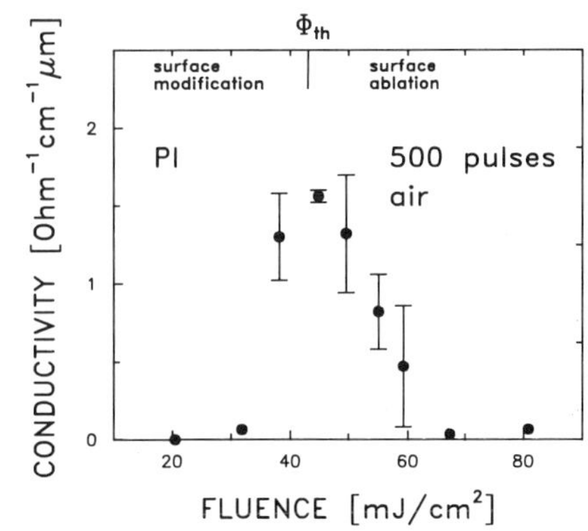

Fig. 3. Conductivity of PI foils as a function of laser fluence, ϕ(248 nm). The number of laser pulses was $N = 500$. The ambient medium was air.

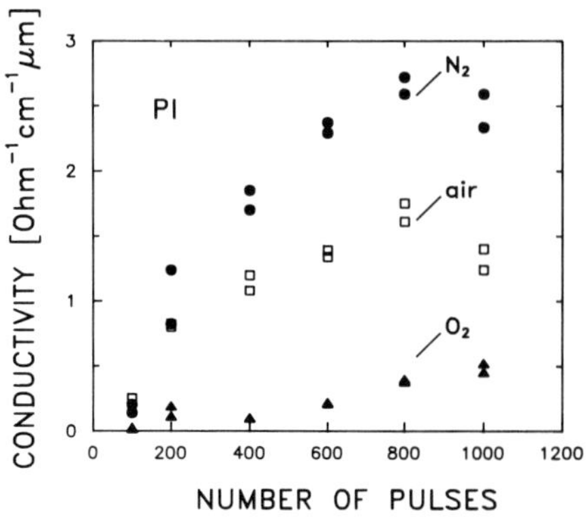

Fig. 4. Conductivity of PI foils versus the number of pulses for different atmospheres. The gas pressure was always 1 bar. 248 nm KrF laser radiation and a fluence of 50 mJ/cm^2 have been employed.

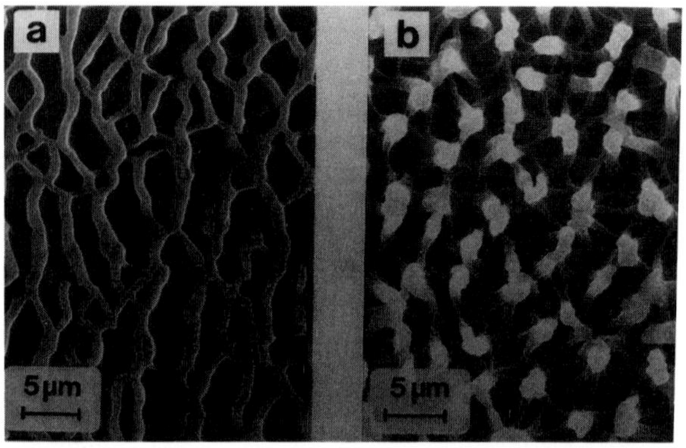

Fig. 5. SEM pictures of wall- and nap-type structures observed on PET foils after ablation with 248 nm KrF-laser radiation. (a) Foil uniaxially stretched before laser ablation (10 pulses, ϕ(248 nm) ≈ 60 mJ/cm^2). (b) Foil biaxially stretched before ablation (10 pulses, ϕ(248 nm) ≈ 90 mJ/cm^2).

ity is shown in fig. 4. The gas pressure was in each case 1 bar and the laser fluence 50 mJ/cm^2. The figure shows that the conductivity with respect to air is suppressed by pure O_2 while it is enhanced by pure N_2. Clearly, this is expected with this process.

4. Structure formation

It is well known that laser irradiation of untreated PET foils results in the formation of wall-type or nap-type structures as shown in figs. 5a and 5b. This structure formation has been related to the presence of spherulites [6], to selective etching of amorphous and crystalline material [7], and to convection [8]. In a recent paper we have demonstrated that the formation of these structures can uniquely be assigned to the mechanical and thermal pretreatment of the PET foils [9]. The structures are neither related to the presence of spherulites nor to selective etching. In order to test the effect of convection, we have extended our recent experiments to PI [10]. In contrast to PET, PI shows no molten phase, at least under quasi-stationary conditions. Nevertheless, uniaxial and biaxial stretching of the foils prior to laser ablation results in wall-type and nap-type structures, respectively. For these reasons we believe that structure formation is not related to a molten surface layer.

The mean distance between walls or naps, d_A, increases with the number of laser pulses, N. This is shown in fig. 6 for the case of naps observed on PET after 248 nm KrF-laser irradiation with a fluence of 80 mJ/cm^2. With $N < 100$ pulses, the measured dependence for both wall-type and nap-type structures on PET can be fitted by

$$d_A(N) = C_1 \ln N + C_2.$$

This relation holds for fluences $\phi \leq 180$ mJ/cm^2 and different magnitudes of frozen-in stresses.

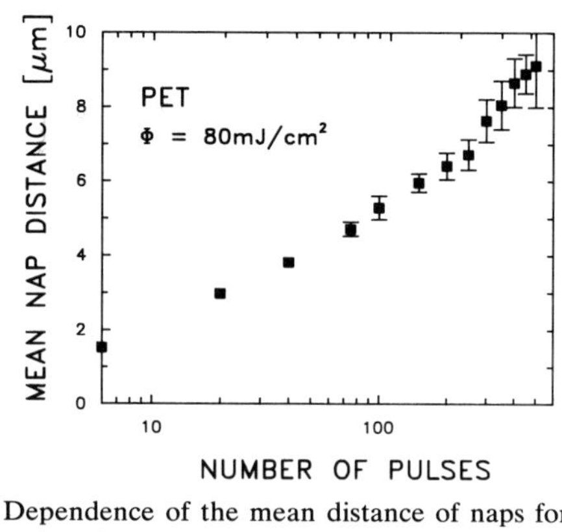

Fig. 6. Dependence of the mean distance of naps formed on PET as a function of the number of laser pulses. 248 nm laser radiation and a fluence of 80 mJ/cm^2 have been employed.

The same dependence was reported for fibres in ref. [11]. With pulse numbers $N > 100$, however, the distance d_A increases about linearly with N. The results can tentatively be described by a stress release model [12].

Acknowledgement

We wish to thank the Jubiläumsfonds der Österreichischen Nationalbank for financial support.

References

[1] D. Bäuerle, Chemical Processing with Lasers, Vol. 1 of Springer Series in Materials Science (Springer, Berlin, 1986).

[2] R. Srinivasan, in: Interfaces under Laser Irradiation, NATO ASI Series, Eds. L.D. Laude, D. Bäuerle and M. Wautelet (Nijhoff, Dordrecht, 1987) p. 359.

[3] Y. De Puydt, P. Bertrand and P. Lutgen, Surf. Interface Anal. 12 (1988) 486.

[4] M. Schuhmann, A. Sauerbrey and M.C. Smayling, Appl. Phys. Lett. 58 (1991) 428.

[5] T.H. Kefer, E. Arenholz and J. Heitz, unpublished data; determination of Φ_{th} by the method described in: J.H. Brannon et al., J. Appl. Phys. 58 (1985) 2036.

[6] J.E. Andrew, P.E. Dyer, D. Forster and P.H. Key, Appl. Phys. Lett. 43 (1983) 717.

[7] Y. Novis, J.J. Pireaux, A. Brenzini, E. Petit, R. Caudano, P. Lutgen, G. Feyder and S. Lazare, J. Appl. Phys. 64 (1988) 365.

[8] T. Bahners and E. Schollmeyer, J. Appl. Phys. 66 (1989) 1884.

[9] E. Arenholz, V. Svorcik, T. Kefer, J. Heitz and D. Bäuerle, Appl. Phys. A 53 (1991) 330.

[10] E. Arenholz, M. Wagner, J. Heitz and D. Bäuerle, Appl. Phys. A 55 (1992) 119.

[11] W. Kerting and E. Schollmeyer, Angew. Makromol. Chem. 193 (1991) 169.

[12] E. Arenholz, J. Heitz and D. Bäuerle, to be published.

Applied Surface Science 69 (1993) 20–26
North-Holland

applied
surface science

Mechanisms of synchrotron radiation-excited etching reactions of semiconductor materials

Haruhiko Ohashi, Akira Yoshida, Kiyohiko Tabayashi and Kosuke Shobatake
Institute for Molecular Science, Myodaiji, Okazaki 444, Japan

Received 2 June 1992; accepted for publication 30 October 1992

Synchrotron radiation-excited etching of semiconductor materials including SiO_2 and SiC is studied to clarify the mechanisms of the etching processes. Synchrotron radiation (SR) was shone upon SiO_2 and SiC surfaces in SF_6 atmosphere, and the etch rates against SF_6 pressure and the cross-sectional profiles of the etched surfaces for different light spectra were measured. From the obtained results it is concluded that excitation of the surface by SR is essential and is the rate determining step when the etchant pressure is high enough. We have proposed a simple reaction model for the SR-excited etching of the insulator surface that the excitation of the outermost layer is essential in the etching reaction.

1. Introduction

Recently, studies of synchrotron radiation-excited surface reaction processes such as etching reactions of semiconductor materials [1–8] and thin film formation using photochemical vapor deposition (photo-CVD) [7,9–11] have attracted considerable attention in the hope that they would provide us with unique features owing to many advantages of using SR; (1) cleanliness, (2) damagelessness, (3) lowering of the process temperatures, and (4) material selectivity owing to strong absorption cross sections in the extreme ultraviolet region.

Surface science aspects of etching reactions have recently been reviewed by Winters and Coburn [12], who have shown that much knowledge has been accumulated but SR-excited etching has not been well studied. In the last five years we have been working on the study of the SR-excited etching processes of semiconductor material surfaces [2,5,8]. In order to control the reactions by photoexcitation it is essential to clarify the mechanisms of the photochemical surface reactions. Since the photoexcitation process plays an important role in the SR-excited etching and the etch rate is expected to strongly depend upon wavelength, we have studied the wavelength dependence upon etching rates [5]. The ideal experiment should be done by using monochromatized light. Because of the experimental limitations, however, we have taken two approaches to meet the condition; (1) the use of quasi-monochromatic undulator radiation with or without a thin metal filter or (2) to carry out the pressure dependence experiment, by which the spectrum of the irradiated light upon the surface can be varied by changing the concentration of the etchant layer between the light source and the surface to be etched. A reaction model in which surface excitation is essential in SR-etching has been proposed to explain the absolute magnitude of the etch rates [5].

2. Experimental

2.1. Spectra of synchrotron radiation light

The used SR came from the 750 MeV electron storage ring of the ultraviolet synchrotron orbital radiation facility at IMS (UVSOR). Two kinds of synchrotron radiation sources were used for the present experiments. First, fig. 1 illustrates the

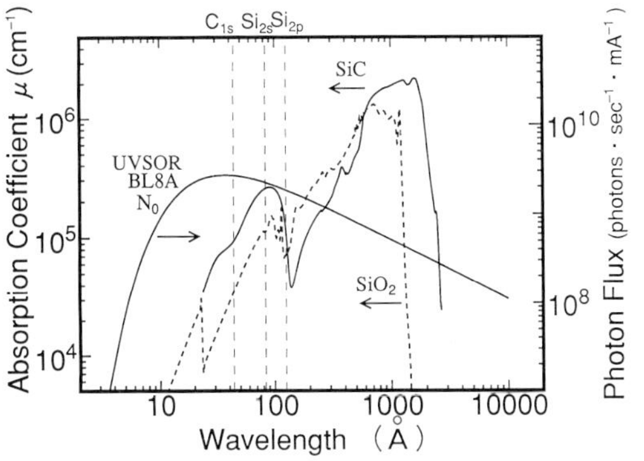

Fig. 1. Theoretical spectrum of SR from the bending magnet at BL8A of UVSOR facility at IMS when shone upon the substrate at 5.85 m from the source point through a channel of 3 mm diameter. Also shown are the absorption coefficients of SiO_2 [13] and SiC [14].

theoretical spectrum [13] of SR irradiated upon the substrate surface of 3 mm in diameter located 5.85 m from the source point of the bending magnet on the beamline BL8A. Fig. 1 also shows the absorption spectra of SiO_2 and SiC [14]. The photon flux integrated over the wavelength from 4 Å to 1 μm is 8.6×10^{12} photons/mA · s.

Second, pseudo-monochromatic undulator radiation from beamline BL3A1 was used, whose theoretical spectral profiles are illustrated in fig. 2 for the two gap lengths, $L_G = 70$ and 33 mm. The spectrum of the undulator light for $L_G = 70$ mm essentially contains the first order light

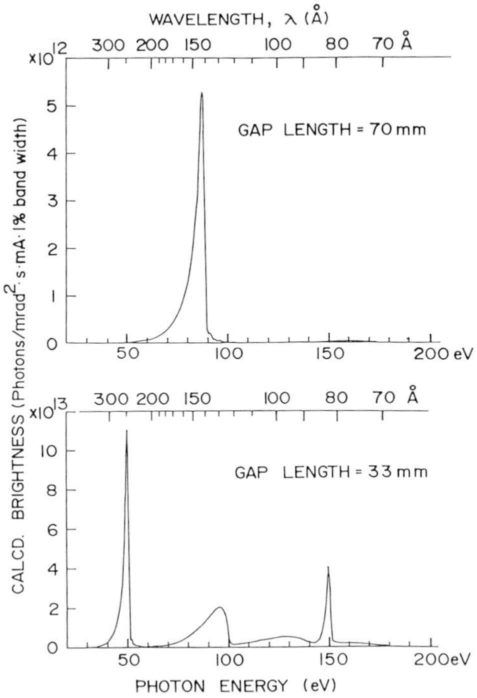

Fig. 2. Calculated spectra of undulator radiation for two gap lengths $L_G = 70$ and 33 mm.

peaked at 86.7 eV (143 Å). For $L_G = 33$ mm the theoretical undulator light consists of first ($\lambda_1 =$ 251 Å) to fourth order light below 200 eV. The experiments for gap length $L_G = 33$ mm described here were carried out using a thin (1060

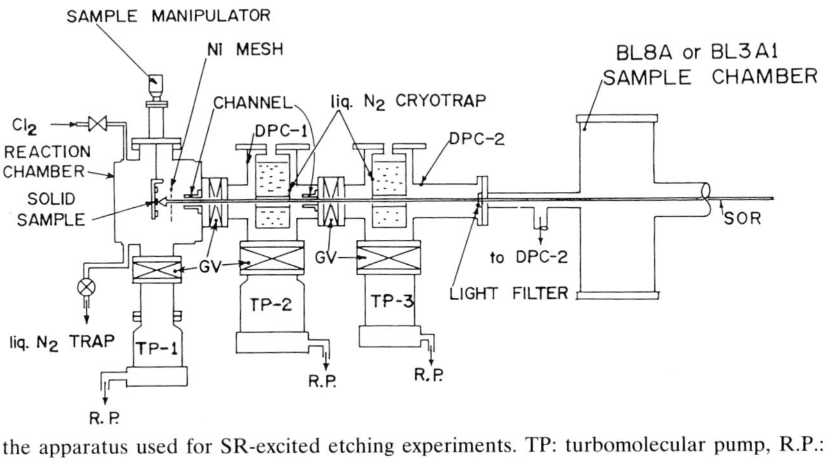

Fig. 3. Schematic of the apparatus used for SR-excited etching experiments. TP: turbomolecular pump, R.P.: rotary pump, DPC: differential pumping chamber.

Å thick) Al filter and the spectrum of the irradiated light essentially consists of first order light peaked at 251 Å. The light intensities measured using an Al diode were determined as 1.47×10^{14} photons/s at 100 mA of stored current for $L_G = 70$ mm ($\lambda_1 = 143$ Å) and 1.21×10^{14} photons/s at 100 mA of stored current for $L_G = 33$ mm with an Al filter ($\lambda_1 = 251$ Å), respectively. The undulator light was irradiated upon an elliptical spot on the SiO_2 surface with axes of 0.8 and 0.6 mm for 143 Å light and 2.0 and 1.2 mm for 251 Å, respectively.

2.2. Apparatus and experimental procedures

The irradiation experiments were carried out using an apparatus schematically shown in fig. 3. After going through five stages of differential pumping, SR enters the reaction chamber through a channel (3 mm ∅ and 30 mm long) and irradiated upon the substrate perpendicularly to the surface. The sample chamber is filled with an etchant gas at pressures ranging from 1.0×10^{-3} to 1.0 Torr.

In the SR-excited etching reactions, SR plays two roles: (i) production of radical species such as F atoms which attack the sample surface, and (ii) excitation of surface atoms, if the process is important. In order to minimize the first effect, experiments were also carried out by flowing atoms formed by microwave discharge.

The SiO_2 sample used was a commercially obtained, thermally oxidized SiO_2 film (about 8000 Å thick) on crystalline Si. The SiC sample used was β-SiC(220) obtained from Nippon Pillar Packing Co. [15]. The cross-sectional profiles of the etched surfaces of the thin SiO_2 were obtained by measuring the interference pattern as a function of wavelength using a film thickness measurement system (Nikon Model NFM-5). The etched depth of the β-SiC surface was measured using a stylus monitor. The etch rate defined in the present paper is the etched depth (in Å) divided by the accumulated photon dose in units of stored current (in mA) per minute (Å/mA · min) or the number of SiO_2 or SiC moieties removed from the $(3\ \text{mm})^2$ area per mA per min.

3. Results and discussion

3.1. Etching of the SiO_2 surface with SF_6

SR-excited etching of SiO_2 with SF_6 was first carried out by Urisu et al. [4] who have found that surface excitation by SR is important in overcoming the thermal barrier (the apparent activation energy from thermal experiment $F + SiO_2$ is determined as about 0.20 eV [16]). The information on the wavelength dependences of the etch rates plays an important role in clarifying the etching reaction mechanism. We have carried out the SR-excited etching reaction of SiO_2 with SF_6 using both pseudo-monochromatic undulator light at 143 and 251 Å as well as the SR from the bending magnet [5]. In order to make it easier to discuss a reaction model the results on the wavelength dependence of the etch rate will be described.

3.1.1. Undulator radiation experiments (at BL3A1)

Fig. 4 illustrates the SF_6 pressure dependences of etch rates for pseudo-monochromatic light at 143 and 251 Å. The results show that, at the two wavelengths studied, the etch rate quickly decreases with increasing SF_6 pressure in the pressure range of 0.016 to 0.5 Torr. The curves shown in fig. 4 represent the calculated transparent light intensity impinging upon the SiO_2 surface through

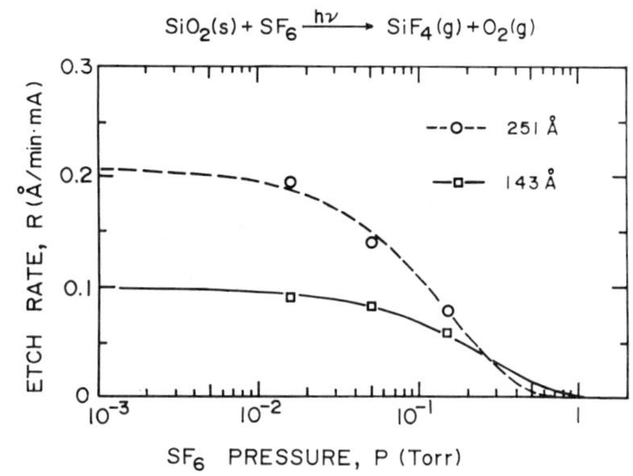

Fig. 4. SF_6 pressure dependence of the SiO_2 etch rate obtained for two irradiation wavelengths, 251 Å (○------○) and 143 Å (□——□). The curves are calculated transparent light intensities fitted to the experiment.

an SF_6 layer between the channel exit and the sample surface ($L = 4.9$ cm long). Although not shown, in the low pressure region below 0.016 Torr, the etch rate decreases with SF_6 pressure, since it is limited by the F atom supply-rate. Etching experiments have been also carried out by flowing microwave discharged SF_6 gas to see whether the etch rate increases drastically or not by changing the F concentration, but it has been found that the rates increase by factors of at most two. The results indicate that the etch rate is not strongly dependent upon the F atom concentration in the high pressure region.

3.1.2. Experiments using SR from the bending magnet (at BL8A)

Synchrotron radiation from beamline BL8A was irradiated upon the SiO_2 sample in the SF_6 atmosphere at room temperature. Fig. 5 illustrates the etch rate against SF_6 pressure along with the data obtained by Takahashi et al. [17]. The distance between the defining channel exit and the sample surface was $L = 1.0$ cm. As shown in fig. 5 the etch rate decreases as the pressure is lowered below 0.1 Torr and it also gradually decreases with pressure in the pressure region above 0.1 Torr. In the lower pressure region the etch rate is limited by the F atom supply rate by photodissociation and in the higher pressure region it is limited by the number of photons irradiated upon the sample surface.

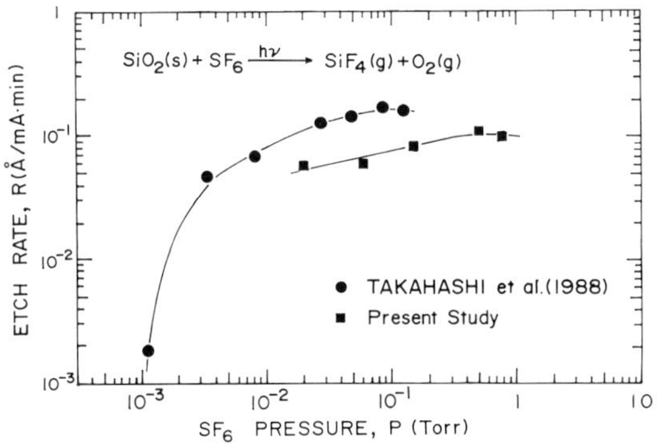

Fig. 5. Etch rate against the SF_6 pressure. Solid squares are present results and solid circles are due to measurements of Takahashi et al. [17]. The curves are drawn to guide the eyes. The distance between the channel exit and the sample surface was $L = 1.0$ cm.

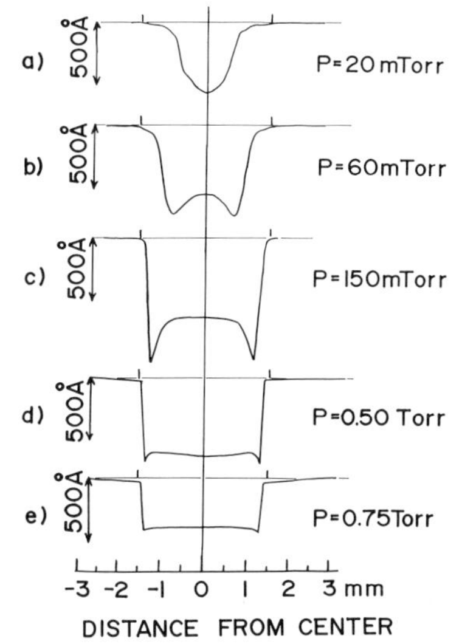

Fig. 6. Cross-sectional profiles of the etched surfaces of SiO_2 with SF_6 for various pressures using SR from the BL8A bending magnet.

Fig. 6 shows the cross-sectional profiles of the etched surface for different SF_6 pressures (0.02 to 0.75 Torr). The profiles for SF_6 pressures at 0.5 and 0.75 Torr show that the irradiated area is strongly etched and its edge is only slightly etched. These results indicate that surface excitation by SR is very important for the etching reaction to proceed. In the intermediate pressure region the central part is less etched than the rim region. Why this occurs has not been well understood.

Now we compare the absolute etch rates obtained for different undulator lights. The intensity for 143 Å (1.47×10^{14} photons/100 mA) is comparable to that for 251 Å light (1.21×10^{14} photons/s 100 mA). However, the maximum etch rate for 143 Å light is about half as large as that for 251 Å light. This fact can be explained by the following arguments: since we do not exactly know which kinds of species on the surface absorb SR to stimulate the etching reaction, we simply assume that the surface species absorb the light with the absorption coefficient μ of glassy SiO_2.

The μ values are estimated as 3.3×10^5 and 1.29×10^5 cm^{-1} at 251 and 143 Å, respectively.

3.1.3. Reaction model

Suppose that the etch rate is proportional to both (i) the number of photons impinging upon the surface and (ii) the absorption coefficient μ estimated above, then the ratio of the maximum etch rate at 0.016 Torr for 251 Å light to that for 143 Å light radiation is 2.1. This value is in excellent agreement with the observed ratio 2.2.

It is thus concluded that the etch rate is proportional to (i) the number of photons impinging upon the surface and (ii) the absorption coefficient of surface species, as long as the concentration of F atoms formed by photodissociation is high enough such that the surface is saturated by F atoms. The thickness of the effective absorption layer, d, which is effective to induce etching reaction, was estimated as 4.1 Å, *assuming that the (desorption) reaction is induced with a unit probability when the SiO_2 layer absorbs an undulator photon* and also taking into account the above-mentioned photon intensity, etch rate, and irradiated area. This value ($d = 4.1$ Å) agrees well with the size of one SiO_2 moiety in the glass.

A reaction model has been proposed based on the results for the SR-excited etching of the SiO_2 surface at two wavelength 143 and 251 Å. Excitation at these wavelengths is rather soft and valence electrons are excited (Menzel–Gomer–Redhead mechanism [18]). However, excitation at shorter wavelengths than 115 Å is expected to excite inner core electrons and thus more efficient in promoting etching reaction than valence excitation. The modified version of the reaction model proposed above was applied to explain the pressure dependence of the etch rate obtained for SR from the bending magnet at BL8A. The analyzed data were the ones obtained for the SF_6 layer of $L = 4.4$ cm, as shown in fig. 7. The unit of etch rate (Å/mA · min) was converted to the number of SiO_2 moieties removed from the irradiated area, $(0.5\ \text{mrad})^2$. I_s (solid curve) in fig. 7 is the theoretical total number of photons impinging upon the sample surface as a function of SF_6 pressure, which is about 200 times larger than the observed etch rates. This means that the quantum yield defined by the number of SiO_2 moieties removed per total number of photons irradiated is about 0.5%. Since in the present experiment, however, "white light" is irradiated, the number of photons absorbed by the outermost layer of SiO_2 with a thickness $d = 4.1$ Å for the photons for valence excitation (1 $\mu m > \lambda > 115$ Å) and for the photons for inner core excitation (115 Å $> \lambda > 4$ Å) have been estimated using the absorption spectra of SF_6 and SiO_2 as well as the thickness of the gas absorption layer of $L = 4.4$ cm assuming the outermost SiO_2 layer of $d = 4.1$ Å can be applied for another wavelength. These

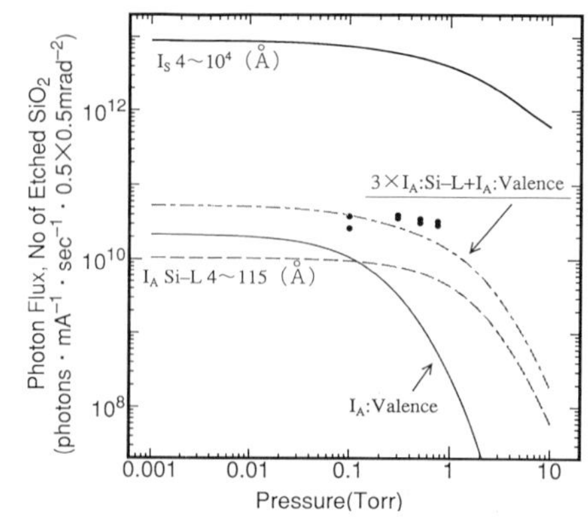

Fig. 7. Etch rate of SiO_2 versus SF_6 pressure. The curves are the calculated photon flux irradiated upon the sample (I_s), the number of photons absorbed by the outermost layer (thickness $d = 4.1$ Å) in the wavelength regions from 4 to 10^4 Å (I_A: Valence) and from 4 to 115 Å (I_A Si–L 4 ~ 115 Å).

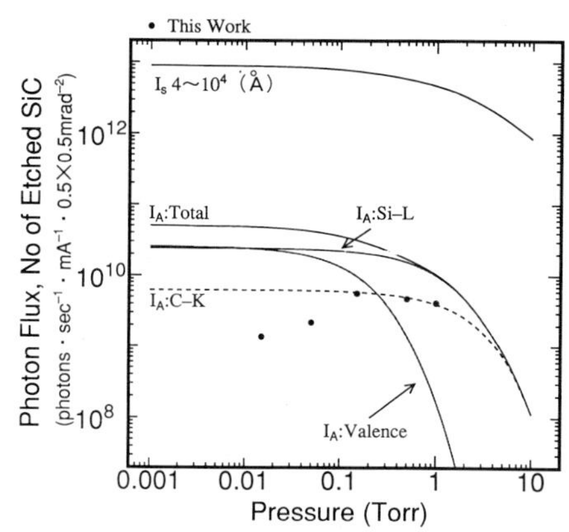

Fig. 8. Etch rate of β-SiC(220) against SF_6 pressure. Solid circles represent experiment. See the text for explanations.

curves show that valence excitation alone cannot explain the pressure dependence of the etch rate nor the inner core excitation alone. If one assumes that the contributions of both valence excitation and three times inner core excitation, the data can be fairly well explained. The fact that the inner core excitation is three times more effective than the valence excitation may be due to the fact that it is followed by an Auger process and then more violent movement of atoms excited (Knotek–Feibelman mechanism [19]). Therefore, the effective thickness of the outermost layer is thicker for inner core excitation than the valence excitation.

At the moment we cannot say if neutral or ionic species are desorbed immediately after the surface excitation. However, considering the fact that the apparent thermal barrier for the etching reaction is low (about 0.20 eV [16]), surface excitation prepares the surface state which easily undergoes thermal reaction with F radicals formed by photodissociation of the SF_6 molecule in the gaseous phase. We speculate that the inner core excitation at wavelengths below 115 Å simply affect the atomic motion on the surface more efficiently than the valence excitation.

3.2. Etching of the β-SiC(220) surface with SF_6

SiC is an important semiconductor material with a wide band gap, and is also expected to be used as a thermal resisting optical material for high brightness synchrotron radiation sources to be constructed in the future. However, since SiC is a very hard material and chemically very stable, its fabrication is difficult and thus synchrotron radiation was used to etch it.

The SR-excited etching of the β-SiC(220) surface with a resist pattern of 1.0 μm line width (500 lines/mm) was carried out at room temperature using SF_6 gas as an etchant. Although not shown in the present paper a micrograph of the etched area clearly shows that the irradiated area is strongly etched, but the edge of the irradiated area is also etched. Since the contrast between the irradiated and the non-irradiated area is large, it is concluded that the surface excitation by SR is also important in the etching process. The etch rate is plotted against SF_6 pressure as illustrated in fig. 7 which also shows the theoretical total photon numbers impinging upon the substrate surface, I_s. One finds that the etch rate in units of the number of SiC moieties removed per mA and per second absorbed is 1200 times smaller than I_s, the numbers of photons impinging upon the surface and thus about six times smaller than the etch rate of SiO_2.

One finds that the etch rate increases with pressure below 0.15 Torr, peaks at around 0.15 Torr, and then decreases at higher pressures. From these results one may conclude that in the low pressure region (below 0.15 Torr) the etch rate is limited by the F atom supply. A similar trend was observed in the SR-excited etching of the SiO_2 surface with SF_6 gas. However, at pressures higher than 0.15 Torr the rate seems to be limited by the photon intensity irradiated upon surface.

In order to explain the approximate magnitude of the etching rate similar calculations have been made to that of SiO_2 etching. It is noted that carbon, especially diamond, is hard to etch with SF_6 gas. Therefore, the low etch rate for the SiC surface may be due to the difficulty in etching of C atoms. It is thus expected that the inner core excitation of SiC, in particular C(1s), is expected to promote the etching reaction. It is thus assumed that excitation of the carbon 1s electron is the rate determining step, that is, excitation by the photons in the wavelength region between 23.6 and 44.7 Å is essential in promoting the etching reaction.

In order to fit the experimental etch rates in the higher pressure region (above 0.15 Torr) to the calculated values represented by the dashed curve, the thickness of the outermost layer was determined as $d = 4.36$ Å [20], which is in exact agreement with the unit size of the SiC cell. I_A's are also calculated assuming that the reaction proceeds when the outermost layer ($d = 4.36$ Å) absorbs photons in the wavelength region (i) from 4 Å to 1 μm (represented by (I_A: total)), (ii) from 115 Å to 1 μm (represented by (I_A: valence)), and (iii) from 4 to 115 Å. The pressure dependence of the calculated I_A for the valence excitation does not fit to the experiment and thus the inner core excitation seems to be essential in the SR-excited etching reaction of the SiC surface. Although the apparent activation energy for the etching reaction is not known, we believe that, except for radical formation by photodissociation, surface photoexcitation simply assists in a process, i.e. overcoming a thermal barrier, out of many steps of the thermal etching reaction, such as adsorption of F atoms, diffusion of adsorbed F atoms under the outermost layer, and desorption of the product molecules. This is probably the reason why the observed rates of the SR-excited etching reactions are roughly estimated from the photoexcitation probability of the surface layer.

Acknowledgements

The present research was partially funded by a Grant-in-Aid on Priority Area Research "Photo-Excited Process" from the Ministry of Education, Science and Culture of Japan. The authors thank Professor K. Kitamura of Photon factory, KEK for providing us with a program to calculate the undulator radiation spectra used for the present experiment. The authors acknowledge a great help of K. Sano, M. Koeda and T. Nagano of Shimadzu Corporation for supplying SiC specimens and the measurements of the cross-sectional profiles of the etched SiC surfaces. Thanks are due to Professor E. Ishiguro of Osaka City University for his suggestion of SiC experiments.

References

[1] H. Kyuragi and T. Urisu, Appl. Phys. Lett. 50 (1987) 1254.
[2] N. Hayasaka, A. Hiraya and K. Shobatake, Jpn. J. Appl. Phys. 26 (1987) L1110.
[3] T. Urisu and H. Kyuragi, J. Vac. Sci. Technol. B 5 (1987) 1436.
[4] T. Urisu, H. Kyuragi, Y. Utsumi, J. Takahashi and M. Kitamura, Rev. Sci. Instrum. 60 (1989) 2157.
[5] K. Shobatake, H. Ohashi, K. Fukui, A. Hiraya, N. Hayasaka, H. Okano, A. Yoshida and H. Kume, Appl. Phys. Lett. 56 (1990) 2189.
[6] H. Akazawa, Y. Utsumi, J. Takahashi and T. Urisu, Appl. Phys. Lett. 57 (1990) 2302.
[7] See references in: T. Urisu, J. Takahashi, Y. Utsumi and H. Akazawa, Appl. Organomet. Chem. 5 (1991) 229.
[8] K. Shobatake, J. Soc. Syn. Radiat. Res. 3 (1990) 27.
[9] H. Kyuragi and T. Urisu, J. Appl. Phys. 61 (1987) 2035.
[10] H. Hochst and M.A. Engelhardt, J. Vac. Sci. Technol. B 8 (1990) 686.
[11] H. Ohashi, K. Inoue, Y. Saito, A. Yoshida, H. Ogawa and K. Shobatake, Appl. Phys. Lett. 55 (1989) 1644.
[12] H.F. Winters and J.W. Coburn, Surf. Sci. Rep. 14 (1992) 161.
[13] H. Winick, in: Synchrotron Radiation Research, Eds. H. Winick and S. Doniach (Plenum, New York, 1980) p. 11.
[14] H.R. Philipp, Handbook of Optical Constants of Solids, Ed. E.D. Polik (Academic Press, New York, 1985) pp. 749, 587;
D.L. Windt, W.C. Cash, Jr., M. Scott, P. Arendt, B. Newman, P.F. Fisher, A.B. Swartzlander, P.Z. Takacs and J.M. Pinneo, Appl. Opt. 27 (1988) 279.
[15] Nippon Pilar Packing Co., 11-48, 2-Chome, Nonaka Minami, Yodogawa-ku, Osaka 532, Japan.
[16] D.L. Flamm, V.M. Donnelly and J.A. Mucha, J. Appl. Phys. 52 (1981) 3633.
[17] J. Takahashi, Y. Utsumi and T. Urisu, Proceedings of 1988 Conference on Solid State Devices and Materials, Tokyo (Publication Office, Business Center for Academic Societies of Japan, Tokyo, 1988) p. 73.
[18] D. Menzel and R.J. Gomer, J. Chem. Phys. 41 (1964) 3311; P.A. Redhead, J. Can. Phys. 42 (1964) 886.
[19] M.L. Knotek and P.J. Feibelman, Phys. Rev. Lett. 40 (1978) 964.
[20] A. Taylor and R.M. Jones, in: Silicon Carbide, Eds. J.R. O'Conners and J. Smittens (Pergamon, New York, 1960) p. 147.

Applied Surface Science 69 (1993) 27–30
North-Holland

applied
surface science

Temperature dependence of synchrotron radiation induced growth initiation in the molecular beam chemical vapor deposition of Al on a SiO_2 surface

Fumihiko Uesugi and Iwao Nishiyama

Opto-Electronics Research Laboratories, NEC Corporation, 34 Miyukigaoka, Tsukuba, Ibaraki 305, Japan

Received 2 June 1992; accepted for publication 30 October 1992

Synchrotron radiation (SR) irradiation effect in the Al thermal growth on a SiO_2 surface is studied by using a molecular-beam chemical vapor deposition (MB-CVD) technique. Al growth is observed only on the SR-irradiated area at 200°C. The substrate temperature dependence of the SR-assisted Al growth rate is found to be non-monotonous: it rises from 150°C to reach a peak at 250°C, then falls to reach a minimum at 300°C, and rises again over 300°C. The origins of this non-monotonous temperature dependence are investigated by chemical shifts analysis in Auger electron spectroscopy (AES). We find that the peculiar temperature dependence is attributed to the change of the chemical bonding nature of the initiation layers formed by SR irradiation.

1. Introduction

Aluminum is one of the important metallization materials in the LSI circuit. The Al growth has been studied extensively using the CVD technique, since this technique has the advantages of a good step coverage and selective growth. It already demonstrated a high purity and good selective Al growth on Si/SiO_2, and a blanket CVD on SiO_2 using dimethylaluminum hydride (DMAH) [1–3]. Selecting the growth mode, the selective CVD or the blanket one, the surface preparation plays a very important role. We tried to control the surface condition by SR irradiation and making a spatial selective CVD of Al. Up to this study, we have found two characteristic SR irradiation effects: the suppression effect observed on a clean Si surface [4,5], and the initiation effect on a SiO_2 surface [5,6].

In the case of a clean Si surface, thermal CVD is possible near 200°C and pure Al growth is confirmed by Auger electron spectroscopy (AES). We found that this thermal CVD reaction was suppressed almost completely by SR irradiation. By using this suppression effect, we succeeded in making inverse projection patterning. On the other hand, pure thermal CVD is not possible on a SiO_2 surface near 200°C. However, Al growth was found to be initiated by the SR irradiation. Consequently, direct normal projection patterning of Al was successfully demonstrated.

The cause of these two particular SR irradiation effects, suppression and initiation, has been studied by AES. As a result, the continuance of these effects after stopping of the irradiation was confirmed [4,6]. Then, we have revealed that these SR irradiation effects were caused by the modified layers with an atomic level thickness formed from DMAH by the SR photochemical effects [5]. In the case of the Si surface, an aluminium carbide (AlC) layer inhibits the thermal reaction. On the other hand, metallic Al initiates the thermal reaction on the SiO_2 surface.

These reaction control effects by SR irradiation are strongly dependent on the substrate temperature. Therefore, the temperature dependence of the initiation effect on a SiO_2 surface was examined in detail. In this paper, the temper-

0169-4332/93/$06.00

ature dependence of the Al growth rate and its origin will be discussed on the basis of AES and chemical shifts analysis of both the initiation layer and the thermally grown films.

2. Experiment

Experiments were done using the SR beam line, BL-9B constructed in a synchrotron radiation facility of the Photon Factory at the National Laboratory for High Energy Physics. The details of the experimental apparatus are described in ref. [4]. A brief experimental procedure is mentioned. Fig. 1 shows a schematic diagram of the experimental set-up used for SR irradiation. The photon energy of the used SR covered the full range under 300 eV. Its beam flux was estimated to be on the order of 10^{16} photons/cm^2 s. The incident angle of SR to the substrate was 25°. The DMAH was supplied as the supersonic molecular beam seeded in He gas. The beam flux was on the order of 10^{14} molecules/cm^2 s. Its beam was exposed to the substrate with an incident angle of 65°. The substrate temperature was ranging from 100 to 600°C.

The 200 nm SiO_2 film was made by thermal oxidation of a Si(100) surface. The SiO_2 surface was cleaned chemically by dipping in dilute HF ($HF/H_2O = 1.6\%$) solution. After a 15 min rinsing in de-ionized water, it was cleaned thermally for 10 min at 1000°C in the vacuum chamber under 2×10^{-10} Torr just before Al CVD.

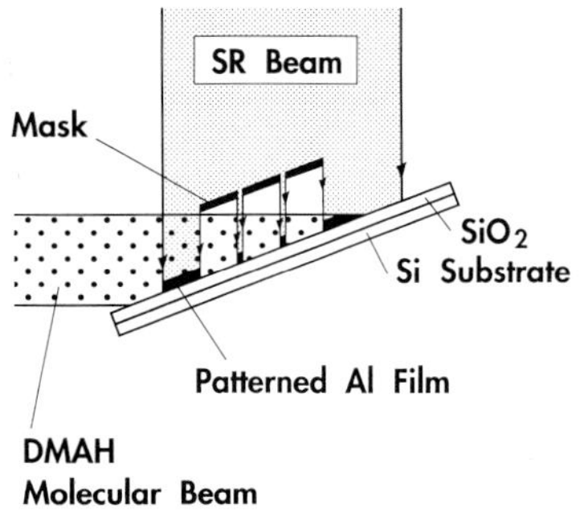

Fig. 1. Schematic diagram of SR-induced direct patterning of an Al film.

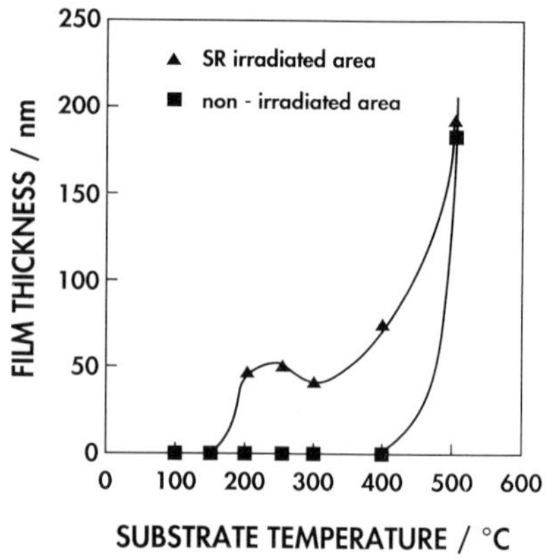

Fig. 2. Temperature dependence of the Al film thickness on the SR-irradiated area and non-irradiated area.

The procedure of the Al direct patterning consists of two sequential steps: the first one is about 10 min SR irradiation in CVD during which an initiation layer of atomic layer level thickness is formed on SiO_2. The next one is 80 min of thermal CVD without SR irradiation. Al films on the order of 100 nm thickness are grown during this step. The 10 min irradiation time in the first step was optimized in the previous experiment.

The thickness of films grown by thermal CVD was measured with the step-height meter, α-step 200 (Tencor Co.). The composition of the initiation layer and the CVD films was analyzed by in situ Auger electron spectroscopy (Ulvac-Phi Co.). The chemical bonding nature was evaluated by chemical shifts analyses of Auger electron spectra.

3. Results and discussion

Fig. 2 shows the temperature dependence of the growth rate both on the SR irradiated and non-irradiated SiO_2 surfaces. The growth rate on the SR-irradiated surface shows the up-and-down temperature dependence: it rises from 150°C to reach a peak at 250°C, then falls to reach a minimum at 300°C and rises again over 300°C. On the other hand, the growth rate on the non-irradiated SiO_2 surface increases monotonously above 400°C.

The increasing and decreasing temperature dependence of the SR assisted CVD, especially the negative temperature dependence observed in the range from 250 to 300°C, is not explainable with the simple Arrhenius equation. The change in activation energy for a CVD reaction is necessary to be considered for such a non-monotonous temperature dependence. As the cause of the change in activation energy, we postulated the variation of the surface composition with the temperature. In this case, we must consider the composition changes of the initiation layer and the thermally grown film on its layer. Therefore, the compositions of them were measured by in situ AES.

The initiation layer consists of Al and carbon (C) with a constant ratio of 6/4 over the whole temperature range. On the other hand, the thermally grown film consisted of pure Al below 300°C, but above 400°C its composition changed to Al and C. However, no composition change was observed corresponding to the decrease of the growth rate from 250 to 300°C, since the pure Al was observed in this temperature range. The cause of the negative temperature dependence could not be found in the composition changes both in the initiation layer and the thermally grown film.

Since no change was observed in the composition, we tried to measure the chemical bonding nature in order to search the cause of the up-and-down temperature dependence of the growth rate. However, below 300°C no change was observed in the chemical shifts of thermally grown films: the Al film was pure and metallic. Therefore, we measured the variation in the chemical shifts of the initiation layers with the temperature.

Figs. 3a and 3b show the typical AES spectra for the initiation layers obtained at 200 and 500°C, respectively. At the former temperature, clear spatial selectivity induced by SR irradiation was observed; no selectivity was observed at the latter temperature. As shown in fig. 3a, Al LVV has a peak at 67.5 eV near the pure metallic Al state and Si LVV has a peak at the 92 eV pure Si elementary state [7]. On the other hand, fig. 3b shows that Al LVV has a peak at the 64 eV

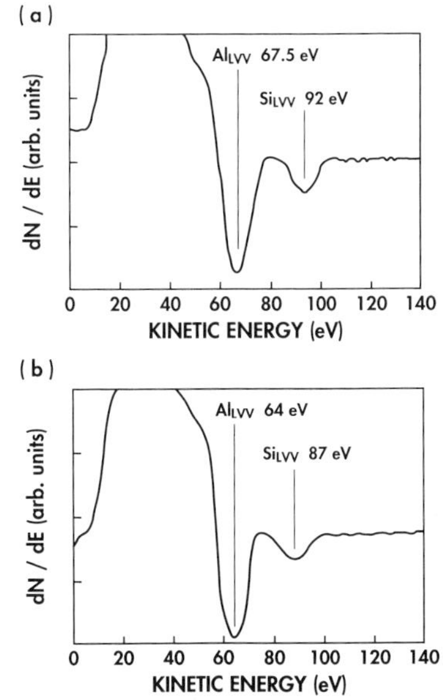

Fig. 3. AES spectra of the initiation layers obtained at (a) 200 and (b) 500°C.

carbide state and Si LVV has a peak at the 87 eV carbide state [8,9]. In this way, the chemical bonding states of Al and Si were found to change with the temperature. So, we measured the temperature dependence of the chemical shifts in detail, which is shown in fig. 4.

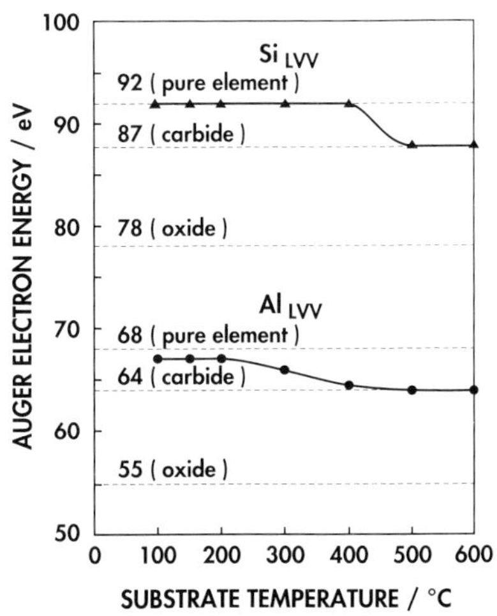

Fig. 4. Temperature dependence of the chemical shifts of the initiation layers.

As shown in the figure, Al LVV has a value of 67.5 eV below 200°C, decreases gradually near 300°C, and then reaches the constant value of 64 eV. On the other hand, Si LVV remains constant at 92 eV below 400°C, and decreases to 87 eV at 500°C. When comparing these temperature dependences of chemical shifts with that of the growth rate in fig. 2, two relationships are noticed: (1) the change of the Al LVV chemical shift of the initiation layer corresponds to the decrease of the thermal growth rate occurring at the same temperature range, (2) the change of the Si LVV chemical shift corresponds to the disappearance of the spatial selectivity between the irradiated and non-irradiated area.

From the temperature dependence of the Al LVV chemical shift, the Al chemical bonding state was found to change from the pure elementary state to the carbide state. Since aluminum carbide has a covalent bond structure and the bonds are closed to each other, its reactivity might be lower than that of the metallic aluminum. The suppression effect caused by AlC is found indeed on the Si surface [4,5]. Therefore, we concluded that the decrease of the Al growth rate from 250 to 300°C was caused by the change of the initiation layer's bonding nature of Al. The second increase of the growth rate, above 300°C, is caused by thermal AlC growth on the carbide state initiation layer.

On the other hand, the Si LVV chemical shift of the initiation layer remains, below 400°C, in the pure elementary state. This means that the SiO_2 surface is reduced to elementary Si by the SR-induced reaction as described in a previous paper [5]. In this paper the Si state was found to be changed from the elementary to the carbide state near 400–500°C. This temperature range corresponds to the temperature at which thermal AlC growth begins even on the non-irradiated SiO_2 surface. Therefore, we conclude that the pure thermal reaction forming AlC is essential and conceals the SR irradiation effect over 400°C, which results in the loss of SR-induced spatial selectivity.

4. Summary

The temperature dependence of the SR-induced thermal Al-CVD rate on a SiO_2 surface is studied. The origin of its increasing and decreasing temperature dependence is the change of the Al chemical bonding nature of the initiation layer formed due to the change of the metallic to the carbide state induced by SR irradiation. On the other hand, the cause of the disappearance of the SR-induced spatial selectivity is a high-speed thermal AlC growth reaction both on SR-irradiated and non-irradiated SiO_2 surfaces.

Acknowledgements

The authors are grateful to T. Shinzawa, K. Sugai, Y. Teraoka, Y. Morishige, Y. Seki and B. Haba for useful discussions. They wish to thank K. Asakawa, S. Kishida and K. Kobayashi for their continuous support and encouragement.

References

[1] T. Shinzawa, K. Sugai, S. Kishida and H. Okabayashi, Mater. Res. Soc. Symp. Proc. VLSI V, Tokyo, 1990 (Materials Research Society, Pittsburgh, PA, 1990) p. 377.
[2] K. Tsubouchi, K. Masu, N. Shigeeda, T. Matano, Y. Hiura and N. Mikoshiba, Appl. Phys. Lett. 57 (1991) 1221.
[3] K. Sugai, T. Shinzawa, S. Kishida and H. Okabayashi, Digest of Papers 4th Micro Process Conference, Kanazawa, 1991 (The Japan Society of Applied Physics, Tokyo, 1991) p. 206.
[4] F. Uesugi and I. Nishiyama, Appl. Surf. Sci. 54 (1992) 284.
[5] F. Uesugi and I. Nishiyama, Appl. Surf. Sci. 60/61 (1992) 587.
[6] F. Uesugi and I. Nishiyama, Appl. Surf. Sci. 62 (1992) 151.
[7] L.E. Davis, N.C. Mcdonald, P.W. Palmberg, G.E. Riach and R.E. Weber, Handbook of Auger Electron Spectroscopy (Perkin-Elmer, Eden Prairie, MN, 1978).
[8] P.G. Lurie and J.M. Wilson, Surf. Sci. 65 (1977) 453.
[9] Y. Katayama, K.L.I. Kobayashi and Y. Shiraki, Surf. Sci. 86 (1979) 549.

Applied Surface Science 69 (1993) 31–37
North-Holland

applied
surface science

Large scale excimer laser production of submicron periodic structures on polymer surfaces

Matthias Bolle and Sylvain Lazare

Laboratoire de Photophysique et Photochimie Moléculaire, URA CNRS 348, Université de Bordeaux 1, 351, Cours de la Libération, F-33405 Talence, France

Received 2 June 1992; accepted for publication 30 October 1992

By absorption of a polarized excimer laser beam of low fluence a new class of submicron laser induced structures is produced on various polymer surfaces. Parallel ripples of less than 0.2 μm spacing in the direction of the electric field are obtained with the ArF or KrF radiation when it is strongly absorbed by the polymer. The amplitude of the roughening increases with the number of pulses used, with a maximum obtained after one thousand. Several polymers in thin or thick films of commercial or laboratory origin can be textured in air or in vacuum or both depending on the chemical structure. Fluence is below the ablation threshold and must be chosen in a narrow window, which depends on the polymer and the wavelength used. The obtained relief is observed by SEM, TEM and characterized by ellipsometry. Structures are obtained also by irradiation of thin film (2000 Å) coated on silicon wafers. In many applications, a controlled submicron level roughening of surfaces is desired. Owing to the simplicity of this irradiation procedure large scale surface treatments are possible.

1. Introduction

Topography of surfaces strongly influences their properties. A surface roughness or structure not only increases the surface area and thus gives more space for chemical interaction but also changes tribological and optical properties. A well dimensioned surface structure decreases for example the reflectivity of a surface. Further possible applications can be to facilitate glueing or deposition of surface films in composite materials and to trap or orient molecules or clusters on the surface. In this work we have explored the possibilities of creating a given periodic structure on polymer surfaces by irradiation with the beam of an excimer laser at moderate fluences, that is to say below the ablation threshold.

Recently, using fluences above the ablation threshold, periodic structures on polymers were obtained, which were not caused by a spacial modulation of the irradiating beam as done by Ilcisin and Fedosejevs [1] or Sauerbrey and coworkers [2] or by relaxation of internal stress of the polymer [3,4]. Niino et al. [6] reported self-organizing periodic ripple structures on polyethersulphone occurring with ablative photodecomposition but without polarization of the beam. The ripple spacing that this group found was between 0.7 and 1.2 μm and varied strongly with the applied fluence. Unless an inclination of the sample to the beam the pattern was regular on the scale of some ripples but without a global orientation direction. Dyer and Farley [5] showed that self-organizing periodic structures can accompany photoablative processing of polyimide (PI), polyethersulphone (PES), polyetherketone (PEEK) and poly(ethylene terephthalate) (PET). For perpendicular incidence the ripple spacing is equal to the irradiating wavelength in the ambient and the ripples are running perpendicular to the polarization of the irradiating laser beam. Under an inclination larger than 40° of the sample surface towards the beam also ripples running parallel to the polarization direction occur, but their spacing is larger then the wavelength of the laser.

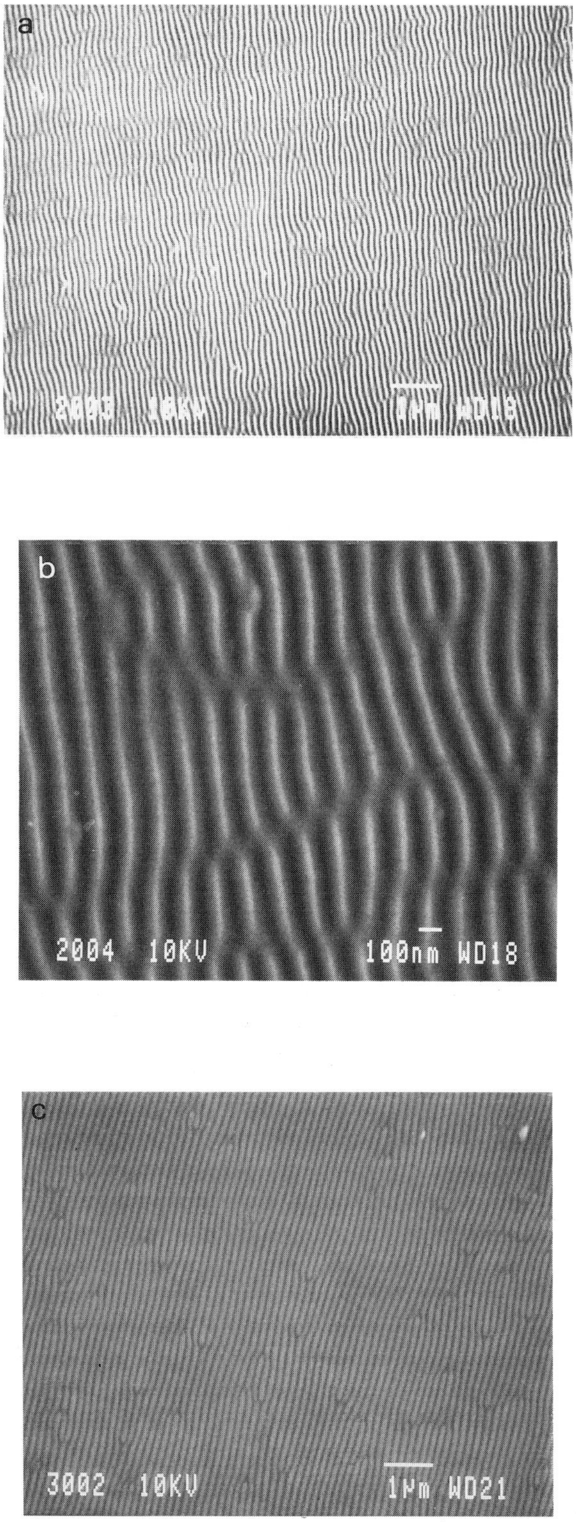

As we reported in an earlier communication [8] submicron laser induced surface structures can be observed even without ablative processes on large surface areas of various polymers. The structures that we observe, differ from conventional ones [7] not just by their smaller periodicity, smaller than the laser wavelength used, but also by its direction which is parallel to the polarization of the laser beam.

In this paper we specify the conditions under which these phenomena occur and we discuss the mechanisms which lead to the ripple growth. Since the size and distance of the ripples are much lower than the wavelength of visible light, scattering methods for fast investigation of the surface fail. We present a way to detect and to classify the periodic structure by monochromatic ellipsometry.

2. Experimental

The polymers studied in this work were commercial films of poly(ethylene terephthalate) PET (Mylar, DuPont) of different thicknesses, as well as spincoated films (900 to 2000 Å) of PET, poly(butylene terephthalate) (PBT), polystyrene (PS) and polycarbonate (PC) on silicon wafers.

Irradiation was done with a Lambda Physik EMG 200EF excimer laser, using its 193 nm ArF and its 248 nm KrF emission. A polarized beam was obtained by using the reflection of a fused silica plate which was inclined by 60°. The degree of polarization that can be reached by this set-up is up to 80%. The samples were mounted onto a positioning stage to allow displacement during irradiation or placed into a small vacuum cham-

Fig. 1. SEM photographs of submicron structured polymer surfaces. (a) PET thin film on a silicium wafer. The film was irradiated with 1000 pulses of 193 nm at a fluence of 3 mJ/cm^2. (b) Close-up of (a). (c) PS spin coated onto silicium wafer. The film was irradiated with 2000 pulses of 193 nm at a fluence of 3.9 mJ/cm^2.

Table 1
Ripple spacing for different materials and wavelength for ripples observed after 1000 pulses 3 Hz irradiation

Material	193 nm		248 nm	
	Ripple spacing in nm	Wavelength / Ripple spacing	Ripple spacing in nm	Wavelength / Ripple spacing
Mylar, PET, PBT	145	1.33	185	1.34
PC	132	1.46	No structures observed	
PS	139	1.39	No structures observed	

ber to do irradiation in vacuum (10^{-2} mbar). The fluence at which the irradiation took place was measured with a pyroelectric joule meter (Gentec). The irradiations were performed at a repetition rate between 1 and 7 Hz.

Some samples were metallized with an Al layer after irradiation. This was done by vacuum evaporation of Al wire and condensation at 10^{-5} mbar.

Ellipsometric mesurements were performed with a monochromatic (632.8 nm) ellipsometer (Gaertner LB 116 B). A set of measurements was done with the same sample but the ellipsometer table turned around its vertical axes. In this way the angle between the intersection of the plane of incidence with the surface and a specified direction on the surface was varied (sample rotating ellipsometry SAREL).

The samples were also studied by SEM and TEM. Therefore, its surface was metallized by a 50 Å film of Au/Pd (plasma discharge).

3. Results

Different forms of submicron structures were observed as there are ripple structures (fig. 1) in different directions and dot patterns. The most often seen pattern is a periodic ripple structure that developed on the surface with its ripple direction parallel to the main polarization of the laser beam. An example of such a pattern on a PET film is shown in figs. 1a and 1b. The pattern is characterized by zones of parallel ripples. At the matching line of those zones sometimes mismatching of the ripples occurs: a ripple may abruptly end or two ripples may join to a single one. The size of such a zones ranges from about 1 μm to a couple of μm in direction parallel to the ripples and is somewhat larger in the perpendicular direction. On smoother samples like a PS film on a silicon wafer the mismatching is found less often (fig. 1c).

The distance between the ripples for different materials and wavelengths is given in table 1. The spacing of the ripples for PET and PBT treated with 248 nm is with 185 nm the same for both materials. The ripple spacing for PS is with 139 nm and PC with 132 nm smaller than the 145 nm spacing for PET. No difference in the ripple spacing was found doing irradiation in vacuum or in air.

Table 2
Smallest and largest fluence for which we observed periodic structures after 1000 pulses 3 Hz

Material	193 nm			248 nm		
	Min. fluence in mJ/cm^2	Max. fluence in mJ/cm^2	Abl. thresh. in mJ/cm^2	Min. fluence in mJ/cm^2	Max. fluence in mJ/cm^2	Abl. thresh. in mJ/cm^2
Mylar	3.5	5.5	17	< 10	14.8	28
PET	3	5	17	4	10.5	28
PC	3	5.2	11	No structures observed		39
PS	3	5	19	No structures observed		40

PET and PS are thin films of 1000 to 1700 Å on silicon wafers, Mylar indicates commercial films made by DuPont. For sake of comparison the table gives also the ablation threshold. As can be seen, the fluence window for producing the observed structures is much below the ablation threshold.

TEM investigation of pictures taken structured and perpendicular to the structure direction sliced Mylar films show that hill to valley modulation of the surface can be up to 140 nm, which is about the same as the distance of the ripples [8].

The ripples occur in a narrow fluence window applying a large number of pulses (> 400). The smallest and largest fluences for which we observed a fully developed ripple structure with the SEM are shown in table 2. No difference was found between irradiation in air and in vacuum for PET, PBT and Mylar films. PS did not show ripples after being irradiated in vacuum. On PS and PC using 248 nm radiation neither in air nor in vacuum structures were found.

Inclination of the sample around an axis parallel to the polarization direction of the laser beam produces the ripple distances larger than that for perpendicular incidence but develops a less regular pattern.

Ellipsometric measurements using the SAREL method were done on thin films on silicon wafers and on metallized thin films. Films with a ripple pattern show a modulation of the ellipsometric angles Δ and Ψ when the sample is turned around its vertical axis. Fig. 2 shows the graph of the ellipsometric angles for a PS film on a silicon wafer irradiated with 240 to 10000 pulses of 3.9 mJ/cm^2 at 193 nm. The table of the ellipsometer was turned in steps of 10° from 0° to 180°. The periodic structure gives an ellipse in the plot of Δ and Ψ like an anisotropic film does. The centers of gravity of the ellipses follow nearly a curve of constant index of refraction. For comparison such a curve for $n = 1.6$ is also drawn in fig. 2.

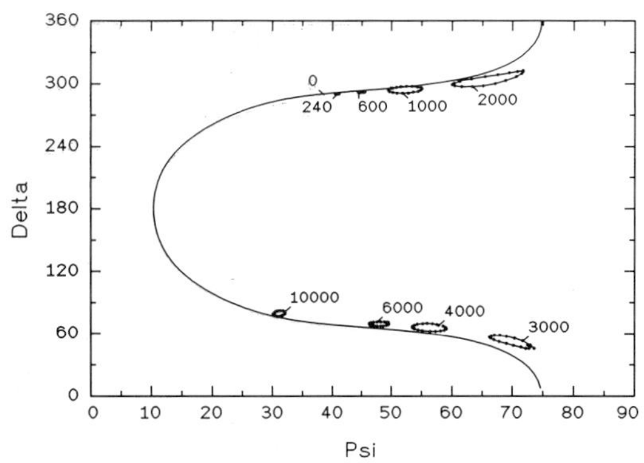

Fig. 2. Ellipsometric angles measured on a film of PS on a silicon wafer. The film was treated with 240 to 10000 pulses of 3.9 mJ/cm^2 at 193 nm. SAREL measurements were performed on the samples for angles between the plane of incidence and the ripple direction from 0° and 180° in steps of 10°. The number of applied pulses is indicated at the corresponding ellipsometric results. A global decrease in the film thickness can be followed by the position of the center of gravity of the obtained curves.

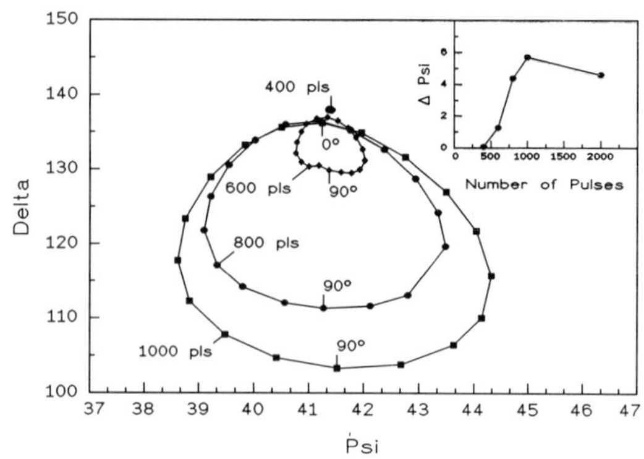

Fig. 3. Ellipsometric angles measured on metallized (Al) PS samples structured with 400, 600, 800 and 1000 pulses of 3.9 mJ/cm^2 at 193 nm. The angle of incidence of the analyzing beam was 70°. SAREL were performed on the samples for angles between the plane of incidence and the ripple direction from 0° to 180° in steps of 10°. The structure growth can be followed by the increasing size and the displacement of the center of gravity of the obtained curves. The small figure shows the development of the modulation of Ψ in dependance of the number of the applied pulses. The slow onset of the structure growth and the fast increase when a small structure is established (at about 500 pulses) can clearly be seen.

Structured and afterwards metallized films also show the modulation of the ellipsometric angles when applying the SAREL method. Fig. 3 shows the obtained values for Δ and Ψ for Al covered PS thin film on a glass substrate, irradiated with 400, 600, 800 and 1000 pulses before metallization. The samples were measured every 10°. Like in the case of a non-metallized film the obtained modulation of the ellipsometric angles grows with the amplitude of the structure, that is with the number of the applied pulses. Another common feature of both kinds of samples is the shift of the center of gravity of the curves towards values of smaller index of refraction than those of a sample without surface structure.

4. Discussion

4.1. Discussion of the ripple forming mechanism

The mechanism of the ripple development has two major aspects: (i) the optical effect, which is the mechanism causing the laser light to have a spatial intensity modulation in the first polymer layers and (ii) the physical structure development, which accounts for the conversion of an intensity modulation into a surface relief. These two aspects may be regarded separately, but for structure development both are essential. The result of one process is the preliminary of the other one. A surface structure causes light intensity modulation which causes the surface structure to grow. Once this feedback system is established (after about 500 pulses), the structure growing mechanism is very efficient.

The ripple distance we observe is shorter than the wavelength in the ambient medium and longer than the wavelength in the material. Due to the large aspect ratio of the structure, the structure has to be considered as a layer composed of these two media. The resulting index of refraction is in between those of the ambient and of the polymer. The length of a surface wave in this layer determines the ripple distance by interference with an incident plane wave.

The fluence that we use to produce the surface structure may be at the limit for melting the whole surface layer. Experiments on PET show that irradiation with fluences above the value for structure generation (4.5 mJ/cm^2 for PET, 193 nm) but below the ablation threshold (18 mJ/cm^2 for PET, 193 nm) lowers the crystallinity of the surface layer [9]. This can be explained by a fast melting and cooling of the surface and reorganisation of the polymer chains. In the fluence region, where the structures develop, melting occurs partially, controlled by the modulation in the light intensity. This causes a viscosity gradient lateral as well as axial and, forced by surface tension, gathers the material in the colder zones. Also photofragmentation of the polymer chains will take place. In the more liquid parts of the surface the smaller fragments can move more freely and evaporate more easily.

The importance of these two factors, melting of the polymer and shortening of the polymer chains can be seen in the case of PS. At 248 nm the absorption coefficient is about a factor 15 lower as for 193 nm. Even though photo-oxidation takes place when doing 248 nm irradiation in air [10] and a decrease of thickness can be observed, no structure develops. Below the ablation threshold there is not enough energy absorbed to melt the polymer layer on the surface. At higher fluence the amount of photo-oxidation increases and the etchrate is too high for a pulse to pulse feedback, which controls the structure growth. On the other hand, using 193 nm irradiation a structure develops when the irradiation is done in air, but in vacuum the structure does not develop. The surface is molten, but broken polymer chains tend to recombine or to crosslink with other chains in lack of a reactive partner like oxygen. Until the onset of photoablation no structuring occurs.

For PET and PC oxygen in the ambient is not necessary and the structures occur also in the vacuum. Both polymers contain already oxygen in form of C=O groups in their chains.

4.2. Discussion and interpretation of ellipsometric results

Until today there is no model to simulate the complicated film surface system that we have in our case. In the case of a non-metallized sample the simplest model to use is a single layer on a substrate on a substrate model [11]. Fig. 2 shows a constant index of refraction curve for a film with an index of refraction $n = 1.6$ on a silicon substrate, together with the results of a structured PS film on a silicon wafer. Using this model and the center of gravity values of the angles obtained for each irradiation, the thickness of the film and the index of refraction can be calculated. The etch rate can be determined to be 210 ng/J, which agrees with etch-rate measurements done with a quartz microbalance. But this loss of mass is not sufficient to generate an aspect ratio as observed in the TEM experiment.

Using the single layer model a decrease of the index of refraction is observed with decreasing thickness. This is an error caused by the models insufficency. The film surface is covered with a structure, a layer in which the polymer and air are present. The index of refraction of this anisotropic layer will thus be a value between that of air and that of the polymer. With decreasing film thickness this layer has an increasing effects in the total result and the index of refraction seems to decrease.

On metallized as well as not-metallized samples a plot of Ψ and Δ reveals to be ellipses. To analyze the results for the metallized sample the 860 Å layer of Al is thick enough to be treated as a rough aluminium substrate. Therefore, the change of Ψ and Δ can be regarded as resulting from one or several flat overlayers which have the same ellipsometric properties as the structure of the surface. The data to be interpreted are (a) the shifted position of the center of mass of the obtained curves and (b) the modulation in Ψ and Δ. A model is going to be developed.

As can be remarked, the center of gravity of the obtained curves obtained with the SAREL method for non-oriented rough samples shows also a shift in the ellipsometric angles. It can be concluded that the offset of the center of gravity with regard to a non-treated surface is related to the degree of surface roughness and the modulation in Ψ and Δ to its degree of orientation.

5. Conclusion

It was shown that on a variety of polymers submicron structures can be produced with polarized excimer laser radiation. The periodicity of the structure is smaller than the wavelength of the irradiating light in air and is dependent on the polymer, the wavelength used and the angle of incidence.

The fluence which is necessary for structuring the polymer has to be in a narrow window far below the ablation threshold at the level where melting of the surface starts. Thus the material loss when doing the irradiation is much smaller than in ablative processes. This low fluence level allows to treat large surface areas with a single excimer laser beam.

The structure develops after about 500 pulses with a low amplitude and reaches after about 1000 pulses a stable region of constant amplitude. The structure is self-organizing and does not need an external spacial fluence modulation of the laser beam as for example an interference set-up. Translational displacement of the laser beam on the sample does not affect the development of the structure and large areas can be treated. On the other hand the structure is quite sensitive to rotation of the polarisation direction, which causes a disturbance of the regular pattern.

Photo-oxidation and local successive melting are besides possible other phenomena the main steps in the structure development, as shown for PS.

Monochromatic ellipsometry has proven to be a useful tool to study the structure development on polymer samples. Even though there is not yet a model available to simulate the obtained data in a way to fully characterize the structure, ellipsometry succeeds in showing whether a structure has developed and to identify its direction. This is of great importance since optical scattering techniques in the visible do not give acceptable results because of the size of the structure a couple of times smaller than the wavelength of visible light. Ellipsometry is non-destructive, fast and the size of the sample is not limited. These advantages make it a perfect complement of classical microscopic techniques like SEM.

Acknowledgements

Financial contributions of the ECC, BRITE and ESPRIT programmes, as well as from DuPont Luxembourg are gratefully acknowledged. Monique Leblanc is thanked for her skillful assistance in electron microscopy.

References

[1] K.J. Ilcisin and R. Fedosejevs, Appl. Opt. 26 (1987) 396.
[2] H.M. Phillips, D.L. Callahan, R. Sauerbrey, G. Szabó and Z. Bor, Appl. Phys. Lett. 58 (1991) 2761.
[3] S. Lazare and R. Srinivasan, J. Phys. Chem. 86 (1986) 2124.

[4] T. Bahners and E. Schollmeyer, Angew. Makromol. Chem. 151 (1987) 1.
[5] P.E. Dyer and R.J. Farley, Appl. Phys. Lett. 57 (1990) 765.
[6] H. Niino, M. Nakano, S. Nagano, A. Yabe and T. Miki, Appl. Phys. Lett. 55 (1989) 510.
[7] J.E. Sipe, J.F. Young, J.S. Preston and H.M. van Driel, Phys. Rev. B 27 (1983) 1141.
[8] M. Bolle, S. Lazare, M. Leblanc and A. Wilmes, Appl. Phys. Lett. 60 (1992) 674.
[9] P. Darles-Benet, Thesis, Université de Bordeaux I, 1991.
[10] M. Bolle and S. Lazare, Appl. Surf. Sci. 54 (1992) 471.
[11] R.M.A. Azzam and N.M. Bashara, Ellipsometry and Polarized Light (North-Holland, Amsterdam, 1977).

Applied Surface Science 69 (1993) 38–45
North-Holland

applied
surface science

Maskless excimer laser induced projection patterning of InP in Cl_2 etch gas

R. Heydel, R. Matz
Research Laboratories of Siemens AG, Otto-Hahn-Ring 6, D-8000 München 83, Germany

and

W. Göpel
Universität Tübingen, Institut für Physikalische und Theoretische Chemie, Auf der Morgenstelle 8, D-7400 Tübingen 1, Germany

Received 2 June 1992; accepted for publication 30 October 1992

A chlorine-based excimer laser (KrF, 248 nm) induced etch process has been developed with focus on etch rate, spatial resolution capability and morphology of etched surfaces. The thermal desorption of spontaneously formed chemical reaction products of Cl_2 on InP during the short laser pulse permits excellent spatial resolution of about 1 μm. We demonstrate for the first time maskless micropatterning of a III–V semiconductor by projection etching of micron-wide waveguide structures in InP. Etch rates can exceed those of conventional dry etching processes by a factor of 10 to 20, resulting in processing times which are conveniently short but well controllable due to the pulsed nature of the present process. Regarding surface morphology, the process is apparently sensitive to the laser speckle pattern and etch gas or surface contaminations. However, under proper processing conditions, the influence of both can be sufficiently reduced so that a surface smoothness is achieved comparable to conventional dry etching.

1. Introduction

Photoinduced etching processes have been studied during the recent decade worldwide by several research and development groups with various goals. The feature to determine the etch rate by the intensity of light being projected onto the surface to be etched, is considered most attractive. It allows one e.g. to pattern a suitable etch mask in situ (similar to some ion beam writing approaches [1]) or to abandon any etch mask usually in contact with the surface, so that integrated processing without atmospheric interference between processing steps appears possible. Another conceivable attractive application is trimming of single critical devices on a more or less complete integrated circuit to improve yield. Furthermore, the light intensity can be varied continuously across the surface processed so that the resulting smooth etch rate variations can be used to form tapered structures (e.g. vertical waveguide tapers [2]) and integrated lenses [3] which would be difficult to realize by other means).

An especially promising and flexible approach is pattern transfer etching by optical projection of a reticle onto an unpatterned wafer surface in a processing chamber filled with a suitable etch gas. For pattern etching of III–V semiconductors the formation of high vapour pressure halide reaction products is attractive. Therefore, hydrogen halides were used initially as etch gases [4]. However, for etching, photodissociation of these gases is required and spatial resolution was found to be limited by gas-phase diffusion of the radicals generated [5]. To circumvent the diffusion problem we suggested the use of a spontaneously reacting etch gas with subsequent laser-induced product desorption [5]. At elevated temperature, Cl_2 is able to etch InP spontaneously [1,6]. This is in contrast to Si where Cl_2 photodissociation plays an important role [7]. Consequently, Cl_2 has been used early to photothermally etch InP with an Ar ion laser [8]. Similar spatial resolution of some 10 μm has been demonstrated by Maki and Ehrlich [9] with a pulsed excimer laser and Cl_2 on GaAs. Excimer laser induced etching of InP at 193 nm in Cl_2 has been discussed recently by Donnelly and Hayes [13]. They used SiO_2 making of the substrate to study etch patterns. The work described here is aimed at the demonstration of the original predictions of inherently high etch rate and high spatial resolution [5] when gas-phase photodissociation and diffusion are negligible, so that device-relevant microstructures can be etched even without masking layers at the more convenient wavelength of 248 nm.

2. Experimental

As a first approach to process development the dependence of the etch rate on various experimental parameters like gas pressure, laser fluence, pulse repetition rate and laser resonator conditions was investigated using an ultra-high vacuum (UHV) capable etch chamber (base pressure 1×10^{-7} mbar without baking) with airlock and low numerical aperture (NA) projection optics for moderate spatial resolution (20 μm) but large working distance (5 cm). This set-up has been described before [5,10]. In order to apply the process studied such to micropatterning, a second set-up was added with a cylindrical flat etch chamber (stainless steel, 6 cm diameter) allowing for a small working distance, increased NA and improved spatial resolution (fig. 1). A vacuum of 2×10^{-4} mbar is obtained with a turbomolecular pump in a few minutes (base pressure 1×10^{-5} mbar). The etch gas is supplied to and removed from the chamber through two valves located on opposite sides of the chamber.

The gas pressure is adjusted under flow conditions between 10^{-4} and 1 mbar by varying the pump rate. At higher pressures no gas flow is possible. Substrate temperatures up to 250°C can be reached by resistive heating of the substrate holder integrated into the bottom of the chamber. Laser light passes through a Viton-sealed quartz window of 1 mm thickness. The excimer laser (Lambda Physik 150 TMSC) comprises injection-locked oscillator and amplifier tubes. It is

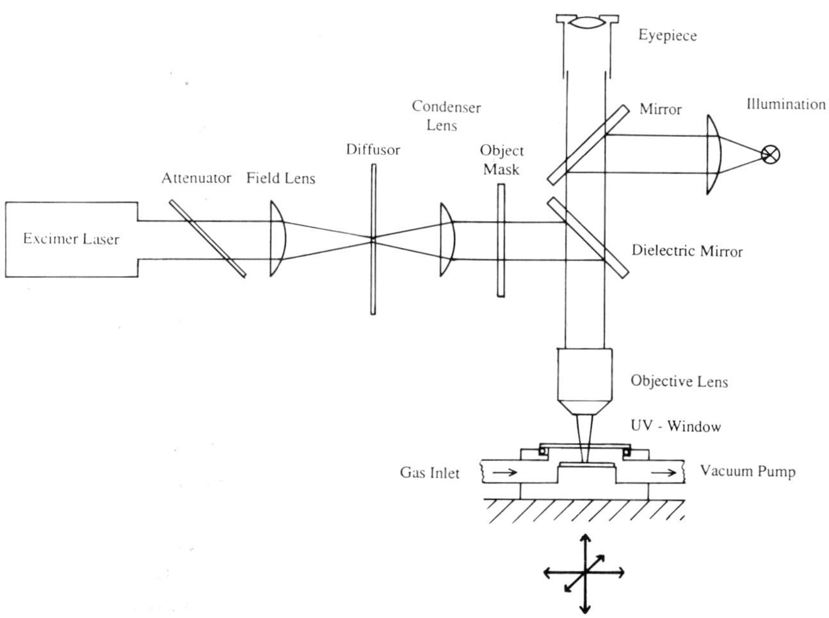

Fig. 1. Experimental set-up for maskless projection patterning.

operated here at a wavelength of 248 nm (KrF) with a nominal pulse duration of 18 ns in three different resonator modes.

First, the amplifier can be operated as an isolated laser with stable optics resulting in a broad spatial line width ("stable mode"). To narrow the line width to 2×10^{-3} nm for minimum chromatic aberration injection-locking between oscillator (prism optics) and amplifier (unstable resonator) is used ("unstable mode"). The pulse energy is about 200 mJ in both modes. A third modified laser mode is obtained from the unstable mode when the rear amplifier mirror is replaced by a quartz diffusor plate and the front mirror is removed. As described in ref. [11], the consequence is only little loss in spectral line width (4×10^{-3} nm) and pulse energy (135 mJ) but a strongly reduced speckle contrast (from 74% to 8%) due to reduced coherence length. Behind the variable attenuator the beam is collimated onto a rotating quartz diffusor disc (grinded and etched in HF), which is placed slightly ahead of the focus. The divergent Gaussian beam leaving this set-up is collected by the field lens to achieve homogeneous illumination of the mask (chromium lines on quartz) and maximum brightness at the objective lens (Zeiss Ultrafluar). The projection system, which is similar to one described in ref. [12] is designed for a 10:1 image size reduction. The usable field size on the sample amounts to about 500 μm with fluences up to 1 J/cm^2 in the center for any of the three laser resonator modes. InP(100) substrates (n-type: 6×10^{18} cm^{-3}) polished on both sides were covered with a 100 nm thick evaporated Cr film on the back side for reproducible resistive heating. For cleaning and removal of native oxide and surface damage ethanol, HF (40%) and a solution of $H_2SO_4/H_2O_2/H_2O$ (3:1:1) were applied prior to mounting in the processing chamber. Etched profiles were analyzed with a stylus profilometer, light microscope and scanning electron microscope (SEM).

3. Results

The observed dependences of the InP etch rate on laser fluence, Cl_2 gas pressure and pulse repetition rate as measured in the low-pressure chamber are shown in figs. 2a–2f. No significant differences were found in large-area etch rates between the stable (low divergence, wide spectrum) and the modified (increased divergence, narrow spectrum) laser mode. Fig. 2, therefore, shows differences only between the stable and unstable (low divergence, narrow spectrum) modes.

In the unstable mode, a threshold fluence for etching at ambient temperature of 100 mJ/cm^2 is found (fig. 2a), above which the etch rate increases linearly. Above about 200 mJ/cm^2 this increase becomes steeper and a rate of more than 1 nm/pulse is reached before a saturation is observed (300–400 mJ/cm^2, fig. 2b). Etching at elevated wafer temperatures starts from a lower threshold fluence but exhibits otherwise similar characteristics, e.g. the rate versus fluence curve appears shifted down by 20–30 mJ/cm^2 at 140°C wafer temperature (fig. 2a). In the stable laser mode (fig. 2b), the behaviour is also similar. However, a higher fluence is required to obtain the same rate as in the unstable mode.

With regard to etch gas pressure (figs. 2c and 2d), etching requires at least 10^{-5} mbar Cl_2. Above this pressure, the rate increases logarithmically to a fluence-dependent saturation value of up to 0.15 nm/pulse (in the range studied) for both illumination modes. In this high-pressure range of almost constant etch rate we also observe the beginning of product redeposition around the irradiated surface area. The details in the range below depend on the laser mode: a simple logarithmic increase is observed for the unstable mode between 10^{-5} mbar and saturation at 10^{-3} mbar. Under stable mode conditions (fig. 2d), in contrast, this range is shifted by an order of magnitude to higher pressures (saturation at 10^{-2} mbar) and an additional range appears below 10^{-3} mbar in which the slope of the logarithmic increase is reduced. Figs. 2e and 2f demonstrate the decrease in etch rate per pulse with increasing pulse repetition rate f or decreasing time t between pulses at low pressure (10^{-3} mbar). The best simple fit is log t as shown. At higher pressures within the saturation range of figs. 2c and 2d, e.g. 1 mbar, the etch rate per pulse becomes practically independent of pulse rate.

For device application also the quality of etched surfaces is important. Within this work the surface morphology is found to depend mainly on the laser resonator conditions, i.e. spatial and temporal coherence of the laser light, but negligibly on other processing conditions like gas pressure, laser fluence and pulse repetition rate (except for very high fluences above 350 mJ/cm^2 which generally produce rough surfaces). Fig. 3 demonstrates typical rough and smooth surfaces etched under unstable and stable resonator conditions, respectively. These morphologies are a consequence of the different illumination and are not characteristic of InP. In fact, they are similarly reproducible in PMMA.

To apply the process to pattern etching of high-resolution waveguide-like structures the modified laser mode is used since it avoids speck-

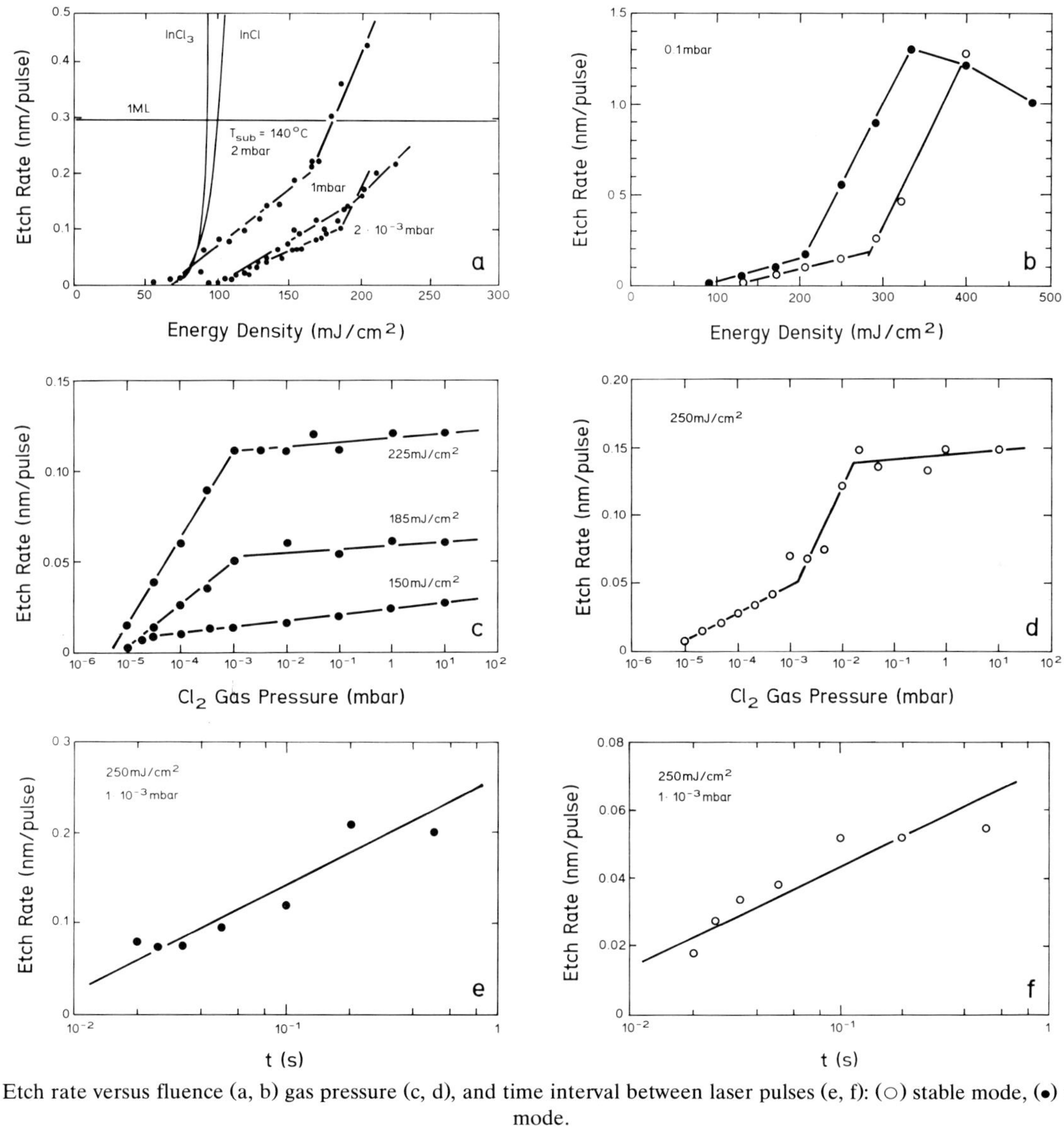

Fig. 2. Etch rate versus fluence (a, b) gas pressure (c, d), and time interval between laser pulses (e, f): (○) stable mode, (•) unstable mode.

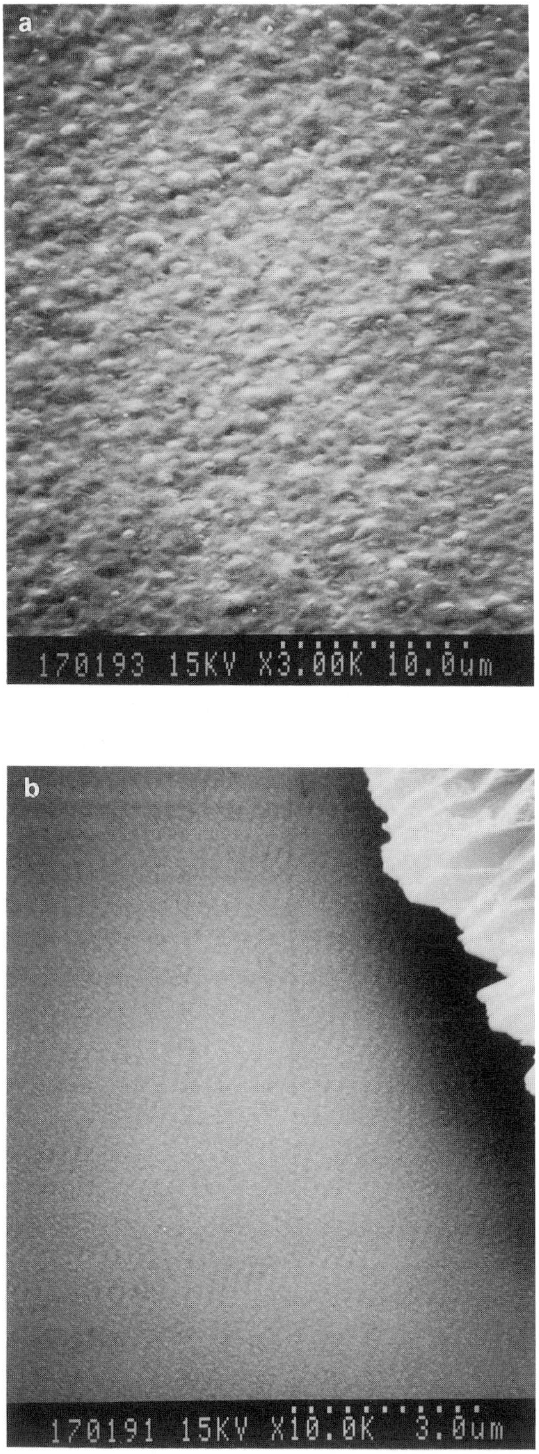

Fig. 3. Comparison of Cl_2 etched InP surfaces, 300 mJ/cm^2, 20 Hz, 0.1 mbar: (a) unstable resonator mode, 1.3 nm/pulse, 2000 pulses; (b) stable resonator mode, 0.17 nm/pulse, 3000 pulses.

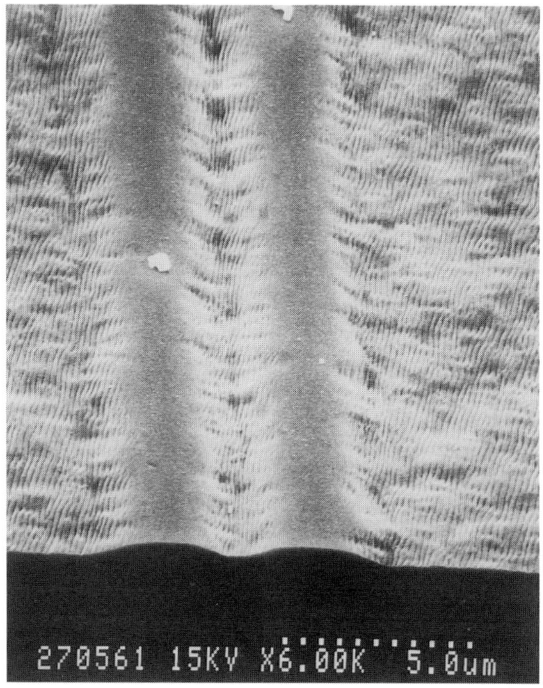

Fig. 4. SEM micrograph of a waveguide-like structure on InP formed by maskless projection etching in 10 mbar Cl_2 (120 mJ/cm^2, 10 mbar, 20 Hz, 70 000 pulses; etch rate = 0.005 nm/pulse).

les (like the stable mode) and chromatic aberration (like the unstable mode). It, therefore, produces also smooth surfaces under conditions of fig. 3, i.e. in a clean UHV environment. However, these results were not transferable into the high-resolution etch chamber which has, due to the short working distance, less favourable figures of volume-to-wall surface ratio, base pressure and residual contamination level. Therefore, rough surfaces are formed (as shown in fig. 4) unless further counter-measures are taken.

The raised lines on the surface were formed without an etch mask by projection of corresponding chromium lines from the reticle onto the surface. It was observed and is evident from

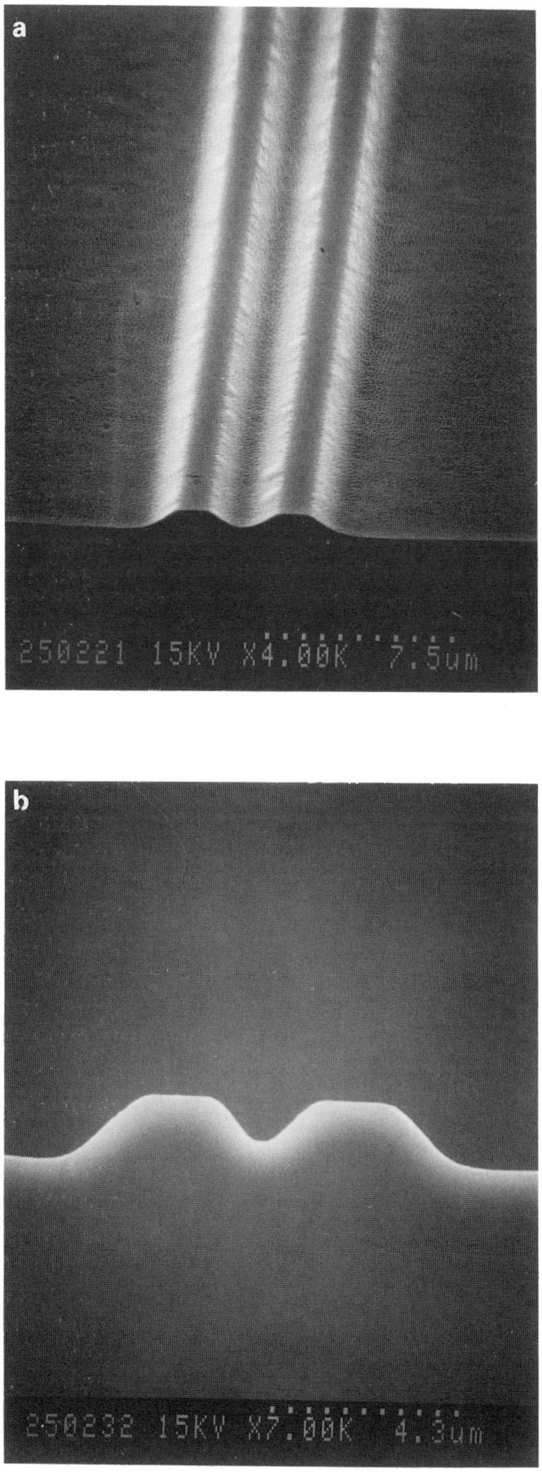

Fig. 5. SEM micrographs: (a) top view, (b) cross section of waveguide-like structures on InP etched in 10 mbar Cl_2 + 590 mbar Ar (320 mJ/cm^2, 20 Hz, 100 s; etch rate: 0.45 nm/pulse, 540 nm/min; depth: 0.9 μm.

fig. 5 that the detrimental effects in this simple vacuum chamber can be completely eliminated when a buffer gas like Ar is used to raise the total pressure to an almost atmospheric level and a sufficiently high fluence between 250 and 350 mJ/cm^2 is applied. The SEM micrographs present top view and cross section through a structure resembling a planar integrated optical directional coupler on InP.

4. Discussion and interpretation

In principle, the laser-induced desorption of reaction products (chlorides) from the InP surface may occur along two alternate reaction paths, namely photochemically or photothermally. However, the absence of a photochemical component is suggested here by the distinct threshold fluence of 100 mJ/cm^2 (figs. 2a and 2b). The photothermal reaction path has been observed for the InP/Cl_2 system also by Donnelly and Hayes [13] at a laser wavelength of 193 nm. Hence, InP/Cl_2 is in contrast to GaAs/Cl_2 where a photochemical process was considered likely [9]. For the present interpretation, the calculated maximum surface temperature during the laser pulse is plotted as a function of incident laser fluence in fig. 6 (wafer at room temperature). The calculated threshold for surface melting near 200 mJ/cm^2 coincides well with the position of the kink in figs. 2a and 2b. In hydrogen halide etch gases (e.g. HCl) this threshold represents the threshold for notable etching [10] which was understood as a phosphorus depletion and reactive site formation process for photodissociated halogen atoms. Also here, the steeper rise of etch rate with fluence reflects enhanced surface reactivity in this range. Below 200 mJ/cm^2 the maximum surface temperature increases at a rate of about 5 K per mJ/cm^2. A bias wafer temperature of 140°C should therefore permit an equivalent flu-

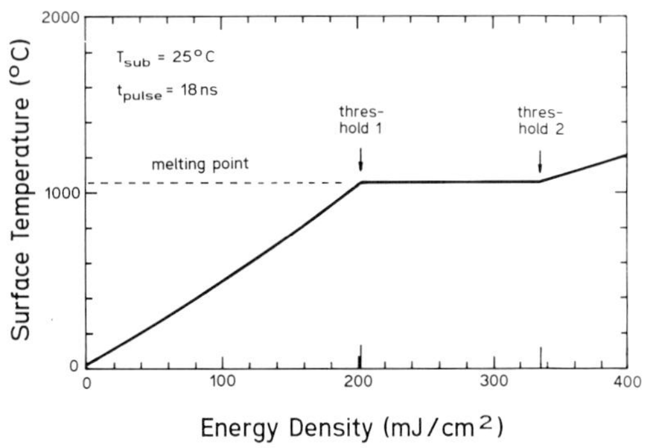

Fig. 6. Maximum surface temperature of InP as a function of laser fluence at 248 nm, $T_{melt} = 1062°C$.

ence reduction of approximately 28 mJ/cm^2. This agrees very well with the measured shift in fig. 2a. Above 340 mJ/cm^2 the model predicts higher surface temperatures and a melt depth exceeding the thermal diffusion length (0.5 μm). In this range generally a poor surface morphology was observed. With regard to the melt threshold at 200 mJ/cm^2 etching is possible here with Cl_2 in contrast to HCl also at lower fluences between 100 and 200 mJ/cm^2. This is due to the fast and complete chlorination of the InP surface between laser pulses (0.1 s) at pressures above 10^{-3} mbar [5] and sufficient surface temperature during the laser pulse to photothermally desorb even the least volatile $InCl_x$. The theoretical curves for the maximum expectable etch rate designated by $InCl_3$ and InCl in fig. 2a were calculated by integration of the Langmuir-free evaporation of $InCl_3$ and InCl, respectively, over the time interval of the laser pulse, assuming vapour pressures from ref. [14]. The result reveals notable desorption is possible above about 100 mJ/cm^2 in agreement with the experiment. The model overestimates, however, the increase of etch rate with fluence, since pulsed laser induced desorption is far from slow Langmuir-free evaporation near thermal equilibrium. Comparing theory and experiment in fig. 2a one would conclude that only a fraction of the chloride layer is laser desorbed and that the process is desorption rate limited. This agrees well with the observation that at high pressures ($> 10^{-3}$ mbar) the etch rate per pulse is almost independent of the pulse repetition rate or the pressure (figs. 2c and 2d). Since the chloride layer reaches a final thickness of 1–2 nm [5] within 0.1 s for 0.1–10 mbar Cl_2, complete removal of this layer would be indicated by a fluence-independent rate of this amount. The saturation of etch rate near 1.5 nm/pulse above 400 mJ/cm^2 (fig. 2b) is therefore interpreted as limited by the maximum chloride layer thickness. At pressures below 10^{-3} mbar (unstable resonator) or 10^{-2} mbar (stable resonator) the etch rate depends logarithmically on both the gas pressure p and the time interval t between laser pulses. It therefore seems to be limited by the gas exposure (pt) between pulses in this range. The logarithmic dependence on time differs essentially from $\sqrt{t}$ laws found earlier for HCl on InP [10] and for Cl_2 on GaAs [9] which could be explained by diffusion-limited surface chlorination. The present situation may be described by a sticking coefficient s of gas-phase chlorine on the surface which decreases exponentially with chlorine coverage c:

$$s/s_0 = \exp(-c/c_0), \qquad (1)$$

where s_0 and c_0 are for normalization. The coverage increases according to

$$\mathrm{d}c/\mathrm{d}t = s(p/w), \qquad (2)$$

where $w^2 = 2\pi m k_B T$ and p/w denotes the surface collision frequency. Insertion of eq. (1) into eq. (2) and integration yields

$$c(t) = c_0 \ln(1 + s_0 pt/c_0 w). \qquad (3)$$

This would explain a logarithmic dependence of the etch rate on pressure and inverse pulse repetition rate, supposing that the desorption efficiency, i.e. the fraction of x which is desorbed by a laser pulse, is constant. Etch rate differences between stable and unstable resonator optics are attributed to a considerably larger speckle contrast in the latter case. Under the spectrally broader illumination of the stable mode the surface remains smooth (fig. 3b) and generally exhibits lower etch rates (figs. 2b, 2d and 2f) or requires higher fluence or higher gas pressure for a comparable etch rate. his may result from changes in surface area, chemical reactivity or optical constants related to the smooth-to-rough transition.

Apart from inhomogeneous illumination also process contaminations can lead to rough surfaces (fig. 4). This is even more likely in small etch chambers with a relatively large area of wall surface, on which exchange reactions take place. The effect has been suppressed here by adding Ar as buffer gas, thus reducing the gas-phase diffusion length of contaminations. Furthermore, a high fluence above the melting threshold of InP (200 mJ/cm^2) helps to keep the surface planar as demonstrated in fig. 5. Also, high etch rates above 0.5 $\mu m/min$ and short processing times result. These rates exceed those of conventional dry etching by a factor of 10 to 20. The sidewalls of etched features are determined by the photothermal spatial resolution of the process (thermal diffusion length 0.5 μm [10]) when the etched depth is small (≤ 1 μm).

5. Conclusion

Chlorine gas was shown to be a useful etch gas for excimer laser induced dry etching of InP. Related compounds like InGaAsP are expected to behave similarly. Stimulated etching occurs via photothermal and, hence, doping-independent desorption of spontaneously formed surface chlorides above 100 mJ/cm^2. Except for very low Cl_2 pressures, the etch rate is limited by this desorption step. The surface morphology obtained depends generally on the speckle contrast within the illuminated wafer surface. If not a low base pressure processing chamber is used, optimum performance requires the addition of Ar buffer gas and laser fluences above the threshold for transient surface melting. The buffer gas also serves to compensate for atmospheric pressure on the thin chamber window thus allowing a more favourable window design. Under these conditions micron-wide ribs with smooth surface resembling integrated optical waveguides and directional couplers were successfully etched with high etch rate (0.5 $\mu m/min$) by 10:1 projection of a quartz reticle.

Acknowledgements

The authors alone are responsible for the contents. Valuable discussions with U. Wolff are gratefully acknowledged.

References

[1] Y.L. Wang, H. Temkin, L.R. Harriott, R.A. Logan and T. Tanbun-Ek, Appl. Phys. Lett. 57 (1990) 1864.
[2] G. Müller, B. Stegmüller, H. Westermeier and G. Wenger, Electron. Lett. 27 (1991) 1836.
[3] F.W. Ostermayer, Jr., P.A. Kohl and R.H. Burton, Appl. Phys. Lett. 43 (1983) 642.
[4] P.D. Brewer, D. McClure and R.M. Osgood, Jr., Appl. Phys. Lett. 47 (1985) 310.
[5] R. Matz, J. Meiler and D. Haarer, Mater. Res. Soc. Symp. Proc. 158 (1990) 307.
[6] S.C. McNevin, J. Vac. Sci. Technol. B 4 (1986) 1216.
[7] R. Kullmer and D. Bäuerle, Appl. Phys. A 47 (1988) 377.
[8] L. Ding, Q. Mingxin and Kuang Zhong, J. Electron. Mater. 17 (1987) 29.
[9] P.A. Maki and D.J. Ehrlich, Appl. Phys. Lett. 55 (1989) 91.
[10] J. Meiler, R. Matz and D. Haarer, Appl. Surf. Sci. 43 (1989) 416.
[11] G. Solaro, E. Keckeis and F. Bachmann, Proc. SPIE 1264 (1990) 350.
[12] J.H. Brannon, J. Vac. Sci. Technol. B 7 (1989) 1064.
[13] V.M. Donnelly and T.R. Hayes, Appl. Phys. Lett. 57 (1990) 701.
[14] Gmelins Handbuch der Chemie, Indium, Ed. Deutsche Chemische Gesellschaft (Verlag Chemie, Weinheim, 1969).

Applied Surface Science 69 (1993) 46–51
North-Holland

applied
surface science

Reflectance anisotropy from (001) GaAs surfaces during pseudo-ALE growth of GaAs

S.R. Armstrong, R.D. Hoare, I.M. Povey, M.E. Pemble *, A. Stafford, A.G. Taylor

Department of Chemistry and Centre for Electronic Materials, University of Manchester Institute of Science and Technology, P.O. Box 88, Sackville Street, Manchester M60 1QD, UK

and

D.R. Klug

Centre for Photomolecular Sciences, Imperial College of Science, Technology and Medicine, Prince Consort Road, South Kensington, London, UK

Received 2 June 1992; accepted for publication 30 October 1992

We demonstrate that the technique of reflectance anisotropy (RA) has now evolved to the level of sophistication required to be of use in direct MOVPE process control. Comparison with data obtained under high vacuum GaAs MBE growth and vacuum chemical epitaxy growth conditions permits the interpretation of RA changes that occur under higher pressure growth conditions. Inflections in RA responses can indicate changes in surface reconstruction that occur again both under high vacuum conditions and at higher pressures. Under pseudo-ALE growth of GaAs on (001) GaAs from triethylgallium (TEGa) and arsine in a conventional hot-wall MOVPE reactor maintained at a constant pressure of 200 mbar, the onset of Ga incorporation via TEGa decomposition has been monitored as a function of substrate temperature. For this reaction, where the rate-determining step may occur either homogeneously in the gas phase or heterogeneously on the substrate surface, changes in RA intensity have been observed at temperatures as low as 350°C. Under these conditions, while the surface changes resulting from TEGa decomposition occur quite slowly, the subsequent surface recovery upon exposure to arsine gas in the second half of the growth cycle occurs very rapidly, the data suggesting that the surface initially relaxes beyond the initial As-stabilised surface towards a very As-rich surface in response to an oversaturation of the surface with Ga atoms during the first half of the growth cycle. This surface then slowly loses As to return to the stable initial condition. In order to observe the arsine recovery kinetics in more detail, the flow rates were reduced. Under these conditions, while rapid RA variations were again observed, inflections in the RA responses became increasingly prominent, which are discussed in terms of possible surface reconstructions that may occur during recovery back to the initial surface condition.

1. Introduction

The need for greater control in the growth of layers of materials for use in the fabrication of electronic devices has led to the application of many photon-based in-situ diagnostic methods to the study of epitaxial growth systems, and the subsequent appearance of a new field of study, namely that of the interaction of probe electromagnetic radiation with surface layers. This field has become known as "epioptics", a term first used by McGilp [1], and which describes the combination of the Greek word "epi" meaning upon, with optical methods. Many methods fall into this category and are thus currently under investigation, including ellipsometry [2], second harmonic generation [3,4], Raman spectroscopy [5], IR spectroscopy [6,7] and the technique known as reflectance anisotropy (RA) or reflectance dif-

* To whom correspondence should be addressed.

ference (spectroscopy) (RD(S)) [8–19]. This method, first demonstrated in this context by Aspnes and coworkers [8] is now an established means of monitoring layer-by-layer growth of compound semiconductors having the zinc blende structure. The sensitivity of the technique towards these epitaxial growth processes arises from the behaviour of the growing (001) surfaces of these materials, which typically undergo reconstructions involving formation of Group III atom dimers aligned along the ⟨110⟩ azimuth and Group V dimers aligned along the orthogonal $\langle 1\bar{1}0 \rangle$ azimuth. While the exact nature of the processes leading to a difference in reflectivity for visible radiation polarised parallel to these two azimuths is not clear, the generally accepted view as proposed by Aspnes and coworkers is that selective excitation of surface dimer resonance states via a suitable choice of excitation wavelength may be responsible [10,18], although little direct evidence in support of this proposal exists. To date, most applications of this technique to the study of the epitaxial growth of III–V materials have been limited to systems where the deposition reactions take place under some form of high vacuum, such as MBE growth [11,12] and higher pressure vacuum chemical epitaxy (VCE) [13,14], although in the literature there are reports of the application of RA to atomic layer epitaxy (ALE) systems [15] and even near atmospheric pressure metalorganic chemical vapour deposition (MOCVD) systems [10,16–18], notably for heteroepitaxial systems where large RA responses are observed [16,17] and recently for clean (001) GaAs surfaces prepared under MOCVD conditions [18]. In a previous publication, we demonstrated the sensitivity of the RA method to the study in *real-time* of surface reconstruction changes occurring at a (001) GaAs substrate in high vacuum by considering the response of an As-rich c(4 × 4) (001) GaAs surface under conditions of progressive As depletion [12]. RA transients and reflection high energy electron diffraction (RHEED) intensity oscillations recorded *simultaneously* during single experimental runs demonstrated that the RA response contained well-defined plateaus which correlated with the appearance of a new RHEED pattern.

The purpose of this present communication is to demonstrate that the sensitivity of the RA technique is now sufficient to enable the technique to be applied to the real-time monitoring of layer-by-layer growth of GaAs on (001) GaAs under relatively high pressure conditions (200 mbar). Under these high pressure conditions, where the Group III and Group V reagents are introduced into the reactor sequentially, RA is used to monitor the surface kinetics leading to the formation of Ga dimers from triethylgallium and their subsequent destruction upon exposure to arsine gas.

2. Experimental

The optical and electronic layout of the RA measurements has been described in detail in a previous publication [19]. The laser used in these experiments, a continuous-wave Kr^+ (Coherent Innova 90), impinged on the substrate surface (001) GaAs (approximately 1 cm^2, semi-insulating, Cr-doped, MCP Wafer Technology Ltd.) at near-normal incidence through the unmodified walls of the MOCVD reactor (MOCVD 4000, CVT Ltd.), the reflected intensity being measured on a silicon diode detector. The polarisation of the laser beam was controlled by a high-frequency electro-optic modulating device and was switched at MHz frequencies, such that detection of RA changes on a time scale of 10 ms was possible. The experimental arrangement is sensitive to relative changes in surface reflectivity, $\Delta R/R$, on the order of 10^{-6}. As noted earlier, all data shown were recorded in real time, with no signal averaging.

Triethylgallium (Epichem Ltd.) was transported to the reactor in a flow of dihydrogen. Arsine gas (Matheson Phoenix Plus grade) was used without further purification. During the experiments the reactor pressure was maintained at a constant value of 200 mbar by pressure balancing. This was necessary since pressure changes, like temperature changes can cause major fluctuations in the RA response due to changes in the birefringence characteristics of the reactor walls. For this reason the RA response was stabilised at each temperature selected prior to introducing the growth precursors.

3. Results and discussion

Fig. 1 depicts the normalised RA response, expressed as $(R_{\langle 110 \rangle} - R_{\langle 1\bar{1}0 \rangle})/R_{\text{total}}$ (arbitrary units) from an (001) GaAs substrate mounted in the MOCVD reactor, as a function of time, at a series of substrate temperatures. Prior to recording these responses the substrate was cleaned in the reactor by heating to ~ 900 K in an AsH_3/H_2 overpressure – a routine procedure prior to the MOCVD growth of GaAs which is believed to leave the surface in an As-rich condition. Similar

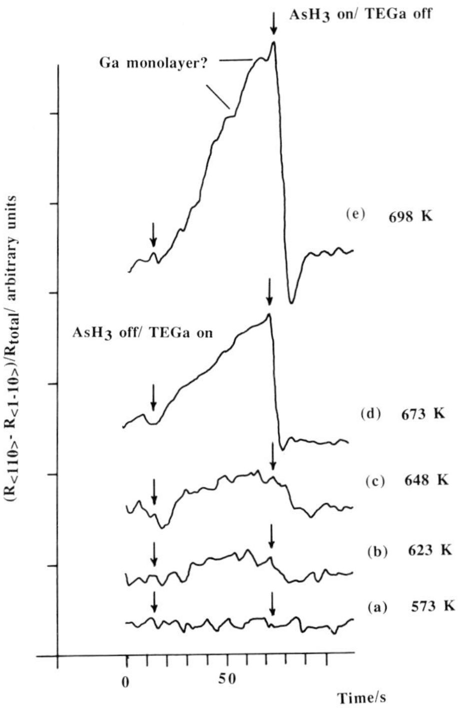

Fig. 1. RA transients recorded as a function of time from a (001) GaAs substrate mounted in an MOCVD reactor, at a series of susceptor temperatures. Indicated on the figure are the times at which TEGa and arsine were introduced into the reactor and inflections which may correspond to the Ga monolayer transition point, see text. The total reactor pressure remained constant at 200 mbar throughout each experiment. The molar flow rates of TEGa and arsine were 3.5×10^{-5} and 2.4×10^{-4} mol min^{-1}, respectively.

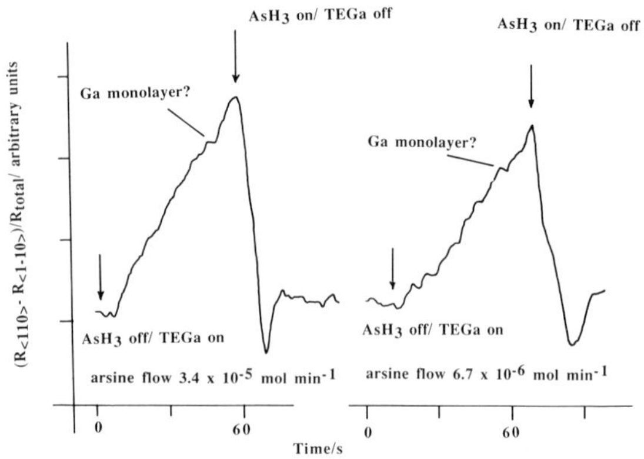

Fig. 2. RA transients recorded as a function of time from a (001) GaAs substrate mounted in an MOCVD reactor, at 673 K, depicting the response upon introduction of TEGa as in fig. 1 but with the recovery transient under arsine recorded for two discrete reduced flow rates of arsine, (a) 3.4×10^{-5} and (b) 6.7×10^{-6} mol min^{-1}.

surfaces have recently been studied by Aspnes and coworkers using RA. By comparison with the RA responses from (001) GaAs surfaces produced under controlled MBE conditions, these workers have concluded that the clean "MOCVD" surfaces adopt a "(2 × 4)-like" surface reconstruction [18] despite the high overpressures employed. Fig. 2 depicts similar experiments obtained under conditions of reduced arsine flow rates in an attempt to reveal structure inherent in the arsine recovery transients.

At the times indicated TEGa was introduced into the reactor, then, at a later time, the TEGa flow was switched off and AsH_3 introduced to complete the sequential Group III/Group V cycle. From fig. 1 it may be seen that introduction of TEGa at 573 K produces little or no change in the RA response. Indeed little change is observed until the susceptor temperature reaches 623 K. At this temperature and above, clear surface changes are indicated from the RA responses during each half of the "ALE" growth cycle.

Consider first the lack of response during either TEGa or arsine exposure below 623 K: the lack of change which could be associated with Ga dimer formation is of great interest since surface chemical studies performed in our laboratories and elsewhere [20,21] have shown that at lower temperatures, ~ 300 K, TEGa adsorbs strongly and furthermore undergoes at least partial decomposition by 573 K on such surfaces. Under the conditions of molar flow rates employed, the partial pressure of TEGa in the reactor was approximately 0.03 mbar (as determined from the mole fractions of TEGa and dihydrogen). It is possible that under these conditions TEGa adsorption did not occur but given the surface science data and the relatively high overpressures employed this seems unlikely. We suggest that it is more likely that TEGa adsorption and decomposition was occurring but that the RA method could not *observe* these processes since they did not result in the formation of Ga dimers. Possible reasons for the apparent lack of dimer formation at temperatures below 623 K may be summarised as, (a) insufficient coverage of Ga atoms, (b) lack of complete dissociation of TEGa molecules such that Ga atoms adsorbed at lattice sites do not have a spare electron for dimerisation, or (c) a lack of surface mobility required in order to place adsorbed Ga atoms on lattice sites. All of these phenomena would be temperature-dependent, i.e. the subsequent observation of a real change in the RA response at temperatures > 623 K would be consistent with points (a) and (b) noted above providing the chemisorption and decomposition processes were significantly activated. Furthermore, at higher temperatures one would expect to observe enhanced surface mobility of the adsorbed Ga atoms (point (c) above). At present we favour the model which involves enhanced decomposition of the adsorbed TEGa species since enhanced decomposition will also lead to enhanced Ga coverage which will necessarily increase the likelihood of dimer formation. Also we note that in previous work using ex-situ FTIR measurements under MOCVD growth conditions, we have demonstrated that the production of ethene from TEGa and dihydrogen in the absence of arsine is quite noticeable even at temperatures as low as 600 K [22]. While the exact conditions of these experiments do not match with those used in this present work, they demonstrated that TEGa decomposition can occur at an appreciable rate at 623 K, where the first changes in the RA response are observed.

Of further interest is the observation that at temperatures of 673 K and above, the recovery under arsine introduction is both rapid and furthermore results in a transient which dips below the initial RA value prior to recovering back to this level, figs. 1d, 1e and 2. To the best of our knowledge this is the first time that such an effect has been reported. Structure in this recovery trace is evident from the experiments performed with reduced arsine flow rates, fig. 2. The kinetics of this recovery are almost certainly heterogeneous in nature since the temperatures at which they occur are well below the usual temperatures required for efficient cracking of arsine in the gas phase. This observation is believed to be the first *direct* evidence of a heterogeneous arsine decomposition mechanism obtained under high-pressure "pseudo-MOCVD" conditions. As yet we do not have a satisfactory explanation for the dips below the starting "baseline" observed in the arsine recovery transients at temperatures in excess of 650 K, although some suggestions can be proposed following consideration of other relevant data: the work of Maa and Dapkus on ALE-type experiments using trimethylgallium (TMGa) and arsine at 773 K [15] and the VCE experiments of Paulsson et al. using TEGa and arsine at 683 K [14] are relevant to this present study, although these experiments were carried out under much lower pressures. In these experiments, dips and shoulders have been observed not in the arsine recovery transients but the Group III metalorganic exposure transients. The interpretation of these dips is at present not clear, but two explanations have been proposed: firstly, Maa and Dapkus attribute the inflections noted for TMGa exposure over (001) GaAs at 773 K to kinetic limitations imposed by As-site blocking from methyl radicals preventing further TMGa adsorption and reaction [15]; secondly and perhaps the most pertinent to this present study is the argument of Paulsson et al., who suggest that for TEGa exposure over (001) GaAs at 683 K oversaturation of surface Ga occurs via a mechanism in which TEGa adsorption and decomposition results in Ga coverages in excess of one monolayer [14]. The points on the RA transients for TEGa exposure corresponding to the completion of the first monolayer of Ga and the onset of the oversaturation process were marked by sharp dips in the RA responses. Furthermore, a two-component arsine recovery transient was observed, with one component assigned to "fast" processes involving the reaction of excess surface Ga and a second more gradual change which was assigned to the reaction of the first chemisorbed monolayer of Ga [14].

With these proposals in mind, if one re-examines the data depicted in figs. 1 and 2 one observes inflections in the TEGa exposure transients (as indicated) which may correspond to the Ga monolayer transition point. If this were the case, then a possible explanation for the dip in the arsine recovery traces depicted in figs. 1 and 2 is that the coverage of As required in order to completely react with the Ga overlayer was greater than the initial As coverage, i.e. that the Ga coverage following exposure to TEGa but prior to the introduction of the arsine exceeded one monolayer. The "dip" may thus arise simply because of a large excess of Ga resulting from adsorption and decomposition at higher pressures, which occurs prior to the surface returning to the stable initial condition. The relaxation of such surfaces back to a (2×4) reconstruction following reaction of excess Ga is not unusual and has been observed previously for the chemical beam epitaxy growth of GaAs by Chiu et al. [23].

Obviously, more data are required in order to test the validity of this proposal but the implication is that under conditions which approximate to those used in GaAs MOCVD, oversaturation of the surface with excess Ga may occur when using TEGa as a precursor.

Turning now to the arsine recovery traces depicted in fig. 2, again no satisfactory explanation exists for these data at the present time. However, under As recovery conditions starting from a Ga-rich $(4 \times 2)/(4 \times 1)$ surface in an MBE system we have observed inflections having a similar appearance although on a much longer time scale [12]. These inflections were shown to correlate well with changes in surface reconstruction as determined by RHEED. We therefore speculate here that the inflections depicted in fig. 2 represent a relaxation of a Ga-rich surface to an As-rich surface via an intermediate reconstruction.

4. Conclusions

We have demonstrated for the first time that RA transients can be obtained in real time, from (001) GaAs surfaces in an MOCVD reactor at 200 mbar, during exposure to TEGa and arsine. The RA response clearly shows the onset of the complete decomposition of TEGa at temperatures of ~ 623 K and the incorporation of Ga atoms onto lattice sites at these high overpressures. The heterogeneous nature of the subsequent reaction of arsine gas with the Ga-rich surface is indicated directly from the RA measurements. Via a comparison with VCE data, structure observed in the TEGa exposure and arsine recovery transients is tentatively attributed to oversaturation of the surface with Ga atoms during the first half of the growth cycle.

Acknowledgements

The authors gratefully acknowledge the support of the Science and Engineering Research Council in the UK for provision of capital equipment, research assistantships for R.D.H. and A.G.T. and postgraduate studentships for S.R.A., I.M.P. and A.S. In addition the support of Epichem Ltd. is gratefully acknowledged.

References

[1] J.F. McGilp, J. Phys. (Condens. Matter) 1 (1989) SB85.
[2] S. Andrieu and F. Arnaud d'Avitaya, Surf. Sci. 219 (1989) 277.
[3] Y.R. Shen, Nature 337 (1989) 519.
[4] D.S. Buhaenko, S.M. Francis, P.A. Goulding and M.E. Pemble, J. Phys. (Condens. Matter) 1 (1989) SB251.
[5] J. Geurts and W. Richter, in: Springer Proceedings in Physics, Vol. 22, Eds. G. Lelay, J. Derrien and N. Boccara (Springer, Berlin, 1987) p. 328.
[6] A.V. Annapragada, S. Salim and K.F. Jensen, in: Atomic Layer Growth and Processing, Eds. T.F. Kuech, P.D. Dapkus and Y. Aoyagi, Mater. Res. Soc. Symp. Proc., Vol. 222 (1991).

[7] H. Patel and M.E. Pemble, J. Cryst. Growth 116 (1992) 511.

[8] D.E. Aspnes, J. Vac. Sci. Technol. B 3 (1985) 1498.

[9] D.E. Aspnes, J.P. Harbison, A.A. Studna and L.T. Florez, J. Vac. Sci. Technol. A 6 (1988) 1327.

[10] E. Colas, D.E. Aspnes, R. Bhat, A.A. Studna, J.P. Harbison, L.T. Florez, M.A. Koza and V.G. Keramidas, J. Cryst. Growth 107 (1991) 47.

[11] D.E. Aspnes, J.P. Harbison, A.A. Studna, L.T. Florez and M.K. Kelly, J. Vac. Sci. Technol. B 6 (1988) 1127.

[12] S.R. Armstrong, M.E. Pemble, A.G. Taylor, B.A. Joyce, J.H. Neave, J. Zhang and D.A. Klug, Appl. Surf. Sci. 54 (1992) 493.

[13] L. Samuelson, K. Deppert, S. Jeppesen, J. Jönsson, G. Paulsson and P. Schmidt, J. Cryst. Growth 107 (1991) 68.

[14] G. Paulsson, K. Deppert, S. Jeppesen, J. Jönsson, L. Samuelson and P. Schmidt, J. Cryst. Growth 111 (1991) 115.

[15] B.Y. Maa and P.D. Dapkus, Appl. Phys. Lett. 58 (1991) 2261.

[16] S.M. Koch, O. Acher, F. Omnes, M. Defour and M. Razeghi, J. Appl. Phys. 68 (1990) 3364.

[17] O. Acher, S.M. Koch, F. Omnes, M. Defour, M. Razeghi and B. Drevillon, J. Appl. Phys. 68 (1990) 3564.

[18] D.E. Aspnes, R. Bhat, C. Caneau, E. Colas, L.T. Florez, S. Gregory, J.P. Harbison, I. Kamiya, V.G. Keramidas, M.A. Koza, M.A.A. Pudensi, W.E. Quinn, S.A. Schwarz, M.C. Tamargo and H. Tanaka, J. Cryst. Growth 120 (1992) 71.

[19] S.R. Armstrong, A.G. Taylor and M.E. Pemble, J. Phys. (Condens. Matter) 3 (1991) S85.

[20] D.S. Buhaenko, S.M. Francis, P.A. Goulding and M.E. Pemble, Vacuum 41 (1990) 972.

[21] J.A. McCaulley, V.R. McCrary and V.M. Donnelly, J. Phys. Chem. 93 (1989) 1148.

[22] R.D. Hoare, PhD Thesis, UMIST, Manchester, UK, 1990.

[23] T.H. Chiu, W.T. Tsang, J.E. Cunningham and A. Robertson, Jr., J. Appl. Phys. 62 (1987) 2302.

Applied Surface Science 69 (1993) 52–58
North-Holland

applied
surface science

Laser etching of silicon by chlorine: effect of post-desorption collisions and chlorine in-diffusion on the laser desorption yield

A. Aliouchouche [a], J. Boulmer [a], B. Bourguignon [b], J.-P. Budin [a], D. Débarre [a] and A. Desmur [a]

[a] *Institut d'Electronique Fondamentale, CNRS, Bât. 220, Université Paris-Sud, 91405 Orsay-Cedex, France*
[b] *Laboratoire de Photophysique Moléculaire, CNRS, Institut de Physico-Chimie Moléculaire d'Orsay, Bât. 213, Université Paris-Sud, 91405 Orsay-Cedex, France*

Received 2 June 1992; accepted for publication 30 October 1992

The yield of SiCl laser induced desorption in laser chemical etching of silicon by chlorine is found to never exceed 80%, even under conditions where the desorption rate is saturated. We show that this incomplete desorption is not a consequence of SiCl redeposition due to post-desorption collisions. Instead it is accounted for by in-diffusion and segregation of chlorine during the superficial melting of silicon induced by the pulsed laser.

1. Introduction

Laser chemical etching is potentially useful in the fabrication of microelectronic circuits [1–3]. To improve its practical applications, the physical mechanisms involved in this process have to be identified and well characterized. Laser etching of silicon by chlorine is one of the most extensively studied systems. Experiments have been conducted in very different conditions: in cells at chlorine pressures P ranging from 0.1 to 100 mbar [4–12], or at very low pressures in static conditions ($P < 10^{-4}$ mbar) [13–15] or with a pulsed molecular beam ($P < 10^{-6}$ mbar) [16–18]. Moreover, CW lasers (Ar^+ and Kr^+), pulsed lasers (excimer, nitrogen and YAG), or a combination of both, have been used. Depending on the experimental conditions, a number of basic processes occurring in laser etching of silicon have been identified: chemisorption of chlorine molecules or of photogenerated atoms, generation of photocarriers, influence of dopant impurities, laser induced desorption and melting of the silicon surface. Despite the large amount of available data, disagreement between various results or with model calculations still persist [21], showing that laser etching of silicon is not yet fully understood.

This paper deals with the laser desorption yield of the silicon chlorides which are formed by chlorine chemisorption between laser pulses. It was observed previously that the desorption yield is saturated at approximately 0.5 ML per pulse, even as the laser fluence is increased well above the melting threshold. Under such conditions, it is very unlikely that SiCl molecules may stay on the surface for tens of ns (the duration of surface melting). However, we do find that additional SiCl molecules may be desorbed at the next laser pulse even if the surface is not rechlorinated. We investigate experimentally the possible role of SiCl redeposition due to post-desorption collisions. It is found that post-desorption collisions cannot account for the observed desorption yields, and Cl diffusion to the bulk due to surface melting is invoked to explain the results.

2. Experimental

The experimental set-up and procedures have been described previously in more detail [16–18].

Laser etching of B doped Si(100) samples is studied in a vacuum chamber equipped with a conventional pumping system allowing a base pressure of 10^{-7} mbar. A pulsed molecular beam of chlorine (pulse duration ≈ 0.5 ms) directed towards the silicon surface delivers chlorine doses adjustable from 10^{12} to 10^{18} $Cl_2/cm^2/pulse$. This pulsed injection mode, the continuous pumping and liquid nitrogen trapping of chlorine leaving the silicon surface allow us to maintain the background pressure in the reactor chamber in the 10^{-7} mbar range and to control the dynamics of the chlorination process.

Samples are generally laser etched at 308 nm (Questek model 2040). A homogeneous part of the laser beam (dimension: 2×8 mm^2, inhomogeneity $< 5\%$) is selected and imaged onto the sample. The laser flux incident onto the surface is adjusted up to 1 J/cm^2 by varying the magnification of the imaging system and by adding Spectrosil plates in the optical path.

The etching process is followed pulse-by-pulse by a double time-of-flight (TOF) mass spectrometer [17] which provides, at every laser pulse, the mass spectrum of the etched species leaving the surface in a geometrically defined solid angle and possessing the velocity determined by the time delay between the laser pulse and the TOF operation. In preceding studies, neutral species desorbed from the surface were ionized by a pulsed electron beam before being mass analyzed. It was shown that SiCl is the major laser-desorbed species so that the amplitude of the SiCl TOF signal is used as a measure of the etch rate and calibrated in Å/pulse by comparison with ex-situ profilometer measurements. For the present study, desorbed SiCl molecules are ionized by a second laser beam at 308 nm (Lambda Physik model LPX105i) according to the following two-photon resonant process [19,20]:

$$SiCl(X^2\Pi) + h\nu(308\ nm) \rightarrow SiCl(B^2\Sigma^+),$$

$$SiCl(B^2\Sigma^+) + h\nu(308\ nm) \rightarrow SiCl^+ + e^-.$$

Unlike ionization by electron collisions, laser induced resonant ionization is a selective and very efficient process. As a consequence, SiCl is the only detected mass peak, so that a simplified low resolution TOF mass spectrometer could be used. The detection sensitivity is improved by about two orders of magnitude, by comparison with electron impact ionization, and is about 10^{-5} ML per pulse. The velocity distribution of SiCl molecules desorbed from the surface is measured by scanning the time delay between the etching laser pulse and the ionization laser pulse.

3. Experimental results

Different pulsed lasers (pulse duration from 7 to 25 ns) were used: KrF (248 nm), XeCl (308 nm), N_2 (337 nm), Nd:YAG (355 or 532 nm). No significant laser wavelength dependence of the process was observed, but it was shown that only excimer lasers exhibit beam properties adequate for a practical etching process. However, an N_2 laser was successfully used for some specific measurements where an XeCl laser was necessary for the detection of the etching products and when only one excimer laser was available.

Under our experimental conditions, laser etching proceeds in two steps: spontaneous chemisorption of chlorine molecules that saturates when all available dangling bonds at the surface are occupied by chlorine atoms (as illustrated by fig. 1), followed by laser desorption of the chlorinated

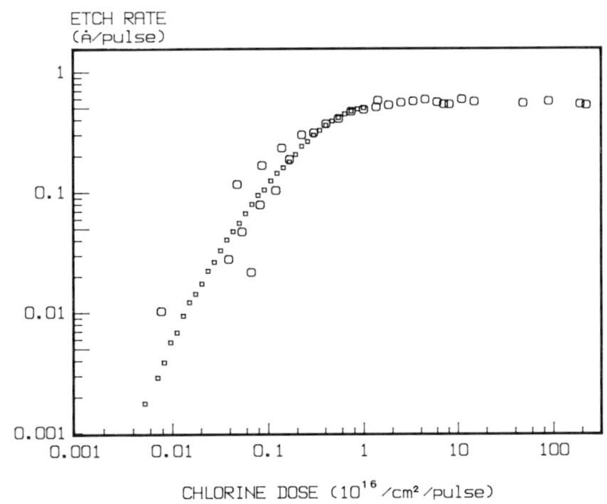

Fig. 1. Etch rate as a function of the incident chlorine dose injected onto the surface by a pulsed molecular beam. The laser wavelength is 308 nm (□) or 337 nm (○) and the laser fluence is above the saturation threshold (480 mJ/cm^2).

layer that requires laser induced melting of the silicon surface [4,5,13,14,18].

The etch rate dependence on laser fluence (fig. 2) exhibits two main features: a laser fluence threshold, at ~ 340 mJ/cm^2, which coincides with the measured laser induced melting threshold of silicon [18], and a saturation plateau, above 480 mJ/cm^2, corresponding to the removal of about half a monolayer of silicon at each laser pulse. The existence of this plateau suggests that the whole chlorinated layer, created during the chemisorption step, is desorbed when a laser fluence of 480 mJ/cm^2 is reached. The surface remains melted for a duration that increases rapidly with laser fluence: we measured a melting duration of about 20 ns at 500 mJ/cm^2 and 60 ns at 650 mJ/cm^2 [21,22]. SiCl desorption is expected to continue as long as the surface remains melted and as long as SiCl is present at the surface. Therefore it might be concluded that the whole chlorinated layer is totally desorbed provided that surface melting of the silicon surface lasts for ~ 20 ns or more.

This commonly accepted hypothesis is in contradiction with the experimental result shown on fig. 3. In this experiment, the silicon surface is saturated by one single chlorine dose (above 2×10^{16} Cl_2/cm^2, see fig. 1); then a series of laser

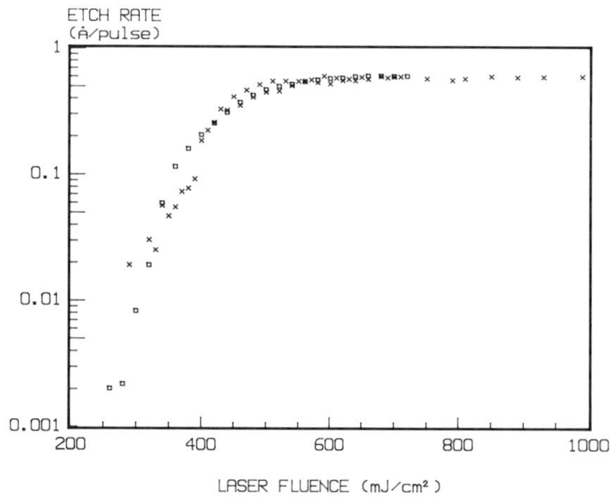

Fig. 2. Etch rate as a function of laser fluence at 248 nm (□) and 308 nm (×) when chlorine is injected onto the surface by a pulsed molecular beam. The chlorine dose is above the saturation threshold (2×10^{16} Cl_2/cm^2/pulse).

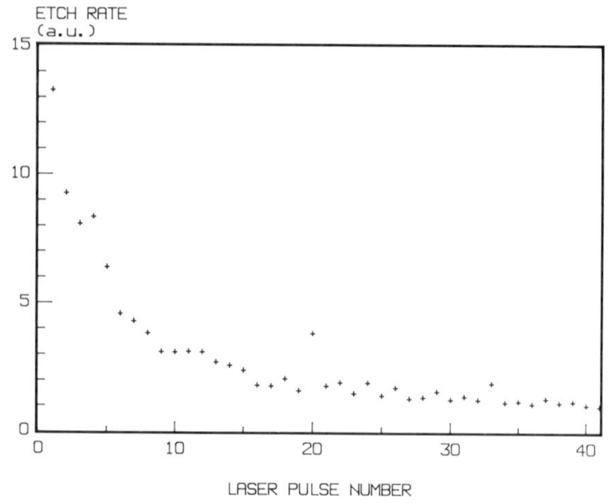

Fig. 3. Unaveraged record of the etch rate at 308 nm as a function of the laser pulse number after the silicon surface was submitted to a single chlorine dose large enough to saturate the chemisorption process. The laser repetition rate is 10 Hz.

pulses irradiate the surface while the amount of SiCl molecules desorbed at each laser pulse is recorded. The hypothesis of total desorption in one laser pulse would imply that only the first laser pulse induces significant SiCl desorption. On the contrary, fig. 3 (recorded without any averaging) shows that such is not the case: the SiCl signal induced by at least the first ten 600 mJ/cm^2 pulses is quite significant.

This partial desorption is also directly observed by profilometer measurements when samples are etched by 2^{14} laser pulses in three different experiments: the chlorine dose is injected before each laser pulse or only before a series of 2 or 3 laser pulses. The measured etch depths are respectively 9000, 5800 and 4100 Å, which implies that the etch rates for the first, the second and the third laser pulse following one chlorine dose injection are, respectively, 0.55, 0.16 and 0.04 Å/pulse.

To improve the signal-to-noise ratio, data similar to those of fig. 3 can be accumulated over many sequences of one chlorine dose injection followed by a given number of laser pulses. For such measurements, possible spurious effects due to the experimental procedure must be carefully examined. Most of these effects are the consequences of chlorination from both the pulsed chlorine beam and the residual vacuum:

(1) the dynamic range of the molecular beam intensity modulation by the pulsed valve is found to be of the order of 2×10^4.

(2) the time dependence of the molecular flux on the surface shows a main pulse (time duration ~ 0.5 ms), followed by a weak tail corresponding to the time required for the molecules to leave the valve nozzle (1 mm in diameter, 30 mm long) and to be further pumped by the liquid nitrogen trap and the diffusion pump.

(3) the residual Cl_2 pressure ($P < 5 \times 10^{-7}$ mbar for the largest chlorine dose) may induce a measurable chlorination of the surface.

Since very low chlorine coverages are used, these three spurious effects are difficult to characterize directly but may play a major role in these experiments. Therefore they were checked by changing the characteristics of the sequences of measurements: chlorine injection and laser repetition rates, and time delay between the chlorine injection and the first laser pulse of the sequence. We found that the chlorination from the molecular beam becomes negligible 70 ms after the main chlorine pulse, and that the sample may be significantly chlorinated by the residual gases for delays exceeding 350 ms. Then, for the experiment to be valid, the first and the last pulses in the sequence to be taken into account should occur between 70 and 350 ms after the pulsed injection of chlorine. Provided this condition is fulfilled, the measured SiCl desorption rates induced by the first three laser pulses do not depend on the sequence characteristics by more than 15%. The asymptotic behavior of the desorption rates of fig. 3 cannot be clearly interpreted since the upper delay limit is exceeded. To account for the above-mentioned or any other unknown effects responsible for this asymptotic behavior, the asymptote is substracted from the data, and, in the following, the only 4 or 5 first laser pulses are used.

The partial desorption process could also be attributed to surface diffusion of chemisorbed chlorine from unirradiated areas in the time between two successive laser pulses. However, we do not observe any significant modification in the partial desorption yields when the laser spot diameter is changed by an order of magnitude (from 0.3 to 3 mm). Moreover, when the laser repetition rate is changed from 5 to 50 Hz, the laser shot number dependence of the desorption yield does not vary by more than 20%. Since partial desorption depends neither on the spot size nor on the repetition rate, surface diffusion is certainly not involved in this process.

On the other hand, two other processes could explain the observations of fig. 3: bulk diffusion of chlorine when laser induces silicon melting, or redeposition of etched products due to post-desorption collisions. This latter hypothesis was examined in more detail.

Post-desorption collisions were first observed by Cowin et al. [23] in laser induced desorption experiments. Further calculations [14,24,25] and experiments [13–15] showed that these collisions lead to significantly modified angular and velocity distributions of desorbed molecules if the average number of collisions N_c per desorbed molecule is not much smaller than 1. In our experimental conditions, for laser desorption of one SiCl monolayer, a rough evaluation leads to $N_c \approx 10$ [21,23], so that post-desorption collisions would become negligible only for etch rates of the order of 0.01 Å/pulse or less. Kelly and Dreyfus [25] calculated that, due to post-desorption collisions, 20 to 25% of the desorbed molecules can return to the surface and redeposit on it. The redeposition rate depends on the average number of post-desorption collisions and should decrease with the etch rate.

This hypothesis was tested by experiments similar to those of fig. 3, during which the etch rate was scanned over five orders of magnitude by varying the chlorine dose (see fig. 1). Data were accumulated over many sequences consisting of one chlorine dose followed by 16 laser pulses. Such results are shown on fig. 4, where the pulsed molecular beam is operated at a 0.5 Hz repetition rate and the laser at 8 Hz. For etch rates ranging from 10^{-5} to 0.6 Å/pulse, only about 70% of the chlorinated layer is desorbed by the first laser pulse. Should redeposition be involved in these experiments, it would imply that post-desorption collisions are present for etch rates as low as

10^{-5} Å/pulse, which is supported neither by Monte Carlo calculations [14,24] nor by rough evaluations [21,23].

The etch rate threshold above which post-desorption collisions do occur, and are therefore able to induce redeposition, can be determined by measuring the etch rate above which the velocity distribution of desorbed molecules is etch-rate dependent. For this purpose, we have measured the velocity distributions of desorbed SiCl molecules along the surface normal, where they are known to be very sensitive to such collisions [15,23], and we scanned the etch rate over more than 3 orders of magnitude by varying the chlorine dose. Fig. 5 shows the measured SiCl velocity distributions defined by $f(v) = \mathrm{d}N(v)/\mathrm{d}v$ where $\mathrm{d}N(v)$ is the number of detected SiCl molecules with a velocity ranging from v to $v + \mathrm{d}v$. The velocity resolution was $\mathrm{d}v/v = 0.12$. All distributions are normalized to unity, but the total number N_{tot} of detected molecules in the solid angle defined by the TOF spectrometer and the ionization laser beam ($\mathrm{d}\Theta = 8°$, $\mathrm{d}\Phi = 1°$) is indicated in the figure caption. N_{tot} is expressed in arbitrary units; the highest value shown on fig. 5 ($N_{tot} = 10^4$) corresponds to the saturation etch rate (0.6 Å/pulse). As N_{tot} decreases from its maximum value, $f(v)$ shifts to lower velocities until $N_{tot} \approx 100$, and then does not depend any more on N_{tot}. This effect is clearly shown in fig. 6, where the mean energy $\langle E \rangle$ of the desorbing molecules, as deduced directly from experimental points of fig. 5, is plotted as a function of N_{tot}. It is observed that $\langle E \rangle$ first decreases with N_{tot} and then stays at approximately 0.22 eV for N_{tot} values smaller than 100. Therefore we conclude that post-desorption collisions do occur only for $N_{tot} > 100$, or etch rates larger than 0.01 Å/pulse.

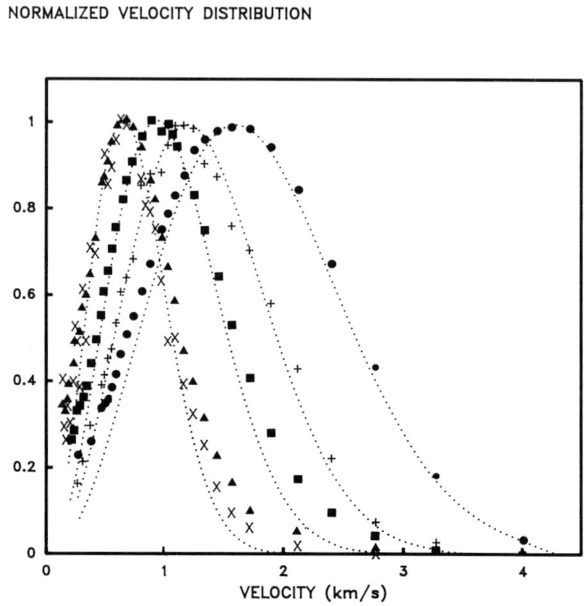

Fig. 5. Normalized velocity distribution functions of SiCl desorbed along the silicon surface normal for different etch rates. Scaling factors and total numbers of desorbed molecules within the detector solid angle are respectively: 2820 and 10000 (●), 560 and 1387 (+), 417 and 871 (■), 45 and 71 (▲), 10 and 13 (×). Fits by Maxwell distributions are for temperatures of 9850, 5500, 3400 and 1650 K.

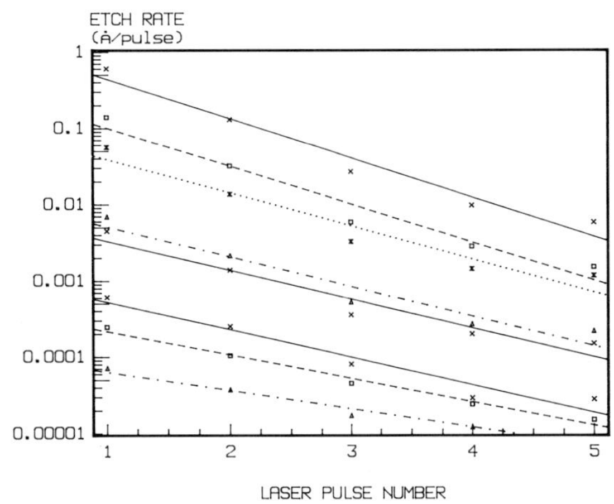

Fig. 4. Etch rate at 308 nm as a function of the laser pulse number following one single pulsed injection of chlorine. The etch rate for the first laser pulse is scanned over 4 orders of magnitude by changing the chlorine dose, as shown on fig. 1. The laser fluence is 600 mJ/cm^2. The straight lines are exponential fits.

We made no assumption so far about the velocity distribution. We observe that the measured distributions are well fitted by the following Maxwellian distribution:

$$f(v) \propto v^2 \exp(-mv^2/2kT)$$

(k is the Boltzmann constant). With this distribution, T is related to the mean energy by: $\langle E \rangle = 3kT/2$. When plotted in terms of temperature,

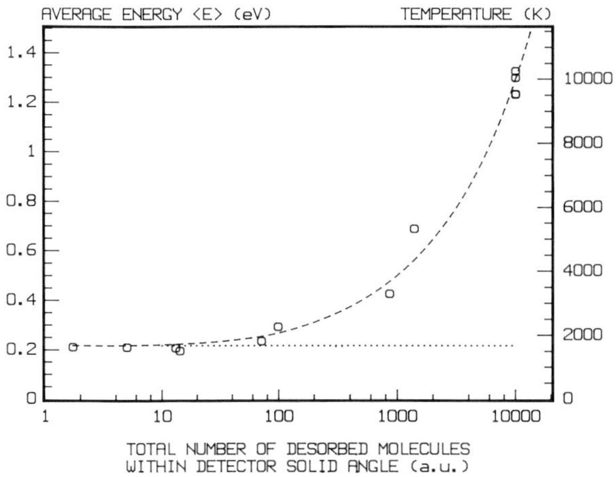

Fig. 6. Average energy $\langle E \rangle$ of SiCl molecules desorbed along the silicon surface normal as a function of the total number of SiCl measured within the detection solid angle and for different etch rates. $\langle E \rangle$ is also scaled in terms of the temperature of the corresponding Maxwell distribution. The melting temperature of silicon is indicated by the horizontal dashed line (1680 K).

the data of fig. 6 show that, when velocity distributions are not perturbed by post-desorption collisions, the temperature of the laser desorbed molecules is very close to the melting temperature (1680 K) of the silicon surface. These results strongly support our previous conclusions [18] that the mechanism of laser etching under our experimental conditions is evaporation from the melted surface.

4. Discussion

From the results shown in figs. 5 and 6, we conclude that post-desorption collisions may be considered negligible provided that the etch rate is maintained under 1% of its largest value, that is to say under 6×10^{-3} Å/pulse. Referring again to the data of fig. 4, it is now clear that SiCl molecules desorbed after the first laser pulse of a sequence cannot originate from redeposition induced by post-desorption collisions, since this effect is still observed for etch rates as low as 10^{-5} Å/pulse. As a consequence, the only physical process capable of inducing this effect is bulk diffusion of chlorine during laser induced melting of the silicon surface.

During surface melting, two processes compete: desorption and diffusion. Our results are all consistent with the hypothesis that the desorption mechanism is evaporation from the melted surface; in such a case, the desorption rate mainly depends on surface coverage. On the other hand, diffusion mostly depends on the chlorine concentration gradient, which is very strong at the beginning of the laser pulse. The diffusion constant D of chlorine in liquid silicon is not known. However, the diffusion constant in liquid silicon depends little on the type of atom; it varies in the range 1.5×10^{-4} (for Sb) to 7.8×10^{-4} cm^2 s^{-1} (for Th) [26]. Assuming $D = 2 \times 10^{-4}$ cm^2 s^{-1}, the diffusion length in 20 ns (the melting duration at a laser fluence of 480 mJ/cm^2 [21,22]) would be 200 Å. Therefore, it appears quite reasonable to estimate that 20 to 30% of the chlorine atoms are removed from the surface by diffusion.

Why most chlorine atoms that have diffused are again available for desorption at the next laser shot is probably related to segregation at the solid/liquid interface during recrystallization. The segregation coefficient k_i (the ratio of the impurity concentrations in the solid and the liquid) of chlorine in silicon is not known. Moreover, it depends on the velocity of the solid/liquid interface due to non-equilibrium effects when surface melting is induced by laser pulses [27,28]. However k_i values for In and Bi in Si of the order of 10^{-4} in equilibrium (zero melt-front velocity) and of 0.1 in our experimental conditions (melt-front velocities between 1 and 5 m/s) have been reported [26–28]. Therefore, it is quite possible that the segregation coefficient would also be small for chlorine, even though this argument is to be taken with care, since the segregation coefficient sometimes varies erratically from one element to another. In the hypothesis of a small segregation coefficient for chlorine in silicon, diffusion strongly reduces the surface coverage, thus limits the etch rate, most of the desorption probably occurring at the beginning of the laser pulse. Segregation during solidification brings back to the surface most of the chlorine atoms at the end of surface melting; these atoms become available again for desorption at the next laser pulse.

More experiments are underway in our laboratory to investigate chlorine diffusion and segregation in silicon.

Acknowledgements

This work was supported by the Centre National d'Etudes des Télécommunications, the Groupement Circuits Intégrés Silicium, and the Groupement Dynamique des Réactions Moléculaires of the CNRS. The authors are indebted to Photon Science Instruments Corp. for lending a Lambda Physik excimer laser.

References

[1] D. Bäuerle, Chemical Processing with Lasers, Vol. 1 of Springer Series in Material Science (Springer, Berlin, 1987).
[2] I.W. Boyd, Laser Processing of Thin Films and Microstructures, Vol. 3 of Springer Series in Material Science (Springer, Berlin, 1987).
[3] T.J. Chuang, in: Interfaces under Laser Irradiation, Eds. L.D. Laude, D. Bäuerle and M. Wautelet (Nijhoff, Dordrecht, 1987) p. 235.
[4] W. Sesselmann, H. Hudeczek and F. Bachmann, J. Vac. Sci. Technol. B 7 (1989) 1284.
[5] R. Kullmer and D. Bäuerle, Appl. Phys. A 43 (1987) 227.
[6] P. Mogyorosi, K. Piglmayer, R. Kullmer and D. Bäuerle, Appl. Phys. A 45 (1988) 293.
[7] R. Kullmer and D. Bäuerle, Appl. Phys. A 47 (1988) 377.
[8] W. Sesselmann and T.J. Chuang, Surf. Sci. 176 (1986) 67.
[9] J.H. Brannon, J. Vac. Sci. Technol. B 7 (1989) 1064.
[10] J.L. Peyre, C. Vannier, D. Rivière and G. Villela, Appl. Surf. Sci. 36 (1989) 313.
[11] Y. Horiike, N. Hayasaka, M. Sekine, T. Arikado, M. Nakase and H. Okano, Appl. Phys. A 44 (1987) 313.
[12] T. Arikado, M. Sekine, H. Okano and Y. Horiike, Mater. Res. Symp. Proc. 29 (1984) 167.
[13] J.C. Kools, T.S. Baller and J. Dieleman, in: Photochemical Processing of Electronic Materials, Eds. I.W. Boyd and R.B. Jackman (Academic Press, New York, 1991) p. 339.
[14] T.S. Baller, Thesis, 1990.
[15] T.S. Baller and J. Dieleman, Appl. Surf. Sci. 43 (1989) 409.
[16] J. Boulmer, B. Bourguignon, J.P. Budin and D. Débarre, Appl. Surf. Sci. 43 (1989) 424.
[17] J. Boulmer, B. Bourguignon, J.P. Budin and D. Débarre, Chemtronics 4 (1989) 165.
[18] J. Boulmer, B. Bourguignon, J.P. Budin, D. Débarre and A. Desmur, J. Vac. Sci. Technol. A 9 (1991) 2923.
[19] K.P. Huber and G. Herzberg, Molecular Spectra and Molecular Structure, IV, Constants of Diatomic Molecules (Van Nostrand Reinhold, New York, 1979) p. 596.
[20] H. Bredohl, Ph. Demoulin, Y. Houbrechts and F. Mélen, J. Phys. B 14 (1981) 1771.
[21] J. Boulmer, B. Bourguignon, J.P. Budin, D. Débarre and A. Desmur, in: Laser Ablation of Electronic Materials: Basic Mechanisms and Applications, Eds. E. Fogarassy and S. Lazare (Elsevier, Amsterdam, 1992) p. 239.
[22] A. Desmur, Thesis, Orsay, 1992.
[23] J.P. Cowin, D.J. Auerbach, C. Becker and L. Wharton, Surf. Sci. 78 (1978) 545.
[24] I. NoorBatcha, R.R. Lucchese and Y. Zeiri, J. Chem. Phys. 86 (1987) 5816.
[25] R. Kelly and R.W. Dreyfus, Surf. Sci. 198 (1988) 263.
[26] B. Leroy, Rev. Phys. Appl. 21 (1986) 467.
[27] R.F. Wood, J.R. Kirkpatrick and G.E. Giles, Phys. Rev. B 23 (1981) 5555.
[28] R.F. Wood, Phys. Rev. B 25 (1982) 2786.

Applied Surface Science 69 (1993) 59–64
North-Holland

applied
surface science

ArF excimer laser deposited tin oxide films studied by "in situ" surface diagnostics and by synchrotron radiation induced UV photoemission

R. Larciprete [a], E. Borsella [a], P. De Padova [b], M. Fanfoni [c], M. Mangiantini [b] and P. Perfetti [b]

[a] *Dipartimento INN-SVIL, ENEA, P.O. Box 65, 00044 Frascati (RM), Italy*

[b] *Istituto di Struttura della Materia del CNR, Via E. Fermi 34, 00044 Frascati (RM), Italy*

[c] *Dipartimento di Fisica dell'Università di Roma "Tor Vergata", Via O. Raimondo, 00173 Rome, Italy*

Received 2 June 1992; accepted for publication 30 October 1992

Tin oxide films were deposited by ArF excimer laser induced photolysis of $Sn(CH_3)_4/O_2$ mixtures. The "in situ" Auger analysis was used to determine the effect of the gas-phase oxygen concentration on the film composition. The surface properties of the films were investigated by synchrotron radiation UV photoemission spectroscopy. To this aim samples were transferred to a synchrotron radiation facility, without any exposure to air and kept in dynamical high vacuum. The results obtained showed that the valence-band spectrum was characteristic for SnO_2 and that the Sn 4d core peak did not have any contribution due to metallic tin. Furthermore, by tuning the synchrotron radiation photon energy, in order to vary the photoelectron escape depth, we got the indication that the oxygen vacancies responsible for the weak emission in the band-gap region were localized at the sample surface.

1. Introduction

Tin oxide SnO_2 exhibits a rather uncommon combination of optical (it is transparent in the visible and reflecting in a wide region of the IR spectrum) and electronic properties (it can be made highly conductive when sufficiently doped) which makes it a very interesting material for a variety of technological fields ranging from optoelectronics, solar cell manufacture, heating element fabrication and gas monitoring.

During the last years tin oxide films have been achieved with various techniques such as chemical vapour deposition (CVD) [1], spray pyrolysis [2–4] or oxidation of evaporated tin films [5], which however very often require high temperatures. Since usually applications in optoelectronics device manufacture demand low substrate temperatures, significant attention has been paid to growth methods which are promising in this direction. Interesting results have been achieved using photochemical deposition processes enhanced or induced by UV photons: SnO_2 films were achieved by irradiating $SnCl_4/SbCl_5/O_2$ mixtures with a Hg lamp [6] or driving the reaction between $SnCl_4$ and N_2O by ArF laser radiation [7].

In the present work we have studied the deposition of tin oxide films induced by ArF laser photolysis of tetramethyltin $Sn(CH_3)_4$ (TMT) diluted in O_2. Recently the UV gas-phase photochemistry of TMT was investigated in our laboratory using spectroscopic diagnostics such as multi-photon ionization (MPI) [8], resonance enhanced MPI [8] and analysis of visible spontaneous emission [9,10]. The evidence that TMT could be efficiently dissociated when excited at 193 nm [8] indicated the suitability of this organometallic as precursor for the ArF excimer laser-CVD of tin and tin oxide films.

Surface analysis of the films made by "in situ" Auger electron spectroscopy (AES) showed that the photolysis of pure TMT produced only polymeric organotin films containing carbon, whereas

0169-4332/93/$06.00

the irradiation of TMT in excess of oxygen resulted in the occurrence of oxidation and in the complete elimination of the organic contamination. AES, however, does not permit the oxidation state of the metal to be determined, since, as it is well known [11], the two stable oxides of tin, SnO and SnO_2, present identical Auger spectra. It has been shown [12,13] that the valence band (VB) of SnO_2 presents distinctive features which result in close agreement with the calculated band structure of the material [13–15], whereas a different shape is exhibited by the VB of SnO [16]. Analysis of the VB of the tin oxide laser-CVD samples appeared suitable, therefore, for the identification of the oxidation state of tin in the films.

To this aim, UV photoemission spectroscopy (UPS) being a surface-sensitive technique, it was compulsory to minimize the surface modification caused by placing the sample in a poor vacuum environment. The samples were then transferred to the synchrotron radiation beam line without breaking high vacuum, by use of a removable chamber equipped with an ion pump/portable battery system. The UPS analysis of the VB and of the 4d core levels of Sn clearly showed that the laser-CVD films consisted of SnO_2, whereas the presence of oxygen vacancies localized at the surface was indicated by the weak emission originated in the band-gap region.

2. Experimental

The photodeposition apparatus consists of a growth chamber and a surface analysis chamber separated by a gate valve. The analysis chamber (background pressure 5×10^{-10} Torr) is equipped with an ion gun and an Auger spectrometer consisting of a 3-grid reverse view LEED (VG RVL-640) system, operated as a retarding-field analyzer electron spectrometer, coupled with an additional electron gun (VG LEG 31).

Substrates were Ar^+ sputtered, moved into the deposition chamber and irradiated at room temperature, in perpendicular geometry, in the presence of the gas mixture. The mildly focused ArF excimer laser beam (Lambda Physik 100, 12–15 ns FWHM) entered the growth chamber through a sapphire window. Typical laser fluence values on the substrate were 10–20 mJ/cm^2. After deposition the sample was transferred to the other chamber and analyzed by AES.

For the UPS study an additional chamber, necessary to contain the sample during the transfer to the synchrotron radiation beam line, was connected to the deposition section. Vacuum was maintained in the removable chamber by an ion pump which was kept "on" during the transfer, power being supplied by a portable battery. The whole transfer procedure to the Grasshopper (PULS) laboratory of the Adone (Frascati) storage ring lasted almost 30 min, during which dynamical pumping kept the sample at a pressure around 2×10^{-8} Torr. The UHV analysis chamber ($p \approx 10^{-11}$ Torr) attached to the synchrotron beam line is equipped with an ion gun and a double-pass cylindrical mirror analyzer (CMA) (Perkin-Elmer PHI 255G), which allow, respectively, Ar^+ ion sputtering and Auger analysis of the surface. For the UPS study the CMA was used to measure the energy distribution curve of the photoelectrons emitted when the sample was excited with UV photons.

3. Results and discussion

The first stage of this work consisted in the characterization of films deposited by dissociating pure TMT at 193 nm. Films were deposited irradiating the organometallic at a pressure in the range 10–80 mTorr in the vicinity of a substrate at room temperature and were analyzed by "in situ" AES. In agreement with results obtained with other deposition techniques such as plasma deposition [17] or glow discharge [18], TMT produced carbon-rich polymeric organotin films containing Sn and C. A typical AES spectrum is shown in fig. 1a.

Investigation of the TMT photolysis at 193 nm [8–10] showed the occurrence of complete dissociation producing ground state and excited free Sn atoms and hydrocarbon radicals, such as CH_3 [8] or excited CH [9,10]. The presence of a parallel channel forming ligand carrying photofrag-

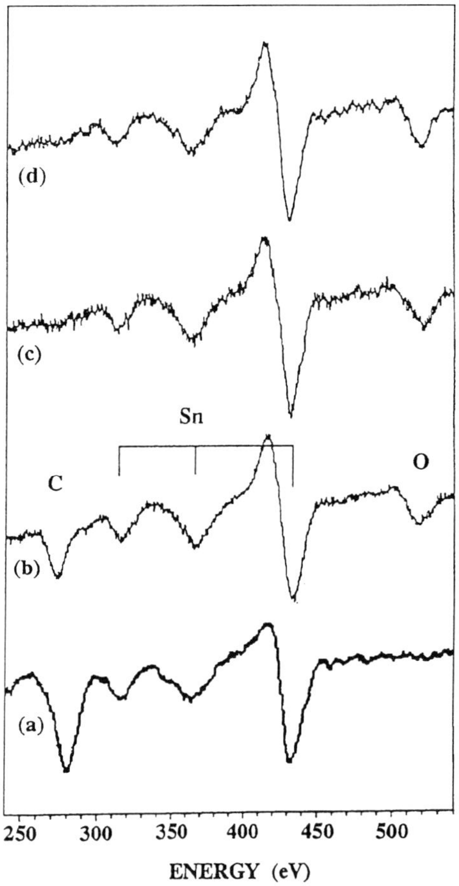

Fig. 1. "In situ" Auger spectra of samples deposited by ArF laser-CVD dissociating TMT ($p = 3$ mTorr) diluted in O_2 at the following TMT:O_2 concentrations, (a) 1:0; (b) 1:0.1; (c) 1:1; (d) 1:50.

ments is, however, strongly supported by the high organic contamination presented by the films.

It has been shown that the level of carbon incorporated in films deposited from organic tin precursors could be successfully reduced by performing the reactions in oxygen. In the case of glow discharge deposition of films from TMT/O_2 mixtures [19], the role of oxygen as non-polymerizing species was successfully utilized to lower the carbon content of the films, by producing CO and CO_2.

A similar behaviour took place when TMT was irradiated at 193 nm in excess of O_2 which allowed the reaction of the organometallic photoproducts with oxygen atoms [20]. Figs. 1b–1d show the AES spectra relative to films grown by ArF laser photolysis of TMT:O_2 mixtures having, respectively, 1:0.1, 1:1 and 1:50 concentrations and a fixed TMT pressure of 3 mTorr. The occurrence of oxidation is indicated by the appearance of the oxygen peak at 512 eV, whereas only a slight change in the Sn/O peak-to-peak height ratio is observed.

Since, at variance with the Auger spectrum, the valence band is sensitive to the change from SnO to SnO_2 [12,13,16], we used UPS spectroscopy to investigate the composition of the laser-CVD samples. To this aim three films deposited in large excess of oxygen (TMT:O_2 = 1:100) on Si and Au substrates were separately brought to the synchrotron radiation beam line and analyzed [21]. Since the results achieved were completely reproducible, in the following we report the results relative to a single sample deposited on a Si(100) wafer. Synchrotron radiation photons having energies in the 50–150 eV range were used as excitation source, whereas Auger spectroscopy was used to check the composition of the surface.

The AES spectrum of the "as-received" sample showed a trace of C (the Sn (433 eV)/C(275 eV) peak-to-peak height ratio was about 13) caused by the transfer of the sample from the laser-CVD apparatus to the synchrotron radiation facility, which disappeared completely after 3 min of Ar^+ sputtering (500 eV, 10 $\mu A/cm^2$). No significant change in the Sn/O peak-to-peak height ratio was observed after a sputtering of 7 min.

The UPS valence-band profiles, induced by 50 eV photons, measured after 0, 3 and 7 min of Ar ion sputtering are reported in fig. 2. The band measured after ion cleaning clearly shows the spectral features A, B and C, which can be taken as fingerprints of SnO_2 [12–15]. In fact, the prominent peak A, appearing at a binding energy of 5.2 eV below the Fermi level, derives primarily from O(2p) lone-pair-like states, the structure B, located about 8 eV below the Fermi level, has been attributed to the hybridization between Sn 5p and O 2p orbitals, whereas the lower part of the valence band (feature C) to the strong interaction between Sn 5s and O 2p orbitals.

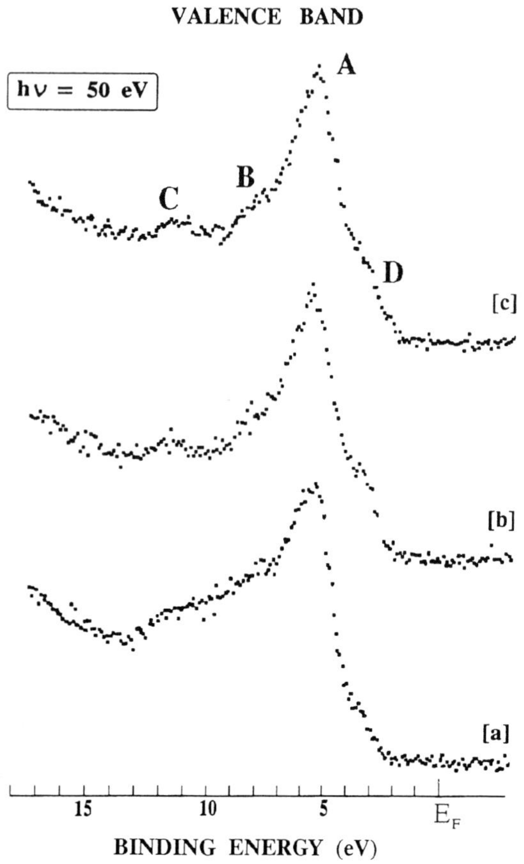

Fig. 2. Valence-band photoemission spectra of ArF laser deposited tin oxide films induced by 50 eV synchrotron radiation: "as-received" sample (curve (a)); after 3 (curve (b)) and 7 (curve (c)) min of Ar^+ sputtering (500 eV, 10 $\mu A/cm^2$).

The attribution of the observed VB to SnO_2 can be further proved by the comparison with the VB of a SnO sample. The latter can be prepared by exposing a clean tin slab to molecular oxygen. In fact, Powell [11] demonstrated, by electron energy loss spectroscopy, that the oxide layer formed after room temperature oxidation of tin is made essentially of SnO. In agreement with this result, the VB we have measured in a Sn sample dosed with 45000 L of O_2, which is displayed in fig. 3a, results similar to the VB curve reported by Lau and Wertheim [12], and more recently by Themlin and co-workers [16], for SnO samples. The profile of this curve results clearly different from the VB of the ArF laser-CVD sample (the VB curves are compared in fig. 3), in agreement with the theoretical calculations of the density of states for SnO_2 [13–15] and for SnO [16].

Further information is given by the shoulder just at the edge of the valence band indicated as D in fig. 2. This feature has been widely studied [22–24] and attributed to the presence of oxygen vacancies. It is quite clear from the spectra that the effect of the Ar^+ ion bombardment is that of increasing the oxygen deficiency of the SnO_2 surface. As fig. 2a shows, this emission, although very weak, is present in the "as-received" sample, too. This is not surprising since in our sample the structure of the surface is strongly determined by the sudden interruption of the laser irradiation, which is quite likely to result in the presence of oxygen vacancies.

The contribution of the subsurface region to the total emission can be enhanced with respect to the surface contribution, by increasing the photon energy from 50 to 150 eV, which results in a variation of the escape depth of the photoelec-

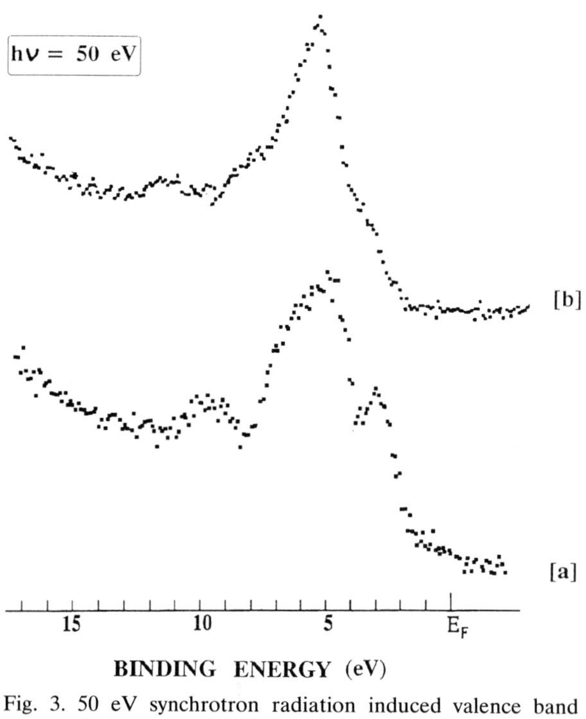

Fig. 3. 50 eV synchrotron radiation induced valence band photoemission spectra of: (a) a slab of metallic tin exposed to 45000 L of O_2; (b) ArF laser-CVD tin oxide film.

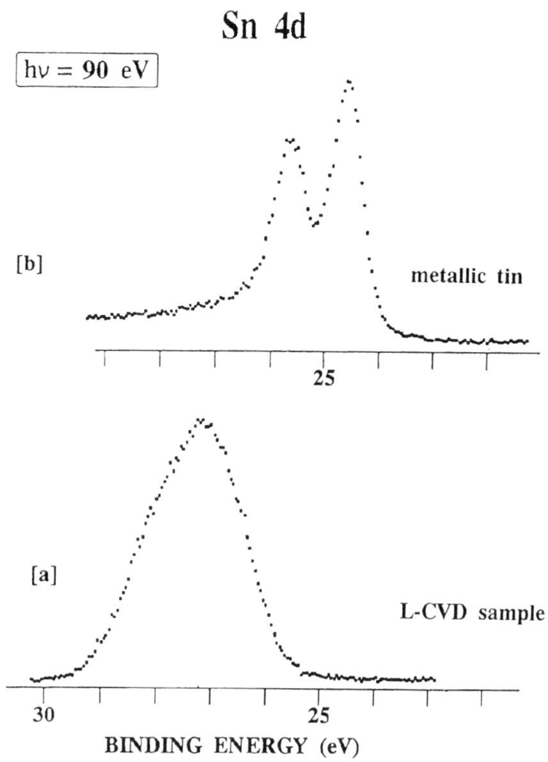

Fig. 4. Sn 4d core-level photoemission spectra induced by 90 eV synchrotron radiation in an ArF laser-CVD tin oxide film (curve (a)). For comparison curve (b) shows the corresponding spectrum relative to a slab of metallic Sn.

trons. The VB of the "as-received" sample induced by 150 eV photons showed a lower contribution due to the emission of the band-gap region (feature D), in agreement with a surface localization of the oxygen vacancies.

The information given by the analysis of the Sn core level was consistent with the former results. Fig. 4 shows the Sn 4d core peak relative to the laser-CVD sample, centred at a binding energy of 27.2 eV, together with the $4d_{3/2}$ and $4d_{5/2}$ components, respectively located at 25.7 and 24.6 eV, of the corresponding 4d peak of a clean Sn slab. The profile of the laser CVD sample did not show any contribution due to the presence of metallic tin, demonstrating that all Sn atoms were oxidized.

4. Conclusions

Tin dioxide films were deposited by ArF laser photolysis of TMT diluted in oxygen. The "in situ" Auger analysis was used to evidence the effect of the gas-phase composition in removing organic contamination and determining the [O]/[Sn] ratio in the films. The possibility of transferring the samples under vacuum and the availability of a synchrotron radiation beam line allowed us to analyze the sample by UPS.

The results obtained in the case of films deposited irradiating a mixture having the $p(O_2)/p(TMT)$ ratio equal to 100 showed that the VB was characteristic of SnO_2 and that the Sn 4d core peak did not have any contribution due to metallic tin. Furthermore, by tuning the synchrotron radiation photon energy in order to vary the photoelectron escape depth, we got the indication that the oxygen vacancies responsible for the weak emission in the band-gap region were localized at the sample surface.

Therefore, we have demonstrated that tin oxide films having a bulk composition close to stoichiometric SnO_2 can be deposited by ArF excimer laser photolysis of TMT/O_2 mixtures.

Acknowledgements

The authors wish to thank M. Nardelli and S. Priori for their precious technical assistance. The UHV transfer system was supported by the finalized project "Materiali Speciali per Tecnologie Avanzate" of CNR.

References

[1] C.G. Borman and R.G. Gordon, J. Electrochem. Soc. 136 (1989) 3820.
[2] J. Vetrone and Y. Chung, J. Vac. Sci. Technol. A 9 (1991) 3041.
[3] W.M. Sears and M.A. Gee, Thin Solid Films 165 (1988) 265.
[4] B.J. Baliga and S.K. Ghandi, J. Electrochem. Soc. 123 (1976) 941.
[5] W.P. Sihna and M. Minsra, Thin Solid Films 101 (1983) L33.

[6] T. Tabuchi, K. Yamagishi and Y. Tarui, Jpn. J. Appl. Phys. 26 (1987) L186.
[7] R.R. Kunz, M. Rothschild and D. Ehrlich, Appl. Phys. Lett. 54 (1989) 1631.
[8] R. Larciprete, S. Fontana and E. Borsella, Chemtronics 4 (1989) 184.
[9] E. Borsella, P. De Padova and R. Larciprete, in: Excimer Laser and Applications III, Eds. T. Letardi and L.D. Laude, Proc. SPIE 1503 (1991) 312.
[10] E. Borsella and R. Larciprete, in: Conf. Proc., Vol. 29, Quantum Electronics and Plasma Physics, Ed. G.C. Righini (SIF, Bologna, 1991).
[11] R.A. Powell, Appl. Surf. Sci. 2 (1979) 397.
[12] C.L. Lau and G.K. Wertheim, J. Vac. Sci. Technol. 15 (1978) 622.
[13] J.M. Themlin, R. Sporken, J. Darville, R. Caudano, J.M. Gilles and R. Johnson, Phys. Rev. B 42 (1990) 11914.
[14] J. Robertson, J. Phys. C 12 (1979) 4767.
[15] S. Munnix and M. Schmeits, Phys. Rev. B 27 (1983) 7624.
[16] J.M. Themlin, M. Chtaib, L. Henrard, L. Lambin, J. Darville and M. Gilles, Phys. Rev. B 46 (1992) 2460.
[17] C. Oehr and H. Suhr, Thin Solid Films 155 (1987) 65.
[18] E. Kny, L.L. Levenson, W.J. James and R.A. Auerback, Thin Solid Films 85 (1981) 23.
[19] R.K. Sadhir, W.J. James and R.A. Auerback, Thin Solid Films 97 (1982) 17.
[20] A.S.C. Cheung, Y. Yoshino, W.H. Parkinson and D.E. Freeman, Can. J. Phys. 62 (1984) 1752.
[21] R. Larciprete, E. Borsella, P. De Padova, M. Fanfoni, M. Mangiantini and P. Perfetti, J. Vac. Sci. Technol. A, to be published.
[22] R.G. Edgell, S. Eriksen and W.R. Flavell, Solid State Commun. 60 (1986) 835.
[23] D.F. Cox, T.B. Fryberger and S. Semancik, Phys. Rev. B 38 (1988) 2072.
[24] C.S. Fang, F.M. Pan, W.S. Tse and S.R. Horng, Surf. Sci. 211/212 (1989) 279.

Applied Surface Science 69 (1993) 65–68
North-Holland

applied
surface science

Nondestructive evaluation of solids and deposited films by thermal-wave interferometry

E. Oesterschulze, M. Stopka, M. Tochtrop-Mayr, K. Masseli and R. Kassing
University of Kassel, Heinrich-Plettstrasse 40, D-3500 Kassel, Germany

Received 2 June 1992; accepted for publication 30 October 1992

In order to investigate heat diffusion processes in dielectric or metallic materials, a differential interferometer was constructed allowing phase-sensitive measurement of the thermal expansion of periodically excited solid surfaces. To get insight into the physical process of heat propagation, a three-dimensional model of temperature diffusion will be introduced. This theoretical model may be considered as a first step for the understanding of signal formation in case of investigating the homogeneity or adhesion properties of deposited films by means of thermal-wave interferometry. First measurements will be presented and thermal parameters evaluated using the model mentioned above.

1. Introduction

The spatially resolved investigation of thermal parameters of materials is of technical interest, especially in the case of structures of small dimensions. The thermal behaviour of dielectric or metallic mirrors under laser illumination, the investigation of deposited films or the heat dissipation in microelectronic devices may serve as example. Thus a method for monitoring thermal parameters is desirable.

For nondestructive and remote-sensing material investigation, photo-thermal measurement techniques have been established in recent years [1–3]. Thermal-wave interferometry is one of the most sensitive techniques; moreover, it is a noncontact measuring system [4,5]. In combination with a three-dimensional model of heat propagation in solids it is possible to extract thermal parameters.

2. Experimental arrangement

The instrumental set-up consists of three main components which are shown in fig. 1.

The sample surface is excited by a focussed Ar-ion laser with a maximum output power of about 1 W. The laser beam is periodically modulated by an acousto-optical modulator (AOM) up to frequencies of 1 MHz. A scanning system allows the computer-controlled variation of the distance between excitation source and the point of measurement of the interferometer as described below.

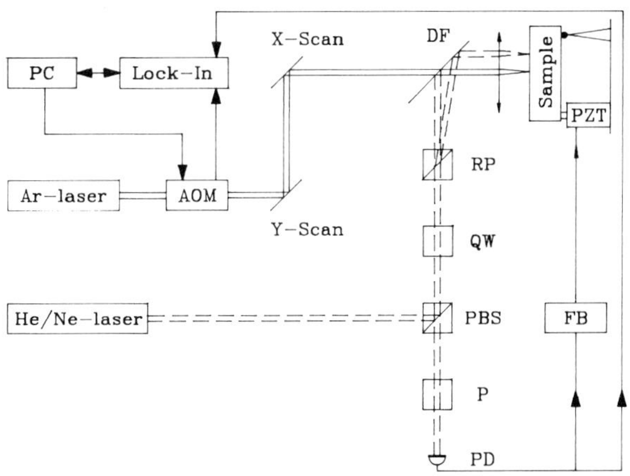

Fig. 1. Schematic experimental set-up of the thermal wave interferometer.

The thermal expansion induced by this laser excitation is detected by a differential interferometer. In this means linear polarized light of a 10 mW He/Ne laser is reflected by a polarizing beam splitter (PBS). The following quarter-wave plate (QW) yields circularly polarized light. Passing through a Rochon prisma (RP) the beam is spatially separated into two orthogonal polarized beams. After reflection at a dichroic filter (DF) both beams are focussed onto the sample, one acting as reference and the other as measurement beam. On their way back through the optical system the polarized beams are recombined by the Rochon prisma. Leaving the quarter-wave plate the beam passes the polarizing beam splitter, and the zero-order interference fringe occurring at the polarizer (P) is detected by a photodiode (PD).

To increase the detection sensitivity, the measured interferometer signal is electronically fed back (FB) to the sample holder using a piezoelectric transducer (PZT) which counteracts incidental vibrations of the measurement system. This procedure reduces the influence of mechanical noise signals on the interferometer output which is given by:

$$I(\delta) = \frac{I_{max} + I_{min}}{2} + \frac{I_{max} - I_{min}}{2} \cos\left(\frac{4\pi\delta}{\lambda}\right). \quad (1)$$

I_{max} and I_{min} denote the minimum and maximum amplitudes of the interferogram, λ is the He/Ne laser wavelength and δ describes the total path difference between the polarized beams. The latter consists of the mechanical and optical static path differences δ_s and the thermal-induced surface expansion $\delta_t(t) = \Delta\cos(\omega t)$, Δ being the amplitude of the thermal expansion. Adjusting the static phase lag to $\delta_s = \pi/2$ by tilting the same holder by means of the computer-controlled piezo-transducer, and considering that Δ will be small in comparison to the wavelength, I may be approximated to:

$$I(t) = \frac{I_{max} + I_{min}}{2} + \frac{I_{max} - I_{min}}{2} \frac{4\pi\Delta}{\lambda} \cos(\omega t). \quad (2)$$

This is a linear dependence between the expansion amplitude Δ and the measured intensity I of frequency ω. $I(\omega)$ is phase-sensitively detected by a lock-in amplifier.

3. Theory of heat propagation in solids

The heat diffusion process in solids may be described by the three-dimensional heat-conduction equation [6]:

$$-\nabla \cdot (\kappa T(\boldsymbol{r}, t)) + \rho c \frac{\partial T(\boldsymbol{r}, t)}{\partial t} = S(\boldsymbol{r}, t), \quad (3)$$

where $T(\boldsymbol{r}, t)$ is the temperature, κ the thermal conductivity, ρ the density, c the specific heat and $S(\boldsymbol{r}, t)$ are the source and sink terms. Assuming an isotropic, homogeneous medium with temperature-independent thermal constants κ and c the conduction equation reduces to:

$$\Delta T(\boldsymbol{r}, t) - \frac{1}{\alpha} \frac{\partial T(\boldsymbol{r}, t)}{\partial t} = -\frac{S(\boldsymbol{r}, t)}{\kappa}. \quad (4)$$

$\alpha = \kappa/\rho c$ denotes the thermal diffusivity. This differential equation must be applied to the experimental situation of fig. 2 showing a solid half-space in contact with a gas half-space. The solid surface is illuminated with a laser beam inducing a heat source with a Gaussian distribution which is periodically modulated and exponentially damped while propagating into the solid in z-direction. Assuming that no absorption takes place in the gas phase, the conduction equation can be written as:

$$z < 0: \quad \left(\Delta - \frac{1}{\alpha_g} \frac{\partial}{\partial t}\right) T_g(\boldsymbol{r}, t) = 0, \quad (5)$$

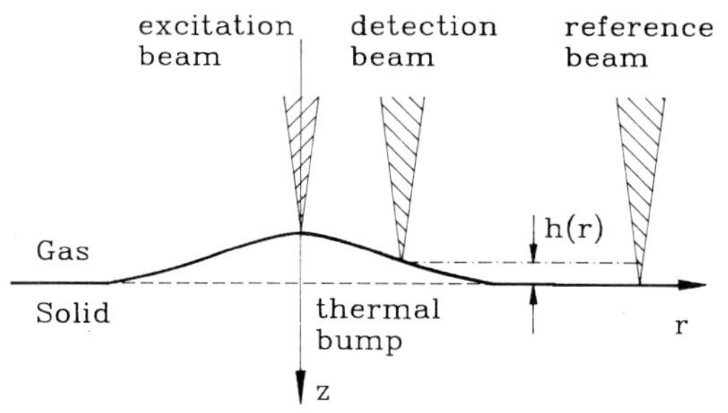

Fig. 2. Schematic description of the experimental geometry of laser illumination of a solid half-space in contact with a gas half-space.

$$z \geqslant 0: \quad \left(\Delta - \frac{1}{\alpha_s}\frac{\partial}{\partial t}\right) T_s(\boldsymbol{r}, t) = -\frac{I_0}{\kappa_s}(1-R)\beta\, e^{-\beta z}\, e^{-r^2/(2\sigma^2)}\, e^{i\omega t}, \tag{6}$$

where the indices g and s denote gas and solid half-space, respectively, β is the absorption coefficient, R the reflection coefficient, σ the waist of the Gaussian distributed laser beam (TEM_{00} mode), I_0 the laser intensity, and ω the modulation frequency.

These differential equations must be treated considering the boundary conditions that both temperature and thermal flux must be continuous functions at the discontinuity ($z = 0$):

$$T_g\,|_{z=0} = T_s\,|_{z=0} \text{ and } \kappa_g \frac{\partial T_g}{\partial z}\bigg|_{z=0} = \kappa_s \frac{\partial T_s}{\partial z}\bigg|_{z=0}. \tag{7}$$

This mathematical problem was solved using a special integral transform technique, the so-called Hankel transform of zero order. Applying the Hankel transform to the differential eqs. (6) and (5) and taking the boundary conditions into account, the resultant temperature field will be given by the Hankel back transform:

$z < 0$:

$$T_g(r, z) = \frac{I_0\beta}{2\pi\kappa_s}\int_0^\infty d\rho\rho J_0(r\rho) \times \frac{\kappa_s q_s - \kappa_s\beta}{\kappa_s q_s + \kappa_g q_g}\, e^{q_g z}\, e^{-\rho^2\sigma^2/2}, \tag{8}$$

$z \geqslant 0$:

$$T_s(r, z) = \frac{I_0\beta}{2\pi\kappa_s}\int_0^\infty d\rho\rho J_0(r\rho) \times \left(e^{-\beta z} - \frac{\kappa_g q_g - \kappa_s\beta}{\kappa_g q_g + \kappa_s q_s}\, e^{-q_s z}\right), \tag{9}$$

where $q_{g,s}^2 = \rho^2 + i\omega/\alpha_{g,s}$ are the quadratic forms of the complex propagation vector in the (ρ, z)-space.

Without explaining any mathematical details, this way of treating heat propagation in solids based on the Hankel-transform technique can be extended to the investigation of layer/substrate systems as has been shown by Iravani and Wickramasinghe [7].

Neglecting elastic deformations, the temperature distribution $T_s(r, z, t)$ considering a solid as shown in fig. 2, leads to the simple linear thermal expansion expression:

$$h(r) = \int_0^\infty dz b T_s(r, z, t), \tag{10}$$

with b denoting the linear thermal expansion coefficient.

4. Results

First measurements were carried out to confirm the above assumptions that thermal parameters may show no measurable temperature dependence in the considered temperature region. Therefore, keeping the distance between excitation laser beam and the point of interferometer measurement constant, the output power of the Ar-ion laser (modulated with 1 kHz) was linearly varied. The detected expansion signal of a brass sample shows an excellent linear behaviour as a function of power as can be seen in fig. 3.

Various metal substrates were investigated to measure the respective thermal diffusivity. As an example, the thermal diffusivity of a Cu sample was determined by measuring the phase lag be-

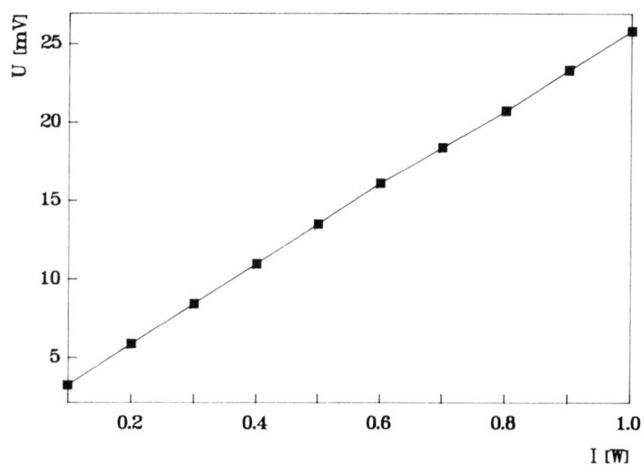

Fig. 3. Amplitude U of the detector output of a brass sample excited with a frequency of 1 kHz as a function of Ar-ion laser output power at a fixed measurement location.

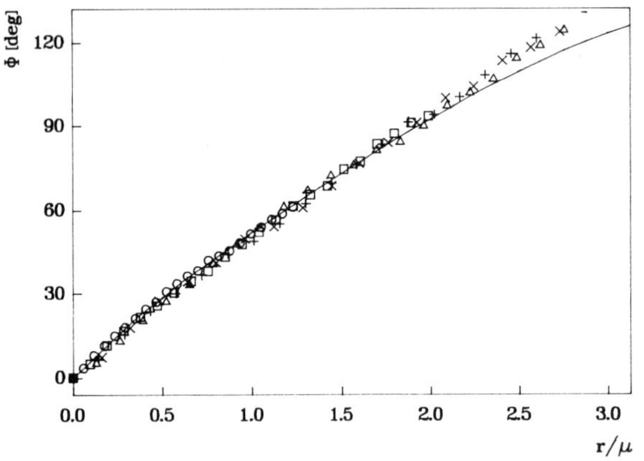

Fig. 4. Phase-lag signal Φ of a Cu sample as a function of the distance between the excitation source and measurement point for various frequencies: (○) 1066 Hz, (□) 3190 Hz, (△) = 5260 Hz, (+) = 7220 Hz, (×) = 9070 Hz. Distances were normalized with respect to the one-dimensional diffusion length μ.

tween excitation signal and the thermal expansion of the point of the interferometer on the sample surface as a function of the distance between them. Fig. 4 shows the normalized phase-lag dependence for various laser modulation frequencies for the Cu sample. The mean value of the thermal diffusivity of this Cu sample was evaluated to be $(111.0 \pm 5.7) \times 10^{-6}$ m^2/s, differing only slightly from the literature value of 117.0 $\times 10^{-6}$ m^2/s.

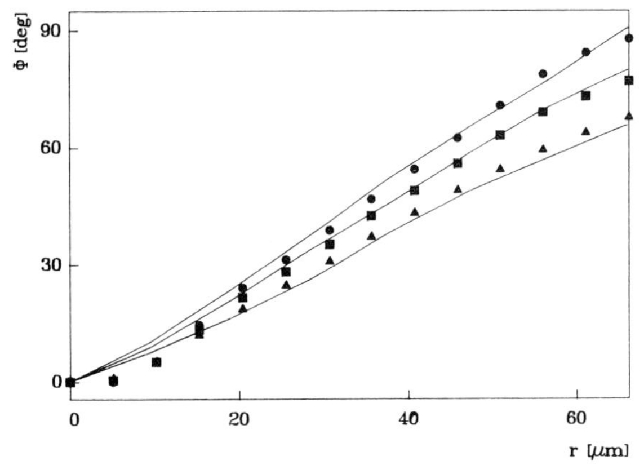

Fig. 5. Phase-lag signal Φ of a 80 nm Au film on a BaK50-glass substrate as a function of frequency ((●) = 3030 Hz, (■) = 1880 Hz, (▲) = 965 Hz) in comparison with theoretical calculations (solid lines). Distances were normalized with respect to the one-dimensional diffusion length μ.

A more complicated situation arises with investigation of layer/substrate systems. Fig. 5 shows the phase-lag measurements of a layer/substrate system consisting of a 80 nm Au film deposited on a BaK50-glass substrate. The solid lines represent theoretical results calculated by means of the extended model mentioned above.

5. Conclusion

It has been shown that thermal-wave interferometry is a powerful tool for noncontact and nondestructive material analysis. This technique allows the spatially resolved investigation of laser-induced thermal expansion of sample surfaces. In combination with a three-dimensional temperature diffusion model, a quantitative evaluation of the thermal diffusivity of various samples was possible.

Acknowledgement

This work is supported by the Department of Research and Technology of the Federal Republic of Germany (BMFT Project 13N6008).

References

[1] F.A. McDonald, Can. J. Phys. 64 (1986) 1023.
[2] J.C. Murphy, L.C. Aamodt, F.G. Satkiewicz, R.B. Givens and P.R. Zarriello, Johns Hopkins APL Tech. Dig. 7 (1986) 187.
[3] R.L. Thomas, L.D. Favro and P.K. Kuo, Can. J. Phys. 64 (1986) 1234.
[4] Z. Sodnik and H.J. Tiziani, in: Photoacoustic and Photothermal Phenomena, Eds. P. Hess and J. Pelzl (Springer, Heidelberg, 1988) pp. 400–403.
[5] V. Kurzmann, J. Stöhr, M. Tochtrop and R. Kassing, in: Acoustic, Thermal Wave and Optical Characterization of Materials, Proc. Eur. Mater. Res. Soc. Spring Meeting 1989, Strasbourg, 1989, Eds. G.M. Crean, M. Locatelli and J. McGilp, pp. 117–120.
[6] H.S. Carslaw and J.C. Jaeger, 2nd ed. (Clarendon, Oxford, 1959).
[7] M.V. Iravani and H.K. Wickramasinghe, J. Appl. Phys. 58 (1985) 122.

Applied Surface Science 69 (1993) 69–74
North-Holland

applied
surface science

Laser direct writing of gold to repair defective lines in thin-film metallizations

D. Metzger and H. Reichl

Forschungsschwerpunkt Technologien der Mikroperipherik, TU Berlin, Sekr. TIB 4 / 2-1, Geb. 17, Gustav Meyer Allee 25, D-1000 Berlin 65, Germany

Received 2 June 1992; accepted for publication 30 October 1992

Laser chemical vapour deposition (LCVD) with dimethylgold hexafluoroacetylacetonate was used to repair defective lines in thin-film metallizations on alumina ceramic and polyimide. A process is described to deposit gold on polyimide and various substrates. Different patterns were deposited. Resistivities on polyimide have been achieved which are lower than 4.7 $\mu\Omega \cdot$cm, which is twice the value of bulk gold. Bond strengths of 10 N/mm^2 were measured on polyimide. With this method contacts have been produced to thin-film gold metallizations. Contact resistances have been measured in the range of 5–10 mΩ.

The experimental set-up has already been described elsewere [1]. It comprised a modified ESI laser trimmer model 4455, a Spectra Physics CW argon-ion laser model 2025, and a HP computer model 9000/300. The substrate samples were placed inside a deposition chamber which was mounted on a computer-controlled z-translation and θ-adjustment stage. Wafers up to 4 inch have been processed. After evacuating the deposition chamber to a base pressure of 1×10^{-6} mbar dimethylgold hexafluoroacetylacetonate was introduced into the chamber with a gas delivery system. The 514 nm TEM_{00} laser line was expanded by a factor of seven and then directed, by means of a mirror system, into the objective lens. After passing through this lens the laser beam is focused through a quartz window into the stainless steel chamber. The $1/e^2$-laser focus spot diameter measured with the knife-edge method was 10 μm on the sample surface. The precise location of the gold deposition was controlled by moving the laser focus in x- and y-direction with the HP computer. Laser power and laser exposure time were also controlled by this computer.

Laser deposition of Au lines has been reported from organometallic films, plating solutions and the vapour phase [3–9].

A number of lines and patterns were laser deposited on various substrates like glass (Corning 7059), alumina ceramics (Coors ADS995), polyimide (Pyralin 2611, Probimide 348) and ORMOCERs (ORganically MOdified CERamics). All experiments were carried out at a vapour pressure of 0.6 mbar corresponding to a temperature of 25°C of the gold precursor measured with a MKS Baratron Type 221A.

Before laser deposition the sample surface was cleaned in acetone, alcohol and rinsed in deionized water. To remove adsorbed hydrocarbon on the surface the samples were etched for 2 min in a reactive O_2-plasma. Typical etching rates were 180 nm/min. Further etching of polyimide led to surface roughness. The surface roughness on polyimide resulted in a homogeneous grain size of the deposited gold crystals and a smoothness of line shape. Fig. 1 shows a meander deposited on polyimide surface treated in this way.

It has been recognized that the preconditioning of the surface plays an important role in laser deposition.

Fig. 2 shows a gold line deposited on glass with a laser power of 800 mW, writing speed 10 μm/s, line width 80 μm/s and line height 20 μm.

After the laser deposition process the samples

0169-4332/93/$06.00

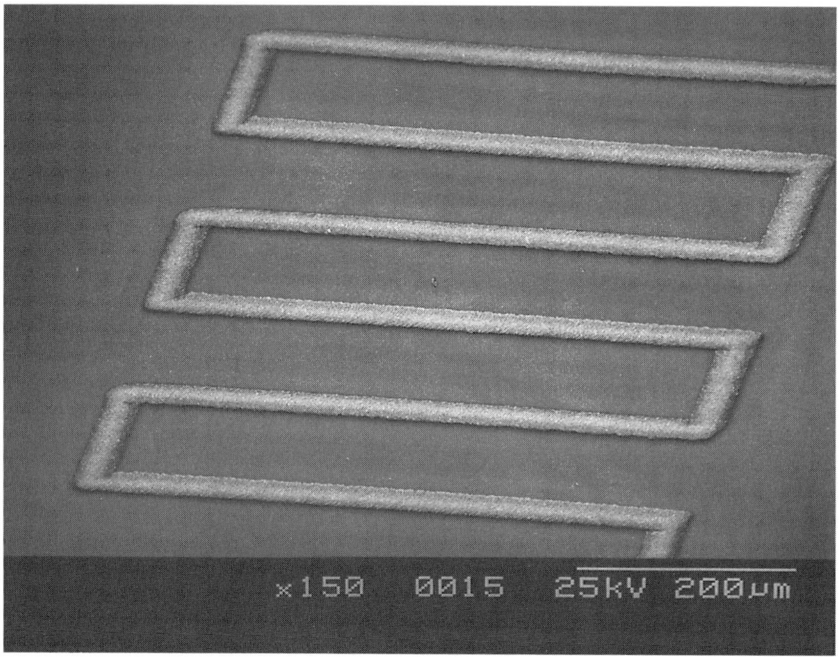

Fig. 1. SEM micrograph of a meander deposited on polyimide 2611. The polyimide surface was treated with an O_2 plasma. The meander was deposited with a laser power of 150 mW and a writing speed of 10 μm/s. Line width and height were 50 and 5 μm, respectively.

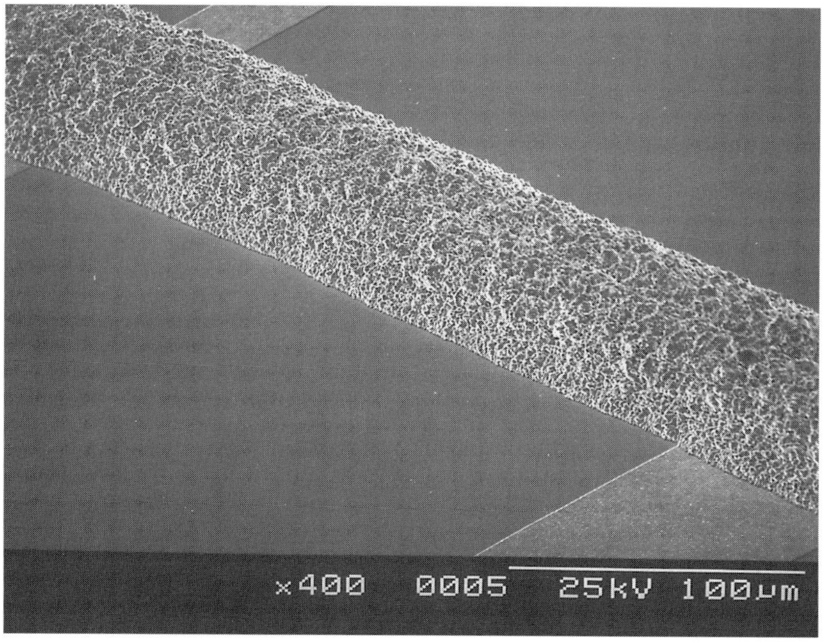

Fig. 2. SEM micrograph of a gold line deposited on glass substrate with laser power of 800 mW and writing speed of 10 μm/s. Line width and height were 80 and 20 μm, respectively.

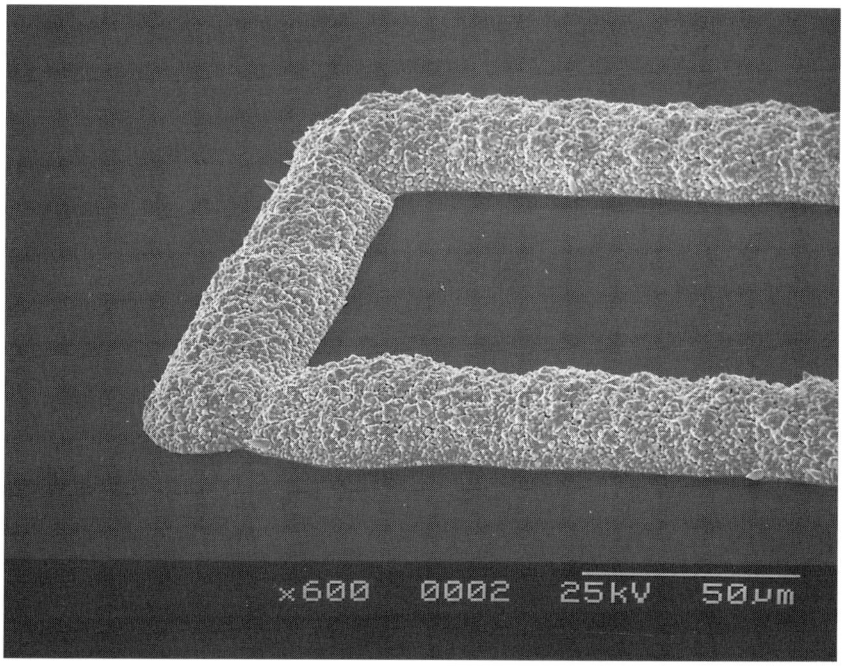

Fig. 3. SEM micrograph of open rectangle deposited on ORMOCERs with writing speed of 10 μm/s and laser power of 400 mW.

were removed from the deposition chamber and cleaned with a gold etching solution. This cleaning removed small particles of gold which had nucleated in the vicinity of the deposited lines.

Scotch-tape tests indicated good adhesion of the laser-deposited gold on polyimide. Pads of 1×1 mm^2 were deposited on Probimide 348. Direct pull tests were used for further quantitative examination of bond strengths. After deposition the pads were immersed in an electroless chemical copper solution (Schering Printoganth ML). At a laser power of 100 mW a bond strength of 10 N/mm^2 were measured. An explanation for this phenomenon is mechanical interlocking. During the deposition process gold crystals are formed and buried into the polyimide surface.

On ORMOCERs laser-induced depositon was carried out with a writing speed of 10 μm/s and a laser power of 300–400 mW, resulting in line widths of 30 μm and heights of 20 μm. Fig. 3 shows an open rectangle deposited on ORMOCERs with the parameters mentioned above.

The resistivities of the deposited gold on various substrates were found to be best at a writing speed of 10 μm/s. Fig. 4 shows resistivities of

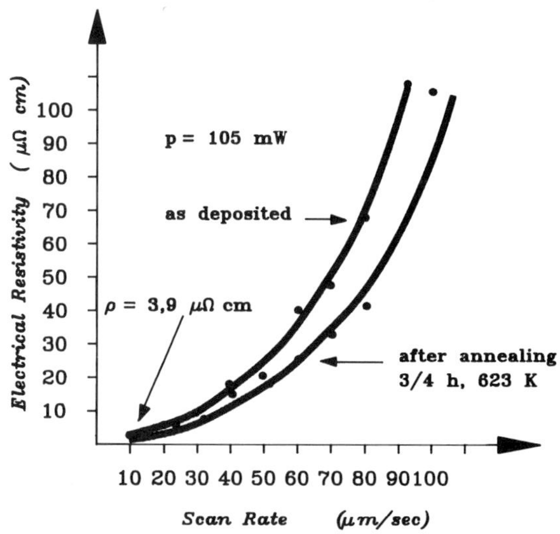

Fig. 4. Electrical resistivity as a function of writing speed deposited on Probimide 348 at a laser power of 105 mW.

deposited gold lines on polyimide as a function of writing speed. Lines were produced with resistivities in the range of 4 $\mu\Omega \cdot$ cm. These results were lower than twice the value of bulk gold (2.35 $\mu\Omega \cdot$ cm). With Auger analysis no carbon and no other impurities were detected after depth profile sputtering. These revealed that the impurities were smaller than 0.1 at%. The difference between the measured resistivities and the values for bulk gold may be due to microporous morphology and distinct grain sizes. Nevertheless, the measurements of resistivities and Auger analysis demonstrated that the quality of gold deposited by this method was excellent.

An important application of the direct writing process is the repair of defects in multilayer metallizations on alumina ceramics with polyimide as the dielectric [2,10]. To accomplish this, gold must be deposited onto alumina ceramics, polyimide and gold conductors. These materials have different thermal and optical properties. Polyimide, for example, has a low thermal conductivity and a greater optical absorption. At a given laser power this will result in a greater temperature rise on polyimide, compared with alumina ceramics or gold interconnect metallizations. These different material parameters were partly accommodated by pre-heating the samples. For the repair technique the laser power must be kept low enough to prevent ablation of polyimide and must be high enough to deposit gold on the patterned gold metallization.

To repair a defective part between two gold conductor lines, with line widths of 50 μm and heights of 5 μm, the laser writing process is started on the first conductor at a distance of typically 20–40 μm from the defect. A writing speed of 10 μm/s thus results in a time interval of 2–4 s, during which the conductor line is heated. Additional heating is necessary to induce laser deposition onto a conductor. This additional heating resulted in cauliflower growth of gold deposition. Controlled growth was achieved by adding 15 mbar argon to decrease the mean free

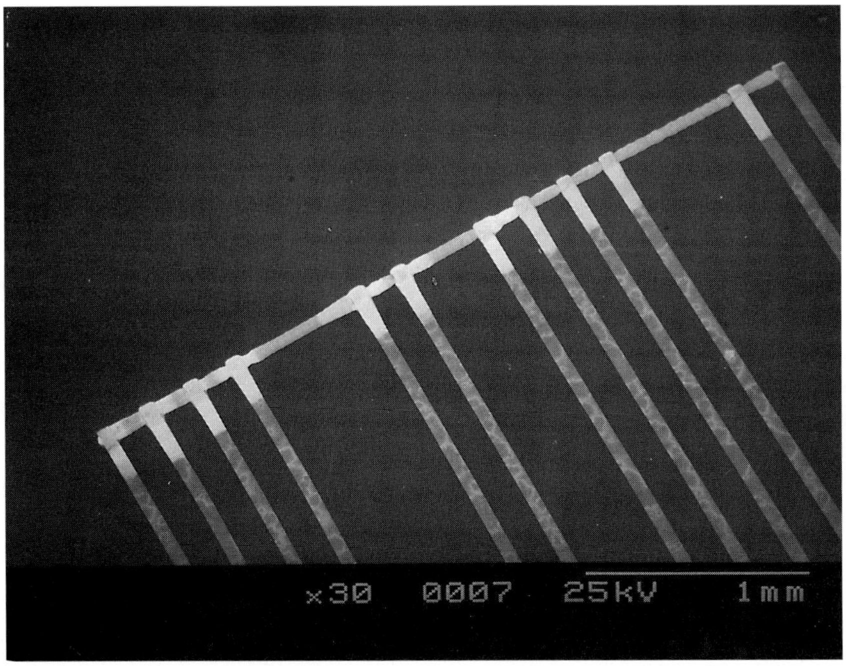

Fig. 5. SEM micrograph of a contact area. Gold was deposited on polyimide and on galvanic gold conductors with resistance of ~ 5 mΩ, sample temperature 100°C, laser power on polyimide 200 mW, laser power on gold conductors 300 mW, and a writing speed of 10 μm/s.

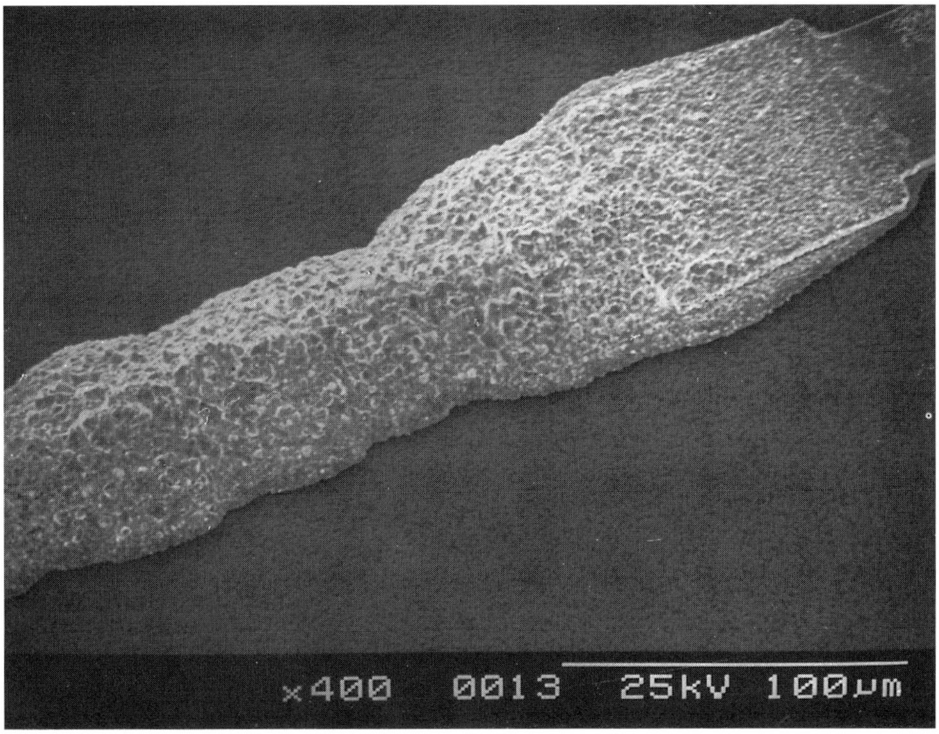

Fig. 6. SEM micrograph with a close-up of a contact area.

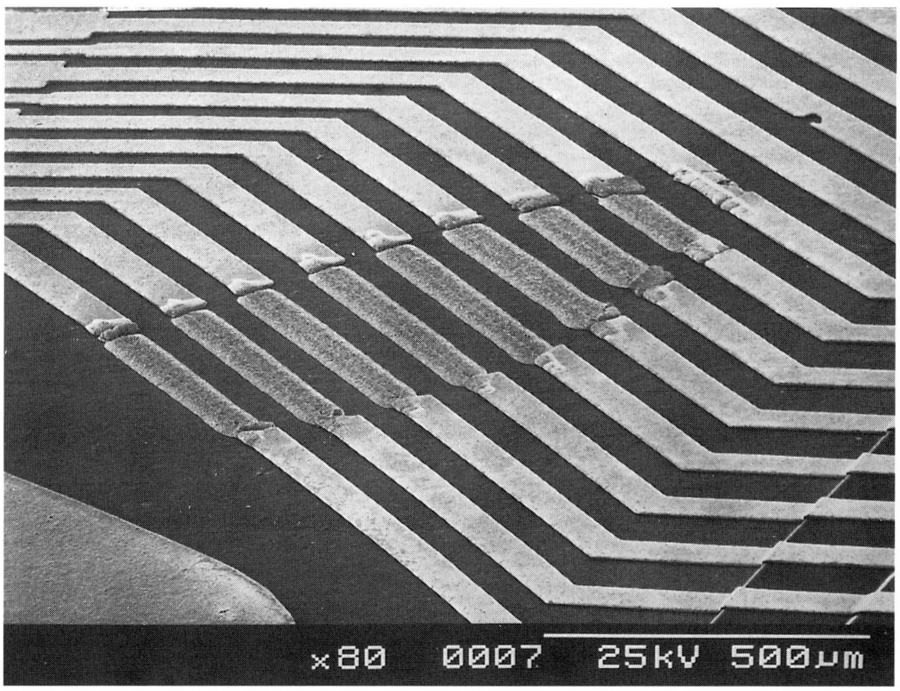

Fig. 7. SEM micrograph of the repair of a defective metallization on alumina ceramic, with laser power of 700 mW, writing speed of 10 μm/s, and line height of 7 μm/s.

path of the gold precursor. Pre-heating the sample surface in the range of 80–100°C increases the surface reaction kinetics. In this range of temperature no dissociation of the organometallic gold compound was measured with a Balzers quadrupole mass spectrometer model QMG 420. Dissociation did occur at temperatures higher than 150°C. At a laser power of 250 mW and a pre-heating of 100°C on polyimide a contact resistance of 10 mΩ to a galvanic thin-film gold metallization was measured. By increasing the laser power on the gold metallization to 300 mW and decreasing the laser power on the polyimide surface to 200 mW to avoid ablation of polyimide, the resistance decreased slightly from 10 to 5 mΩ. In this way we obtained a contact resistance of 25×10^{-8} $\Omega \cdot cm^2$. Contacts formed by these process parameters are shown in fig. 5. Fig. 6 shows a close-up of the contact area.

The repair of a galvanic defect in a thin-film gold metallization on alumina ceramic is shown in fig. 7. The positions of the ends of the conductor lines were detected by a computer program and then linked using a writing speed of 10 μm/s and a laser power of 700 mW.

In fig. 7 it is demonstrated that the width and height of laser-deposited lines is well within the width, height and pitch of the thin-film gold metallization.

In summary, the application of laser deposition to thin-film metallizations has been demonstrated, especially for gold interconnects on polyimide and Al_2O_3. A laser process was described to produce low-resistance contacts. The laser procedure presented in this paper has proved to be practical for the repair of thin-film interconnects and has opened the way for customization of interconnects on multichip modules with polyimide as dielectric. The use of this gold precursor is attractive because of the excellent deposition properties and the low resistivity of the deposited gold.

References

[1] D. Metzger, A. Paredes, Th. Kruck and H. Reichl, in: Micro System Technologies 90, Ed. H. Reichl (Springer, Berlin, 1990) p. 325.

[2] D.J. Ehrlich and J.Y. Tsao, Laser Microfabrication (Academic Press, New York, 1989) p. 51.

[3] T.H. Baum and C.R. Jones, Appl. Phys. Lett. 47 (1985) 538.

[4] T.H. Baum and C.R. Jones, J. Vac. Sci. Technol. B 4 (1986) 1187.

[5] G. Arjavalingam, M.M. Oprysko and J.E. Hurst, Jr., Mater. Res. Soc. Symp. Proc. 101 (1988) 81.

[6] M.E. Gross, G.J. Fisanick, P.K. Gallagher, K.J. Schones and M.D. Fennell, Appl. Phys. Lett. 47 (1985) 923.

[7] V.H. Holding, N.S. Clements, K.W. Beeson and G.A. West, Mater. Res. Soc. Symp. Proc. 75 (1987) 165.

[8] M.E. Gross, A. Appelbaum and K.J. Schones, J. Appl. Phys. 60 (1986) 529.

[9] M. Jubber, J.I.B. Wilson, J.L. Davidson, P.A. Fernie and P. John, Appl. Phys. Lett. 55 (1989) 1477.

[10] R.F. Miracky, Mater. Res. Soc. Symp. Proc. 192 (1989) 547.

Applied Surface Science 69 (1993) 75–78
North-Holland

applied
surface science

High speed laser writing of gold lines from organic solutions

K. Bali [a], T. Szörényi [b], M.R. Brook [c] and G.A. Shafeev [c]

[a] *Department of Experimental Physics, JATE University, Dóm tér 9, H-6720 Szeged, Hungary*
[b] *Research Group on Laser Physics, Dóm tér 9, H-6720 Szeged, Hungary*
[c] *General Physics Institute of the Russian Academy of Sciences, Vavilov Street 38, Moscow, Russian Federation*

Received 2 June 1992; accepted for publication 30 October 1992

A systematic study on pyrolytic laser writing of gold from organic solutions of triphenylphosphine gold (I) complexes reveals that essentially structureless tracks of electrical resistivities between 9 and 22 $\mu\Omega \cdot$cm can be deposited at 3–50 mW power and writing speeds of the order of mm/s. It is shown that under optimized conditions laser deposition from solutions is a competitive method with economic writing speeds and material properties appropriate for practical applications.

1. Introduction

Laser direct writing of gold has received considerable attention in the last decade for the obvious reason that gold is an ideal interconnect material. Inherent to LCVD of gold is its appropriate purity, and, consequently, a resistivity comparable to that of the bulk material with relatively low writing speeds (up to 100 μm/s) [1–6]. Much higher speeds can be attained by laser decomposition of solid films. However, as a rule the features written are considerably larger than the dimension of the laser spot and the appearance of pronounced periodic structures may lead to uneven morphology [7–9]. Furthermore, the lowest resistivities reported are about ten times higher than the bulk value [9,10]. Although the possibility of gold deposition from liquid phase with or without external electron source has been demonstrated [11–16], the limits of gold direct writing are far from being thoroughly investigated.

In this contribution we report pyrolytic laser writing of gold from organic solutions of triphenylphosphine gold complexes with particular emphasis on the unusual properties characterizing deposition initiated by absorption in the solution. It is shown that, under optimized conditions, well-conducting tracks of height 0.04 to 1.6 μm and width 1 to 30 μm can be written with speeds significantly higher than those reported for LCVD [1–6] at powers ranging from 3 to 50 mW.

2. Experimental

Tris-(paranitrophenyl)methyl-triphenylphosphine gold (I) and 8-oxychinolynate-triphenylphosphine gold (I) are dissolved in dimethylformamide. The saturated solutions are strongly absorbing in the visible; the measured absorption coefficients at 514.5 nm are 96.7 and 16.9 cm^{-1}, respectively. Substrates to be patterned are 1 mm thick microscope slides (DIN 58884) covering a solution layer of 1 mm thickness. The Gaussian beam of the Ar^+ laser (Spectra Physics 164) is focused onto the substrate–solution interface ($2w_0 = 4.4$ μm) through the glass plate. Tracks are deposited by translating the sample holder perpendicularly to the laser beam with scanning speeds between 10 μm/s and 1 mm/s. Deposits are characterized by profilometry (TALYSTEP), optical and electron microscopy (SEM, EDX), and electrical resistivity measurements.

3. Results and discussion

Qualitatively identical dependencies were obtained for both solutions and so all the results will be illustrated with the case of tris-(paranitrophenyl)methyl-triphenylphosphine gold. The width and height of the deposited tracks as a function of laser power are shown at four scanning speeds in figs. 1 and 2, respectively. The threshold power for continuous line deposition is ~ 3 mW with a minimum width of ~ 1 μm. Up to 25–30 mW the track widths increase linearly, the slopes being less for higher writing speeds. Above 30 mW the track widths further increase in the case of slow writing, while a plateau value is reached when applying speeds exceeding 500 μm/s. Throughout the power range studied the track width is inversely related to the writing speed.

Unlike the width which monotonously increases, in the case of slow writing the height versus power functions exhibit a biphasic dependence. First, the height rapidly increases up to maximum 1.6 μm. On reaching the range of 20–25 mW a sharp decrease is observed and then a renewed height increase sets on with significantly increased scatter, as indicated with the error bars in fig. 2. In the case of fast writing no

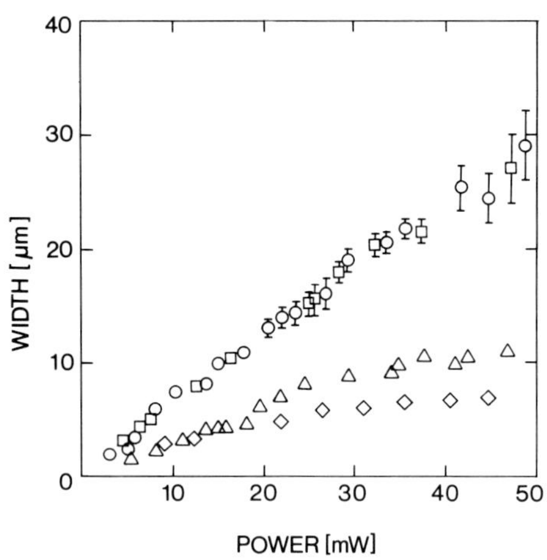

Fig. 1. track width as a function of the laser power at four writing speeds: (○): 80 μm/s, (□): 135 μm/s, (△): 500 μm/s and (◇): 1 mm/s. The error bars represent the 70% confidence intervals on minimum 7 measurements.

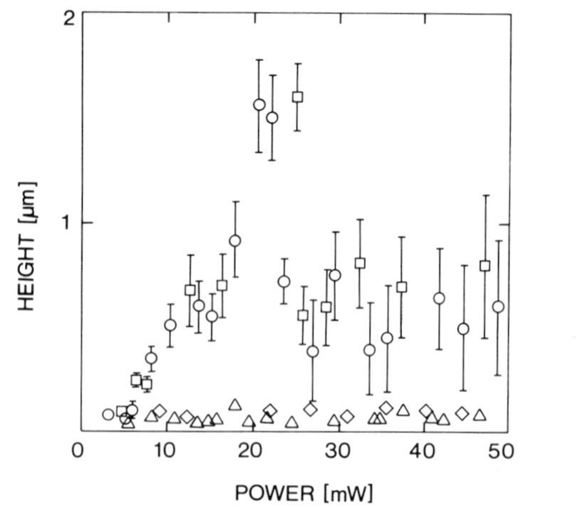

Fig. 2. Track height as a function of the laser power at four writing speeds: (○): 80 μm/s, (□): 135 μm/s, (△): 500 μm/s and (◇): 1 mm/s. The error bars represent the 70% confidence intervals on minimum 7 measurements.

such maxima are detected; all the values lie in the range of 40–120 nm with width-to-height ratios between 50 and 200.

Light microscopic examination of the tracks written at slow speed reveals two distinct types. Tracks generated with low power are continuous with well defined edges and practically without macrostructure. Beyond ~ 20 mW the edges become indefinite and a narrow transparent channel appears in the center of the track. Contrary to this no channel formation is observed at the highest writing speed applied (1 mm/s) up to 50 mW.

The SEM pictures surprisingly reveal exactly the same trends as those reported for copper lines written from copper formate by Hoffmann et al. [17]. Even the well-defined tracks possess a microstructure (fig. 3a) with period lengths decreasing with increasing writing speed. With increasing power the periodic nodular structure formed replaces the straight edges with a less well-defined ripple (fig. 3b). A further increase in power leads to a rough surface with period lengths exceeding 10 μm and clearly seen also at the edges (fig. 3c). The width of the central channel increases but it remains always less than the $(1/e^2)$ beam diameter, 4.4 μm. With increasing writing speed the structure formation becomes

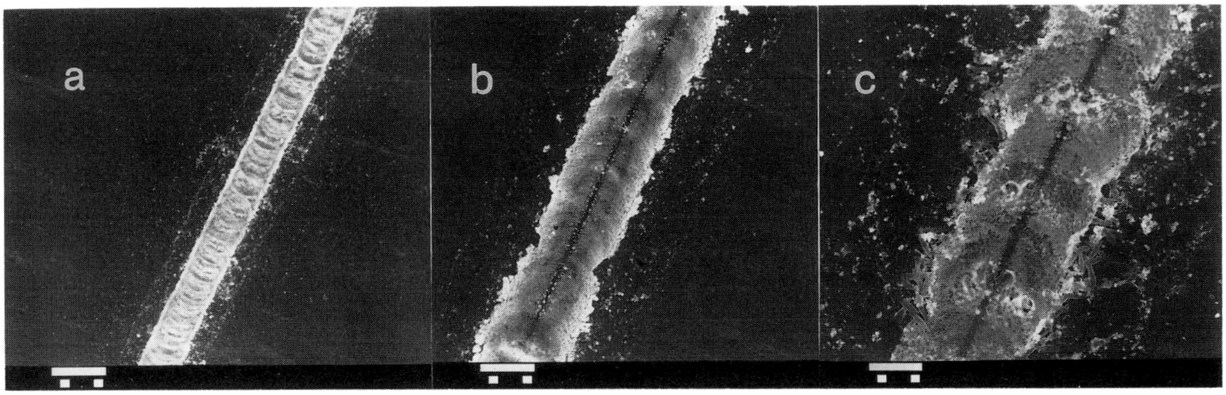

Fig. 3. SEM micrographs of gold lines deposited from a saturated solution of tris-(paranitrophenyl)methyl-triphenylphosphine gold (I) in dimethylformamide using a scanning speed of 80 μm/s. The laser powers are (a) 13 mW, (b) 28 mW and (c) 49 mW. The scaling bars denote 10 μm.

less pronounced. At 1 mm/s smooth, mirror-like tracks are deposited without macrostructure and channel formation.

Deposits are of $90 \pm 2\%$ pure gold as determined by energy dispersive X-ray analysis, with electrical resistivities of 9–22 $\mu\Omega \cdot$cm measured on essentially structureless tracks.

Based on these results a consistent description of the events can be derived. Due to the advantageously low activation energy decomposition commences at a fairly low laser power ($\sim$ 3 mW) on the substrate–solution interface. As deposition proceeds and the film becomes more absorbing the laser energy is more effectively coupled into the system. Since heat loss is minimal the rising temperature further enhances the deposition rate. This positive feedback explains the dramatic increase in the amount of the deposited material: using a writing speed of 80 μm/s and a processing power of 19 mW a deposition rate of $\sim$ 1600 $\mu m^3/s$ is attained. On increasing the speed to 1 mm/s, this rate decreases to $\sim$ 400 $\mu m^3/s$, but even this lesser value competitively compares with those reported for laser direct writing of gold [3–6].

This effective layer growth goes on until the peak power of the Gaussian beam reaches a value sufficient to remove the material from the center of the track. (In control experiments ablation of vacuum evaporated gold layers of thicknesses between 20 and 200 nm required power levels in excess of 200 mW. Unlike for these layers the tape test of the tracks indicated poor adhesion and so we assume that the mechanism of removal is blow-off of the (solid) gold layer in an early stage of its growth.) As the central channel is formed the peak power of the beam is not absorbed in the growing track anymore but penetrates into the adjacent solution layer. The decreasing amount of deposited material (as seen in fig. 2) indicates that light-induced deposition is less effective than the thermally induced process. Another disadvantage of light-induced decomposition is spreading deposition (as seen in figs. 3b and 3c) due to a higher absorption depth and enhanced light scattering.

4. Conclusions

Our systematic study of pyrolytic gold deposition from organic solutions of triphenylphosphine gold (I) complexes onto transparent substrates has demonstrated that tracks of good quality and morphology can be obtained at a low power and a high writing speed. The only drawback of this technique is poor adhesion to the surface as expected for deposition onto bare glass at relatively low temperatures. The present study provides novel evidence in favour of the notion that under optimized conditions gold deposition from solutions can offer economic writing speeds and acceptable material properties.

Acknowledgements

The skilled assistance of A. Tóth in material characterization by SEM is gratefully acknowledged. This work was supported by the Hungarian National Science Foundation (OTKA I/3 3060).

References

[1] T.H. Baum and C.R. Jones, Appl. Phys. Lett. 47 (1985) 538.
[2] T.H. Baum and C.R. Jones, J. Vac. Sci. Technol. B 4 (1986) 1187.
[3] T.H. Baum, J. Electrochem. Soc. 134 (1987) 2616.
[4] M. Jubber, J.I.B. Wilson, J.L. Davidson, P.A. Fernie and P. John, Appl. Phys. Lett. 55 (1989) 1477.
[5] M. Jubber, J.I.B. Wilson, J.L. Davidson, P.A. Fernie and P. John, Appl. Surf. Sci. 43 (1989) 74.
[6] J. Messelhauser, E.B. Flint and H. Suhr, Appl. Surf. Sci. 54 (1992) 64.
[7] G.J. Fisanick, M.E. Gross, J.B. Hopkins, M.D. Fennell, K.J. Schnoes and A. Katzir, J. Appl. Phys. 57 (1985) 1139.
[8] G.J. Fisanick, J.B. Hopkins, M.E. Gross, M.D. Fennell and K.J. Schnoes, Appl. Phys. Lett. 46 (1985) 1184.
[9] M.E. Gross, G.J. Fisanick, P.K. Gallagher, K.J. Schnoes and M.D. Fennell, Appl. Phys. Lett. 47 (1985) 923.
[10] M. Widmer, P. Hoffmann, B. Lecohier, H. Solka, J.-M. Philippoz and H. van den Bergh, in: Photochemical Processing of Electronic Materials (Academic Press, New York, 1992) p. 433.
[11] R.F. Karlicek, V.M. Donnelly and G.J. Collins, J. Appl. Phys. 53 (1982) 1084.
[12] H.K. Kuiken, F.E.P. Mikkers and P.E. Wierenga, J. Electrochem. Soc. 130 (1983) 554.
[13] R.J. von Gutfeld, E.E. Tynan, R.L. Melcher and S.E. Blum, Appl. Phys. Lett. 35 (1979) 651.
[14] R.J. von Gutfeld, M.H. Gelchinski, L.T. Romankiw and D.R. Vigliotti, Appl. Phys. Lett. 43 (1983) 876.
[15] S. Tamir and J. Zahavi, J. Vac. Sci. Technol. A 3 (1985) 2312.
[16] M.R. Brook, K.I. Grandberg and G.A. Shafeev, Appl. Phys. A 52 (1991) 78.
[17] P. Hoffmann, B. Lecohier, S. Goldoni and H. van den Bergh, Appl. Surf. Sci. 43 (1989) 54.

Applied Surface Science 69 (1993) 79–86
North-Holland

applied
surface science

Deposition of micron-size nickel lines by argon-ion laser-assisted decomposition of nickel tetracarbonyl

S. Boughaba and G. Auvert

Centre National d'Etudes des Télécommunications, BP 98, F-38243 Meylan, France

Received 2 June 1992; accepted for publication 30 October 1992

Nickel lines were deposited from the decomposition of $Ni(CO)_4$ on silicon/silicon-dioxide/silicon substrates locally heated by means of a focused CW argon-ion laser operating at several wavelengths around 0.5 μm. Growth kinetics, morphology and electrical resistivity of the nickel microstructures were investigated at various scanning speeds of the laser spot, laser beam powers and reactant gas pressures. Nickel lines as small as 1 μm in width were written. From the growth kinetics, the vertical deposition rate of nickel can be calculated. At low $Ni(CO)_4$ pressures, typically below 0.3 mbar, deposition of flat-topped nickel lines occurs. The deposition rate is proportional to the $Ni(CO)_4$ pressure and independent of the laser-induced temperature. At reactant gas pressures above 0.3 mbar, Gaussian nickel lines are obtained. The deposition rate is found to be independent of the $Ni(CO)_4$ pressure and exhibits an activation energy of about 11.5 kcal mol^{-1}. The measured electrical resistivity of the flat-topped lines is about 10 times higher than the bulk resistivity (7 $\mu\Omega \cdot$ cm) while in the Gaussian regime this ratio falls below 1.5. The morphology and roughness of the deposited lines were investigated using both a scanning electron microscope (SEM) and an atomic force microscope (AFM). In the flat-topped regime, grains of about 100 nm are observed at the top of the nickel lines while in the Gaussian regime the surface is smooth.

1. Introduction

In recent years laser-induced chemical vapour deposition (LCVD) has become of increasing interest because of its potential applications in microelectronics, electro-optics and hybrid technology. Depending on the wavelength of the laser light used, the LCVD of materials may be accomplished photochemically, thermally or through a combination of both. Pyrolytic LCVD is an adaptation of conventional CVD. The main difference between LCVD and conventional CVD lies in the use of a focused laser beam on absorbing substrates as heat source, leaving the reactant gas at room temperature.

Various metals have been deposited on various substrates by thermal LCVD [1,2]. Laser-induced pyrolytic deposition of metal appears to be well-suited to the production of fine interconnect conductor lines in prototype integrated circuits (IC's) and therefore to their remodelling [3–5]. For this purpose, the thermal decomposition of nickel tetracarbonyl, $Ni(CO)_4$, which occurs at low temperatures (about 200°C) with high rates appears to be a very attractive reaction for use in LCVD of nickel interconnects. Furthermore, this reaction is known to provide metal films of high purity and low resistivity [6] (the bulk resistivity of nickel is 7 $\mu\Omega \cdot$ cm).

Localized deposition of nickel by thermal decomposition of $Ni(CO)_4$ on various substrates was first demonstrated with the use of the 10.6 μm radiation of a CO_2 laser beam [7–10]. In order to reduce the size of the deposited structures, CW Ar^+ and CW Kr^+ lasers operating at wavelengths of around 0.5 μm were used and nickel lines as narrow as 1 μm were deposited [6,11–13]. However, very few results have been reported on the deposition kinetics of nickel for the case of micrometer-size laser spots which are of interest in the laser direct-writing technique.

The aim of this work is to report on LCVD of micrometer-size nickel lines from $Ni(CO)_4$ using the visible radiation of a CW Ar^+ laser. The

0169-4332/93/$06.00

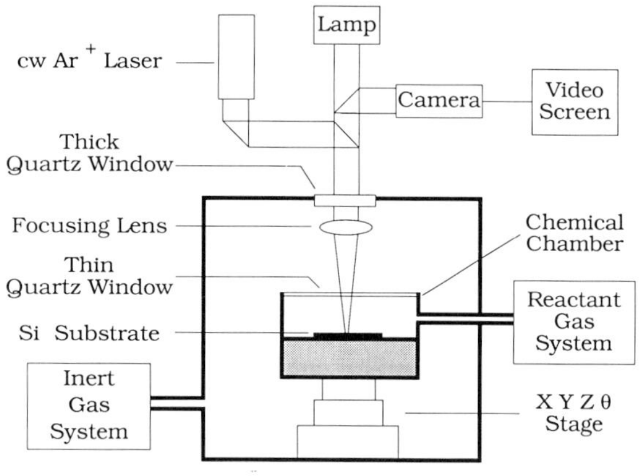

Fig. 1. Experimental set-up to deposit very narrow lines using the laser-induced decomposition of gaseous species.

growth kinetics, morphology and electrical properties of the deposited lines have been investigated. The effects of the reactant gas pressure and deposition temperature on the vertical growth rate and the electrical resistivity are reported.

2. Experiment

The experimental set-up is schematically represented in fig. 1 [14]. The TEM_{00} beam of a 5 W CW argon-ion laser operating in the range 488–514 nm passes through a fixed optical system and is focused by means of microscope objectives of ×5, ×20 or ×50 magnifications. A substrate placed in a chemical chamber filled with a reactant gas at a pressure ranging between 0.1 and 100 mbar, is moved by a $XYZ\theta$ table system until the absorbing surface is in the focal plane of the objective and then laterally displaced at a scan speed of up to 3 mm/s. A conventional fibre optic source is used to form an image of the substrate surface on a video screen allowing an operator to control the deposition process. The laser spot diameter at half-height has been measured [15]. It has been found to be 1.1, 2 and 4.8 μm when using the ×50, ×20 and ×5 objectives, respectively.

The substrates used in our experiments were ⟨100⟩ silicon substrates covered with two coating layers. The first is a 1 or 2 μm thick thermally grown silicon dioxide layer and the second is a 0.1, 0.2 or 0.3 μm thick polysilicon film. The role of the polysilicon film is to absorb the laser radiation in order to locally increase the surface temperature. The silicon dioxide layer thermally insulates the silicon coating from the silicon substrate which is at room temperature.

After pumping down to about 10^{-3} mbar, the reaction chamber is filled with nickel tetracarbonyl. A nickel line is deposited when the substrate is laterally moved in the focal plane of the laser beam which is turned on at a high enough power to locally heat the substrate surface to a high temperature, i.e., between 300 and 1400°C.

The morphology of the deposited nickel lines was investigated using an optical microscope, a scanning electron microscope (SEM) and an atomic force microscope (AFM) while the buried interfaces were controlled with a focused ion beam system. The thickness and width of the deposits were measured using a mechanical stylus and a SEM, respectively. The resistivity of the nickel lines was deduced from resistance measurements which were achieved by a two-point test method.

3. Results and discussion

Accurate determination of the surface temperature induced by absorption of the laser radiation in the substrate is a major factor for relevant investigations of the growth kinetics. A point of reference for such a determination is the output laser power P_m required for melting the surface polysilicon layer. This laser power threshold was determined at the beginning of each experiment. After pumping down to about 10^{-3} mbar in the chemical chamber, the laser is turned on and its focused beam interacts with the substrate surface. Locally, the temperature increases with the laser power and no modification occurs as long as the laser-induced temperature remains below the silicon melting point (1410°C). The laser beam power threshold P_m is obtained when a modification, consisting of a molten line, is optically detected on the silicon film.

Using the laser power threshold P_m and considering the modification of the reflectivity during the deposition of nickel on the silicon surface, the laser-induced surface temperature can be expressed as a function of the output laser power P by the following equation:

$$T(°C) = (T_m - T_r)\left(\frac{P}{P_m}\right)\left(\frac{R_{Si}}{R_{Ni}}\right) + T_r, \qquad (1)$$

where T_m is the melting temperature of the polysilicon coating (1410°C) and T_r the room temperature usually taken as equal to 20°C. R_{Si} and R_{Ni} are the reflectivities of silicon and nickel in the range of wavelength emitted by the Ar^+ laser. R_{Si} and R_{Ni} have been measured previously [6] and have been found to be 0.38 and 0.48, respectively.

Eq. (1) has been derived from laser-induced surface temperature calculations [16,17] and its validity has been substantiated by the kinetic data obtained for LCVD of nickel dots [6]. The results of these calculations can be easily extended to our experimental configuration [18] and eq. (1) is well-suited to determining the laser-induced surface temperature of our substrates.

Furthermore, for a given laser output power P, one can expect that the laser-induced surface temperature depends on the substrate structure [15]. Since no difference in growth kinetics was detected using the various substrates available for our experiments, the results presented in this paper were obtained using the 0.1 μm poly-Si/2 μm $SiO_2/\langle 100 \rangle$ Si structure.

In fig. 2, the laser-induced deposition temperature threshold (T_d) is represented versus the nickel tetracarbonyl pressure at a scanning speed of 30 μm/s. The decomposition temperature threshold for $Ni(CO)_4$ molecules decreases with increasing reactant gas pressure and is in the range of 240–320°C. This occurs at all scanning speeds investigated between 5 and 100 μm/s. This laser-induced deposition temperature threshold is slightly higher than for the conventional CVD technique or for that of the large spot LCVD (100 and 200°C, respectively) [6,19]. This can be explained by the fact that in micrometer-size spot LCVD, the gas molecules remain "cold" until reaching the hot zone of the laser spot at the substrate surface.

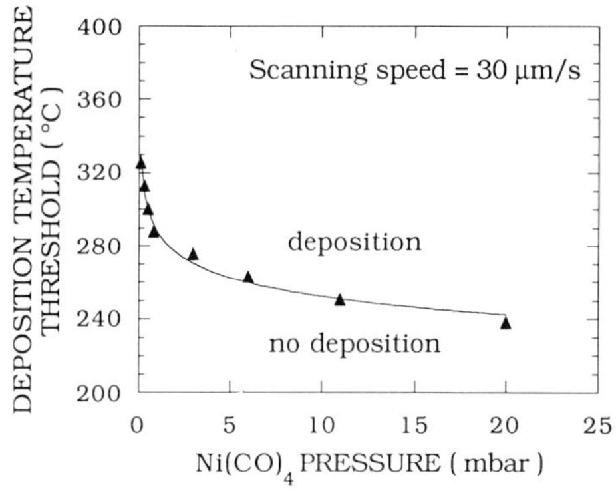

Fig. 2. Behaviour of the laser-induced deposition temperature threshold, T_d, versus the $Ni(CO)_4$ gas pressure.

Fig. 3 shows interference contrast micrographs of two nickel lines deposited at two different $Ni(CO)_4$ pressures, 0.1 and 1 mbar. Both lines were deposited at a laser-induced temperature of 600°C and a scanning speed of 20 μm/s. The line deposited at 0.1 mbar appears to be flat-topped (fig. 3a) while the line deposited at 1 mbar has a Gaussian profile (fig. 3b). Apart from this difference in shape, it appears that the flat-topped line is slightly wider than the Gaussian one. The width of both lines was measured and is equal to 4.8 μm for the flat-topped line and 3 μm for the Gaussian one. This particular feature was observed for all the scanning speeds considered (5–100 μm/s). In the power range investigated, it was found that the flat-topped nickel lines are deposited for reactant gas pressures typically below 0.3 mbar, whereas for $Ni(CO)_4$ pressures above 0.3 mbar, lines with a Gaussian profile are obtained.

Fig. 4 shows scanning electron micrographs of two nickel lines deposited at two different gas pressures, 0.1 and 0.8 mbar. Both lines were deposited at a laser-induced temperature of 600°C and a scanning speed of 30 μm/s. The lines are observed laterally at a tilt angle of about 70°. The structure of the deposits is polycrystalline in both the flat-topped and Gaussian regimes. The grain size appears to be larger in the flat-topped line than in the Gaussian one and is, in both regimes, larger than in the polysilicon film. The latter fact can be due either to a low nucleation rate or an

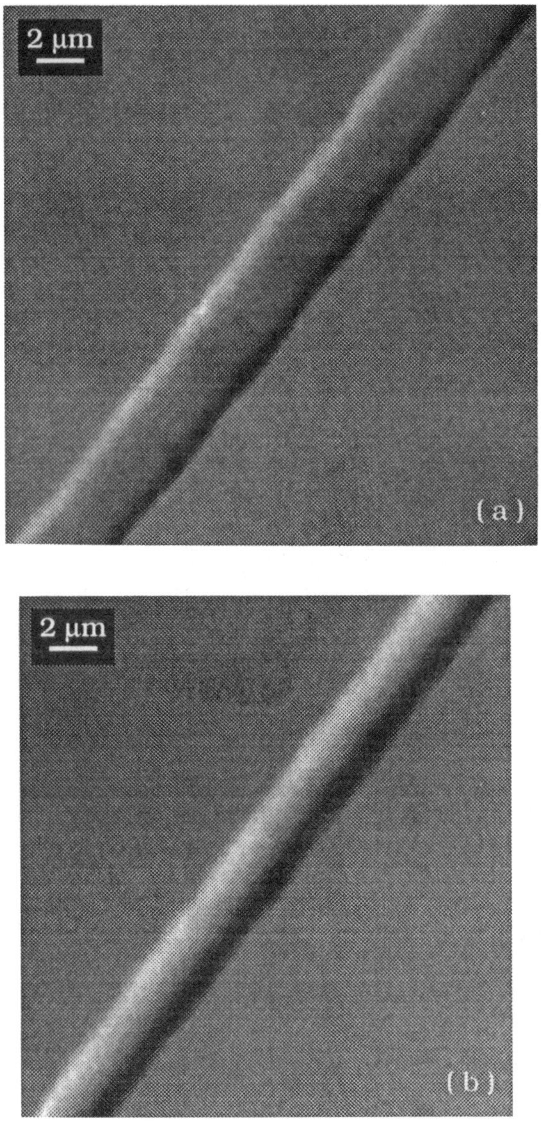

Fig. 3. Interference contrast micrographs of nickel lines deposited on a 0.1 μm poly-Si/2 μm SiO_2/⟨100⟩Si substrate using the ×20 objective at a laser-induced deposition temperature of 600°C and a scanning speed of 20 μm/s. $Ni(CO)_4$ gas pressure: (a) 0.1 mbar, (b) 1 mbar.

anisotropic growth of the nickel grains. Moreover, the flat-topped line shows a more pronounced roughness than the Gaussian one whose surface is relatively smooth.

The shape of the deposited nickel lines and their roughness were investigated more accurately using an atomic force microscope. Fig. 5 shows atomic force micrographs of two nickel lines deposited at 0.1 and 0.8 mbar of $Ni(CO)_4$. Both lines were deposited at a laser-induced temperature of 600°C and a scanning speed of 20 μm/s and have a thickness of 0.14 and 0.25 μm, respectively. The difference in shape is striking. The line deposited at 0.1 mbar of $Ni(CO)_4$ (fig. 5a) shows a flat-topped profile with abrupt sides which lead to a well-defined lateral width. The line deposited at 0.8 mbar of $Ni(CO)_4$ (fig. 5b) shows a Gaussian profile with an ill-defined lateral width. Furthermore, the difference in grain size is confirmed. In the flat-topped line, grains of about 100 nm are observed while in the Gaussian one, the grain size is about 50 nm.

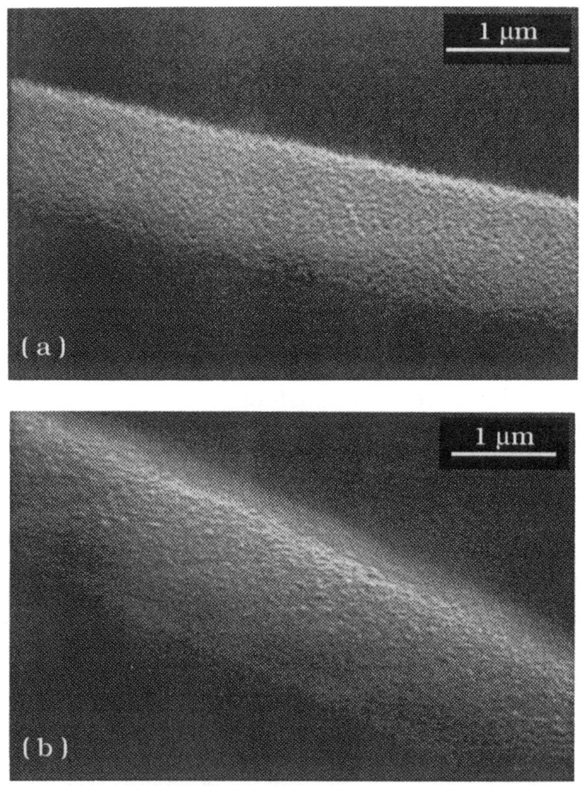

Fig. 4. Scanning electron micrographs of nickel lines deposited on a 0.1 μm poly-Si/2 μm SiO_2/⟨100⟩Si substrate using the ×20 objective at a laser-induced deposition temperature of 600°C and a scanning speed of 30 μm/s. $Ni(CO)_4$ gas pressure: (a) 0.1 mbar, (b) 0.8 mbar.

In order to investigate the quality of the buried interfaces, cross sections of the deposited nickel lines were performed using a focused ion beam system. Fig. 6 shows such a cross section of a Gaussian shape nickel line deposited at 0.8 mbar of $Ni(CO)_4$. The laser-induced deposition temperature is 600°C and the scanning speed is 20 μm/s. The section is achieved at a high current density and then observed at a tilt angle of 60° at a low current density. This cross section shows clearly that the deposition process occurring at 600°C does not damage the two underlying interfaces. This is the case for all nickel lines we deposited in the pressure and temperature range investigated, namely 0.1–20 mbar and 300–1000°C, respectively.

The vertical growth rate was deduced from the following equation [20]:

$$G_R = (h/\Phi)V_S, \tag{2}$$

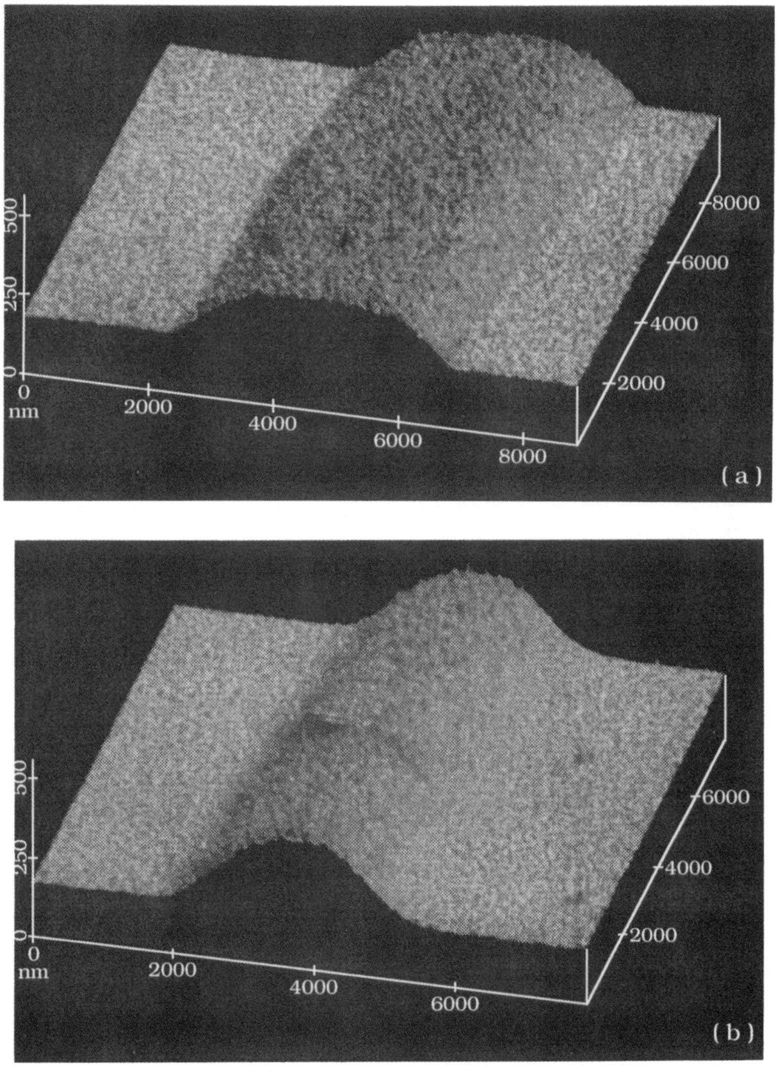

Fig. 5. Atomic force micrographs of nickel lines deposited on a 0.1 μm poly-Si/2 μm SiO_2/⟨100⟩Si substrate using the ×20 objective at a laser-induced deposition temperature of 600°C and a scanning speed of 20 μm/s. $Ni(CO)_4$ gas pressure: (a) 0.1 mbar, (b) 0.8 mbar.

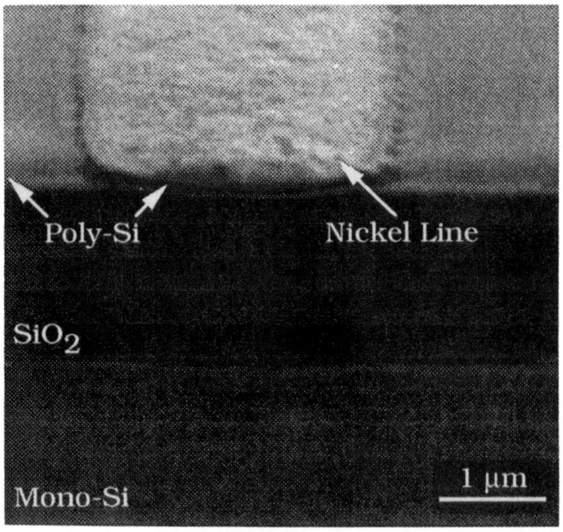

Fig. 6. Ion beam picture of a cross section of a Gaussian nickel line deposited on a 0.1 μm poly-Si/2 μm SiO_2/⟨100⟩Si substrate using the ×20 objective at a laser-induced deposition temperature of 600°C and a scanning speed of 20 μm/s. $p(Ni(CO)_4) = 0.8$ mbar.

where h is the line thickness, Φ the measured laser spot diameter and V_S the scanning speed. Fig. 7 shows the dependence of the vertical nickel deposition rate on the $Ni(CO)_4$ gas pressure for two temperatures, 750 and 850°C. At low reactant gas pressures (below 0.3 mbar), the deposition rate increases rapidly as the $Ni(CO)_4$ pressure increases. This corresponds to the flat-topped regime in which the deposition rate is found to be independent of the output laser power. At higher $Ni(CO)_4$ pressures, the deposition rate is independent of the reactant gas pressure and increases with increasing output laser power (i.e. laser-induced temperature). This corresponds to the Gaussian regime. No difference in growth kinetics is observed using the ×5, ×20 or ×50 objectives.

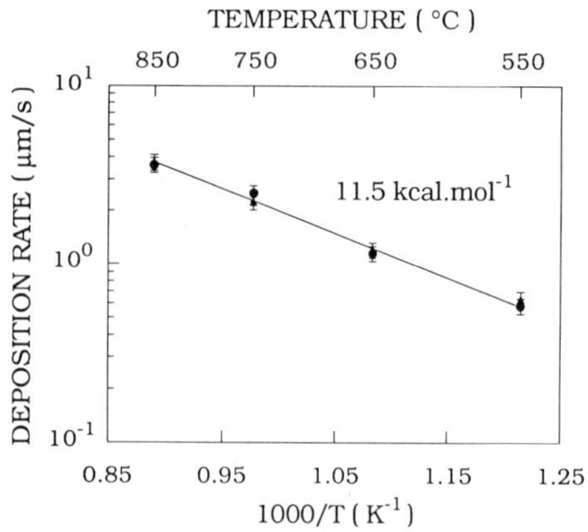

Fig. 8. Arrhenius plot of the Gaussian shape nickel lines for two different $Ni(CO)_4$ gas pressures: (▲) 1 mbar, (●) 3 mbar.

Fig. 8 shows an Arrhenius plot of the Gaussian shape nickel lines for two different $Ni(CO)_4$ pressures, 1 and 3 mbar. The relationship between the deposition rate and deposition temperature is in agreement with an Arrhenius law, and the

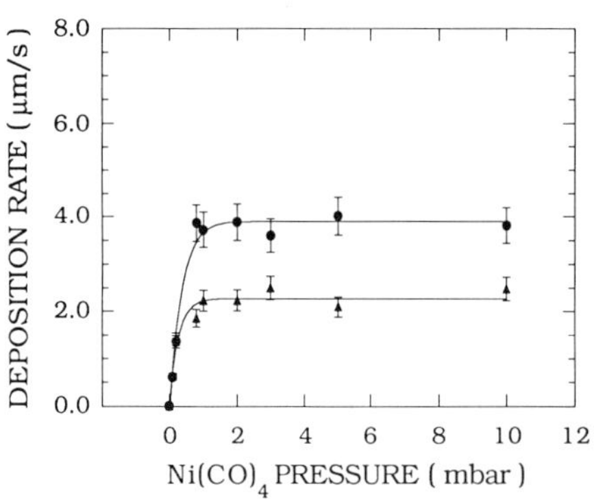

Fig. 7. Nickel deposition rate versus $Ni(CO)_4$ gas pressure for two different laser-induced deposition temperatures: (▲) 750°C, (●) 850°C.

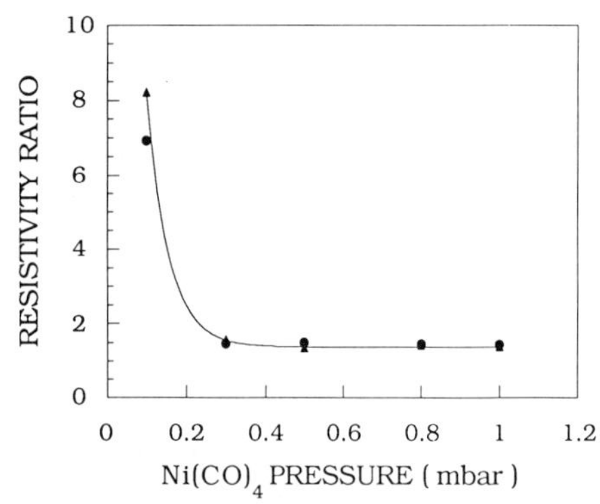

Fig. 9. Behaviour of the ratio of measured electrical resistivity of deposited nickel lines to bulk resistivity (7 $\mu\Omega \cdot$cm) versus the $Ni(CO)_4$ gas pressure for two different laser-induced deposition temperatures: (▲) 550°C, (●) 650°C.

apparent activation energy is calculated to be 11.5 kcal mol^{-1}. This value is close to that obtained by other techniques or with a large spot size [6,19], showing that no optical laser effects are present in the decomposition mechanism of the $Ni(CO)_4$ molecules. As a result of both, (i) the existence of an activation energy and (ii) a deposition rate independent of pressure, one can suggest that the limiting step of the decomposition process lies in the adsorbed layer which is saturated by some intermediate species.

The electrical resistivity of the deposited nickel lines was investigated. In fig. 9, the ratio of measured resistivity of deposited nickel lines to bulk resistivity is plotted versus $Ni(CO)_4$ pressure for two different temperatures, 550 and 650°C. It was found that the electrical resistivity of the flat-topped lines (0.1 mbar of $Ni(CO)_4$) is about 10 times higher than the bulk resistivity while in the Gaussian regime this ratio falls below 1.5. Furthermore, in the flat-topped regime the electrical resistivity slightly decreases when increasing the deposition temperature, whereas no dependence was observed in the Gaussian regime.

4. Conclusion

We have reported on LCVD of micrometer-size nickel lines from the decomposition of nickel tetracarbonyl. At low $Ni(CO)_4$ pressures, typically below 0.3 mbar, deposition of flat-topped nickel lines occurs. The deposition rate is proportional to the $Ni(CO)_4$ pressure and is independent of the laser-induced temperature, and the electrical resistivity is about 10 times higher than that of the bulk. At reactant gas pressures of above 0.3 mbar, Gaussian nickel lines are obtained. The deposition rate is found to be independent of $Ni(CO)_4$ pressure. The decomposition process is thermally activated and is limited by surface reactions. The electrical resistivity of the Gaussian lines is less than 1.5 times that of the bulk.

This highly localized laser-induced deposition of metal as well as deposition of polysilicon [15] is widely used in our laboratory to repair and remodel very large scale integrated (VLSI) and hybrid circuits.

Acknowledgements

The authors are indebted to E. André for technical assistance in performing the AFM investigations. The focused ion beam cross sections were carried out by J.P. Gonchond to whom we are very grateful.

References

[1] For a review, see: D. Bäuerle, Chemical Processing with Lasers, Vol. 1 of Spinger Series in Materials Science (Spinger, Berlin, 1986).
[2] For a review, see: J. Haigh and M.R. Aylett, in: Laser Microfabrication: Thin Film Processes and Lithography, Eds. D.J. Ehrlich and J.Y. Tsao (Academic Press, New York, 1989) p. 453.
[3] D.J. Ehrlich and J.Y. Tsao, in: VLSI Electronics: Microstructure Science, Vol. 7, Ed. N.G. Einspruch (Academic Press, Orlando, FL, 1983) p. 129.
[4] G. Auvert, Y. Pauleau and D. Tonneau, in: Emerging Technologies for in situ Processing, Eds. D.J. Ehrlich and V.T. N'Guyen, NATO ASI Ser. E 139 (1988) 201.
[5] G. Auvert, in: Reduced Thermal Processing for ULSI, Ed. R.A. Levy, NATO ASI Ser. B 207 (1989) 227.
[6] D. Tonneau, G. Auvert and Y. Pauleau, J. Appl. Phys. 64 (1988) 5189.
[7] S.D. Allen, J. Appl. Phys. 52 (1981) 6501.
[8] S.D. Allen, A.B. Tribugo and R.Y. Jan, in: Laser Diagnostics and Photochemical Processing for Semiconductor Devices, Mater. Res. Soc. Symp. Proc., Vol. 17, Eds. R.M. Osgood, Jr., S.R.J. Brueck and H.R. Schlossberg (North-Holland, Amsterdam, 1983) p. 207.
[9] S.D. Allen, in: Physical Processes in Laser–Materials Interactions, Ed. M. Bertolotti (Plenum, New York, 1983) p. 455.
[10] S.D. Allen, R.Y. Jan, S.M. Mazuk and S.D. Vernon, J. Appl. Phys. 58 (1985) 327.
[11] D. Bäuerle, in: Laser Diagnostics and Photochemical Processing for Semiconductor Devices, Mater. Res. Soc. Symp. Proc., Vol. 17, Eds. R.M. Osgood, Jr., S.R.J. Brueck and H.R. Schlossberg (North-Holland, Amsterdam, 1983) p. 19.
[12] W. Kräuter, D. Bäuerle and F. Fimberger, Appl. Phys. A 31 (1983) 13.
[13] F. Pedzoldt, K. Piglmayer, W. Kräuter and D. Bäuerle, Appl. Phys. A 35 (1984) 155.
[14] D. Tonneau, Y. Guern and G. Pelous, in: Optical Microlithography and Metrology for Microcircuit Fabrication, SPIE Proc., Vol. 1138, Eds. M.J. Lacombat and S. Wittekoek (SPIE, Bellingham, WA, 1989) p. 184.
[15] S. Boughaba and G. Auvert, Appl. Surf. Sci. 54 (1992) 25.
[16] D. Tonneau and G. Auvert, in: Photon, Beam and Plasma Enhanced Processing, Eur. Mater. Res. Soc. Symp. Proc.,

Vol. 15, Eds. A. Golanski, V.T. N'Guyen and E.F. Krimmel (Editions de Physique, Paris, 1987) p. 169.

[17] D. Tonneau and G. Auvert, in: Laser and Particle-Beam Chemical Processing for Microelectronics, Mater. Res. Soc. Symp. Proc., Vol. 101, Eds. D.J. Ehrlich, G.S. Higashi and M.M. Oprysko (Materials Research Society, Pittsburgh, PA, 1988) p. 131.

[18] S. Boughaba, G. Auvert and Y. Pauleau, J. Electrochem. Soc., submitted.

[19] H.E. Carlton and J.H. Oxley, AIChE J. 13 (1967) 86.

[20] Y.S. Liu, in: Tungsten and Other Refractory Metals for VLSI Applications, Ed. R.S. Blewer (Materials Research Society, Pittsburgh, PA, 1986) p. 43.

Applied Surface Science 69 (1993) 87–93
North-Holland

applied
surface science

Laser direct-write Al deposition on Si, GaAs and diamond from trialkylamine-alane precursors

F. Foulon, O. Lehmann and M. Stuke

Max-Planck-Institut für biophysikalische Chemie, P.O. Box 2841, D-3400 Göttingen, Germany

Received 2 June 1992; accepted for publication 30 October 1992

Laser direct writing of aluminium on silicon, gallium arsenide and diamond film substrates using the Al precursors $(C_2H_5)_3N \cdot AlH_3$ (TEAA) and $(CH_3)_3N \cdot AlH_3$ (TMAA) is reported. Deposition, carried out with a continuous wave Ar^+ laser working at 514 nm, results from the thermal decomposition of the precursor at the laser-heated surface. Estimated temperature thresholds for Al deposition from TEAA and TMAA are similar: about 370 K on Si(111) and 385 K on GaAs(100) for both precursors. The deposited height increases with laser power density but decreases with scanning speed. Uniform lines are produced at scanning speeds up to 250 and 400 $\mu m/s$, using TEAA and TMAA precursors, respectively. The height of the deposited line is also strongly dependent on the precursor pressure in the processing chamber. Larger deposited heights are obtained with TMAA due to its higher vapour pressure compared to TEAA. 7 μm wide, 0.7 μm thick Al lines are generated by a single surface scan at a scanning speed of 100 $\mu m/s$ and a high power density (1.55 MW/cm^2). Deposited Al lines deposited from TEAA and TMAA are polycrystalline. The rate of crystallite growth is greater than the rate of nucleation, leading to an increase of the surface roughness with the line thickness. Adhesion of the aluminium lines to the substrates is good and their resistivities range from 4.5 to 7 $\mu\Omega \cdot cm$ for both precursors (bulk: 2.8 $\mu\Omega \cdot cm$).

1. Introduction

Aluminium (Al) is the most common conductor metal used in the microelectronics industry, both for device and interconnect fabrication. Deposition of high-purity Al is therefore under intensive investigation. Trimethylamine alane (TMAA) [1–8] and triethylamine alane (TEAA) [9–11] are suitable precursors to generate high-quality aluminium films. TEAA is a liquid at room temperature and has a vapour pressure of about 0.65 mbar at 25°C. TMAA has a higher vapour pressure of 1.3 mbar at 25°C but is a solid. In the trialkylamine alane molecules ($AlH_3 \cdot NR_3$, R = alkyl), the amines form Lewis acid–base adducts with the alane. These complexes contain no Al–C bonds, in contrast to trimethyl and triethyl aluminium precursors that were found to produce carbon-contaminated films due to incorporation of alkyl radicals [12,13]. The weak Al–N bond thermally dissociates at low temperature ($T \approx 320$ K on Al substrate) in the adsorbed or gase phase to form adsorbed AlH_3. Further decomposition of AlH_3 through the recombinative desorption of hydrogen from the surface provides Al atoms for the deposition process [3,4,9].

Al deposition from TMAA and TEAA by CVD [1–3] and focused ion beam stimulated deposition [5] has been reported. High-quality films, having C and N contamination levels about 1.5 times higher than sputtered films, were produced [5]. These films had electrical resistivity down to 3.2 $\mu\Omega \cdot cm$, close to the bulk value of 2.8 $\mu\Omega \cdot cm$. Higher deposition rates were obtained with TMAA because of its higher vapour pressure. Laser direct writing of Al lines from TMAA has also been reported using an argon-ion laser (Ar^+, $\lambda = 514.5$ nm) [6,7]. Al lines having a minimum linewidth of 2 μm and resistivities in the range 7 to 11 $\mu\Omega \cdot cm$ were produced. Auger analysis of

0169-4332/93/$06.00

the lines showed that the purity of Al was 95%–97% with 3%–5% of oxygen being present in the lines. Generation of three-dimensional free-standing metal objects was achieved by this technique [7]. Very recently, aluminium lines have been deposited from the TEAA precursor by a two-step laser-assisted technique [11]. Aluminium oxide substrates covered with a spin-coated film of palladium acetate were irradiated through a contact mask with a KrF excimer laser ($\lambda = 248$ nm). Irradiation resulted in the formation of a thin palladium film in the irradiated areas. Untreated palladium acetate was then removed before Al films were formed on predeposited palladium through catalytic thermal decomposition ($T \approx 370$ K) of a thin liquid film of TEAA sprayed onto the substrate.

This paper reports Ar^+ laser direct-write of Al lines on Si, GaAs and diamond/Si substrates by thermal decomposition of TEAA and TMAA. The threshold for Al deposition and' the deposited height per scan are given for each substrate/precursor system. The influence of power density and scanning speed on the deposited height and line morphology is discussed. The adhesive property and electrical resistivity of the deposited lines are reported.

2. Experiment

Laser direct writing was carried out with an Ar^+ laser working at 514.5 nm in the TEM_{00} mode. The beam was focussed with a microscope objective (8 ×, 0.2 NA) into a vacuum chamber through a quartz window. The beam diameter at the sample surface is approximately 7 μm. The laser power was measured using a detector external to the vacuum system. The power value was corrected taking into account the transmission losses in the objective and in the window of the processing chamber in order to give the laser power density at the sample surface; values in the range 0.01 to 2.6 Mw/cm2 were obtained. The processing chamber was fixed on a microprocessor-controlled *XYZ* stage. Irradiation was performed at scanning speeds in the range 25 to 1000 $\mu m\ s^{-1}$. Samples used were: (111)-oriented-n-type Si, (100)-oriented semi-insulating GaAs and polycrystalline diamond film (1 μm thick) deposited on Si by a CVD technique (P. Bachmann). Prior to their introduction into the processing chamber, the samples were degreased by successive immersion in each of the following baths, trichloroethylene, acetone, isopropanol-alcohol and de-ionized water, for 2 min in an

Table 1
Power-density threshold and deposited height per scan for Ar^+ laser-induced Al deposition from TEAA and TMAA on Si, GaAs and diamond/Si substrates at a scanning speed of 25 μm/s

		Material		
		Si	GaAs	Diamond/Si
Deposition threshold (MW/cm^2); $v = 25$ μm/s	TEAA 0.2 mbar	0.55	0.18	0.6
	TMAA 0.5 mbar	0.52	0.16	0.57
Deposited height (μm/scan); $v = 25$ μm/s	TEAA 0.2 mbar	0.6 [1.65]	0.65 [0.54]	1 [1.8]
	TEAA + N_2 1 mbar	1.1 [1.65]	0.9 [0.54]	1.6 [1.8]
	TMAA 0.5 mbar	1.9 [1.65]	1.7 [0.54]	2.9 [1.8]

The deposited height is given for a power density (in MW/cm^2, indicated in brackets) equal to three times the deposition threshold for a given material in a TEAA atmosphere.

ultrasonic tank. Immersion in dip HF (5% HF) after degreasing was not found to result in significant modification of the deposition process. The chamber was pumped down to 5×10^{-5} mbar using a turbo-molecular pump before being filled with the precursor. TEAA and TMAA were passed through the chamber at pressures of 0.2 and 0.5 mbar, respectively. In case of TEAA, the precursor molecules were also transported to the processing chamber by passing ultrahigh-purity N_2 through a TEAA bubbler at room temperature. Total TEAA + N_2 pressures were in the range 0.8 to 15 mbar. Prior to any deposition experiment, the precursor was left to pass through the processing chamber for 5 min to allow precursor degassing and system equilibration. The thickness, width and shape of the deposited lines were measured by surface profilometry. The morphology and the width of the lines were obtained using scanning electron microscopy (SEM). The volume resistivity of the deposited lines was calculated after measuring the resistance, the length and the cross section of lines written between gold pads.

3. Experimental results and discussion

We report in table 1 the power density threshold for Al deposition from TEAA and TMAA on Si, GaAs and diamond/Si substrates, at a scanning speed of 25 μm/s. For a given material, the deposition thresholds for the two precursors are almost identical. This is in good agreement with previous results reported by Dubois et al. [4,9], showing that TEAA and TMAA have similar temperature thresholds for Al deposition on Al substrates (310 and 320 K for TEAA and TMAA, respectively). In contrast, there is a strong dependence of the deposition threshold on the type of material. The deposition threshold on Si, either from TEAA or TMAA, is about three times higher than the deposition threshold on GaAs. This behaviour is attributed to the different thermal and optical properties of the two materials. According to heat-flow calculations for a Gaussian beam source of constant power, heating a semi-infinite solid, whose temperature dependence on the thermal and optical parameters is not taken into account, the surface temperature increase ΔT in the centre of the beam is given by [14]:

$$\Delta T = P(1-R)w\sqrt{\pi}/2K_T, \tag{1}$$

where P, w, R and K_T are the incident power density, the beam radius ($w = 3.5$ μm), the surface reflectivity and the thermal conductivity of the substrate, respectively. Given the values of R, 0.36 and 0.39 at 514 nm, and K_T, 1.4 [15] and 0.37 W/cm · K [16] at 300 K, for Si and GaAs, respectively, it is clear that in order to produce a similar temperature increase ΔT, the laser power has to be about 3.5 times higher for Si than for GaAs. This is in good agreement with the reported experimental results. The activation energy of the decomposition process on Si and GaAs is therefore believed to be identical for these two materials. According to eq. (1), the calculated thresholds for Al deposition on Si and GaAs from TEAA are about 370 and 385 K, respectively. These values are higher than those previously reported for Al deposition from TEAA on Al substrates (310 K) which was shown to be less activated than deposition on Si substrates [4,9]. The interpretation of the threshold values in case of diamond/Si is more complicated since the thermal conductivity of our polycrystalline diamond film is not known and heat loss in the film is influenced by the underlying Si. We have observed for all materials a slight increase of the deposition threshold with the scanning speed. In case of Si for example, at 200 μm/s, the deposition threshold under a TEAA atmosphere increases up to 0.65 MW/cm^2. This may result from both a reduction of the maximum surface temperature reached at higher scanning speed and a decrease of the surface heating time, which can strongly influence the nucleation process. Al deposition from TEAA and TMAA at scanning speeds above 250 and 400 μm/s, respectively, was found to produce broken lines even at high power densities. Moreover, deposition on GaAs at moderate power densities (> 0.7 MW/cm^2) and high scanning speeds (> 150 μm/s) resulted in significant surface damage, which may be attributed to loss of arsenic atoms, and only in small and nonuniform deposition. In this case the As loss is believed to occur more rapidly than nucleation and film growth.

Fig. 1 shows the deposited height per scan as a function of the power density for Si samples irradiated at different scanning speeds in 0.5 mbar TMAA. For energy densities above 0.5 MW/cm^2, the deposited height continuously increases with power density. This results from an increase of the surface temperature, as expected from eq. (1), and a longer surface cooling time after surface irradiation, when increasing the power density. After surface scanning at 25 $\mu m/s$ and 1.28 MW/cm^2, an Al line having a thickness of 1.7 μm and a width of 14 μm is deposited. Fig. 1 also shows that an increase in the scanning speed by a factor of two, which corresponds roughly to a decrease of surface heating time by a factor of two, results in deposited heights about two times smaller. Thus, at 0.89 MW/cm^2, the deposited height at 25, 50, 100, and 200 $\mu m/s$ are respectively equal to 1.1, 0.62, 0.38 and 0.15 μm. The decrease of the deposited height is therefore strongly correlated to the surface heating time, i.e. the deposition time. Similar evolution of the deposited height on the power density and scanning speed were observed for deposition from

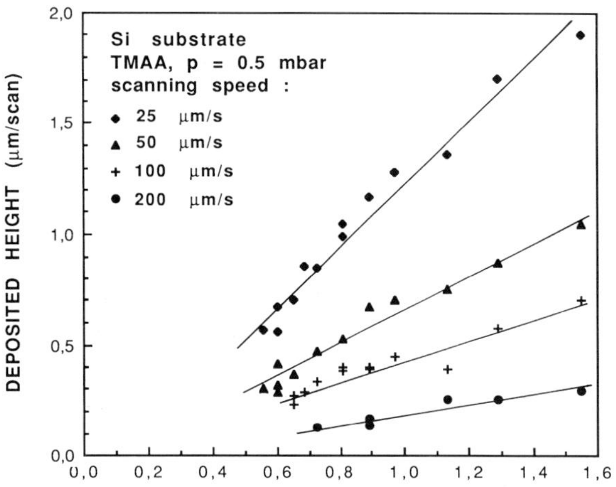

Fig. 1. Height of the deposited Al lines as a function of the laser power density for laser direct writing on Si(111) in 0.5 mbar TMAA at the scanning speeds: 25, 50, 100 and 200 $\mu m/s$.

TEAA and TMAA on Si, GaAs and diamond/Si substrates.

Table 1 reports the height of Al lines deposited from TEAA and TMAA on Si, GaAs and diamond/Si substrates, at a scanning speed of 25 $\mu m/s$ and at power densities equal to three times the deposition threshold measured for each material in a TEAA atmosphere (this was done because of the different thermal conductivities of the materials in order to give the deposited height at comparable surface temperatures). Al deposition from TEAA was achieved with and without a bubbler. Investigation on the influence of the total TEAA + N_2 pressure on the height of the deposited lines showed that the deposited height sharply increases up to a maximum for a total pressure of about 1 mbar before it slowly decreases when increasing the flow of N_2 passing through the bubbler. At the maximum, the deposited heights are about 1.6 times larger than without the bubbler due to an increase of the partial pressure of TEAA in the chamber when using the bubbler. Table 1 also shows that Al deposition on Si, GaAs and diamond/Si substrates under 0.5 mbar TMAA, results in deposited heights about 3 times larger than under 0.2 mbar TEAA. These larger deposition rates are also attributed to the larger flux of precursor molecules striking the hot surface during laser irradiation under TMAA atmosphere. One can point out that the ratio of the height deposited under TEAA + N_2 ($p = 1$ mbar) and TMAA ($p_{tot} = 0.5$ mbar) atmosphere to the height deposited under TEAA ($p = 0.2$ mbar) atmosphere shows no significant dependence on the substrate nature. Therefore, the kinetics of the deposition process does not seem to depend on the substrate nature for the materials used in this study.

Fig. 2a is a SEM photograph of Al lines deposited from TEAA + N_2, $p_{tot} = 1$ mbar, at a power density of 1.05 MW/cm^2 and at different scanning speeds: 25, 50, 100 and 150 $\mu m/s$ successively, from the right to the left of the photograph (vertical lines). The lines deposited at scanning speeds of 25, 50, 100 and 150 $\mu m/s$ are 9.6, 5.5, 4.4 and 3.8 μm wide and 0.55, 0.22, 0.12 and 0.09 μm thick, respectively. The line width and thickness decrease with the scanning speed results from a decrease of the surface area heated above the deposition threshold and a decrease of the time available for the lateral growth of the deposit when increasing the scanning speed. Figs. 2b and 2c show the morphology of Al lines deposited at 25 $\mu m/s$ from TEAA + N_2 ($p_{tot} = 1$ mbar) on Si and diamond/Si substrates at power densities of 1.05 and 1.21 MW/cm^2, respectively. The width, thickness and morphology of the lines are uniform. Al lines deposited on Si and diamond/Si are 0.55 and 1.1 μm high, respectively. They are rough and formed of grains having a diameter of about 0.5 and 1.1 μm, respectively. Al lines deposited on diamond/Si were not found to be significantly wider than lines deposited on Si. This may be due to the poor thermal conductivity of the deposited diamond films or to their small thickness. Al lines having identical morphology were deposited from TEAA on GaAs substrates. The roughness is found to increase

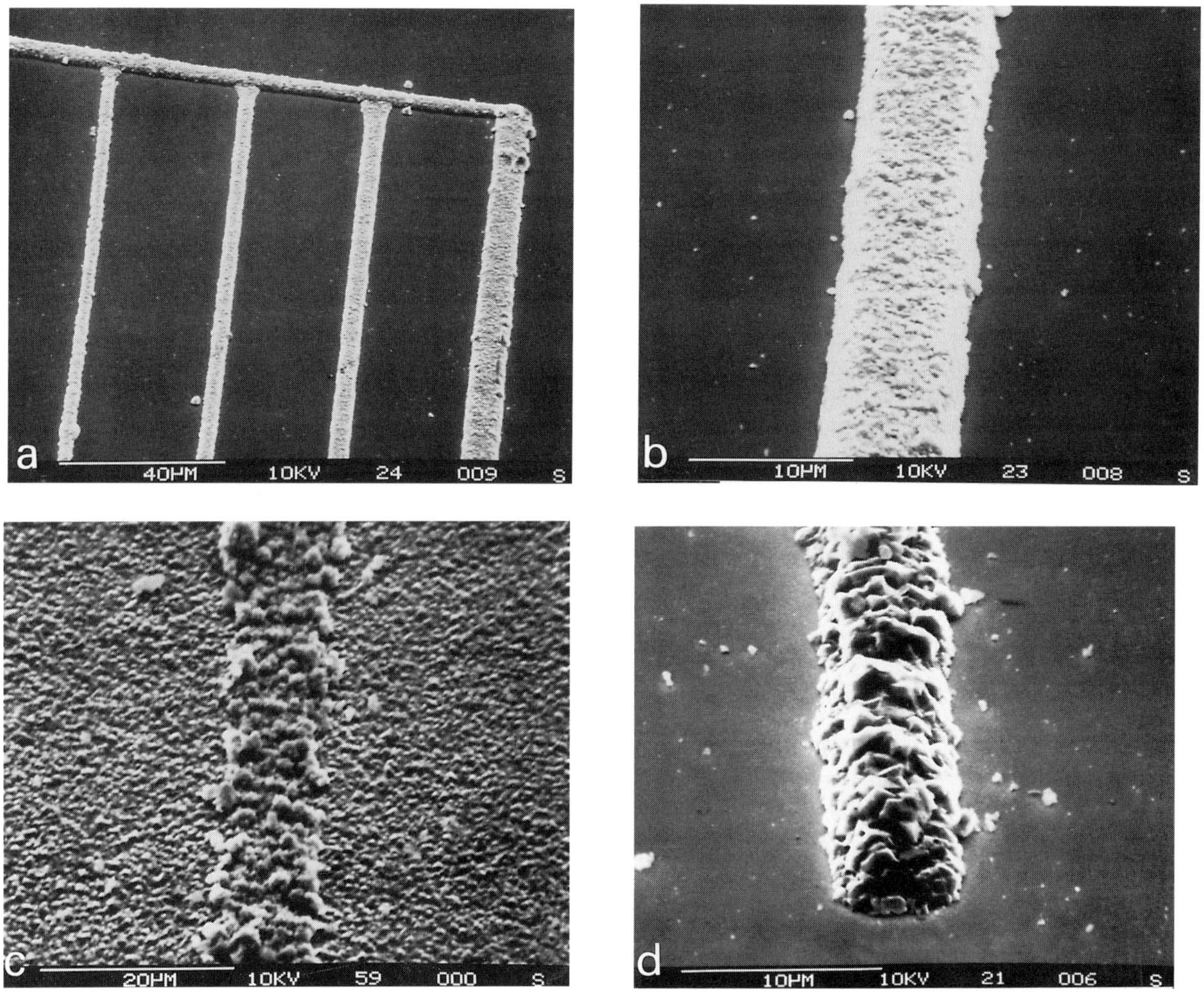

Fig. 2. SEM photographs of deposited Al lines. (a) Deposition on Si, at 1.05 MW/cm^2 and at different scanning speeds: 25, 50, 100 and 150 $\mu m/s$ successively, from the right to the left of the photograph (vertical lines), in 1 mbar TEAA + N_2. (b) Deposition on Si, at 1.05 MW/cm^2 and 25 $\mu m/s$, in 1 mbar TEAA + N_2. (c) Deposition on diamond/Si, at 1.21 MW/cm^2 and 25 $\mu m/s$, in 1 mbar TEAA + N_2. (d) Deposition on GaAs, at 0.64 MW/cm^2 and 25 $\mu m/s$, in 0.5 mbar TMAA.

with the deposited height, i.e. when increasing the power density at low scanning speeds (typically ≤ 50 μm/s). Al deposition from TMAA at low scanning speeds (≤ 75 μm/s) was found to produce highly polycrystalline films. Fig. 2d shows an Al line deposited from TMAA, $p = 0.5$ mbar, at 0.64 MW/cm^2 and 25 μm/s. The crystallite size of about 1 μm at 25 μm/s was found to decrease down to 0.2 μm at 200 μm/s. SEM observation of the morphology of the lines showed an increase in the crystallite size rather than an increase in the number of small crystallites when the thickness of the film increases. This suggests that the rate of crystallite growth is larger than the rate of nucleation, leading to an increase of the surface roughness with the film thickness. Such a phenomenon was already observed for Al deposition from TMAA by CVD [1].

The adhesive properties of the Al lines were qualitatively investigated by Scotch-tape test. All the films grown from TEAA and TMAA on Si, GaAs and diamond/Si substrates remained intact as the tape was peeled away from the deposit. The electrical resistivity of the Al lines was found to be in the range 4.5 to 7 μΩ · cm, i.e. about 1.5 to 2.5 the bulk value. These low-resistivity values confirm that high-purity Al lines can be deposited from TEAA and TMAA by the laser-induced CVD technique.

4. Conclusion

Ar^+ laser direct-write of Al lines on Si, GaAs and diamond/Si from TEAA and TMAA is shown to produce uniform lines at scanning speeds up to 250 and 400 μm/s, respectively. Deposition results from the thermal decomposition of the precursor. The temperature threshold for deposition from both TEAA and TMAA on Si and GaAs substrates are about 370 and 385 K, respectively. Deposited heights depend strongly on laser power density, scanning speed and gas pressure conditions. The deposited height strongly depends on laser power density, scanning speed and gas pressure conditions. Higher deposited heights are obtained with TMAA due to its higher vapour pressure when compared to TEAA. Passing N_2 gas through a TEAA bubbler increases the deposited height for the same power density and scanning speed conditions by about 60%. 7 μm wide and 0.7 μm thick Al lines are produced after a single surface scan at a scanning speed of 100 μm/s and a power density of 1.55 MW/cm^2. The deposited Al lines are polycrystalline. The increase in the crystallite size rather than an increase in the number of small crystallites, when the thickness of the film increases, suggests that the rate of crystallite growth is greater than the rate of nucleation, leading to an increase of the surface roughness with the film thickness. The adhesion of the Al lines to the substrates is good. Resistivities of deposited aluminium lines range from 4.5 to 7 μΩ · cm, i.e. 1.5 to 2.5 times the bulk value.

Acknowledgements

We would like to thank K. Müller for technical assistance, P. Bachmann (Philips) for providing the diamond/Si substrates, F.P. Schäfer for support and SFB 93 for financial support.

References

[1] W.L. Gladfelter, D.C. Boyd and K.F. Jensen, Chem. Mater. 1 (1989) 339.
[2] D.B. Beach, S.E. Blum and F.K. LeGoues, J. Vac. Sci. Technol. A 7 (1989) 3117.
[3] M.E. Gross, K.P. Cheung, C.G. Fleming, J. Kovalchick and L.A. Heimbrook, J. Vac. Sci. Technol. A 9 (1991) 57.
[4] L.H. Dubois, B.R. Zegarski, C. Kao and R.G. Nuzzo, Surf. Sci. 236 (1990) 77.
[5] M.E. Gross, L.R. Harriot and R.L. Opila, J. Appl. Phys. 68 (1990) 4820.
[6] T.H. Baum, C.E. Larson and R.L. Jackson, Appl. Phys. Lett. 55 (1989) 1264.
[7] O. Lehmann and M. Stuke, Appl. Phys. A 53 (1991) 343.
[8] O. Gottsleben, H.W. Roesky and M. Stuke, Adv. Mater. 3 (1991) 201.
[9] L.H. Dubois, B.R. Zegarski, M.E. Gross and R.G. Nuzzo, Surf. Sci. 244 (1991) 89.
[10] M.E. Gross, C.G. Fleming, K.P. Cheung and L.A. Heimbrook, J. Appl. Phys. 69 (1991) 2589.
[11] O. Lehmann and M. Stuke, Appl. Phys. Lett. (1992), in press.
[12] D.W. Squire, C.S. Dulcey and M.C. Lin, J. Vac. Sci. Technol. B 3 (1985) 1513.

[13] G.S. Higashi, L.J. Rothberg and C.G. Fleming, Chem. Phys. Lett. 115 (1985) 2771.
[14] M. Von Allmen, Laser–Beam Interactions with Materials, Vol. 1 of Springer Series in Materials Science (Springer, Berlin, 1987) p. 49.
[15] J.W. Amos and T.B. Wolfe, IEEE Trans. Electron Devices ED-25 (1978) 1160.
[16] S. De Unamuno and E. Fogarassy, Appl. Surf. Sci. 36 (1989) 1.

Applied Surface Science 69 (1993) 94–100
North-Holland

applied
surface science

Direct deposition of patterned copper films on Teflon

W.L. Perry [a], K.M. Chi [b], T. Kodas [a,*], M. Hampden-Smith [b,*] and R. Rye [c]

[a] *Chemical Engineering Department,* [b] *Chemistry Department, Albuquerque, NM 87131, USA*
[c] *Sandia National Laboratory, Albuquerque, NM 87185, USA*

Received 2 June 1992; accepted for publication 30 October 1992

Current subtractive methods yielding patterned Cu features on PTFE substrates rely on wet Cu etching processes. We have developed three variations of a new *dry, additive* patterning process. The mechanisms for patterning include Mg Kα X-ray-induced cross-linking, e-beam-induced cross-linking, or laser patterning. The X-ray and e-beam patterning processes rely on irradiation followed by selective etching of the non-irradiated areas. The laser patterning begins by chemically etching PTFE which leaves a rough surface with good adhesion characteristics. An argon-ion laser beam is then used to selectively remove the etched layer, revealing the underlying surface which has physical properties closely resembling unmodified Teflon. Typical laser patterning conditions are scan rates of 0.005–5.5 mm/s, incident powers of 40–380 mW at 514 nm, and base pressures of 10^{-2} Torr and at atmospheric pressure in air. In all cases, CVD from (β-diketonate)CuL compounds is used to deposit copper only on the etched regions of the sample, leaving the irradiated regions copper-free. The advantages of this procedure are: (1) subtractive Cu wet etching of copper is avoided, so no masking techniques are necessary and no liquid waste is generated; (2) ten-micron-sized features can be produced; (3) excellent adhesion is obtained.

1. Introduction

Copper deposition on poly(tetrafluoroethylene) (PTFE), better known by the DuPont trade name, Teflon, is of great interest and value to the electronics industry. One attractive physical property of Teflon is its low dielectric constant of approximately 2.1. Copper is a highly conductive material with a very low resistivity of approximately 1.7 $\mu\Omega \cdot$cm. The combination of these properties makes these materials ideal for developing new printed wiring boards and integrated circuit packaging which require small feature sizes and high frequencies of operation. As feature sizes decrease, the problem of capacitive coupling becomes appreciable, but can be minimized by a material with a low dielectric constant. In addition, a measure of a circuit's delay time is the product of its overall resistance and capacitance. Delay time can be reduced, relative to standard materials, by using copper on Teflon. Thus, combining these materials can result in fast circuits with relatively little capacitive coupling.

* To whom correspondence should be addressed.

Although the low dielectric constant of Teflon is a beneficial property, the extreme inertness of PTFE makes metallization a difficult process. Direct deposition of copper onto fresh Teflon yields unsatisfactory adhesion. Several researchers have developed methods to modify the surface of Teflon [1] in a manner that improves adhesion [2]. These methods include O_2 ion beam presputtering, excimer laser irradiation, and thermal treatments. The most successful of these methods, ion presputtering, yields an adhesion strength of approximately 53 g/mm [3,4]. In addition to these methods, sodium naphthalenide (Poly Etch, Matheson Products) etches Teflon, leaving a rough, discolored surface with good adhesion qualities [2].

In addition to the problem of adhesion, methods must be explored to pattern copper features on Teflon. Standard wet etch methods rely on patterning after deposition. However, there are several problems associated with this method.

0169-4332/93/$06.00

These include the generation of copper-containing liquid waste, complex masking procedures and large feature sizes. We have developed new methods to circumvent these problems. These methods, laser patterning and X-ray- or e-beam-induced cross-linking, both require chemical modification of the Teflon surface before CVD. The processes differ in the order of process steps. As shown in fig. 1, the Teflon surface can be patterned prior to chemical modification, as with the e-beam and X-ray techniques, or after, as with the laser process.

X-rays and electron beams incident on unmodified Teflon can protect the surface from significant subsequent reaction with sodium naphthalenide [5,6]. The e-beam method generated patterns on the Teflon surface by using computer-controlled deflection plates at the emitter. The X-ray system required a line-of-sight mask to generate patterns. Experimental results exist for both of these methods [7,8]. This protection mechanism is believed to result from radiation-induced cross-linking in the region near the surface of the polymer [2]. Subsequent selective chemical vapor deposition (CVD) occurs only on chemically etched regions and not those areas protected by irradiation as shown on the left side of fig. 1 [7,8]. These beam processes have advantages over the subtractive wet etch procedure. Firstly, they are dry patterning processes which require only three steps. Also, small feature sizes are possible. Unfortunately, since both of these methods require an ultra-high vacuum system, the expense and complexity of the equipment, and also the time required to pattern make these methods less attractive at this point.

Several variations on this basic process for patterned adherent Cu on Teflon have been developed. For example, selective Cu CVD can be replaced with electroless Cu deposition [9], a process widely used in the electronics industry. However, instead of the selective nucleation growth on the non-irradiated and etched areas, as observed in the CVD method, electroless deposition is uniform over the entire surface with strong adhesion being selective to the non-irradiated regions. On the irradiated areas, Scotch tape cleanly removes the Cu film along with all nucleation sites. While an extra step of film peeling is required with electroless Cu, the process is general to any metal that can be electrolessly deposited. This includes metals such as Au, Ag, Ni, Pd, Co, etc.

As a further alternative, we have developed a procedure that uses a laser to pattern substrates, after chemical modification (etching), as shown on the right side of fig. 1. This paper will concentrate on describing this patterning method, which

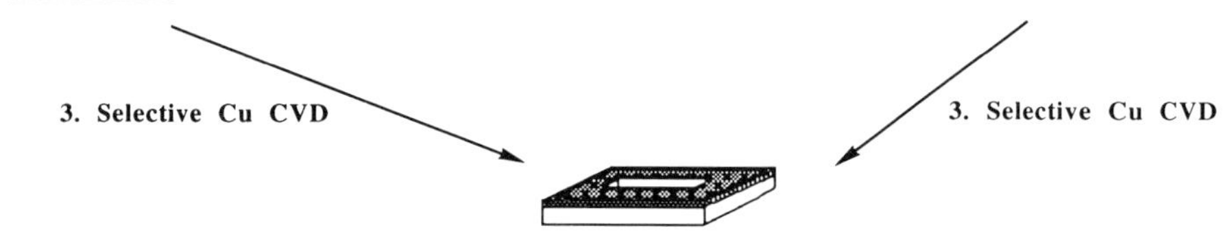

Fig. 1. Routes to patterned metallization of Teflon. In one approach, the Teflon is etched first, a laser beam is used to selectively remove the etched layer, and copper is selectively deposited. In the other route, the sample is selectively etched after X-ray or e-beam irradiation, and selective CVD metallizes the surface.

employs a focused laser beam to selectively remove the modified Teflon surface region which has been etched by sodium naphthalenide. The material removal is self-limiting and stops when the laser removes the entire etched layer and exposes underlying Teflon. This procedure has several advantages over the previously mentioned processes. First, as with the other methods, this is a dry procedure. Also, the cost and complexity are significantly lower than the other methods. Most importantly, however, the laser process can rapidly produce small features, without requiring a high vacuum chamber. In addition, the procedure works over a wide range of operating parameters.

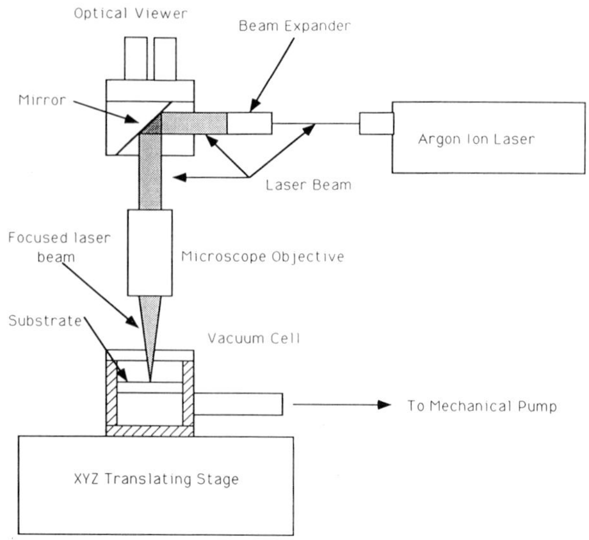

Fig. 2. Argon-ion laser system used for patterning Teflon substrates.

2. Experimental

The experimental procedures include etching, laser patterning methods, chemical vapor deposition and analysis. The X-ray and electron beam methods have been described elsewhere [5,6].

After patterning with e-beams or X-rays, or before patterning with the laser, the Teflon surface must be etched. The etching process begins with a 30 s immersion in sodium naphthalenide (Poly Etch). The immersion is followed with a water rinse, and ultrasonic cleaning. This cleaning is done first in water, then acetone. Had the sample been exposed to X-rays or an electron beam, a visible pattern would appear during this procedure as in fig. 1.

Fig. 2 shows a schematic of the laser patterning system. The laser was a Lexel Model 88 argon-ion laser and produced most of its power at 514 nm wavelength operating in the TEM_{00} mode with 0.5 W total power. The power was attenuated to ~350 mW through various optical components, including a 1-to-5 beam expander, a beam splitter, an attenuator, and a 10× microscope objective. The combination of the objective and expander located the beam waist in the optical focal plane. This configuration allowed the operator to view the substrate in the same plane as the beam waist. The beam diameter and intensity profile were measured using a sharp stainless steel blade attached to a micrometer stage. The blade was incremented across the beam, and the reduction from full power was measured using an optical pyrometer. The beam was roughly Gaussian with a $1/e$ beam diameter of 10 μm.

The substrate was positioned using an *XYZ* translation stage. Both horizontal axes were driven by high-resolution stepper motors. The range of speed provided was 0.005–5.5 mm/s. The substrate was located inside a stainless steel vacuum cell, sealed with a Pyrex window. A 0.5 HP mechanical roughing pump brought the base pressure of the system to 10^{-2} Torr. Experiments were performed at the base pressure and at atmospheric pressure.

Using this system, we have generated three types of samples for analysis. One type was simply to determine the conditions where ablation would occur. The second was to determine the surface characteristics of the laser-patterned region. The third type was to produce small features on the Teflon surface. All of the samples began as unmodified, commercial Teflon and were etched, as described earlier.

The first type of samples generated were used

to determine the conditions where removal of the etched layer would occur, and also to observe the selectivity of the CVD process to these ablated areas. To do so, scan rates were varied from 0 to 5.5 mm/s and the power at the substrate was varied from 10 to 385 mW. In addition, experiments were performed at atmospheric pressure and at the base pressure of the system. Finally, copper CVD was carried out on the sample, using synthesized [10–13] copper precursors (hfac)-Cu(1,5-COD), (hfac)Cu(butyne) and (hfac)-Cu(PMe_3). This was done in a hot-wall reactor at 200°C using a procedure previously described [14–18].

The second type of sample was generated to determine the characteristics of the patterned surface by XPS. To generate samples for characterization, the beam was rastered over a 20 mm^2 area at a variable scan rate. The sample was held under vacuum and the power at the substrate was approximately 0.3 W.

The third type of sample was prepared to determine the feature resolution of this approach. This was studied by generating increasingly narrow, continuous lines of non-ablated material. Lines were generated using scan rates of 0.05 mm/s and a power variation of 40 to 350 mW. To create the lines, a 50 μm by 1 mm (approximately) area was rastered under high laser power of ~ 350 mW. One edge, which was the edge of the small feature, was trimmed using a low power, approximately 40 to 50 mW. Finally, another rectangle was ablated in the same manner. The distance between the two trimmed edges (the feature) was varied from approximately 50 to approximately 5 μm. The features were then examined using an optical microscope and SEM.

The methods used to characterize the substrates included X-ray photoelectron spectroscopy (XPS), scanning electron microscopy (SEM), and optical microscopy. The XPS analysis system included an Mg Kα X-ray source and a cylindrical mirror analyzer (CMA). XPS analysis was carried out with minimum X-ray exposure to avoid surface modification. SEM was used to examine CVD morphology, feature size, and patterned features after CVD. An optical microscope was also used to view patterned features.

3. Results and discussion

We begin by presenting a summary of the results for the chemically etched Teflon. We will then discuss our observations and results concerning the laser patterning of etched Teflon.

The etched Teflon surface has been examined previously, by XPS [2], SEM, optical microscopy, and RBS (Rutherford back scattering) [19]. Visually, a sample of Teflon turns dark brown after immersion in a sodium naphthalenide solution for 30 s. Observing an etched sample by SEM and optical microscopy, one can see a grainy structure, as in the unpatterned region of fig. 3. The "grains" appear to vary in size between 5 and 10 μm. (At higher magnification, a dendritic or columnar structure is revealed and this structure has a relatively high surface area.) RBS has shown that etching extensively defluorinates the surface to approximately 4000 ångströms [19]. This chem-

Fig. 3. Optical micrograph of a laser-patterned substrate showing surface features varying in width from ~ 40 to 10 μm. Close examination reveals the grainy nature of the surface. At 10 μm the line size approaches the grain size.

ically modified surface is a high energy, reactive structure and we have observed that an etched sample left standing at room temperature will slowly turn to a lighter color, indicating a change in the surface region.

The etched surface, relative to unmodified Teflon, provides vastly improved adhesion characteristics for copper metallization. Copper-coated unetched Teflon consistently fails the "Scotch tape" peel test for copper deposited by CVD and electroless deposition. Copper deposited onto Teflon that has been etched with sodium naphthalenide has good adhesion properties. Adhesion tests consistently show failure in the bulk material, indicating the adhesion strength exceeds the ultimate tensile strength of Teflon.

The laser-patterning process was self-limiting and the etched layer of Teflon could be removed without harming the unetched bulk material below. The etched Teflon has a high absorption coefficient, and low reflectivity in the visible region. This is in contrast to the fresh Teflon which has a high reflectivity and low absorption coefficient at these wavelengths. The optical power absorbance of etched Teflon is 300 times greater than virgin Teflon [20]. As a result, the laser light was absorbed near the surface of the etched region, and this, coupled with the low thermal conductivity of Teflon, caused localized high temperatures. Removal of etched Teflon proceeds until the beam encounters the unmodified Teflon underneath. When the laser penetrates the etched layer, the underlying Teflon reflects most of the light, lowering the temperature, and as a result, material removal ceases. Thus, the process is self-limiting.

Laser patterning of etched Teflon is a process which works over a wide range of operating conditions and these conditions were examined in a series of experiments. Most importantly, removal of a narrow line of the etched layer occurred for scanning velocities ranging from 0 to 5.5 mm/s at a beam power of approximately 350 mW. We have also examined the influence of laser power on the patterning process. Patterning occurred at laser powers ranging from 30 to 380 mW. For laser power greater than 200 mW, at a scan speed of 0.05 mm/s, blistering of the Teflon and/or the etched layer occurred. The blistering effect may also occur at higher scan speeds, but scan rates above 0.05 mm/s have not been examined. Interestingly, the effect seems to be temporary and does not seem to have any detrimental effects on deposition or adhesion. The influence of gas pressure on the patterning process was examined at 10^{-2} Torr and atmospheric pressure. The low-pressure patterning process left cleaner lines and less debris, whereas the high-pressure case resulted in observable black flecks inside and outside the ablated features. Also, we observed differences in the variation of feature characteristics with laser power. No attempt at quantifying these differences has been made so far. However, the debris may be products of the patterning process which were confined to the surface region by the gas acting as a barrier.

We also used laser patterning to remove a large area, 20 mm^2, of the etched layer to allow characterization of the surface. XPS showed the region ablated by the laser was slightly defluorinated as compared to the fresh Teflon (fig. 4). The carbon to fluorine ratio in this region was approximately 1 : 1.3 as shown by the relative intensity levels, while the carbon-to-fluorine ratio in unmodified Teflon is approximately 1 : 1.87. In contrast, as shown in fig. 4, the upper 30 Å (XPS sampling depth) of etched Teflon is almost completely composed of carbon as shown by XPS; most of the fluorine was removed by the etching process. Thus, these results indicate that the patterning process removes the etched surface of the substrate, leaving a surface resembling unmodified Teflon.

The laser process was also used to generate a variety of line widths (fig. 3). Features as small as 10 μm were formed. This feature size limit approached the "grain" size on the surface. At these dimensions, the unevenness of the Teflon surface became a limiting factor, and discontinuities appeared in the lines. Fig. 3 shows this result in the circled area. As indicated in the experimental section, the bulk of the material was removed at high laser power. To complete the process, low laser power, 35 mW, was used for slower material removal in the areas which would form the boundaries of the feature. As shown in the optical micrographs of fig. 3, this lower power beam, scanned at 0.05 mm/s produced reasonably smooth and consistent edge characteristics considering the relative dimensions present. These results show that we can produce feature sizes in the 10 μm range. However, smoother Teflon surfaces will allow even narrower features to be formed.

Finally, the selective deposition of copper onto the laser-patterned substrates was examined using several copper precursors. Patterned substrates were put into a hot-wall CVD reactor [14–18] to determine if copper would deposit selectively onto the areas not exposed to laser irradiation. Experiments were performed using (hfac)Cu(butyne), (hfac)Cu(PMe_3) and (hfac)-Cu(1,5-COD). Characterization of this process is in progress. Qualitatively, the best selectivity was obtained with (hfac)Cu(PMe_3). With this precursor, copper selectively deposited on the chemically etched areas while the ablated regions remained copper-free. The selectivity was also a function of the time that had elapsed since the etching process which may be due to degradation of the etched Teflon surface with time. The results of a typical deposition experiment are shown in fig. 5. The selective deposition process is the most difficult part of the overall process to control, and we are currently studying this process in more detail.

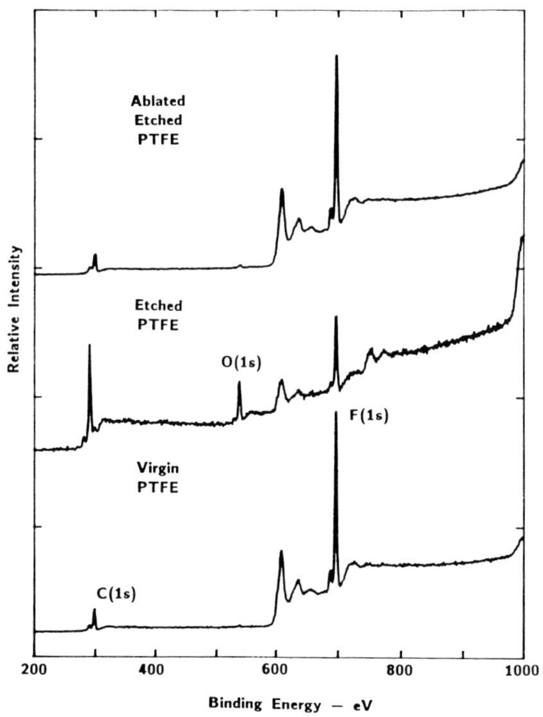

Fig. 4. XPS results for three samples. Top: chemically etched Teflon after laser patterning. Middle: Chemically etched Teflon with no other modification. Bottom: completely unmodified Teflon. Comparison of the top and bottom spectra reveals the similarity of the ablated area to the unmodified surface.

Fig. 5. SEM micrographs of copper deposited selectively on etched and patterned Teflon.

4. Summary and conclusions

The combination of chemical etching with sodium naphthalenide, laser patterning and CVD, as discussed above, provides an excellent method for creating small (10 μm) copper features on Teflon. The laser process requires three steps. (1) Etch in sodium naphthalenide, which is a method to chemically modify the Teflon surface to render it suitable for metallization by CVD with good adhesion. (2) Generation of a pattern using a laser to selectively remove the etched surface layer of the Teflon. (3) Chemical vapor deposition to selectively metallize the surface in the areas that were not exposed to the laser beam. This process works over a wide range of scanning speeds and laser powers, and allows rapid patterning without requiring complex, expensive equipment as with other methods [2,6]. The process is also dry, avoiding the problems associated with wet patterning methods.

CVD of copper occurred selectively on the non-irradiated areas of substrates using the (hfac)Cu(PMe_3) precursor and work is in progress to better understand this process. Work is also in progress to quantify characteristics of the laser-induced removal process as a function of the various operating parameters.

Acknowledgement

The authors acknowledge support by the National Science Foundation Presidential Young Investigator Award #CTS-9059538 (T.T.K.).

References

[1] A.A. Benderly, J. Appl. Polym. Sci. 20 (1962) 221.
[2] R.R. Rye and R.J. Martinez, J. Appl. Polym. Sci. 37 (1989) 2529.
[3] A. Clerier and J. Machet, Thin Solid Films 323 (1987) 148.
[4] Y.K. Kim, C.A. Chang, A.G. Schrott, J. Andreshak and M. Cali, Mater. Res. Soc. Symp. Proc. 153 (1989) 279.
[5] R.R. Rye, Langmuir 6 (1990) 338.
[6] R.R. Rye and N.D. Shinn, Langmuir 6 (1990) 142.
[7] R.R. Rye, K.M. Chi, M.J. Hampden-Smith and T.T. Kodas, Mater. Res. Soc. Proc. ULSI-VIII (1992).
[8] M.J. Hampden-Smith, T.T. Kodas and R.R. Rye, Adv. Mater. 4 (1992) 524.
[9] R.R. Rye and J.A. Ricco, J. Electrochem. Soc., submitted.
[10] H.K. Shin, M.J. Hampden-Smith, T.T. Kodas and E.N. Duesler, Polyhedron 6 (1991) 645.
[11] H.K. Shin, K.M. Chi, J. Farkas, M.J. Hampden-Smith, T.T. Kodas and E.N. Duesler, Inorg. Chem. 31 (1992) 424.
[12] K.M. Chi, H.K. Shin, M.J. Hampden-Smith, T.T. Kodas and E.N. Duesler, Polyhedron 10 (1991) 2293.
[13] K.M. Chi, H.K. Shin, M.J. Hampden-Smith, T.T. Kodas and E.N. Duesler, Inorg. Chem. 30 (1991) 4293.
[14] H.K. Shin, K.M. Chi, M.J. Hampden-Smith, T.T. Kodas, M.F. Paffet and J.D. Farr, Angew. Chem. Adv. Mater. 3 (1991) 246.
[15] H.K. Shin, K.M. Chi, M.J. Hampden-Smith, T.T. Kodas, M.F. Paffet and J.D. Farr, Chem. Mater. 4 (1992) 788.
[16] A. Jain, K.M. Chi, M.J. Hampden-Smith, T.T. Kodas, M.F. Paffett and J.D. Farr, J. Mater. Res. 7 (1992) 261.
[17] A. Jain, K.M. Chi, H.K. Shin, M.J. Hampden-Smith, T.T. Kodas, M.F. Paffett and J.D. Farr. Metallization: Performance and Reliability Issues for VLSI and ULSI SPIE Proc. 1596 (1991) 23.
[18] A. Jain, K.M. Chi, M.J. Hampden-Smith, T.T. Kodas, M.F. Paffett and J.D. Farr, Chem. Mater. 3 (1991) 995.
[19] R.R. Rye and G.W. Arnold, Langmuir 5 (1989) 1331.
[20] C. Seager, private communication.

Applied Surface Science 69 (1993) 101–107
North-Holland

applied
surface science

The ohmic contact formation mechanism in laser beam mixed Au/Te/Au/GaAs structures

J. Watté [a], R.E. Silverans [a], M. Van Hove [b] and K. Wuyts [c]

[a] *Laboratorium voor Vaste Stof-Fysika en Magnetisme, KU Leuven, Celestijnenlaan 200D, B-3001 Leuven, Belgium*
[b] *Interuniversitair Microelectronica Centrum, Kapeldreef 75, B-3001 Leuven, Belgium*
[c] *Instituut voor Kern- en Stralingsfysika, KU Leuven, Celestijnenlaan 200D, B-3001 Leuven, Belgium*

Received 2 June 1992; accepted for publication 30 October 1992

The electronic and structural properties of the intermixed interfaces of Au/Te/Au/n-GaAs structures subjected to continuous wave and to pulsed laser irradiations were studied. The transition from Schottky-type to ohmic contact was examined to identify the ohmic contact formation mechanism for which three types of conflicting models exist. The electrical characteristics have been correlated to the structural properties investigated by the combined application of X-ray diffraction, laser Raman spectroscopy and Mössbauer spectroscopy. In summary of the results, first the ohmic contact model based on the formation of a highly Te-doped GaAs surface layer could be questioned. For the low fluence pulsed laser irradiations, the formation of a Ga_2Te_3 interface layer is crucial for the ohmic character of the contact, corresponding to the model based on the formation of a crystalline heterojunction. Ohmic contact structures formed by CW and by high energy fluence laser irradiations were found to be correlated to the formation of a high density of defect complexes in the GaAs top layers, corresponding to the highly defective/amorphous heterojunction model.

1. Introduction

Laser processing has emerged to be an attractive alternative for the conventional furnace alloying in GaAs device technology, e.g. to produce satisfactory ohmic contacts [1]. The main technological advantage of laser beam mixing lies in the ability to heat localized areas while the bulk of the wafer is kept at room temperature. However, as even for furnace alloying the origin of ohmic behavior in several GaAs metallic contacts has remained controversial [2], the ultra-short heating and cooling during laser beam mixing makes the ohmic contact formation mechanism even more difficult to predict. Nevertheless, a thorough understanding of these contacts is interesting from both a fundamental and a technological viewpoint.

The rapid freezing of dopants may result in a degenerate doping of the semiconductor allowing tunneling through a narrowed metal/semiconductor potential barrier. The model based on this physical mechanism is widely accepted in ohmic contact research [3,4]. Alternatively, the short heat treatment can lead to the formation of a heterojunction by the growth of an intermediate semiconducting layer during the rapid recrystallization process. If the composition of that layer gradually varies, reducing the heterointerface band discontinuity, ohmic conductivity results [5]. Furthermore, laser beam mixing is also known to cause deep levels in the substrate surface region. Low resistive electron transport is then believed to occur by a phonon-assisted hopping or a multi-step tunneling process mediated by these localized gap states in the contact zone. These ohmic conduction mechanisms form the basis of respectively the graded crystalline and the highly defective heterojunction model, proposed as alternatives for the high-doping model [6]. Finally, modification of the Fermi level pinning position has also been taken into consideration [7].

We developed a Te-based metallization scheme [8] which allows characterization of the alloyed structures by Mössbauer spectroscopy as a complement to other characterization techniques. In previous work we reported the study of furnace mixed Au/Te/Au/n-GaAs ohmic contacts. The low resistance conductivity of these contacts was found to be related to the formation of an n^+-type Ga_2Te_3/n-GaAs heterojunction [9]. In this paper we report the study of the ohmic conduction mechanism in pulsed and CW laser beam mixed contacts by X-ray diffraction (XRD), Mössbauer and Raman spectroscopy. The discussion centers on the applicability of the above given conflicting models to explain the low resistive conductivity in laser beam mixed structures at particular irradiation conditions.

2. Metallization and experimental procedures

The metallization of the n-GaAs substrates was performed by successive deposition of 50 Å Au, 500 Å Te and 1200 Å Au on (100)-oriented GaAs by resistive heating evaporation in a vacuum of the order of 10^{-6} Torr. Uniformly Si (3×10^{17} atoms/cm^3) and Zn (2×10^{18} atoms/cm^3) doped wafers were used for the current–voltage (I–V) tests and semi-insulating GaAs was taken for the other measurements. Mössbauer samples were prepared by implanting ^{129m}Te in the evaporated Te layer after which the Au top layer was deposited. For the pulsed irradiations, frequency-doubled ($\lambda = 532$ nm, pulse duration $t_p = 7$ ns) Nd:YAG laser light was used to alloy the contact structures; single pulse and multiple pulse alloying was performed with energy fluences ranging from 0.20(5) to 2.5(5) J/cm^2. CW laser irradiations were performed by scanning the beam of a CW Ar^+-ion laser over the contact structures with a galvano-controlled mirror (scan speed ~1 mm/s), corresponding with an irradiation time of ~10^{-2} s. Normalized power values $p = (1 - r)P/w$ (with r the reflection coefficient of the Au top layer, P the laser power applied and w the beam waist) ranged from 0.040(2) W/μm to 0.070(2) W/μm. $\theta:2\theta$ XRD scans and rocking curve measurements were carried out on a computer-controlled Rigaku Dmax II rotating anode spectrometer. Cu K_α radiation ($\lambda = 1.54184$ Å) was selected with a flat pyrolytic graphite monochromator. Mössbauer spectra were recorded at 4.2 K using a CuI absorber. Raman measurements were performed at room temperature using the 514.5 nm line of an Ar^+-ion laser. The scattered light was analyzed by a tripple spectrometer equipped with multichannel detection. The diameter of the laser focus was 1–2 μm and in order to avoid heating effects the spectra were recorded at an input power < 2 mW.

2.1. I–V measurements

Figs. 1a and 1b show the resistance values of pulsed and CW laser beam mixed contacts deduced from room-temperature I–V measurements by a linear fit through the data points, both for contacts exhibiting a linear as well as rectifying character. Ohmic contacts are those with a resistance < 25 Ω [10]. For comparison, laser irradiations on identically metallized p-type GaAs were also carried out. Most strikingly, for contacts CW laser beam mixed and contacts pulsed laser beam mixed with an energy fluence > 0.35 J/cm^2 similar resistance values on n- and p-type GaAs can be discerned. However, the data obtained for the 0.20 J/cm^2 condition stands definitely in contrast with this behavior and is found to be more closely related to the electrical behavior of furnace mixed identical contacts [9] (see fig. 1).

2.2. XRD measurements

Two examples of XRD spectra, taken of ohmic contact structures obtained after pulsed low energy fluence (0.20 J/cm^2, 20 pulses) and CW ($p = 0.067$ W/μm) laser beam mixing, are presented in fig. 2. Besides AuGa phases (see fig. 2 for the identification) $(As)_x(Ga_2Te_3)_{1-x}$ phases ($0 \leq x \leq 0.01$) [16] were found to be present in both samples. Only the reflections of the (200) and (400) planes of Ga_2Te_3 are apparent, which is an indication of a preferential growth with respect to GaAs. In addition, ω-scans were performed around the (400) reflection. For the pulsed low energy fluence condition the Ga_2Te_3 was found to be highly textured since the FWHM of the rocking curve was equal to 5.63°. For the CW laser beam mixed contact a FWHM of 0.33° was measured, indicative for a pseudomorphic growth of the latter compound. No traces of Ga_2Te_3 could be found in contacts mixed with high energy fluences. For those structures the major

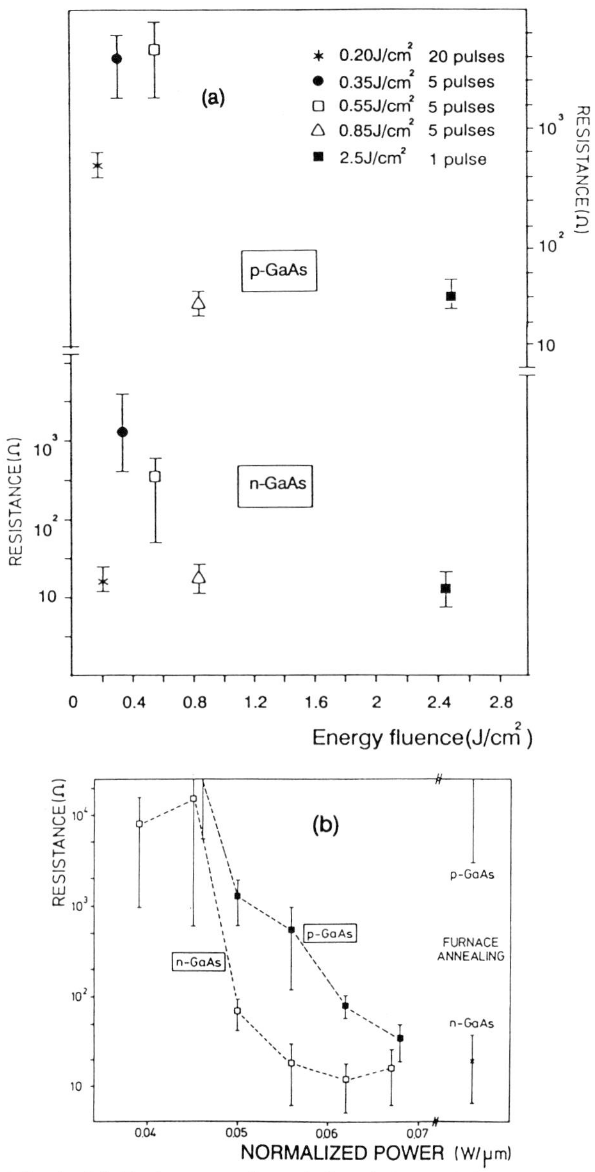

Fig. 1. (a) Resistance values deduced from $I-V$ measurements on pulsed laser beam mixed Au/Te/Au/n-GaAs and p-GaAs contacts. (b) Idem for CW laser beam mixed and furnace mixed identical structures.

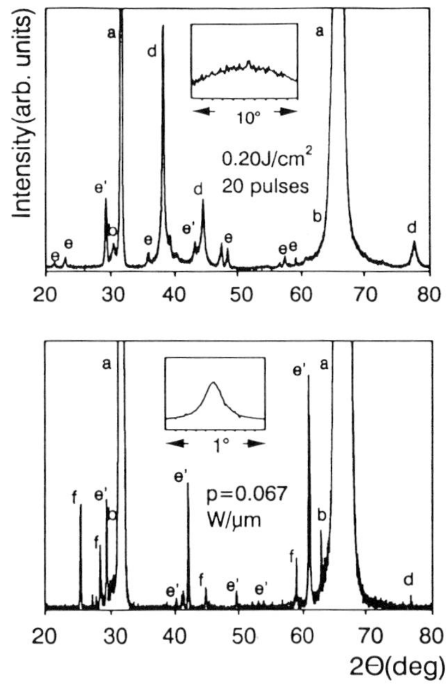

Fig. 2. X-ray profiles of laser irradiated Au/Te/Au/GaAs structures: 0.20 J/cm², 20 pulses (above) and CW $p = 0.067$ W/μm (below): peaks labeled by a correspond to GaAs, b = Ga_2Te_3, d = Au, e′/e = $Au_{0.79}Ga_{0.21}$/Au_2Ga and f = As. In the insets, ω-scans around the Ga_2Te_3(400) reflections are shown.

diffraction peaks originated from elemental As, $Au_{0.79}Ga_{0.21}$, AuGa and Au_2Ga.

2.3. Raman measurements

Raman measurements were performed in backscattering geometry from the (100) face of GaAs with the polarization of the incident light and scattered light along (010) and (001) crystallographic directions, respectively. Making use of the Porto notation this will be denoted by $z(xy)\bar{z}$. The penetration depth in GaAs of the laser line being used is ~ 100 nm. Furthermore, the spatial resolution of the micro-Raman technique (1 μm) is particularly useful to reveal the interface inhomogeneity. A selection of Raman spectra of laser beam mixed samples is presented in fig. 3. In spectra 3b and 3d (see figure caption of the mixing condition) the symmetry-allowed LO model of GaAs at ~ 293 cm^{-1} is accompanied by a mode at ~ 269 cm^{-1}. There are two possible explanations for the appearance of this mode. First, this peak can correspond to the Ω^- mode of the coupled phonon–plasmon modes which arise when doping of the underlying GaAs is involved [11]. A reference study on MBE-grown n^+-GaAs has proven our Raman analysis technique to be very sensitive in detecting free carrier layer thicknesses of the order of 50 Å [9]. Secondly, the mode at ~ 269 cm^{-1} can originate from the presence of a high density of defect complexes in the GaAs top layers. This relaxes the $\boldsymbol{k}$-selection rule enhancing the scattering contributions from TO phonons at 269 cm^{-1} [12]. By using different orientations for the polarization of the incident and scattered light the symmetry of the mode at 269 cm^{-1} can be probed. If we assume that it is an Ω^- mode it must obey the same selection rules as the LO phonon and consequently vanish in the forbidden $z(yy)\bar{z}$ configuration. A check of the selection rules has been performed (see fig. 3e) yielding appealing evidence that the mode at 269 cm^{-1} should be labeled as a disorder-activated $\boldsymbol{k}$-induced TO mode.

Applying higher energy fluences in pulsed laser beam mixing resulted in a broadening of the

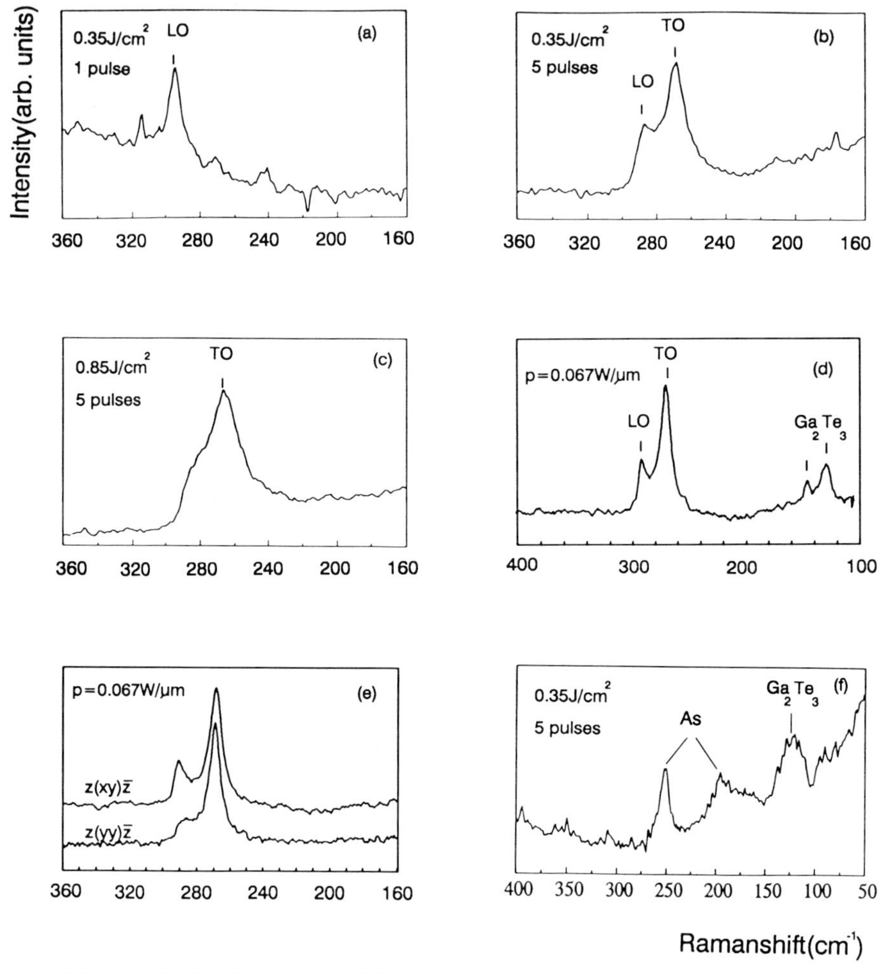

Fig. 3. Raman spectra of Au/Te/Au/GaAs structures: (a) GaAs signals of pulsed laser beam mixed 0.35 J/cm^2, 1 pulse; (b) 0.35 J/cm^2, 5 pulses; (c) 0.85 J/cm^2, 5 pulses; (d) CW laser beam mixed $p = 0.067$ W/μm; (e) GaAs signals of $p = 0.067$ W/μm in allowed and forbidden scattering configurations; (f) 0.35 J/cm^2, 5 pulses.

GaAs modes (see figs. 3a–3c) which points to an increase of the disorder in the regrown GaAs region. A downshift of the LO phonon frequency can be discerned which is more pronounced in pulsed laser beam mixed structures (5 cm^{-1}) compared to the CW mixed structures (2 cm^{-1}). This shift can be explained by a change in the ionic plasma frequency and also accounts for a high density of defects [13].

By probing different positions on the samples non-uniformly dispersed Ga_2Te_3 was identified by its vibrational modes at 128 and 144 cm^{-1} (see figs. 3d and 3f). For the pulsed laser beam mixed structures the phonon peaks of the Ga_2Te_3 phase are rather broadened, which might be indicative for highly disordered Ga_2Te_3 crystallites. Additionally, in pulsed laser beam mixed structures (high energy fluence regime) peaks at 195 and 253 cm^{-1} are present, corresponding respectively to the E_g and A_g mode of elemental As [14].

2.4. Mössbauer spectroscopy results

^{129}I Mössbauer spectroscopy (in the decay of ^{129m}Te) was applied to CW and pulsed laser beam mixed structures. The radioactive probes are found to occupy different sites after recrystallization, the corresponding Mössbauer parameters (Δ quadrupole splitting, IS isomer shift, η asymmetry parameter, F fractional resonance area) are listed in table 1. For comparison, reference parameters of Ga_2Te_3 and Te-implanted GaAs are included. When the Te atoms are substitutionally incorporated on vacant As sites, shallow donors are created and a single line in the spectra will arise due to the absence of the quadrupolar interaction [15]. No traces of an occupation of the Te_{As} site were found for the 0.20 J/cm^2 multiple pulse condition since the Mössbauer data could be fitted with two components the parameters of which are indicative for a Ga_2Te_3 regrowth [16] (see table 1).

The spectra of the non-ohmic structures obtained after alloying with an energy fluence 0.35 J/cm^2, 10 pulses and 0.55 J/cm^2, 5 pulses could be fitted with a single line and a multiplet. For the former mixing condition the Te site, yielding the multiplet contribution to the spectrum, is characterized by $\Delta = 290(10)$ MHz and can be interpreted as the ^{129}I Mössbauer trace of a donor-related deep level, the Te DX center in GaAs [17]. The Te site occupied in the latter mixing condition corresponding with the quadrupole splitting $\Delta = 330(10)$ MHz is identified as a $Te_{As}V_{Ga}$ defect complex [18], where the Te atom is covalently bound to three gallium atoms and a vacancy in the first-neighbour shell.

Table 1
Mössbauer parameters of pulsed and CW laser beam mixed 1200 Å Au/500 Å (^{129m}Te)Te/50 Å Au/GaAs; reference parameters of Ga_2Te_3 and Te implanted GaAs (5×10^{14} atoms/cm^2) are included for comparison

Sample	Parameters of spectral components			
	Δ (MHz)	δ (mm/s)	η	F (%)
0.20(5) J/cm^2 50 pulses	361(10)	+1.14(5)	0	66(5)
	584(10)	1.19(5)	1	34(5)
0.35(5) J/cm^2 10 pulses	0	+0.60(5)		45(1)
	290(10)	+0.86(5)	0	55(1)
0.55(5) J/cm^2 5 pulses	0	+0.68(5)		28(1)
	330(10)	+0.91(5)	0	72(1)
2.5 J/cm^2 1 pulse	343(10)	+0.91(5)	0	100
$p = 0.045$ W/μm	366(10)	+0.82(5)	0	70(5)
	523(10)	+0.94(5)	0.73	30(5)
$p = 0.067$ W/μm	335(10)	+0.93(5)	0	84(5)
	504(10)	+0.87(5)	0.73	16(5)
Ga_2Te_3	369(10)	+0.82(5)	0	67(5)
	550(10)	+0.94(5)	0.74	33(5)
GaAs(Te) 5×10^{14} atoms/cm^2	0	+0.70(5)	0	10(1)
	290(10)	+0.90(5)	0	90(1)

Strikingly, for the 2.5 J/cm^2, 1 pulse condition which resulted in low resistive contacts the single line contribution to the spectra vanishes which implies a dominant formation of $Te_{As}V_{Ga}$ defect complexes. A similar behavior of the Te atoms is found for CW laser beam mixed contacts (see table 1). For those structures the transition from Schottky to ohmic behaviour is also marked by a preferential population of the $Te_{As}V_{Ga}$ defect site. Notable is the fact that attempts to fit single-line contributions in these spectra always yielded a zero occupation of that component.

3. Discussion and conclusions

From the data different regimes in the ohmic contact behavior of laser beam mixed structures can be discerned. For the lowest energy fluence (0.20 J/cm^2 multiple pulses) the conclusion that can be drawn from the XRD analysis is that the low resistive electron transport can be explained by the formation of a $(As)Ga_2Te_3/GaAs$ heterojunction. The same explanation was given to identical furnace mixed contacts; however for the latter the FWHM of the rocking curve was found to be 0.16° [9]. The fact that I–V measurements performed at 77 K on the lowest energy fluence mixed contacts yielded Schottky behavior is consistent with the broad rocking curve, indicating a higher number of dislocations. This results in a larger band bending and higher barriers at the $Ga_2Te_3/GaAs$ interface and can explain the temperature dependence of the contact resistivities [19].

Although Ga_2Te_3 is present in structures alloyed with 0.35 J/cm^2, 5 pulses Schottky behaviour was observed. The same feature is observed in CW laser beam mixed structures ($p =$ 0.045 $W/\mu m$). The Raman data can solve this anomaly since the Ga_2Te_3 precipitates were found to segregate as locally crystallized structures, only implying a degradation of the Schottky characteristics. The single line contributions in the Mössbauer spectra of non-ohmic pulsed laser beam mixed contacts should according to the doping model constitute an ohmic contact structure on n-GaAs. However, due to the accompanying Te DX- and $Te_{As}V_{Ga}$ defect complexes a trapping [20] and compensation [21] of the free electrons can be expected. This is consistent with the fact that no traces of coupled phonon–plasmon modes are observed in the corresponding Raman spectra.

From Mössbauer analysis it seems that the low contact resistance values obtained after CW and pulsed laser beam mixing (high energy fluence regime) can be attributed to a dominant presence of $Te_{As}V_{Ga}$ defects in the GaAs top layers. Unfortunately, Mössbauer spectroscopy cannot probe directly the presence of non-Te-related point defects like Ga and As vacancies and anti-site defects. The enhanced population of the $Te_{As}V_{Ga}$ defect site can however be regarded as an indication of an increased formation of the latter type of defects which can provide a non-negligible contribution to the transport mechanism in the contact zone [22]. Furthermore, for those structures the similar electrical behaviour on n- and p-GaAs is indicative for low resistive electron transport mediated by localized defect states. The shape of the GaAs signals found in the Raman measurements, indicative for a defective regrown GaAs layer, is another argument in favour of the highly defective heterojunction model. Temperature-dependent conductivity measurements are currently in preparation to elucidate whether a multistep tunneling, a hopping or a recombination process, assisted localized defect states is the dominant transport mode.

Acknowledgements

This work was supported by the Interuniversitair Instituut voor Kernwetenschappen (IIKW) under contract No. 4.0010.90. J.W. is a research fellow of the Instituut tot Aanmoediging van het Wetenschappelijk Onderzoek in de Nijverheid en Landbouw (IWONL).

References

[1] S. Imanaga, H. Kawai, K. Kajikawa, K. Kaneko and N. Watanabe, J. Appl. Phys. 62 (1987) 2381.
[2] C.J. Palmström, S.A. Schwarz, E. Yablonovitch, J.P. Harbison, C.L. Schwartz, L.T. Florez, T.J. Gmitter, E.D. Marshall and S.S. Lau, J. Appl. Phys. 67 (1990) 334.
[3] R.S. Popovic, Solid-State Electron. 21 (1978) 1133.
[4] R.K. Kupka and W.A. Anderson, J. Appl. Phys. 69 (1991) 3622.
[5] D.T. Cheung, S.Y. Chiang and G.L. Pearson, Solid State Electron. 18 (1975) 263.
[6] T. Sebestyen, Solid State Electron. 25 (1982) 543.
[7] P.L. Meissner, J.C. Bravman, T. Kendelewicz, K. Miyano, W.E. Spicer, J.C. Woicik and C. Bouldin, Mater. Res. Soc. Proc. 181 (1990) 265.
[8] K. Wuyts, A. Vantomme, R.E. Silverans, M. Van Hove and M. Van Rossum, Mater. Res. Soc. Proc. 144 (1989) 545.

[9] K. Wuyts, J. Watté, H. Vanderstraeten, G. Langouche, R.E. Silverans, H. Münder, M.G. Berger, M. Van Hove, H. Bender and M. Van Rossum, Phys. Rev. B 45 (1992) 11863.

[10] K. Wuyts, J. Watté, R.E. Silverans, M. Van Hove and M. Van Rossum, Appl. Phys. Lett. 59 (1991) 1779.

[11] G. Abstreiter, E. Bauser, A. Fisher and K. Ploog, Appl. Phys. 16 (1976) 345.

[12] K.K. Tiong, P.M. Amirtharaj, F.H. Pollak and D.E. Aspnes, Appl. Phys. Lett. 44 (1984) 122.

[13] G. Burns, F.H. Dacol, C.R. Wie, E. Burnstein and M. Cardona, Solid State Commun. 62 (1987) 449.

[14] W. Richter, T. Fjeldy, J. Renucci and M. Cardona, in: Lattice Dynamics, Ed. M. Balkanski (Flammerion, Paris, 1975).

[15] G. Langouche, D. Schroyen, H. Bemelmans, W. De Raedt, M. de Potter and M. Van Rossum, Mater. Res. Soc. Proc. 104 (1988) 527.

[16] K. Wuyts, J. Watté, G. Langouche, R.E. Silverans, G. Zégbé and J.C. Jumas, J. Appl. Phys. 71 (1992) 744.

[17] H. Bemelmans, G. Borghs and G. Langouche, Nucl. Instr. Meth. B 63 (1992) 151.

[18] K. Wuyts, G. Langouche, M. Van Rossum and R.E. Silverans, Phys. Rev. B 45 (1992) 6297.

[19] K. Wuyts, PhD Thesis, KU Leuven, 1991.

[20] D.J. Chadi and K.J. Chang, Phys. Rev. Lett. 61 (1988) 873.

[21] C.J. Hwang, J. Appl. Phys. 40 (1969) 4584.

[22] D.C. Look, D.C. Walters, M.O. Manasreh, J.R. Sizelove, C.E. Stutz and K.R. Evans, Phys. Rev. B 42 (1989) 3578.

Applied Surface Science 69 (1993) 108–114
North-Holland

applied
surface science

High-intensity KrF excimer laser processing of metal surfaces

J.E. Montagne, Th. Sarnet, Ch. Prat, G. Inglesakis and M. Autric

Laser–Matter Interaction Group, Institute of Fluid Mechanics, Pôle Scientifique et Technologique de Luminy, 163 Avenue de Luminy, 13288 Marseille, France

Received 2 June 1992; accepted for publication 30 October 1992

The interaction of high-intensity pulsed UV radiation with solid materials involves complex phenomena. Generally, they depend on different factors including laser wavelength, pulse duration, energy and power density, laser spot size, nature of the material and surface characteristics. The experiments presented in this paper have been performed using a 248 nm KrF laser which allowed aluminium and titanium alloys to be irradiated with an energy density from 0.3 up to 100 J/cm^2 and a power density from 1 up to 300 MW/cm^2.

1. Introduction

Metal–laser interactions have become increasingly important due to advances in machining processes. The advent of excimer lasers, which emit in the UV with good efficiency, has allowed one to consider an important development in short-wavelength treatment. Effectively, the sharpness with which UV radiation can be focused, as well as the increased absorptivity of metals at such wavelengths, makes them interesting for machining applications.

The work carried out by the Laser–Matter Interaction Group (EILM) of the Institute of Fluid Mechanics of Marseilles (IMFM) in recent years has permitted a characterization of the different hydrodynamic and thermomechanical phenomena that rule the interaction at 10.6 μm. The construction of a high-energy KrF source ($\lambda = 248$ nm) by the team allows a study of the interaction between metal alloys and UV radiation, leading to a characterization of mechanisms involved in laser–material processing. First results have already been obtained and are presented in this paper.

2. Laser–metal interaction

When a pulsed laser beam irradiates a metallic sample, we generally observe a melting and vaporization of the solid surface. At high irradiance, the vapor above the sample is ionized, leading to the formation of a plasma. The dynamics of the generated plasma is a function of the incident flux level, ambient gas pressure around the surface, laser pulse duration and shape.

For low pressure ($P < 1$ Pa) and low flux, material vapor expansion in the vacuum is the main process. For higher intensity ($\Phi >$ plasma threshold), plasma appears close to the surface and absorbs energy.

When the interaction occurs in a rarefied atmosphere, the formation of a spherical shock wave, completely transparent to the incident radiation (blast wave) may be observed. The wave is a result of ambient air compression due to the ejection of particles from the sample surface. It occurs at sufficiently high pressure (≥ 1 Pa) permitting a collisional transfer of the energy from the sample to the ambient gas.

At high pressure ($> 10^3$ Pa), laser absorption

0169-4332/93/$06.00

waves can be ignited. Two regimes dependent upon the laser intensity can be observed. At a flux level slightly in excess of the plasma threshold, a laser-supported combustion wave (LSC) is generated [1]. As a consequence of the ignition process and the presence of the target surface, a precursor shock preceeds the LSC wave. At higher intensity the short absorption length of the radiation allows a laser-supported wave (LSD) to form. Most of the laser energy is absorbed in the leading edge of this surrounding gas plasma [2].

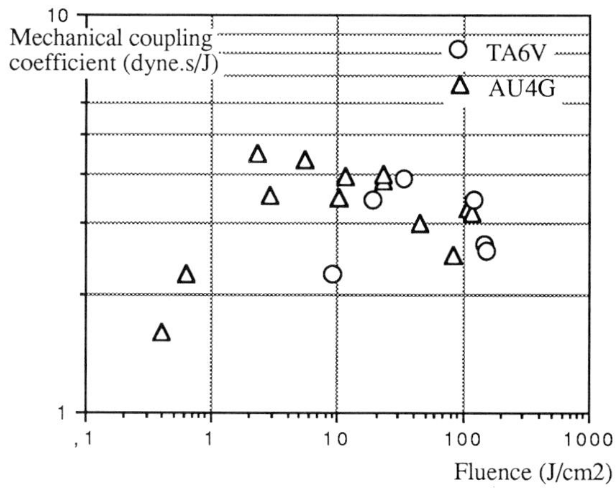

Fig. 1. Mechanical coupling versus fluence for AU4G and TA6V (ambient pressure = 2×10^{-2} Pa).

3. Experimental set-up

The experiments have been performed with a KrF laser (KIR), built at IMFM by EILM. This laser delivers pulses of 400 ns duration with a maximum energy of about 200 J at $\lambda = 248$ nm. A double-sided pumping scheme has been adopted using two cold-cathode diodes with "cheese grater emitters", each one powered by a six-stage, 600 kV, 36 kJ, fast-rise Marx generator.

For our experiments the laser was configured with a 80 mm diameter stable resonator which produced a pulse width of 300 ns and an output energy of 20 J on the target. With focusing optics, the spot size varied from 3 to 28 mm, leading to a fluence of 0.3 up to 100 J/cm^2 with a good spot uniformity.

The materials studied were aluminum (AU4G/2017A) and titanium (TA6V/Ti–6Al–4V) alloys. The workpiece samples were mechanically polished to obtain a roughness $R_a < 0.1$ μm; the irradiance was normal to the surface.

The surrounding gas was air with a pressure varying from 10^{-2} up to 10^5 Pa.

4. Results and discussion

We have first measured induced impulse and mass loss from exposures of AU4G and TA6V to KrF laser pulses as a function of incident laser fluence and ambient atmospheric pressure. The results gave us a first indication of the pressure at the sample surface and the ejection velocity. Then scanning electron microscopy (SEM), X-ray diffraction (XRD) and energy-dispersive X-ray spectroscopy (EDS) were used to characterize surface micro-structure changes.

Fig. 1 presents the impulse coupling coefficient versus fluence for both AU4G and TA6V. The coupling threshold is found at 0.3 J/cm^2 (1 MW/cm^2) for AU4G and 7 J/cm^2 (20 MW/cm^2) for TA6V. This difference may be explained by the highest vaporization temperature of titanium and the high absorption of copper, an element of AU4G alloy. The plasma seems to become important above 3 J/cm^2 for AU4G and 30 J/cm^2 for TA6V when the impulse coupling begins to decrease.

Fig. 2 presents the mechanical coupling versus ambient pressure. The coupling is found to be two times greater at atmospheric pressure than at pressures below 10 Pa. The surface sample pressure may be estimated by

$$I = \int\int P_{\text{sample}} \, \mathrm{d}S \, \mathrm{d}t.$$

If we consider a uniform intensity distribution over the spot area S and a stationary phenomenon during the pulse duration τ, we have:

$$\bar{P} = I/S\tau.$$

We find $P = 2.5 \times 10^8$ Pa for $F = 100$ J/cm^2 with AU4G in vacuum.

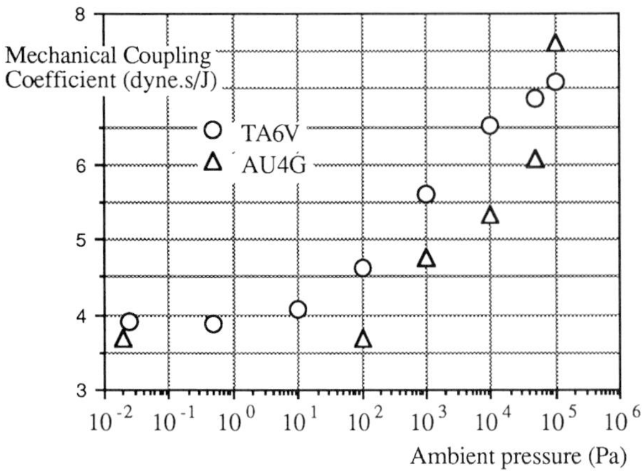

Fig. 2. Mechanical coupling versus ambient pressure for AU4G and TA6V ($F = 30$ J/cm^2).

We can also estimate the particle ejection velocity with motion conservation:

$$\bar{V} = I/\Delta M,$$

where ΔM is the ablated mass. We measured an ablated mass per shot of 22 μg with 31 dyne · s for 50 J/cm^2, leading to a particle ejection mean velocity of 1.4×10^4 m/s.

A large amount of energy is deposited in a very thin surface layer; this heat is rapidly conducted away from the layer towards the sample. If τ is the pulse duration, the heat will penetrate to a diffusion depth z,

$$z \approx (4k\tau)^{1/2},$$

where k is the thermal diffusivity $= K/\rho C$, with K the thermal conductivity, C the specific heat and ρ the density of the material. For AU4G, $k = 0.68$ cm^2 s^{-1} (i.e. for $\tau = 300$ ns, $z \approx 9$ μm) and for TA6V, $k = 0.062$ cm^2 s^{-1} (i.e. for $\tau = 300$ ns, $z \approx 3$ μm).

Laser irradiation may produce several topographic alterations (surface damage) at an initially "flat surface" [3]. We can generally distinguish three different zones on the sample [4]: the irradiated zone with an arched, rippled or smoothened surface, the non-irradiated outer region sometimes covered with laser-ablated debris and the transition zone that delimits the border between the ablated and non-ablated area.

The plasma exerts a pressure on the liquid displacing it laterally: fig. 3 shows the border of the ablated zone (AU4G) and the displacement of the liquid towards the non-affected zone; be-

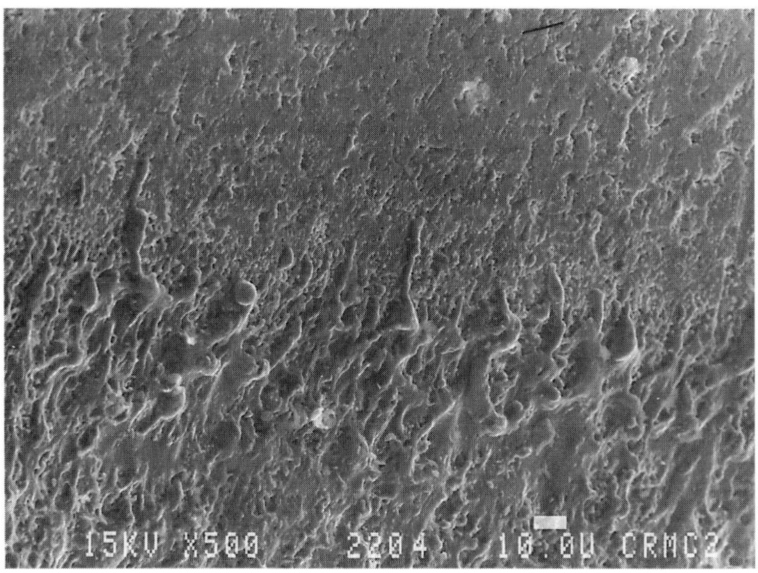

Fig. 3. Detail of the border of the ablated zone (AU4G). Displacement of the liquid from the irradiated zone (right) towards the non-affected zone (left) ($F = 22$ J/cm^2).

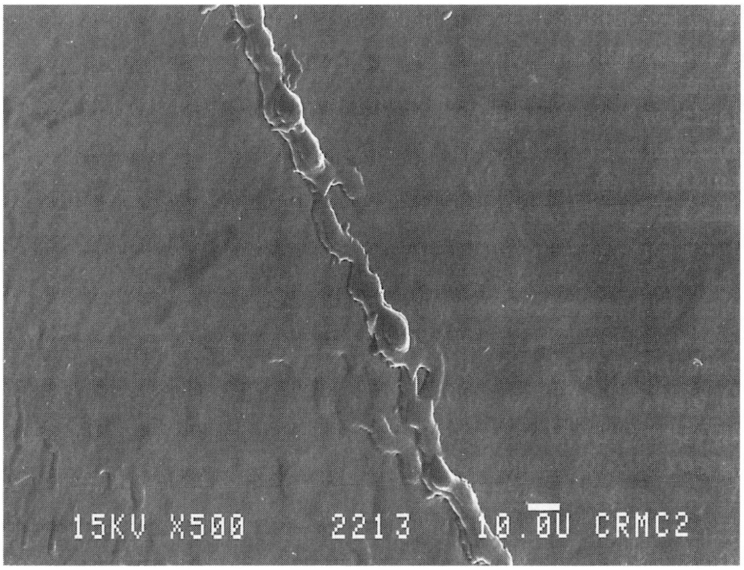

Fig. 4. Detail of the border (TA6V) between the ablated zone (up) and the non-affected zone.

cause of the high cooling rates, the displaced melt subsequently freezes in that position, forming a kind of burr.

The aspect of this border is different for TA6V (fig. 4), where no projections have been evidenced, leading to a smaller size of the transition zone (< 50 μm for AU4G and < 10 μ for TA6V).

The ablated zone shows arched and rippled surfaces but the amount of melt is higher for AU4G than TA6V. A single laser shot on the

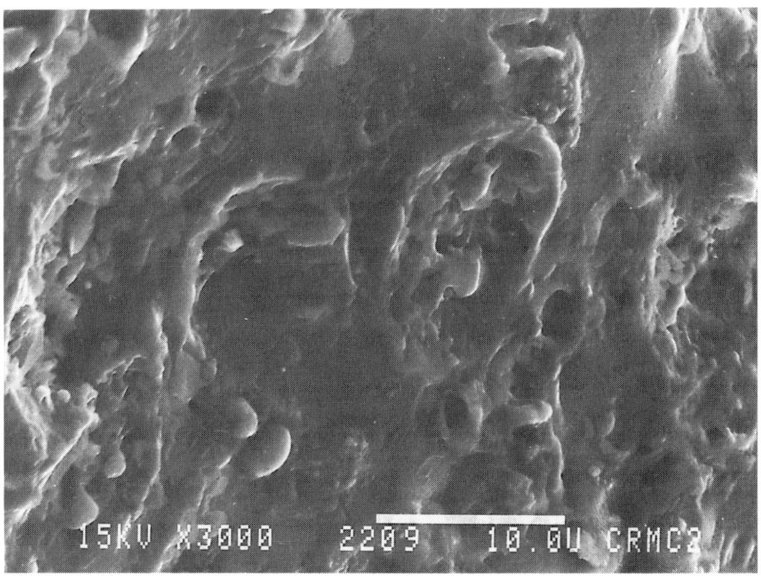

Fig. 5. Detail of AU4G surface after 1 shot at 22 J/cm^2.

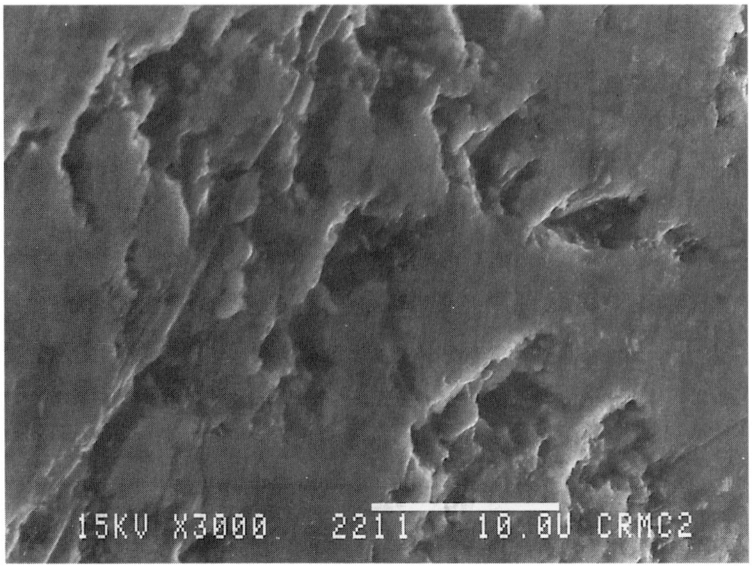

Fig. 6. Surface discontinuities on a non-treated sample of AU4G.

AU4G sample enhances the melting at surface discontinuities like scratches, inclusions and impurities [5], and the onset of miniature eruptions is seen in fig. 5. This has to be compared to a non-treated surface like that shown in fig. 6.

After 3 shots (13 J/cm^2 per shot) on this sample, the scale of the eruptions grows considerably: fig. 7 shows a typical crater (average size is 200 μm) observed on the AU4G surface; the depth of this crater, from the top of the rim to the bottom, is about 17 μm.

The energy was probably insufficient to melt the central island that still remains. This could be attributed to impurities or to different alloying

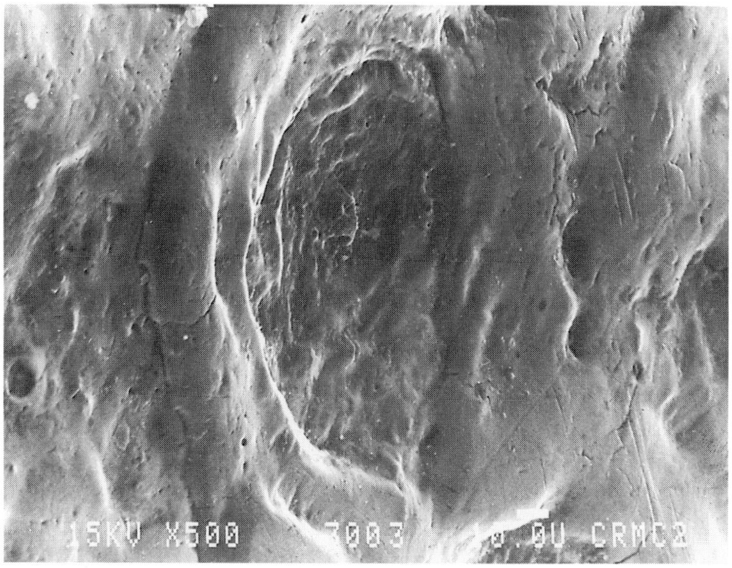

Fig. 7. Detail of a crater on AU4G sample after 3 shots at 13 J/cm^2. Depth: 17 μm. Diameter: 200 μm.

Table 1
Relative proportions (wt%)

	Ti	Al	V
TA6V (typical)	90	6	4
Non-irradiated zone	87.8	8.5	3.7
Irradiated zone	88.2	7.4	4.4

elements of AU4G with different thermal properties and has to be confirmed by energy dispersive spectroscopy (EDS) in further experiments.

The heating and evaporation of the material has been enhanced around this remnant particle, leading to a consequent disruption of the surroundings to form the final crater.

The laser-induced surface of AU4G has been considerably roughened; fig. 8 is a 3D-surfascan visualization of the surface roughness. This roughening can be an interesting behavior for the CO_2 laser-assisted machining of AU4G; it has been shown that the 10.6 μm absorptivity of aluminum can be considerably improved by pretreating the surface with a series of overlapping excimer laser pulses which oxidize and roughen the metal surface, leading to a high enhancement of the drilling efficiency [6].

This roughening has not been seen on the TA6V sample which still presents after irradiation a bright and shiny aspect.

X-ray diffraction (XRD) measurements using Cu Kα radiation have been carried out on the TA6V sample (irradiated and non-irradiated); fig. 9 shows that there were no important changes in the crystallographic structure, there was no shift in the d-spacing value and one relative intensity,

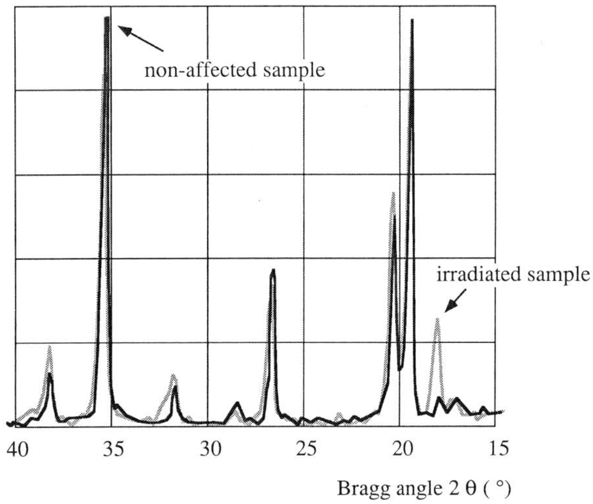

Fig. 9. XRD measurement on a TA6V sample.

corresponding to a Bragg angle of reflection, has been increased while another one decreased. The different phases have not been identified yet; however, an analysis using energy-dispersive spectroscopy (EDS) has revealed little changes in the relative proportions of the elements constituting the TA6V alloy, and could be correlated to the XRD measurements (table 1).

EDS has been used on the AU4G sample and revealed a high percentage of iron on the surface, probably due to the machining of the sample. It has also shown that inside a crater region the amount of copper, magnesium and iron is higher, compared to the non-affected zone (table 2).

These results reveal that the original percentage of copper, magnesium and iron (12.8%) tends to increase in the crater zone (17.1%), whereas

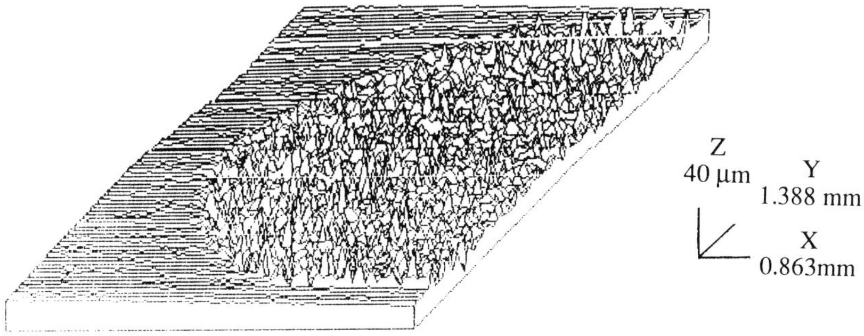

Fig. 8. 3D-surfascan visualization of the roughness (AU4G). $R_a < 0.1$ μm on the non-treated zone, $R_a > 0.5$ μm on the irradiated zone.

Table 2
Relative proportions (wt%)

	Al	Cu	Mg	Fe
AU4G (typical)	94.5	4.5	0.5	0.5
Non-irradiated zone	87.2	3.8	1.0	8.0
Irradiated zone	82.9	5.8	1.3	10.0

the amount of aluminum decreases from 87.2% to 82.9% in the same area.

5. Conclusions

Because of its controllability and reproducibility, a pulsed laser is an important tool for studies of basic mechanisms involved in materials treatment (change in the mechanical properties, surface planarization, microstructural characteristics, etc.).

This paper presents preliminary experiments carried out in order to characterize the surface modification of materials.

Future investigations such as measurements of reflected and scattered radiation using an integrating sphere, measurements of absorptivity by means of thermocouples, studies of the plasma by spectroscopic methods, will allow an understanding of the phenomena related to the high-intensity laser material processing.

Acknowledgements

The authors would like to thank J. Bouesc for technical assistance, W. Marine for XRD measurements and S. Nitsche for SEM and EDS experiments.

This work has been supported by the "Direction des Recherches Etudes et Techniques".

References

[1] J.P. Reilly, A. Ballantyne and J.A. Woodroffe, AIAA J. 17 (1979) 1098.
[2] G. Weyl, A. Pirri and R. Roost, AIAA J. 19 (1981) 460.
[3] A.J. Pedraza, J. Met. 39 (1987) 14.
[4] J. Arnold, Structuring of Metals by Ablation with Excimer Laser, in: Proc. ICALEO Conf., San Jose, CA, 1991.
[5] A.S. Bransden, J.H.P.C. Hegaw, P.H. Balkwill and C. Westcott, J. Mod. Opt. 37 (1990) 813.
[6] G. Kinsman and W.W. Duley, Appl. Phys. Lett. 56 (1990) 996.

Applied Surface Science 69 (1993) 115–120
North-Holland

applied
surface science

Critical issues for single-chamber manufacturing: the role of laser technology

D.J. Ehrlich

Lincoln Laboratory, Massachusetts Institute of Technology, Lexington, MA 02173, USA

Received 2 June 1992; accepted for publication 30 October 1992

Some critical issues confronting single-chamber, in situ processing are reviewed with an eye toward the role laser technologies are likely to play in enabling this emerging trend in microfabrication.

1. Introduction

The development of single-chamber or in situ processing is among the most important current trends in microfabrication. This development is motivated by disparate forces all pointing toward some common themes. Three of the strongest forces are the evolution toward larger wafers, the scaling toward smaller device size, and an explosion in the variety of specialized (and lower-volume) circuits. These influences are combined with intensive competitive pressures associated with product development efficiency and time-to-market.

It is probable that wafer sizes will reach 300 mm diameter by the turn of the century. Even with current 200 mm dimensions many process steps are now becoming impractical in batch processing machines. Single-wafer handling, in turn, favors a new factory architecture based on cluster tooling. Similarly the scaling of devices (particularly toward thinner layers), the greater importance of interfaces, and the incorporation of new materials (e.g., III–IV semiconductors) stress the importance of controlling semiconductors throughout their fabrication history without the random degradation from chemical reaction with air. The time-to-market issues are separable from the other forces. These issues are intense and will in all probability result in a new class of fast-turn-around fabrication tools, distinctly different from the tools of the production line.

Three single-chamber machine architectures are idealized in fig. 1. The cluster-tool geometry is now becoming prevalent in silicon manufacturing, and the trolley tool has emerged first as the

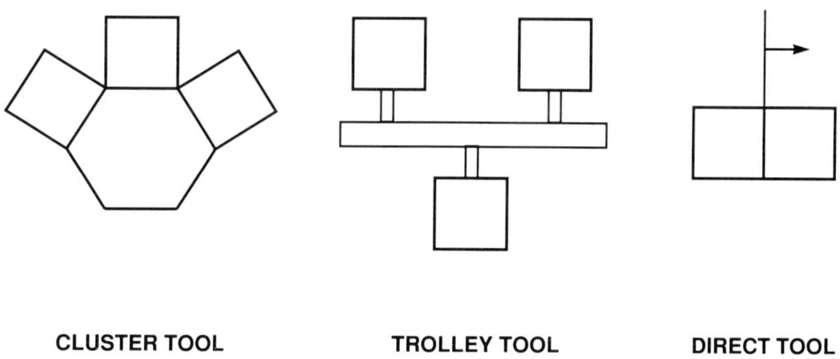

Fig. 1. Three machine architectures for in situ processing of semiconductors (see text).

geometry for III–V materials/device engineering, particularly where ultrahigh vacuum is required. (Neither of these is really single-chamber, in that each processing step for both cases probably requires an individual process cavity separated from a robotic, or manual, handler by a gate valve.) In the direct-tool architecture the wafer is held stationary for a number of steps, and several processes are brought sequentially to it, controlled for example with a beam. In all three machine architectures, the wafer remains in a machine envelope for a series of steps.

Much current process technology is entirely consistent with single-chamber processing and is being transported directly into the new machine architectures without difficulty. In this paper we briefly outline several more difficult issues, emphasizing those for which laser processes may provide the leading potential solutions.

2. Lithography and patterning for wafer manufacturing

Typical wafer fabrication sequences now involve about 20 mask levels. This number will probably increase by ~ 50% in the next ten years. Each lithography cycle entails five to eight wafer-handling steps. Since such a dominant fraction of processing time is bound up in these steps, the motivations for single-chamber tools carry over very strongly to lithography. The main problems are replacing the wet chemical steps such as resist spin-on and development. The lithography cluster almost certainly must be capable of carrying out these steps dry. Although a lamp-based exposure tool might be consistent with this requirement, the emergence of laser steppers seems to greatly enhance the prospect. It is our view that an all-dry lithography cluster will, in fact,

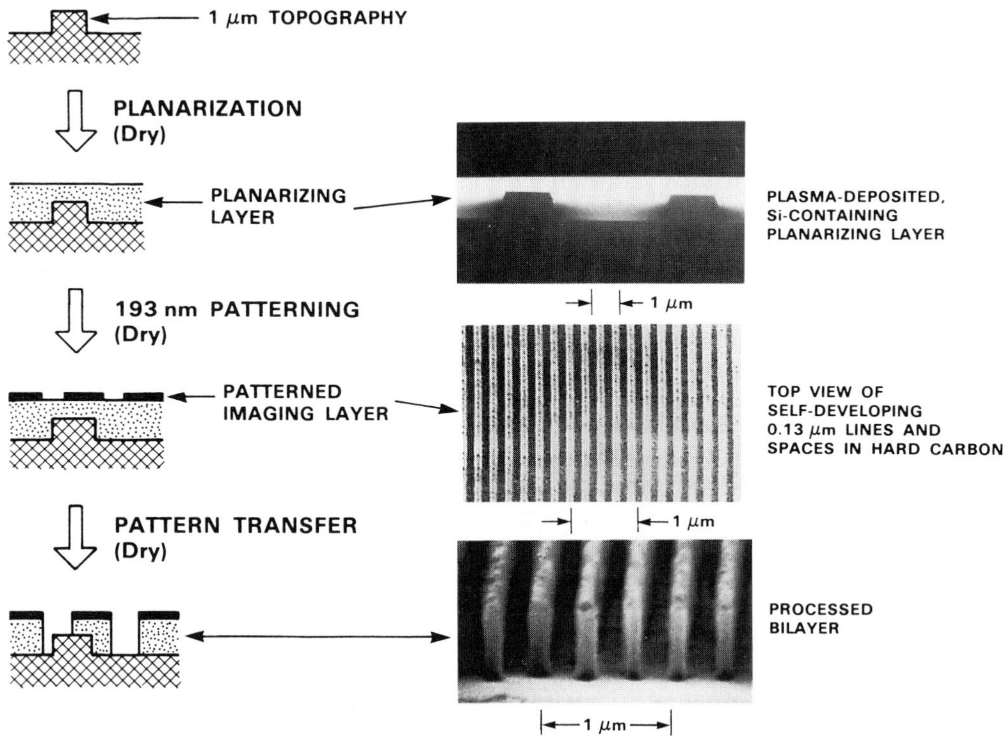

Fig. 2. All-dry lithography process (after Horn et al., see ref. [1]). The process flow is first the plasma deposition of planarization and imaging layers, laser exposure and two-step pattern transfer by reactive ion etching.

become available with the 193 nm (ArF laser) stepper in about five years time. The optical engineering of the stepper is non-trivial but is straightforward and, in fact, is not significantly more difficult than the engineering of a more conventional 193 nm tool. The all-dry lithographic process to go along with the tool is the primary concern, since it must meet all the specifications of a state-of-the-art process. In our own research we have identified planarization (normally achieved reasonably well by spin-on resist) and process perfection (particulate/pinhole intolerance) as two of the most difficult issues for a dry lithographic process [1]. An all-dry processing sequence is illustrated in fig. 2. The conceptual configuration of a single-enclosure fabrication line that might execute such a sequence along with other silicon processing steps is indicated in fig. 3.

3. Wafer cleaning

Wafer cleaning will be a major issue not just for single-chamber processes but for all future VLSI technologies. For example, the current sputter cleaning of silicon contacts is almost certain to be too damaging as a process for future scaled devices. Particulate removal is more controversial. Current yield loss is more than 50% attributable to particulates. As killing defect size shrinks, the issue becomes worse. There is little debate among process designers that the best strategy is to stringently eliminate as many sources of particulates as possible. However, few argue with merits of any process that can efficiently recover circuits.

The unusual efficiency of pulsed laser particulate removal has been demonstrated by Allen et al. [2] and Zapka et al. [3]. On the assumption that blanket-irradiation cleans are not adopted, the cost-effective location of randomly occurring defects is the primary issue (see discussion of monitors below). There is a very strong motivation to solve this latter problem. In addition, a specific issue for single-chamber processing is the elimination of wet cleaning steps currently used to control trace metallic and organic impurities. UV-pulsed laser cleaning will be fairly effective at this. The critical experiments, actual cleans in a factory environment, remain to be done.

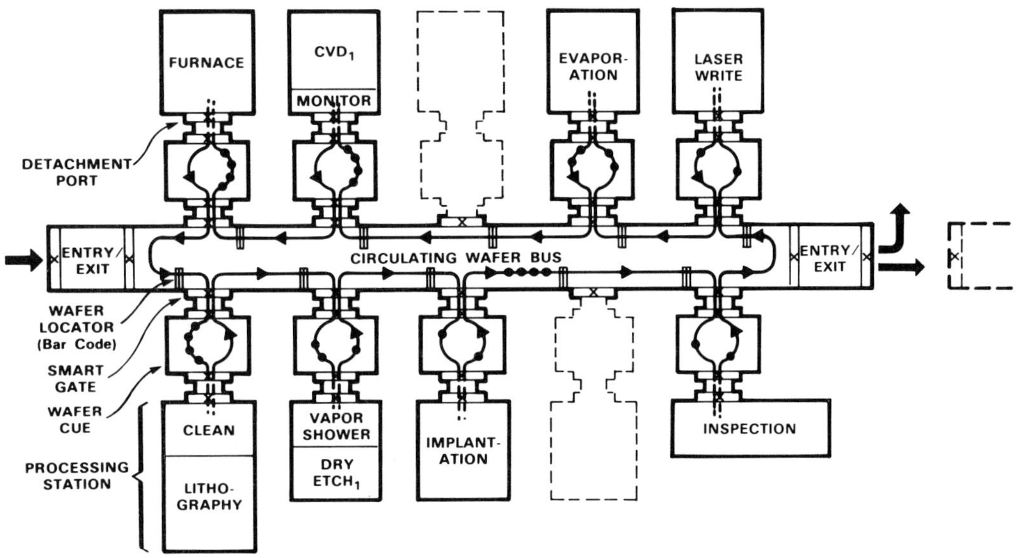

Fig. 3. Conceptual scheme for a single-enclosure fabrication line. Cluster and trolley architectural from fig. 1 are combined around a recirculating wafer buss.

4. Process drivers

The use of lasers as surface-chemical process drivers for numerous deposition, doping and etching steps has been under development for some time. The topic is far too large to treat here, and the reader is referred to ref. [4]. Particularly encouraging results in this area are the announcements by several semiconductor companies of manufacturing using direct (ablative) etching of polyimide. One of these processes is also a good example of in situ patterning using a direct geometry.

5. Monitors

Real-time monitors and controls are intimately related to the ideas of single-wafer and single-chamber handling. Monitors/controls are essential to future factories and, fortuitously, the architectures for the new machines, with a single-wafer surface to look at, are conducive to monitors/controls. It is our view that the dominant monitor techniques will be laser based. Compared to charged-particle probes, optical beams are unique in their ability to monitor through process gases. A single laser probe beam can probably read and deconvolve three to six process parameters (temperature, profile, thickness, etc.). This will be an important application area for lasers in future factories.

6. Fast prototyping

One of the strongest trends of the 1990's is the rapid increase in the number of new IC designs. This increase is driven, for example, by the need of suppliers to offer a massive variety of slightly differentiated consumer electronics. The explosion in new IC designs currently far outpaces trends toward shrinking linewidth or larger wafer size. Manufacturing tools are poorly equipped to address this fundamental change in the industry. Every IC must go through one and often several prototyping cycles. Multiple cycles, despite sophisticated simulation, are currently frequent and

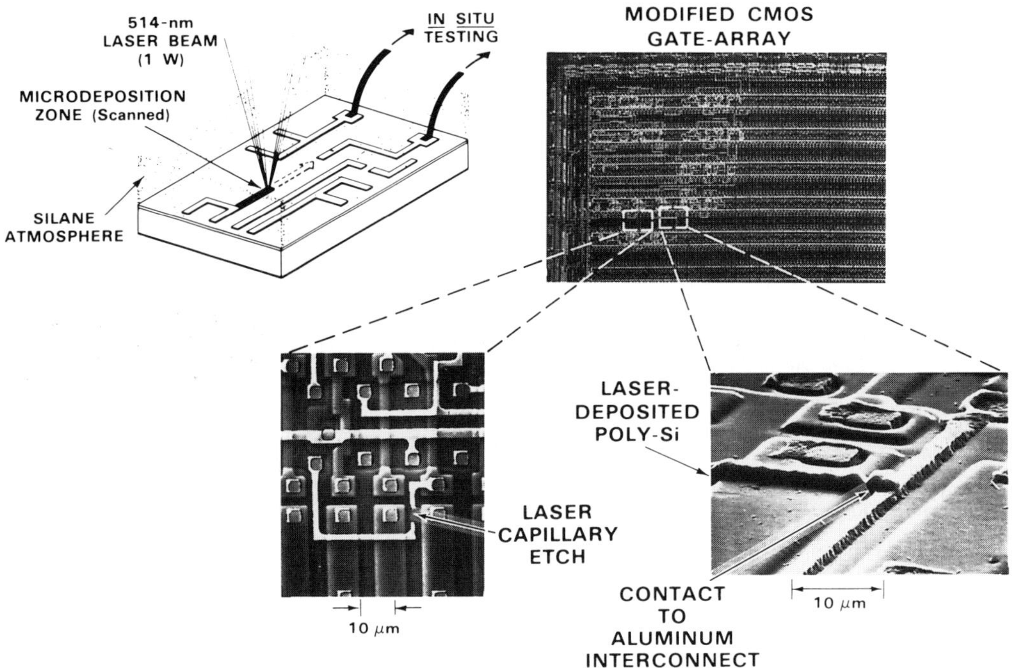

Fig. 4. Laser direct writing processes as applied to rapid implimentation of engineering changes on circuit prototypes.

will remain so. Laser and ion-beam restructuring tools will be used increasingly to cut the time of multiple cycles. Compared to ion-beam systems a commercial laser restructuring tool would offer significant advantage in process speed (~ 100 ×), conductivity (10–100 ×), and through-dielectric viewing. The typical procedures of laser restructuring using deposition in the direct geometry are shown in fig. 4. Restructuring of prototypes will also be important at the next higher level of system integration, notably for multi-chip modules.

7. Tools for micromechanics

The field of silicon-based microelectromechanical devices is about to make the leap from single devices to integrated functioning systems. An important need is for primary three-dimensional (3-D) patterning tools and for system integration and assembly tools. A laser micromachining method has recently been developed [5] for part and mold making in a size regime one to two orders of magnitude smaller than that of the finest conventional machine tools. With further development of associated processes, this primary patterning tool will enable low-cost replication of 3D parts in analogy with projection printing of integrated circuits from electron-beam-fabricated masks.

The approach makes use of fast laser microchemical etching, combined with a digitally addressable acoustooptic laser scanner and computer-aided-design/computer-aided-manufacturing (CAD/CAM) software. In order to obtain the required degrees of dimensional control and surface cleanliness the technique uses precision laser

Fig. 5. Scanning electron micrograph (SEM) of cone inside dish laser etched into silicon using a chlorine ambient of 100 Torr. The beam spot size is 1 μm and the objective is lowered in 1 μm increments after each plane is scanned at 7.5 mm/s. The dish has an upper radius of 80 μm and a lower radius of 35 μm with a radius of curvature of 82 μm and spacer layers extending 8 μm inward. The clipped cone has an upper radius of 4 μm and a lower radius of 16 μm. The laser was swept in a circular scan in each plane. The white bar is a 10 μm scale marker.

microreactions of silicon in a chlorine ambient. The chlorine ambient reacts with the silicon at temperatures above the melting point to form volatile silicon chlorides, which are pumped away from the surface. The digitally controlled laser scanning machine employes a CW argon-ion laser operating at 488 nm wavelength. A typical result is shown in fig. 5.

Acknowledgements

This work was supported by the Defense Advanced Research Projects Agency and by the Department of the Air Force under a specific program sponsored by the Air Force Office of Scientific Research.

References

[1] D.J. Ehrlich, R.R. Kunz, M.A. Hartney, M.W. Horn and J. Melngailis, New Photoresist Processes at UV Wavelengths Less than 200 nm (American Chemical Society, Washington), to be published;
M.W. Horn, S.W. Pang and M. Rothschild, J. Vac. Sci. Technol. B 8 (1990) 1463;
M.W. Horn, J. Vac. Sci. Technol. B 8 (1990) 1980.

[2] S.D. Allen, J.O. Porteus, W.N. Faith and J.B. Franck, Appl. Phys. Lett. 45 (1984) 997.

[3] W. Zapka, W. Ziemlich and A.C. Tam, Appl. Phys. Lett. 58 (1991) 2217.

[4] D.J. Ehrlich and J.Y. Tsao, Laser Microfabrication: Thin-Film Processes and Lithography (Academic Press, Boston, 1989).

[5] T. Bloomstein and D.J. Ehrlich, Appl. Phys. Lett. 61 (1992) 708.

Applied Surface Science 69 (1993) 121–126
North-Holland

applied
surface science

Fabrication and characterization of selectively grown $Si_{1-x}Ge_x/Si$ p^+/n heterojunctions using pulsed laser induced epitaxy and gas immersion laser doping

K.-J. Kramer [a], S. Talwar [a], E. Ishida [a], K.H. Weiner [b] and T.W. Sigmon [1]

[a] *Stanford Electronics Laboratories and Department of Electrical Engineering, Stanford University, Stanford, CA 94305, USA.*
[b] *Lawrence Livermore National Laboratory, Livermore, CA 94550, USA*

Received 2 June 1992; accepted for publication 30 October 1992

A pulsed XeCl excimer laser is used to grow ideally strained heteroepitaxial $Si_{1-x}Ge_x/Si$ layers with Ge fractions up to 21% by intermixing a structure of electron beam evaporated a-Ge on Si(100). The rapid regrowth process induces an interfacial grading of the Ge fraction, which results in unusual stability of the layer strain upon heat treatment, as confirmed by MeV-ion channeling along ⟨100⟩ and ⟨110⟩. Boron dopant is incorporated during the melt process by using a BF_3 gas ambient. Hall/van der Pauw and SIMS analysis reveal that the incorporated dopant dose scales with the number of laser pulses. The junction depth is controlled by the incident laser fluence. The melt time is monitored in-situ utilizing the transient reflectance of the sample during the phase transformations. A patterned reflective aluminum mask is used to obtain spatially selective melting. In-plane Hall mobilities are found to be lower for the heteroepitaxial junctions than for Si homojunctions. We believe this is due to different transport behavior for holes in the observed doping regime of 10^{18}–10^{20} cm^{-3}. Quasiplanar p^+/n heterojunction diodes are fabricated and exhibit near-ideal forward I–V characteristics. Heterojunction diodes exhibit lower turn-on voltages than equivalent Si homojunction diodes, indicative of a lowered bandgap. The turn-on voltages also depend on the B junction depth with respect to the $Si_{1-x}Ge_x/Si$ interface. Both quantities are controlled independently by separating the epitaxy from the doping step.

1. Introduction

In earlier work we demonstrated that pulsed laser induced epitaxy (PLIE) can be used as a means for creating high quality $Si_{1-x}Ge_x$ heteroepitaxial layers on Si [1,2]. In recent time similar methods have gained popularity [3,4]. What makes the method interesting for both research and fabrication is (i) the ultra-short time scale (< 1 μs) that a sample is exposed to elevated temperatures, (ii) the spatially selective nature of the process, (iii) its flexibility of allowing several different processing conditions on one wafer, and (iv) the availability of reliable process control [5]. In this paper we examine the crystallinity, strain and thermal stability for layers formed in this manner.

[1] Presently at the Oregon Graduate Institute, Beaverton, OR 97006, USA.

Gas immersion laser doping (GILD) has been shown to be a powerful tool for producing ultra-shallow junctions of excellent quality in Si [5–7]. Precise control of junction depths and doping levels are achievable using this method. The process consists of supplying enough energy to melt the surface of a semiconductor in the presence of a doping gas, such as BF_3 or AsF_5 for the case of Si. The use of spatially homogenized pulsed UV-laser beams has many advantages in different applications of laser processing [8–10], particularly since the total amount of deposited energy needed to melt the surface is low, resulting in a short thermal budget for the wafer. The dopant dose incorporated is determined by the number of laser pulses in the dopant ambient, whereas the junction depth is controlled by varying the intensity of the incident laser beam. In our apparatus, the melting process is monitored using a HeNe-laser beam to probe the processed region.

0169-4332/93/$06.00

As the area melts, the reflectivity changes. By measuring the melt-time with a fast photodiode, the relation between laser energy fluence and melt time is established as a calibration curve. Furthermore, the melting can be made spatially selective simply by using a reflective metal mask over regions where doping/heating is not desired. This mask is patterned using standard photolithographic techniques. Doping occurs when the surface-adsorbed dopants incorporate into the molten layers. Since the layers are molten, diffusion is very fast. The doping in the molten layer is uniform and essentially no solid phase diffusion occurs beneath this layer, a result of the short thermal cycle ($< 1\ \mu s$) to which the material is subjected.

The laser light is chosen to be highly absorptive in the layer to be melted. We use a 308 nm XeCl excimer laser with a pulse width of 27 ns which has an absorption depth of about 70 Å in Si and roughly 100 Å for Ge [11,12]. The energy of the pulse is deposited only in the topmost layers of these materials. Other parts of the device/circuit are protected from doping thermal cycles. Since the dopant is immediately active after incorporation, no post-GILD annealing is necessary. This differs from ion implantation, where the entire wafer is subjected to a high temperature anneal to activate the implanted dopant.

2. Experimental

For our application of the PLIE/GILD process, we first deposit a thin layer of a-Ge on the surface of a Si wafer using an electron beam evaporator. During the rapid melt and regrowth process induced by the energetic laser pulse, the Ge and the underlying Si intermix and form a single crystalline layer. The difference in the lattice parameter of the $Si_{1-x}Ge_x$ alloy layer and the Si substrate is accommodated by a compressive tetragonal strain in the alloy layer, rather than by the growth of dislocations [1,2].

To fabricate diodes, additional processing is required. Starting with n-type Si(100) wafers, SiO_2 is thermally grown, then 1 μm of pure Al is sputter-deposited. The wafers are patterned using photolithography. The Al and the underlying SiO_2 are etched off in the active areas. The photoresist is kept on the wafers during the following deposition of Ge by electron beam evaporation. After this step, the wafers are subjected to an organic resist strip solution, resulting in the Ge on top of the photoresist being lifted off, Ge remaining only in the opened active areas.

After this initial processing the wafer is transferred to the laser processing apparatus. It is placed in a gas cell with a quartz window allowing laser irradiation of the wafer surface. The cell is pumped out and refilled with dry nitrogen gas to 70 Torr. To form the heteroepitaxial $Si_{1-x}Ge_x$ structure, 2–10 laser pulses at an energy fluence ranging between 0.8 and 1.1 J/cm^2, are used to alloy the Ge with the Si. Following this step, the cell is then evacuated and refilled with BF_3 gas to 60 Torr. The doping process used to form the p-doped fabricated $Si_{1-x}Ge_x$ consists of two different schemes; (i) certain areas of the wafer are doped with laser pulses of the same energy as used to produce the alloy layer, (ii) remaining areas are doped at lower energies resulting in confinement of the B completely within the alloy layer.

Further processing includes removal of the Al mask, CVD SiO_2 deposition and contact metallization, which consists of Al on Ti.

3. Results

Fig. 1 shows random and ⟨100⟩-aligned Rutherford backscattering spectra taken using 2.2 MeV $^4He^+$ together with a RUMP simulation for an undoped $Si_{1-x}Ge_x$ on Si(100) sample [13]. Ideal epitaxial growth is achieved with no significant dechanneling by dislocations observed. This sample was irradiated with 2 pulses at the same energy fluence of 0.92 J/cm^2. The minimum yield is calculated to be 4% for this sample. From the simulation, the layer stoichiometry is best described as a 350 Å thick layer of constant Ge content of 21%, followed by a region of about 350 Å thickness, in which the Ge is linearly graded from 21% to 0% towards the substrate.

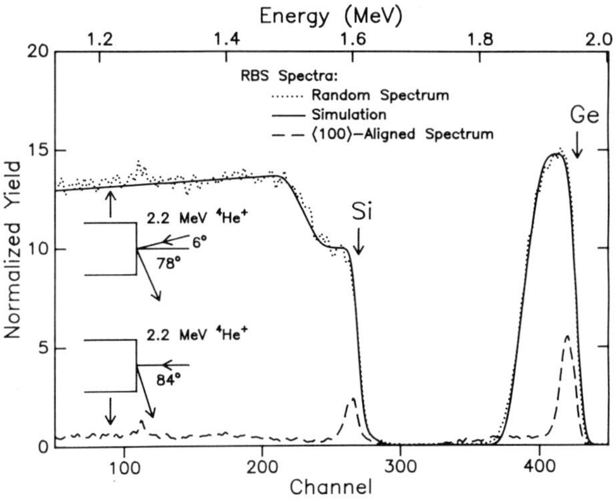

Fig. 1. RBS random and channeling spectra together with a RUMP computer simulation for a $Si_{1-x}Ge_x$/Si sample created using PLIE. The minimum yield, corrected for the different scattering geometries, is 4%. The simulation yields a film of 350 Å of 21% Ge at the top followed by a layer of 350 Å in which the Ge content is linearly graded from 21% to 0% towards the interface.

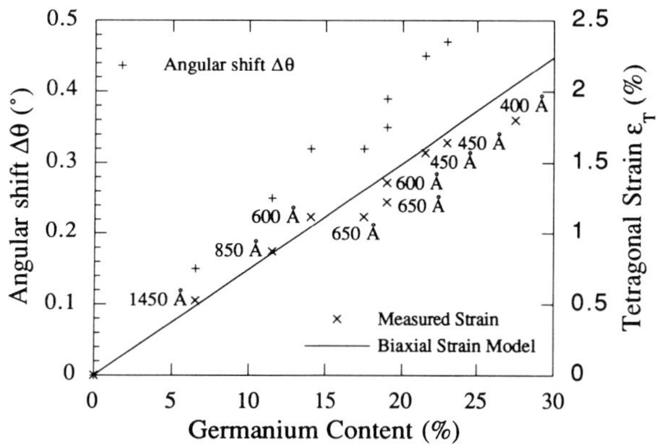

Fig. 2. Shift of the angle of the minimum yield in the alloy layer compared to the substrate as recorded by MeV-ion channeling along ⟨110⟩-directions and the resulting tetragonal strain ϵ_T as a function of Ge content. The solid line represents the predicted relation between strain and Ge content for commensurate heteroepitaxy. For each sample, the thickness of the alloy film is labeled adjacent to its strain value.

We believe our data indicates that the ultra-short time scale of the process results in epitaxy without significant development of dislocations. The melt times are about 50 nanoseconds with the total time the surface layer remains at an elevated temperature after resolidification being on the order of a microsecond. For very high Ge fractions, and thus higher misfit strain, increased stress at the moving liquid–solid interface inhibits

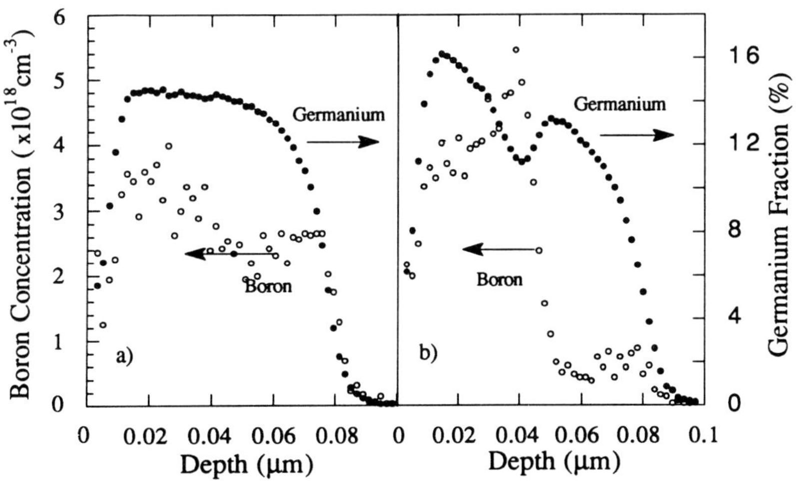

Fig. 3. Boron concentration and Ge fraction versus depth for $Si_{1-x}Ge_x$ heterojunction layers obtained using secondary ion mass spectroscopy. The data is from samples fabricated using two-step laser driven epitaxial mixing–doping processes. (a) These data are for samples receiving B doping using the same laser energy fluence used for the epitaxial layer formation. (b) These data are for identically processed samples, however, the doping is performed using a lower laser energy fluence. The B tail in (a) between 500 and 850 Å results from residual BF_3 gas present during the $Si_{1-x}Ge_x$ epitaxial formation.

perfect regrowth, resulting in the fabrication of defected layers.

Strain in epitaxial $Si_{1-x}Ge_x$ on Si(100) layers can be measured by ⟨110⟩ MeV-ion channeling [14,15]. The tetragonal strain is calculated directly from the angular shift $\Delta\theta$ of the ⟨110⟩ direction in the alloy layer with respect to the substrate. The obtained angular shifts $\Delta\theta$ and the calculated strain ϵ_T are given in fig. 2 as a function of Ge content. Also shown is the predicted elastic strain for ideal commensurate growth. Within $\pm 0.05°$, our estimated experimental error, the strain values obtained match with the ideal, calculated values. Even for a Ge content of 28%, where perfect epitaxy is not achieved, the crystalline alloy film is in a highly strained state.

To examine thermal relaxation behavior, we examined identical samples of 19% Ge and a total layer thickness of 650 Å after exposure to different thermal cycles. The as-grown layer shows a minimum channeling yield of 3.8% for both the Si and Ge signals and is strained to 1.22%. Substantial relaxation of the layers only occurs after heating to 950°C or above for one hour. There the remaining elastic strain is found to be 0.98%.

The high level of residual elastic strain after heat treatment stands in contrast to observations made on $Si_{1-x}Ge_x$/Si films grown by other techniques [15,16]. The unusual thermal stability of the compressed state of these layers is most likely due to two features of the PLIE process; (i) the lattice mismatch is adapted gradually at the interface, thereby reducing the strain energy per bond in this region; (ii) the ultra-rapid nature of the process, resulting in a low total thermal budget during growth.

For B-doped layers, secondary ion mass spectroscopy (SIMS) is performed using 3 keV O_2^+-ions. The depth scale is calculated using profilometry of the analytical crater. The depth values are confirmed by Rutherford backscattering spectroscopy (RBS). We also use the latter to determine the Ge fraction in our alloy layers. Fig. 3 shows the results of SIMS and RBS measurements for two samples processed with pulses of identical laser energy fluences for the fabrication of the $Si_{1-x}Ge_x$ alloy layers. The average Ge fraction is found to be ~ 14.5% for these layers. The first sample of this group is then B-doped, using three laser pulses of the same energy fluence as used before, to form the alloy, whereas for the second sample, the energy fluence is reduced, resulting in a shallower B profile.

From figs. 1 and 3 we see that the Ge signal is graded toward lower Ge content as the substrate is approached, whereas in the center part of the alloy layer it appears flat. This indicates that the Ge tends to segregate into the melt during resolidification. It is for this reason, that the second sample in fig. 3 exhibits a pronounced dip in the Ge signal precisely at the melt depth chosen for the B-doping pulses.

The position of the B profile for sample one in fig. 3a indicates that the electrical junction depth, the metallurgical junction depth and the melt depth are equal. For the second sample, fig. 3b, the B profile falls off at a shallower depth than the Ge and the electrical junction lies within a region of higher Ge fraction. In this sample, a tail is visible in the B signal up to the metallurgical interface. From SIMS measurements on control samples, we conclude that this tail results from doping from residual BF_3 gas in the cell during the alloy formation. This problem is easily overcome by proper purging of the gas transfer system to avoid gas condensation.

A shift towards lower turn-on voltages is expected for fabricated diodes with junctions placed within the $Si_{1-x}Ge_x$ alloy. This is a consequence of the significant energy gap reduction for strained $Si_{1-x}Ge_x$ alloy layers on Si [17].

In fig. 4 we show forward I–V characteristics for a shallow ($x_j = 400$ Å) silicon diode and two sets of alloy layer diodes consisting of 10% and 28% Ge fractions. For each pair of diodes the one exhibiting the lower turn-on voltage has been B-doped at the reduced laser energy fluence. The ideality factors of the diodes are between 1.05 and 1.12. The effect of the location of the junction within the alloy layer on the electrical properties of the diode can be seen in this figure. Although the diodes fabricated in 28% Ge fraction material contain many defects, they remain highly strained, as found using MeV-ion channeling (fig. 2). In summary, our I–V measurements indicate that bandgap reductions are achieved in

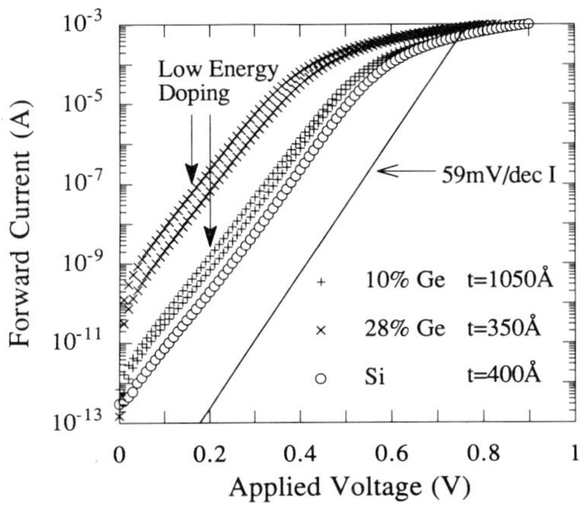

Fig. 4. Forward current–voltage characteristics for p^+/n $Si_{1-x}Ge_x/Si$ diodes. (a) $x = 0.28$, $t = 350$ Å; (b) $x = 0.1$, $t = 1050$ Å; (c) $x = 0$, $t = 400$ Å. For the 10% and 28% Ge layers, B doping was performed at a lower laser pulse energy than used for the epitaxial layer formation. This results in a lower turn-on voltage than those that received B doping at the same laser energy as the epitaxial layer formation. The ideality factor is closer to 1 for the lower Ge percentage diodes since they are fabricated in virtually defect-free material.

these devices. Reverse leakage currents for the diodes are found to lie between 1 and 50 $\mu A/cm^2$ at −5 V. Refinement of the CVD-SiO_2 passivation layer is believed to improve upon this value [7].

We have also measured the electrically active dopant dose incorporated and the room temperature Hall mobility using a standard Hall configuration setup to perform Hall-effect and van der Pauw measurements of sheet resistance, active dose and mobility. From Hall measurements we find that the active dopant dose is nearly proportional to the number of laser pulses at each laser energy. This is interpreted in that dopant incorporation conditions are similar for each pulse. The dopant dose per pulse is found to lie between 1×10^{13} and 2×10^{13} cm^{-2}. It is, however, also dependent on the laser energy fluence in such a way that for higher energy pulses generally higher doses are incorporated.

Mobility values are obtained using a combination of Hall and van der Pauw measurements. The "in-plane" hole mobility is a determining factor for bipolar transistor performance, as carrier transport from the base contact to the intrinsic base is strongly affected by its value.

We find that the mobility values for $Si_{1-x}Ge_x$ heterojunction layers are generally 15% to 25% lower than those for Si homojunction layers that are comparable in doping level and layer thickness. We believe that there is no simple explanation to this result. The physical phenomena which affect the hole mobility in $Si_{1-x}Ge_x$, as compared to Si, are strain, alloy composition and heavy doping. At this time we have not found calculations of the hole mobility which include all these effects simultaneously. We believe, however, that due to the strain-induced split of the lowest energy hole bands, most holes flow in the heavy hole band [18]. Its separation from the light hole band, which reduces interband scattering, may not be significant for the high doping levels under investigation.

4. Conclusion

We demonstrated that pulsed laser induced epitaxy and gas immersion laser doping combined are able to form ideally strained boron-doped p^+/n $Si_{1-x}Ge_x/Si$ heterojunction layers of excellent crystal quality and heterojunction diodes with lowered turn-on voltages. The epitaxial process is spatially selective and the diodes are quasi-planar. Dopant dose and electrical junction depth can be controlled independent of the $Si_{1-x}Ge_x/Si$ epitaxial layer thickness. In-plane hole mobilities are found to be lower for $Si_{1-x}Ge_x$ than for similarly doped Si layers. We believe that this results from increased carrier transport occurring in the heavy hole band, which is heavily modified by the misfit strain in the film.

Acknowledgements

This work was supported by Semiconductor Research Corporation Contract No. 91-SJ-102, and IBM Corporation Contract No. 1577. Part of this work was performed under the auspices of the United States Department of Energy by

Lawrence Livermore National Laboratory under Contract No. W-7405-Eng-48.

References

[1] J.R. Abelson, T.W. Sigmon, K.B. Kim, and K.H. Weiner, Appl. Phys. Lett. 52 (1988) 230.
[2] Y. Chang, S.Y. Chou, K.-J. Kramer, T.W. Sigmon, A.F. Marshall and K.H. Weiner, Appl. Phys. Lett. 58 (1991) 2150.
[3] S. Lombardo, K. Kramer, M.O. Thompson and D.R. Smith, Appl. Phys. Lett. 59 (1991) 3455.
[4] M. Berti, G. Mazzi, L. Calcagnile, A.V. Drigo, P.G. Merli and A. Migliori, J. Mater. Res. 6 (1991) 2120.
[5] P.G. Carey, T.W. Sigmon, R.L. Press and T.S. Fahlen, IEEE Electron Device Lett. 6 (1985) 291.
[6] F. Foulon, A. Slaoui, E. Fogarassy, R. Stuck, C. Fuchs and P. Siffert, Appl. Surf. Sci. 36 (1989) 384.
[7] K.H. Weiner, P.G. Carey, A.M. McCarthy and T.W. Sigmon, IEEE Electron Device Lett. 13 (1992) 369.
[8] S. Küper and M. Stuke, Appl. Phys. A 49 (1939) 211.
[9] T. Licata, R. Scarmozzino, B.J. Garcia and R.M. Osgood, J. Vac. Sci. Technol. B 10 (1992) 77.
[10] W. Zapka and W. Ziemlich, Appl. Phys. Lett. 58 (1991) 2217.
[11] G.E.. Ellison and F.A. Modine, Appl. Phys. Lett. 41 (1982) 180.
[12] J. Kolodzey, S. Aljishi, R. Schwarz, D. Slobodin and S. Wagner, J. Vac. Sci. Technol. A 4 (1986) 2499.
[13] L.R. Doolittle, Nucl. Instr. Meth. B 15 (1986) 277.
[14] J.C. Bean, L.C. Feldman, A.T. Fiory, S. Nakahara and I.K. Robinson, J. Vac. Sci. Technol. A 2 (1984) 436.
[15] B. Hollaender, S. Mantl, B. Stritzker, H. Jorke and E. Kasper, J. Mater. Res. 4 (1989) 163.
[16] J.-M. Baribeau, S. Kechang and K. Munro, Appl. Phys. Lett. 54 (1989) 323.
[17] R. People, Phys. Rev. B 32 (1985) 1405.
[18] G.C. Osbourn, Bull. Am. Phys. Soc. 30 (1985) 594.

Applied Surface Science 69 (1993) 127–132
North-Holland

applied surface science

A novel VUV photochemical deposition apparatus

C. Manfredotti, F. Fizzotti, M. Boero and G. Piatti
Experimental Physics Department, University of Torino, Via Giuria 1, 10125 Torino, Italy

Received 2 June 1992; accepted for publication 30 October 1992

A new photo-CVD apparatus has been built in order to deposit a-Si:H films and other kinds of amorphous thin films by a technique which is both simple and versatile. This apparatus is composed of three chambers connected together: a load-lock chamber, a process chamber and a third chamber for in-situ analysis of deposited films. A peculiarity of the lamp, a dielectric discharge lamp which can work with noble gases like Xe or Kr, is that it can be completely dismounted without breaking the vacuum in order to clean the optical MgF_2 window. By this method, the deposition chamber can be kept very clean. In this apparatus, we started to deposit a-Si_xC_{1-x}:H of very good quality. These films have been completely characterized by chemical (RBS, ERDA) and optical (PDS) methods. Their quality can be compared with that of a-Si:H samples of the same thickness, obtained by PECVD.

1. Introduction

Photochemical vapor deposition has been recognized since long as a very suitable technique for depositing a-Si:H and related films of good electronic quality at a relatively low temperature [1]. The lack of electron and ion bombardment on the substrate and on growing films allows a real optimization of interfaces, in particular p/i and ITO/p-Si:H interfaces of solar cells, improving their conversion efficiency [1]. As a consequence, this technique seems to be the best one for depositing a-Si:H based superstructures like a-Si/a-SiC, giving conversion efficiencies as high as 10.5% over 1 cm^2 surface area [2]. Another advantage of photo-CVD technique is the possibility to achieve large deposition rates over large areas, by keeping excellent electrical and photoelectric properties of deposited films. In general, this goal is reached by direct photo-CVD by using a rare gas discharge lamp emitting below 200 nm with good efficiency and by flushing Si_2H_6 gas, which maintains a fairly large absorption cross section between 150 and 200 nm, while SiH_4 practically has a threshold for photoionization at 150 nm. By a Xe discharge lamp, driven by a magnetron, a deposition rate of 7.5 nm/min was obtained for a-Si:H at 100 W input microwave power over areas of the order of 1 cm^2 [3].

Direct photo-CVD was also used to obtain high quality p-Si:H films for solar cells over areas of 10×10 cm^2 [4]. However, Hg lamps emit with very low efficiency in the VUV range and cannot reach normally either high powers or large areas. Therefore, the possibility of obtaining larger efficiencies in the VUV range was investigated [5], by concluding that dielectric-barrier (silent) discharges are ideally suited for efficient excitation of high-intensity UV radiation from excimers, particularly of noble gases. Since the emission is due to transitions between a weakly bound upper state and a repulsive ground state, no self-absorption occurs as in Hg or similar lamps. By using this kind of lamp, a-Si:H and similar films of good quality were deposited over 4 cm^2 areas with growth rates as high as 50 Å/min [6]. However, the large absorption of VUV light in the deposition chamber, which runs at 0.5 mbar, limits the window-to-substrate distance to 1 cm. Since Fomblin oil must be deposited onto the internal face of the window in order to avoid the progressive darkening due to a-Si:H deposition onto it, the consequence is that films do contain carbon. Moreover, the deposition run cannot be pushed beyond a few hours because Fomblin oil evaporates and darkening problems start again.

A possible way to overcome this difficulty is to use an internal lamp [7] but, in this case, even if deposition rates as large as 10–20 Å/s can be obtained, direct thermal decomposition due to high filament temperature cannot be avoided and, in practice, the method is not a true photo-CVD one, but more likely a hot-filament method, which seems to have some drawbacks at least for a-Si:H deposition. Finally, VUV photo-CVD methods should not be restricted to a-Si:H and similar films, since other applications are possible, like low temperature growth of thin SiO_2 films [8], of paramount importance in InP and related technologies.

In this paper, we present a relatively new photo-deposition apparatus which displays several facilities not generally available in systems of this kind. Another novelty is represented by the lamp which, even if based on principles reported in refs. [5,6], presents technical innovations that make the entire system practically unique.

2. The system

Photochemical vapor deposition has been proved in the past to be a really suitable method to obtain a-Si:H films of good quality, particularly as far as a-Si:H p/i interfaces are concerned. In general, this technique suffers from some problems, connected to the surface area and the homogeneity of the deposited film and to the darkening of the optical window which is used for light input because of deposition on it. The periodic window cleaning implies vacuum breaking, a procedure which must be avoided if one wants to keep the deposition chamber as clean as possible.

In the present work, we describe a deposition apparatus, which has been designed in order to overcome these problems: the sample holder, on the top of the chamber, can be pushed from the outside downwards against the front of the lamp, which is approximately in the center of the chamber and which is equipped with a particular flange on its periphery with an O-ring groove. The same flange has many holes in its inner periphery, which are used in order to flow nitrogen or argon onto the window to keep it clean. This simple trick allows the complete extraction of the lamp together with the window, which can be easily cleaned, if necessary, in the laboratory.

The deposition chamber is linked by a load-lock to a sample introduction chamber: both chambers are equipped with turbomolecular pumps and therefore the deposition chamber can be maintained under high vacuum conditions continuously. The lamp designed for the deposition apparatus is a dielectric discharge lamp in xenon and in other gases, which will be described in detail in the next paragraph. This two-chamber system is linearly load-lock connected with a Varian VT14 chamber equipped with XPS and Auger apparatus. Therefore, surface chemical analysis can be carried out directly in-situ on growing layers.

The deposition chamber was designed large enough to host also other deposition systems like hot-filament or PECVD. Both the sample heater-holder, which can reach 400°C, and the annular gas outlet around the sample holder, can be vertically displaced from the top manually. A big optical window (∅ = 15 cm) allows an easy monitoring of what is going on. The sample holder can host 10×10 cm^2 substrates. The lamp is fixed with the optical window at a distance of 5 cm from the comber bottom. The active volume thickness of the lamp can be varied from the bottom in the range 1–3 mm.

3. The lamp

A schematic diagram of the lamp pushed against the sample holder, in order to extract the lamp itself and the magnesium fluoride window, is shown in fig. 1. Starting from the bottom, we find the knob for varying the gap between the dielectric (quartz) and MgF_2 window, the external body of the lamp, in stainless steel, the central part with a high-voltage electrical feedthrough,

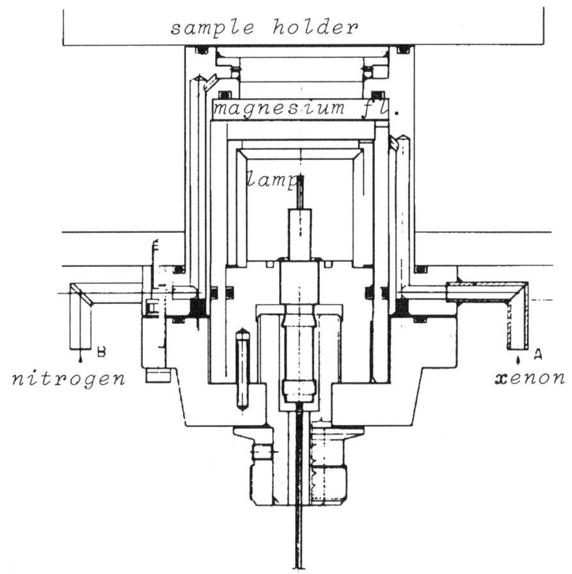

Fig. 1. Xenon dielectric discharge lamp pushed against the sample holder in order to maintain vacuum. Xenon and nitrogen inlets are also shown: nitrogen can be used in order to keep the window clean. The central wire applies AC power to the Cu electrode. Dielectric is quartz. The anode is a thin Pt wire mesh stretched over the window (not shown).

(on the periphery) xenon and nitrogen gas inlets, the former one flushing the gas below the MgF_2 window and the latter one in the gap above the same window, the Cu electrode soldered to the Cu disk-shaped electrode glued to dielectric, which has the shape of an inverted cylindrical glass and it is laterally protected by teflon, the lamp active volume, the Pt wire mesh (grounded electrode), MgF_2-window, the flange used for radial N_2 or Ar blowing onto the window and, finally, the flange pushed against the mirror-finished sample holder surface. The sample is shown just in the gap below it. During deposition, the sample is maintained at a distance of 1.6 cm from the window, in order to obtain a gap of about 0.5 cm on the periphery and to allow Si_2H_6 flushing from the external annular gas outlet (not shown). Xenon has been chosen taking into account that 6.2 mm thick MgF_2 presents a transmittance of only 10% around 150 nm and, consequently, severely reduces the radiation emitted from the krypton, which is centered at 144 nm. Fomblin oil does not constitute a problem, since it starts to absorb just at this wavelength. Xe emission (see fig. 2) is fairly well matched with the Si_2H_6 absorption cross section and, consequently, the use of Xe seems to be the best compromise.

In order to get the optimum geometrical and physical working conditions, the behaviour of the photon absorption rate has been simulated as a function of temperature, pressure and distance from the window. This absorption rate is given by the relationship

$$B(x) = K\int_{\Delta\Omega}\int_{\Delta E} \frac{1}{\vartheta KT} S(\hbar\omega) p\sigma(\hbar\omega)(1-2\rho\vartheta) \times \exp\left(-\frac{p\sigma\vartheta x}{KT}\right) \mathrm{d}\vartheta \ \mathrm{d}(\hbar\omega), \qquad (1)$$

where $S(\hbar\omega)$ is the emission intensity per unit solid angle, $\sigma(\hbar\omega)$ is the optical absorption cross section in Si_2H_6, T and p are the deposition temperature and pressure, ρ is the (Pt wire radius)/(mesh size of the grid) ratio, $\vartheta = 1/\cos\Omega$, where Ω is the angle of diffused radiation and K the Boltzmann constant.

In our case, the light source has been considered as isotropic and 2ρ is equal to 0.05. By looking at fig. 3, one notices that the light intensity is strongly absorbed in the gas and, consequently, the minimum window-to-substrate distance, still compatible with geometrical constraints, must be chosen. $B(x)$ clearly shows a

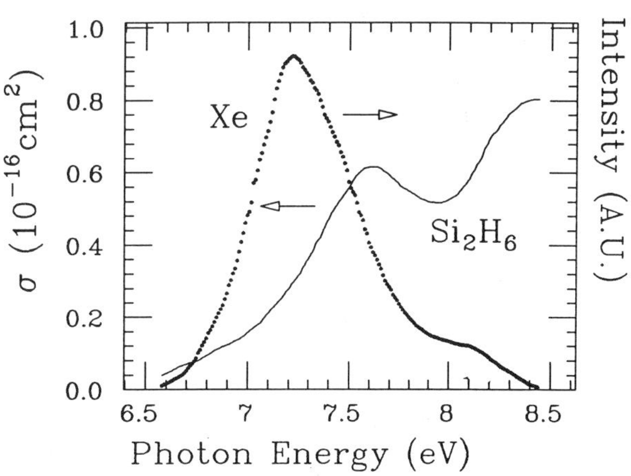

Fig. 2. Xenon excimer emission spectrum and absorption cross section of disilane as a function of photon energy.

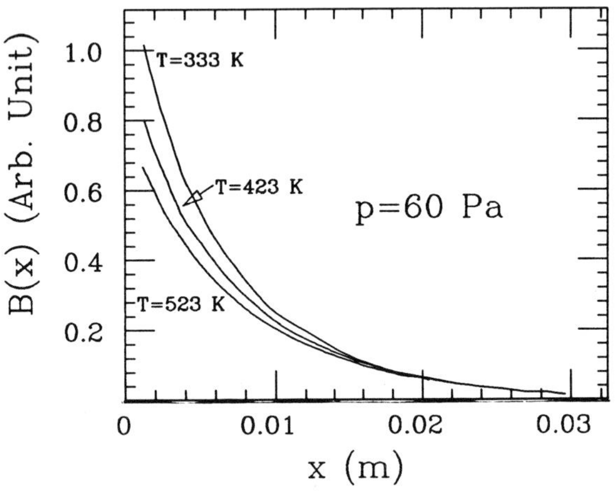

Fig. 3. Behaviour of photon absorption rate $B(x)$ in $cm^{-3}\ s^{-1}$ as a function of source-to-substrate distance at three different temperatures. This behaviour should be related to the radical generation rate in the gas and for a process gas pressure of 60 Pa.

maximum as a function of pressure. For the chosen minimum window-to-substrate distance the corresponding optimum pressure was 60 Pa.

The discharge which is used in the lamp is a special non-equilibrium discharge that can be operated at about atmospheric pressure [5]. It is characterized by the presence of a dielectric layer between the discharge gap and at least one electrode. At atmospheric pressures, a large number of microdischarges are observed in the discharge gap. The function of the dielectric is to avoid the formation of a single discharge by limiting the total current, to keep the energy input in the microdischarges reasonably low and to distribute them evenly over large areas. The microdischarges have diameters of a few hundred micrometers, durations of about 200 ns and can carry several hundred amperes per square centimeter. The electron energies and densities in the microdischarges are 1–10 eV and 10^{14} cm^{-3}, respectively, i.e. very similar to those encountered in discharge pumped excimer lasers. The collisions between electrons and rare gas atoms lead to excited atoms which, at these high pressures, react to form excimers before being quenched by other processes. The major advantage of dielectric-barrier discharge is that the conditions of microdischarges, i.e. electron energy and density, can be influenced by external parameters and by geometry and, therefore, can be optimized.

In our case, gap spacing and dielectric thickness were 1.5 mm, while xenon pressure was 1.75 bar and the total electrical power delivered to the lamp was 30 W at a frequency of 230 kHz. With increasing the electrical power, heating produced by discharges lowered the lamp efficiency, giving rise to a light intensity slightly decaying as a function of time.

The emission spectrum of the lamp was measured by using a vacuum UV monochromator of 0.5 m and a sodium salicilate as a wavelength converter [5]. The electrical power could be varied up to 150 W, with a maximum peak-to-peak voltage of 10 kV, while frequency could be tuned in the 125–375 kHz interval.

The measured emission spectrum of the lamp is shown in fig. 4, as obtained under the conditions mentioned above. The excimer peak is slightly displaced with respect to 172 nm, reported in the literature, and a continuum due to atomic Xe excitation increasing towards 400 nm can be sometimes observed. Traces of air or of nitrogen in the lamp chamber are sometimes responsible for some lines from 220 to 320 nm. Small displacements of the spectrum are also observed, depending on the discharge pressure. It

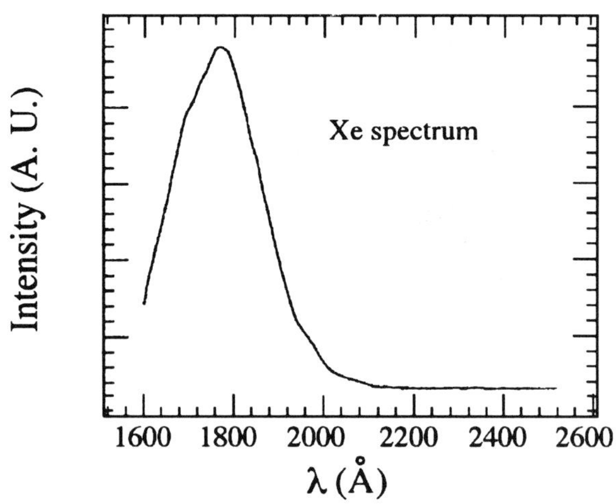

Fig. 4. Measured xenon excimer emission spectrum as a function of wavelength. See the text for details.

is quite difficult to evaluate exactly the energy conversion efficiency of the lamp in terms of (optical power output)/(electrical power input), referred only to the excimer peak, because of the many constraints and indeterminations used during emission spectrum measurements. What can be roughly concluded is that excimer light power onto the substrate should be of the order of 0.95 W, corresponding to an external efficiency of about 3%, which is not too far from literature data [5]. The internal conversion efficiency, taking into account that solid angle, MgF_2 window, and wire mesh losses amounting to 90%, should be of the order of 30% which is near to the maximum of 40% quoted in the literature.

4. Film deposition

Only preliminary results concerning a-Si:H thin film deposition and characterization will be presented here. A more exhaustive report will be published elsewhere. In general, deposition runs were carried out at a pressure of 70 Pa, with temperatures in the range 150–250°C by using very pure disilane from Voltaix at a flux between 5 and 20 sccm. Best deposition conditions, in our case, are met at 15 sccm and $T = 250°C$. Thin films of thicknesses from 500–1500 Å were deposited on 7059 Corning Glass substrates over an area of 7.5 cm^2. Deposition rates were of the order of 5 Å/min without any kind of light focusing onto the sample. The samples have been characterized by PDS and optical absorption for the determination of optical pseudogap, defect density of states (DOS) and tail characteristic energy. Chemical analysis was performed with nuclear techniques such as RBS, ERD and NRA. The hydrogen content depended on the deposition temperature and was in the range 8–12 at%. In the sample deposited by using Fomblin oil, the carbon atomic percentage was around 12%.

Thin samples displayed good electronic properties: an example is shown in fig. 5, in which PDS data for three samples are reported, one of which was obtained by our apparatus equipped with a Xe excimer lamp, the other one by the same apparatus equipped with a D_2 lamp and, finally, the third one by a standard PECVD apparatus. All the films have a thickness of about 700 Å. The DOS value is 6×10^{16} cm^{-3} eV^{-1} for a Xe-photo-CVD sample, a result that can be considered quite well, if one takes the sample thickness into account.

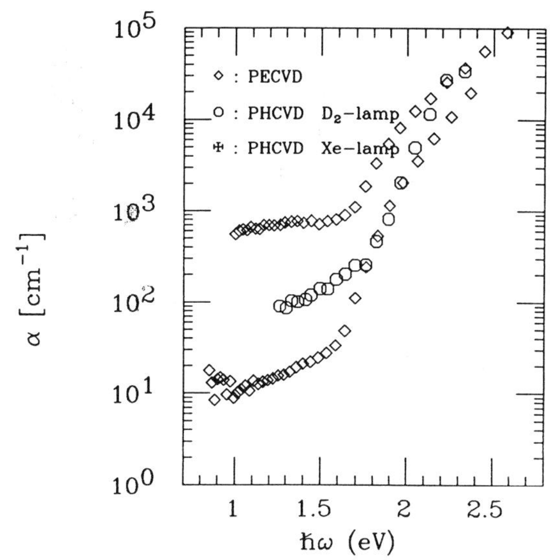

Fig. 5. PDS spectra for samples of the same thickness (500 Å) as obtained by PECVD, photo-CVD with a D_2 lamp and with a Xe excimer lamp. Note that in the last two cases, the obtained films are a-Si_xC_{1-x}:H with roughly 10 at% of carbon.

5. Conclusions

Preliminary results have been reported concerning a relatively new photo-CVD apparatus which makes use of a Xe excimer lamp. Particular care has been taken to reach very low pumping down pressures both in the deposition and in the load-lock chambers. The apparatus can run continuously without vacuum- or other problems, generally connected with window cleaning, and deposition onto the window cleaning can be avoided by an inert gas flushing channel obtained in the body of the lamp.

More complete results concerning both routine running of the apparatus and films characterization will be presented in the near future.

References

[1] M. Konagai, Tech. Dig. Int. PVSEC-3, Tokyo (1987) p. 15.
[2] S. Nakano, K. Wakisaka, M. Kameda, M. Isomura, T. Matsuyama, N. Nakamura, S. Tsuda, M. Ohnishi and Y. Kuwano, Jpn. J. Appl. Phys. 26 (1987) 28.
[3] T. Fuyuki, K. Du, S. Okamoto, S. Yasuda, T. Kimoto, M. Yoshimoto and H. Matsunami, Tech. Dig. Int. PVSEC-3, Tokyo (1987) p. 21.
[4] S. Nakano, K. Wakisaka, M. Kameda, M. Isomura, T. Matsuyama, N. Nakamura, S. Tsuda, M. Ohnishi and Y. Kuwano, Mater. Res. Soc. Symp. Proc. 149 (1989) 417.
[5] B. Elliasson and U. Kogelschatz, Appl. Phys. B 46 (1988) 299.
[6] F. Kessler, H.D. Mohring and G.H. Bauer, Mater. Res. Soc. Symp. Proc. 192 (1990) 99.
[7] P.A. Robertson and W.I. Milne, Mater. Res. Soc. Symp. Proc. 70 (1986) 31.
[8] P. Patel and I.W. Boyd, Appl. Surf. Sci. 46 (1990) 352.

Applied Surface Science 69 (1993) 133–139
North-Holland

applied surface science

A simple formalism for the prediction of angular distributions in laser ablation deposition

J.C.S. Kools, E. van de Riet and J. Dieleman [1]
Philips Research Laboratories, P.O. Box 80.000, 5600 JA Eindhoven, Netherlands

Received 2 June 1992; accepted for publication 30 October 1992

An analytical model allowing the quantitative description of gas clouds generated by near-threshold laser ablation is described. The model is based on a continuum description of the expansion of an idealized laser-generated gas cloud. This analysis yields the time- and space-dependent gas density and the angular intensity distribution. Dependencies of the angular intensity distribution on experimental variables (laser fluence, atom mass and laser spot size) are determined. It is found that the laser spot dimensions are most important parameters in determining the angular distribution. For a circular spot, a simple formula is derived which allows quantitative prediction of angular distributions. These predictions are in good agreement with the angular distributions of ablation plumes reported in the literature. Finally, recommendations towards deposition practice are given.

1. Introduction

Laser ablation deposition (LAD), also known as pulsed laser deposition (PLD), is a thin-film deposition technique of growing importance [1]. The versatility and power of the technique have been demonstrated [2] convincingly by the deposition of high-quality thin films of high-T_c superconductor materials. This has resulted in the introduction of the technique in nearly every major materials science laboratory as a tool for development of complex, high-quality thin-film material stuctures on a laboratory scale with a short optimalization procedure. Extending the possibilities of LAD/PLD to industrial applications still requires solution of some problems, a most important one being the scale-up of the technique. Although homogeneous deposition over 2 and 3 inch wafers has been reported [3,4], most present deposition machines are in general able to deposit homogeneously over surfaces of only a few square centimeters. This is related to the angular distribution of the atoms and molecules vaporized by the laser ablation process. Due to expansion of the laser-created gas cloud, this angular distribution is found to be peaked along the target normal. An understanding of this expansion effect and, if possible, quantitative description would enable one to tackle the scale-up problem with judgment, enlarging the opportunities to use LAD as an industrial technique. In this paper, the results of a model [5,6] describing gas clouds generated by laser ablation of solids are used to describe the relationship between (a) angular distributions obtained as in LAD and (b) experimental parameters such as laser spot geometry, laser fluence and particle mass. This leads to a number of practical recipes relevant to deposition practice.

2. Hydrodynamical model: physical assumptions

The physical picture on which the model is based will be discussed first. It is shown schematically in fig. 1. The gas cloud history is divided in three stages:

[1] Present address: DSA Consultants, Geraniumlaan 11, 5582 GD Waalre, Netherlands.

0169-4332/93/$06.00

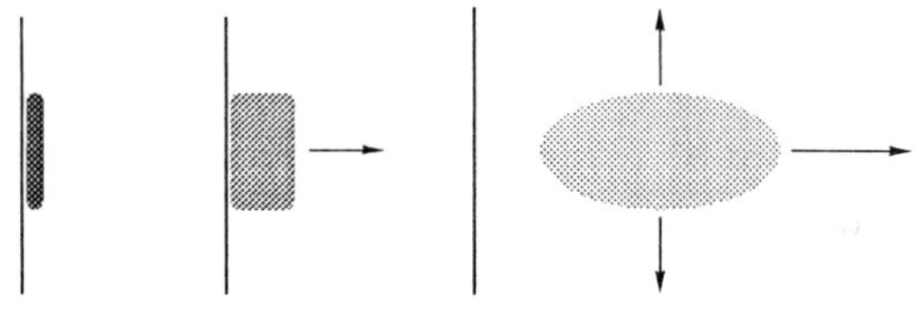

Fig. 1. Physical picture on which the hydrodynamical model is based. A gas cloud is generated by ultrafast evaporation from a solid. The cloud will at first expand one-dimensionally (1D) and gain a stream velocity. The cloud will then expand three-dimensionally (3D) in its center-of-mass (COM) system.

(1) A gas cloud is created by ultrafast desorption from a solid or liquid surface.
(2) Immediately after desorption, the kinetic degrees of freedom of the gas cloud equilibrate to a full MB-distribution in its center-of-mass (COM) system, followed by a one-dimensional (1D) expansion of the cloud along the target normal.
(3) This is followed by adiabatic three-dimensional (3D) expansion of the gas cloud in its COM reference frame.

The two expansion processes (1D and 3D) are described in a continuum hydrodynamical formalism (Navier–Stokes equations).

This complicated behavior is caused by an anisotropy induced by the presence of the target. For times shortly after desorption, the gas cloud is in contact with the target, and considerable momentum exchange with the target is thus possible. This results in a pressure of the gas cloud on the target surface, causing the gas cloud to gain momentum along the target normal, thus being accelerated away from the target (1D expansion). The gas cloud also expands into the surrounding vacuum (3D expansion) due to the pressure gradient induced by its density profile. The gas cloud does not change its overall momentum in this process. Although both expansion processes overlap, it is reasonable to make the simplification of a 1D expansion stage followed by a 3D expansion stage, since the pressure on the target is a strongly decreasing function of time. This picture is supported by ultrafast photography of laser ablation plumes. It is found there that, under typical LAD conditions, for the shortest time-scales ($t \leqslant 100$ ns) the expansion of the laser-generated gas cloud is essentially one-dimensional [7], whereas for longer time-scales ($t \geqslant 300$ ns) the expansion is essentially three-dimensional [8]. The model should apply to all systems of transient gas desorption from a solid or liquid target in a vacuum where the following conditions are met:

(1) The number of post-desorption collisions is sufficient for hydrodynamics to apply. To get some estimate of this condition, consider the following numerical example: five monolayers of copper ($\sim 1 \times 10^{16}$ atoms cm^{-2}) are evaporated from a 5000 K surface in 20 ns. The atoms will have typical velocities of about 1000 ms^{-1}, resulting in a cloud which has a length of about 20 μm along the surface normal. The gas density is then 5×10^{18} atoms cm^{-3} resulting in a pressure of 3.5×10^5 Pa. Taking a typical collision cross section of 5×10^{-15} cm^{-2}, the mean free path becomes 0.3 μm. This is significantly smaller than the gas cloud dimensions. As long as the expansion process is mainly 1D, both mean free path and smallest gas cloud dimension are approximately linear in time and the ratio cloud dimensions/mean free path does not change. This motivates the treatment of the gas as a continuum.
(2) The majority of the atoms are in a neutral state. The flow of charged particles is not considered here. This condition is not fulfilled for laser vaporization phenomena in the higher fluence range such as in the case of laser-generated X-ray sources, ion sources or controlled-fusion devices. For LAD circumstances near the ablation threshold, the ionized fraction is sufficiently low. However, deviations may occur in the case of atoms with very low ionization energies. Since the depositing flux in LAD is in general found to consist mainly of single atoms, only the results for atoms are given here.
(3) The particles gain energy as long as momentum exchange with the target is possible. Absorption of laser radiation after considerable 1D expansion has occurred, is not considered here.
(4) The expansion is happening in a vacuum. More quantitatively: the mean free path of the ablated particles with respect to collisions with background gas atoms or molecules is large compared to the distances in the deposition geome-

try. Since these distances are typically on the order of a few cm, the model applies for ablation events in pressures typically below 10^{-1} Pa. This implies that an important class of LAD processes, i.e. reactive LAD processes cannot be described properly by the present model. Nevertheless, it can be concluded that the model applies to a wide class of laser vaporization phenomena, including LAD situations.

3. Hydrodynamical model: results

The primary quantity of the model is the time- and position-dependent particle density distribution $n(x_i, t)$ (or equivalently a velocity distribution $f(v, \vartheta)$) which comes as a solution of the Navier–Stokes equations. From this quantity, the angular intensity distribution $I(\vartheta)$ (ϑ is the polar angle) can be calculated by integration of the total number of particles moving in one direction:

$$I(\vartheta)\ \mathrm{d}\Omega = \left(\int_{v \geqslant 0} f(v, \vartheta)\ \mathrm{d}v \right) \mathrm{d}\Omega.$$

This yields for small angles:

$$I(\vartheta) \propto \cos^p(\vartheta),$$

where

$$p = 1.75 \times \frac{T_z}{T_\varphi} \left(\frac{mu^2}{kT_z} \right)^{3/4}, \tag{1}$$

where m is the atomic mass (in amu) and the parameters u and T_z, T_φ are called stream velocity and elliptical temperatures, respectively. The meaning of these parameters and their dependence on experimental conditions will be discussed now.

The *stream velocity* u is the velocity of the center-of-mass of the gas cloud after 1D expansion. It is related to the temperature after 1D expansion (T_0) by the relation:

$$u = M(\gamma k T_0/m)^{1/2}, \tag{2}$$

where M is the dimensionless Mach number of this 1D expansion. It is found [5] experimentally, and motivated theoretically, that $M = 2.33$ for atoms in a well-developed expansion.

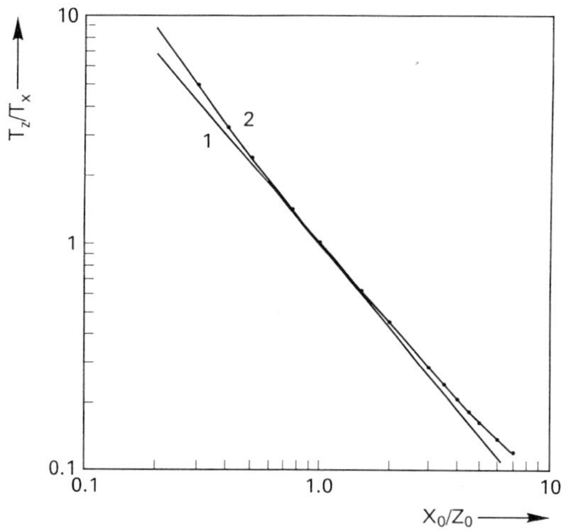

Fig. 2. Dependence of the ratio T_z/T_x on the ratio X_0/Z_0 for a circular laser spot as determined numerically.

The *elliptical temperatures* are parameters describing the direction-dependent expansion velocity of the cloud for the 3D expansion. *It must be emphasized that they should not be interpreted as real physical temperatures.* It is found that T_{xi} is linear in T_0 and independent of m. Furthermore, it can be proven that

$$\Sigma T_{xi} = [2/(\gamma - 1)] T_0. \tag{3}$$

The elliptical temperatures are found to be strongly dependent on the dimensions of the gas cloud prior to 3D expansion (called X_i: $i = 1, 2, 3$). Fig. 2 shows T_z/T_x as a function of X_0/Z_0 for a circular laser spot. For dimensions of the same order of magnitude, this ratio scales approximately as:

$$T_z/T_x \simeq (X_0/Z_0)^{1.2}. \tag{4}$$

For a non-circular laser spot, the ratio T_y/T_x is plotted as a function of the aspect ratio $A = Y_0/X_0$ for a few values of X_0 in fig. 3. The behavior shown in fig. 3 is typical. The parameters X_0 and Y_0 can be evaluated straightforwardly as the halves of the widths of a gaussian laser spot intensity profile on the surface, and can be measured directly.

The evaluation of Z_0 is less simple, Z_0 is a measure for the dimensions in the z-direction of the gas cloud at the onset of 3D expansion.

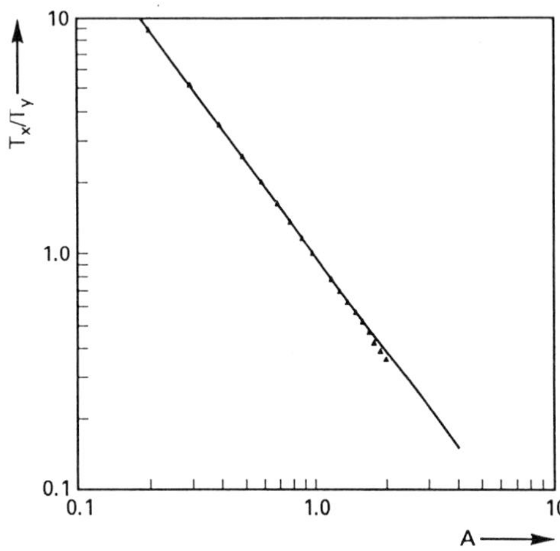

Fig. 3. Dependence of the ratio T_y/T_x on the aspect ratio A for an asymmetric laser spot as determined numerically.

Values of about 400–500 μm are found to give a reasonable prediction of the experimental data presented [5]. A gas cloud with a typical expansion velocity of 10^3 m/s will have such dimensions after 400 ns. This is typically the time-scale on which 3D expansion has become important [8]. Further work, both experimental and theoretical, is required to investigate the dependence of Z_0 on experimental parameters such as the surface temperature, the particle mass, or the density of the particles.

For a circular spot, it is possible to combine eqs. (1)–(4) to

$$p = 3.9\left(1 + 2\left(\frac{Z_0}{X_0}\right)^{1.2}\right)^{3/4}\left(\frac{X_0}{Z_0}\right)^{1.2}. \tag{5}$$

The dependence of angular intensity distributions $I(\vartheta)$ on the experimental conditions in a deposition experiment as predicted by the model will be discussed now.

(1) Atomic mass: The formalism described above applies in principle to the expansion of a gas consisting of one type of atom. If Z_0 has no dependency on the atomic mass, it follows from eq. (4) that T_φ/T_z is independent of the mass. It is then found from eq. (3) that T_z/T_0 must be independent of the mass, and thus from eq. (2) that mu^2/T_z is also independent of the mass and finally that $I(\vartheta)$ is independent of m.

(2) Laser fluence: Provided the etch yield is sufficiently high for a well developed expansion to occur, and the laser fluence F is not so high that plasma effects disturb the neutral atom flow, the only parameter that depends on the fluence is the surface temperature T_s. From eq. (2) it can be seen that u^2 is linear in T_0, just like T_{xi}. Once again, this condition is only fulfilled if Z_0 is independent of T_0. Their respective ratios are then independent of T_0 and thus of T_s. In conclusion, it is found (because of eq. (1)) that the functional form of $I(\vartheta)$ is independent of F.

(3) Spot dimensions: The elliptical temperatures are sensitive functions of the initial lateral dimensions of the gas cloud (X_0, Y_0). These dimensions are dictated by the dimensions of the laser spot. For a circular spot with $X_0 = Y_0$ of the same order as Z_0, it is possible to calculate $I(\vartheta)$ from eq. (5). For an asymmetric spot, it is possible to predict the ratio of the exponents p in the cosine power law from fig. 3.

4. Comparison to experimental deposition patterns

In this section, the predictions of the hydrodynamical model formulated above will be compared to angular intensity distributions $I(\vartheta)$ reported in the literature. These angular intensity distributions are measured by evaluation of the thickness profile of the deposited film. The sticking coefficient of the atoms arriving on the substrate is assumed to be constant for all species under all conditions. Since (1) the sticking coefficient is in general a function of velocity and polar angle of impact, and (2) the velocity of the incoming particles is strongly dependent on the polar angle this simplification might a priori be a source of discrepancies.

(1) Atom mass: The formalism applies in principle to the expansion of a cloud of a gas consisting of one type of atom. If one assumes Z_0 to be independent of the mass of the atom, it follows that $I(\vartheta)$ is independent of the mass. Gorbunov and Konov [9] found indeed that ablation of gold and graphite yields the same angular distribution. Akhsakhalyan et al. [10] investigated a large se-

ries of elemental solids for a much higher power regime ($\geqslant 10^9$ W cm^{-2}). In contradiction to the previous authors, it was claimed there that the angular distribution has a dependence on the mass of the atom, although the scatter between data points is considerable. Elements with a low ionization energy (Bi, Pb) are found to give different angular distributions, probably due to plasma formation effects at these high fluences.

The situation is somewhat more complex for the ablation of a multicomponent solid. The stream velocities of several components of such a target are somewhere in between equality (all types of atoms move in a single gas cloud) and scaling with the square root of the mass $u \propto m^{-1/2}$ (all types of atoms are in different gas clouds). A scaling behavior of the stream velocity implies independency of the angular distribution on the mass, as has been described for the ablation of CdS [11] and $SmBa_2Cu_3O_7$ [12].

(2) Laser fluence: The angular distribution is predicted to be independent of the fluence. A small second-order effect is expected, since a higher fluence causes the ablation threshold to be exceeded for a larger fraction of the spot. Such a weak dependency of the angular distribution on the laser fluence is expected, and indeed found experimentally [9,12,13].

(3) Laser spot dimensions: As mentioned above, the elliptical temperatures, and therefore $I(\vartheta)$ are sensitive functions of the initial dimensions of the cloud, i.e. the geometry of the laser spot. For a circular spot, it is possible to predict the angular intensity distribution from the spot radius X_0, using eq. (5), provided Z_0 is known. If the angular distribution is measured for a series of radii, Z_0 can be fitted by plotting p versus X_0/Z_0. This is done in fig. 4 for a series of angular distributions measured for W-ablation by Akhsakhalyan et al. [10]. The experimental data can be fitted fairly well. The value of Z_0 is 350 μm, comparable to the values obtained earlier [5] in ARTOF measurements (400 and 500 μm). Since the three fit values obtained for Z_0 are comparable (350, 400 and 500 μm), it is reasonable to suppose Z_0 to be a constant quantity. This allows the prediction of angular distributions of atoms ablated by circular laser spots. Table 1

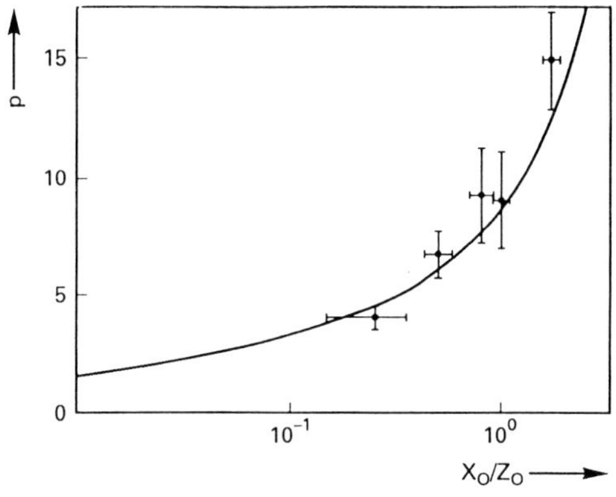

Fig. 4. Comparison between measured [33] powers p in cosine power law angular distributions for Nd:YAG ablation of W with a varying laser spot size with the predictions based on eq. (5) and a value $Z_0 = 350$ μm.

compares this prediction to a few experimental values of p, taking $Z_0 = 400$ μm. Considering the (lack of) accuracy of the input data, the agreement is found to be quite good for these experiments. Cheenne et al. [13] determined the dependence of the angular distribution on the laser spot size at fixed fluence for the case of excimer ablation of high-T_c superconductors. Increasing the laser spot from 1 to 8 mm^2 caused a change from $p = 2.5$ to $p = 5.5$. Substituting $Z_0 = 2.2$ mm in eq. (5) gives $p = 2.6$ and $p = 6.0$. It is thus found that this case requires a somewhat larger value for Z_0. It is not clear what is responsible for this effect.

For an asymmetric laser spot, the behavior is more complex: the azimuth-dependent angular intensity distribution is determined by the aspect ratio A. As was noticed experimentally [16], there

Table 1
Comparison between the angular distribution (5) and some literature results

Ref.	Target	X_0 (mm)	p_{cal}	p_{meas}
[9]	Graphite	0.22	6.4	8
	Gold	0.22	6.4	8
[14]	YBaCuO	1.2	16	$\geqslant 11$
[15]	TiN	0.3	7.0	6

The exponent in the cosine power law is calculated from the spot dimensions as given in the paper, and the assumptions $M = 2.33$ and $Z_0 = 0.4$ mm.

is a "flip-over": the longest direction of the laser spot results in the shortest direction of the deposition pattern, and vice versa. Afonso et al. [16] have measured the azimuthal dependence of the peaking of a Ge ablation cloud ablated by a spot with $A = 2.5$. They find a ratio of 7.5 for the power factor in the cosine law for x and y directions. Fig. 3, combined with eq. (1), predicts a ratio of 4.5.

5. Conclusions and predictions towards deposition practice

From the extensive discussion of the deposition pattern given above, it must be clear that the formalism presented above gives a fairly good description of angular intensity distributions $I(\vartheta)$. A number of "rules of thumb" relevant to deposition practice will be presented now:

Laser spot geometry: The analysis given above is for a Gaussian elliptical gas cloud generated by an elliptical laser spot. In practice significant inhomogeneities in the laser spot may occur. The analysis given above does no longer apply then, but it is clear that the same gas expansion effects play a role. The homogeneity of the deposition geometry is thus directly related to the homogeneity of the laser spot. This is clearly illustrated in fig. 5. This figure shows the thickness profiles (visible as interference rings) of two BN films deposited on Si(100). Film a is deposited using an elliptical Gaussian laser spot, resulting in a smooth elliptical deposition pattern. Film b is deposited using an asymmetric laser spot, resulting in an asymmetric deposition pattern.

For a homogeneous laser spot, it is possible to manipulate the distribution pattern in a predictable way using the spot dimensions. For a given set of dimensions $(2X_0, 2Y_0)$, the deposition pattern can then be estimated using eqs. (1) and (5) and figs. 2 and 3.

Other experimental parameters: The fact that the angular intensity distribution is in principle independent of the other experimental parameters (laser fluence and atomic mass), makes the deposition uniformity stable towards small instabilities in laser fluence. Another conclusion is that changing the target does not result in another deposition uniformity, a feature which is important when multilayers have to be deposited.

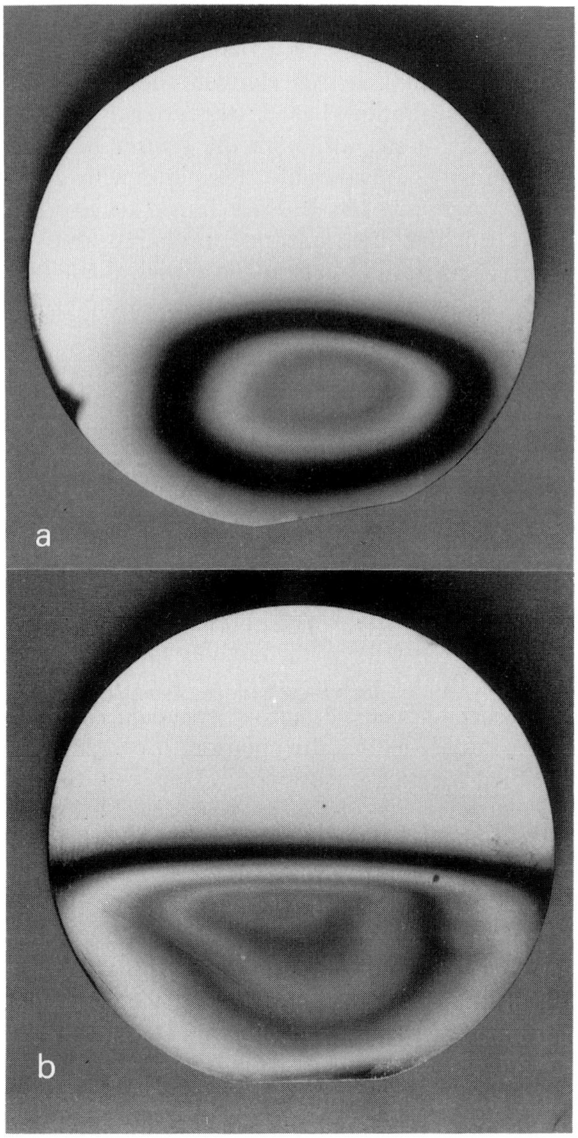

Fig. 5. Comparison between the deposition uniformity (visible as interference rings) of two BN films deposited on Si(100) using (a) an elliptical laser spot and (b) an irregularly shaped laser spot.

As a last remark, it is important to note that the present analysis does not include the effects of laser-induced target surface morphology, since the target is assumed to be flat. The development of laser-induced surface roughness upon prolonged irradiation will cause the depositing flux to be directed along the local target normal, which does not have to coincide with the macroscopic target normal.

References

[1] MRS Bull. XVII (2) (February 1992). Special issue on PLD/LAD.

[2] T.S. Baller, G.N.A. van Veen and H. van Hal, Appl. Phys. A 46 (1988) 215;
D. Dijkkamp, T. Venkatesan, X.D. Wu, S.A. Shaheen, N. Jisrawi, Y.H. Min-Lee, W.L. McLean and M. Croft, Appl. Phys. Lett. 51 (1987) 619.

[3] S.R. Foltyn, R.E. Muenchausen, R.C. Dye, X.D. Wu, L. Luo, D.W. Cooke and R.C. Taber, Appl. Phys. Lett. 59 (1991) 1374.

[4] J. Greer and H.J. van Hook, MRS Proc. 191 (1990) 171.

[5] J.C.S. Kools, PhD Thesis, Eindhoven University of Technology, Eindhoven, 1992.

[6] J.C.S. Kools, T.S. Baller, S.T. De Zwart and J. Dieleman, J. Appl. Phys. 71 (1992) 4547.

[7] A. Gupta, B. Braren, K.G. Casey, B.W. Husey and R. Kelly, Appl. Phys. Lett. 59 (1991) 1302.

[8] R. Gilgenbach and P.L. Ventzek, Appl. Phys. Lett. 58 (1991) 1597.

[9] A.A. Gorbunov and V.I. Konov, Sov. Phys. Tech. Phys. 34 (1989) 1271.

[10] A.D. Akhsakhalyan, S.V. Gaponov, V.I. Luchin and A.P. Chirimanov, Sov. Phys. Tech. Phys. 33 (1988) 1146.

[11] A. Namiki, T. Kawai and K. Ichige, Surf. Sci. 166 (1986) 129.

[12] R.A. Neifeld, S. Gunapala, C. Liang, S.A. Shaheen, M. Croft, J. Price, D. Simons and W.T. Hill, Appl. Phys. Lett. 54 (1988) 954.

[13] A. Cheenne, J. Perriere, F. Kerherve, G. Hauchecorne, E. Fogarassy and C. Fuchs, MRS Proc. 191 (1990) 229.

[14] T. Venkatesan, X.D. Wu, A. Inam and J.B. Wachtman, Appl. Phys. Lett. 52 (1988) 1193.

[15] J.C.S Kools, C.J.C. Nillesen, S.H. Brongersma, E. van de Riet and J. Dieleman, J. Vac. Sci. Technol. A 10 (1992) 1809.

[16] C.N. Afonso, R. Serna, F. Catalina and D. Bermejo, Appl. Surf. Sci. 46 (1990) 249.

Applied Surface Science 69 (1993) 140–148
North-Holland

applied
surface science

Pulsed laser deposition and processing of biocompatible hydroxylapatite thin films

C.M. Cotell

Surface Modification Branch, Naval Research Laboratory, Washington, DC 20375-5000, USA

Received 2 June 1992; accepted for publication 30 October 1992

Calcium phosphate-based ceramics were deposited on Ti–6Al–4V and Si substrates by pulsed laser deposition. The calcium phosphate phase (amorphous, α- or β-tricalcium phosphate, tetracalcium phosphate, or hydroxylapatite), Ca/P ratio and surface morphology of the films were related to deposition conditions. Deposition of crystalline films required both elevated substrate temperature and a reactive gas environment. At substrate temperatures < 400°C, films were amorphous. Above 400°C, the gaseous environment during deposition determined the crystalline phase of the films. In Ar or in vacuum at 500–600°C, films were amorphous. In O_2, deposition of α-tricalcium phosphate was favored between 400 and 700°C and β-tricalcium phosphate was favored above 700°C. Films deposited in low pressures of O_2 (< 0.05 Torr) were P-deficient. The correct stoichiometry was approached as the pressure of O_2 in the chamber during deposition increased. In water vapor-enriched inert gas environments, deposition of hydroxylapatite was observed at temperatures between 400 and 700°C and tetracalcium phosphate was observed at temperatures above 700°C. Adhesion of crystalline HA films deposited at 600°C in Ar/H_2O was excellent, but films that were amorphous (deposited either at room temperature or in a non-reactive gas environment) and tricalcium phosphate films did not adhere well to the substrates.

1. Introduction

Calcium phosphate-based bioceramics have been used in non-load-bearing medical and dental implant devices for applications in which the characteristics of natural bone are desirable. However, use of these materials for orthopedic devices that must withstand substantial forces during their expected lifetimes has been limited by the low tensile strength of these ceramics. Consequently most orthopedic implant devices comprise metals, usually corrosion resistant alloys such as Ti–6Al–4V. While the mechanical integrity of metal implant devices exceeds that of ceramics, corrosion products from these devices give rise to tissue irritation and inflammation. One suggested method to overcome this deficiency is to coat a high strength metal with a biocompatible calcium phosphate-based ceramic. Calcium hydroxylapatite ($Ca_{10}(PO_4)_6(OH)_2$), or HA, is the most stable calcium phosphate in contact with body fluids; therefore, research directed toward thin film deposition of biocompatible ceramics has focused on HA. Deposition methods for HA that have been reported in the literature include sputtering [1], plasma-spraying [2], electrophoretic deposition [3] and combinations of these methods with post-deposition treatments to reconstitute crystalline HA from as-deposited films that are frequently partially or wholly amorphous and/or dehydrated. Recently, pulsed laser deposition (PLD) has been reported to produce films of pure, crystalline HA that are adherent to Ti–6Al–4V substrates at loads in excess of the yield strength of Ti–6Al–4V [4]. It is the purpose of this paper to review what is known about PLD of HA including the relationship between PLD processing parameters and the characteristics of calcium phosphate-based films deposited.

Elsevier Science Publishers B.V.

2. Experimental procedures

Targets of HA were produced from a suspension of HA in a phosphate buffer supplied by the Sigma Chemical Company. The suspension was dried overnight in a vacuum oven at 100°C and the resulting powder was pressed at 45 000 psi into 0.75 inch diameter pellets. Each pellet was sintered in flowing oxygen at 550°C for one hour and slowly cooled (1°/min) to room temperature. Chemical analysis by RBS showed a P-excess in the pressed, sintered pellets; the atomic ratio of Ca/P was 1.5, less than the expected 1.7. The P-excess may have resulted from contamination of the dried powder with residual phosphate salt from the buffer. Some films were deposited from pellets pressed and sintered in the same way using HA supplied by Integrated Separation Systems. This HA was in the form of dry powder of two different particle sizes: 40 μm and 60–100 μm.

The PLD apparatus has been described in detail previously [5]. Briefly, as shown in fig. 1, the deposition chamber was a stainless steel vacuum chamber, which could be evacuated to ~ 10^{-6} Torr. Prior to each deposition, the chamber was equilibrated with the chosen ambient gas at the desired pressure. Depositions were conducted in pure oxygen at pressures between 10^{-6} and 0.3 Torr or in water vapor-enriched gaseous environments produced by bubbling O_2, He, Ar, or Kr gas at a rate of 10 sccm through a water bath at room temperature. The total pressure in the chamber was held at a dynamic equilibrium pressure of 0.3 Torr. The substrate was positioned on a stage normal to the target pellet at a distance of ~ 4 cm. The stage was heated by means of a quartz halogen projector lamp and substrates were heated either conductively in the case of Si substrates by affixing the substrate to the holder with silver paint, or radiatively in holders machined to hold the Ti–6A1–4V substrates. The substrate temperature was monitored with a thermocouple embedded in the heated substrate stage mount for Si substrates or inserted into a hole drilled through the side of Ti–6Al–4V substrates. Films were deposited at substrate temperatures from room temperature to 800°C. The equilibration procedure resulted in the growth of a thin layer of oxide on the surfaces of the substrates prior to deposition of the laser ablated film. Once the temperature and gas environment had been equilibrated (~ 5 min), the beam from a KrF (248 nm) excimer laser operating at 20 Hz with an energy of ~ 200 mJ/pulse was focused on the pellet which was rotated such that the laser beam was effectively rastered over its surface to an energy density of ~ 2 J/cm^2 at a 45° angle of incidence. The number of laser pulses determined the desired film thickness which ranged

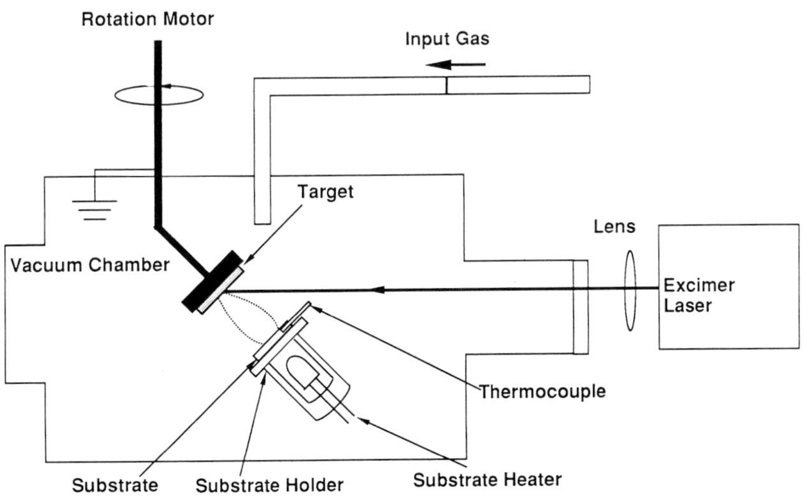

Fig. 1. Schematic diagram of the pulsed laser deposition equipment used to deposit thin films of hydroxylapatite.

from 0.05–5 μm deposited at a rate of $\sim$0.02 nm/laser shot.

Films were characterized by X-ray diffraction (XRD), scanning electron microscopy (SEM), and Rutherford backscattering spectroscopy (RBS). The adhesion of films to substrates was studied by means of scratch tests.

3. Results

As has been reported earlier [4,6], the temperature and the gas environment during deposition control the chemical and phase composition of films grown by PLD from HA targets. Deposition at temperatures less than 400°C in all the gas environments studied to date has resulted in amorphous films [4]. At temperatures greater than 400°C, films were either crystalline or amorphous according to the gas environment. Fig. 2 shows the effect of gaseous environment at constant temperature on the crystal phase of the film. The XRD patterns in fig. 2 were recorded from films that were deposited on Ti–6Al–4V at 500°C in pure Ar (fig. 2a), pure O_2 (fig. 2b), Ar/H_2O (fig. 2c) and O_2/H_2O (fig. 2d). The films are identified as amorphous, crystalline α-tricalcium phosphate, crystalline HA and crystalline α-tricalcium phosphate mixed with crystalline HA and some amorphous material, respectively. Fig. 3 is an

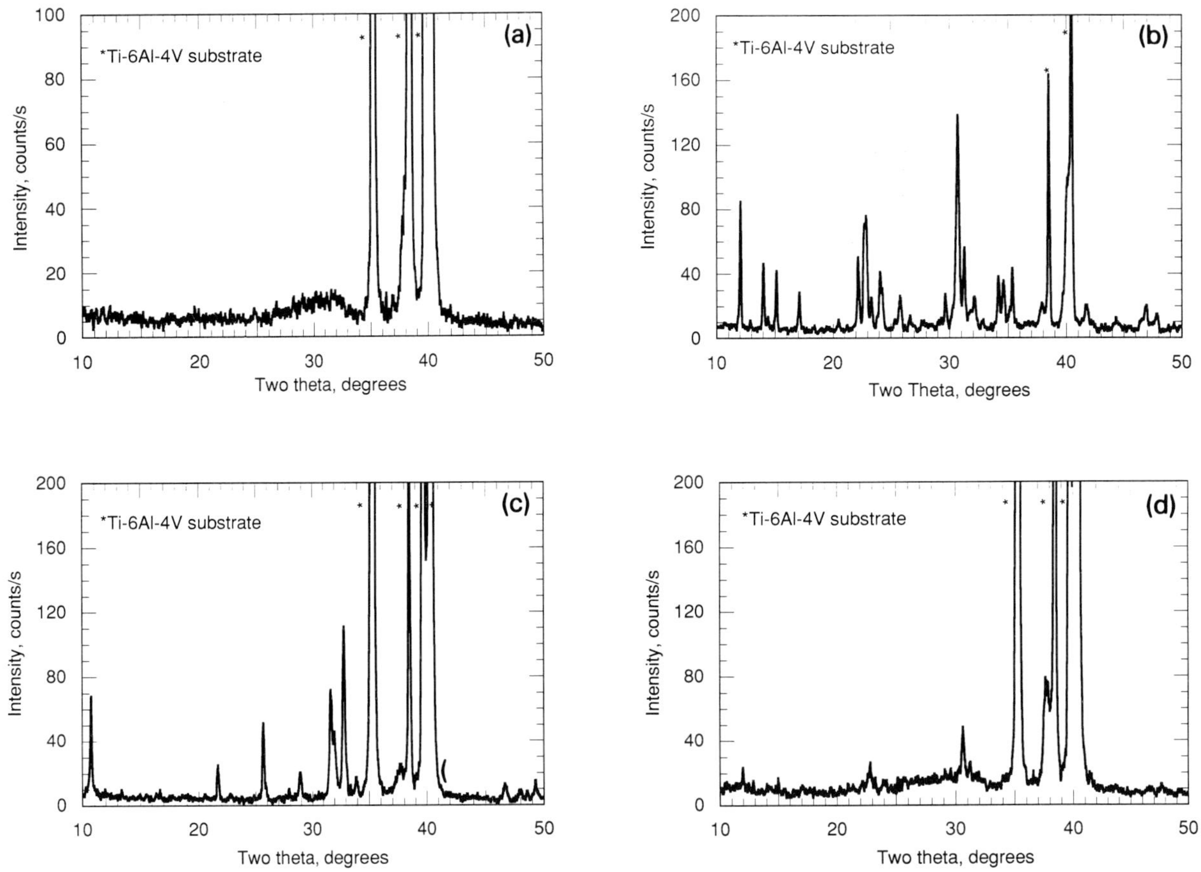

Fig. 2. X-ray diffraction patterns from thin films deposited by PLD on Ti–6Al–4V substrates at 500°C from a target of HA in (a) Ar, (b) O_2, (c) Ar/H_2O, and (d) O_2/H_2O. The film in (a) was amorphous; that in (b) was crystalline α-TCP; that in (c) was crystalline HA; and that in (d) was a mixture of crystalline α-TCP, crystalline HA and amorphous material.

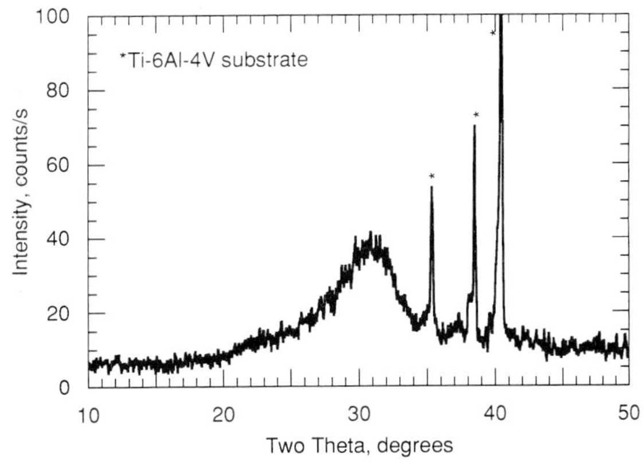

Fig. 3. X-ray diffraction pattern from a film deposited in vacuum by PLD on Ti–6Al–4V at 600°C from a target of HA. The film was amorphous.

XRD pattern from a film deposited on Ti–6Al–4V at 600°C under vacuum ($\sim 10^{-6}$ Torr). This film was amorphous.

The effect of temperature on the crystal phase of films grown in Ar/H_2O and O_2 is shown in fig. 4. Figs. 4a–4c are XRD patterns from films deposited in Ar/H_2O at 400, 600, and 800°C, respectively, and figs. 4d–4f are XRD patterns from films deposited in pure O_2 at these same temperatures. The films in figs. 4a–4c were identified as amorphous material mixed with crystalline HA, crystalline HA, and crystalline HA mixed with tetracalcium phosphate, respectively, while those in figs. 4d–4f were identified as amorphous, α-tricalcium phosphate, and α-tricalcium phosphate mixed with β-tricalcium phosphate and HA, respectively. The intensity and signal-to-noise ratios of the XRD patterns in figs. 2 and 4 are low because the films were quite thin ($\sim$ 100 nm). Table 1 summarizes the conditions that favor the deposition of different calcium phosphate phases by PLD, based upon the results presented here and in refs. [4,5].

X-ray diffraction patterns from crystalline HA films deposited by PLD (e.g., figs. 2b and 4b) showed a strong preferential (002) orientation of the films. Characterization by SEM and TEM of films deposited at 500°C in Ar/H_2O has shown that the microstructure consisted of porous particulates, 1–3 μm in size, on the surface of a columnar film [4]. Figs. 5 and 6 show the relationship between the density of particulates on the surface and the degree of (002) orientation of the films. In fig. 5, XRD patterns from two films deposited at 500°C in He/H_2O and Kr/H_2O are compared. The (002) orientation is much stronger in the film deposited in Kr/H_2O. Fig. 6 shows the surface morphology of these films; the film deposited in Kr/H_2O showed fewer particulates. It is concluded, therefore, that the (002) orientation results from the columnar part of the film while the other other peaks in the XRD pattern result from the random orientation of the particulates on the surface.

The Ca/P ratio in the films increased as the oxygen content of the gas decreased [6]. The Ca/P ratios of films deposited in Ar/H_2O varied considerably from the 1.7 ratio of Ca_{10}-$(PO_4)_6(OH)_2$, even for those films that were identified as crystalline HA by XRD. Amorphous films deposited at room temperature were generally Ca-deficient ($1.2 < Ca/P < 1.6$). Crystalline HA films deposited between 400 and 700°C had Ca/P ratios ranging from 1.6 to 3.6. A very general trend toward higher Ca/P ratios at higher substrate temperatures was observed, but even at

Table 1
Favorable conditions of substrate temperature and gaseous environment during PLD from a target of HA for the deposition of specific phases of calcium phosphate-based ceramics

Desired phase	Substrate temperature and gas environment
Amorphous "HA"	$T_{sub} < 400°C$, any gas; $T_{sub} \leq 600°C$, inert gases or vacuum
Crystalline HA	$400°C < T_{sub} < 700°C$, inert gas/water vapor
α-Tricalcium phosphate	$400°C < T_{sub} < 700°C$, O_2
β-Tricalcium phosphate	$T_{sub} > 700°C$, O_2
Tetracalcium phosphate	$T_{sub} > 700°C$, inert gas/water vapor

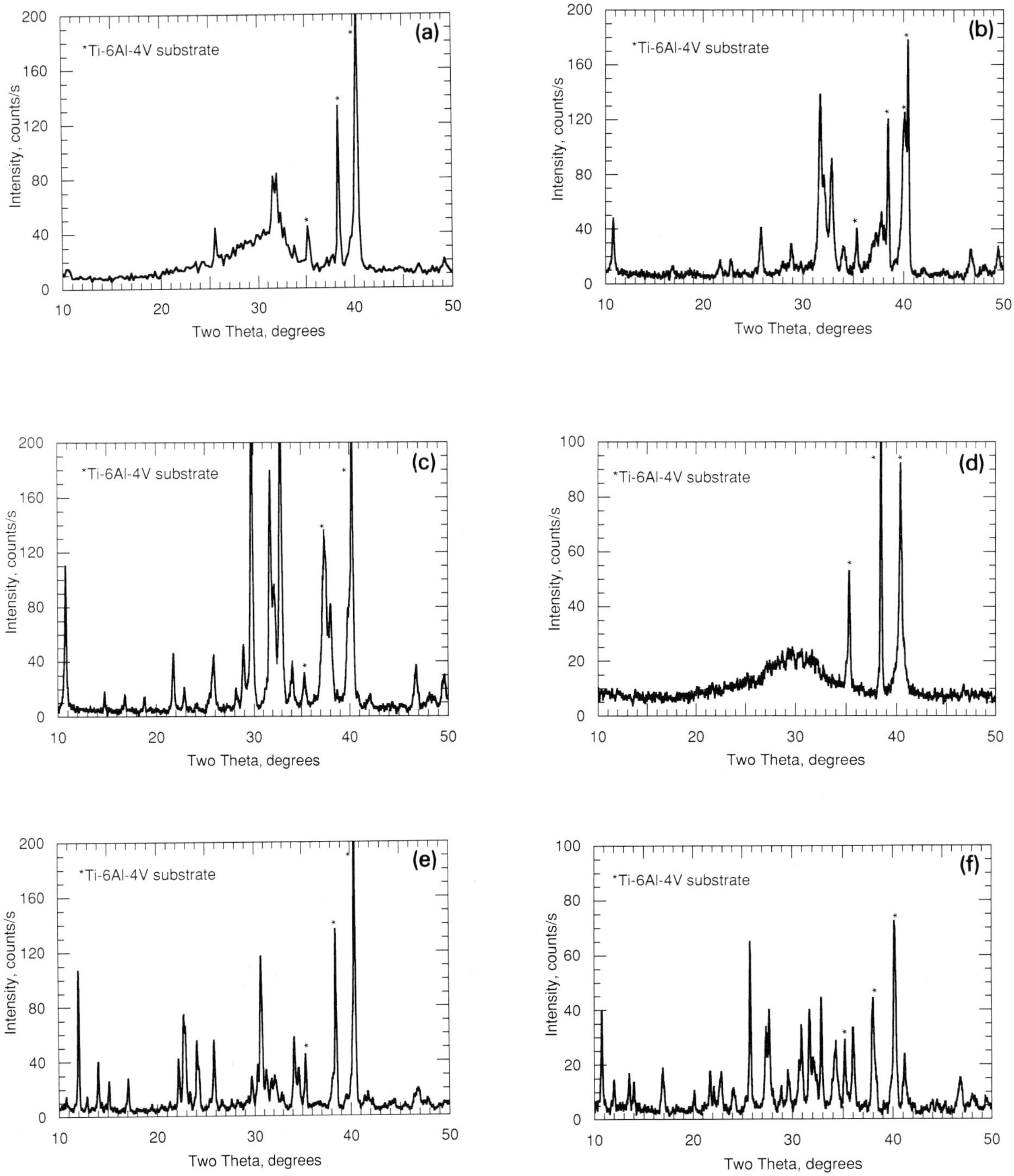

Fig. 4. X-ray diffraction patterns from thin films deposited by PLD from a target of HA on Ti–6Al–4V substrates in Ar/H_2O at (a) 400, (b) 600 and (c) 800°C, and in O_2 at (d) 400, (e) 600 and (e) 800°C. The film in (a) was a mixture of amorphous material and crystalline HA; that in (b) was crystalline HA; and that in (c) was a mixture of crystalline HA and crystalline tetracalcium phosphate. The film in (d) was amorphous; that in (e) was crystalline α-TCP; and that in (f) was a crystalline mixture of α-TCP, β-TCP and HA.

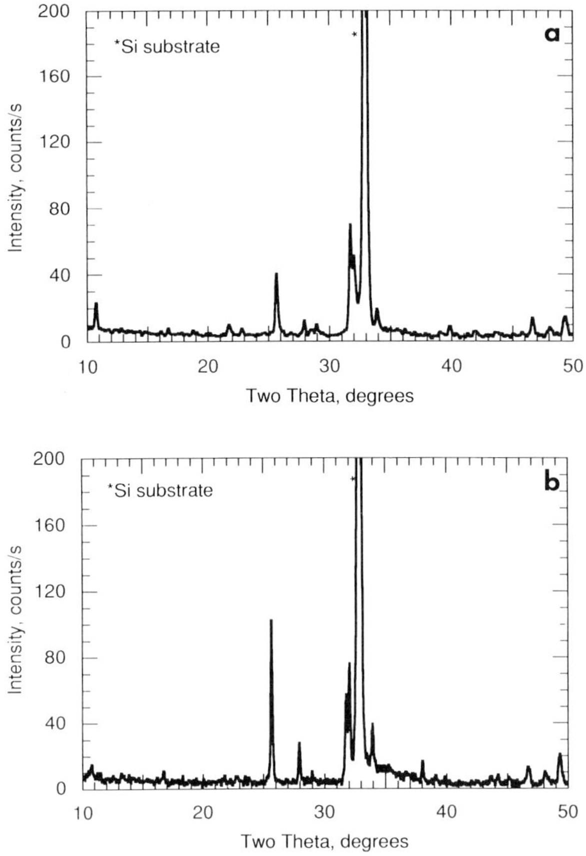

Fig. 5. X-ray diffraction patterns from thin films deposited by PLD from a target of HA on Si substrates at 500°C in (a) He/H_2O and (b) Kr/H_2O. The film in (b) shows a stronger (002) orientation.

one substrate temperature, variations in Ca/P ratio of as much as 60% were observed.

The adhesion of crystalline HA films deposited on Ti–6A1–AV substrates in Ar/H_2O at 600°C has been shown by means of scratch tests to exceed the yield strength of Ti–6Al–4V [4]. However, HA films deposited at 800°C in Ar/H_2O were not as adherent, presumably due to the presence of a titanium oxide at the interface between the substrate and the HA film [4]. The same scratch tests were performed on films deposited under conditions that resulted in calcium phosphate phases other than crystalline HA. The load on a 0.125 inch spherical stylus was linearly increased from 0–200 N over a scratch length of 5 mm. Fig. 7 shows SEM photomicrographs of the scratch marks left on the surface of an amorphous film deposited from a target of HA on Ti–6Al–4V at room temperature in Ar/H_2O. Fig. 7a shows the entire scratch length of two scratches while fig. 7b shows the beginning of one of the scratches. It is clear from fig. 7b that the

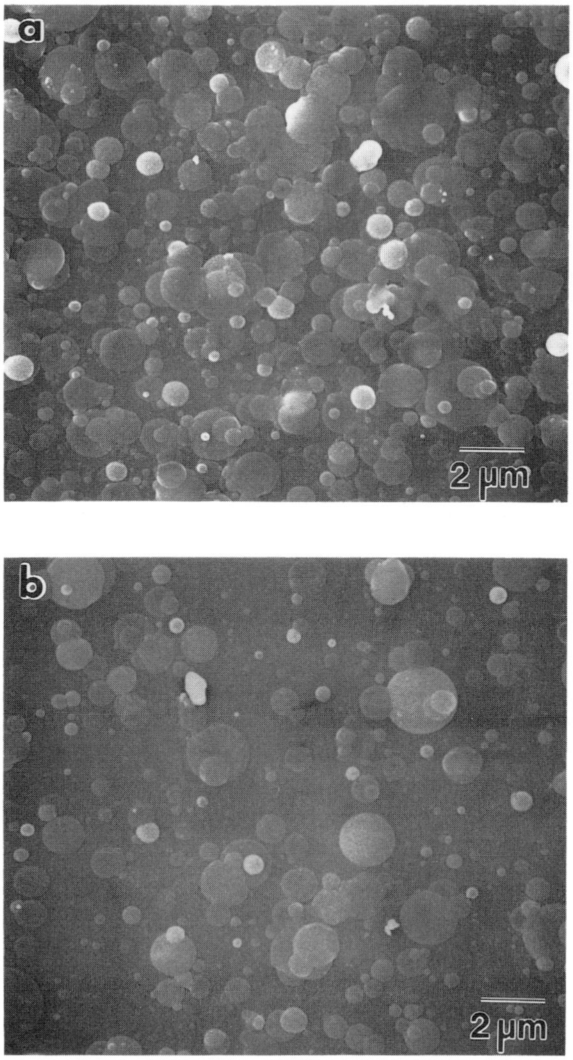

Fig. 6. SEM photomicrographs of the surfaces of thin films of HA deposited by PLD on Si substrates at 500°C in (a) He/H_2O and (b) Kr/H_2O. The XRD patterns from these films are shown in fig. 5. The film in (b) has fewer particulates on its surface.

film failed immediately. Very similar results, shown in fig. 8, were found for amorphous films deposited on Ti–6Al–4V at 600°C in Ar (fig. 8a) and at room temperature in O_2 (fig. 8b). The films failed immediately. Films deposited at elevated temperatures in O_2 also showed poor adhesion to Ti–6Al–4V substrates.

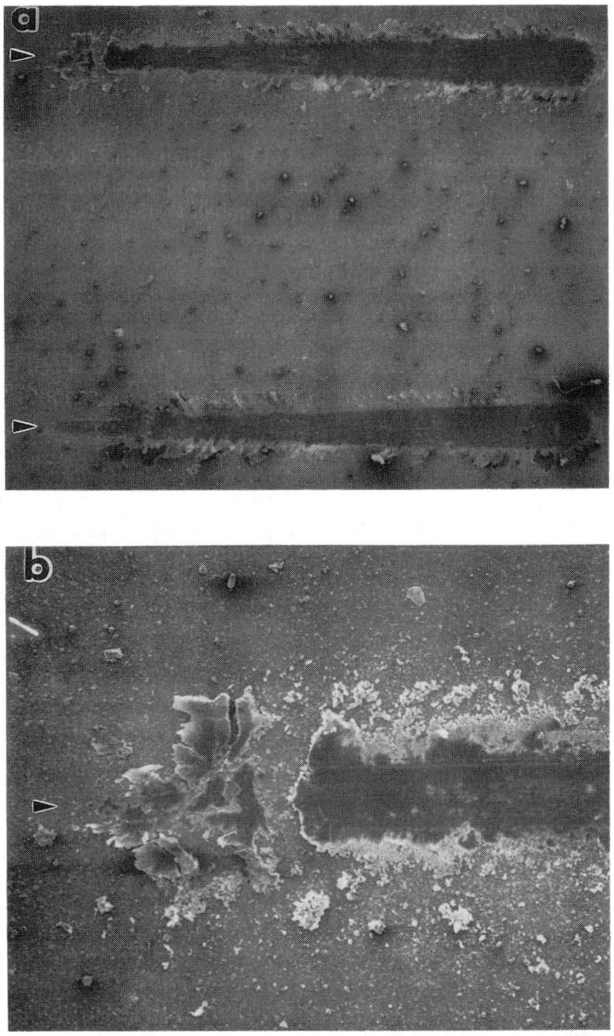

Fig. 7. SEM photomicrographs of the scratches left on the surface of an amorphous film deposited from a target of HA on Ti–6Al–4V at room temperature in Ar/H_2O. The scratches were produced by linearly increasing the load on a 0.125 inch spherical stylus from 0–200 N over a scratch length of 5 mm. Arrows indicate the direction of motion of the stylus. The entire length of the scratches is shown in (a) while the beginning of one of the scratches is shown in (b). Failure was immediate.

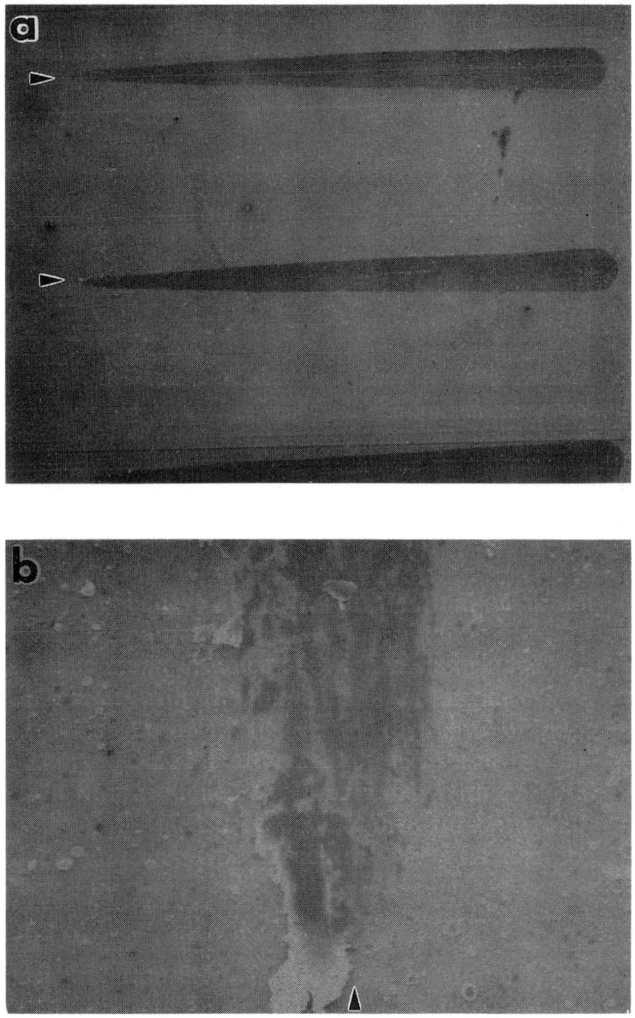

Fig. 8. SEM photomicrographs of the scratches left on the surface of an amorphous film deposited from a target of HA on Ti–6A1–4V at 600°C in Ar. The scratches were produced by linearly increasing the load on a 0.125 inch spherical stylus from 0–200 N over a scratch length of 5 mm. Arrows indicate the direction of motion of the stylus. The entire length of the scratches is shown in (a) while the beginning of one of the scratches is shown in (b). Failure was immediate.

4. Discussion

These results show that there are recognizable relationships between PLD processing parameters and film characteristics. The phase deposited depends on the substrate temperature and the gas environment during deposition according to the data in table 1. Pulsed laser deposition of crystalline calcium phosphate phases requires both an elevated substrate temperature and a reactive gas environment. Deposition in reactive gases (O_2, water vapor) at temperatures less than 400°C resulted in amorphous films as did deposition at elevated temperatures in the absence of a reactive gas species (e.g., deposition in Ar at 500°C or deposition under vacuum at 600°C). This combination of necessary conditions implies that chemical reactions are taking place on the substrate surface during the deposition process. In particular, the deposition of HA in water vapor and dehydrated HA (tricalcium phosphate) in O_2 suggests that the interaction of the laser with the HA target may produce fragments of the HA molecules that are reconstituted to HA or TCP by reaction with water vapor or oxygen on the substrate surface. Time-of-flight mass spectroscopy is presently being used to identify the species in the plume ablated from a target of HA [7]. The temperature of the substrate on which the chemical reactions with the gas take place determines the phase deposited in a given gas environment. In Ar/H_2O, HA was favored at temperatures between 400 and 700°C, and tetracalcium phosphate was found at temperatures > 700°C. In O_2, α-TCP was found between 400 and 700°C and β-TCP between 700 and 800°C.

The ability to vary the calcium phosphate phases deposited, as well as the Ca/P ratio, has significant implications for the application of these films in implanted devices. The degree of bioresorbability of calcium phosphate-based ceramics reportedly depends on the Ca/P ratio; higher Ca/P ratios lead to lower rates of resorbability [8]. Likewise, among the five phases deposited by PLD there are differences in rates of resorbability. The ability to change the phase and the Ca/P ratio in situ during a deposition by changing the substrate temperature or the oxygen content of the gas environment suggests the possibility of design of films to take advantage of the differential rates of resorbability of different calcium phosphates.

It is believed that the particulates observed on the surfaces of films deposited at lower substrate temperatures (400–500°C) result from particles of HA ablated from the target intact. The smoother films beneath the particulates (see fig. 3b) are oriented crystallographically in the (002) direction as indicated by XRD and the particulates give rise to the other diffraction peaks in the XRD patterns. The particles ablated from the target intact are much smaller than the particle size in the powders used to press the targets. These particles must result from laser/solid interactions since neither the density of the targets [4] nor the particle size of the powders used to fabricate the targets influences the size or fractional coverage of the particulates found on the film surfaces. Some influence of gas environment on the fractional coverage by particulates was observed. Heavier inert gas/water vapor mixtures produced films with lower fractional coverage by particulates.

Among the most puzzling of these results is the variation in Ca/P ratio among films that were all identified as HA by XRD. One possible explanation is the substitution of carbonate groups for phosphate groups on the HA molecule. Such a substitution has been observed in HA ceramic retrieved after implantation in animals [9,10]. After implantation, diffraction from the HA ceramic implants was not significantly different from preimplantation diffraction, but infrared spectra showed the presence of C–O functional groups. The IR bands reported were characteristic for CO_3 substituted (CO_3-for-PO_4 substitution) synthetic and biological apatites [10]. Infrared spectra from PLD HA films have shown weak C–O absorptions of somewhat greater intensity than those present in the starting powders [4]. Ratios of Ca/P in excess of that expected for HA could be due to substitution of CO_3 for PO_4 on some of the molecules comprising the films.

Adhesion of crystalline pulsed laser deposited calcium phosphate-based films on Ti–6A1–4V substrates appears to be related to the existence of an oxide at the interface between the film and the substrate. Adhesion was less good when films were deposited under more oxidizing conditions (e.g., higher temperature or higher partial pressure of oxygen in the gas environment). Amorphous films were not adherent regardless of the temperature or gas environment. This suggests there may be some effect of epitaxy, which is presently under investigation.

Acknowledgements

The contributions to this work by D.B. Chrisey, C.A. Carosella, C.C. Sheridan, H. Funkhouser and J. Fields are gratefully acknowledged.

References

[1] P. Ducheyne, W. Van Raemdonck, J.C. Heughebaert and M. Hueghebaert, Biomaterials 7 (1986) 97.

[2] S.D. Cook, J.F. Kay, K.A. Thomas, R.C. Anderson and J. Jarcho, J. Dental Res. 65 (1986) 222.

[3] E. Ruckenstein, S. Gourisanker and R.E. Baier, J. Coll. Interf. Sci. 96 (1983) 245.

[4] C.M. Cotell, D.B. Chrisey, K.S. Grabowski, J.A. Sprague and C.R. Gossett, J. Appl. Biomat. 3 (1992) 87.

[5] D.B. Chrisey, J.S. Horwitz and K.S. Grabowski, in: Laser Ablation for Materials Synthesis, Eds. D.C. Paine and J.C. Bravman, Mater. Res. Soc. Proc. 191 (1990) 25.

[6] C.M. Cotell, D.B. Chrisey and K.S. Grabowski, in: Tissue-Inducing Biomaterials, Eds. L.G. Cima, M. Flanagan and E. Ron, Mater. Res. Soc. Proc. 252 (1992) 3.

[7] C.M. Cotell, R.E. Leuchtner and D.B. Chrisey, unpublished research.

[8] K. Koster, E. Karbe, H. Heide and R. Konig, Langenbecks Arch. Chir. 341 (1976) 77.

[9] G. Daculsi, R.Z. LeGeros, M. Heughebaert and I. Barbieux, Calcif. Tissue Int. 46 (1990) 20.

[10] M. Heughebaert, R.Z. LeGeros, M. Gineste, A. Guilhem and G. Bonel, J. Biomed. Mater. Res. (Appl. Biomater.) 22 (1988) 257.

Applied Surface Science 69 (1993) 149–155
North-Holland

applied surface science

Lithium niobate films grown by excimer laser deposition

C.N. Afonso [a], F. Vega [a], J. Gonzalo [a] and C. Zaldo [b]

[a] Instituto de Optica, CSIC, Serrano 121, 28006 Madrid, Spain
[b] Instituto de Ciencia de Materiales, CSIC, Cantoblanco, C-IV, 28049 Madrid, Spain

Received 2 June 1992; accepted for publication 30 October 1992

Laser ablation of single-crystal $LiNbO_3$ targets is performed by means of an ArF laser. Films are grown onto glass, quartz and carbon-coated mica substrates held at room temperature. Amorphous and particulate-free films with the refractive index and optical energy gap close to those of bulk single-crystal $LiNbO_3$ are obtained depending on the oxygen pressure applied during deposition and the laser fluence. Optical diagnostics of the laser induced plume show that the complex dynamics followed by lithium species in the gas phase are responsible for the variation of the film refractive index as a function of the laser fluence.

1. Introduction

Lithium niobate ($LiNbO_3$) single crystals exhibit several interesting optical properties including acousto-optic, electro-optic and non-linear optical effects [1]. It is, therefore, a material very appreciated in optoelectronics technology to implement integrated optics devices [2]. The fabrication of such devices requires the deposition of high quality films on a substrate with a lower refractive index and a small lattice mismatch. Several attempts have been reported in the literature to form $LiNbO_3$ films by liquid phase epitaxy [3], molecular beam epitaxy [4], sol–gel process [5] and mainly by RF sputtering [6–10]. Although waveguiding [11] and second-harmonic generation [12] have been reported in single-crystal films deposited by RF sputtering, studies of $LiNbO_3$ sputtering [8] pointed out that a very precise control of the source material and the deposition conditions is needed in order to obtain reproducible film properties.

The successful results obtained on high T_c superconducting films prepared by pulsed laser deposition [13] accelerated the development of this field and its application to other kinds of materials. The possibility to deposit films in a reactive atmosphere and the fact that film composition reproduced approximately that of the target made this technique very suitable for deposition of complex oxide materials. The latest trends in the field include semiconductor oxides [14,15], ferroelectric [16–18] and photorefractive materials [19]. Laser ablation of solid targets is in general accompanied by a bright "plume" on the target surface. Its diagnostics has been the subject of many studies [20–24] in part because the high quality of the as-deposited superconducting films was related to the existence of gas phase reactions between the plume species and the foreign oxygen atoms [20,21]. The plume dynamics itself are important since it is known that the nature, speed and charge of the species present in the gas phase during deposition can substantially affect the film quality and properties [25].

The present work for the first time reports results of films deposited by pulsed laser ablation of lithium niobate single-crystal targets in vacuum and in an oxygen atmosphere. It aims to correlate the optical properties of the films with the dynamics of the ablation process. "In situ" measurements of the optical properties of the films, as they grow, allow us both to monitor the optical quality of the deposited material and to determine the optical constants and the thickness of the films. We have analyzed the role of the target stoichiometry ([Li]/[Nb] ratio), the oxygen pressure during deposition and the laser fluence on the film optical properties. The last two parameters may have a direct influence on the dynamics of the ablated species before they reach the substrate. We have analyzed this influence by studying the nature and velocity of the species in the gas phase by real time spectroscopy of the light emitted by the plume with space resolution.

2. Experimental

Laser ablation is carried out on $LiNbO_3$ single-crystal targets grown from melts with [Li]/[Nb] ratios between 0.94 and 1.2 which has been reported [26] to lead to crystal stoichiometries in the 0.93–0.94 range. An ArF laser (12 ns, 193 nm) working at 10 Hz is focused onto the target at 45°. Glass, quartz and carbon-coated mica substrates held at room temperature are located at 28 mm along the normal to the target surface. The experiments are carried out in a turbo-pumped system at a background pressure of 8×10^{-6} mbar where oxygen is dynamically allowed up to pressures of 10^{-1} mbar. A He–Ne laser beam ($\lambda = 632.8$ nm) is used to record in real time the evolution of the reflectivity and transmission of the substrate as the film grows.

In order to characterize the light emitted by the plume, the excimer laser beam is focused onto the target at normal incidence and the image of the plume is formed onto the entrance slit of a SPEX spectrometer (0.05 nm resolution). The emission intensity transients are recorded in a 500 MHz transient digitizer with a 15 ns rise-time photomultiplier. The plume is scanned along the target normal with 0.18 mm spatial resolution. The average energy density is calculated to an accuracy within 15% by integrating the laser energy profile over the range from its maximum to the point at which it decays to 1/10 of its maximum value. Further details of the experimental setup used for film growth and plume analysis can be found elsewhere [27,28].

The optical absorption of the films grown onto quartz substrates is determined after deposition in the 210–1300 nm wavelength interval in Perkin–Elmer (Lambda 9) and Varian (Cary 17) spectrophotometers. The optical energy gap of the films is determined from the extrapolation of the linear dependence of $(\alpha(\omega)\hbar\omega)^{1/2}$ versus $\hbar\omega$, where $\hbar\omega$ is the photon energy [29]. Some samples are annealed after deposition either in vacuum (10^{-4} mbar) or in an oxygen atmosphere (500 mbar) at temperatures ranging from 400 to 800°C. Above 700°C, a degradation of the films is observed.

The structure of the films deposited on carbon-coated mica substrates is analyzed by transmission electron microscopy (TEM). The results show that the films are amorphous, very homogeneous and particulate-free in all the cases.

3. Results

Fig. 1 shows the evolution of the light transmitted through the substrate as the film grows at different oxygen pressures. It is clearly seen that the second maximum is attenuated as the films are grown in lower oxygen pressures, which clearly shows that the optical absorption of the films increases. Pressures above 10^{-1} mbar lead to a

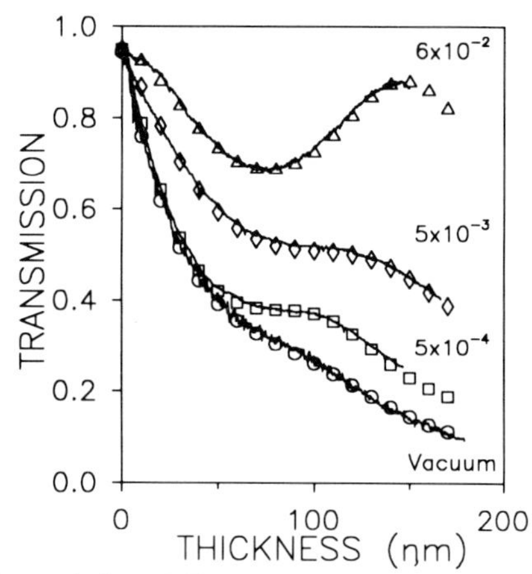

Fig. 1. Transmission of films grown in vacuum and in different oxygen pressures (in mbar) as a function of thickness. The symbols are simulations of the experimental data recorded during film growth (full lines). The plotted results correspond to films grown from a single-crystal $LiNbO_3$ target grown from a melt with a [Li]/[Nb] = 1.1 ratio and at 1.6 J/cm^2 laser fluence.

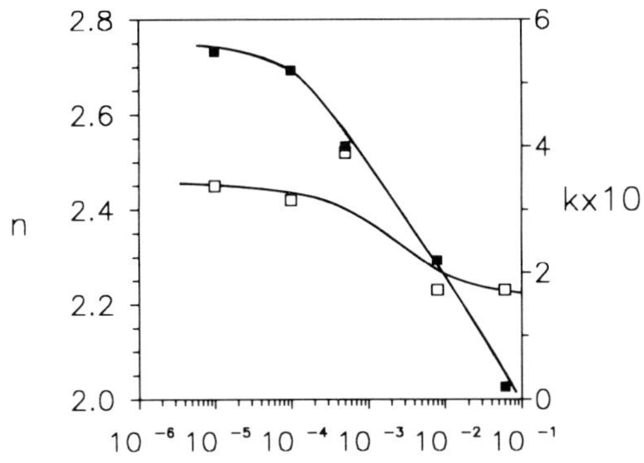

Fig. 2. Real (n,□) and imaginary (k,■) parts of the refractive index as a function of the oxygen pressure used to grow the films. The plotted results correspond to films grown from a single-crystal $LiNbO_3$ target grown from a melt with a [Li]/[Nb] = 1.1 ratio and at 1.6 J/cm^2 laser fluence.

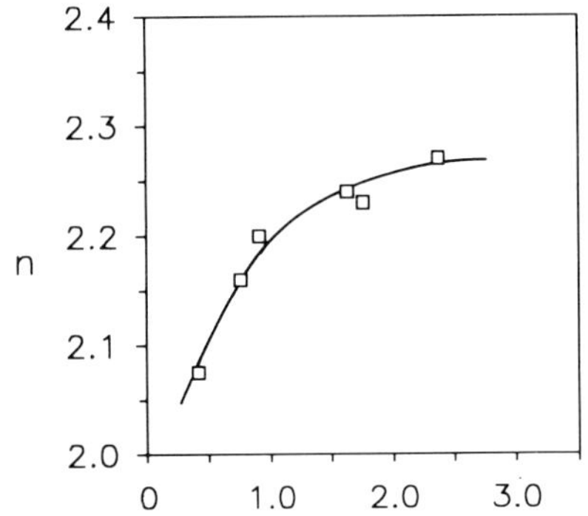

Fig. 3. Real (n) part of the refractive index as a function of the laser fluence for films grown from a single-crystal $LiNbO_3$ target grown from a melt with a [Li]/[Nb] = 1.1 ratio at an oxygen pressure of 6×10^{-2} mbar.

strong reduction in the deposition rate. Simulations of the experimental data, using a home developed computer program based on the theory of Abeles for calculating the reflectivity and transmission of an isotropic planar multilayer system [30], are also included. The simulations which are in excellent agreement with experimental results provide the film optical constants as well as the film thickness. The horizontal scale in fig. 1, which was originally a temporal scale, has been normalized to the thickness provided by the simulations in order to ease the comparison. The film optical constants obtained as a function of the oxygen pressure used during deposition are plotted in fig. 2. It is clearly seen that both real (n) and imaginary (k) parts of the refractive index decrease as the oxygen pressure is increased. Films grown at 6×10^{-2} mbar have $n = 2.23 \pm 0.02$ which is close to the ordinary refractive index of bulk single crystal $LiNbO_3$ ($n = 2.286$ [1]) and $k < 0.02$. This oxygen pressure will be hereafter selected to analyze the influence of other deposition parameters. The dependence of n on the laser fluence is shown in fig. 3. The values obtained at low fluences are below 2.20, whereas the k values remain below 0.02 in the whole studied range. Finally, films grown at a constant fluence from different targets show no variation of n within the experimental error as

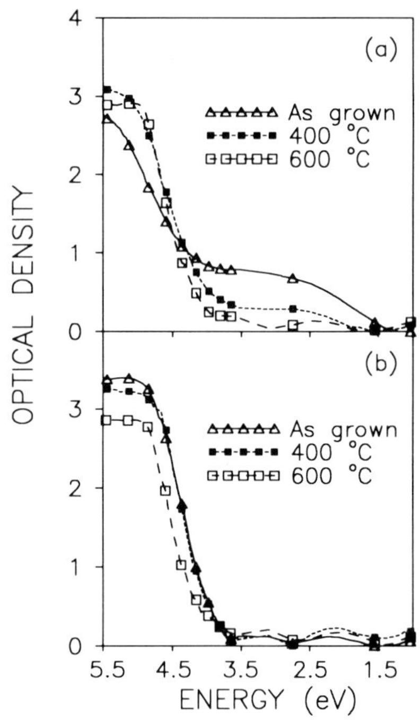

Fig. 4. Optical density as a function of the photon energy measured in a film, grown in vacuum (a) and in 6×10^{-2} mbar of oxygen pressure (b), from a single-crystal $LiNbO_3$ target grown from a melt with a [Li]/[Nb] = 1.2 ratio at 1.6 J/cm^2 laser fluence. Results for as-grown films and films after thermal annealing in 500 mbar of oxygen pressure during 2 hours at each temperature are included.

Table 1
Optical energy gap measured in as-grown films deposited in 6×10^{-2} mbar of oxygen pressure at different laser fluences, and from single-crystal $LiNbO_3$ targets grown from melts with different [Li]/[Nb] ratios

[Li]/[Nb] ratio in the melt	Fluence (J/cm^2)	Gap energy (eV)
1.1	0.4	3.60
1.1	1.6	3.58
1.1	2.4	3.60
0.94	1.63	3.63
1.0	1.63	3.62
1.1	1.63	3.58
1.2	1.63	3.60

Table 2
Optical energy gap (in eV) determined in films grown in vacuum and in an oxygen pressure of 6×10^{-2} mbar from the optical absorption plotted in fig. 4 after thermal treatment in 500 mbar of oxygen pressure

	Films grown in	
	Vacuum	Oxygen
As-grown	–	3.60
400°C	3.50	3.60
600°C	3.78	3.75

the Li/Nb melt ratio used to grow the single-crystal targets increases.

Fig. 4 shows the optical absorption of as-grown films prepared in vacuum and in oxygen atmosphere. In all cases, the absorption edge observed in the UV is sharper in the latter than in the former films. Films grown in vacuum exhibit an intense optical absorption band in the visible (fig. 3a). Upon thermal annealing to 400° C both in an oxygen atmosphere and in vacuum this band decreases and the absorption edge becomes sharper. In fig. 3b it is seen that films grown in an oxygen atmosphere show instead two weak optical absorption bands in the visible (at ~2.22 and 3.4 eV) which are very similar to those observed in the films grown in vacuum after thermal treatment. Upon thermal treatment to 600°C a shift to the UV of the optical absorption edge is also observed.

The optical energy gap values determined from the optical absorption curves following the procedure described earlier in this paper, are summarized in tables 1 and 2. The energy gap of as-deposited films grown in oxygen atmosphere is similar to that of bulk $LiNbO_3$ [31] and no significant variations are observed which depend either on the targets used or on the laser fluence (table 1). The dependence of the optical energy gap on the annealing temperature is shown in table 2.

The analysis of the plume emission in the 400–700 nm range show several lines which are mostly related to Nb and Li neutrals (NbI, LiI) [32]. For the present study, we have selected the 407.97 nm (NbI) and 607.79 nm (LiI) lines which exhibit a transient emission with a sharp peak shortly delayed with respect to the laser pulse. The delay of the maximum emission with respect to the laser pulse maximum and its intensity, as we move away from the target surface along its normal, are plotted in fig. 5. It is worth noting

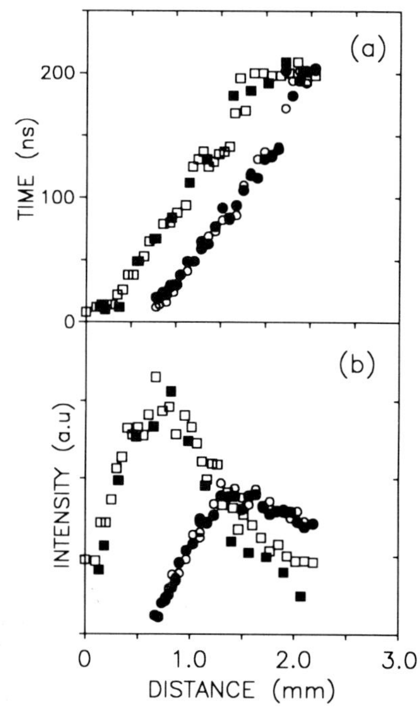

Fig. 5. Delay of the maximum transient emission with respect to the laser pulse maximum (a) and its intensity (b) as a function of the distance to the target surface. Emission from NbI (□,■) and LiI (○,●) species both in vacuum (□,○) and in 6×10^{-2} mbar of oxygen pressure (■,●) are included. The experiments were carried out on a single-crystal $LiNbO_3$ target grown from a melt with a [Li]/[Nb] = 1.1 ratio at 2.6 J/cm^2 laser fluence.

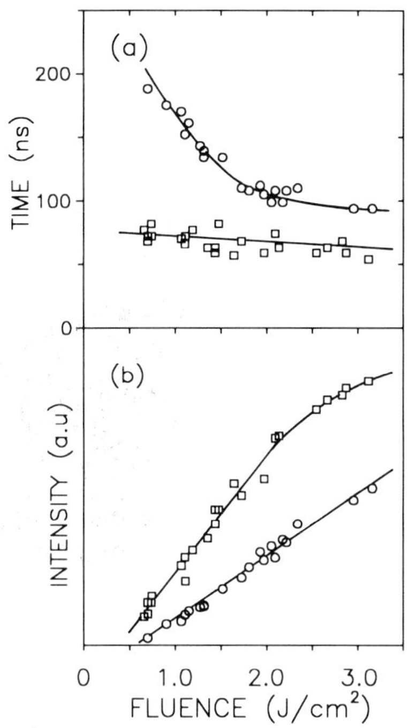

Fig. 6. Delay of the maximum transient emission in respect to the laser pulse maximum (a) and its intensity (b) as a function of the laser fluence. The NbI (□) and LiI (○) emission lines are recorded at 0.7 and at 1.5 mm from the target surface, respectively. The experiments were carried out on a single-crystal $LiNbO_3$ target grown from a melt with a [Li]/[Nb] = 1.1 ratio.

that no emission is recorded for LiI species in the neighbourhood of the target surface and that their emission occurs earlier than that of NbI species at the same distance. The mean velocities of the species, as calculated from the inverse of the slope of the data in fig. 5a, are $(9.1 \pm 0.1) \times 10^5$ cm/s and $(7.8 \pm 0.1) \times 10^5$ cm/s for NbI and LiI species, respectively. Results recorded in vacuum and at 6×10^{-2} mbar of oxygen pressure are both included and no variations are observed either in the emission maxima or in their delay with respect to the laser pulse.

The delay and intensity maximum of the transient emission of NbI and LiI species as a function of laser fluence are shown in fig. 6. The transients were recorded at those distances from the target where the maximum emission from each species occurs (see fig. 5). The delay of the emission of the NbI species decreases slightly as the fluence is increased, whereas that of the LiI species exhibits a sharp decrease as the fluence is increased (fig. 6a). The intensity of both lines (fig. 6b) increases as the fluence is increased, but at different rates, following a nearly linear behaviour in the case of LiI, whereas that of NbI tends to saturate within the studied experimental range.

4. Discussion

The refractive index and optical absorption of lithium niobate films deposited by reactive magnetron sputtering have been reported [6,8] to depend strongly both on the oxygen pressure applied during deposition and on the RF power which determines the particle flux impinging the substrate. Values of n higher than that of bulk single-crystal $LiNbO_3$ were reported in films grown under certain conditions from $LiNbO_3$ targets or from Nb targets [8]. LiO_2 enriched $LiNbO_3$ targets lead instead to films with a refractive index close to that of bulk $LiNbO_3$ [8] and to films with an [O]/[Nb] ratio equal to 3 [10]. The comparison of these results suggested that high n values were associated with a "poor" lithium content in the films. Increasing amounts of LiO_2 on $LiNbO_3$ targets [8] lead to films with n values lower than that of bulk single-crystal $LiNbO_3$ which was related to an increase of the [Li]/[Nb] ratio in the film.

From the results plotted in figs. 2–4 it can be seen that the optical properties of laser-deposited films depend strongly on the oxygen pressure applied during film deposition, where 6×10^{-2} mbar of oxygen is a critical pressure for the growth of low absorbing films. The results also show that the target stoichiometry does not significantly influence the film optical properties in the studied range. This result is not too surprising since the [Li]/[Nb] ratios involved in previous experiments [10] were higher than the ones expected [26] in the targets we have used in this work. The most significant variation observed in n depends on the laser fluence. If we take into account what was deduced from the results reported in ref. [8], the decrease of n in the low fluence regime may be related to an increase in the [Li]/[Nb] ratio. From the analysis of the dependence of the emission of LiI and NbI species on the laser fluence, it is clearly seen that the dynamics of LiI and NbI species are different. In particular, the velocity of LiI species is strongly dependent on the laser fluence. We can then conclude that the observed low *n* values must be related to the dynamics of lithium species in the gas phase which affects the incorporation of Li to the films.

The fact that no emission was experimentally recorded from LiI species for distances to the target shorter than 0.6 mm supports the idea that the observed emission comes from species which are created during the plasma expansion rather than directly ejected from the target. Several authors have reported the existence of slow velocity species or species which present two velocity populations [20,28,33]. The slow velocity component is usually related to species which are produced by recombination/collision processes in the neighbourhood of the target where the density of energetic charged species and, therefore, the collision probability is very high. In addition, molecules with a low dissociation energy were found to dissociate either by electron collisions [33] or photon excitation [34]. If Li is ejected from the target, mainly in the form of oxidized molecules or atomic metallic clusters, these species can be efficiently dissociated producing excited LiI. This process should then strongly depend on the plasma density and/or the number of photons and, therefore, on the laser fluence. Our experimental results show that, whereas species directly ejected from the target (i.e. NbI) slow down their velocity slightly (less than 15%), a 50% decrease in velocity is observed for LiI species when the fluence decreases. Further support for this picture is given by ablation experiments of Li solid targets [35] where the anomalously low density of Li atoms measured in the gas phase was related to a surface oxidation of the Li target and to a preferential ablation of molecular species.

This reasoning explains the spatial features observed in the emission of LiI species and the abrupt reduction observed in their velocity for low fluences. It is not straightforward to determine which is the real influence of this complex dynamics of lithium species in the gas phase on the final lithium content of the films. Nevertheless, some quantitative arguments can be given. The decrease in velocity of LiI species as the fluence decreases suggests that the arrival rate ratio of Nb to Li atoms at the substrate should increase and, therefore, the measured low *n* values could be related to a low lithium content. There is another effect which can also contribute to decrease the refractive index. It is very well known that ion bombardment during the film growth improves its packing density [18,25,36] and, therefore, increases the refractive index of the films [36]. The observed decrease in the kinetic energy of LiI species as the fluence decreases (a factor of 4) can also contribute to the decrease observed in *n*.

The presence of oxide molecules in ablation processes involving complex targets has been widely reported [20,21,23,24,33,34]. Although we have not detected clearly the presence of these molecules, it is very likely that they are involved in the ablation process, most probably in the ground state. The oxygen content of films deposited in vacuum should be mainly determined by the contribution of these oxidized species. When the ablation process is performed in the oxygen pressure necessary to grow optimum films, no new lines are observed in the emission spectra in the studied wavelength range and no variation is observed in the velocity of NbI and LiI species. This result allows us to conclude that the oxygen enrichment, of the film, as the oxygen pressure is increased should, occur mainly because of the enhancement in the flux ratio of oxygen to Li and Nb atoms which arrive at the substrate, rather than because of reactions in the gas phase, as has been reported in the case of YBaCuO films [21,23,33]. We have earlier observed similar behaviour in ablation processes of Ge in an oxygen atmosphere [15].

5. Conclusions

Amorphous and particulate-free films with an optical energy gap and optical constants similar to those of bulk single-crystal $LiNbO_3$ can be prepared by laser ablation of single-crystal targets in 6×10^{-2} mbar of oxygen pressure, but a very precise control of the laser fluence is needed. The complex dynamics of Li species in the gas phase is found responsible for the variation of the film refractive index as a function of the laser fluence.

Acknowledgements

This work was partially supported by CICYT under the TIC-90 program. We would like to thank Dr. J. Solis (Instituto Optica, Spain) for helpful discussions and Dr. E. Dieguez ad Mr. R. Esquevillas (UAM, Spain) for providing some of the crystals and experimental assistance, respectively.

References

[1] See, for instance, Properties of Lithium Niobate, EMIS Datareviews Series No. 5 (1989).
[2] E. Voger, in: Electro-optic and Photorefractive Materials, Ed. P. Gunter (Springer, Berlin, 1987) ch. 3.
[3] S. Miyazawa, S. Fushimi and S. Kondo, Appl. Phys. Lett. 26 (1975) 8.
[4] R.A. Betts and C.W. Pitt, Electron. Lett. 21 (1985) 960.
[5] K. Nashimoto and M.J. Cima, Mater. Lett. 10 (1991) 348.
[6] G.H. Hewig, K. Jain, F.O. Sequeda, R. Tom and P. Wang, Thin Solid Films 88 (1982) 67.
[7] G. Griffel, S. Ruschin, A. Hardy, M. Itzkovitz and N. Croitoru, Thin Solid Films 126 (1985) 185.
[8] P.R. Meek, L. Holland and P.D. Townsend, Thin Solid Films 141 (1986) 251.
[9] T. Kanata, Y. Kobayashi and K. Kubota, J. Appl. Phys. 62 (1987) 2989.
[10] R.C. Baumann, T.A. Roast and T.A. Rabson, J. Appl. Phys. 68 (1990) 2989.
[11] S. Takada, M. Ohnishi, H. Hayakawa and N. Mikoshiba, Appl. Phys. Lett. 24 (1974) 490.
[12] G.H. Hewig and K. Jain, J. Appl. Phys. 54 (1983) 57.
[13] J. Cheung and J. Horwitz, MRS Bull. (1992) 30.
[14] A. Slaoui, E. Fogarassy, C. Fuchs and P. Siffert, J. Appl. Phys. 71 (1992) 590.
[15] F. Vega, C.N. Afonso and J. Solis, Appl. Surf. Sci. 69 (1993) 403.
[16] H. Tabata, T. Kawai, S. Kawai, O. Murata, J. Fujioka and S. Minakata, Appl. Phys. Lett. 59 (1991) 2354.
[17] D. Roy, B. Krupanidhi and J.P. Dougherty, J. Appl. Phys. 69 (1991) 7930.
[18] N. Maffei and S.B. Krupanidhi, Appl. Phys. Lett. 60 (1992) 781.
[19] F.E. Youden, R.W. Eason, M.C. Gower and N.A. Vainos, Appl. Phys. Lett. 59 (1991) 1929.
[20] X.D. Wu, B. Dutta, M.S. Hedge, A. Inam, T. Venkatesan, E.W. Chase, C.C. Chang and R. Howard, Appl. Phys. Lett. 54 (1989) 179.
[21] C. Girault, D. Damiani, J. Aubreton and A. Catherinot, Appl. Phys. Lett. 54 (1989) 2035.
[22] K. Scott, J.M. Huntley, W.A. Phillips, J. Clarke and J.E. Field, Appl. Phys. Lett. 57 (1990) 922.
[23] D. Fried, G.P. Rech, T. Kushida and E.W. Rothe, J. Appl. Phys. 70 (1991) 2337.
[24] H. Izumi, K. Ohata, T. Sawada, T. Morishita and S. Tanaka, Appl. Phys. Lett. 59 (1991) 597.
[25] J.K. Hirvonen, Mater. Sci. Rep. 6 (1991) 215.
[26] B.C. Grabmaier and F. Otto, J. Cryst. Growth 79 (1986) 682.
[27] C.N. Afonso, F. Vega, J. Solis, F. Catalina, C. Ortega and J. Siejka, Appl. Surf. Sci. 54 (1992) 175.
[28] F. Vega, C.N. Afonso and J. Solis, J. Appl. Phys., March 1993.
[29] J. Tauc, in: Optical Properties of Solids, Ed. F. Abeles (North-Holland, Amsterdam, 1970) p. 277.
[30] M. Born and E. Wolf, Eds., Principles of Optics (Pergamon, New York, 1980) pp. 51–70.
[31] D. Redfield and W.J. Burke, J. Appl. Phys. 45 (1974) 4566.
[32] J. Reader, C. Corliss, W.L. Wiese and G.A. Martin, Wavelengths and Transitions Probabilities for Atoms and Atomic Ions, NSRDS, December 1980.
[33] W. Marine, M. Gerri, J.M. Scotto d'Aniello, M. Sentis, Ph. Delaporte, B. Forestier and B. Fontaine, Appl. Surf. Sci. 54 (1992) 264.
[34] C.E. Otis and R.W. Dreyfus, Phys. Rev. Lett. 67 (1991) 2102.
[35] M. Harnafi and B. Dubreil, J. Appl. Phys. 69 (1991) 7565.
[36] J.C. de Sande, C.N. Afonso, J.L. Escudero, R. Serna, F. Catalina and E. Bernabeu, Appl. Opt. 31 (1992) 6133.

Applied Surface Science 69 (1993) 156–160
North-Holland

applied
surface science

Pulsed laser deposition growth of ZnSe, MnSe, and ZnSe/MnSe epilayers

J. Misiewicz [1], C. Huber [2], D. Heiman
Francis Bitter National Magnet Laboratory, MIT, Cambridge, MA 02139, USA

and

T.Q. Vu
Physics Department, Tufts University, Medford, MA 02155, USA

Received 2 June 1992; accepted for publication 30 October 1992

Epitaxial layers of ZnSe and MnSe as well as heterostructures of ZnSe/MnSe were grown on GaAs(100) substrates by pulsed-laser deposition (PLD). The X-ray diffraction patterns for different layers are presented as well as the low temperature magnetization for ultrathin MnSe/ZnSe structures.

1. Introduction

The growth of epitaxial layers and superlattices of II–VI compounds has attracted much attention. Recently, a new system ZnSe/MnSe was proposed [1,2]. Because of their wide band gap (~ 2.7 eV at 300 K), the materials of this system may prove to be of use in the development of blue light emitters and display devices.

The presence of paramagnetic Mn^{2+} ions is also a promising factor both from a basic as well as an applied point of view.

The (PLD) growth method has recently been applied to a number of new materials, see reviews of refs. [3–9]. This technique has some advantages over the MBE and MOCVD methods for fabricating layered semiconductors and microstructures. Instead of using an effusion cell or metalorganic gas as a source of growth material PLD uses solid targets that are heated locally to a few thousand Kelvin with a pulsed laser beam. To produce an atomic beam source which is expected to be more reactive on the substrate than a molecular beam, changes in composition are made during the epitaxial layer growth by directing the laser beam to different targets using a computer controlled mirror. This has been employed to layer different materials in superlattices and quantum wells [5]. In addition, since a single laser shot produces only a part of a monolayer (about 1/10 in our case), materials from different targets can be evaporated and directed to the substrate during the deposition of a single monolayer. Therefore, one might anticipate that monolayer growth with any reasonable composition can be obtained. If single monolayers can be deposited without intermixing, ordered ternary materials such as $ZnMnSe_2$ might be grown by this technique [10].

In fig. 1, our PLD growth apparatus is shown. A pulsed Nd:YAG laser operating at 1.06 μm, with a repetition rate of 10 Hz and a pulse of 10

[1] Present and permanent address: Institute of Physics, Technical University of Wrocław, Wybrzeze Wyspianskiego 27, 50-370 Wrocław, Poland.

[2] Present address: Naval Surface Warfare Center, 10901 New Hampshire Ave., Silver Spring, NH 20903-5000, USA.

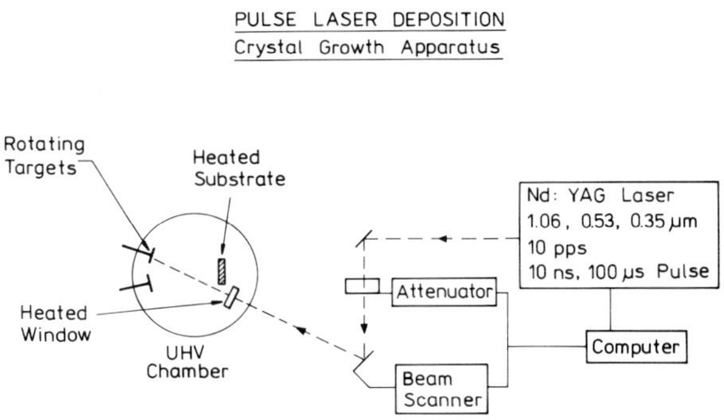

Fig. 1. Diagram of the pulsed laser deposition system used to grow ZnSe, MnSe and MnSe/ZnSe structures.

ns was used for the target ablation. A computer controlled reflecting mirror allowed scanning of the focused laser beam (0.5 mm in diameter) over each rotating solid target to ensure a uniform consumption of the target material. The target-to-substrate distance was ~ 7 cm. Using an average laser power of 0.25 W (25 mJ/pulse) we obtain a growth rate of 0.1 μm/h. The energy density per pulse was 20 J/cm^2. An ultrahigh-vacuum chamber with a base pressure of 1×10^{-10} Torr was used. The approximate growth rate was monitored by a quartz-crystal thickness monitor positioned near the substrate. For reflection high-energy diffraction (RHEED) measurements an electron gun was operated at 10 kV with an emission current of 1 mA. A phosphor screen was used to image the RHEED diffraction patterns. The substrate growth temperature was controlled within a temperature range of 300–400°C, with accuracy of ± 1°C.

ZnSe, MnSe and ZnSe/MnSe epilayers were grown on GaAs substrates made by Sumitomo Inc., with (100) orientation and 2° off towards (110) direction. GaAs substrates were used for their technological importance and because of the lattice match with ZnSe (0.25% mismatch). The GaAs substrates were cut into plates of 1 cm $\times$ 1 cm by cleaving along the [110] faces. Prior to the growth, substrate was cleaned and finally baked at 600°C for 5 min in the vacuum chamber to remove atmospheric contaminations and oxides from the surface. The quality of the GaAs surface was controlled by observation of RHEED patterns. These patterns were used also to characterize the growth of the layer during the process. The quality of the layer and its orientation were determined by two-theta diffraction spectra measured by a Rigaku diffractometer for the whole sample surface and by scanning electron

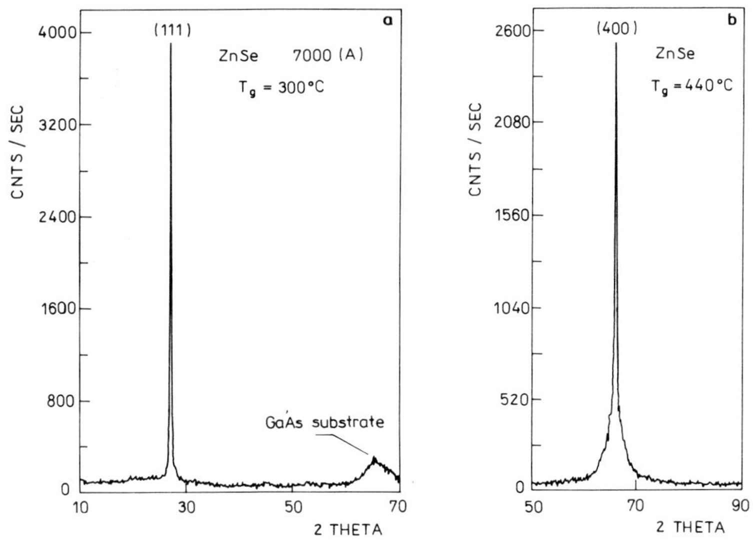

Fig. 2. Two-theta X-ray diffraction spectra of ZnSe layers: (a) grown at 300°C showing preferred orientation (111); and (b) grown at 440°C showing epitaxial layer oriented at (100).

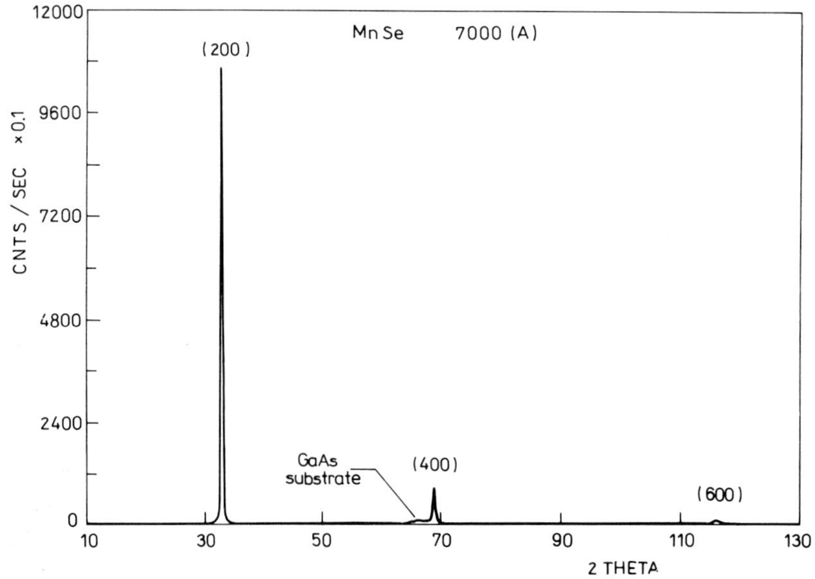

Fig. 3. Two-theta X-ray diffraction spectra of MnSe layer grown at 400°C on the a 100 Å ZnSe buffer layer. Besides the main (200) peak, the (400) and (600) peaks are also visible.

microscopy. The stoichiometry of layer was determined by Auger spectroscopy and Rutherford backscattering (RBS) spectroscopy. In addition to the thickness monitor in the chamber, more accurate layer thicknesses were determined from a combination of sputtering Auger profiles, Dektak

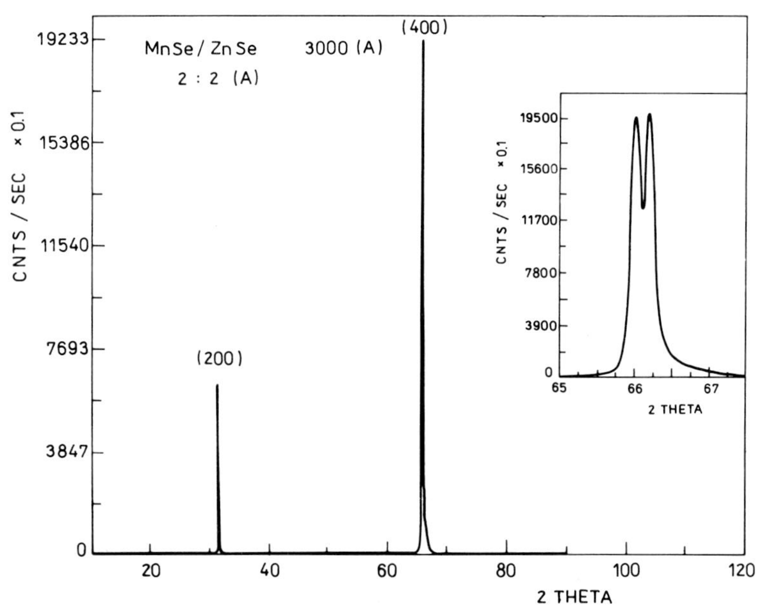

Fig. 4. Two-theta X-ray diffraction spectra of the 850 period MnSe/ZnSe multilayer containing 1.5 Å/2.2 Å of each compound. Inset shows splitting of the main diffraction maximum arising from the Cu X-ray doublet.

thickness scans, and RBS spectra. Details will be published elsewhere [11].

2. Results

2.1. ZnSe growth

A high purity (99.999%) ZnSe crystal obtained from Eagle-Pitcher-Industries was used as a target for ZnSe films. Layers of 10^2 to 10^3 Å thickness were grown at different substrate temperatures. For temperatures below 400°C preferred layer orientation was (111), but layers were mostly polycrystalline. For substrate temperatures higher than 400°C epitaxial layers were obtained with orientation (100). Fig. 2 shows the two-theta spectra for ZnSe layers grown at 300 and 440°C.

2.2. MnSe growth

MnSe powder (99.9%) from Aesar Inc. was melted into a pellet and used as a MnSe target. Layer thicknesses of 10^2 to 10^3 Å were grown. Layers grown directly on GaAs were sometimes polycrystallines. The best MnSe epitaxial layers were grown at 400°C and higher on a 100 Å thick ZnSe buffer layer. Thick MnSe epilayers indicated (100) orientation of the rock salt structure – see fig. 3.

2.3. MnSe / ZnSe structures

Various MnSe/ZnSe structures were grown, including multilayer MnSe/ZnSe structures with different ratios of MnSe to ZnSe layer thicknesses. Despite the relatively large lattice constant mismatch between ZnSe and zincblende MnSe (~ 4.6%) such structures have exhibited a rather low dislocation density [12]. The X-ray diffraction of these multilayers was usually consistent with a zincblende structure.

Fig. 4 shows diffraction spectra of the multilayer consisting of 850-periods of MnSe/ZnSe with 1.5 Å of MnSe and 2.2 Å of ZnSe. This layer seems to be well oriented in the [100] direction of a zincblende crystal structure. The unusually high diffraction intensity, appearance of the (200) diffraction peak, and resolution-limited line widths are indications of the excellent layer quality.

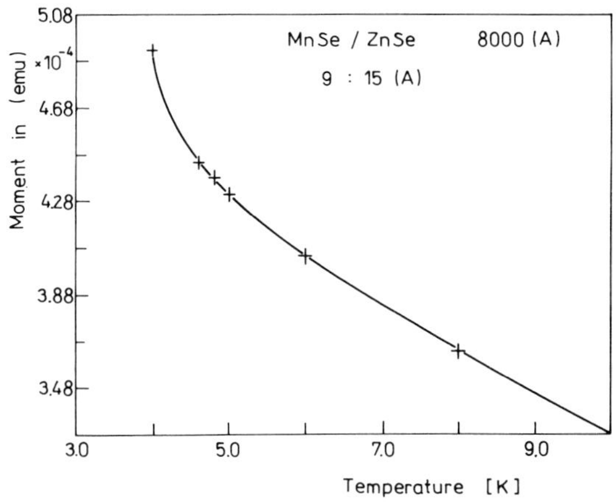

Fig. 5. The low-temperature magnetic moment of ultrathin MnSe/ZnSe structure (9 Å:15 Å) at a magnetic field of 5 tesla.

Bulk grown and thick films of MnSe have the rock-salt crystal structure and are antiferromagnetic semiconductor. In our case, ultrathin molecular monolayers of MnSe separated by ZnSe were grown. This structure indicates different magnetic behaviour. Fig. 5 presents the low temperature part of magnetic moment as a function of temperature for a structure of 200 periods of 3 monolayers (9 Å) of MnSe and 5 monolayers (15 Å) of ZnSe. In contrast to bulk-like MnSe, where the magnetization increases with increasing temperature, we observe a decrease of magnetization with increasing temperature. This dependence may indicate paramagnetic behaviour of the Mn–Mn spin interactions in our layer.

3. Conclusions

Epitaxial layers of ZnSe, MnSe and multilayered structures of ZnSe/MnSe were succesfully grown on GaAs(100) substrates by the pulsed laser deposition. X-ray diffraction spectra as well as the results of other experimental techniques confirmed high quality of the films. The degree of intermixing of thin-layers structures is not known at this time, but work is in progress to determine how the growth parameters will affect the intermixing between layers.

Acknowledgements

This work was supported by NSF Grants DMR-8807419 and DMR-890419. One of us (J.M.) was partially supported by the contract 34124-4 from the Technical University of Wrocław.

References

[1] L.A. Kolodziejski, R.L. Gunshor, T.C. Bounsett, R. Venkatasubramanian, S. Datta, R.B. Bijlsma, W.M. Becker and N. Otsuka, Appl. Phys. Lett. 47 (1985) 169.

[2] L.A. Kolodziejski, R.L. Gunshor, N. Otsuka, B.P. Gu, Y. Hefetz and A.V. Nurmikko, Appl. Phys. Lett. 48 (1986) 1482.

[3] J.J. Dubowski, Chemtronics 3 (1988) 66; J. Cryst. Growth 101 (1990) 105.

[4] J.T. Cheung and H. Sankur, CRC Crit. Rev. Solid State Mater. Sci. 15 (1988) 63.

[5] J.J. Dubowski, J.J. Thompson, S.J. Rolfe and J.P. Mc Caffrey, Superlatt. Microstruct. 9 (1990) 327.

[6] X.L. Zheng, C.A. Huber, P. Becla, M. Shih and D. Heiman, Mater. Res. Soc. Proc. 161 (1990) 465.

[7] X.L. Zheng, C.A. Huber, M. Shih, P. Becla, A.M. Rao and D. Heiman, ICPS-20, Thessaloniki, Greece, 1990, p. 260.

[8] J.T. Cheung, M. Khoshnevisan and T. Magee, Appl. Phys. Lett. 43 (1983) 462.

[9] A. Aydinli, G. Contreras Puente, A. Bhat and A. Campaan, J. Vac. Sci. Technol. A 9 (1991) 3031.

[10] D. Heiman, E.D. Isaacs, P. Becla and S. Foner, in: Diluted Magnetic (Semimagnetic) Semiconductors, Eds. R.L. Aggarwal, J.F. Furdyna and S. von Molnar, Mater. Res. Soc. Proc. 89 (1987) 21.

[11] J. Misiewicz, C.A. Huber, D. Heiman and T.Q. Vu, to be published.

[12] S.K. Chang, D. Lee, H. Nakata, A.V. Nurmikko, L.A. Kolodziejski and R.L. Gunshor, J. Appl. Phys. 62 (1987) 4835.

Applied Surface Science 69 (1993) 161–168
North-Holland

applied
surface science

Pulsed laser ablation and deposition of semiconducting thin films: characterization of transient species

A. Giardini Guidoni, A. Mele, G. Pizzella

Dipartimento di Chimica, Università degli Studi di Roma "La Sapienza", P.le A. Moro 5, 00185 Roma, Italy

and

R. Teghil

Dipartimento di Chimica, Università degli Studi della Basilicata, Via N. Sauro 85, 85100 Potenza, Italy

Received 2 June 1992; accepted for publication 30 October 1992

A study of laser-induced production and mass spectrometric analysis of ionized clusters from powder mixtures of the Group VI elements S, Se and Te and Cr, Fe, Ni, Cu, Ag, Zn, In and Cd is discussed. The binary systems Li_2S, FeS, Ni_2S, Ag_2S, Ag_2Te, ZnSe, CdS, CdSe and CdTe have also been investigated. The results show the formation of sequences of ionized products from laser ablation. Their relative stabilities are correlated to structure and reaction kinetics parameters.

1. Introduction

The production of refractory semiconductor materials has undergone a very large development in the last few years [1]. The field of interest appears to involve many branches of science and technology. Thin layer deposition of these materials for applications in optics and microelectronics is extensively reviewed [2]. Deposition of thin films by pulsed laser ablation of a solid target has been demonstrated to be a useful technique which may compete with other procedures. High T_c superconducting films were obtained by laser evaporation of a suitable target [3]. Diamond-like carbon films were also grown by laser ablation of graphite and some organic compounds [4]. This technique is an alternative method of semiconductor materials growth similar to chemical vapor deposition [5]. Thus far we have used laser mass spectrometry to study aggregation phenomena on pure compounds and mixed systems. A knowledge of the type, concentration and distribution of the laser-induced species as well as an understanding of the mechanism involved are of great interest not only for physics and chemistry aspects, but also for the application of laser ablation techniques in thin film processing. The investigation on oxides and on carbon and organic material is a typical example [6,7]. The experimental results have shown some of the chemical pathways which follow laser evaporation and lead to nucleation and growth processes for thin film formation. Metal sulphides, selenides and tellurides have been indicated as products which may find application in the field of semiconductivity [8]. In the present study we have produced clusters by chemical combination of various metals with Group VI elements by laser evaporation of suitable mixtures. Crystals of binary systems of these mixtures have also been laser-ablated. The reaction products have been analyzed by using a time-of-flight mass spectrometer. The intensity distribution of the observed peaks allows the identification of clusters of high and low stability

0169-4332/93/$06.00

which can be correlated with their geometrical structure.

2. Experimental

The laser mass spectrometer consists of a Nd:YAG laser system quadrupled in frequency ($\lambda = 266$ nm $\tau_{FWHM} = 6$ ns) combined with a reflectron time-of-flight mass spectrometer (TOF). When the laser is focused on the sample to a spot of 10 μm in diameter the power is ranging from 10^8 to 10^{10} W/cm^2. The procedures of clusters production and measurements have been previously described [6].

Reagent grade sulphur, selenium and tellurium were used as powder sample. Sulphides, selenides and tellurides of various transition metals were pure grade products. Mixtures of these Group VI elements with various metals were obtained combining finely grinded powders of the products and inserting them into the sample stage, covered by a quartz slide. The spectra of positive and negative ions were recorded and averaged over several runs.

3. Results and discussion

TOF mass spectra of laser-ablated samples of pure sulphur, selenium and tellurium powders show positive and negative cluster ion formation of formula S_n^+, Se_n^+ and Te_n^+ or S_n^-, Se_n^- and Te_n^- with n up to 32. The ion intensity of the clusters decreases with increasing size and is accompanied by discontinuities of the mass distribution of positive and negative cluster ions. The positive and negative cluster size is larger for sulphur than for selenium and tellurium. In our experimental conditions, that is under direct laser-induced desorption, the temperature at the solid–gas interface is high enough for all three elements to produce mainly atoms and ionized atoms. The process of aggregation and growth of the observed clusters appears to depend likely only on the binding energy that is, in the order S > Se > Te [9]. The analysis of mass spectra shows that the distribution presents some alternation. It may be noted that in the spectra of the three pure elements, higher intensities at $n = 2$, 5, 7, 8 for sulphur, at $n = 2$, 5, 7 for selenium and at $n = 2$, 3 for tellurium are observed. It may be tempting to correlate these numbers with the

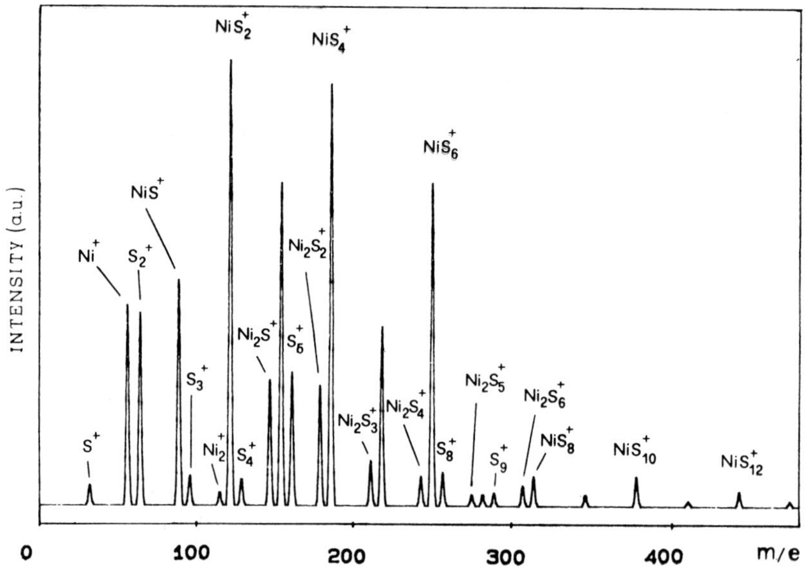

Fig. 1. Laser time-of-flight mass spectrum of a mixture of nickel and sulphur powders.

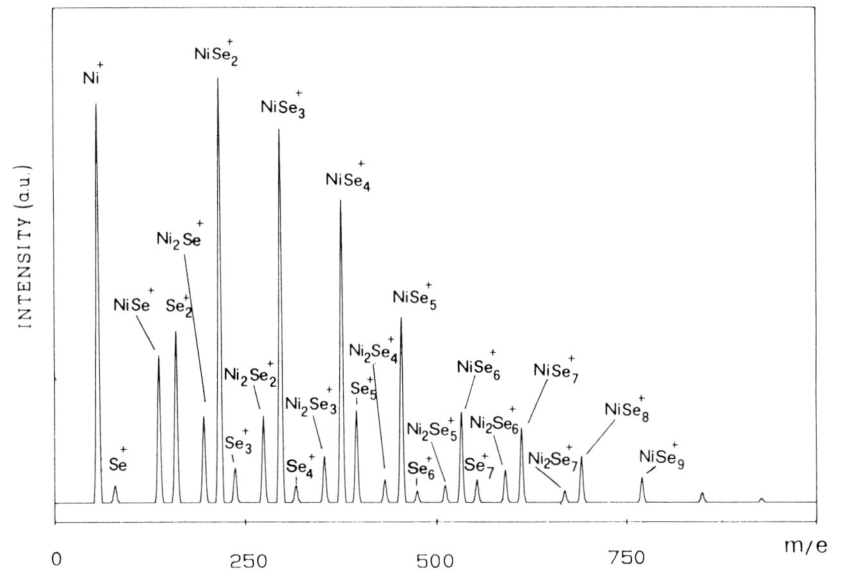

Fig. 2. Laser time-of-flight mass spectrum of a mixture of nickel and selenium.

experimental findings of the UV photoelectron studies of Se_n clusters that show characteristic differences between the spectra of Se_2, Se_5 and Se_7 with respect to Se_6 and Se_8. These differences have been attributed to the different geometrical structure [10]. Photodissociation of tellurium clusters also suggests the existence of stable two, five and seven atom ion clusters which

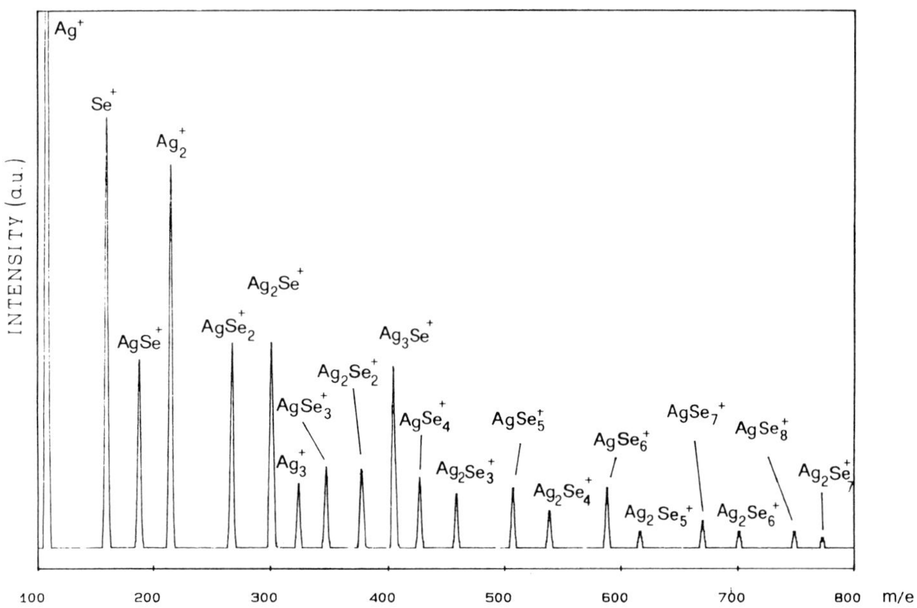

Fig. 3. Laser time-of-flight mass spectrum of a mixture of silver and selenium.

are preferentially produced [8]. Sulphur cluster stability may probably be correlated to the one that is found for selenium and tellurium. The S_8 structure is also particularly stable in a simple laser evaporation experiment. The geometrical structures and stability of sulphur and some selenium and tellurium clusters have been extensively investigated. The most interesting aspect of these studies is that many of these clusters exhibit open or ring forms, whose energy has been calculated [10]. On the contrary, the stability among clusters containing different numbers of atoms has not been determined.

Laser ablation of mixtures of Ni–S, Ni–Se and Ni–Te produces positive and negative cluster ions. Sequences of the formula NiS_n^+, $NiSe_n^+$, $NiTe_n^+$ are predominant; n is ranging up to 22 for sulphur and up to 11 for selenium and tellurium. In fig. 1 the low resolution spectrum of a mixture of nickel and sulphur is reported. S_n^+ cluster ions can be seen together with some less intense structures attributed to $Ni_2S_n^+$ ions. Concerning the predominant series NiS_n^+, it may be noted that higher intensities are observed at $n = 2$, 4, 6, 8, 10, 12, 14. This higher stability for even values of n is not found in the spectra of the bare sulphur clusters. The difference between the spectra of the bare elements and the mixtures cannot be interpreted in terms of a simple molecular orbital theory as in the case of carbon clusters [6].

Similar sequences of $NiSe_n$ and $NiTe_n$ of positive and negative cluster ions are found in the mass spectra of nickel selenium and nickel tellurium mixtures. The peak intensity continuously decreases by increasing n, as shown in fig. 2 for the Ni–Se mixture. Other less-intense sequences containing two nickel atoms $Ni_2Se_n^+$ are observed in the spectra up to $n = 8$. Very weak peaks at m/e, corresponding at $Ni_3Se_n^+$ with n ranging up to 4, were also observed. This trend is typical of an ionic crystal structure building up. Similar structures for this type of compounds have been previously discussed [11,12].

Cu–S, Cu–Se and Cu–Te mixtures also produce positive and negative cluster ion sequences of $CuSe_n$, Cu_2Se_n and $CuTe_n$.

A list of the examined mixtures is summarized in table 1, together with a rough estimate of the trend of the clusters formed by laser ablation. In the spectra of mixture, with S, Se and Te, the bare metal abundance is much higher than that of the Group VI elements or of their clusters. In some cases unreacted S, Se and Te or their clusters are completely lacking. On the other hand, a large amount of these elements is combined with a few metal atoms in the mixed clusters. The initial composition of the mixtures affects the cluster abundance distribution. Formation of mixed clusters occurs only when large quantities of metal are present. The mass spectra

Table 1
Cluster sequences formed from laser ablation of mixtures of Group VI elements and metals

Group VI elements	Metals							
S	Cr	Fe	Ni	Cu	Ag	–	In	–
Se	Cr	–	Ni	Cu	Ag	Zn	–	Cd
Te	Cr	–	Ni	Cu	Ag	–	–	Cd
	Cluster sequence							
M	M_m	M_m	M_m	M_m	M_m	M_m	M_m	M_m
X	X_n	X_n	X_n	[a]	–	X_n	–	X_n
MX	MX_n	MX_n	MX_n	MX_n	MX_n	–	–	–
					M_mX_n	M_mX_n	M_mX_n	M_mX_n

[a] The assignment is uncertain being $m/e = 64$, for S_2^+ and $m/e = 63$, for Cu^+.
M_m = metal; X = Group VI element, $m = 1, \cdots$;
MX_n = polysulphides etc., $n = 1, \cdots$;
M_mX_n = subsulphides etc., $m = 1, \cdots$.

Table 2
Cluster sequences formed from laser ablation of some sulphide, selenide and telluride crystals

	Crystals					
	Li_2S	FeS	Ni_2S	Ag_2S Ag_2Se Ag_2Te	ZnSe	CdS CdSe CdTe
	Cluster sequence					
M	M_m	M_m	M_m	M_m	M_m	M_m
X	X	–	–	–	X_n	X_n
MX	–	–	–	MX_n	MX_n	MX_n
	M_mX_n	M_mX_n	M_mX_n	M_mX_n	M_mX_n	–

See table 1 for symbols.

of mixtures with a high content of Group VI elements mainly show the bare metal and S, Se and Te clusters in a high abundance, but no mixed clusters. In our experimental condition a metal excess in the reaction region seems to have a stabilizing effect on aggregation.

In fig. 3 the TOF mass spectra of a mixture of Ag and S show that AgS_n cluster and other metal rich clusters of higher intensity are present. Mixtures of Zn, Cd and In with S, Se and Te produce analogously only metal rich clusters.

In summary, irradiation of mixtures of a metal and Group VI elements produces clusters containing polyatomic anions. Polysulphide, polyselenide and polytelluride clusters are constructed with the metal cation M^{2+} and the polyanion X^{2-}. This process is favoured by a large concentration of sulphur microaggregates. Metal rich subsulphide clusters which are formed only in mixtures of certain metals are commonly observed in the spectra of all laser-ablated metal crystals.

Several sulphide, selenide and telluride crystals have been irradiated and analysed by TOF mass spectrometry. These compounds represent the final stage of the laser ablation process, leading to thin film deposition. A comparison of the mass spectra of these compounds with those of mixtures of the pure elements may be indicative of the mechanisms of evaporation and formation of mixed clusters. In fig. 4 the TOF mass spectrum of Ni_2S crystals is reported. Laser ablation of Ni_2S crystals produces bare Ni clusters and mixed Ni rich sulphur subsulphide clusters.

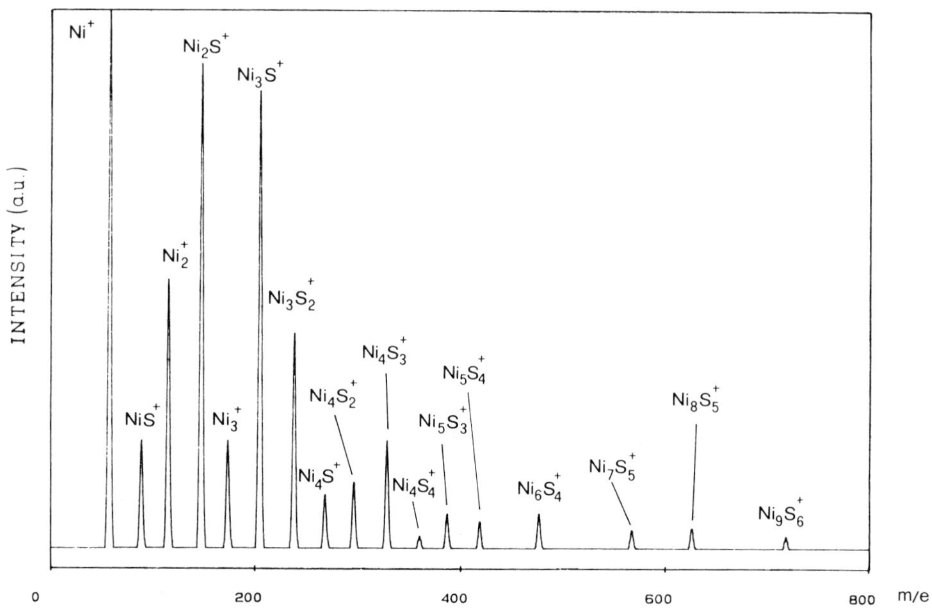

Fig. 4. Laser time-of-flight mass spectrum of Ni_2S.

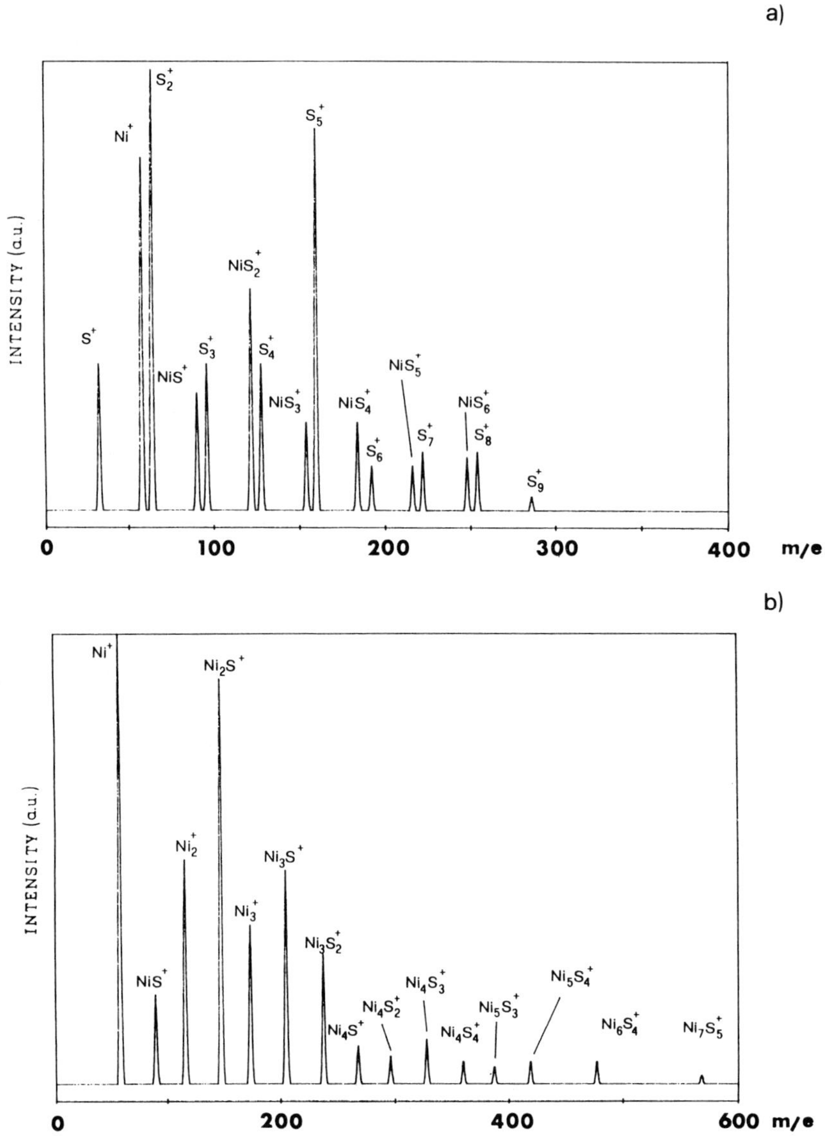

Fig. 5. Laser time-of-flight mass spectrum of a thin film obtained from a Ni and S mixture. (a) Polysulphides, (b) subsulphides.

Sulphur clusters and polysulphide clusters are completely lacking. Other sulphides such as Li_2S, FeS, Ag_2S and CdS produce mainly subsulphide clusters. In table 2 the cluster sequences formed from several crystals are reported.

Thin film deposits from Ni + S and Ni_2S targets yield laser mass spectra which confirm the tendency of mixtures to provide polysulphides and that of sulphides to give subsulphides. The optical microscope observation of the deposits from the Ni–S mixture shows that a gray deposit with a rough structure is present together with a thin film. The laser mass spectra obtained by hitting the thin film are typical of sulphides, while the rough structure yields polysulphides as shown in fig. 5.

4. Conclusions

Laser ablation of mixtures of transition metals and Group VI elements may lead to competing processes of cluster formation. Homonuclear cluster ions built up of pure metal or constructed of bare sulphur, selenium and tellurium are two possible aggregation patways. The mixed aggregation of metal and of Group VI elements is the other process. The models which depict the structures and stability of these three types of clusters are reported in recent literature [12–16]. The stability of metal clusters has been described in terms of an electronic closing shell model which describes the cluster as a structureless ionic jelly [14]. The hypothesis is that in the metal the electrons move rather free in an almost spherically symmetrical environment, so that the shell structure appears to be of electronic origin. Experimental abundance spectra correlate with this shell model for several of the metals shown in table 1. Group VI elements have properties which depend on their position in the periodic table. Sulphur typically is an insulator with an S–S bond energy around 101.6 kcal/mol. Selenium and tellurium are semiconductors and their bond energies are 79.5 and 63 kcal/mol, respectively. Bond energies are higher in S_n, Se_n and Te_n with $n > 2$ [9]. By laser ablation these elements form small covalent clusters with well defined relative abundance. Mixed clusters formed by metal and Group VI elements have a dominant ionic character. This is shown in a calculation of cluster structures and stabilities made with an ionic model with weak covalent contribution [15].

Competition between two reaction pathways corresponding to homonuclear cluster growth or mixed Group VI elements and metal clusters depends on several parameters not always identified. Clusters strongly deviating from normal stoichiometry such as subsulfides, selenides and tellurides as well polysulphides, polyselenides and polytellurides have been observed. Polysulphides, polyselenides and polytellurides are energetically favoured with respect to the formation of metal clusters and mixed clusters on a simple thermodynamic basis.

The basic considerations in play during the preparation of the important metal–Group VI element semiconductors are related to chemical and physical parameters of the target material, transient cluster nucleation, plasma cloud and, finally, to thin film growth conditions such as growth-rate and temperature on the substrate. The stoichiometry in the solid material starting either from pure elements or from more complex chemical substances as well as the impurity content play a relevant role in the distribution of the laser-induced species in the reaction region. This, in turn, determines the thermodynamic equilibrium composition of the transient species and ultimately will characterize the semiconductor crystal growth. The physical and chemical state of the solid target will also affect the stoichiometry of the intermediate products, as found by comparing the results obtained by laser ablation of powders of the two components and of crystals of pure sulphide selenide or telluride. In the vapour phase the metal and sulphur selenium and tellurium vapours are found in monatomic and polyatomic form as clusters. One of these species usually predominates and an estimate of the relative abundance provides basic information on the complex solid–vapour equilibrium. In the experimental conditions adopted in laser ablation the concentrations of the reactants in the vapour phase cannot be very easily adjusted. The composition of the vapour essentially reflects their precursors in the solid as far as their components and depends on their physical properties. The assumption that the aggregation phenomena in the vapour phase predicts what is going to occur in the process of epitaxial growth of the crystal appears to be only partly consistent with the experimental data. This process is also affected by factors related to the deposition rate and to the temperature of the substrate. Ablation has a typical deposition rate in the range of a few Å/pulse which depends on the power of the laser and the geometry between the target and the substrate.

Acknowledgement

This work has been partially supported by CNR P.F. Chimica Fine.

References

[1] R. Teghil, A. Giardini Guidoni, S. Piccirillo, A. Mele and F. Polla-Mattiot, Appl. Surf. Sci. 64 (1990) 220.

[2] Deposition Technologies for Films and Coatings (Noves, New Jersey, 1982).

[3] (a) T. Venkatesan, H.D. Wu, A. Inam and J.B. Wachtman, Appl. Phys. Lett. 52 (1988) 1193;
(b) M. Balsoch, D.R. Olander and R.E. Russo, Appl. Phys. Lett. 52 (1989) 197;
(c) A. Giardini Guidoni, A. Morone and M. Ambrico, High Temperature Superconducting Thin Films, Ed. L. Correra, Proc. Symp. Al Int. Conf. Advanced Materials, ICEM 91, Strasbourg, May 27–31, 1991, p. 451.

[4] (a) I.S. Athwal, A. Mele and E.A. Ogryzlo, Diamond Relat. Mater. 1 (1992) 731;
(b) B.R. Mehta and E.A. Ogryzlo, Surf. Coating Technol. 43 (1990) 80.

[5] P. John, Chemistry of the Semiconductor Industry, Eds. S.J. May and A. Ladwith (Champman & Hall, New York, 1987) p. 98.

[6] (a) A. Mele, D. Consalvo, D. Stranges, A Giardini Guidoni and R. Teghil, Int. J. Mass Spectrom. Ion Proc. 95 (1990) 359;
(b) A. Giardini Guidoni, A. Mele, G. Pizzella and R. Teghil, Z. Phys. D 20 (1991) 89.

[7] D.N. Lineman. K.V. Somayajula, A.G. Sharkey and D.M. Hercules, J. Phys. Chem. 93 (1989) 5026.

[8] S.P. Keller, Handbook on Semiconductors (North-Holland, Amsterdam, 1980).

[9] K.C. Mills, Thermodynamic Data for inorganic Sulphides, Selenides and Tellurides (Butterworths, London, 1974).

[10] J. Becker, K. Rademan and F. Hensel, Z. Phys. D 19 (1991) 229.

[11] K.F. Willey, P.Y. Cheng, T.G. Taylor, M.B. Bishop and M.A. Duncan, J. Phys. Chem. 94 (1990) 1544.

[12] T.P. Martin, J. Chem. Phys. 81 (1984) 4426.

[13] K. Raghavachari, C.M. Rohlfing and J.S. Binkley, J. Chem. Phys. 93 (1990) 5862.

[14] W.A. de Heer, W.D. Knight, M.Y. Chon and M.L. Cohen, Solid State Phys. 40 (1987) 93.

[15] P.J. Ziemann and A.W. Castlemann, J. Chem. Phys. 94 (1991) 718.

[16] A. Giardini Guidoni and A. Mele, Laser Chem. 11 (1991) 205.

Applied Surface Science 69 (1993) 169–173
North-Holland

applied
surface science

Mass spectrometric investigation of the KrF laser-induced plasma plume created above an YBaCuO superconducting target: correlation with thickness distribution of deposited thin films

C. Champeaux [a], D. Damiani [b], J. Aubreton [a] and A. Catherinot [a]

[a] *URA 320 CNRS, Faculté des Sciences, 123 Avenue A. Thomas, 87060 Limoges Cedex, France*
[b] *Industrial Laser Partners, 87130 St. Laurent-sur-Gorre, France*

Received 2 June 1992; accepted for publication 30 October 1992

The plasma plume created during the ablation of YBaCuO superconducting targets by a KrF laser beam (248 nm) is studied by mass spectrometry so as to understand one of the key-points of this thin films elaboration technique, that is to say the transport of yttrium, barium, copper and oxygen species to the substrate. The evolutions of the relative concentrations of the detected species are reported as functions of the laser fluence on the target surface. The angular distribution of ejected species is measured and fitted by $\cos^n$ curves. The results, correlated with thickness, allow an estimation of the relative contributions of neutral and ionic species in deposited thin films.

1. Introduction

The ability of UV lasers to transfer the stoichiometry from the target to the substrate explains why pulsed laser deposition has quickly emerged as the leading technique to deposit in situ high quality epitaxial superconducting thin films. However, only few studies are devoted to the fundamental aspects of the mechanism of laser ablation and to the transport phenomena of vaporized species from the target to the substrate.

This paper deals with mass spectrometric investigations of the plasma plume produced by KrF excimer laser ablation of bulk YBaCuO superconducting targets, under vacuum conditions. The evolutions of the relative concentrations of the detected species are reported as functions of the laser fluence on the target. Their angular distributions are measured and these data, correlated with the thickness distribution, lead to an estimation of the relative contributions of neutral and ionic species in the deposited thin films.

2. Experimental device

Experiments are performed in an ultrahigh vacuum cell with an introduction chamber for target and substrate manipulation. The laser source is an excimer laser (Lambda Physik EMG 101) operating at 248 nm (KrF), with a repetition rate of 3 Hz, a pulse duration of 15 ns and an energy per pulse in the range 20–300 mJ. After spatial filtering, the laser beam is focused onto a rotating bulk superconducting YBaCuO target, home-prepared by a usual sintering technique. According to the 35° incidence angle, the laser spot is elliptical and the fluence can be changed by varying the laser energy. The vaporized material normally ejected to the target surface is investigated using a quadrupolar mass spectrometer (QMG 420 Balzers) equipped with a differential turbomolecular pumping system (sampling hole of 50 μm in diameter). This device is substituted to the substrate holder. Measurements of the angular distribution of the ionic species concentrations have been realized by tilting the target.

3. Mass spectrometric investigations

3.1. Detected species

All the experiments are performed under a residual pressure of about 10^{-7} mbar within the cell. According to the poor signal-to-noise ratio observed for the detection of neutral particles, this work is limited to the study of ionic species. Each measurement corresponds to an average over, at least, 500 laser shots. Table 1 lists the different ions detected in the laser-induced plasma plume created above an YBaCuO superconducting target. In addition to monatomic (Y^+, Ba^+, Cu^+, O^+) and diatomic (YO^+, BaO^+, CuO^+) species previously observed by optical spectroscopic investigations [1,2], oxide clusters like YO_2^+, $Y_2O_2^+$, BaO_2^+ and $Ba_2O_2^+$ are detected. However, no mixed metal oxide ions are observed.

The relative concentrations of the detected species are compared to those given by Dietze and Becker [3] for a YAG (1.06 μm) laser-induced plasma plume on YBaCuO targets (fig. 1). The results are normalized to the measured Ba^+ concentration. We notice that detection of O^+ and CuO^+ ions is not reported by Dietze and Becker, and that some significant differences are observed between the measured relative concentrations of oxide molecular ions such as YO^+, Y_2O^+, Ba_2O^+ and BaO_2^+. This indicates, as expected, that the interactions between 1.06 and 0.248 μm laser beams and YBaCuO targets are significantly different.

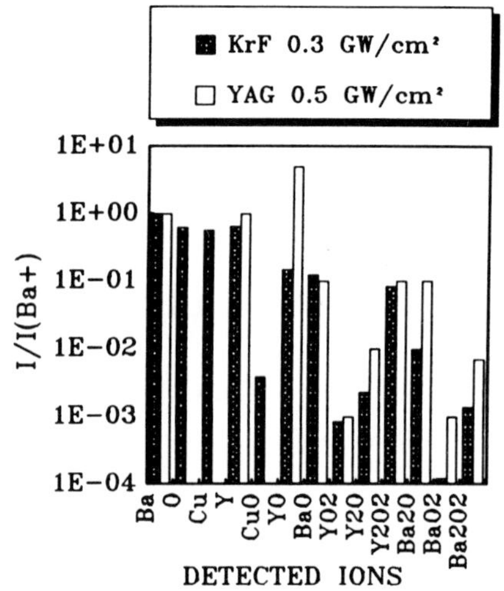

Fig. 1. Comparative relative concentrations of detected positive ionic species using mass spectrometry in the plasma plume on an YBaCuO target created by a YAG laser (1.06 μm, 5×10^8 W/cm^2; from Dietze and Becker [3]) and a KrF laser (0.248 μm, 3×10^8 W/cm^2; our work). Results are normalized to the measured concentration in Ba^+.

3.2. Influence of the laser fluence F

The measured intensities, which are proportional to the concentrations of the detected ionic species, are studied as functions of the KrF laser fluence.

• For all monatomic (Y^+, Ba^+, Cu^+, O^+) and diatomic (YO^+, BaO^+, CuO^+) ions, the evolution of the detected signal I_i as a function of the laser

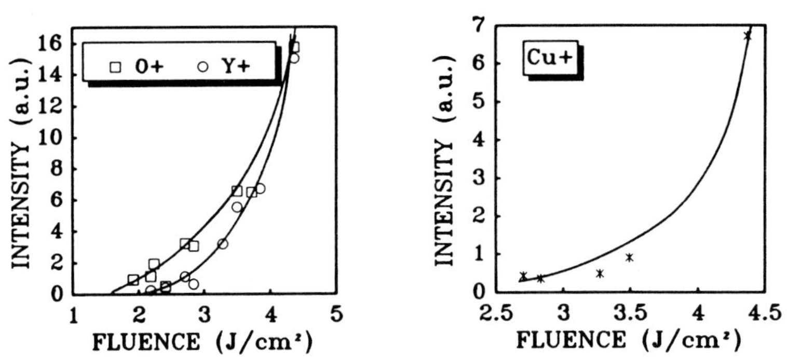

Fig. 2. Influence of the fluence on the mass spectrometric signal intensities of monatomic ionic species (examples given for Y^+, Cu^+, O^+). The background pressure is about 10^{-7} mbar and the measurement accuracy is estimated to be 20%.

Table 1
Ionic species observed in the KrF laser ablation of YBaCuO bulk superconducting samples

Detected species	Mass number (amu)
O^+	16
Cu^+	63
Cu^+	65
CuO^+	79
CuO^+	81
Y^+	89
YO^+	105
YO_2^+	121
Ba^+ isotopes	130–138
BaO^+ isotopes	146–154
BaO_2^+	170
Y_2O^+	194
$Y_2O_2^+$	210
Ba_2O^+	292
$Ba_2O_2^+$	308

fluence is well described by a monoexponential relation:

$$I_i = I_{0i} \exp(F/F_{0i}),$$

where F_{0i} is the detection threshold fluence for the ion i. The obtained results are summarized in figs. 2 and 3.

• For the oxide clusters, some results are given in fig. 4. Except for $Y_2O_2^+$, which exhibits a concentration evolution similar to those of monatomic or diatomic ions, the detected intensities for ionic aggregates increase up to a maximum reached for a laser fluence of about 3 J/cm^2. It is important to notice that this value corresponds to the laser fluence that we currently use to deposit thin films with "good" superconducting characteristics.

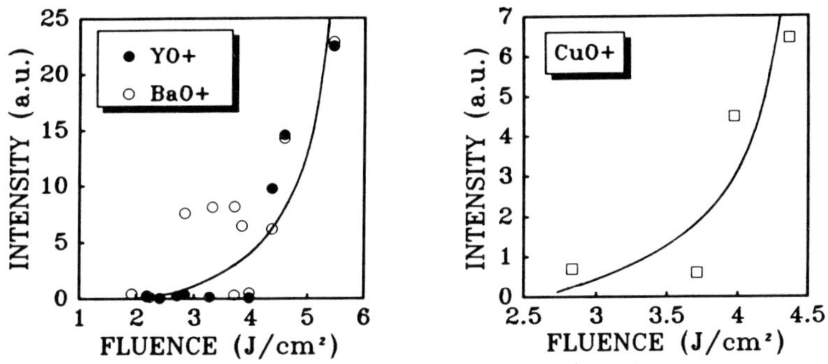

Fig. 3. Influence of the fluence on the mass spectrometric signal intensities of diatomic oxide ionic species (YO^+, BaO^+, CuO^+). The background pressure is about 10^{-7} mbar and the measurement accuracy is estimated to be 20%.

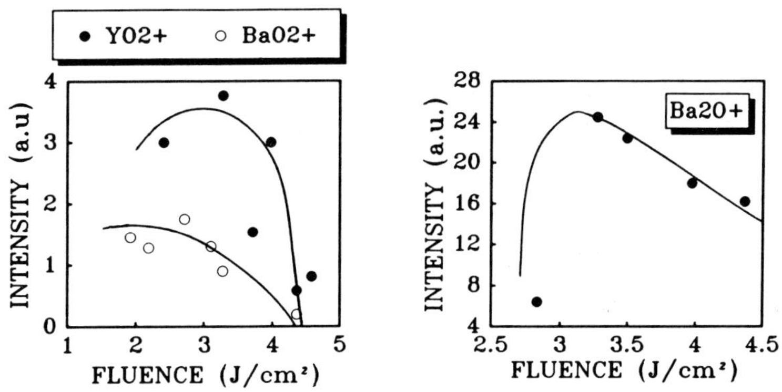

Fig. 4. Influence of the fluence on the mass spectrometric signal intensities of oxide ionic clusters species (YO_2^+, BaO_2^+, Ba_2O^+). The background pressure is about 10^{-7} mbar and the measurement accuracy is estimated to be 20%.

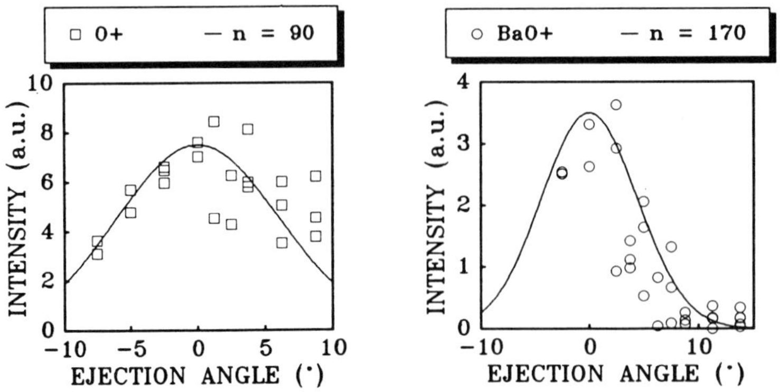

Fig. 5. Examples of angular concentration distributions for some ejected ionic species. The continuous curve is the best fit by $\cos^n\phi$.

3.3. *Angular distributions of ejected ionic species*

Angular distribution measurements are performed by tilting the target. They exhibit a very strong peak, that can be fitted by $\cos^n\phi$ curves, where ϕ is the angle relative to the direction normal to the target surface. Values of the exponent n as large as 90 (for Y^+, Ba^+, O^+) or 170 (for Cu^+, BaO^+ and Ba_2O^+) are found, as shown in fig. 5. So this indicates that ionic species are approximately ejected within the plume in a cone, which has a vertex half-angle of about 10°. This result has to be correlated with observations of other authors [4] which show that, beyond an angle of 10°, the "good" stoichiometry is not obtained and that the films are not superconducting. Consequently, we can mention that ionic species have probably an important role and contribute largely to film growth and quality.

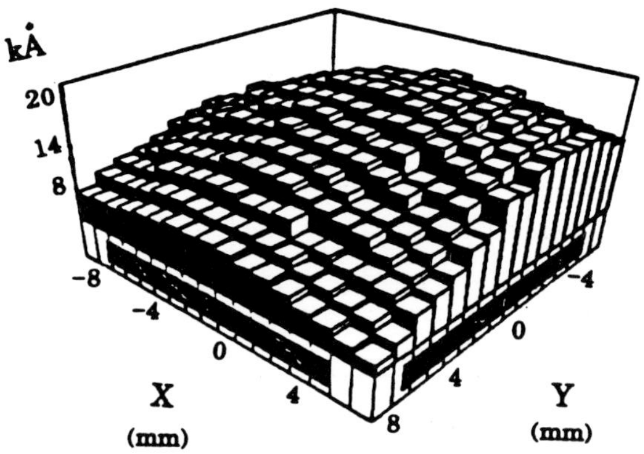

Fig. 6. Measured thickness distribution of an YBaCuO film deposited through a grid-shaped mask on a silicon wafer.

4. Correlation between mass spectrometric investigation and thickness distribution of deposited thin films

In order to correlate mass spectrometric studies with thickness distribution, YBaCuO thin films are deposited on silicon wafers, through a coupled grid-shaped mask. The film thickness, shown in fig. 6, is measured using a DEKTAK profilometer, and its distribution can be fitted by $\cos^p$ curves where $p = 16$ for the central part of the film (disk of about 1 cm in diameter), and $p = 12$ for the periphery.

As suggested by Sajjadi et al. [5], we may assume a similarity of the growth of films by laser ablation technique to a photometric flux-transfer problem. So, the ejected material flux originated from the target element A_T and received in the x position on the substrate (fig. 7) is given by:

$$\frac{\delta\Phi}{\delta A_T} = L \cos\phi \cos^3\beta \frac{x}{a^2} \delta\alpha \, \delta x,$$

where L is the flux per unit projected area and solid angle.

If we assume $L = L_0 \cos^n \phi$, the deposition thickness $d(x)$ in the x position is:

$$d(x) = \frac{L_0 a}{2\pi(a^2 + x^2)^{3/2}} \times \int_0^{2\pi} \left(\frac{a^2 + xr\cos\alpha}{\sqrt{(a^2 + x^2)(a^2 + r^2)}} \right)^{n+1} d\alpha.$$

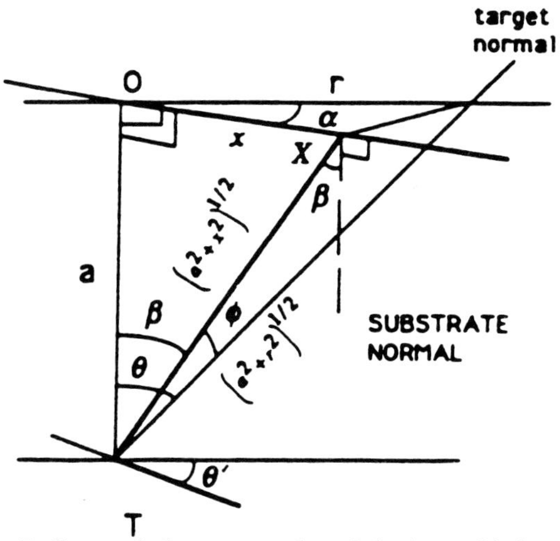

Fig. 7. Geometrical representation of the laser ablation system from Sajjadi et al. [5].

Comparison between the measurements (fig. 6) and the thickness distribution calculated using the previous relation leads to: $L = L_0 \cos^{13}\phi$ in the central region, and $L = L_0 \cos^9\phi$ in the peripherical region of the deposited film.

According to the previous results given by mass spectrometry (see section 3.3), we can assume that the film thickness in the peripherical region is a consequence of an exclusive deposition of neutral particles and that the thickness distribution of the central region results from contributions of both neutral and ionic species. Therefore, the "luminance" L in the central area can be approximated by:

$$L = L_0\left[(1-X)\cos^9\phi + X\cos^{90}\phi\right],$$

where, in the right member, the first term corresponds to the contribution of neutrals and the second to the contribution of ions in relative concentration X. According to the assumptions previously quoted, the comparison with the experimental distribution of thickness indicates that ionic species may contribute up to 10% to the film thickness in the central region (fig. 8).

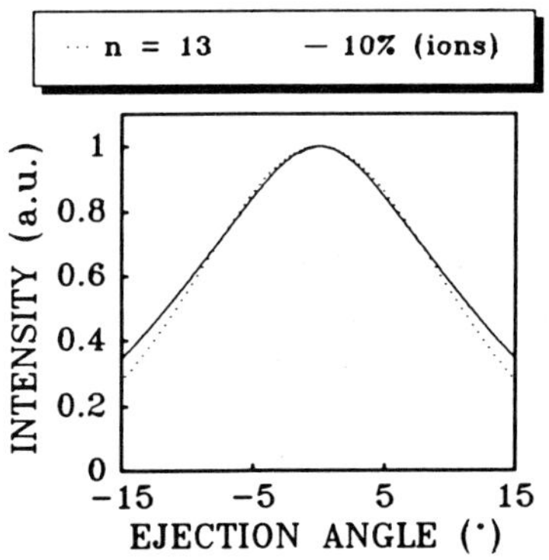

Fig. 8. Contribution of ions to the film thickness (an ejection angle in the range −10° to +10° corresponds to a deposited disk of about 1 cm in diameter).

5. Conclusion

The mass spectrometric investigation of the plasma plume created above an YBaCuO superconducting target shows the presence of atomic, diatomic oxide and clusters of oxide ions. These ions seem to be very important for the transport phenomena of Y, Ba, Cu, O species from the target to the substrate. They contribute up to 10% to the thickness distribution.

Acknowledgements

The authors thank P. Marchet for the preparation of bulk superconducting samples. This work is partly supported by DRET under contract No. 88-084.

References

[1] C. Girault, D. Damiani, C. Champeaux, P. Marchet, C. Germain, J.-P. Mercurio and A. Catherinot, Appl. Surf. Sci. 46 (1990) 78.
[2] C. Girault, D. Damiani, J. Aubreton and A. Catherinot, Appl. Phys. Lett. 55 (1989) 182.
[3] H.J. Dietze and S. Becker, Int. J. Mass. Spectrom. Ion Processes 82 (1988) R1.
[4] T. Venkatesan, X.D. Wu, A. Inam and J.B. Watchman, Appl. Phys. Lett. 52 (1988) 14.
[5] A. Sajjadi, K. Kuen-Lau, F. Saba, F. Beech and I.W. Boyd, Appl. Surf. Sci. 46 (1990) 84.

Applied Surface Science 69 (1993) 174–179
North-Holland

applied
surface science

Time-resolved diagnosis of processes in PLD of ceramics

A. Voss, E.W. Kreutz, J. Funken, M. Alunovic and H. Sung

Lehrstuhl für Lasertechnik der Rheinisch-Westfälischen Technischen Hochschule Aachen, Steinbachstrasse 15, D-5100 Aachen, Germany

Received 2 June 1992; accepted for publication 30 October 1992

The material transfer of ceramics (Al_2O_3, ZrO_2) in pulsed laser deposition (PLD) is studied as a function of pulsed CO_2 laser radiation parameters and processing variables. Time-resolved measurements of the composition and the electron-temperature as well as the geometry and dynamics of the vapour/plasma were obtained by emission spectroscopy and high-speed photography. The vapour/plasma states are related to the structures of ceramic films grown on metallic substrates.

1. Introduction

Thin films elaborated by pulsed laser deposition (PLD) are used in a variety of applications [1,2]. Several novel deposition techniques (e.g. ion-assisted deposition or pulsed laser deposition) have the advantage to produce energetic and excited species in the vapour/plasma plume, which may originate in various structural film properties at moderate substrate temperatures [3,4] depending on the parameters and processing variables of the deposition technique used. Excimer, Nd:YAG and CO_2 laser radiation are mainly used in PLD for thin film deposition of different structures [5–8]. Films deposited at various target–substrate distances on cold substrates [9] show characteristic differences in growth and film structure known from films achieved by other deposition techniques on substrates at high temperatures. As the material transfer plays a major role in PLD, the composition, development and electron temperature of the generated material vapour/plasma have been studied as a function of laser parameters and processing variables by time and space-resolved emission spectroscopy as well as high-speed photography.

2. Experimental

The PLD is performed in a vacuum chamber at a base pressure of 6×10^{-5} mbar with the laser radiation provided by a CO_2 TEA laser (spike of 100 ns (FWHM) and tail of about 3.5 μs, pulse energy 2–4.6 J). The beam is focussed onto the plane surface of the cylindrically rotating target under an inclination ($\simeq$ 30–45°) to the laser beam in that the incident laser radiation is absorbed on the surface effectively because of the incidence-angle-dependent variation in reflectance [4,9]. The films were deposited onto stainless steel substrates, which are oriented parallel to the plane surface of the target [4,9]. The composition and ionization state of the vapour/plasma were measured by emission spectroscopy (fig. 1) with a spectral resolution of 0.065 nm. Emission spectra of the plume were recorded in the 380–800 nm region using a monochromator equipped with an OSMA (optical spectroscopy multichannel analyser) detector system, which is triggered by a pulse generator for the time-resolved measurements allowing a minimum gate width of 200 ns. For time- and space-averaged measurements as well as time- and space-re-

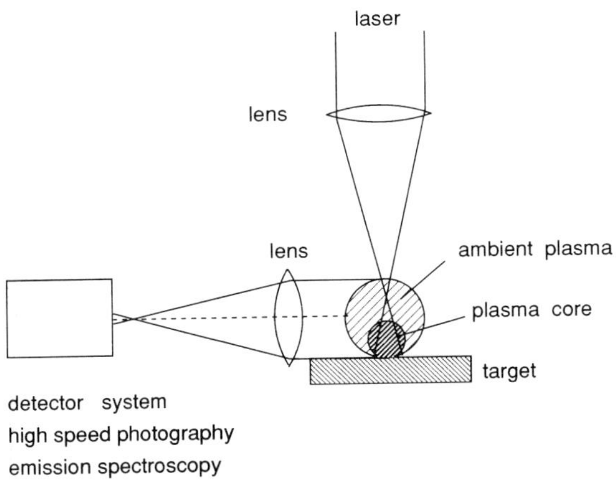

Fig. 1. Experimental set-up for high speed photography and emission spectroscopy.

solved measurements have been performed. During space-averaged measuring the overall luminous vapour/plasma (20 mm in extension) or the vapour/plasma core (2 mm in extension) was imaged onto the fiber optic. The time-averaged measuring was done by using an integration time of 1 μs either for the overall vapour/plasma or the vapour plasma core. In any case, the integration time is larger than the generation time of the vapour/plasma, which is smaller ($\simeq 0.5\ \mu$s) for the given vapour/plasma geometry. The emitted light of the core and the overall vapour/plasma plume was focussed by a lens onto a fiber optic, which guided the light to the entrance slit of the monochromator using a shielding plate to avoid the detection of light emissions of vapour/plasma regions not to be measured and a pinhole (1 mm in diameter) in front of the fiber optic to improve the spatial resolution. The temporal resolution is given by the minimum gate width. The spatial resolution in the range of 1 mm is determined by imaging magnification ~ 1 of the luminous vapour/plasma plume onto the movable fiber optic taking into account the extension of overall vapour/plasma or of the vapour/plasma core by moving the fiber optic perpendicular to the surface of the ceramic target. The spatial measurements do not consider the movement of the vapour/plasma particles, which takes about 0.1 ns to pass geometrically the overall vapour/plasma. High speed photography was used in framing (one image each 50 ns) and streak (highest resolution 10 ns/mm) mode of operation for time and space-resolved analysis of the formation and development of the vapour/plasma plume in the direction perpendicular to the target surface (fig. 1). Laser parameters and processing variables were varied in the same ranges as used for film deposition [4,9] in order to relate results from material transfer investigations to different growth and structure of films deposited on metallic substrates.

3. Results

At base pressures, high-speed photography (fig. 2) in the framing mode shows only the small plasma core at the target surface, which increases in the first 100 ns. At processing gas pressures (Ar) above 10^{-2} mbar a relatively sharp defined spherical luminous front detaches from the target surface 75 ns after the beginning of the laser pulse, which decelerates during expansion. This front is mainly responsible for the plasma emission in the region of the ambient plasma. At Ar pressures above 10 mbar the vapour/plasma expands mainly into the direction of the incident laser radiation and at even higher pressures of about 100 mbar the laser radiation is partly absorbed in the surrounding processing gas above the target surface giving evidence for a laser-induced breakdown in the ambient gas atmosphere. The velocity of the luminous front as a function of time [10] was derived by high-speed photography in the streak mode of operation for Al_2O_3 and ZrO_2 targets at processing gas pressures (Ar) normally used for film deposition (10^{-3}–10 mbar). The velocity decreases with time and with increasing processing gas pressure as well. The highest velocities ($2–6 \times 10^4$ m/s) are measured in the beginning of the material removal.

Typical time-averaged plasma emission spectra of the vapour/plasma plume from Al_2O_3 and ZrO_2 [10] targets show that the vapour/plasma consists of neutral and ionized atoms [11]. For Al_2O_3 target, spectral emission lines from single and double ionized Al atoms and from neutral

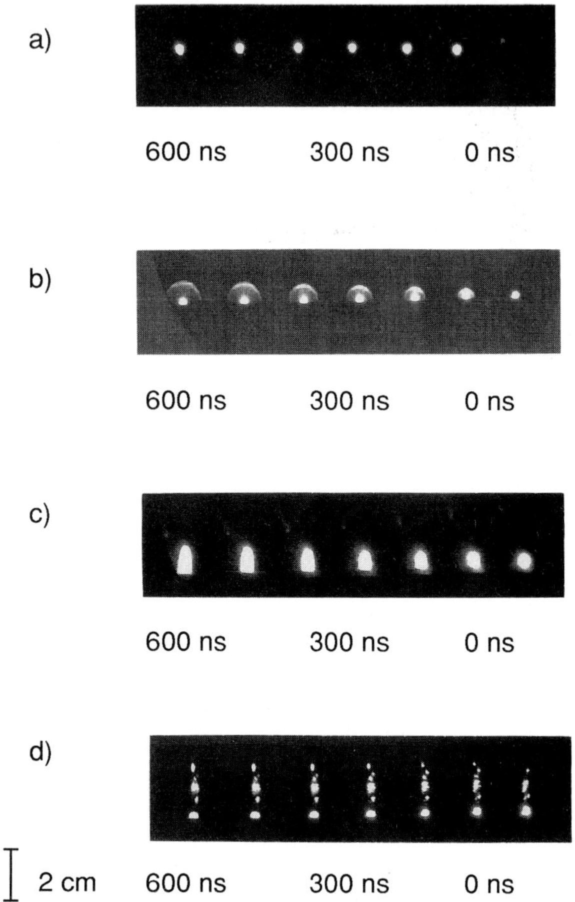

Fig. 2. High speed photography (framing mode) of the vapour/plasma plume of ZrO_2 (laser fluence $\epsilon = 22$ J/cm^2, laser pulse length $t_p = 3$ μs) as a function of processing gas (Ar) pressure ((a) 6×10^{-3}, (b) 6×10^{-1}, (c) 10, (d) 100 mbar).

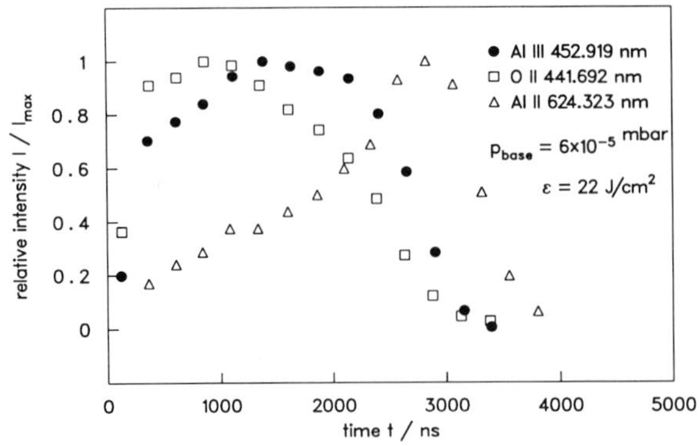

Fig. 3. Temporal development of the relative intensities of the emission lines in the vapour/plasma plume of Al_2O_3.

and single ionized O atoms were detected as well as emission bands of AlO complexes [12]. For ZrO_2 target spectral emission lines from neutral and single ionized Zr atoms as well as single ionized O atoms are observed. For Ar as processing gas emission lines of neutral Ar atoms additionally are detected. The overall intensity of the emission lines is influenced by the pressure and type of processing gas [13] and depends strongly on the laser fluence. Time-resolved measurements of the relative intensities of the emission lines show that the maximum intensities of the higher ionized species (Ar III and O II) occur earlier after the beginning of the laser pulse as those of the lower ionized and neutral species (fig. 3). The emission of Ar species (ambient plasma) is observed up to 150 μs depending on

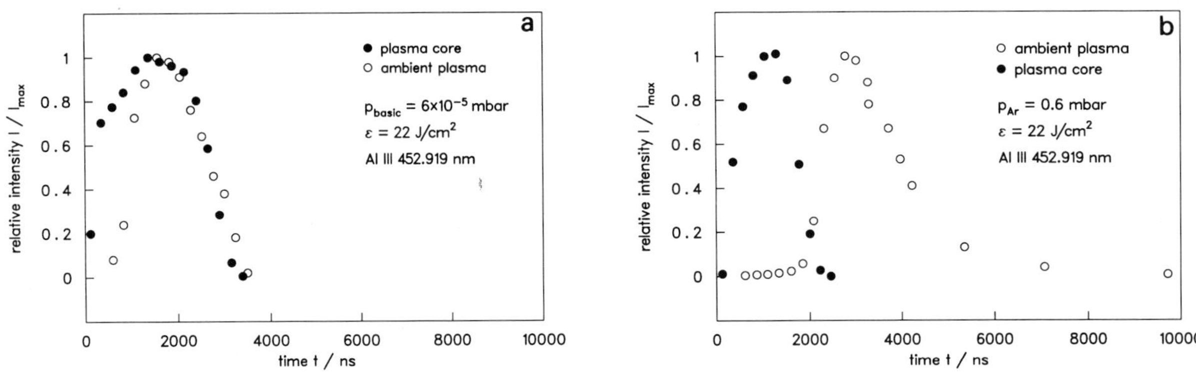

Fig. 4. Temporal development of the relative intensities of the Al III emission line in the plasma core (●) and the ambient plasma (○) as a function of pressure ((a) $p_{base} = 6 \times 10^{-5}$ mbar, (b) $p_{Ar} = 0.6$ mbar).

the Ar pressure. The emission of AlO complexes occurs after the end of the laser radiation (3–5 μs) with a duration up to 30 μs. Time- and space-resolved emission spectroscopy shows the different temporal behaviour of the relative intensity of an Al III emission line (452.919 nm) in the plasma core and in the ambient plasma at different processing gas pressures (fig. 4). At base pressure the emission is increasing faster in the plasma core than in the ambient plasma, but its maximum is reached at the same time after the beginning of the laser radiation (fig. 4a). At the end of the radiation pulse the emission signal decreases. At Ar pressures of 0.6 mbar the maximum intensity in the plasma core is reached earlier than at base pressure, but in the ambient plasma the signal is shifted towards higher times, that the maximum intensity occurs at 3 μs (fig. 4b). The weight loss at the target [4] and the intensity of the emission of the Al III triplet (452.919 nm) near the target surface ($\simeq$ 1 mm) measured as a function of fluence 1 μs after the beginning of the laser radiation qualitatively show the same behaviour (fig. 5): a strong increase at low laser fluences, a maximum at a fluence of 13 J/cm^2 and a decrease at higher fluences. Following well-known methods [11,14,15] the corresponding plasma temperatures are derived from the time- and space-resolved measurement of emission spectroscopy of the Al III triplet showing strong fluctuations in the range 40 000–140 000 K. These methods [11,14,15] also allow to estimate the spatial and temporal densities of particles within the vapour/plasma.

4. Discussion

The initial stage of PLD generally involves the coupling of the optical energy into the target. At laser fluences above the evaporation threshold the dominant mechanism is vaporization, at higher laser fluences the plasma formation plays the major role in governing the material transfer from the target to the substrate [4]. To get time- and space-resolved information about the physical processes concerning the material transfer different optical detection methods were used (section 2). The interaction between vapour/plasma and surrounding processing gas was confirmed by high-speed photography (fig. 2). The luminous spherical front of the vapour/plasma plume shows velocities in the range of weak detonation waves. The relatively high pressure of the evaporated material acts to sweep-up and drive the background gas at supersonic velocities during outward expansion [16]. In the pressure range of 10^{-2}–10 mbar the highest intensity in the luminous front is always detected in the direction normal to the target. For pressures above 10 mbar a laser supported detonation wave [17] is detected, which expands with nearly spherical

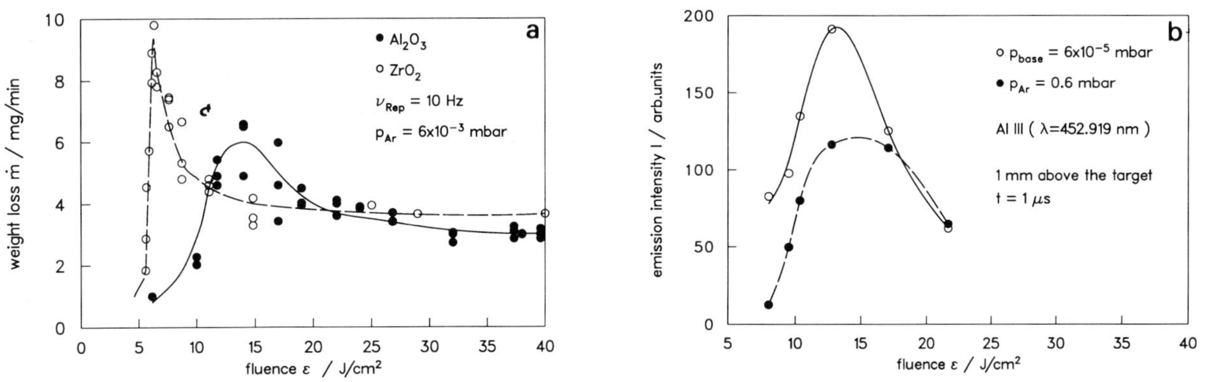

Fig. 5. (a) Weight loss of the target as a function of fluence of the laser radiation. (b) Relative intensity of the Al III emsission line as a function of fluence of the laser radiation at the target surface for base pressure and processing pressure.

geometry into the ambient atmosphere above the target surface and, thus, also in the direction of the incident laser beam, for even higher pressures of about 100 mbar a breakdown in the surrounding processing gas atmosphere takes place (fig. 2).

Although a further increase of fluence does not result in an increase of weight loss of the target material at high laser fluence (fig. 5), higher emission intensities of all lines are observed and simultaneously the overall number of activated and ionized particles and hence the degree of ionization of the vapour/plasma is increasing. The appearance of Ar lines supports the assumption that the surrounding processing gas is activated by the evaporated material and by the absorption of laser radiation. Due to the higher absorption coefficient the resulting vapour/plasma state absorbs the incident laser radiation, which reduces thereby the material removal as supported by the decreasing weight loss at high fluences (fig. 5). Time-resolved emission spectroscopy shows that at the beginning of the laser radiation highly ionized species are formed (fig. 3) which change with time into single ionized and neutral species by recombination. The relative emission intensities are also depending on the processing gas pressure and the distance from the target (fig. 4). At high expansion velocities of the vapour/plasma plume at base pressures the maximum intensity occurs nearly at the same time in the core and in the ambient plasma. At low expansion velocities of the vapour/plasma plume at processing gas pressures of 0.6 mbar the maximum intensity is detected about 2 μs later in the ambient plasma than in the plasma core (fig. 3).

Films were deposited in the pressure range of 10^{-2} mbar, where the detonation wave (fig. 2) only absorbs a part of the incident laser energy, which is supported by the decrease of the emission intensity of the Al III triplet measured at the target surface 1 μs after the beginning of the laser radiation (fig. 5b). A part of the laser energy is absorbed in the vapour/plasma above the target surface, thus not contributing to the thermal vaporization and non-thermal excitation processes in the material removal at the surface. The absorption of the laser radiation in the vapour/plasma and the removal of target material is a self-regulating process at higher laser fluence which in equilibrium yields a constant weight loss of the target (fig. 5a). The properties of the vapour/plasma plume, which depend on the processing gas pressure, the laser fluence and the distance from the target, affect strongly the properties of the deposited films [9]. Amorphous films are generated without contact of the luminous plasma zone with the substrate, because of the low energies of the vapour/plasma particles impinging on the substrate surface yielding consequently low substrate temperatures. Crystalline films are obtained with contact of the luminous plasma zone with the substrate because of the high energies of the vapour/plasma particles impinging on the substrate resulting as a consequence in high substrate temperatures. Films deposited at various distances from the target on cold substrates show characteristic differences in growth and film structure known from films deposited by conventional deposition techniques on substrates at high temperatures [18].

5. Conclusion

Investigations of the vapour/plasma plume during PLD by different optical detection methods as time-resolved emission spectroscopy and high-speed photography show highly activated species with high kinetic energies in high ionization states and densities. Depending on laser fluence and properties of processing gas pressure different vapour/plasma states are observed, which are used in combination with the geometrical arrangement of target and substrate to generate various film structures in view of applications.

Acknowledgement

The authors are very indebted to the Ministry for Research and Technology of the Federal Republic of Germany for financial support within the framework "PVD at low temperatures and new coatings (13 N5611)".

References

[1] H. Frey and G. Kienel, Dünnschichttechnologie (VDI, Düsseldorf, 1987).

[2] S. Schiller, G. Beister, U. Heisig and H. Foerster, J. Vac. Sci. Technol. A 5 (1987) 2239.

[3] E.W. Kreutz, M. Krösche, H. Sung, A. Voss, T. Leyendecker, O. Lemmer and G. Erkens, in: Dünnschichttechnologien, Vol. 2, (VDI Technologiezentrum Physikalische Technologien, Düsseldorf, 1990).

[4] J. Funken, E.W. Kreutz, M. Krösche, H. Sung, A. Voss, G. Erkens, O. Lemmer and T. Leyendecker, Surf. Coat. Technol. 52 (1992) 221.

[5] B. Roas, J. Schultz and G. Endres, Appl. Phys. Lett. 53 (1988) 1557.

[6] H. Sankur, W.J. Gunning, J. DeNatale and J.F. Flintoff, J. Appl. Phys. 65 (1989) 2475.

[7] J. Klosowski, H. Mai, G. Oertel, M. Procop and S. Völlmar, Proc. 12th Int. Congr. X-Ray Optics and Microanalysis, Cracow, Poland, 1, 1989, p. 271.

[8] A. Matsunawa, S. Katayama, H. Miyazawa and K. Oka, Trans. Jpn. Weld. Soc. 19 (1990) 113.

[9] E.W. Kreutz, M. Krösche, H. Sung, A. Voss, A. Jürgens and T. Leyendecker, Surf. Coat. Technol. 53 (1992) 57.

[10] H. Sankur, J.G. Nelson, A.T. Pritt, Jr. and W.J. Gunning, J. Vac. Sci. Technol. A 5 (1987) 15.

[11] J. Reader and C.H. Corliss, Wavelengths and Transition Probabilities for Atoms and Atomic Ions (US National Standard Reference Data System, Washington, 1980).

[12] R.W.B. Pearse and A.G. Gaydon, The Identification of Molecular Spectra (Chapman & Hall, London, 1976).

[13] E.W. Kreutz, M. Krösche, H. Sung and A. Voss, Appl. Surf. Sci. 54 (1992) 69.

[14] H.R. Griem, Spectral Line Broadening by Plasmas (Academic Press, New York, 1964).

[15] H.R. Griem, Plasma Spectroscopy (McGraw-Hill, New York, 1964).

[16] P.E. Dyer, A. Issa and P.H. Key, Appl. Surf. Sci. 46 (1990) 89.

[17] Ya.B. Zel'dovich and Yu.P. Raizer, Physics of Shock Waves and High-Temperature Hydrodynamic Phenomena (Academic Press, New York, 1966) p. 343.

[18] J.A. Thornton, Annu. Rev. Mater. Sci. 7 (1977) 239.

Applied Surface Science 69 (1993) 180–184
North-Holland

applied
surface science

Time-resolved absorption studies of excimer laser ablation of CaF_2

R. Mitzner, A. Rosenfeld and R. König

Institute of Nonlinear Optics and Time Resolved Spectroscopy, Rudower Chaussee 5, D-1199 Berlin, Germany

Received 2 June 1992; accepted for publication 30 October 1992

Time-resolved optical absorption spectroscopy has been used to determine the particle densities of Ca, Ca^+, Ca* and relative densities of CaF molecules produced by KrF-laser ablation of CaF_2 in air. The time-resolved spectra show a very strong broadening of the absorption lines at short delay times near the surface. The broadening mechanism has been discussed. The spatial and temporal behaviour of the particles in the ablation plume near the surface was found to be strongly affected by both the air atmosphere and the reaction dynamics in the plasma zone.

1. Introduction

Laser ablation of dielectric oxides and fluorides is of great practical interest. There are several papers dealing with either the deposition of dielectric films [1] or with the damage threshold of these wide-bandgap materials [2].

Important information about the mechanism of ablation processes and the following reactions in the gas phase have been obtained analyzing the plasma plume by means of mass spectroscopic [3] and emission spectroscopic techniques [4,5]. Nevertheless, the established methods have certain limitations concerning e.g. the quantitative time-resolved detection of the ablated species (atoms, molecules, ions or molecular fragments) and the quantitative detection on atmospheric conditions. Geohegan and Mashburn [6] used the time-resolved absorption method to investigate the composition and the dynamics of the plume during laser ablation of a high-temperature superconductor. In this paper we present results of time-resolved absorption spectroscopy investigations of the KrF-laser ablation of CaF_2. These studies demonstrate the ability of the absorption method to determine quantitative particle densities even in the ablation plasma, near the surface and with short delay to the laser pulse. In this way the limitations of the emission and mass spectroscopic methods can be partly overcome.

2. Experimental

In all ablation experiments the mechanically roughened (111) surface (roughness about 1 μm) of optical-grade CaF_2 was used. After every laser shot the target was shifted to provide fresh surface to the next laser shot.

Fig. 1 shows the experimental set-up used to

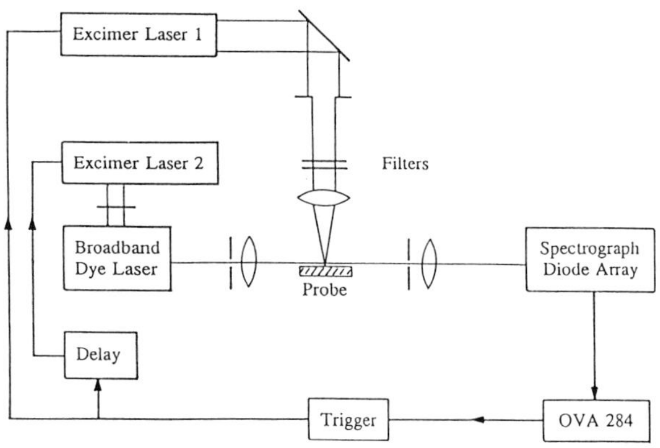

Fig. 1. Experimental set-up for the measurement of the time-resolved absorption spectra of the ablation plume.

obtain the time-resolved absorption spectra of the ablation plume. A Lambda Physik EMG 1003i KrF excimer laser (248 nm, 20 ns) was used as ablating laser. After passing a variable dielectric attenuator and a rectangular aperture the laser beam was imaged by a 120 mm lens onto the target (demagnification factor 20:1). Pulse energies were measured by a pyroelectric detector (ZWG Berlin). The size of the laser spot on the target was determined to using a beam diagnostic system (Laserlaboratorium Göttingen). The second laser which was in a defined and adjustable delay to the ablating laser pumped a broad-band dye laser. The probe laser beam had to pass a small pinhole (250 μm) and was focussed into the ablation plume parallel to the target surface. The transmitted dye laser beam was focussed onto the entrance slit of a 0.25 m spectrograph and detected and processed by an optical multichannel analyzer (OVA 284, ZWG Berlin). The spectra were averaged over three shots. At the target the probe laser beam had a diameter below 100 μm (beam waist about 50 μm). All ablation experiments were performed in air atmosphere.

3. Results and discussion

In the ns laser ablation of CaF_2 both laser fluence fluctuation and surface alteration show a great nonlinear influence on the ablation process. This extreme nonlinear behaviour resulting from the wide bandgap of the CaF_2 [2] is responsible for the relatively strong fluctuation of the measured absorption spectra of the ablation plume. We measured in the range from 4 to 10 J cm^{-2} which is well above the ablation threshold for rough CaF_2 indicated by the bright emitting plume.

Fig. 2 shows examples of the time-resolved absorption spectra of the plume obtained in different spectral areas. Beside the Ca atom, the

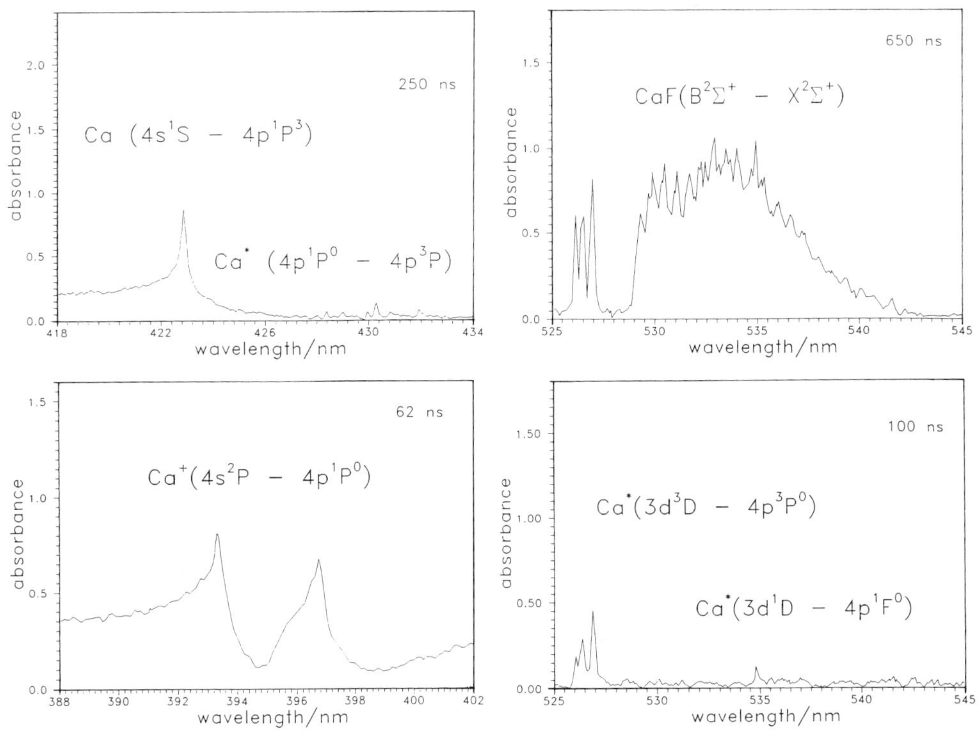

Fig. 2. Absorption spectra of the plume during KrF-laser ablation of CaF_2 (energy density 10 J cm^{-2}, distance to the surface 0.20 mm). The electronic transitions are marked inside the figure.

Ca^+ cation and the CaF molecule, all in the electronic ground state, also excited states of the Ca atom could be detected. The time-resolved absorption spectra show a very strong broadening of the absorption lines up to more than 1 nm at short delay times. This effect is especially marked close to the surface of the target. Three mechanisms could be responsible for this broadening: (1) the Doppler broadening (due to high temperature), (2) the pressure broadening (due to high pressure and high temperature) and the (3) Stark broadening mechanism (due to high electron density).

According to the model of Afanas'ev and Krokhin [7,8] the temperature and the vapour pressure at the vaporation front are estimated to be several thousand Kelvin and above 100 MPa under the conditions of our experiments. The electron density is typically of the order of 10^{16}–10^{18} cm^{-3} [9]. Under these conditions the Doppler broadening can be neglected. Even at a temperature of 10^5 K the Doppler line width is only some picometers [10]. The estimation of the Stark broadening is very complex but tabulated values (10 pm for the Ca II line at 393.37 nm at an electron density of 10^{17} cm^{-3} and 40000 K) [10] suggest that this electron impact broadening mechanism can be responsible for a small fraction of the experimentally observed spectral line width broadening. The pressure broadening can simply be estimated [11]. On the experimental conditions we determined a pressure broadening for Ca of about 1 nm assuming an impact diameter of about 1 nm. Our estimations strongly suggest that the pressure broadening is the main broadening mechanism affecting spectral lines in the ablation plume. The absorption lines of Ca and Ca^+ shown in fig. 2 have unsymmetrical line profiles which are distortions due to refractive index effects in the ablation plume.

An important advantage of the absorption method is the possibility to calculate absolute particle densities N_i according to the following equation:

$$N_i = \frac{g_i 8\pi c}{g_k \lambda^4 A_{ki} l} \int \ln\frac{I_0}{I} \mathrm{d}\lambda .$$

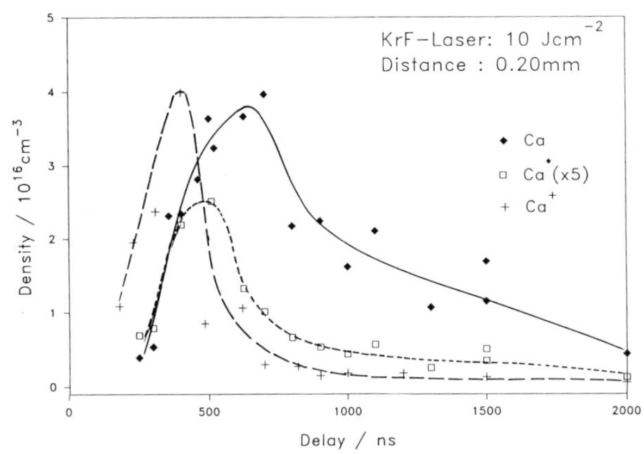

Fig. 3. Temporal evolution of particle densities in the plume at a KrF-laser fluence of 10 J cm^{-2} and a distance to the surface of 0.20 mm.

The speed of light c, the transition probability A_{ki} and the statistical weights g_k and g_i were taken from the literature [12]. The wavelength λ and the integral of the absorption line (I_0 and I are the intensity of the probe beam before and behind the probe at the wavelength λ) were obtained from the experiment. The interaction length l between the probe beam and plume was defined to be constant and comparable to the size of the diameter of the ablated area.

In figs. 3–7 the particle densities of the detected species are plotted versus the delay to the ablating laser for two energy densities and two distances to the surface. At 10 J cm^{-2} and a distance of 0.20 ± 0.04 mm (figs. 3 and 4) all investigated species were found with densities up to 4×10^{16} cm^{-3}. The measurements at smaller distance to the surface could not be evaluated due to the high resonant absorption. It can be clearly distinguished between (1) the Ca^+ cations and the excited Ca atoms peaking at 400 and 500 ns, respectively, with a small temporal distribution, (2) the Ca atoms having a peak density at 600 ns and a remarkable slowed down decrease of their density and (3) the "slow" CaF molecules with their very broad temporal distribution. Decreasing the fluence to 4 J cm^{-2} at 0.20 mm distance (fig. 5) only Ca^+ and Ca were detected. Both the absolute particle density and the ratio of atom-to-cation density had dramatically changed.

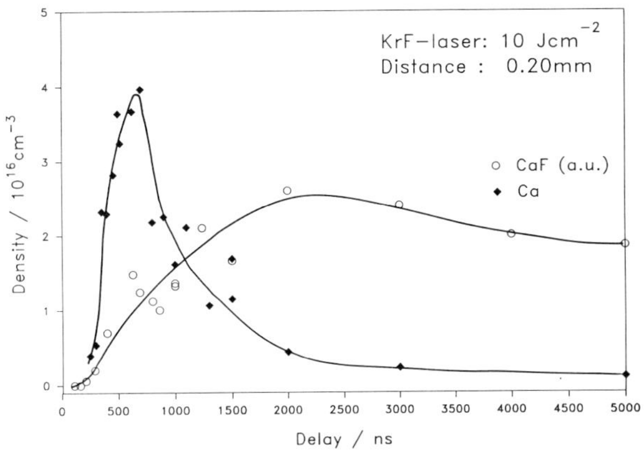

Fig. 4. Temporal evolution of particle densities in the plume at a KrF-laser fluence of 10 J cm^{-2} and a distance to the surface of 0.20 mm.

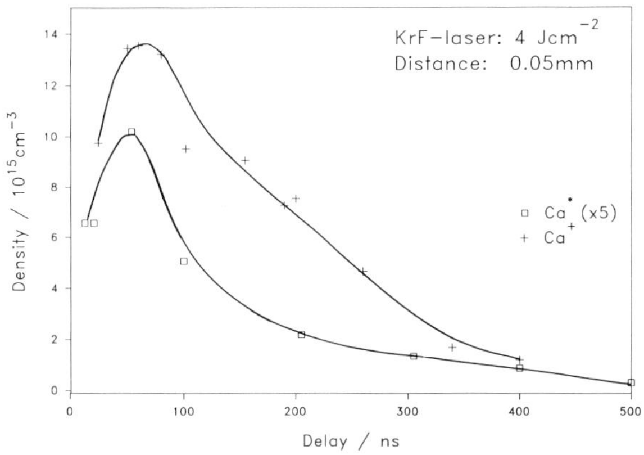

Fig. 6. Temporal evolution of particle densities in the plume at a KrF-laser fluence of 4 J cm^{-2} and a distance to the surface of 0.05 mm.

More close to the surface at 0.05 ± 0.04 mm distance (figs. 6 and 7) the qualitative picture is similar to that found at the high fluence and a distance of 0.20 mm. Naturally, the particle densities were found peaking at shorter delays due to the small distance to the surface.

The conditions in the ablation plume close to the surface are very complex. For a discussion of the dynamics of the particles in the plume several processes must be taken into account: (1) the highly nonlinear ablation process, (2) the expansion of the plume, (3) the effect of the buffer gas (air) resulting in a slowing down and a cooling of the ablated particles and (4) reactions of the particles (especially the recombination of cations and electrons). Especially the air atmosphere and the high density of the particles even in the probe volume alter the temporal distribution to a large extent compared with measurements on vacuum conditions and at a fairly large distance from the surface. Cations are known to have a higher velocity than atoms [13]. But the temporal behaviour of the Ca^+ is also very strongly affected by the electron capture of the Ca^+ producing excited Ca atoms. At a distance of 0.20 mm the

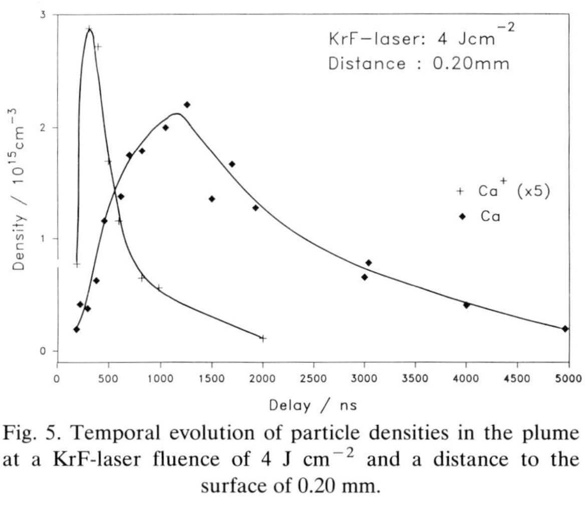

Fig. 5. Temporal evolution of particle densities in the plume at a KrF-laser fluence of 4 J cm^{-2} and a distance to the surface of 0.20 mm.

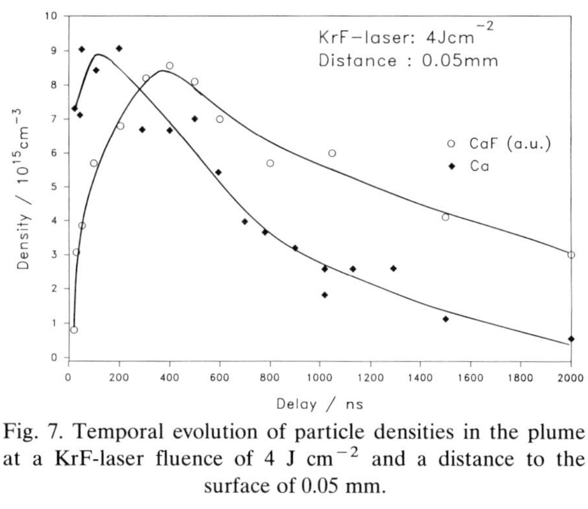

Fig. 7. Temporal evolution of particle densities in the plume at a KrF-laser fluence of 4 J cm^{-2} and a distance to the surface of 0.05 mm.

recombination is the only source of excited atoms because of the low kinetic energy of the particles due to the slowing down by the collision with the molecules of the air atmosphere (mean free path ~ 60 nm). Close to the surface (fig. 6) also inelastic impacts are assumed to play an important role in the population of excited states. The velocity of the ground-state atoms and molecules is strongly affected by the air atmosphere resulting in the slow decrease of their particle density. The low velocity of the CaF molecules compared to the atoms is similar to the results of LIF studies of the excimer laser ablation of Al_2O_3 [3]. In an additional experiment we could detect the CaF molecule even during the pulse length of the ablating laser. This suggested that the CaF molecules are ejected directly from the irradiated surface, but it cannot explain their low kinetic energy.

4. Conclusion

We have used time-resolved absorption spectroscopy to determine the particle densities of Ca and Ca^+ in their electronic ground states, of excited Ca atoms and relative densities of the CaF molecules produced by KrF-laser ablation of CaF_2 in air. The spatial and temporal behaviour of the particles in the ablation plume near the surface is strongly affected by both the air atmosphere and the reaction dynamics in the plasma zone. Measurements on vacuum conditions are planned to obtain a more detailed insight into the reaction processes in the ablation plume. The time-resolved absorption spectroscopy is well suited for the examination of the ablation plume even in the plasma zone.

Acknowledgements

We would like to acknowledge the support from the BMFT.

References

[1] S. Amirhaghi, A. Archer, B. Taguiang, R. McMinn, P. Barnes, S. Tarling and I.W. Boyd, Appl. Surf. Sci. 54 (1992) 205.
[2] J. Reif, S. Petzold, A.P. Elg and E. Matthias, Appl. Phys. A 49 (1989) 199.
[3] S.G. Hansen, J. Appl. Phys. 66 (1989) 1411.
[4] R.W. Dreyfus, R. Kelly and R.E. Walkup, Appl. Phys. Lett. 49 (1986) 1478.
[5] W. Sdorra and K. Niemax, Spectrochim. Acta 45B (1990) 917.
[6] D.B. Geohegan and D.N. Mashburn, Appl. Phys. Lett. 55 (1989) 2345.
[7] Y.V. Afanas'ev and O.N. Krokhin, Sov. Phys. JETP 24 (1967) 639.
[8] R. Poprawe, PhD Thesis, Darmstadt, 1984.
[9] T.P. Hughes, Plasma and Laser Light (Hilger, Bristol, 1975).
[10] H.R. Griem, Spectral Line Broadening by Plasmas, Vol. 39 of Pure and Applied Physics (Academic Press, New York, 1974).
[11] F.K. Kneubühl and M.W. Sigrist, Laser (Teubner, Stuttgart, 1989).
[12] J.R. Fuhr and W.L. Wiese, in: Handbook of Physics and Chemistry, 71st ed. (CRC Press, Boca Raton, FL) pp. 10–136.
[13] W. Marine, M. Gerri, J.M. Scotto d'Aniello, M. Sentis, Ph. Delaporte, B. Forestier and B. Fontaine, Appl. Surf. Sci. 54 (1992) 264.

Applied Surface Science 69 (1993) 185–192
North-Holland

applied
surface science

VUV light production from a formed-ferrite plasma source

M.L. Sentis, Ph. Granier, F. Chazaud
Institut de Mécanique des Fluides de Marseille, UM 34 CNRS, 1, Rue Honnorat, 13003 Marseille, France

W.I. Marine, M. Gerry
Groupe des Etats Condensés, URA 783 CNRS, Faculté des Sciences de Luminy, Case 901, 13288 Marseille Cedex 9, France

and

R.C. Sze
Los Alamos National Laboratory, Los Alamos, NM 87545, USA

Received 2 June 1992; accepted for publication 30 October 1992

Emission spectra from a high-current line plasma produced by means of a formed-ferrite plasma source have been measured in the 1200–2900 Å spectral region. The formed-ferrite plasma source was driven by a 4.6 μF capacitor bank charged up to 25 kV. In order to obtain this intense flash light source, a conducting channel was first made on the ferrite surface which has a measurable high impedance, but orders of magnitudes less than the impedance of the unprocessed ferrite ($> 10^5$ k$\Omega \cdot$cm). Two different methods were successfully used: laser ablation and explosion of a brass wire. A study of the formation and the characteristics of the channel and the surroundings on different ferrite materials has shown chemical and strong microcrystalline structure modifications. We report the electrical and optical properties of such a source by high voltage measurements, time integrated and time resolved VUV optical spectroscopy and frame photographs. An observation of the plasma expansion with the formation and the propagation of a blast wave (1.8 km/s) is presented for the first time. The possibility of a peak intensity shift of VUV radiation to shorter wavelengths is discussed.

1. Introduction

Broadband optical pumping sources have been extensively studied and developed since many years. Among them, open discharge is one of the most suitable sources of optical excitation of gases since in this case a maximum of spectral distribution of photons falls generally within the UV or VUV spectral region where the strongest absorption bands of the overwhelming majority of molecules is located. In open discharges, active gas is directly excited by the pulsed discharge plasma radiation, the discharge device itself taking up a small portion of the total cell volume. Open discharges include exploding wire by an electrical discharge and sliding arc discharges on various types of surfaces and materials [1]. Intense broadband UV and VUV radiations are produced over a wide range of active gas pressures, from vacuum to several atmospheres, with possibilities of a relatively long pulse operation (1 to 10 μs). These characteristics have allowed the open discharges to be attractive optical sources for many applications as preionization or photochemical excitation sources for high power gas lasers.

In this work, we report on an open discharge source called "formed-ferrite plasma source" [2] that creates an intense radiation from the UV to the VUV region and X-ray emission [3]. This

0169-4332/93/$06.00

light source that approximates the exploding wire technique, has the attractive characteristic to be capable of repetitive operation [4] with an efficiency in the order of 20% for the wavelength interval between 140 and 200 nm and 45% for the transmission band of CaF_2 window (140–5000 nm) [5]. We present a study of the formed-ferrite-filament processing that allows to produce intense radiation in the UV and VUV region. The formed channel made on a virgin ferrite ($Ni-ZnFe_2O_4$ or $Li-ZnFe_2O_4$) is obtained by laser ablation or exploding wire. An investigation of the characteristics (impedance, material structure) and the transformation of the filament is performed. The discussion concerning the physical mechanism, which leads to an impedance drop, and a study of the plasma formation and expansion in gas ambient are reported. The corresponding evolution of the spectral emission, measured in the UV and VUV spectral region, 1200–2900 Å, is presented with emphasis on the ambient gas pressure and discharge voltage dependence.

2. Forming procedure and characteristics of the formed-ferrite-filament

A channel called "formed-ferrite-filament" (FFF) has to be formed on the surface of a virgin ferrite in order to initiate an electrical discharge on the ferrite surface. Initiation of this discharge is made possible by dropping the impedance of the written line to $10-10^2$ $k\Omega \cdot cm$, while the impedance of the virgin ferrite exceeds 10^5 $k\Omega \cdot cm$. This line may be formed either by evaporation of an aluminum thin film [2] or by exploding a thin wire [6] onto the ferrite. In the present work the conducting channel on $Ni-ZnFe_2O_4$ or $Li-ZnFe_2O_4$ ferrite has been formed by the technique of exploding a thin brass wire, or by laser ablation [7].

2.1. Laser machining of the ferrite filament

Ferrite filament was ablated by different types of lasers: excimer ArF ($\lambda = 193$ nm), excimer XeCl ($\lambda = 308$ nm) and YAG ($\lambda = 1.06$ μm) lasers. The general conclusion of these experiments was that the use of the excimer laser allows to form the conductive line on the ferrite surface. However, we could not reach (after a reasonable time of irradiation of less than 1000 shots on the same surface) a strong modification of the materials properties and concomitantly induce an important resistivity change of the ablated lines. We explain these results by congruent ablation of the surface and a very shallow melting depth of the ferrite during UV irradiation. Several years ago we showed that using of the pulse of the mode-locked YAG laser induces a non-stoichiometric ablation of multicomponent target [8]. This kind of laser was used in the present work to ablate ferrites.

2.1.1. Experimental of the ferrite filament machining

The ferrite rods were irradiated in air with the pulse train of a mode-locked YAG laser which had 9 ns time separation between 35 ps (FWHM) pulses, the train duration was about 100 ns. The laser beam was focused on a spot of about 120 μm, so that the laser fluence could be varied between 1 and 20 J/cm^2. The ferrite rod was placed on the translation stage with a microcomputer controlled displacement. Usual process parameters were: laser repetition rate 10 Hz and scanning rate 0.8–1 mm/s. During irradiation, the DC resistivity change was controlled by a two points method. The results of the laser processing were characterized by scanning electron microscopy (SEM), energy dispersive X-ray spectroscopy (EDX), and X-ray diffractometry (XRD).

2.1.2. Results and discussion

SEM analysis of the untreated ferrite shows a typical grained structure of the sintered material, with grain sizes of about 1–5 μm. Laser irradiation of the ferrite induces strong changes of its morphology and conduction properties.

Fig. 1 shows the Ni–Zn ferrite surface morphology after laser irradiation. Laser energy density was about 20 J/cm^2. The depth and the width of the ablated channel are about 30 and 120 μm, respectively. After laser processing the microcrystalline morphology of the virgin ferrite was transformed into a more dense structure. Many cracks due to the thermal shock and a more compact structure of the treated zones were observed along the laser-ablated channel. However, the presence of these cracks does not have any crucial influence on the conduction properties of the channel.

Fig. 2 shows the evolution of the FFF impedance as a function of the YAG laser scans number on each 1 cm-line. The four channels were interconnected in parallel and the distance between two adjacent channels was 200 μm. Each channel corresponds to a total number of 80 shots. One shot produces a hole with a diameter of 120 μm. It can be observed that, for a given channel, the overall channel impedance decreases down to a saturation value, reached after 5 scans over the same channel. The lower the number of the channels, the greater this saturation value is. The reason is that several channels behave like several impedances connected in parallel. But we notice that the presence of n channels decreases the overall impedance to more than n times; thus, these conditions are promising to obtain an impedance as low as possible.

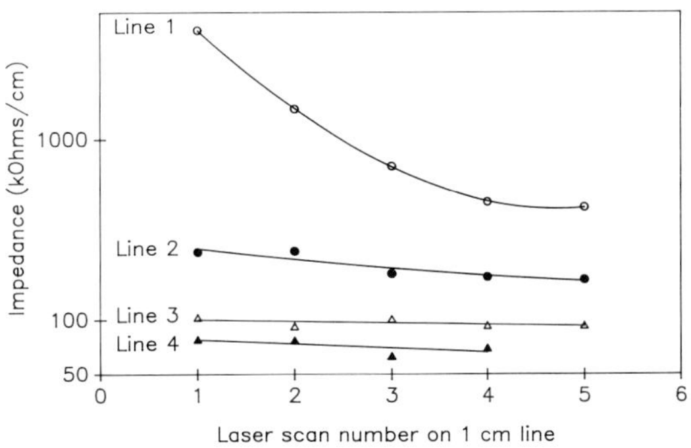

Fig. 2. Evolution of the FFF impedance as a function of the laser scans per cm. The four lines are connected in parallel.

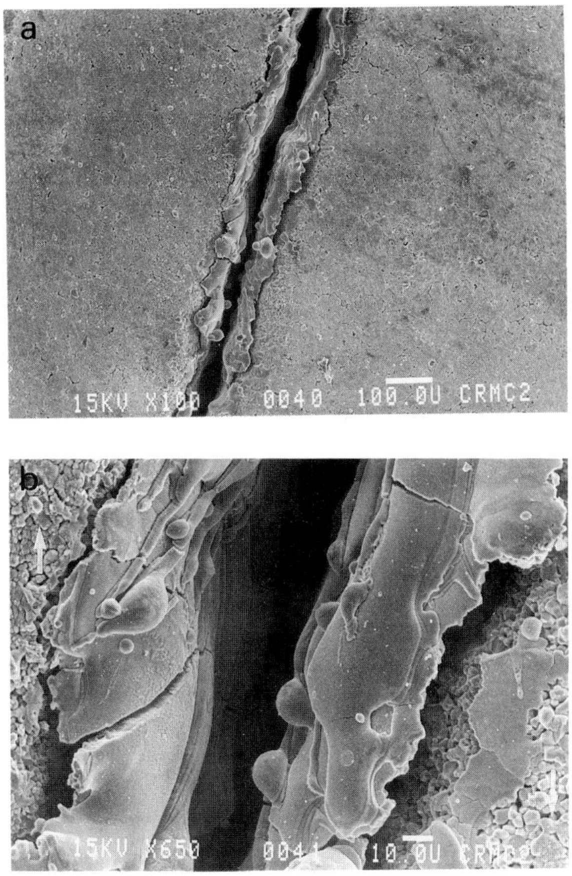

Fig. 1. SEM image of the ferrite filament formed by YAG laser ablation. (a) General view of the channel. Scanning rate and laser energy density were 0.8 mm/s and 20 J/cm^2, respectively. (b) Magnified view of the ablated zone. White arrows show microcrystalline morphology of the virgin ferrite.

The electrical properties of the spinel ferrites are governed by the distribution of the iron and the bivalent metal ions in the spinel lattice. The EDX analysis of the laser-treated zones shows the increase of about 2–4% of the Zn concentration with a corresponding decrease of the Fe concentration down to 86%. The change of the ferrite composition influences the change of the crystalline structure. From X-ray diffraction experiments we found an expansion of the fcc lattice parameters from $a = 8.38$ Å (untreated ferrite) to $a = 8.40$ Å (laser-treated ferrite). In fact, it is well known that incorporation of bivalent ions in tetrahedral sites induces a large lattice expansion [9]. Our observation of conductivity decrease is in agreement with the results of Ravinger and Seshagiri Rao [10], who show that the electrical conductivity of mixed ferrites like $Li_{(0.5-x)/2}Zn_xFe_{(2.5-x)/2}O_4$ decreases with the increase of the Zn content up to $x = 0.6$ and increases if the Zn content is increased above this

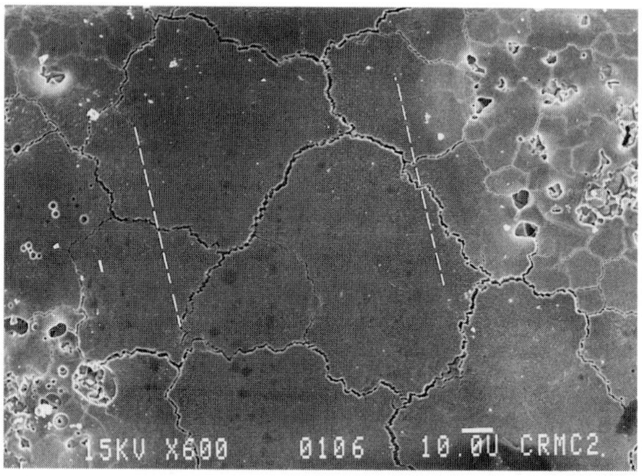

Fig. 3. SEM micrograph of the ferrite filament formed by wire explosion. The dashed line shows a contact zone of the wire and the ferrite.

value. Kishan et al. have observed the same effect but their threshold value was 0.4 [11].

Another effective process in the raising of the ferrites resistivity is oxidation. Our experiments confirm this result. One must note that processed samples in contact with air for a few hours have shown an increased resistivity of about 30–40%. Moreover, as oxygen is the first element displaced by the laser processing, the variation of its concentration in ferrites must be drastic for the decrease of the sample resistivity.

2.2. Ferrite filament formed by wire explosion

A brass wire of 0.2 mm diameter was placed in contact with the ferrite surface and was exploded in 1 atm of argon. Argon was used to prevent oxidation as was discussed above. The stored energy for the wire explosion was about 1 kJ.

Fig. 3 shows the surface structure of the ferrite formed by this method. Two significant zones can be observed. The first one is the zone (delimited by the dashed line) where the wire was in contact with the ferrite surface. This zone is characterized by a flat surface morphology of the homogeneous domain, separated by the cracks formed, probably, after heating and melting with subsequent solidification of the zone which was in contact with the wire during its explosion. The micromorphology of this domain is quite different compared to the virgin ferrite microstructure (fig. 1b).

Characteristic of the second zone (right and left parts of the SEM image) is a formation of a

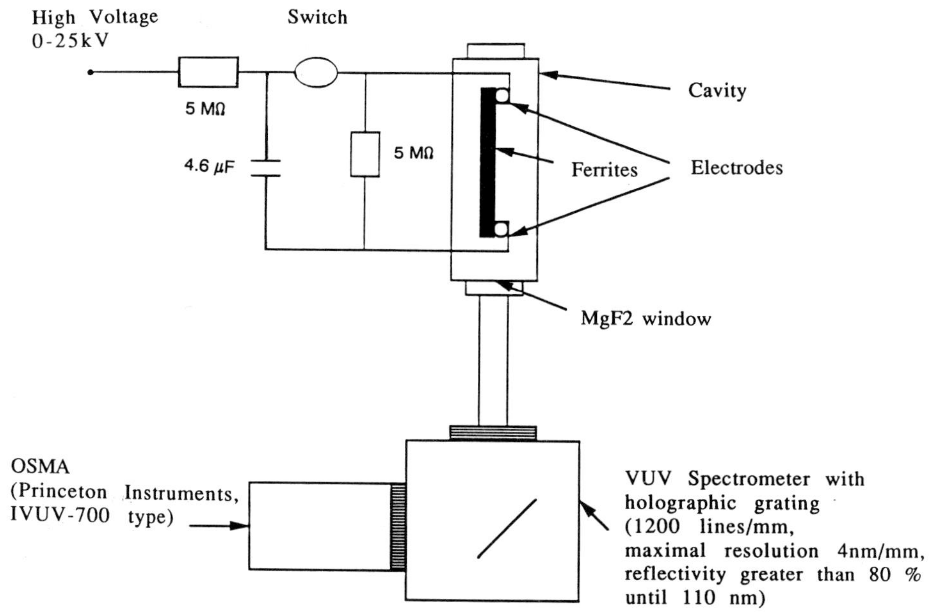

Fig. 4. Schematic of the experimental device including the spectroscopical setup.

smaller flat domain with a typical size of about 10 μm. The dimension of this domain is very close to the size of the microcrystalline zones on the unprocessed ferrites (fig. 1a). However, after the wire exploded, the featureless structure of this domain clearly indicates the strong surface heating. Another peculiar feature, in the second zone, is the formation of numerous large craters and holes with diameters of about 5–20 μm and 2–3 μm, respectively.

The mechanism of the surface modification is not yet known. But we believe that the melting of the ferrite is the dominant process. The following possibility is the doping of the surface by the wire particles or by the droplets generated during shock-wave formation. The impact of these atoms and particles leads to the shift of the surface resistivity and to the local heating, which acts as a local annealing.

After the wire explosion the resistivity of the preformed ferrite is about 50–100 $k\Omega \cdot cm$. After some more discharges (10 to 20 shots) the surface resistivity decreases to a value of about 5–10 $k\Omega \cdot cm$ and the processed ferrite channel shows excellent discharge properties.

3. Experimental apparatus

A schematic of the experimental device, including the setup of the spectroscopic study, is shown in fig. 4. A $(Ni–Zn)Fe_2O_4$ or $(Li–Zn)Fe_2O_4$ ferrite road (10×200 mm) was fixed by two brass electrodes in a PVC cell evacuated to better than 10^{-4} Torr and filled with various gases up to 1.5 atm. The formed-ferrite-filament (FFF) was driven by using an electrical circuit which consisted of an energy storage capacitor of 4.6 μF charged up to 25 kV and a triggered spark switch. Voltage and current waveforms were monitored with a high-voltage probe (Tektronic model P 6015) and a current transformer (Stangenes model 3 – 0.01), respectively.

The optical radiation pulse in the UV and visible spectral region was transversally measured by an ITL model TF 1850 photodiode. The formed-ferrite flash source was connected to a 1/4 meter ARC vacuum spectrometer or to a

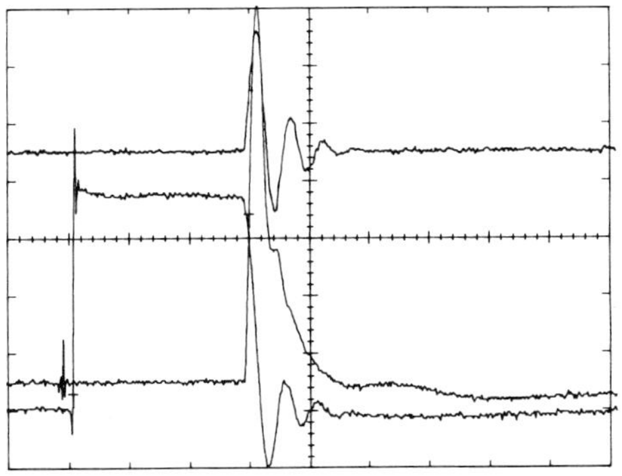

Fig. 5. Typical optical pulse (PD), voltage (V), and current (I) waveforms for FFF plasma operated in 1 atm argon with a charging voltage of 19 kV.

1/2 meter McPherson vacuum spectrometer (diffraction grating of 1200 groves/mm) via a half meter long pipeline. Light in the 120–290 nm spectral region was detected by a solar blind R 1220 Hammamatsu photomultiplier with a CsTe photocathode via a MgF_2 window or by a VUV optical spectrometric multichannel analyser (Princeton Instruments model IVUV 700). Plasma expansion was imaged by a digitally controlled CCD video camera featuring exposure times down to 5 ns (PCO model Dicam-2).

4. Results and discussion on plasma light source

Fig. 5 shows the measured optical pulse (PD) in the UV and visible spectral region (200–800 nm), the voltage (V) and current (I) waveforms for a charging voltage of 19 kV in 1 atm of argon. The voltage on the FFF reaches a steady value equivalent to the charging voltage in a short rise time, negligible as compared to the time scale used (20 μs/division). After a delay time of about 60 μs, the steady-state applied voltage collapses. The delay time is mainly dependent on the length of the FFF and on the charging voltage value. During this interval time, the current (< 1 A) ohmically heats the filament to very high temperatures causing it to emit thermal electrons and ions. The ferrite and gas impedance now collapse

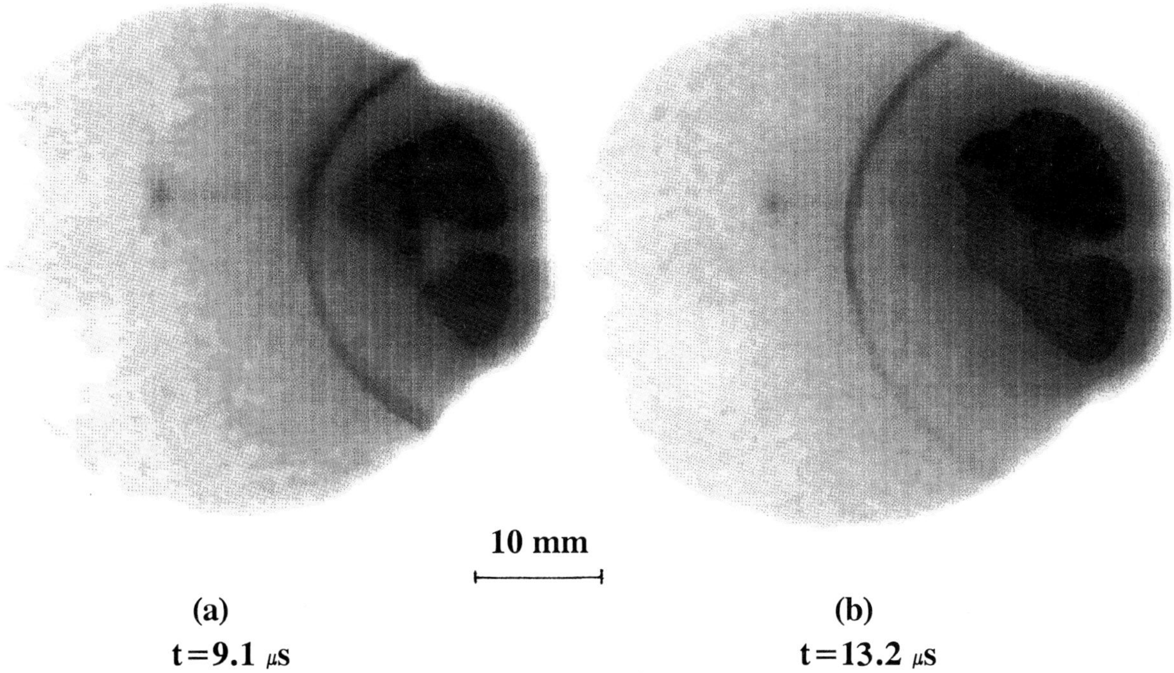

Fig. 6. Transversal plasma expansion in 1 atm argon and a charging voltage of 19 kV after a delay time of 9.1 (a) and 13.2 μs (b). The inhomogeneity in the center of the bright plasma comes from the presence of the anode of the discharge circuit in the CCD camera shot.

and the bulk of the stored energy is deposited in the FFF channel, its surrounding area, and into the ambient gas. The ringing current reaches a maximum value of 25 kA, corresponding to an impedance of 0.1 Ω for a channel of 18 cm long, at a rising time of about 2.5 μF. An intense

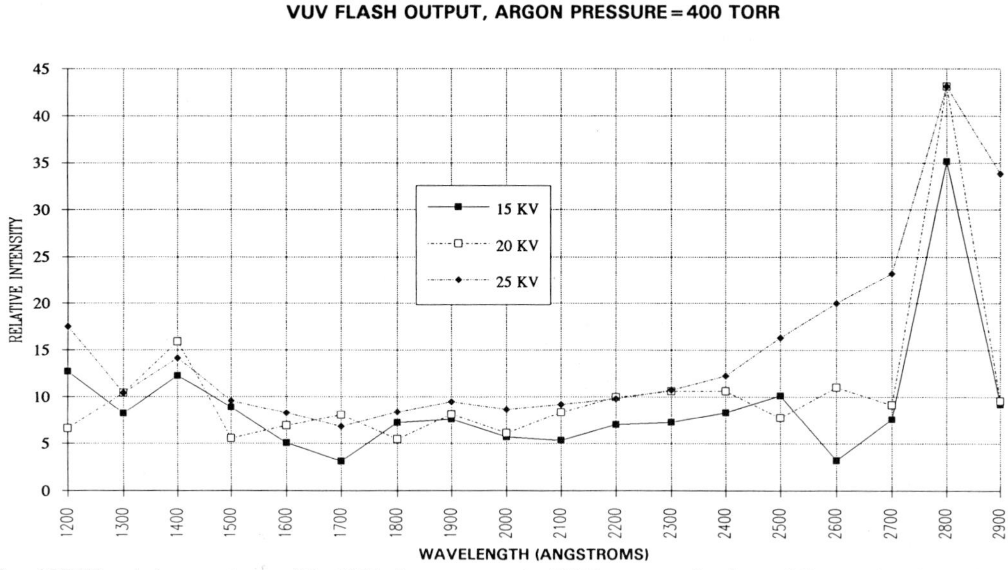

Fig. 7. UV and VUV emission spectrum of the FFF plasma source in 400 Torr argon for three different charging voltage values: 15, 20 and 25 kV.

radiation of approximately 4.5 μs FWHM duration is observed near the ferrite surface with a peak corresponding to the maximum power deposition of about 125 MW.

Fig. 6 demonstrates the transversal plasma expansion initiated by an electrical discharge applied on the formed-ferrite-filament. This experiment was performed in the same experimental conditions as for fig. 5 ($V_c = 19$ kV, 1 atm argon). The two-dimensional digitized photographs of the plasma expansion were recorded with an exposure time of 80 ns at two delay times of 9.1 (fig. 6a) and 13.2 μs (fig. 6b), respectively, after the steady-state voltage had collapsed.

The electrical energy deposition produces a highly emitting plasma column of an approximately half-cylindrical form which expands radially. During a first time interval (0 ~ 5 μs), the transversal velocity of the plasma propagation is about 1.5 km/s. Then the plasma column is surrounded by a layer of a shock-compressed gas, which is seen in figs. 6a and 6b, as being due to the scattered light. The blast wave is at first observed after (approximately) a delay time of 5 μs which corresponds to the time when the discharge current reaches its maximum value (25 kA). The observed blast wave form is a perfect cylindrical layer. Between 5 and 9 μs the mean velocity of the blast wave is ~ 1.8 km/s. As the plasma expands, the mean velocities of the blast wave and plasma decrease to 0.9 and 0.7 km/s, respectively, at a delay time of 17 μs. After around 20 μs the blast energy is too low to still observe the scattered light, while plasma light is still perceptible during more than 50 μs. Photographs have shown very uniform longitudinal plasma along the 18 cm of the FFF.

To our knowledge, the spatial development of the plasma of formed-ferrite-filament is observed for the first time. This spatial development is comparable to a plasma initiated by an exploding wire or during laser ablation of material at a high background gas pressure.

Figs. 7 and 8 give the relative light intensities from the formed-ferrite-filament plasma source for 600 and 400 Torr of argon, respectively, and for various values of charging voltage (15, 20 and 25 kV). Each point of the different curves correspond to the maximum of light intensity for a given wavelength. The arbitrary scale of the relative intensity is the same at 400 and 600 Torr. The spectral resolution was approximately 17 Å

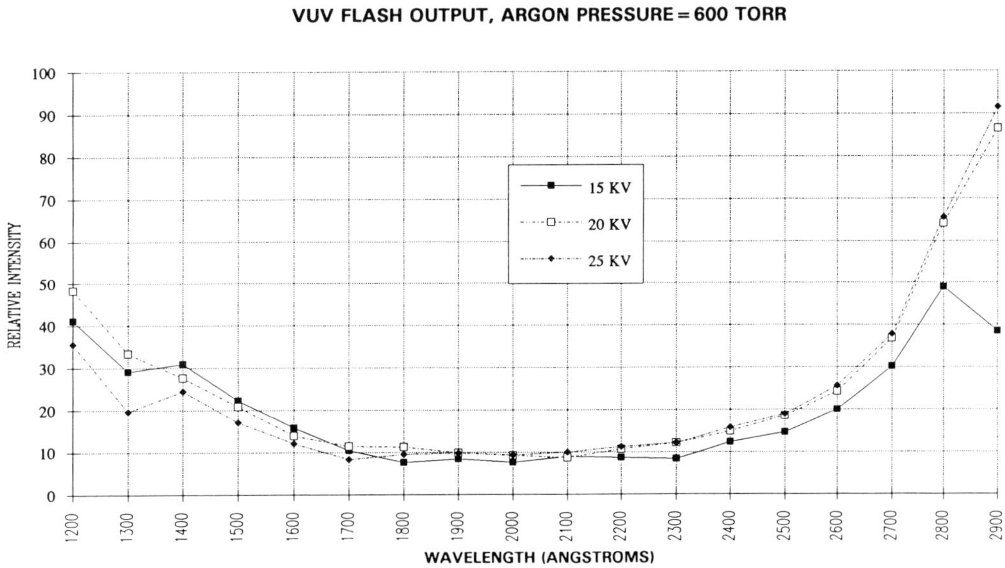

Fig. 8. UV and VUV emission spectrum of the FFF plasma source in 600 Torr argon for three different charging voltage values: 15, 20 and 25 kV.

for an output aperture of the spectrometer of 1 mm. The data have been corrected for the photomultiplier and for holographic grating array responses. The maximum relative intensity is observed in the 2700–2900 Å spectral region. A shift of about 100 Å to shorter wavelengths, which seems to appear when the background gas pressure is decreased, has to be confirmed by a more detailed study in the 2700–3200 spectral region. The light intensity is continuously rising as we go to shorter wavelengths and should peak substantially below 1200 Å.

5. Conclusion

In this work we have reported the experimental study of an intense formed-ferrite-filament plasma source which shows an efficient emission in the UV and VUV region. With the exploding wire technique, we were able to decrease the impedance of the ferrite channel from 10^5 to 5 $k\Omega \cdot cm$. SEM shows high density structures on the formed-ferrite-filament. By YAG laser ablation we demonstrated the possibility to decrease the impedance to 20 $k\Omega \cdot cm$ and to control its value by laser processing. X-ray fluorescence studies show an increase in zinc content and a decrease in iron content. Microcrystalline structure of virgin ferrite is transformed, by YAG laser ablation, to a high density structure due to melting down and rapid cooling of the ferrite surface. Laser ablation technique allows one to control the shape and the impedance value of the formed-ferrite-filament and should dramatically increase the lifetime of FFF.

A formed-ferrite-filament plasma source was built that allows one to inject an electrical energy of 1 kJ in a cell along the FFF of 18 or 36 cm long. A current of up to 25 kA produces an intense plasma flash in argon gas (1 atm) with a pulse duration of 4.5 μs FWHM. A study of plasma expansion allows, one to our knowledge for the first time, to observe a blast wave in such an experiment. This blast wave expands in a transversal direction to the ferrite surface with an average velocity of about 1.8 km/s. The propagation velocities of the blast wave and plasma are related to the rate of electrical energy deposition, background gas pressure. A spectroscopic study of light produced in the spectral region of 1200–2900 Å demonstrates an intensive radiation peak at shorter wavelengths.

Acknowledgements

The authors would like to thank their colleagues for valuable suggestions or technical assistance. We greatly appreciate the collaboration with the Rice University group, especially Professor F.K. Tittel. This work was supported by the DRET.

References

[1] V.S. Zuev and L.D. Mikheev, Photochemical Lasers (Harwood, Switzerland, 1991).
[2] K. Watanabe, S. Kashiwabara and R. Fujimoto, Appl. Phys. Lett. 50 (1987) 629.
[3] K. Watanabe, K. Furusho, S. Kashiwabara and R. Fujimoto, J. Appl. Phys. 68 (1990) 5059.
[4] K. Watanabe, F.A. van Goor, R. Fujimoto and W.J. Witteman, GCL – Seventh International Symposium on Gas Flow and Chemical Lasers, SPIE 1031 (1988) 450.
[5] R.W.F. Gross, L.E. Schneider and S.T. Amimoto, Appl. Phys. Lett. 53 (1989) 2365.
[6] X.F. Han, T. Sakai, G.J. Hirst, R. Sauerbrey and F.K. Tittel, Phys. Lett. A 137 (1989) 34.
[7] R.C. Sze and M.L. Sentis, Int. Proc. Lasers 91, to be published.
[8] W. Marine, M. Peray, Y. Mathey and D. Pailharey, Appl. Surf. Sci. 43 (1989) 377.
[9] B. Viswanathan and V.R.K. Murthy, Ferrite Materials (Springer, Berlin, 1990).
[10] D. Ravinger and T. Seshagiri Rao, Cryst. Res. Technol. 25 (1990) 1079.
[11] P. Kishan, D.R. Sager and P. Swarup, J. Less-Common Met. 108 (1985) 345.

Applied Surface Science 69 (1993) 193–197
North-Holland

applied
surface science

New lamp arrangement for rapid thermal processing

J.-P. Zöllner, K. Ullrich, J. Pezoldt and G. Eichhorn
Institut für Festkörperelektronik, TU Ilmenau, Postfach 327, D-6300 Ilmenau, Germany

Received 2 June 1992; accepted for publication 30 October 1992

Temperature uniformity across the wafer is difficult to obtain in RTP systems. It depends essentially on heat source design and chamber geometry. In this work a lamp arrangement commensurate with wafer symmetry is presented. Numerical and experimental investigations show that the chosen lamp and reactor design is able to compensate the radial temperature decrease due to re-absorbed heat radiation from the wafer.

1. Introduction

For the last few years, equipment for rapid thermal processing (RTP) has been successfully used in microelectronics technology of silicon, for dopant activation and defect annealing after ion implantation [1,2]. Further applications for RTP are oxidation (RTO) [3], nitridation (RTN) [4] and cleaning (RTC) [5]. Recently, RTP has also been used for chemical vapour deposition (RTCVD) [6].

RTP is a kind of single-wafer processing. Compared to conventional batch processing, RTP has the advantage of better wafer-to-wafer uniformity and compatibility with cluster tools used for vacuum and low-pressure processes.

RTP equipment has to guarantee homogeneous and reproducible processes at the wafer in order to achieve a uniform thickness of thin layers and pn-junction depth. This requires first of all exact wafer temperature adjustment with respect to the absolute value and its distribution over the wafer.

Exact adjustment of absolute temperature requires precise temperature measurement and regulation. Especially temperature measurement is difficult – normally it is carried out optically – since wafer radiation intensity considerably depends on its emissivity. Emissivity itself is a function of surface roughness, wafer temperature, type and thickness of existing layers on the wafer, realized microstructures, and wavelength used for temperature measurement. During layer deposition optical properties of the wafer, for example emissivity, may change substantially. This can lead to considerable errors during temperature measurement, as shown in refs. [7,8]. That is why a change of emissivity during layer growth processes must be considered, for example by means of emissivity in situ measurement.

Non-uniformity of temperature distribution results from larger heat losses at the wafer edge by radiation, convection and cooling caused by gas flow. To compensate this effect areas with higher heat losses (wafer edge, part of the wafer near the gas inlet) must be irradiated with higher intensity. This requires appropriate lamp arrangement and reflector shaping, since constructive defects cannot be compensated only by single lamp control.

This paper presents first results of a new rotationally symmetric lamp arrangement. A similar construction was presented by Moslehi et al., who used a complicated reactor arrangement, precisely designed for one wafer size [9]. The purpose of this work is to present a design of a simple lamp arrangement which can be used for several wafer sizes.

0169-4332/93/$06.00

2. Lamp arrangement

A rotationally symmetric lamp arrangement was chosen with respect to wafer symmetry. This arrangement and the cylindrical reactor are shown in fig. 1. A stepped arrangement of lamp rings with each ring consisting of 10, 20 and 30 halogen lamps was chosen. To provide increased illumination on the wafer edge, the outer ring was placed near the wafer. Lamp arrays are located above the reactor. The stainless steel reactor is separated by a quartz glass window from the lamp arrays. The wafer temperature is sensed using an IR pyrometer (5–10 μm) focused on the backside of the wafer. Principally, the reactor arrangement admits wafer irradiation from both sides. The size of the reactor and the lamp field were dimensioned so that wafers up to 6 inches in diameter could be treated. Appropriate scaling up may provide applications for larger diameters of wafers.

3. Simulation of heat transfer

A steady-state temperature distribution over the wafer was determined on the basis of the heat balance. Heat balance consist of (i) radiation emitted by lamps and absorbed by the wafer, (ii) radiation emitted by the wafer, and (iii) heat loss caused by gas flow and gas heat conduction. In this study the gas flow influence on the temperature distribution was neglected. This assumption is correct for processes in a pressure range below 10 Torr and medium gas flow velocities (lower 10 m/s).

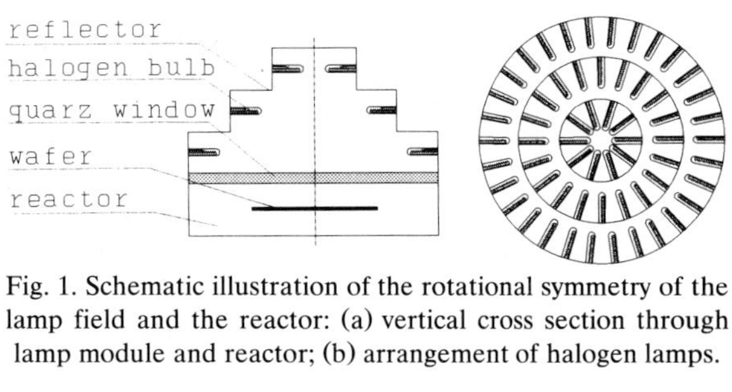

Fig. 1. Schematic illustration of the rotational symmetry of the lamp field and the reactor: (a) vertical cross section through lamp module and reactor; (b) arrangement of halogen lamps.

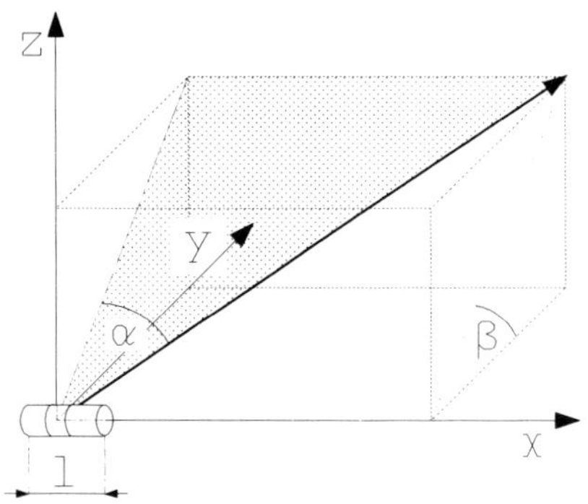

Fig. 2. Characterization of chosen ray with angle α and β for cylindrical light source.

Radiation by lamps was calculated with a ray-tracing method. The lamp radiation characteristic was split into numerous beams. The beam splitting of the cylindrical light source is shown in fig. 2.

The emitted radiation from a small-area source is subject to Lambert's law. The Lambertian emitted radiation described by angles α and β can be expressed by

$$\Delta P(\alpha, \beta) = P_L \cos^2\alpha \frac{\Delta\alpha \, \Delta\beta}{\pi^2} \frac{\Delta l}{l}, \tag{1}$$

where P_L is the power of the lamp, l the helix length, Δl the helix length element, and $\Delta\alpha$ and $\Delta\beta$ are angle elements. The number of rays to be considered for each lamp is determined by choosing $\Delta\alpha$, $\Delta\beta$ and Δl. Each beam is being tracked on its way through the reflecting housing, until its intensity is reduced to a given limit value by absorption ($\Delta P(\alpha, \beta) \leq \Delta P_0(\alpha, \beta)$). The beam can strike the reactor surface, the reflector surface, the wafer or the quartz window. These processes are characterized not only by intensity reduction but also by a change of beam direction. In the case of specular reflection, the beam orientation can be follows from the reflection law under consideration of the reflecting surface

$$\Phi_{in}(\alpha_{in}, \beta_{in}) = \Phi_{out}(\alpha_{out}, \beta_{out})$$

and, in the case of refraction from

$$n_1 \sin(\Phi_{in}(\alpha_{in}, \beta_{in})) = n_2 \sin(\Phi_{out}(\alpha_{out}, \beta_{out})).$$

The incident beam (Φ_{in}) and the reflected beam (Φ_{out}) as well as the surface normal are located in the same plane. Each reflection leads to a reduction of the power of the considered beam $P_{in}(\alpha_{in}, \beta_{in})$ in accordance to

$$\Delta P_{out}(\alpha_{out}, \beta_{out}) = R\ \Delta P_{in}(\alpha_{in}, \beta_{in}), \qquad (2)$$

with $R(x, y, z)$ as the reflection factor of the striking point. The grid used for surface wafer discretization is shown in fig. 3. When the beam strikes the wafer, the surface element is determined where the striking point is located. The radiation absorbed from that beam is added to the already deposited radiation in the given surface element. As a result, we get from one lamp a two-dimensional radiation deposition in terms of power density at the wafer surface. This figure depends on the location of the lamp and the form, type and spectral power distribution. Also, the size, form and optical properties of the reflector and reactor enclosure as well as the geometrical and material properties of the wafer are playing an important role. By superposition of the radiation deposition distributions of all lamps the total irradiation performance absorbed by the wafer is determined as a two-dimensional distribution $I_L(x, y)$.

On the one hand, heat losses of each surface element are described by the Stefan–Boltzmann law:

$$\phi(x, y, t) = \epsilon\sigma\left(T(x, y, t)^4 - T_0^4\right). \qquad (3)$$

They depend on the emissivity ϵ of the Si wafer and the temperature T of the surface element

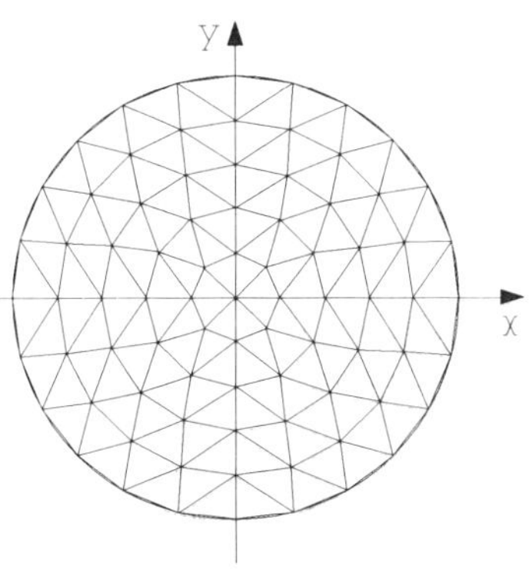

Fig. 3. Wafer discretization in triangle elements.

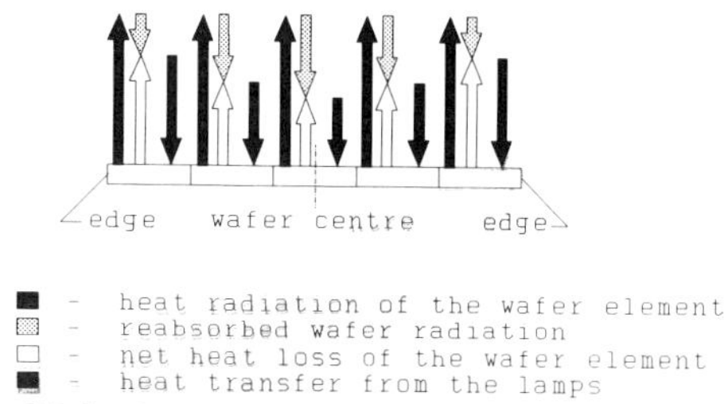

Fig. 4. Wafer heat input–output balance for homogeneous temperature.

concerned as well as on the ambient temperature T_0. On the other hand, the wafer is located in a reflecting housing, so it absorbs part of its own emitted radiation. This part $\Phi_{reab}(x, y, t)$ can be computed in accordance to irradiation by halogen lamps by the ray-tracing method. With a given temperature distribution the emitted radiation distribution on each surface element on the wafer can be determined as (fig. 4)

$$\Phi_{tot}(x, y, t) = \Phi(x, y, t) - \Phi_{reab}(x, y, t).$$

To determine the temperature distribution, the equation of two-dimensional heat conduction

$$\frac{\partial T}{\partial t} = \frac{\lambda(T)}{\varrho_w c_w(T)}\left(\frac{\partial^2 T}{\partial x^2} + \frac{\partial^2 T}{\partial y^2}\right) + \frac{1}{\varrho_w c_w(T)\ d}\left[I_L(x, y, t) - \phi_{tot}(x, y, t)\right] \qquad (4)$$

must be solved. In case of steady state $\partial T/\partial t = 0$. It contains partial derivatives of time and position, thermal and geometrical constants of the materials: heat conductivity λ, density ϱ, specific heat capacity c_w, wafer thickness d, as well as the source term $[I_L(x, y, t) - \Phi_{tot}(x, y, t)]$. It contains also the heat balance at the surface. The differential equation is solved by the finite-element method using the following boundary condition

$$\lambda(T)\left(\frac{\partial T}{\partial x}\boldsymbol{n}_x + \frac{\partial T}{\partial y}\boldsymbol{n}_y\right) = Q(x, y, t), \quad \text{with } x^2 + y^2 = R_w^2, \qquad (5)$$

where R_w is the radius of the wafer.

4. Results

For a constant given wafer temperature (1000°C) the radiation of the wafer was calculated in consideration of re-absorbed wafer radiation for the above-described reactor and lamp array. This is shown in fig. 5 together with the Stefan–Boltzmann radiation and the re-absorbed part. As a consequence of re-absorption of radiation emitted by the wafer, the heat losses increase from the centre to the edge of the wafer.

To achieve a constant temperature across the wafer, the higher radiation of the wafer edge must be compensated by an additional lamp radiation. In our in-house-designed reactor, this was made possible, by using the peripheral lamp ring, with an inner diameter larger then the diameter of the wafer, which is placed near the wafer edge. For two (3 and 6 inch) wafer sizes, the radial irradiation distribution with an additional peripheral lamp ring and the wafer radiation is compared in fig. 6.

For a verification of numerical results calculated and experimentally determined temperature distributions were compared, for a 3 inch wafer, in case of irradiation by the outer ring (fig. 7). By neglecting the effects of convective heat losses, the temperature distribution was determined using low-pressure oxidation kinetics at 5 Torr. The oxide thickness distribution was measured by ellipsometry. A real temperature distribution was obtained on the basis of experimentally determined oxidation kinetics [10].

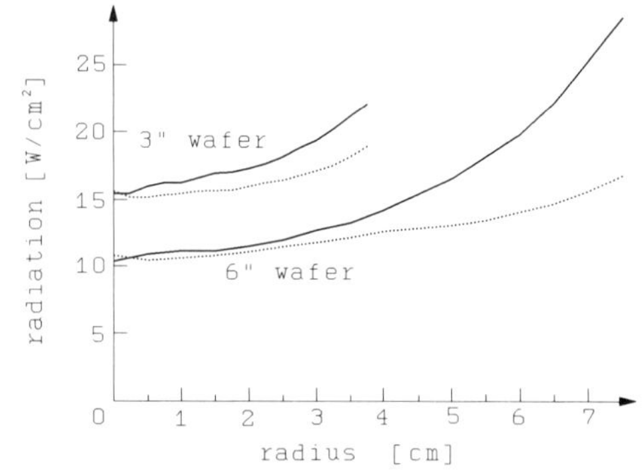

Fig. 6. Radial irradiation distribution of the peripheral lamp ring (solid line) in comparison with the net wafer radiation (dotted line) for 3 and 6 inch wafers.

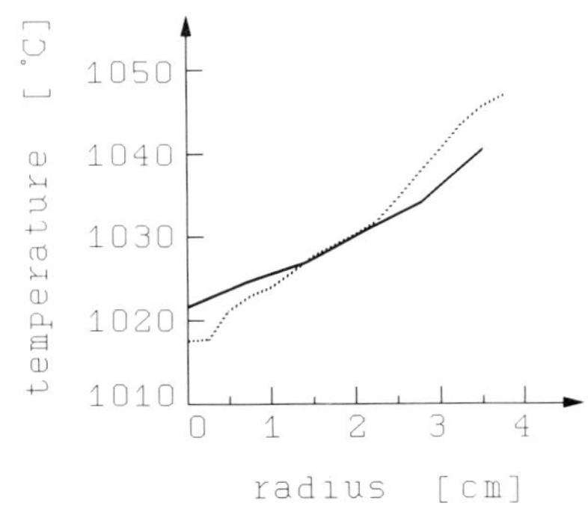

Fig. 7. Realized temperature distribution by using the peripheral lamp ring determined with oxidation experiments (full line) and calculated (dashed line).

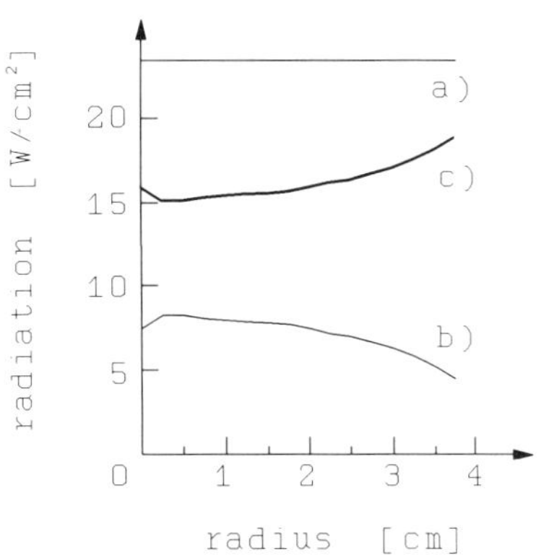

Fig. 5. Wafer radiative losses at constant temperature $T = 1000°C$ versus the Si wafer radius: (a) wafer radiation losses $\Phi(r)$ calculated with Stefan–Boltzmann law only; (b) re-absorbed part $\Phi_{reab}(r)$ of wafer radiation $\Phi(r)$ by itself; (c) total net emission distribution $\Phi_{tot}(r) = \Phi(r) - \Phi_{reab}(r)$.

5. Conclusion

The described rotationally symmetric lamp arrangement located above the wafer is able to compensate heat losses which increase towards the wafer edge. This arrangement provides, therefore, the conditions for the production of homogeneous layers. Suitability of the lamp array was proven by numerical simulation and experiments with low-pressure RTO.

Acknowledgement

The authors acknowledge the support of this research from AST Elektronik GmbH.

References

[1] T.O. Sedwick, J. Electrochem. Soc. 130 (1983) 484.
[2] H.A. Lord, IEEE Trans. Semicond. Manuf. 1 (1988) 105.
[3] J. Nulman, J.P. Krusius and A. Gat, IEEE Electron Device Lett. EDL-6 (1985) 205.
[4] D.R. Shik, A.B. Joshi and D.L. Kwang, J. Appl. Phys. 68 (1990) 5851.
[5] J. Nulman, Mater. Res. Soc. Symp. Proc. 92 (1987) 141.
[6] D.L. Kwong, T.Y. Hsieh, K.H. Jung, W. Ting and S.K. Lee, Rapid Isothermal Processing, Proc. SPIE 1189 (1989) 109.
[7] D.W. Pettibone, J.R. Suarez and A. Gat, Rapid Thermal Processing, Mater. Res. Soc. Symp. Proc. 52 (1986) 209.
[8] F.Y. Sorrell, J.A. Harris and R.S. Gyorisik, IEEE Trans. Semicond. Manuf. 3 (1990) 183.
[9] M.M. Moslehi, J. Kuehne, R. Yeakley, L. Velo, H. Najm, B. Dostalik, D. Yin and C.J. Davis, Mater. Res. Soc. Symp. Proc. 224 (1991) 1.
[10] J.-P. Zöllner and J. Pezoldt, to be published.

Applied Surface Science 69 (1993) 198–203
North-Holland

applied
surface science

Bulk and surface properties of RTCVD Si_3N_4 films for optical device applications

F. Lebland [a], Z.Z. Wang [b], J. Flicstein [a], C. Licoppe [a] and Y.I. Nissim [a]

[a] *CNET Laboratoire de Bagneux, France Telecom, BP 107, 92225 Bagneux Cedex, France*
[b] *CNRS-Laboratoire de Microstructures et de Microélectronique, 196 Avenue Henri Ravera, 92220 Bagneux, France*

Received 2 June 1992; accepted for publication 30 October 1992

Rapid thermal chemical vapour deposition is proposed to obtain optical-quality silicon nitride films on III–V semiconductors. The films have been obtained on InP at high temperature with a very high growth rate. Structural and morphological properties of nearly stoichiometric silicon nitride are presented. Nuclear reaction analysis (NRA), Rutherford backscattering (RBS) and Fourier transform infrared spectroscopy showed that the films are dense and have a uniform thickness. Scanning electron microscopy (SEM) revealed that the surface morphology of the silicon nitride films is dependent on the thickness of the films. Atomic force microscopy (AFM) allowed us to image the dielectric surface and to obtain roughness values as a function of the experimental conditions of deposition. Refractive indices measured at 6328 Å are linearly dependent on the film thickness. We show that the refractive index in this case is also tightly connected to the surface morphology of the deposited thin film.

1. Introduction

The development of micro-optoelectronics in III–V compounds requires the use of high-grade thin-film insulators. Among III–V compound semiconductors, telecommunication systems utilize InP since InP-based alloys have band-gap energies near the minimum loss wavelength in optical fibres. Progress made on optical as well as electronical devices has allowed recently the optoelectronic integration of devices on InP, and dielectric layers have a key role in this perspective. When chemical vapour deposition (CVD) is utilized to deposit a dielectric, both the density and the electronic properties of the film are improved for high-temperature deposition since the pyrolysis of the precursors is complete. Rapid thermal CVD (RTCVD) offers the opportunity to solve the paradox which involves the need for high-temperature processing on the one hand and sublimation of the Group V element for III–V substrates on the other, since the extremely fast growth rates obtained induce instantaneous encapsulation of the substrate. Silicon nitride is an alternative dielectric to silicon dioxide and has shown its capacity to enter as an active or passive material in microelectronic technologies and in optoelectronical devices. The knowledge of optical properties such as density of absorbant species, refractive index, etc. of the material is then necessary. The aim of this work is to show that bulk and surface material structure are determinant for optical device applications. In particular it is shown that the value of the refractive index is not only dependent on the chemical structure of the deposited thin film but also the surface morphology at a nanometer level plays an important role in the determination of optical constants. This has been observed through a detailed study of morphological changes induced on the film by the RTCVD process as a function of the film thickness.

2. Experimental procedure

RTCVD minimizes the thermal exposure of the treated substrate and allows high deposition rates that lead to immediate encapsulation of the sample surface and thus prevents its thermal degradation. In a flow of SiH_4 and NH_3 as reactive gases, silicon nitride layers have been deposited on undoped n-type InP(100) samples deoxidized in a HCl solution. The kinetics of reaction, the composition and the properties of such films are strongly dependent on the conditions under which the deposition is carried out and especially on the NH_3 to SiH_4 ratio, the dilution, the temperature and the total pressure. For a given dilution the rate varies with the $[NH_3]/[SiH_4]$ ratio. A value of 100 Å/s is obtained for a ratio of 10. For this value the electronic properties of the Si_3N_4/InP interface are not degraded [1,2]. The variation of dilution into N_2 induces drastic changes in the deposition rate. The deposition rate varies from 150 Å/s in pure reactive gases to 10 Å/s for a dilution of 1% in the carrier gas. With the following optimal conditions: $NH_3 = 200$ cm^3/min, $SiH_4 = 20$ cm^3/min, $N_2O = 200$ cm^3/min, total pressure 50 Torr, temperature 750°C, a deposition rate of 96 Å/s has been obtained. The thickness of amorphous Si_3N_4 on InP was controlled with the deposition time in a range varying between 500 and 1500 Å.

The contributions of the structural properties and the morphological properties of the films to the value of the index of refraction were analyzed separately. Bulk characterization of silicon nitride layers was obtained through absorption measurements in the transmission mode using a Fourier transform infrared spectrometer (FTIR-Nicolet 750). The width and the position of the Si–H, N–H, and Si–N bonds in the range 4000–400 cm^{-1} are strongly influenced by the stoichiometry, the density and the porosity of the silicon nitride films [3,4]. Thus, the transformations of the structural properties of Si_3N_4 with the thickness were analyzed by variations of the spectral amplitude and position of characteristic bonds in silicon nitride.

Bulk characterization of thin silicon nitride films was performed using Rutherford backscattering (RBS) for silicon and nuclear reaction analysis (NRA) for nitrogen. In a Van de Graaff accelerator we used the reaction $^{14}N(d, \alpha)^{12}C$ at 830 keV. Due to experimental conditions the sensitivity of RBS and NRA was 10%. The refractive index was measured at one wavelength (6328 Å) and angle (equal to 70°) with a Rudolf-Auto ellipsometer. A high-resolution scanning electron microscope (SEM) model Hitachi 800 with a tilt of 45° was used as a first approach to image surface roughness as a function of the thin-film thickness. Finally atomic force microscopy (Nanoscope III) allowed us to image the surface and to obtain roughness values of the deposited thin films. Before surface characterization, in order to avoid artifacts of measurements, samples had been cleaned in boiled trichlorethylene and rinsed with methanol. Such a preparation of the topmost layer allowed us to obtain reproducible results.

3. Bulk characterization

IR absorption peaks of silicon nitride have been intensively studied and are now well known [5]. Between 4000 and 400 cm^{-1} the IR spectrum exhibits three characteristic absorption peaks corresponding to Si–N, Si–H and N–H bonds, respectively located at 835, 2180 and 3350 cm^{-1}. A quantitative determination of the hydrogen concentration in the layers can be obtained using Si–H and N–H absorption peaks [5]. A low value of the hydrogen concentration in the layer of 3×10^{19} atoms/cm^3 has been obtained at 750°C. Due to the deposition temperature, X-ray diffraction measurements showed that the silicon nitride layers are amorphous. The peak position of the Si–N bond is sensitive to the stoichiometry of the films and particularly to the quantity of oxygen in the layer [3]. It has been found that when oxygen is incorporated in the silicon nitride layers the absorption position of Si–N bond shifts toward higher wavenumbers. The IR absorption spectra of silicon nitride layers deposited in the same conditions with different thicknesses are reported in fig. 1. The location of the maximum of the Si–N absorption peak at 835 cm^{-1} shows no significant variation (the increase of the amplitude of the Si–N bond is naturally due to the increase of the thickness of the film). This observation is indicative of the uniformity of the de-

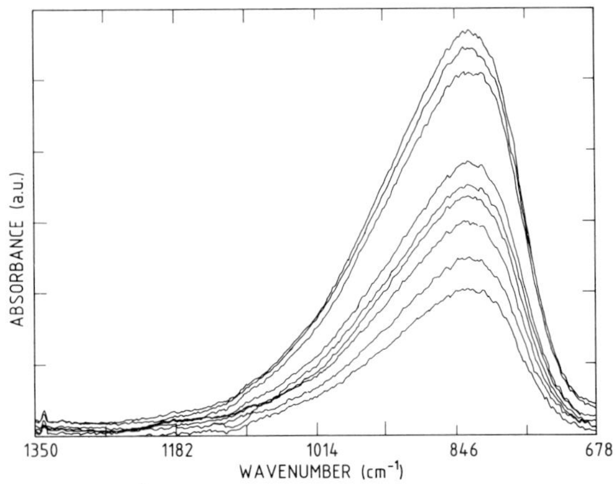

Fig. 1. Infrared absorption spectra of silicon nitride films for several thicknesses. The maximum of the Si–N peak located at 835 cm^{-1} did not vary with the thickness of the deposited films. Precise determination of the stoichiometry (RBS and NRA) confirmed these results.

posited material. A precise analysis of etching rates in buffered HF has been carried out as a function of the thickness. The etching rates are very slow: 8 Å/min. This is a warrant of compactness of the nitride films since films deposited by other techniques have an etching rate larger by an order of magnitude. The silicon nitride layers, due to the high deposition temperature are not porous and no water traces have been observed on the infrared absorption spectra (in the range 3300–3600 cm^{-1}). When dielectrics are used in optical devices such as wave guides, the low loss in the layers is associated with a low density of optical diffusion centres in the guiding films. The compactness, the homogeneity in the bulk, the absence of porosity and water in the films make RTCVD silicon nitride a suitable material for optical device applications.

Nuclear reaction analysis and Rutherford backscattering measurements gave a quantitative confirmation of the structural stability as a function of the thickness. Silicon and nitrogen concentrations were $3.8\text{–}10^{22}$ and 5×10^{22} atoms/cm^3, respectively, and did not vary with the thickness of the thin films in the range 500–1500 Å. With a N/Si ratio of 1.31 the films are nearly Si_3N_4 stoichiometric with an excess of 1.6% of silicon.

4. Ellipsometric measurements

The linear variation of the index of refraction with thickness in the 500–1500 Å range is reported in fig. 2. As can be seen, the variation of the refractive index is important. If one uses the effective dielectric functions for a heterogeneous medium [6], it is then possible to calculate an effective index of refraction of the mixture. A mixture of air and silicon nitride cannot predict such an increase of the refractive index. Furthermore, this important deviation of refractive index from a value of 2, which represents the refractive index of silicon nitride, up to 2.5 cannot be explained by taking into account an excess of silicon in the silicon nitride films only. In any case, such a stoichiometric variation has previously been rejected throughout composition investigations.

5. Surface morphology

High-resolution SEM has been utilized to evaluate the evolution with thickness of the local surface morphology of RTCVD silicon nitride. Fig. 2 shows as-grown silicon nitride films for thicknesses varying from 520 to 1500 Å. The general morphology of the surface exhibits a density of random hillocks of varying size, which grew three-dimensionally. The bigger ones are spaced randomly on the surface (fig. 3). Only the width of these hillocks appeared to increase with the deposition time of the films from 0.1 μm (500 Å)

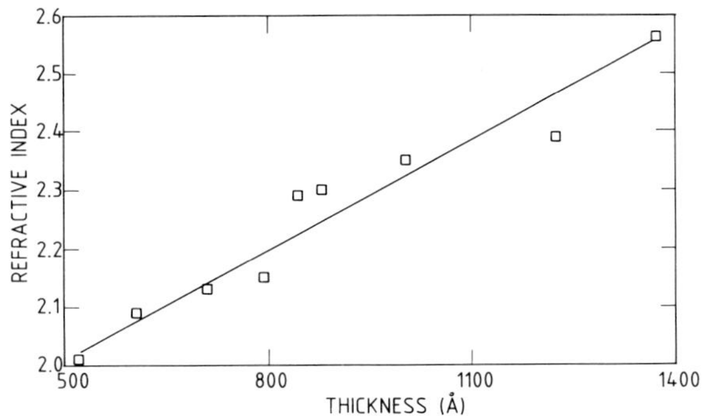

Fig. 2. Ellipsometric refractive index (at $\lambda = 6328$ Å and angle equal to 70°) as a function of the thickness of the deposited silicon nitride film.

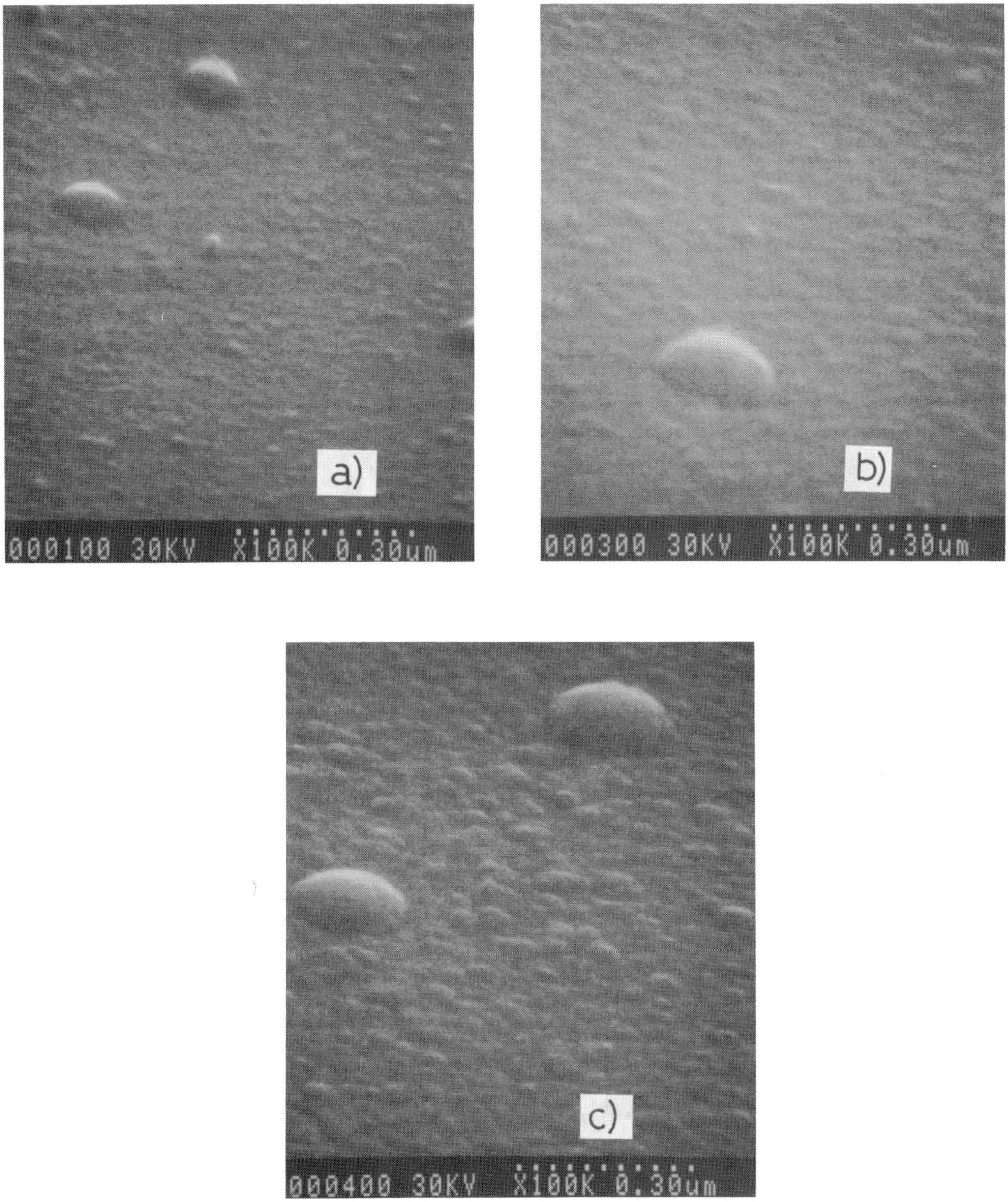

Fig. 3. Typical SEM profile of silicon nitride films. Thickness of the films is (a) 520 Å, (b) 850 Å, (c) 1500 Å. The bigger islands are randomly spaced on the surface. The total roughness increases with the thickness.

to 0.22 μm (1500 Å), the surface density and the height being stable. The origin of these hillocks, which contributed to a micro-roughness, can be attributed to the initial surface state of InP polished substrate prior to the growth and the dielectric nucleation during deposition of the first monolayers [7]. Also, the formation during the growth of the first monolayers of adsorbates and heterogeneous nuclei distributed randomly at the InP surface can be locally responsible for nucleation of islands after only several deposited layers. However, the departure from the smooth surface of the substrate displayed by the SEM picture is quite important. This type of morphology at the microscopic level was next corroborated by AFM and between these bigger hillocks nanoscopic roughness values were determined. Our AFM is operated at room temperature in air using silicon nitride nanotips provided by Digital Instruments. The use of a tiny force ($F < \ln N$) was found to be crucial to obtain AFM images without tip damage. To avoid the artifact due to the tip, different tips were used. The AFM results shown are characteristic of the many images taken of each sample (more than ten for each) at different locations. Direct observations using the AFM profile, which are well reported in fig. 4, have been found in good agreement with SEM profiles. The silicon nitride surface consists of a sum of ellipsoidal hillocks spaced randomly on the surface. The roughness due to these hillocks persists after the islands have joined. Even when,

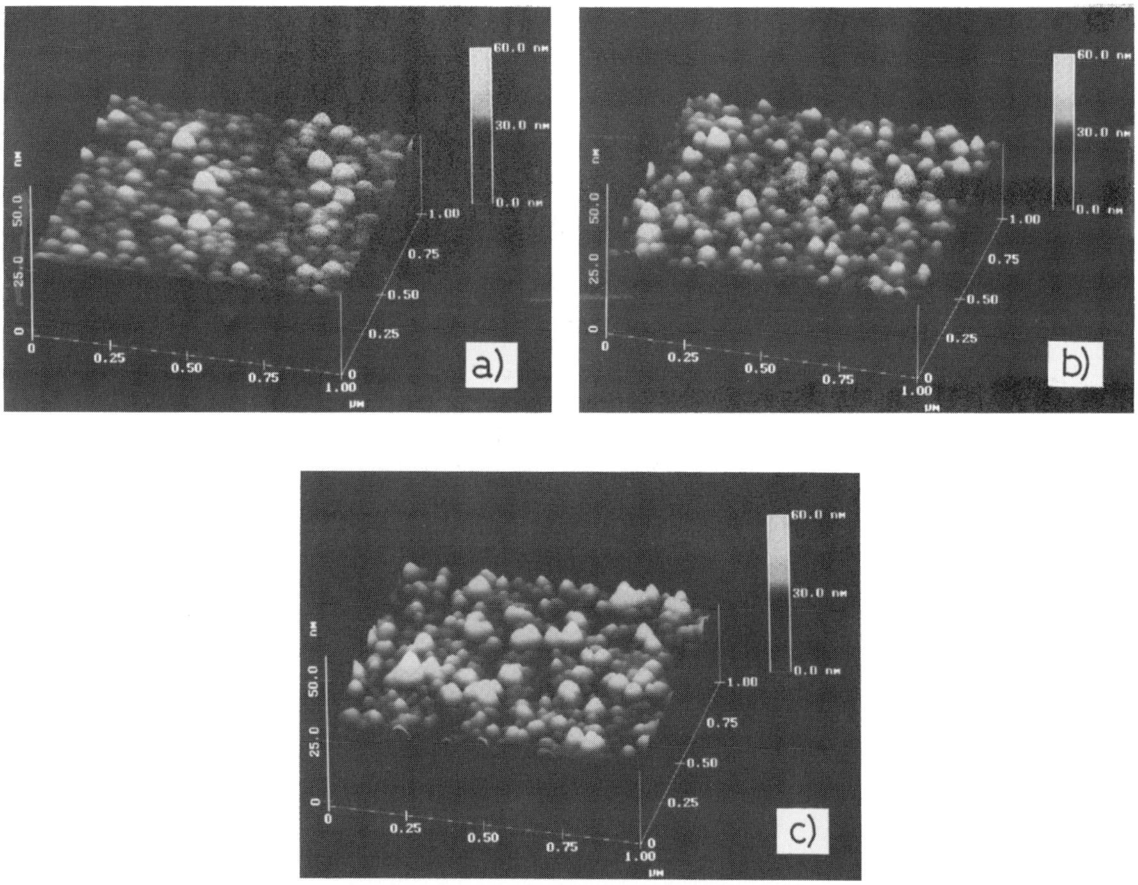

Fig. 4. Typical AFM profile at the nanometer scale. Surface morphology is in good agreement with the analogous SEM profile. Thickness of the films is (a) 520 Å, (b) 850 Å, (c) 1500 Å.

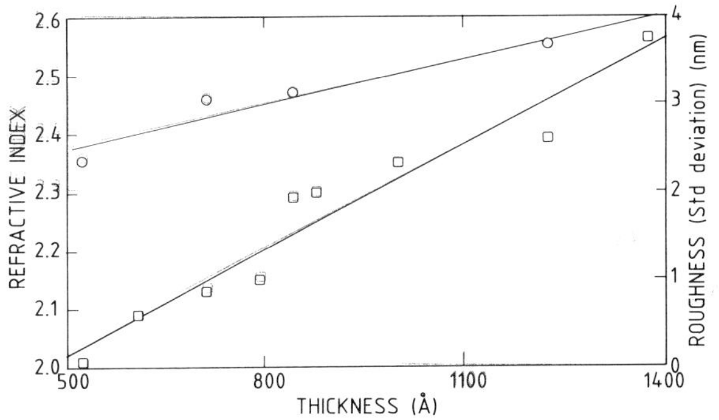

Fig. 5. Index of refraction and nanoscopic roughness as a function of the thickness of the deposited silicon nitride film.

with a very smooth InP substrate (that was utilized here), this peculiar morphology varies with the thickness of the deposited film yielding the roughness. So, the average roughness of the surface increases with the thickness of the deposited film. Nanoscope III software was used to determine these roughness values. To improve the statistics, the roughness was determined in a selected box area instead of line intercepts on the surface. As a roughness parameter we used the standard deviation of the so-called root-mean-square value (RMS). Each image is made with a 512×512 grid of measured points. Refractive index, roughness and thicknesses of the films are reported in fig. 5. Due to the homogeneous structure of the film in its bulk, refractive indices appeared to be related only to film roughness.

6. Discussion

The roughness of the silicon nitride surface appeared to be isotropic. Rayleigh scattering due to increasing film roughness contributed to variations in the intensity and the phase of the reflected beam. It is possible to model this effect by introducing a local dephasing of the reflected beam due to a fluctuation $u(z)$ in surface position.

The phase of the reflected beam will then be: $\Delta_t = \Delta_0 + \delta(z)$ with $\delta(z) = \exp[2ik\,u(z)]$. Thus, if the average value of $\delta(z)$ is taken as 0, we obtain an effective Ψ in the experimental ellipsometric equation:

$$\text{tg } \Psi_{\text{eff}} = \text{tg } \Psi\left[1 - 2k^2\langle u(z)^2\rangle\right].$$

Using this expression and the experimental ellipsometric measurements, the calculated fluctuation in the surface position, $u(z)$, is in good agreement with the determined average roughness (SEM and AFM). A decrease of Ψ leads to an increase in the measured value of the refractive index, in agreement with our experimental data.

7. Conclusion

RTCVD of Si_3N_4 has been shown to result in high-quality films for optical device application. The bulk material is dense and stable in composition. However, it has been shown in this work that the measured value of the refractive index is dependent on the roughness of the surface. Atomic force microscopy allowed us to image the silicon nitride surface and to measure roughness values down to the nanometric level of the films for several thicknesses. With the index–roughness–composition relationship, it is possible to obtain roughness values only with the refractive index measurement. When modulation of the refractive index is used in optical devices, the surface morphology of the dielectric films has to be measured before fabrication of optical devices.

Acknowledgement

The authors are grateful to F. Carcenac from CNRS/L2M for his technical assistance.

References

[1] F. Lebland, C. Licoppe and Y.I. Nissim, Le Vide et Les Couches Minces 251 (1990) 64.
[2] Y.I. Nissim, J.M. Moison, F. Houzay, F. Lebland, C. Licoppe and M. Bensoussan, Appl. Surf. Sci. 46 (1990) 175.
[3] F. Lebland, C. Licoppe, Y. Gao, Y.I. Nissim and S. Rigo, Appl. Surf. Sci. 54 (1992) 125.
[4] W.A. Pliskin, J. Vac. Sci. Technol. 14 (1977) 1064.
[5] W.A. Landford and M.J. Rand, J. Appl. Phys. 49 (1978) 2473.
[6] D.E. Aspnes, Thin Solid Films 89 (1982) 249.
[7] J. Flicstein and J.E. Bouree, in: Photochemical Processing of Electronic Materials, Eds. I.W. Boyd and R.B. Jackman (Academic Press, New York, 1992) p. 106.

Applied Surface Science 69 (1993) 204–211
North-Holland

applied surface science

Pulsed laser deposition – a versatile technique only for high-temperature superconductor thin-film deposition?

H.-U. Habermeier
Max-Planck-Institut für Festkörperforschung, Stuttgart, Germany

Received 2 June 1992; accepted for publication 30 October 1992

One of the most stringent prerequisites for the successful deposition of high-temperature superconductor (HTS) thin films of high quality is the stoichiometry control. This prerequisite will be met easily using the pulsed laser deposition technique (PLD) by the nature of the process. This paper highlights the principles of the preparation of Y-based and Bi-based HTS thin films by PLD, giving some emphasis to the study of the in-plane epitaxy of *c*-axis oriented films by X-ray techniques and Raman scattering. In particular, new results will be reviewed concerning the fabrication and properties of Pr-substituted Y–Ba–Cu–O thin films and Pr–Ba–Cu–O/Y–Ba–Cu–O superlattices with different periodicity. Focussing on the more methodological aspects of these efforts, some possibilities will be explored to prepare new complex materials of technological interest with the pulsed laser deposition technique.

1. Introduction

The discovery of superconductivity above 90 K in the rare-earth–alkaline-earth–copper-oxide system [1,2] initiated an enormous amount of research activities especially in technologies to prepare thin films of these high-temperature superconductors (HTS). All conventional deposition methods capable of multicomponent thin-film preparation have been employed successfully to deposit HTS thin films ranging from equilibrium techniques like molecular beam epitaxy with kinetic energies of the emitted particles, E_p, around 0.1 eV over non-equilibrium techniques such as magnetron sputtering with $E_p \approx 1$ eV to pulsed laser deposition techniques with $10 < E_p < 100$ eV. In a paper by Leskelä et al. [3] the different techniques are reviewed and the common features for HTS thin-film deposition are highlighted. Habermeier et al. [4] recently published a comparative study of pulse laser deposition and RF sputtering discussing the advantages and disadvantages of these two techniques. A review of the achievements of pulsed laser deposition as a general method is given by Cheung and Shankar [5] covering the developments till 1988, the review of Beech and Boyd [6] is focussing on more recent developments. A comprehensive overview of pulsed laser deposition technology can be found additionally in a recent issue of the MRS Bulletin [7].

In this paper a short description of the mechanism of pulsed laser deposition is given and the conditions are derived to fabricate high-quality Y-based and Bi-based HTS thin films. The unique possibilities of the PLD technique are demonstrated for the preparation of Y-based HTS thin films with special cation substitution in order to perform fundamental studies of thin films with high crystallographic perfection. Furthermore, the possibility of heterostructure and superlattice fabrication is explored, giving some technological basis for future work on materials preparation with tailored physical properties.

2. Physical concept of pulsed laser deposition

To illustrate the unique possibilities of pulsed laser deposition in preparing thin films of multicomponent materials some fundamental aspects of the process have to be considered.

0169-4332/93/$06.00

In conventional physical vacuum deposition schemes like thermal evaporation or electron beam evaporation the material to be evaporated is heated continuously either by heat conduction from thermally heated boats or by impact and deceleration of electrons in the keV range. If an intense photon flux is used as energy source there must be a substantial optical absorption in order to couple the energy of the incident photons to the atoms. Athermal elementary interactions of photons which matter coupled by nonlinear mechanisms give rise to heating of the target surface. If CW lasers are used as energy sources, an evaporation characteristics comparable to that of thermal or e^- beam evaporation is expected to occur and in the case of multicomponent targets a congruent removal of material is not likely to be expected. Applying a *pulsed* laser beam as energy source the nature of the evaporation mechanism changes dramatically. The formation of a plasma plume and its expansion from the target surface is the dominant non-equilibrium feature. Four different regimes play a crucial role during the deposition process [8]: (i) the interaction of the pulsed laser beam with the target resulting in evaporation of the surface layers, (ii) the interaction of the evaporated material with the incident photon flux resulting in an isothermal plasma formation and expansion, (iii) the anisotropic adiabatic expansion of the plasma and chemical reactions during the transport, and (iv) the growth of the film at the substrate site. The irradiation of the target surface by a pulsed intense photon flux results in a rapid heating of the surface layer in which the energy absorbed is not diffusing from the surface region but mostly used to excite the atoms. The quantities of interest determining the evaporation process are the optical absorption depth, $1/\alpha$ (α: linear absorption coefficient) and the thermal diffusion length, $L_{th} = (2D\tau_{th})^{1/2}$ (D: thermal diffusivity, τ_{th}: thermal diffusion time which corresponds to the pulse duration, τ_p). If the optical absorption depth is much smaller than the thermal diffusion length – this condition holds especially for materials with absorption coefficients for UV photons in the 10^5 cm^{-1} range –

$$\alpha^{-1} \ll L_{th}, \tag{1}$$

the energy absorbed is not appreciably diffusing away from the surface and is used to excite atoms and particles, consequently surface heating occurs. The standard heat conduction equation gives the following relation between the time-dependent surface temperature, T_s, and the incident laser power density, P [9]:

$$T_S(0, t) = 2P(1-R)/\tau_p\lambda(\lambda t/\rho c_p\pi)^{1/2}, \tag{2}$$

where R is the reflectivity of the target, λ the thermal conductivity, ρ the mass density of the target, and c_p the heat capacity.

With typical values, e.g. for YBCO [10], and standard pulse widths of UV excimer lasers,

$$\langle\lambda\rangle = 2 \text{ W cm}^{-1}\text{ K}^{-1}, \quad R = 0.1,$$
$$\langle c_p\rangle = 0.2 \text{ J/g}\cdot\text{K},$$
$$\rho = 6.68 \text{ g cm}^{-3}, \quad t = \tau = 30 \text{ ns},$$

the surface temperature at pulse termination is 7200 K according to eq. (2) using the simplifying assumption of independence of optical and thermodynamic data from temperature and pulse energy. This value is fairly well above the evaporation temperature of YBCO (2500 K corresponding to a threshold energy for ablation of 0.4 J/cm^2 [8]). The high temperature due to the pulsed laser irradiation gives rise to thermionic emission of positive ions and electrons. Since the time required to heat the YBCO target surface beyond its evaporation temperature is less than 2 ns a laser beam of a typical pulse width of 30 ns duration will interact with the ablated material causing its heating. The plasma absorbs the photons appreciably only in areas where the density of the charged particles is very high, i.e. adjacent to the target surface. The mechanisms of photon absorption and rapid transfer of thermal energy into kinetic energy control the isothermal temperature of the plasma [11] and form a self-regulating regime close to the target. At the target site large density gradients are built up and the high-temperature high-pressure plasma expands into the vacuum. According to the initial dimensions of the plasma (mm in transverse direction corresponding to the size of the laser beam and μm perpendicular to the target) highly anisotropic pressure gradients ($\delta p/\delta z \gg \delta p/\delta x$, $\delta p/\delta y$) develop, giving rise to the forward-directed nature of the laser plume expanding into the vacuum. The emission of photons by the highly excited particles gives the laser plume the characteristic inhomogeneously colored appearance. If this plasma expansion takes place in an ambient enriched with reactive gases chemical processes may occur (in the case of YBCO the metal ions will form metal-oxide ions) and thus the way to tailor reactive processes is paved. Compared to other techniques the growth kinetics of a film at the substrate site is somewhat different due to the pulsed nature of the process and the huge variety of different species arriving at the substrate site (e.g. Cu^+, Ba^+, Y^+, O_2^+, BaO^-, CuO, YO and clusters composed to these components). The outstanding feature of the PLD technique is its ability to precisely reproduce the target composition in the vapor (i.e. not necessarily in the film); this property is basically due to the forward-directed jet-like ejection or in other words due to the dimension of the plasma close to the target and the characteristics of its expansion. The area of reproduction of the target composition in the film, however, is in the case of YBCO limited to an angle of 20° with respect to the target surface normal [12]. For an optimum deposition process parameters like wavelength, pulse duration, pulse energy density at the target site, substrate–target distance, laser spot size at the target, oxygen background pressure and substrate temperature have to be optimized. As laser source for PLD processes usually UV excimer lasers are used with a wavelength of 198 nm (ArF), 248 nm (KrF) or 308 nm (XeCl). The deposition parameters have to be adapted according to the phase diagram of the material to be deposited.

3. Deposition of high-temperature superconductor thin films

3.1. Y–Ba–Cu–O

For the successful growth of high-quality superconducting YBCO thin films the deposition parameters have to be optimized according to the phase stability diagram as given in fig. 1. Deposition conditions beyond the full line in fig. 1 give rise to decomposition of $Y_1Ba_2Cu_3O_6$ into copper oxide, Barium–copper–oxide and the so-called green YBCO phase (Y_2BaCuO_5). To avoid this decomposition, the deposition conditions have to be chosen close to the stability line at the side of the lower temperature for a given oxygen pressure. A typical set of deposition parameters is listed below [13]:

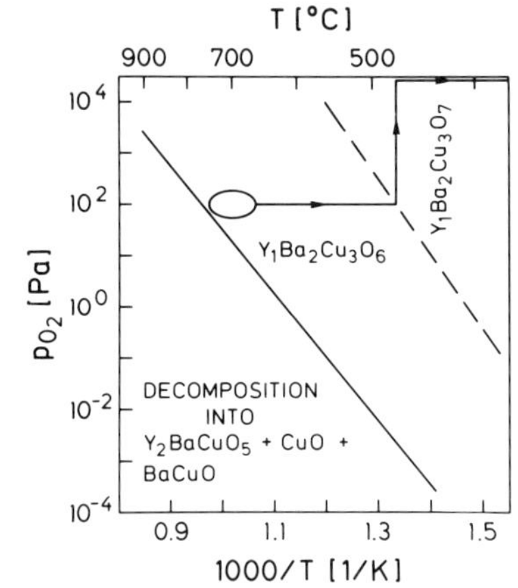

Fig. 1. Oxygen partial pressure versus temperature plot showing the $CuO–Cu_2O$ stability line and the area of single-phase $YBa_2Cu_3O_6$ deposition.

Wavelength: 248 nm (KrF)
Target spot size: 3 mm^2
Oxygen background pressure: 1 mbar
Energy density: 1–2 J/cm^2
Substrate–target distance 3–4 cm
Substrate temperature: 780°C

Using oxide substrates such as $SrTiO_3$, $LaAlO_3$ or $NdGaO_3$ with a fairly good lattice match to the YBCO unit cell the resulting films show a nearly linear metallic behavior of resistance versus temperature which extrapolates to zero resistance at zero temperature, a resistance ratio, $\Gamma = R_{300\,K}/R_{100\,K}$, around 3 and a sharp transition to superconductivity above 88 K with a transition width $\Delta T_c < 0.5$ K. In fig. 2 the characteristics for the transition are represented as a function of the deposition temperature, T_d, for PLD films on $SrTiO_3$ substrates.

There is a broad region around $T_d = 750$°C in which sharp transitions are achieved, the broadening for high T_d is probably due to decomposi-

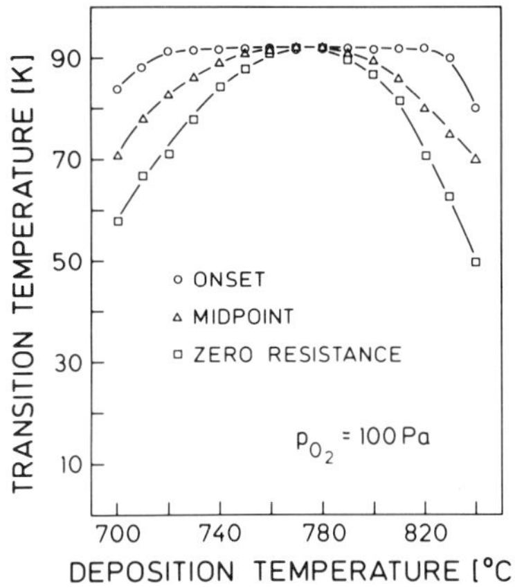

Fig. 2. Optimization of the deposition temperature by the characterization of the transition width to superconductivity.

tion, whereas that for lower T_d is due to non-epitaxial growth and probably incomplete oxygenation. Typical values for j_c measured at 77 K are larger than 10^6 A/cm^2.

Comparing the surface morphology of in-situ PLD deposited films with in-situ evaporated or sputtered films, the striking features are the relative smoothness of sputter films with a typical roughness < 10 nm for 100 nm thick stoichiometric material. Deviations from the stoichiometry lead to outgrowths and precipitate formation leaving the transport properties of the films nearly unchanged. In some contrast to these results are the surfaces of PLD deposited films. Fig. 3 shows the SEM picture of a PLD YBCO thin film in two magnifications, indicating clearly the presence of particles of up to 3/4 μm size on smooth material. These particles are not associated with the outgrowths found in non-stoichiometric sputtered films but are inherent to the PLD process. A series of experiments have been performed aiming to reduce the particle density in PLD films by analyzing the role of pulse energy density, target state, mechanical target density (80% to 97% of theoretical density), deposition temperature and oxygen partial pressure. It appeared that an optimum value for the target density in the range of 94% exists. This combined with a slightly reduced oxygen pressure (30 Pa) results in films with particle densities as low as 400/mm^2 [13].

For the structural analysis of these films standard θ–2θ X-ray scans are conventionally employed resulting in diffraction patterns showing the YBCO (0,0,1) reflection peaks, only. This indicates that the films are at least c-axis textured which means that the c-axis is oriented perpendicular to the film plane. Information about epitaxy, i.e. their alignment of the a–b axes with respect to the substrate a-axis can be obtained if so called X-ray polar scans along a cone with its axis perpendicular to the substrate surface are used. For the (103) pole figure of YBCO a fourfold symmetry is expected as shown in fig. 4a. Quantitative information about the degree of epitaxy cannot be obtained unless a careful analysis of the incoherent scattering amplitudes forming

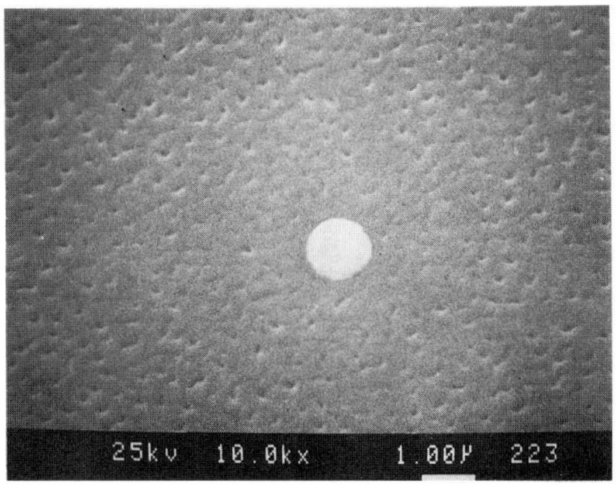

Fig. 3. SEM micrographs of a PLD-deposited YBCO thin film prepared using optimized conditions.

the background intensity is performed. The procedure based on Raman scattering proposed by Thomsen et al. [14] is a more straightforward way to determine the degree of epitaxy. The method uses the selectivity of the appearance of a certain Raman line according to the orientation of the polarization vector of incident and scattered light with respect to the orientation of the a- or b-axis of the unit cell. In fig. 4b the Raman spectra of an YBCO film are given for incident and scattered light polarized along the diagonal of a and b (upper curve) and for crossed diagonal polarization of incident and scattered electrical fields (lower curve). The main feature of interest is the change of the intensity of the 340 cm^{-1} Raman line which corresponds to the in-plane oxygen vibration (B_{1g}-like mode). According to symmetry considerations of the orientation of incident and scattered electrical fields of the light beam with respect to the Raman tensor in perfectly epitaxial film – or single crystal – the B_{1g} Raman intensity for parallel polarization along the a,b diagonal must vanish, whereas for crossed polarization along the diagonal of a,b the B_{1g} intensity must have its maximum value. The degree of epitaxy, η, is calculated from the relative intensities of the B_{1g} mode for the two different polarizations. From the Raman spectra given in fig. 4b the degree of epitaxy is determined as $\eta = 0.91$.

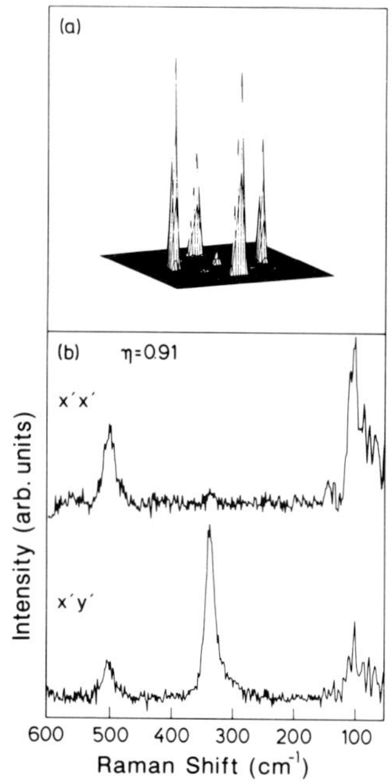

Fig. 4. (103) X-ray pole figure (a) and corresponding Raman spectra of the same film with two different orientations of the polarized light with respect to the diagonal of the unit cell (b).

3.2. Bi–Sr–Ca–Cu–O thin films

In contrast to the YBCO system the Bi-containing HTS material forms a homologous series $Bi_2Sr_2Ca_nCu_{n+1}O_x$ (BSCCO $2:2:n:n+1$) with $n = 0$, 1 and 2; the corresponding superconducting transition temperatures are 20, 80 and 110 K [15]. The structures of these materials differ basically in the number of the CuO_2 planes intercalated by BiO and CaO layers. Due to this complex chemistry compared to YBCO the deposition of pure single-phase thin films is much more difficult. In addition, the different sticking coefficients of the individual elements and suboxides at the substrate site are also a source of problems. According to extensive RBS studies performed by Perriere [16] the Bi concentration in films deposited above 700°C depends strongly on the oxygen pressure during the ablation process. Stoichiometric films of the 2:2:1:2 compound can be obtained if substrate temperatures and oxygen pressures are optimized. Based on results obtained previously [17], Razavi et al. [18] developed an in-situ deposition process of the 2:2:1:2 compound by carefully analysing the influence of substrate temperature, oxygen pressure and thermal history of 2:2:1:2 films on structural transport and morphological properties of the films. Fig. 5 shows the resistivity versus temperature curves for different parameters of substrate temperature and annealing. Corresponding X-ray diffraction patterns reveal that the films are c-axis textured with some broadening of the peaks for in-situ films at low deposition temperatures. It is demonstrated that single-phase 2:2:1:2 films can be grown in situ from stoichiometric targets by the laser ablation technique. For deposition temperatures between 770 and 810°C the critical temperatures of these films are between 60 and 70 K; the critical currents at 4.2 K show values up to 10^6 A/cm^2. The superconducting transport properties can be improved if some post-deposition annealing is applied.

4. Cation substitution in YBCO and YBCO/PrBCO superlattices

The advantage of stoichiometry preservation in the PLD approach for depositing high-quality YBCO thin films allows the application of this technique to other examples of the class of HTS materials and thus easily opens up the possibility to tailor special specimen compositions for basic studies. Fe substitution of Cu gives the possibility to reduce T_c in a controlled way which is inherent to the shift of the energy gap. Investigations of the shift of the gap position with increasing Fe concentration using far-infrared spectroscopy give some insight into gap energies independently from phonon structures [19]. Substituting Y by Pr in YBCO results in a non-superconducting $PrBa_2Cu_3O_7$ which is isostructural to YBCO. Optical investigations of the valency of Pr in $Y_xPr_{1-x}Ba_2Cu_3O_7$ thin films indicate that the Pr is trivalent and the mechanism suppressing superconductivity in PrBaCuO is not hole filling as expected from simple valency arguments [20].

Furthermore, the PLD technique is very attractive to deposit multilayers of structurally similar materials like YBCO/PrBCO or YBCO/SrTiO [21]. The fabrication of YBCO/PrBCO superlattices e.g. is a very attractive tool to investigate the interlayer coupling of the basic structural elements for HTS materials, the CuO_2 planes. In fig. 6 several transition curves of YBCO/PrBCO superlattices are shown, for the $Y_{0.5}Pr_{0.5}Ba_2Cu_3O_7$ solid solution as well as for YBCO/PrBCO superlattices with different periodicity. It is clearly demonstrated that the superlattices have critical temperatures well above the solid solution and

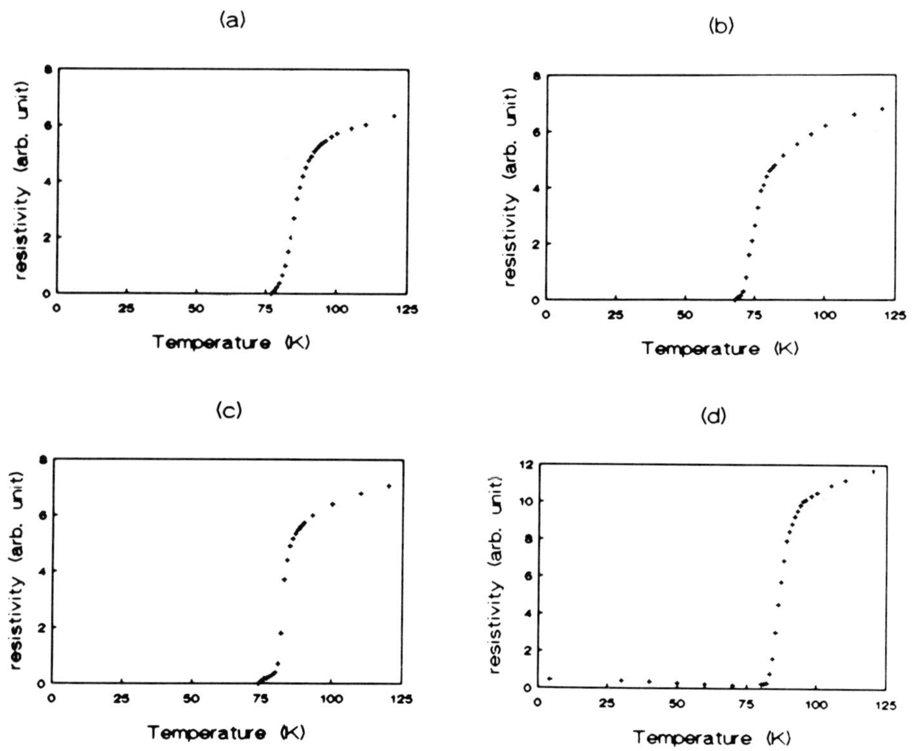

Fig. 5. Resistance versus temperature of 2:2:1:2 BSCCO thin films: (a) substrate temperature of 420°C and annealed for 6 min at 870°C; (b) substrate temperature of 810°C, no annealing; (c) substrate temperature of 770°C and annealed for 18 h at 820°C; (d) substrate temperature of 770°C and annealed for 24 h at 835°C.

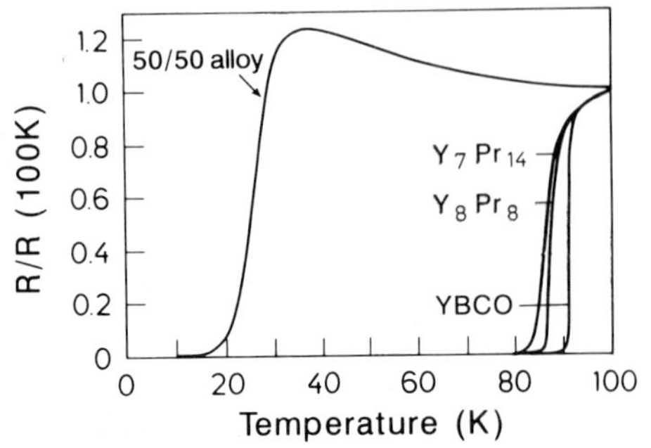

Fig. 6. Transition to superconductivity for a thin film of the $Y_{0.5}Pr_{0.5}Ba_2Cu_3O_7$ solid solution and superlattices with different periodicity.

somewhat below the pure YBCO, indicating a strong interlayer coupling. Raman scattering reveals (cf. fig. 7) that the superlattices of 8 unit cells YBCO and 8 unit cells PrBCO can be treated as to be composed of the two individual materials with no intermixing [22].

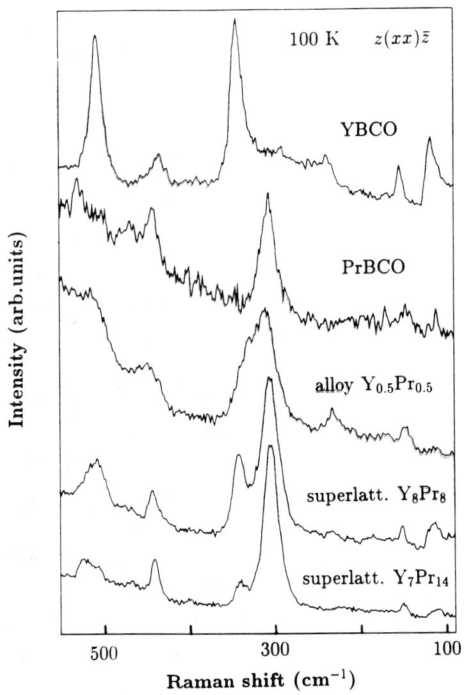

Fig. 7. Raman spectra of YBCO and PrBCO thin films together with YBCO/PrBCO superlattices.

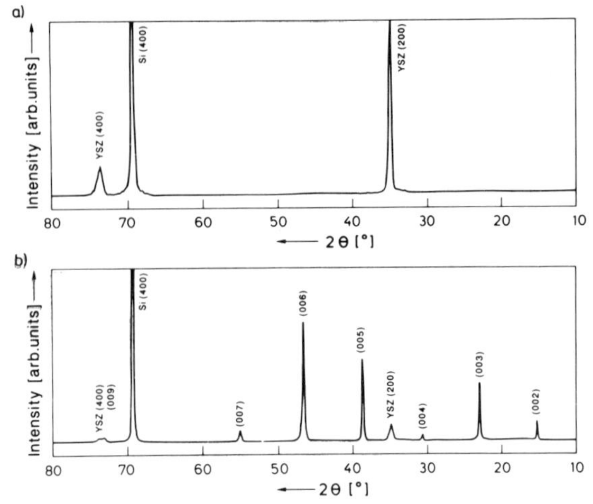

Fig. 8. X-ray diffraction patterns of PLD-grown yttria-stabilized zirconia buffer layer on silicon (a) and the subsequently deposited YBCO thin film, indicating the textured growth of the YBCO (b).

5. Conclusion: pulsed laser deposition as a novel technique for material research

The success of PLD in fabricating HTS thin films has tremendously enhanced the interest using this technique for thin-film deposition of multicomponent materials. Closely connected with the work on HTS materials efforts are made to grow epitaxially yttria-stabilized zirconia (YSZ) on silicon especially as buffer layers between silicon and YBCO [23,24]. In fig. 8 e.g. the X-ray diffraction pattern of a PLD thin film of YSZ on silicon is represented together with the diffraction pattern of an YBCO thin film deposited on top of the YSZ. It is clearly to be seen that the YBCO growth is at least textured, and Raman scattering reveals that 50% of the YBCO material is grown epitaxially. Furthermore, perovskite ferroelectrics, bioceramics and ferrites are other multicomponent oxides which already have been successfully deposited by PLD [7].

The congruent nature of the PLD process which leads to the ability to reproduce the target stoichiometry in conjunction with the possibility to deposit thin films in an environment of reactive gases are the outstanding features of this technique. Since 1987 when Venkatesan et al. [12] first used these properties of PLD to prepare in-situ superconducting YBCO thin films, a dramatic growth in the number of materials deposited by PLD took place. The current activities in using activated reactive gases, substrate biasing, velocity filtering of ejected particles, etc. will certainly enhance the spectrum of applications of the pulsed laser deposition technique to prepare new materials. Furthermore, the possibility of a rather straightforward deposition of superlattices and heterostructures will pave the way for fabricating new types of thin films with tailored physical properties.

Acknowledgements

The author is highly indebted to all members and visiting scientists of the Scientific Service Technology at MPI-FKF, especially to Dipl. Ing. B. Leibold, Dr. G. Beddies, Professor Dr. F. Razavi and Dr. I. Trofimov. The close cooperation with Dr. C. Thomsen, D.A.P. Litvinchuk and other members of Professor Cardona's department is highly acknowledged. The work was funded in parts by the European Community and the German BMFT.

References

[1] J.F. Bednorz and K.A. Müller, Z. Phys. B (Condens. Matter) 64 (1986) 189.
[2] C.W. Chu, P.H. Hor, R.L. Meng, L. Gao, Z.J. Huang and Y.Q. Wang, Phys. Rev. Lett. 58 (1987) 405.
[3] M. Leskelä, J.K. Truman, C.H. Mueller and P.H. Holloway, J. Vac. Sci. Technol. A 7 (1989) 3147.
[4] H.-U. Habermeier, G. Beddies, B. Leibold, G. Lu and G. Wagner, Physica C 180 (1991) 17.
[5] J.T. Cheung and H. Shankar, CRC Crit. Rev. Solid State Mater. Sci. 15 (1988) 63.
[6] F. Beech and I.W. Boyd, in: Photochemical Processing of Electronic Materials, Eds. I.W. Boyd and R.B. Jackman (Academic Press, New York, 1991) p. 387.
[7] MRS Bull., Vol. XVII, No. 2 (February 1992) (Materials Research Society, Pittsburgh, PA, 1992).
[8] R.K. Singh and J. Narayan, Phys. Rev. B 41 (1990) 8843.
[9] J.F. Ready, Effects of High Power Laser Radiation (Academic Press, New York, 1971).
[10] A. Inam, X.D. Wu, T. Venkatesan, S.B. Ogale, C.C. Chang and D. Dijkkamp, Appl. Phys. Lett. 51 (1987) 1112.
[11] A. Caruso and R. Gratton, Plasma Phys. 10 (1968) 867.
[12] T. Venkatesan, X.D. Wu, A. Inam and J.B. Wachtman, Appl. Phys. Lett. 52 (1988) 1193.
[13] H.-U. Habermeier, Eur. J. Solid State Inorg. Chem. 28 (1991) 619.
[14] C. Thomsen, R. Wegerer, H.-U. Habermeier and M. Cardona, Solid State Commun., in press.
[15] C. Michel, M. Hervieu, M.M. Borel, A. Grandin, F. Deslandes, J. Provost and B. Raveau, Z. Phys. B 68 (1987) 421.
[16] J. Perriere, Physica C, in press.
[17] Y. Nakayama, I. Tsukada, A. Maeda and K. Uchinokura, Jpn. J. Appl. Phys. 28 (1989) L1809.
[18] F. Razavi, H.-U. Habermeier and P. Majewski, Mater. Sci. Eng. B 13 (1992) 49.
[19] E. Seider, M. Bauer, L. Genzel and H.-U. Habermeier, Z. Phys. B 83 (1991) 1.
[20] J. Kircher, M. Cardona, S. Gopalan, H.-U. Habermeier and D. Fuchs, Phys. Rev. B 44 (1991) 2410.
[21] U. Poppe, P. Prieto, F. Schubert, H. Saltuer and K. Urban, Solid State Commun. 71 (1989) 569.
[22] A.P. Litvinchuk, C. Thomsen, I.E. Trofimov, H.-U. Habermeier and M. Cardona, Phys. Rev. B, submitted.
[23] D.K. Fork, D.B. Fenner, G.A.N. Connell, J.M. Philips and T.H. Geballe, Appl. Phys. Lett. 57 (1990) 1137.
[24] G. Beddies, B. Leibold, H.-U. Habermeier and G. Lu, in: High T_c Superconductor Thin Films, Ed. L. Correra (Elsevier, Amsterdam, 1992) p. 641.

Applied Surface Science 69 (1993) 212–215
North-Holland

applied
surface science

Hetero-epitaxial oxidic conductor $La_{1-x}Sr_xCoO_3$ prepared by pulsed laser deposition

J.F.M. Cillessen, R.M. Wolf and A.E.M. De Veirman
Philips Research Laboratories, P.O. Box 80 000, 5600 JA Eindhoven, Netherlands

Received 2 June 1992; accepted for publication 30 October 1992

$La_{1-x}Sr_xCoO_3$ films have been grown highly textured on $Si/SiO_2(100)$ and epitaxially on MgO(100) substrates, despite the large lattice mismatch. This is confirmed by XRD pole figures and HR-TEM analyses: misfit corrections are observed in the first atomic layers. Electrical and structural analyses are discussed and show that $La_{1-x}Sr_xCoO_3$ is a promising electrode material, providing a thermally stable and electrically intimate contact with oxidic active layers. Dielectric and ferro-electric materials sandwiched between these oxidic electrodes showed sharp interfaces and excellent electrical properties.

The variety of physical phenomena present in oxidic materials makes this an interesting class of materials for device applications. Especially perovskite and related compounds are serious candidates for future electronics such as integrated electro-optics or non-volatile memories. Single crystalline oxide thin films have always been a wish, however the technical problems encountered during deposition have not been overcome until the discovery of high-temperature superconductors. These materials provided an enormous drive in the development of oxide deposition techniques. Especially the pulsed laser deposition technique proved to be very successful in this area.

In some applications, for example the charging of ferro-electric capacitive structures or the driving of electro-optic devices, electrode layers must be integrated in thin-film stacks. Conventional metallic electrodes, however, often show wetting and adhesion problems. For example, the growth of Pt (an often used electrode material) on MgO proceeds via coalescence of the nucleation sites. We showed this behaviour for pulsed laser deposited epitaxial Pt on MgO(100) [1]. Another disadvantage of metallic electrodes is that they probably form an interface layer between electrode and active layer. This affects material properties and may cause degradation of dielectric or ferroelectric materials by the electrical fields applied.

To prevent these problems, several research groups presented the use of the superconductor $YBa_2Cu_3O_{7-\delta}$ as an alternative electrode material [2]. However, a disadvantage of this material is that it passes a phase transition and shows twinning in the *a–b* plane upon cooling after film growth. This might introduce internal strain in a stack of materials, which is likely to affect the epitaxiality and performance of devices, since several of the phenomena are very sensitive to strain (via electrostriction). Moreover, cuprate superconductor materials require very accurate deposition conditions in order to get the desired conductive properties. To prevent these disadvantages, we introduce the group of $La_{1-x}Sr_xCoO_3$ materials.

Compounds within the $La_{1-x}Sr_xCoO_3$ system have the (pseudo-)cubic perovskite structure. The lattice constant ranges form 0.382 to 0.384 nm, slightly depending on the amount of Sr. The above-mentioned properties make $La_{1-x}Sr_xCoO_3$

an interesting group of compounds for the use as electrodes for perovskite-type materials due to both their structural and lattice match.

Substitution of Sr^{2+} on the La^{3+} site in $LaCoO_3$ introduces a charge compensation resulting in p-type conductivity. Jonker and Van Santen [3] showed that up to about 50% Sr, this charge compensation is dominated by the formal conversion of Co^{3+} to Co^{4+}, whereas at higher Sr content it is also determined by the formation of oxygen vacancies. The overall formula for these materials can therefore be given by:

$$La^{3+}_{1-x}Sr^{2+}_{x}Co^{3+}_{1-x+2y}Co^{4+}_{x-2y}O^{2-}_{3-y}V^{Ox}_{y}.$$

Within this system the electrical resistivity can be tuned over four orders of magnitude by changing the Sr content from 0 to 0.5. The lowest resistivity measured for bulk $La_{0.5}Sr_{0.5}CoO_3$ material is $5 \times 10^{-4}\ \Omega \cdot cm$ [3–5].

$La_{1-x}Sr_xCoO_3$ films were grown using the pulsed laser deposition technique. The turbo-pumped vacuum chamber had a base pressure of 2×10^{-8} mbar. During deposition oxygen system pressures were in the range from 10^{-2} to 0.3 mbar; this parameter did not seem to be very critical, which makes the material compatible with a variety of other oxide materials. An ArF excimer laser was used (193 nm, pulse duration 15 ns, repetition rate 1–5 Hz, 350 mJ/shot) and typical substrate temperatures were between 580 and 650°C. High-density targets of stoichiometric material were prepared by isostatic pressing and sintering. The films (most attention was paid to the $x = 0.5$ composition because of its low resitivity) were deposited on poly-Al_2O_3, Si(100) and MgO(100):

(a) As expected, films grown on poly-Al_2O_3 were polycrystalline. This allowed the accurate refining of the X-ray diffraction (XRD) lines. The cubic lattice parameter of the films with composition $x = 0.5$ was 0.3827 ± 0.0002 nm, comparable with the target material within 0.0005 nm.

(b) On Si(100) substrates (covered by native oxide) a columnar growth morphology was observed, the column width ranging from 20 to 50 nm. XRD mainly shows (00l)-oriented material. The cross-sectional transmission electron microscopy (TEM) image of fig. 1 shows this columnar microstructure for a ferro-electric film of $PbZr_{0.53}Ti_{0.47}O_3$ sandwiched between two $La_{0.5}Sr_{0.5}CoO_3$ layers.

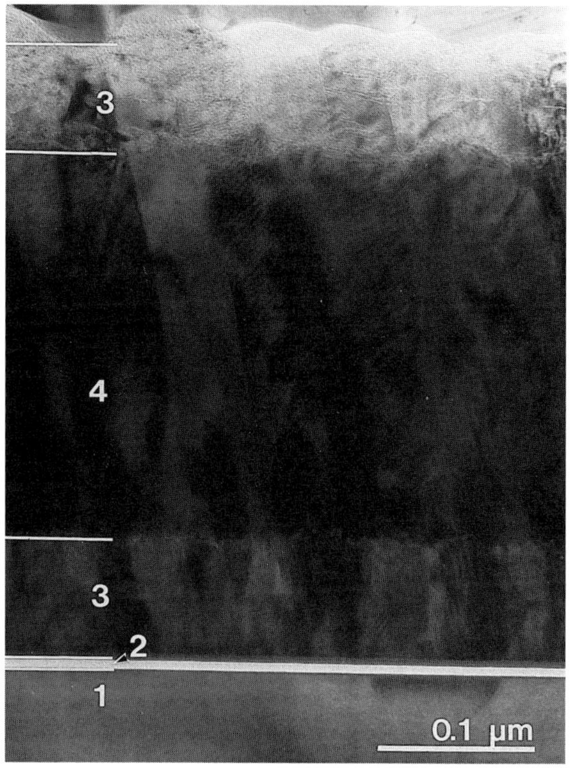

Fig. 1. Cross-sectional TEM micrograph of a stack of materials consisting of $PbZr_{0.53}Ti_{0.47}O_3$ (= 4) sandwiched between two $La_{0.5}Sr_{0.5}CoO_3$ (= 3) layers on a Si(100) (= 1) substrate covered with SiO_2 (= 2) showing a columnar microstructure.

Only a slight interdiffusion between the conductor and the silicon substrate was observed using SIMS depth profiling up to deposition temperatures of 600°C (average steepness 10–15 nm/decade). At temperatures higher than 700°C substantial amounts of silicate compounds were found.

(c) $La_{0.5}Sr_{0.5}CoO_3$ films prepared on single-crystal MgO(100) substrates were epitaxial. XRD only showed sharp (00l) reflections and XRD pole figures confirmed an in-plane alignment of the film and substrate lattice vectors.

Cross-sectional high-resolution TEM (HRTEM) observations of these layers revealed some interesting phenomena: fig. 2 shows a superstruc-

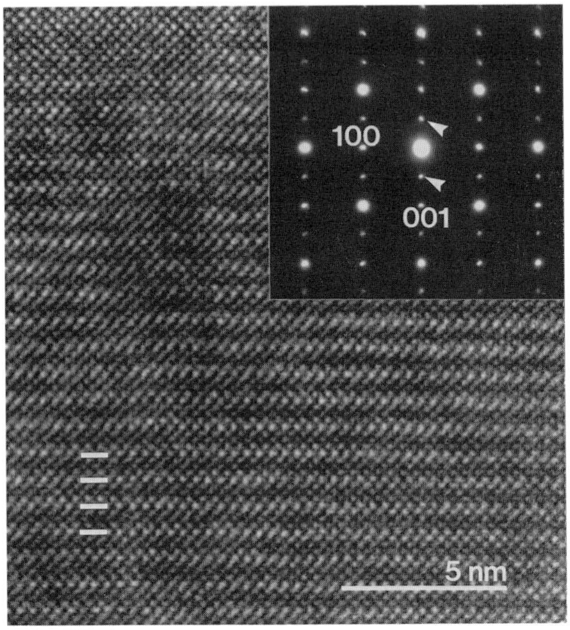

Fig. 2. HR-TEM cross-sectional image of the superstructure in $La_{0.5}Sr_{0.5}CoO_3$ showing a period of twice the XRD lattice parameter (indicated by bars). The inset shows the electron diffraction pattern with extra spots at 0, 0, $\frac{1}{2}$ as indicated by the arrows.

ture in the ⟨001⟩ direction with a period of twice the XRD lattice parameter. Doubling of this parameter is confirmed by extra spots in the electron diffraction pattern (inset) as indicated by the arrows at 0, 0, $\frac{1}{2}$. The superstructure is probably caused by a structural deformation, which is well known in perovskite-type materials [6]: an ordering of the La, Sr or Co ions would introduce extra reflections in the XRD pattern, which in fact was not observed. In some cases (see HR-TEM micrograph of fig. 3) this superstructure was also found in a direction perpendicular to the growth direction, which suggests that it exists in all ⟨100⟩ directions. Fig. 3 also reveals a sharp MgO–$La_{0.5}Sr_{0.5}CoO_3$ transition with misfit dislocations at the interface accommodating the 10% lattice mismatch.

Several films of $La_{1-x}Sr_xCoO_3$ were grown with x ranging from 0 to 0.5. The resistivity of these films was measured using the four-point probe method. They show comparable values with those reported for bulk material [3–5] (see fig. 4).

Fig. 3. HR-TEM cross-sectional image showing a sharp $MgO/La_{0.5}Sr_{0.5}CoO_3$ interface and a superstructure parallel and perpendicular to the interface.

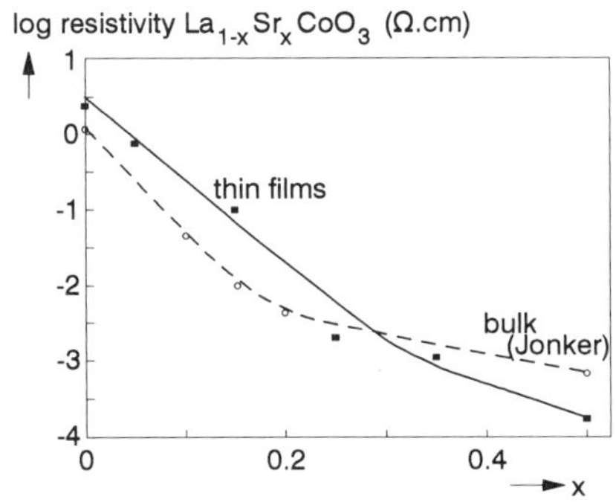

Fig. 4. Resistivity of $La_{1-x}Sr_xCoO_3$ films in comparison with bulk material (after Jonker and Van Santen [3]) as a function of Sr content.

$PbZr_{0.53}Ti_{0.47}O_3$ as well as $SrTiO_3$ films, sandwiched between oxidic bottom and top electrodes, were grown. These layers were structured

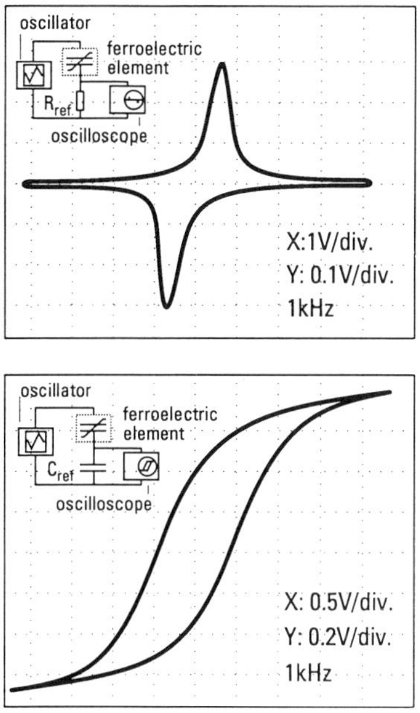

Fig. 5. Ferro-electric switching loop of 450 nm $PbZr_{0.53}Ti_{0.47}O_3$ sandwiched between two $La_{0.5}Sr_{0.5}CoO_3$ electrodes showing a low coercive field strength (E_c) of 10–15 kV/cm, which is comparable with values measured on bulk material.

into small capacitors of sizes ranging from 15 to 2000 μm in diameter. Electrical characterization resulted in high permittivity values for 40 nm $SrTiO_3$ layers ($\epsilon_r = 290$; bulk value 300) and a low coercive field strength (E_c) for $PbZr_{0.53}Ti_{0.47}O_3$ of 10–15 kV/cm, which is also comparable to bulk material values [7] (see fig. 5).

In conclusion, hetero-epitaxial films of $La_{1-x}Sr_xCoO_3$ on MgO(100) as well as films with a columnar growth morphology on Si(100) were prepared using the pulsed laser deposition technique. Deposition conditions with respect to the partial oxygen pressure and substrate temperature were not critical. The films showed electrical resistivities comparable with bulk materials or lower.

Several stacks of perovskite-type oxides sandwiched between two $La_{0.5}Sr_{0.5}CoO_3$ electrodes have been grown successfully, showing sharp interfaces and a stable and intimate electrical contanct between electrode and active layer.

M. Kraan, W. Keur, R. Potze, J. Timmers, B. Giesbers, G. Kampschoër, P. Zalm and F. Hakkens are acknowledged for preparing targets, structuring films, XRD, SIMS TEM preparations and electrical measurements.

References

[1] J.F.M. Cillessen, R.M. Wolf and D.M. de Leeuw, Pulsed laser deposition of hetero-epitaxial thin platinum films on MgO(100), Thin Solid Films, to be published.
[2] R. Ramesh, A. Inam, W.K. Chan, F. Tillerot, B. Wilkens, C.C. Chang, T. Sands, J.M. Tarascon, V.G. Keramidas, Appl. Phys. Lett. 59 (1991) 3542.
[3] G.H. Jonker and J.H. van Santen, Physica 19 (1953) 120.
[4] P.M. Raccah and J.B. Goodenough, J. Appl. Phys. 39 (1968) 1209.
[5] A.N. Vlasov, Inorg. Mater. 26 (1990) 2040.
[6] F.S. Galasso, in: Structure, Properties and Preparation of Perovskite-type Compounds (Pergamon, Oxford, 1969).
[7] R.M. Wolf and J.F.M. Cillessen, to be published.

Applied Surface Science 69 (1993) 216–220
North-Holland

applied
surface science

Laser ablation studies of magnesium oxide

L. Dirnberger, P.E. Dyer, S. Farrar, P.H. Key and P. Monk
Department of Applied Physics, University of Hull, Hull HU6 7RX, UK

Received 2 June 1992; accepted for publication 30 October 1992

The ablation of single crystal magnesium oxide using a UV-excimer KrF laser at 248 nm has been studied using a variety of diagnostic techniques. Ablation depth rate measurements indicate a threshold fluence for significant material removal of ~ 3 J cm^{-2} although surface damage in the form of micro-cracks is observed below this value. The effective absorption coefficient at 248 nm of 4×10^4 cm^{-1}, found from the ablation rate, was much greater than observed from small signal transmission measurements. Scanning electron microscopy reveals evidence of thermal removal in the ablation regime probably mediated by a breakdown plasma, and a degree of exfoliation below the ablation threshold. The velocity of selected ablation species has been determined from time-resolved emission spectroscopy and a charge collector probe used to measure the angular distribution of ions in the ablation plume.

1. Introduction

The pulsed laser ablation-deposition technique is now well established for the growth of thin films of high T_c superconducting (HTS) materials. The technique is also applicable to other oxides (e.g. TiO_2, ZrO_2, CeO_2 and MgO) for optical films [1], substrate buffer layers and HTS-insulator multilayer structures [2,3]. Of these, MgO is of considerable interest as a buffer layer in HTS technology, and the growth of epitaxial films on silicon from magnesium metal ablated in oxygen has recently been reported [4]. A knowledge of the ablation characteristics of single crystal MgO is of intrinsic interest and because of its possible use as a source material for thin film deposition. In addition, laser ablation patterning is a potentially important step for single chamber, in situ, processing of HTS/MgO multilayer structures. It is a requirement of such processing that layers are ablated without damage to underlying material so that the epitaxy into subsequently deposited layers can be maintained.

In this paper we report new studies of the interaction of 248 nm laser radiation with single crystal MgO using a variety of diagnostic techniques. Time-integrated and time-resolved spectroscopy in the UV–visible spectral region are used to provide information on the principal luminous species in the ablation plume and expansion velocities of selected species respectively. Additionally, information on the velocity and angular distribution of ions in the plume is obtained using a charge collector probe. The ablation threshold for MgO has been obtained from ablation depth rate measurements and scanning electron microscopy used to evaluate the irradiated site.

2. Experimental

The ablation rate of single crystal MgO(100) samples was determined as a function of KrF laser fluence from measurements of the depth of the etch crater following multiple pulse exposure and the laser irradiated sites were examined using a scanning electron microscope.

The UV–visible light emission from the ablation plume was examined using a grating monochromator and photomultiplier tube. A CaF_2 lens was used to form an image (×1 magni-

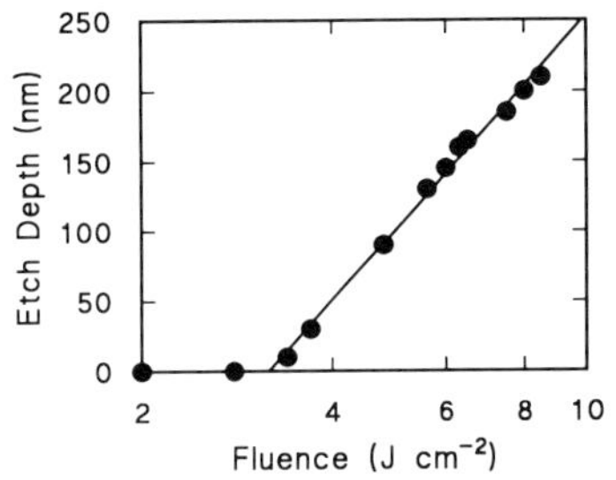

Fig. 1. Etch depth per pulse of single crystal MgO(100) as a function of KrF laser fluence.

fication) of the plume outside of the ablation cell where the end of an optical fibre could be scanned in the image plane in order to direct the radiation to the entrance slit of the monochromator. This arrangement allowed interrogation of the ablation plume along its length with a spatial resolution defined by the fibre core diameter (0.5 mm). The output signal from the photomultiplier was recorded as a function of wavelength using an integrating amplifier and *XY* plotter.

Integrated spectra were obtained at a number of positions along the plume length for various laser fluences and oxygen pressures. The principal spectral features were identified from standard tables of spectral lines for elements and compounds. In addition, time resolved emission from selected species was studied by selecting the appropriate wavelength and recording the photomultiplier output with a fast storage oscilloscope (Tektronix 7834 with 7A15A plug-in amplifier, 50 Ω input impedance). The oscilloscope was triggered by a 1 ns rise time photodiode, allowing velocities to be determined from the emission pulse delay at various distances from the target.

An ion probe consisting of a −80 V biased collector plate sited in a fine mesh cage at earth potential was used to detect ions (non-species resolved) in the ablation plume. The probe could be rotated in a plane containing the ablation site and target normal, allowing ions to be collected within a fixed solid angle at various angles to the target normal. It was also possible to vary the distance between the target and collector.

3. Results

A plot of the etch depth per laser pulse of single crystal MgO(100) as a function of the logarithm of the KrF laser fluence is shown in fig. 1. No differences were observed in the etch rate data between experiments carried out with pulse repetition frequencies of 1 and 10 Hz. It can be seen from fig. 1 that the data are a good fit to a straight line and are thus consistent with Beer's law of absorption so that the etch rate, x, can be represented by;

$$x = \ln(F/F_t)/\alpha,$$

where α is the effective absorption coefficient found from the gradient of the plot to be 4×10^4 cm^{-1} and F_t is the threshold ablation fluence ~3 J cm^{-2}.

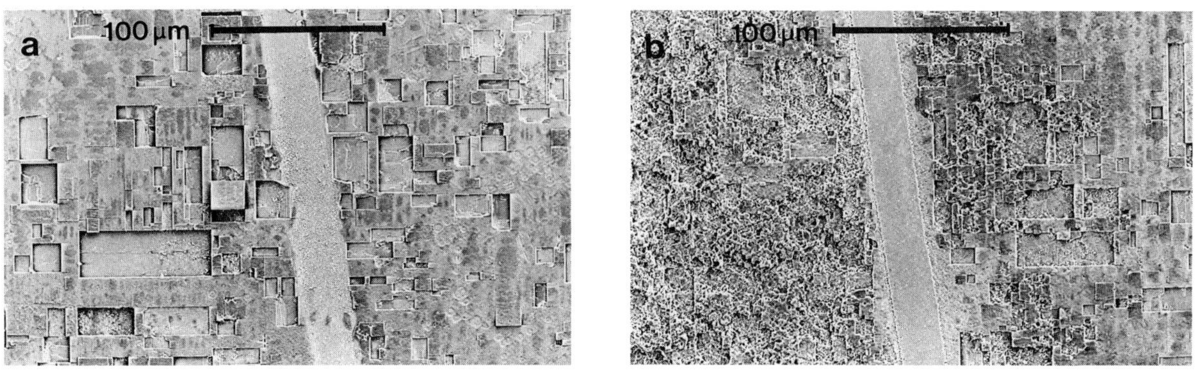

Fig. 2. Scanning electron micrographs of a MgO(100) surface after irradiation with 100 KrF laser pulses at (a) 2.5 and (b) 3.2 J cm^{-2}. (The central line is the image of a focusing bar in the laser beam area selection mask.)

It should be noted that the etch rate per pulse was calculated from the measurement of the depth of the etch crater after irradiation with many ($> 10^3$) laser pulses. It was observed that, for fluences in the range F_t to $2F_t$ there was no characteristic visible or audible indication of ablation for the first ~ 5 laser pulses on fresh target areas. This would indicate that some pre-conditioning of the irradiated zone was necessary in order for the ablation to proceed.

Microscopic examination of target areas irradiated at fluences in the range 1–3 J cm^{-2} revealed surface damage in the form of well ordered micro-cracks along cleavage planes of the cubic crystal. The areal density of these micro-cracks increased with pulse number after the first laser shot. Micro-cracking of the surface was also observed at much lower fluences but only after multi-pulse (> 250) irradiation. The origin of these micro-cracks is thought to be thermally induced stress arising from rapid temperature excursions at the surface of this relatively low thermal diffusivity material.

Examples of surface damaged and of etched targets are shown in the electron micrographs in figs. 2a and 2b. The removal of material by exfoliation of crystallites within the sub-ablation threshold surface damaged region can be observed in some areas of fig. 2a together with areas in which there is a movement of crystallites out of the surface plane. Exfoliation thus would appear to dominate any surface removal in this fluence regime.

The crystal surface in the fluence regime at which significant ablation occurs is illustrated in fig. 2b. In this case the morphological evidence suggests that melting of the surface has occurred and that this is initiated in those regions in which the micro-crack density and embryonic exfoliates are greatest. It is interesting to note that the micro-cracks do not penetrate to any great extent into the unirradiated areas (central line in fig. 2).

At fluences below the surface damage threshold a change of colour extending through the bulk of the laser irradiated region of a 1 mm thick sample was observed. These UV induced colour centres were found to be removed after annealing in an oxygen atmosphere at 1100°C. Small signal transmission measurements at 248 nm using both low fluence KrF laser radiation and, in separate experiments a UV-lamp and monochromator were carried out on a number of crystals from a single source. For unirradiated samples the absorption coefficient was ~ 10 and this increased to a maximum value of ~ 25 cm^{-1} following prolonged exposure to UV irradiation at fluences below the surface damage threshold. This relatively small increase does not appear to explain the high effective absorption coefficient (4×10^4 cm^{-1}) derived from etch-rate measurements (fig. 1) although the possibility that a highly absorbing but shallow layer of near-surface defects is produced which remain unresolved when averaging transmission over the full sample thickness cannot be excluded. It is likely that defect production is responsible for the surface preconditioning effect noted earlier and, as will be argued below, is instrumental in initiating optical breakdown and plasma mediated etching at the surface.

The plume spectra for MgO ablation were relatively simple, (fig. 3), consisting of principally the lines of neutral magnesium, Mg(I), magnesium ions, Mg(II), with weak contributions from oxygen ions, O(II). A comparison of spectra (300–550 nm) for a variety of oxygen pressure conditions (10^{-2}–1.0 mbar) showed an increase in the intensity of lines for Mg(I), Mg(II) and O(II) with increasing laser fluence. Space–time resolved studies of selected species, however, were found to yield more quantitative information.

Figs. 4 and 5 show typical results obtained from measurement of the arrival time of spectral

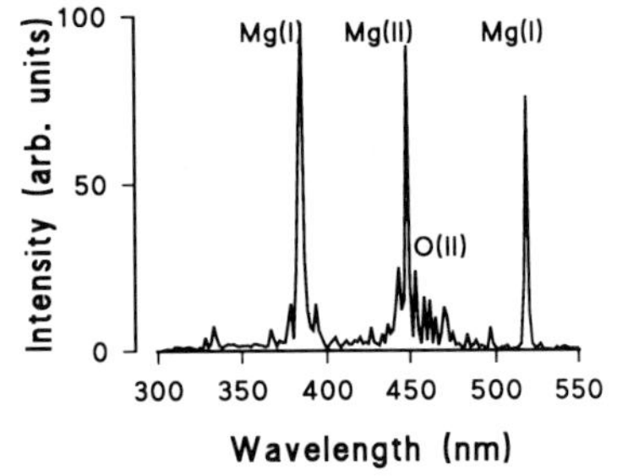

Fig. 3. Emission spectrum of plume from KrF laser ablation of MgO.

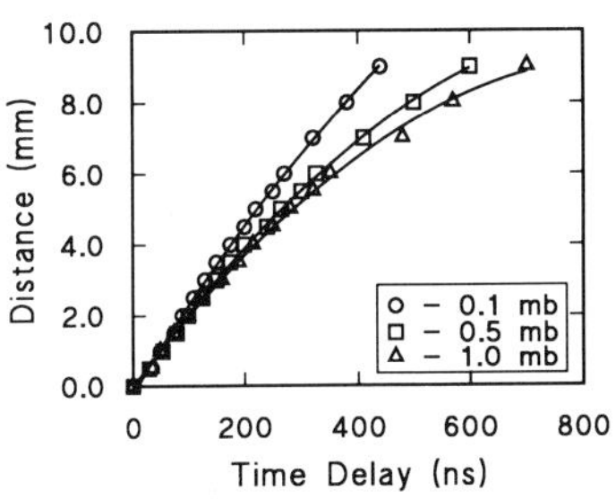

Fig. 4. Distance–time plots for Mg(II) from time-resolved emission spectroscopy of KrF laser ablation of MgO in oxygen.

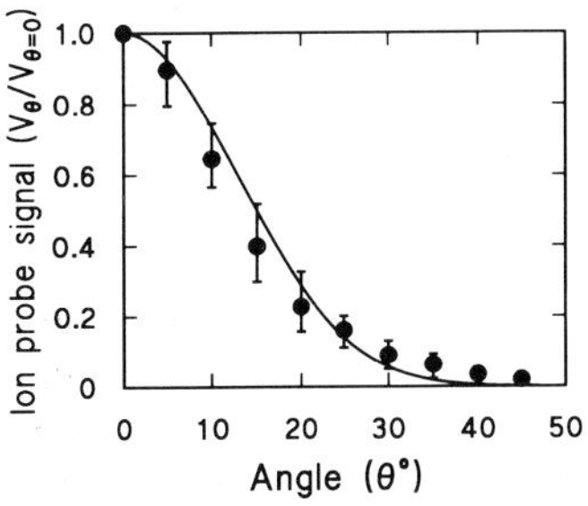

Fig. 6. Angular distribution of non-species resolved ions in the ablation plume from KrF laser irradiation of MgO.

features at various distances from the target surface. These studies show that for a fixed laser fluence, as in fig. 4, the ablated products decelerate more rapidly with increasing background gas pressure. In fig. 5 the velocity for Mg(II) (448 nm emission) measured 1 mm from the target surface is plotted as a function of laser fluence. It can be seen that the velocity increases only slowly with fluence from a value of 1.75×10^4 at threshold to 2.2×10^4 m s^{-1} at 5.2 J cm^{-2}. Similar, but somewhat larger, velocities were determined using the ion probe e.g. 2.1×10^4 m s^{-1} at the ablation threshold.

The angular distribution of ions detected in the ablation plume measured at a range of distances from the target and for fluences in the range F_t to 5 J cm^{-2} is shown in fig. 6. The data have been normalised to the signal corresponding to the forward peak direction ($\theta = 0$, where θ is the angle with respect to the peak) for each fluence and detector probe distance. The error bars represent the full range of measured values and as a guide to the eye a $\cos^n \theta$ with $n = 20$ distribution line is also shown. This sharply peaked distribution can probably be attributed to the relatively large ablation spot size (3 mm) which is known to lead to a strongly directed expansion.

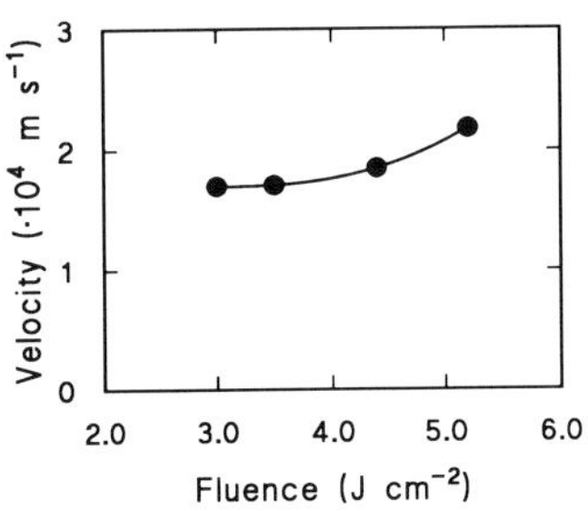

Fig. 5. Velocity of Mg(I) as a function of KrF laser fluence during ablation of MgO from time-resolved spectroscopy at 1 mm from the target.

4. Discussion

On the basis of the small signal absorption coefficient, α_s, at 248 nm only a very small rise in surface temperature is expected for MgO at the fluences involved in these experiments. This can be seen from a simple calculation for the temperature rise, ΔT, obtained from a thermal energy balance:

$$\Delta T = \alpha_s F / C, \quad (1)$$

where F is the absorbed fluence and C is the volume specific heat. For $\alpha_s = 10$–25 cm^{-1} and $F < 3$ J cm^{-2}, $\Delta T < 9$–22 K which is clearly too small to produce vaporisation or to lead to significant surface cracking through thermal stress. It is apparent from fig. 1, however, that relatively high removal rates are obtained at fluences exceeding 3 J cm^{-2} and that the effective absorption coefficient is greater than the small signal value by more than three orders of magnitude. We attribute this dramatic rise in absorption to the onset of laser induced optical breakdown at the crystal surface which leads to a strongly absorbing plasma being formed. With $\alpha = 4 \times 10^4\ \mathrm{cm}^{-1}$ and $F = 3\ \mathrm{J\ cm}^{-2}$ we obtain $\Delta T = 3.5 \times 10^4\ \mathrm{K}$ (~ 3 eV) which would adequately account for the ablative removal of material. This overestimates the temperature rise as we have neglected the enthalpy of ablation and ionisation energy.

From the ion probe and spectroscopic velocity measurements, e.g. of Mg(II), the ion energy, E, near the threshold is found to be 30–45 eV. Using Puells model [5] to relate this to the plasma temperature, T, we obtain:

$$kT = E/5(z+1) = 3\text{–}4.5\ \mathrm{eV},$$

where $z = 1$ is the charge state. This agrees reasonably well with that from the simple energy balance model.

It is likely that the transition from the relatively transparent crystal state to strongly absorbing breakdown plasma is mediated by the surface micro-structural defects, which can enhance the local electric field [6] or volume defects which may be readily photo-ionised and hence provide seed electrons for breakdown. Near the ablation threshold it is observed that several laser pulses are required to suitably "pre-condition" the target before this optical breakdown can take place. Surface micro-cracking may be related to stress associated with volume defect production and further work is under way to investigate this.

5. Conclusions

A description of the ablation mechanism of single crystal MgO using the KrF laser is complicated by the low fluence creation of structural defects and exfoliation and the possibility that these may occur during the early stages of a single high fluence pulse.

The surface cracking and exfoliation at low fluences would suggest that high quality in-situ laser patterning of high T_c superconducting thin films deposited on MgO substrates or buffer layers may be difficult to achieve. This phenomenon may also pose problems if single crystal MgO were to be used as a source material for the laser ablation-deposition of thin films as buffer layers.

References

[1] H.O. Sankur and W. Gunnung, Appl. Opt. 28 (1989) 2806.
[2] H. Koinuma, H. Nagata, T. Tsukahara, S. Gonda and M. Yashimoto, Appl. Phys. Lett. 58 (1991) 2027.
[3] S. Amirhaghi, A. Archer, B. Taguiang, R. McMinn, P. Barnes, S. Tarling and I.W. Boyd, Appl. Surf. Sci. 54 (1992) 205.
[4] D.K. Fork, F.A. Ponce, J.C. Tramontana and T.H. Geballe, Appl. Phys. Lett. 58 (1991) 2294.
[5] H. Puell, Z. Naturforsch. 25a (1970) 1807.
[6] N. Bloembergen, Appl. Opt. 12 (1973) 661.

Applied Surface Science 69 (1993) 221–224
North-Holland

applied
surface science

Deposition of $YBa_2Cu_3O_x$ by laser ablation on Si(100) using different buffer layers

F. Sánchez, M. Varela, X. Queralt, M.V. García-Cuenca, J. Palau, R. Aguiar and J.L. Morenza

Departament de Física Aplicada i Electrònica, Universitat de Barcelona, Avda. Diagonal 647, E-08028 Barcelona, Spain

Received 2 June 1992; accepted for publication 30 October 1992

Superconducting $YBa_2Cu_3O_x$ (YBCO) thin films have been obtained on Si(100) substrates with yttria-stabilized zirconia (YSZ), CeO_2 and $SrTiO_3$ buffer layers. The YBCO and the buffer films were deposited by laser ablation. The films were characterized by X-ray diffraction (XRD), field emission scanning electron microscopy (FESEM), scanning electron microscopy (SEM), secondary ion mass spectrometry (SIMS), and four-contact electrical resistivity. X-ray diffraction spectra show well oriented buffer layers and secondary ion mass spectrometry results indicate low interdiffusion between YBCO, buffer layers and Si. YBCO films with the c-axis perpendicular to the substrate are obtained for the three materials used as buffers. The best films show a normal state metallic behaviour with zero resistance around 84 K. The crystalline and electrical properties of the films are related with the substrate temperature. The substrate temperature range for the best properties depends on the nature of the buffer layer.

1. Introduction

High quality YBCO thin films have been epitaxially grown on oxide single-crystal substrates such as $SrTiO_3$ [1], MgO [2] and $LaAlO_3$ [3]. However, these substrates are not necessarily suitable for device applications because they are expensive and substrate size is restricted. For technological applications silicon is the proper substrate because it is cheap, available in large size and enables the integration of high-temperature superconductor materials with the semiconductor technology. The preparation of YBCO on Si substrates is not as successful as on oxide single crystals because the high chemical reactivity between the film and the substrate degrades the superconducting properties [4]. To overcome this problem it is necessary to introduce intermediate buffer layers between silicon and YBCO. The buffer must be nonreactive with the substrate and the film at the deposition temperature, and have a close match of the lattice constants and thermal expansion coefficients. Moreover, it is very important to deposit well oriented buffer layers with smooth surfaces in order to grow superconducting thin films with the good crystalline quality and flat surfaces required for device applications. Since the buffer layer plays an important role in determining the properties of the YBCO film, it is interesting to study different available materials to use as buffers. In this paper we report the deposition of YBCO films on Si(100) coated with yttria-stabilized zirconia (YSZ), CeO_2 and $SrTiO_3$ layers.

2. Experimental

YBCO films and buffer layers were obtained by laser ablation of YBCO, YSZ, CeO_2 and $SrTiO_3$ targets. Silicon (100) wafers were used as substrates and were chemically and thermally cleaned in order to remove the native oxide. An ArF excimer laser ($\lambda = 193$ nm, 23 ns pulse duration) was used for ablation, and it was focused on the stoichiometric target at an angle of 45°. Conditions for the deposition of the buffer layers were a substrate temperature in the 400–850°C range, an oxygen partial pressure between 10^{-7} and 5×10^{-3} mbar, and 1.5×10^4 pulses of 3–5 J/cm^2 energy density, leading, at a substrate–target distance of 8 cm, to a thickness of about 25, 40 and 50 nm for YSZ, CeO_2 and $SrTiO_3$ films, respectively. To minimize the formation of silicon oxide during the early stages of growth of the buffers obtained in the presence of oxygen, depositions were initiated at the base pressure of the vacuum system (10^{-8}–10^{-7} mbar) and the oxygen pressure was raised after the first 100 pulses.

YBCO superconducting thin films were deposited onto the previously obtained buffer layers using laser pulses with an energy density of 1–2 J/cm^2. The oxygen partial pressure was fixed at 0.3 mbar and the substrate temperature, measured by a thermocouple inside a block of copper, was varied between 700 and 775°C. The thicknesses of the films, at a substrate–target distance of 5 cm, were 0.5 and 1.0 μm for 1×10^4 and 2×10^4 laser pulses, respectively. After deposition, the films were cooled to 400°C in oxygen atmosphere (200 mbar), maintained at this temperature for 15 min and then cooled down to room temperature.

The crystalline properties of the films were investigated by X-ray diffractometry (XRD). The surface morphology was observed by scanning electron microscopy (SEM) and field emission scanning electron microscopy (FESEM). The depth composition profiles of the films have been obtained by secondary ion mass spectrometry (SIMS). The temperature dependence of the electrical resistance $R(T)$ was monitored between 300 and 20 K by an automated measurement system using the standard four-contact technique.

3. Results and discussion

X-ray diffraction spectra indicate that buffer layers have the cubic phase with preferential orientation (hhh) for YSZ and CeO_2 films [5]. $SrTiO_3$ films have preferential orientation ($h00$) or ($hh0$) depending on oxygen pressure and substrate temperature [6], but only films with ($h00$) orientation were used as buffer layers between Si and YBCO. The scanning electron microscope and field emission scanning electron microscope micrographs show, for all the buffer films, a smooth surface without microcracks and which is practically free of particles. This morphology is quite different to that reported for buffers obtained by other deposition methods which show a very granular structure [7]. This is noteworthy because the growth of YBCO films requires smooth surfaces without any particles on top.

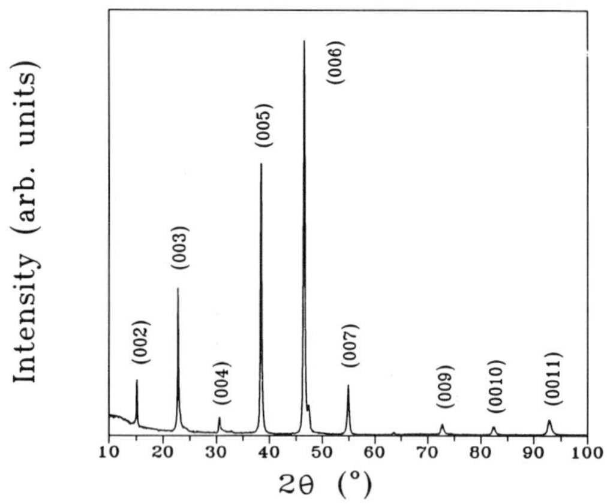

Fig. 1. θ–2θ X-ray diffraction spectrum of an YBCO film on YSZ/Si(100) substrate deposited at 750°C.

XRD spectra of YBCO films show only peaks of silicon, buffer and 123 phase. The films deposited at substrate temperatures between 700 and 750°C with YSZ buffer show preferential orientation ($00h$), corresponding to c-oriented films (fig. 1), whereas YBCO films deposited with CeO_2 buffer are only c-oriented in the 725–750°C substrate temperature range. For YBCO films on $SrTiO_3$ buffers, in the temperature range studied (725–775°C), a c-oriented film has been obtained only at 775°C, but the orientation is partial.

The surface morphology of the YBCO films revealed by the scanning electron microscopy micrographs is quite different from that observed for the buffer layers. YBCO films display a granular structure with globular conglomerates. In some cases the grains are compacted, and the surface is smooth with the typical particles of the ablation process.

Fig. 2 shows a typical SIMS profile corresponding to an YBCO/buffer/Si structure. The main features of the depth composition studies are the good uniformity of Y, Ba, Cu and O in the film, and the low interdiffusion between YBCO, buffer layers and substrate. This result indicates the efficiency of the three materials used as buffer layers.

The results of electrical measurements on YBCO thin films deposited on Si(100) with an YSZ buffer are shown in fig. 3. Curve a corresponds to a film deposited at 700°C, and it can be seen that the film has semiconducting character in the normal state with zero resistance of 63 K. The films deposited at 725 and 750°C (curves b and c) display a metallic behaviour in the normal state ($R_{300}/R_{100} = 1.4$ and 1.7), zero resistance temperature of 76 and 82 K, and a transition width (10%–90%) of 8.8 and 5.5 K, respectively. For the YBCO films grown on CeO_2/Si(100) at substrates temperatures between 725 and 775°C the results are similar, with metallic behaviour ($R_{300}/R_{100} = 1.4$–1.8), whereas the film deposited at 700°C showed semiconducting behaviour. A zero resistance temperature of 83 K was found for films deposited at 775 and 750°C, and 84 and 32 K for films deposited at 725 and 700°C, respectively. YBCO films deposited with a $SrTiO_3$ buffer layer show semiconducting character for a

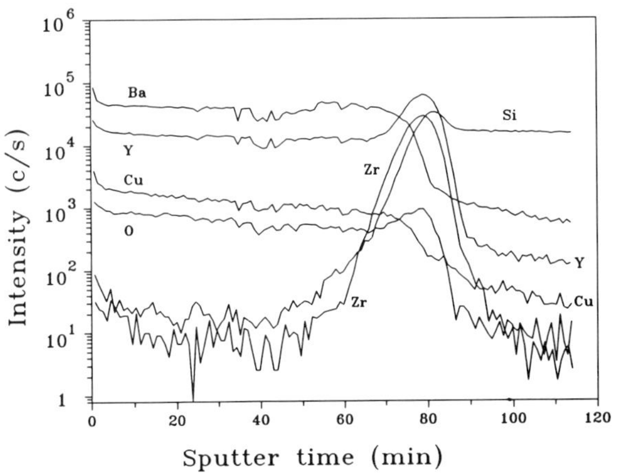

Fig. 2. Secondary ion mass spectrometry profile of the YBCO film deposited on YSZ/Si(100) at 750°C.

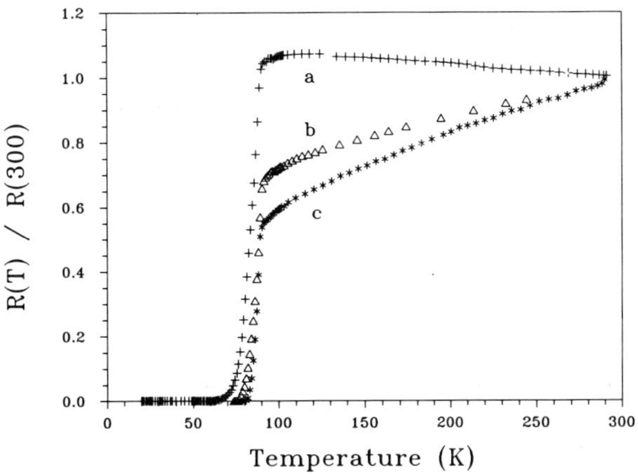

Fig. 3. Normalized resistance versus temperature of the YBCO films obtained on YSZ/Si(100) at temperatures of: (a) 700°C, (b) 725°C and (c) 750°C.

substrate temperature of 725°C, and only a low metallic character ($R_{300}/R_{100} \approx 1$) for a temperature of 775°C. A zero resistance temperature and transition width of 84 and 4.1 K, respectively, were found for the film with metallic behaviour, whereas for the film with semiconducting behaviour these values were 61 and 17 K, respectively.

4. Conclusion

YBCO thin films and YSZ, CeO_2 and $SrTiO_3$ buffer layers were deposited on Si(100) substrates by laser ablation. It should be noted that buffer layers as thin as 25–50 nm prevent diffusion between YBCO and silicon. To obtain good crystalline and electrical properties a minimum substrate temperature is necessary. The minimum substrate temperature for growing *c*-oriented YBCO films depends on the buffer layer used. For YSZ buffer layers *c*-oriented YBCO films are obtained for temperatures higher than 700°C, whereas for CeO_2 buffers the minimum temperature is 725°C, and for $SrTiO_3$ buffers only partially *c*-oriented YBCO film is found at 775°C. For all the buffer layers studied the best resistance–temperature characteristics are obtained for YBCO films deposited in the 725–775°C substrate temperature range. Although good resistance–temperature behaviour is observed with not *c*-oriented films, other electrical properties such as critical current, not studied in this work, require well *c*-oriented YBCO films.

Acknowledgements

The authors acknowledge the collaboration of the Secondary Ion Mass Spectrometry, and Scientific and Technical Services of Universitat de Barcelona.

This work is part of a research programme financed by CICYT of the Spanish Government (project MAT 89-0253) and Red Eléctrica de España, S.A. (MIDAS programme).

References

[1] X.D. Wu, D. Dijkkamp, S.B. Ogale, A. Inam, E.W. Chase, P.F. Miceli, C.C. Chang, J.M. Tarascon and T. Venkatesan, Appl. Phys. Lett. 51 (1987) 861.

[2] R. Ramesh, D. Hwang, T.S. Ravi, A. Inam, J.B. Barner, L. Nazar, S.W. Chan, C.Y. Chen, B. Dutta, T. Venkatesan and X.D. Wu, Appl. Phys. Lett. 56 (1990) 2243.

[3] R.W. Simon, C.E. Platt, A.E. Lee, G.S. Lee, K.P. Daly, M.S. Wire, J.A. Luine and M. Urbanik, Appl. Phys. Lett. 53 (1988) 2677.

[4] H.-U. Habermeier, G. Mertens and G. Wagner, Vacuum 4 (1990) 859.

[5] F. Sánchez, M. Varela, X. Queralt, M.V. García-Cuenca, R. Aguiar and J.L. Morenza, Physica C 195 (1992) 47.

[6] F. Sánchez, M. Varela, X. Queralt, R. Aguiar and J.L. Morenza, Appl. Phys. Lett. 61 (1992) 2228.

[7] W.A. Feil, B.W. Wessels, L.M. Tonge and T.J. Marks, J. Appl. Phys. 67 (1990) 3858.

Applied Surface Science 69 (1993) 225–230
North-Holland

applied
surface science

Dynamics of graphite photoablation: kinetic energy of the precursors to diamond-like carbon

P.T. Murray [a,b] and D.T. Peeler [b,*]

[a] Research Institute, University of Dayton, Dayton, OH 45469-0167, USA
[b] Graduate Materials Engineering, University of Dayton, Dayton, OH 45469-0246, USA

Received 2 June 1992; accepted for publication 30 October 1992

The mechanisms for growth of diamond-like carbon by pulsed laser deposition have been investigated. Two distinct but related issues have been addressed: (1) the nature and speed of the particles ejected from graphite by pulsed UV laser irradiation and (2) the extent to which the kinetic energy of the ablated particles determines the character of the resultant films. The results indicate that C_3^+ and C_3^0 are the predominant ion and neutral particles ejected from graphite by 248 nm laser radiation. These particles are ablated from graphite with mean kinetic energies of 18 and 0.25 eV, respectively. Carbon films were grown in the presence of various collision gases in order to decrease the energy of the ablated particles. Kinetic energy of the particles was found to play a significant role in the generation of diamond-like films.

1. Introduction

Thin films of diamond-like carbon (DLC) can be grown [1–3] by pulsed laser deposition (PLD). PLD is an established thin film growth process possessing a number of advantages over other deposition techniques. In general, these advantages include the ability to evaporate refractory materials, the capability of growing high purity films, and the potential for congruent target removal. Although various techniques have been used to grow DLC films, little of the fundamental growth mechanisms of any of the processes is clearly understood. In contrast to other techniques, the inherent simplicity of PLD, requiring only a graphite target and pulsed laser radiation, allows a more straightforward elucidation of the mechanisms for DLC growth.

Cuomo et al. [4–7] have examined carbon condensation processes capable of producing DLC films. They propose that the mechanism of condensation and growth is dependent upon two critical factors: incident particle energy and a thermal quench. Cuomo et al. hypothesize that the lower energy atoms involved in a thermal process will preferentially condense into the thermodynamically favored, graphite-like structure. In contrast, higher energy particles may form a diamond-like geometry as the particles impact the surface. The tetrahedral diamond-like coordination represents a reduction in average atomic distance from the less dense graphite. The stable low energy solid is compressed into the metastable diamond geometry. Once a particle has established the metastable bond formation, its excess energy will drive it back over the activation barrier into the equilibrium graphitic structure unless the energy is quickly removed.

The primary objective of the work reported here was to investigate the mechanisms of diamond film growth by PLD. Two distinct but related issues have been addressed: (1) the nature and speed of the particles ejected from graphite by pulsed laser irradiation, and (2) the extent to which the kinetic energy of the ablated particles determines the character of the resultant films. The experiments described here provide insight

* Clare Boothe Luce Scholar.

0169-4332/93/$06.00

into the identity of the precursors to diamond-like carbon grown by PLD as well as serving to test directly the hypothesis of Cuomo et al.

2. Experimental

2.1. Plume characterization

These experiments entailed measuring the time of flight (TOF) distributions of positive ions as well as neutral particles ejected from a graphite target by 248 nm laser radiation. The output of a Lumonics (Model EX-400) excimer laser was passed through a defining aperture, a focusing lens ($f = 100$ cm), and a MgF_2 viewport before striking a pyrolytic graphite target (99.99% purity). The temporal width of the laser pulse was reported by the manufacturer to be 15 ns. The target was rotated at 10 rpm, and the laser was operated at a repetition rate of 1 Hz during data acquisition. Fluence at the target during all TOF measurements was 1.0 J/cm^2 and the laser/target angle of incidence was 45°. The experiments were carried out in an ultra high vacuum chamber with a base pressure of 3×10^{-9} Torr, a pressure sufficiently low to ensure a negligible signal from the residual gas in the chamber. Each TOF distribution reported is a summation of 25 individual laser pulses.

The procedure for determining the mass and speed distributions of laser ejected ions has been described previously [8,9]. Positive ions formed by each laser pulse travelled with constant speed in a field-free region of known length before entering a TOF mass spectrometer. The experimental procedure consisted of obtaining a series of mass spectra at selected delay times after the laser pulse. Speed distributions were determined by integrating the ion peak intensities thus obtained, and by converting these delay times to speed ranges.

Measuring the speed distributions of laser-ejected neutrals entailed using a quadrupole mass filter equipped with an electron impact (70 eV) ionizer. The flight path from target to ionizer was 12 cm. These experiments were carried out by recording TOF distributions of species detected with and without the ionizer turned on. Subtraction of the spectra thus obtained yielded the TOF distribution of (post-ionized) neutral species, as described previously by Murray and Peeler [10].

2.2. Film growth and analysis

Films were grown on room temperature Si(100) substrates which had been rinsed with acetone and methanol prior to placement in the deposition chamber. Film growth was carried out in an ultra high vacuum chamber with a base pressure of 3×10^{-9} Torr. Films were grown in vacuum, as well as in the presence of the collision gasses Ar, Ne, and He. The pressure in the deposition chamber was 75 mTorr (monitored by a capacitance manometer) during film growth in the presence of a collision gas. The laser beam was operated at a repetition rate of 20 Hz for 30 min, and the beam was stepped across the pyrolytic graphite target at a rate of 0.5 mm s^{-1}. The rotating (10 rpm) target–substrate distance was 3.5 cm.

After deposition the films were transferred to a surface analysis system (Perkin–Elmer (PHI) Model 550). Because the films were found to be susceptible to electron beam damage, both core and valence EELS spectra were obtained at beam energies not greater than 1 keV and with currents not greater than 25 nA. The EELS spectra were obtained in the derivative as well as in the $N(E)$ mode.

In order to calculate a final kinetic energy for the particles impinging the substrate total scattering cross sections were determined in a beam attenuation experiment [11]. The experiment entailed measuring the carbon film deposition rate on a quartz crystal oscillator as a function of the inert gas pressure.

3. Results and discussion

3.1. Plume characterization

Shown in the inset of fig. 1 is a representative TOF mass spectrum of the positive ions ejected from graphite. The spectrum was acquired with a delay time of 16 μs. The predominant ion detected was C_3^+. Peaks due to C_2^+ and C_4^+ are also present in the spectrum. The speed distribution of C_3^+ is also presented in fig. 1. The circles represent the experimental data. The ion speed distribution exhibits a maximum at 10 km/s with a corresponding kinetic energy of 18 eV. A supersonic expansion distribution, shown as a solid line, is a reasonable representation of the data. The translational temperature of the C_3^+ obtained from such a fit is 71 000 K. An interesting comparison can be made between these and the results of Creasy and Brenna [12]. These authors irradiated graphite with 266 nm laser radiation and, using an ion cyclotron resonance spectrometer, also found C_3^+ to be the predominant ion ablated. Unfortunately, they reported no speed distributions for the ablated ions.

The predominant neutral particle ejected from the graphite target by 248 nm laser radiation was C_3^0. The speed distribution of C_3^0 is presented in fig. 2. The circles represent the experimental data. The distribution exhibits a maximum at 1.1 km/s, with a corresponding kinetic energy of 0.25 eV. The dashed line represents the best fit to the data obtainable using a Maxwell–Boltzmann distribution ($T \approx 3000$ K). A supersonic expansion

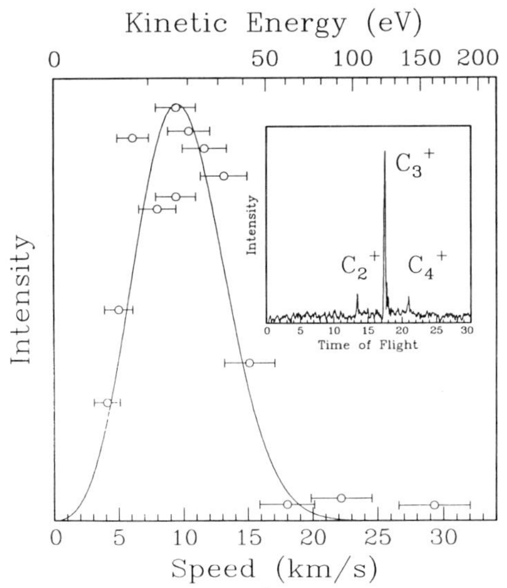

Fig. 1. Speed distribution of laser-ablated C_3^+. Shown in the inset is a TOF mass spectrum of the positive ions.

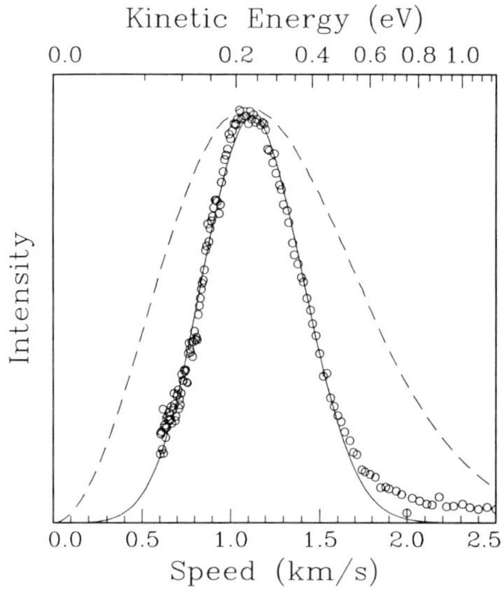

Fig. 2. Speed distribution of laser ablated C_3^0. The circles represent the experimental data. The dashed line represents a Maxwell–Boltzmann fit to the data. The solid line depicts a fit to the data with a supersonic expansion distribution.

distribution, depicted by the solid line, is in considerably better agreement with the experimental data. The translational temperature (~ 390 K) extracted from the supersonic distribution is substantially lower than the sublimation temperature of carbon (~ 3600 K), indicating a significant degree of cooling during adiabatic expansion of the ejected material. The apparent disagreement between the experimental data at higher speeds and the supersonic expansion may be a result of the speed-dependent transmission function of the quadrupole mass filter.

The ion and neutral speed distributions exhibit maxima near 10 and 1 km/s, respectively. The greater ion speed is most likely due to acceleration from Coulombic repulsion of ions in the laser-induced plasma.

3.2. Film analysis

Shown in fig. 3a are the valence EELS results, obtained in the $N(E)$ mode. Core level EELS spectra are presented in fig. 3b. These spectra were obtained in the derivative mode, allowing comparisons of peak-to-peak height ratios among the films. The peaks present in each spectrum are

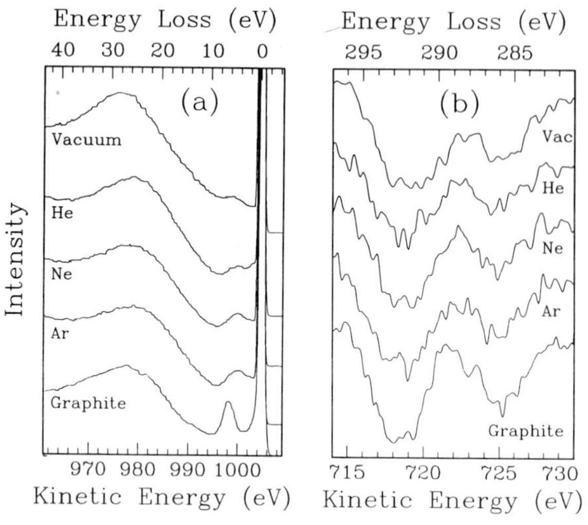

Fig. 3. EELS results from carbon films grown in vacuum, He, Ne, Ar, and from single crystal graphite. (a) Valence EELS spectra ($N(E)$), and (b) core level EELS spectra ($\mathrm{d}N(E)/\mathrm{d}E$).

due to the excitation of 1 s electrons into either π^* or σ^* antibonding states. The peak between 285 and 288 eV loss is due to excitation into the π^* state, while the peak between 290 and 295 eV loss is due to promotion into the σ^* state. The peak-to-peak height ratios were used to determine the corresponding bond ratios ($\pi/(\pi+\sigma)$) by normalizing to the known $\sigma:\pi$ (3:1) bond ratio of graphite.

The valence EELS spectra indicate a general increase in the relative intensity of the π plasmon loss peak with increased mass of the collision gas. The film grown in vacuum exhibits minimal π character when evaluated by either technique. The position of the $\pi+\sigma$ peak (29.5 eV) indicates that this film is near diamond in character [13]. Furthermore, the $\pi+\sigma$ peak shifts to lower loss energy, from 29.5 for the film grown in vacuum, to an average value of 25.7 eV for films grown with a collision gas. This latter value is virtually identical to the $\pi+\sigma$ loss typical of graphite [14].

The core EELS spectra indicate a general increase in $\pi/(\pi+\sigma)$ ratios with an increase in the mass of the collision gas. The film grown in vacuum exhibits the smallest $\pi/(\pi+\sigma)$ ratio, substantiating the conclusion that the film is near diamond in character.

The results of both the valence and core level EELS analyses can be understood by examining the dynamics of the collisions between the ejected carbon species and the various scattering gases. An energetic carbon particle colliding with a scattering gas atom will transfer a well defined fraction of its initial energy to its collision partner with each collision. This fraction, E_s/E_0, is defined by:

$$E_s/E_0 = 4M_1M_2/(M_1+M_2)^2, \qquad (1)$$

and represents the maximum efficiency of energy transfer in each scattering event. Here E_s is the energy of the inert gas species after collision, E_0 is the initial energy of the ejected species, and M_1 and M_2 are the masses of the respective particles. C_3, the most prevalent species ejected by 248 nm light, will transfer energy most efficiently when colliding with Ar, and least efficiently with He. An overall trend of increased energy transfer efficiency with increased mass of the collision gas is predicted by the scattering dynamics. The complimentary result of the greater transfer efficiency of heavier collision partners is lower final energies of the scattered C_3 species.

Total scattering cross sections were determined in a beam attenuation experiment. Cross sections for scattering with He, Ne, and Ar were 1.6×10^{-16}, 5.3×10^{-16} and 7.0×10^{-16} cm^2, respectively. The final energy of the scattered C_3 species was calculated using repeated application of eq. (1) and the absolute cross sections to determine the mean free path between collisions. The final energy, E_f, is that energy which C_3 is expected to have when it reaches the substrate.

Films grown in vacuum were near diamond, exhibiting a negligible π plasmon loss. The ionic portion of the plume, primarily composed of C_3^+ is expected to have incident energies undiminished by collision energy transfer. Thus, C_3^+ ions are anticipated to arrive at the substrate with mean energies of ~ 18 eV. These observations suggest a correlation between incident particle energy and diamond-like character of the film.

Of the films grown with a collision gas, those grown in He exhibit the least intense π plasmon, signifying a small but measurable amount of graphitic character in the film. The final energies

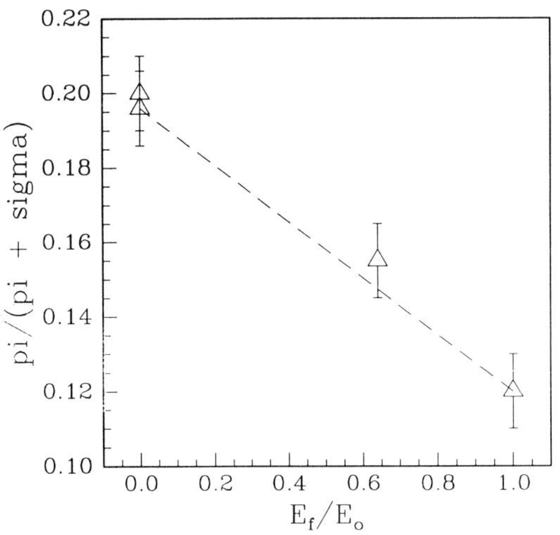

Fig. 4. Calculated $\pi/(\pi+\sigma)$ ratio versus final relative energy (E_f/E_0) of impinging C_3.

of the scattered C_3^+ particles is expected to be 11.5 eV. The fact that a measurable π intensity is observed, indicates that some portion of the distribution did move into a lower energy range, where stable, graphitic product is formed.

Ar and Ne have similar energy transfer efficiencies in collision with C_3. Thus, the character of films grown in these two gases is expected to be comparable. The apparent general trend of increased graphitic character with increased collision gas mass is actually a reflection of transfer efficiencies governed by the relative masses of the colliding species. It is the combination of the particular carbon species being evaluated here, C_3, and gas collision species chosen which creates the apparent dependence upon collision gas mass. The film properties depend upon the kinetic energy of the impinging carbon particles.

The relationship between the incident particle energy and subsequent film properties is clearly demonstrated in fig. 4, showing the correspondence between the $\pi/(\pi+\sigma)$ ratios and the relative energy of C_3 as it reaches the substrate. The horizontal error bars represent the range of kinetic energy for this species. Vertical error bars were obtained from the scatter in the EELS spectra. The ordinate, $\pi/(\pi+\sigma)$, can be considered an indicator of graphitic film character. The abscissa is the relative incident particle kinetic energy. The data clearly indicate that the graphitic character of the film increases with decreasing incident particle energy.

This figure also reveals that kinetic energy is not solely responsible for the ultimate character of the deposited films. The $\pi/(\pi+\sigma)$ ratios achieved for all films were less than the 0.25 value of single crystal graphite. The data demonstrate that films grown with small carbon particles of near zero energy can still result in films with some diamond-like quality. This suggests that small low energy particles form diamond-like carbon through a mechanism other than compaction, possibly by direct formation of a chemical bond between a surface atom and an incident C_n particle arriving in the diamond-like coordination.

4. Summary

Understanding and controlling the mechanisms of diamond thin film growth are prerequisites to the full exploitation of the unique properties this material offers. In this work, the mechanisms of carbon thin film growth by pulsed laser deposition were studied. Carbon film deposition by PLD requires only carbon and laser light; therefore, this process provides a tightly constrained system within which the deposition process can be explored.

The kinetic energy of the species ejected from graphite by 248 nm radiation was studied in detail. Films were grown with 248 nm light in a static pressure environment designed to decrease the energy of the incident particles. Kinetic energy of the incident particles was found to play a significant role in the generation of diamond-like films but was not the sole determining factor. These experimental results suggest that diamond-like films may be produced in PLD by more than a single mechanism, depending upon incident particle energy or size.

Acknowledgements

This work was supported by the Wright Laboratory, Materials Directorate, Wright-Patterson

Air Force Base, Ohio. D.T. Peeler wishes to thank the Clare Boothe Luce Foundation for their support.

References

[1] T. Sato, S. Furuno, S. Iguchi and M. Hanabusa, Jpn. J. Appl. Phys. 26 (1987) L1487.

[2] J. Krishnaswamy, A. Rengan, J. Narayan, K. Vedam and C.J. McHargue, Appl. Phys. Lett. 54 (1989) 2455.

[3] S.S. Wagal, E.M. Juengerman and C.B. Collins, Appl. Phys. Lett. 53 (1988) 187.

[4] J.J. Cuomo, D.L. Pappas, J.P. Doyle, J. Bruley and K.L. Saenger, Proc. TMS, 1991.

[5] J.J. Cuomo, D.L. Pappas, J. Bruley, J.P. Doyle, K.L. Saenger, T. Gu and R.W. Collins, submitted.

[6] J.J. Cuomo, J. Bruley, J.P. Doyle, D.L. Pappas, K.L. Saenger, J.C. Liu and P.E. Batson, Mater. Res. Soc. Symp. Proc. 202 (1991) 247.

[7] J.J. Cuomo, D.L. Pappas, J. Bruley, J.P. Doyle and K.L. Saenger, J. Appl. Phys. 70 (1991) 1706.

[8] E.E.B. Campbell, G. Ulmer, K. Bues and I.V. Hertel, Appl. Phys. A 48 (1989) 543.

[9] P.T. Murray, V.J. Dyhouse, L. Grazulis and D.R. Thomas, Mater. Res. Soc. Proc. 201 (1991) 513.

[10] P.T. Murray, D.T. Peeler and D.V. Dempsey, Mater. Res. Soc. Proc. 236 (1992) 411.

[11] R.D. Levine and R.B. Bernstein, Molecular Reaction Dynamics (Oxford University Press, New York, 1974).

[12] W.R. Creasy and J.T. Brenna, Chem. Phys. 126 (1988) 453.

[13] C. Weissmantel, K. Bewilogua, K. Breuer, D. Dietrich, U. Ebersbach, J.J. Erier, B. Rau and G. Reisse, Thin Solid Films 96 (1982) 31.

[14] L.B. Leder and J.A. Suddeth, J. Appl. Phys. 31 (1960) 1422.

Applied Surface Science 69 (1993) 231–241
North-Holland

applied
surface science

Pulsed laser crystallization and doping for the fabrication of high-quality poly-Si TFTs

E. Fogarassy [a], H. Pattyn [b], M. Elliq [a], A. Slaoui [a], B. Prevot [c], R. Stuck [a], S. de Unamuno [a] and E.L. Mathé [d]

[a] *Laboratoire PHASE (UPR du CNRS no. 292), Centre de Recherches Nucléaires (IN2P3), BP 20, 67037 Strasbourg Cedex 2, France*
[b] *IMEC, Kapeldreef 75, 3030 Leuven, Belgium*
[c] *Laboratoire PHASE–GRPM (UPR du CNRS no. 292), Centre de Recherches Nucléaires (IN2P3), BP 20, 67037 Strasbourg Cedex 2, France*
[d] *Laboratoire de Métallurgie Physique (URA 131 du CNRS), 40 Avenue du Recteur Pineau, 86022 Poitiers Cedex, France*

Received 2 June 1992; accepted for publication 30 October 1992

We review the various applications of pulsed lasers, working in the nanosecond regime, to prepare high-quality poly-Si TFTs. It is shown that the best device performances (field-effect mobilities in excess of 140 $cm^2/V \cdot s$) are achieved by pulsed excimer laser crystallization of unhydrogenated amorphous Si thin films. In addition, for source and drain formation, we demonstrate that the excimer laser induced diffusion of dopant from a solid source (spin-on phosphorus-doped silicate glass) is very attractive to achieve good electrical properties of the n-channel TFTs.

1. Introduction

Polycrystalline silicon thin-film transistors (poly-Si TFTs) have received much attention in recent years for different applications including vertically integrated high-density memories and large-area liquid crystal displays (LCD's) addressed by active matrices [1,2]. For such types of application, which require specific properties like efficient on/off current and high field-effect mobilities in the range of 10 to 100 $cm^2/V \cdot s$, it appears very important to prepare large-grain poly-Si thin films at low temperature, or with a significant reduction in high-temperature processing times, thus permitting the use of various types of amorphous substrates like inexpensive glass, compatible with the manufacturing costs of the market.

Different techniques have been developed, to prepare low-temperature (~ 600°C) poly-Si TFTs, using both thermal [3,4] and continuous [5,6] or pulsed [7–9] laser processing. The use of highly powerful pulsed laser beams, working in the nanosecond duration regime, appears very promising because they allow a large amount of energy to be deposited in short times into the near-surface region. Under suitable conditions, the irradiation leads to the melting of the amorphous silicon (a-Si) layer at a depth not exceeding a few hundred nanometers, and to the crystallization of the molten layer from the underlying substrate with a very high solid–liquid interface velocity of typically several meters per second. Due to the short duration of the overall process, of the order of a few tens of nanoseconds, the average temperature of the substrate can be kept well below 600°C. This greatly reduces any possible degradation and contamination which are likely to occur in the interface region between the active polysilicon layer and the substrate.

Pulsed laser processing also presents the unique advantage to permit selective local crystallization of the starting a-Si layer. Thus it is the most suitable crystallization method to integrate on the same glass substrate both the poly-Si TFTs with high mobility for the peripheral circuits and

0169-4332/93/$06.00

the a-Si TFTs with low leakage current for the matrices.

Poly-Si TFTs also require shallow and high concentration doping for source and drain formation. However, with the conventional ion implantation technology followed by heat treatment in a furnace for times of the order of minutes to hours, it is difficult to achieve a high dose with low-energy doping. In addition, long-term thermal processing could be incompatible with a cheap glass substrate and detrimental for the final quality of the deposited gate insulator (self-alignment technology). To overcome these problems novel techniques such as laser doping from the gas [10] or solid [8] phase and laser annealing of low-energy implanted ions [11] were investigated these last years. In all these experiments pulsed laser irradiation is used to melt locally the amorphous or polycrystalline Si layer and produce liquid-phase incorporation of the dopant and its electrical activation [12–14].

We investigate in this work the possibility to fabricate high-mobility poly-Si thin-film transistors by using either pulsed laser crystallization or doping in the self-aligned TFT process.

The first part of this paper will concentrate on the pulsed ArF excimer laser crystallization step. In these experiments, the use of hydrogen-free amorphous silicon, as starting material, is of paramount importance to prepare high-performance poly-Si TFTs [15–19].

In a second part, we show the possibility to form source and drain regions of the n-channel poly-Si TFT by ArF excimer laser doping from a phosphorus spin-on glass (SOG) solid source [20–22]. This new laser doping technique appears very promising as an alternative to conventional ion implantation.

2. Thermal simulation of the pulsed laser processing

The major difficulty in the interpretation of the pulsed laser processing is the accurate evaluation of the thermal contribution resulting from the interaction between the incoming light and the solid. Surface heating during pulsed laser irradiation is difficult or impossible to measure experimentally. Consequently, the thermal effect has to be simulated by solving the one-dimensional heat flow equation [23]:

$$C_S \rho \frac{\partial T(x,t)}{\partial t} = \frac{\partial}{\partial x}\left(k \frac{\partial T(x,t)}{\partial x} \right) + S(x, t),$$

where C_S is the specific heat, ρ the density and k the thermal conductivity of the sample. The temperature T and heat generation S are functions of space and time. The heat generation:

$$S(x, t) = P(t)(1-R)\ \alpha\ e^{-\alpha x}$$

depends on the optical absorption (α) and surface reflectivity (R) of the irradiated materials. $P(t)$ is the time-dependent power density of the laser beam. The accuracy of the modelling is therefore strongly related to how precisely the optical and thermal properties of the irradiated materials are known both in their solid and liquid phases. The different parameters for amorphous, polycrystalline and liquid Si and for quartz and Corning substrates are reported in table 1.

Fig. 1 shows the time-dependent melting depth of a 200 nm thick a-Si layer deposited onto quartz. Calculations are for a 20 ns pulsed ArF excimer laser irradiation and energy densities between 100 and 600 mJ/cm^2. Above the melting threshold of a-Si layer, which is close to 100 mJ/cm^2, the film is molten deeper and deeper, for a time duration ranging between 50 and 300 ns. Above a laser fluence of about 400 mJ/cm^2, the whole Si layer is fully melted.

As deduced from fig. 2, which gives the distribution profiles of the temperature as a function of depth into the irradiated layer, the melting threshold of quartz substrate is reached at a laser fluence of 600 mJ/cm^2. It is interesting to notice that when using a cheap substrate, such as Corning glass in view of large-area applications, the simulated curves of fig. 3 show that the use of a 500 nm SiO_2 buffer layer can prevent any Corning substrate heating in excess of 600°C.

3. Pulsed excimer laser crystallization

Hydrogen-free (< 1%) a-Si films of 200 nm in thickness have been deposited by low-pressure

Table 1
Optical and thermal parameters of silicon, quartz and Corning glass used in the simulation

	a-Si	μc-Si	Liq-Si	Quartz	Corning
α (cm^{-1}) at 193 nm	1.5×10^6	1.65×10^6 ($T\leqslant1000$ K) 2.8×10^6 ($T>1000$ K)	1.67×10^6	–	–
R at 193 nm	0.55	0.6	0.68	–	–
T_m (K)	1420	1683	–	1920	1113
k (W/cm·K)	0.018	0.318	$0.50+2.9\times10^{-4}(T-T_m)$	8.4×10^{-3} $+1.5\times10^{-5}T$	$1.00\times10^{-2}+1.30\times10^{-5}T$ ($T\leqslant1170$ K)
	$0.18\exp(4.5\times10^{-3}T)$ $-8\times10^{-3}+1.0\times10^{-4}T$ ($T\leqslant300$ K)	$0.18\exp(4.5\times10^{-3}T)$ ($T\leqslant300$ K)		$0.782+5.7\times10^{-4}T$ $-1.88\times10^{4}T^{-2}$ ($T\leqslant856$ K)	
C_s (J/g·K)	$0.69\exp(2.375\times10^{-3}T)$ $-8\times10^{-3}+1.0\times10^{-4}T$ ($T>300$ K)	$0.69\exp(2.375\times10^{-3}T)$ ($T>300$ K)	1.05	$1.0+1.35\times10^{-4}T$ ($T>856$ K)	$0.707+2.99\times10^{-4}T$

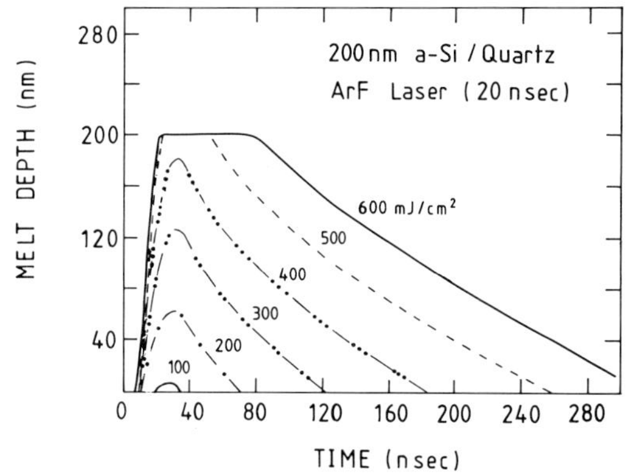

Fig. 1. Time-dependent melting depth of a 200 nm a-Si film onto quartz substrate irradiated with a 20 ns pulsed ArF excimer laser at various fluences.

chemical vapour deposition (LPCVD) at 550°C onto 3 inch diameter quartz wafers and 500 nm SiO_2-coated crystalline Si substrates.

The crystallization step was performed at room temperature and in ambient atmosphere using a pulsed UV excimer (201 MSC Lambda Physik) laser providing at the 193 nm (ArF) wavelength, large area (up to 1 cm^2) and uniform (better than 5%) pulses of 20 ns duration. The laser fluence per pulse was typically in the 60 to 600 mJ/cm^2-range.

The amorphous-to-polycrystalline transition and film crystalline quality were characterized as a function of the laser fluence by various complementary techniques: ultraviolet–visible re-

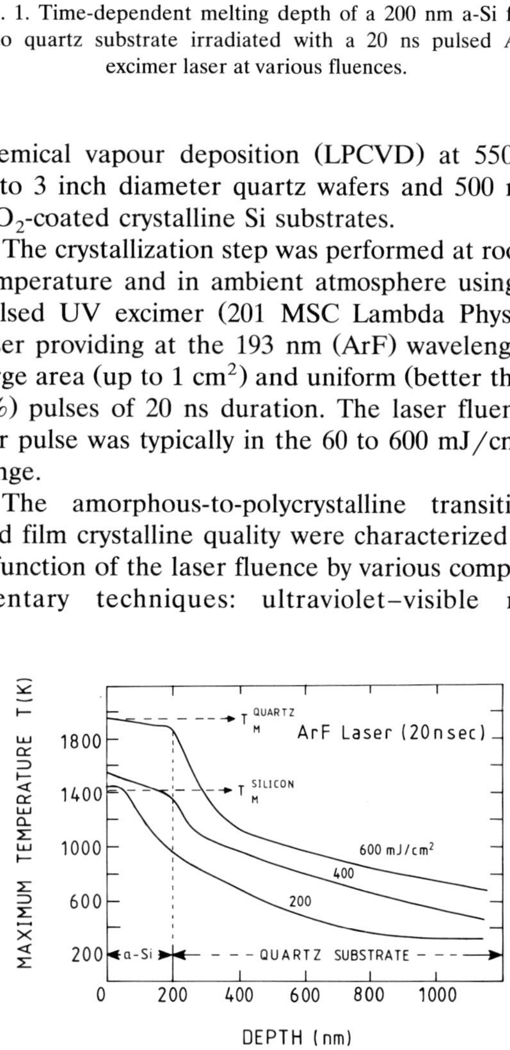

Fig. 2. Temperature distribution profile as a function of depth into a 200 nm a-Si film onto quartz substrate irradiated with a 20 ns pulsed ArF excimer laser at various fluences.

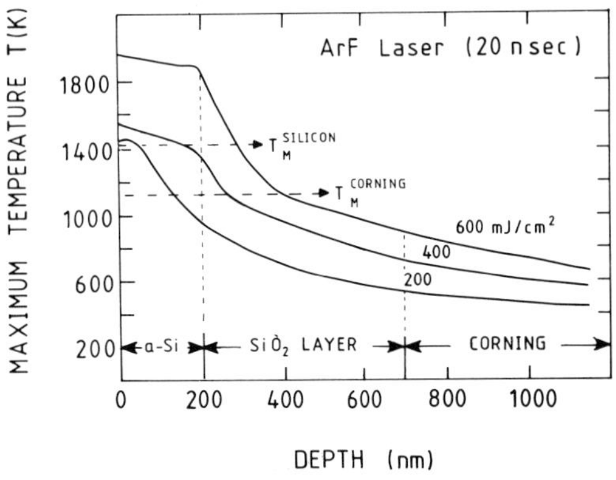

Fig. 3. Temperature distribution profile as a function of depth into a 200 nm a-Si film onto 500 nm SiO_2 coated Corning glass substrate irradiated with a 20 ns pulsed ArF excimer laser at various fluences.

flectance (UVR), Raman spectroscopy (RS) and transmission electron microscropy (TEM). Finally, the surface morphology was controlled by profilometry using a mechanical stylus (Talystep).

We have reported in fig. 4 the UVR spectra of a-Si films onto quartz substrates before and after pulsed ArF laser irradiation at various energy densities. The crystallization of the a-Si layer is characterized by the appearance of two peaks E_1 and E_2 located at 365 (3.4 eV) and 280 (4.6 eV)

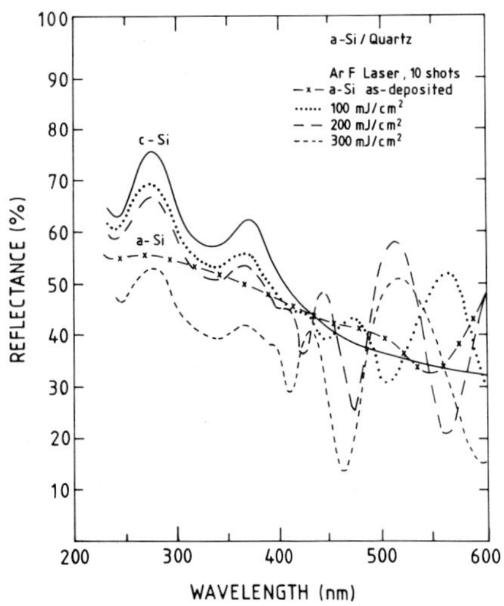

Fig. 4. Reflectance spectra of 200 nm hydrogen-free a-Si films onto quartz substrates before and after pulsed ArF excimer laser irradiation at various fluences.

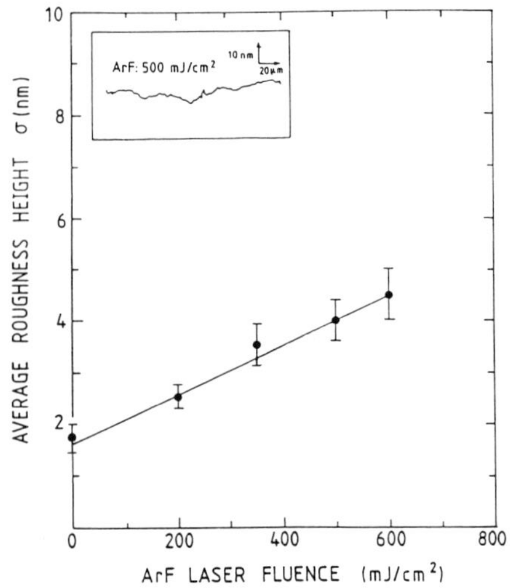

Fig. 5. Average roughness height at surface of a 200 nm a-Si film after pulsed ArF excimer laser irradiation at various fluences.

nm, respectively, and occurs for a laser fluence of about 100 mJ/cm^2. This value is in good agreement with the melting threshold of amorphous silicon deduced from the thermal simulations of fig. 1. Above the crystallization, for increasing values of the laser fluence, a high-reflectivity signal is recorded. However, at fluences higher than 300 mJ/cm^2, a slight decrease of peaks E_1 and E_2 could indicate some small surface degradation, confirmed by profilometry measurements (fig. 5) which show that the surface roughness (σ) increases linearly with the ArF laser fluence but does not exceed 5 nm at 600 mJ/cm^2.

Raman spectroscopy is known to be a particularly useful technique for identifying whether a film exhibits amorphous or crystalline structure [24]. Indeed, the RS spectrum of c-Si is dominated by a sharp feature peaking at 522 cm^{-1}, while the a-Si spectrum shows a very broad ($\sim$ 80 cm^{-1}) and intense structure with a maximum located at an energy about 8% lower than of the crystalline Raman peak. Raman spectra were measured in a quasi-backscattering (Brewster) geometry using the 458 nm line of an Ar^+-ion laser.

Fig. 6 compares the RS spectrum of the as-deposited a-Si film with that observed after a single ArF laser shot performed with an energy fluence

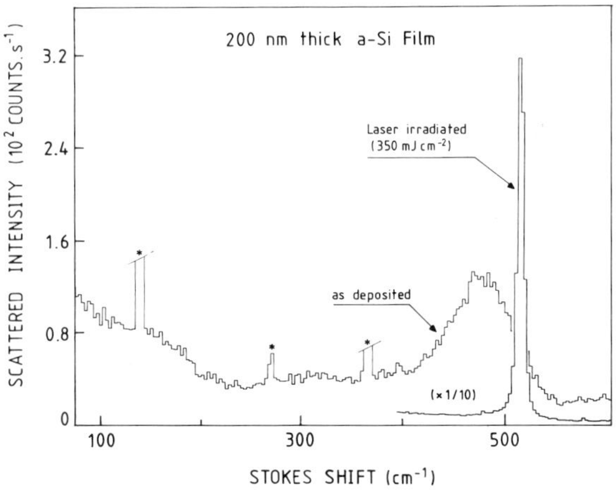

Fig. 6. Raman signature of the as-deposited a-Si film compared with that observed after an ArF irradiation of 350 mJ/cm^2 ($*$ denotes plasma lines originating from the Ar^+ laser used in the Raman experiment).

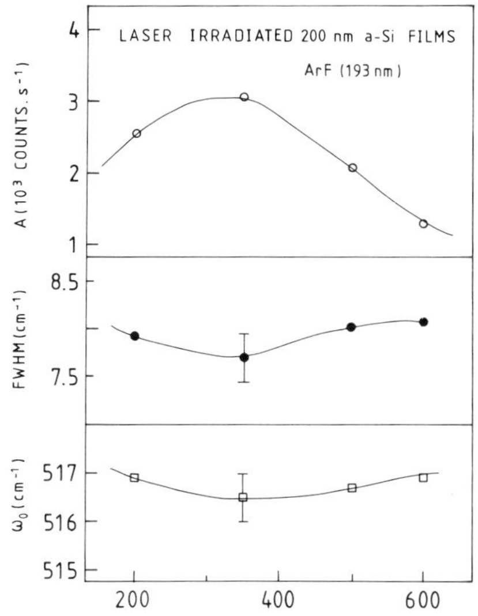

Fig. 7. Raman peak parameters as a function of the ArF laser fluence. From top to bottom: peak magnitude (A), line-width (FWHM) and peak energy (ω_O).

of 350 mJ/cm^2. The former trace is typical of amorphous Si with broad structures located at around 140, 300 and 480 cm^{-1}. They correspond to phonon frequencies which originate from the acoustical (transverse and longitudinal) and optical branches, respectively. After irradiation performed at 350 mJ/cm^2, the amorphous bands have totally disappeared and the Raman signature is now dominated by a very narrow line (8 cm^{-1}) peaking at 516.5 cm^{-1}. As reported in fig. 7, the frequency (ω_0) and the linewidth (FWHM) of this peak do not evolve significantly with the laser fluence ranging between 200 and 600 mJ/cm^2. However, it can be observed that ω_0 shows values which are down shifted by 5 cm^{-1} with respect to that obtained on a single-crystal Si. This behaviour is usually interpreted in terms of phonon-confinement effects which are known to become important in micro-sized structures. From the actual downshift, i.e. 5 cm^{-1}, according to this figure, a value of about 7 nm can be deduced for the typical coherence length of the phonons which propagate in the polycrystalline silicon film [25]. It can be observed that this value is much smaller than the grain size determined by the TEM experiments, i.e. about 100 nm (fig. 8). Such an apparent discrepancy can be explained by looking at the very nature of each technique. Indeed, TEM measurements yield the average

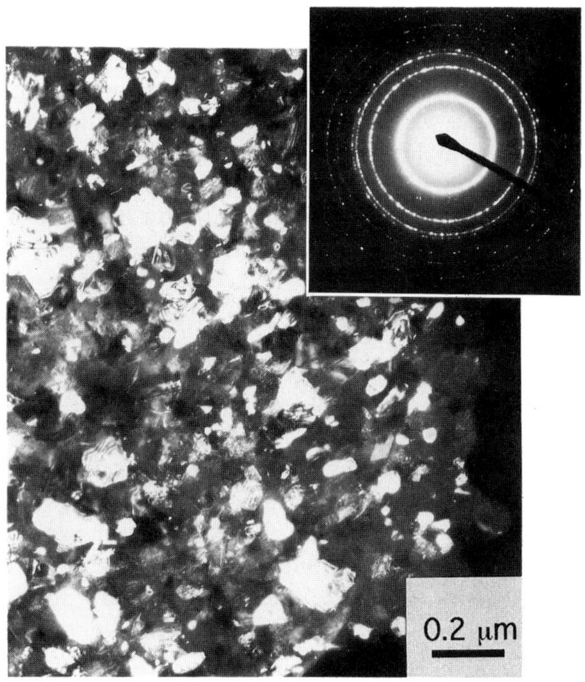

Fig. 8. TEM plan view of a 200 nm a-Si film after ArF laser irradiation at 500 mJ/cm^2. Dark field with part of (111) continuous ring.

size of the crystalline grains irrespective of isolated defects or impurities, whereas the Raman line-shape is affected by such defects through the modification of the electron–phonon system. Since generic defects or impurities are certainly introduced during the laser crystallization of the a-Si layer, it is thus not surprising that typical "crystalline" dimensions determined by TEM are larger than those obtained by Raman scattering [26]. Another interesting observation concerns the scattering level (A) measured as a function of the laser fluence in otherwise strictly identical scattering conditions (see the upper part of fig. 7). As shown, A gets a maximum for E in the range of 350 to 400 mJ/cm^{-2} and then decreases markedly at higher fluences. It is strongly believed that such a behaviour is related to surface degradation phenomena, reported in fig. 5, which reduce the effective photon flux able to excite the Raman effect inside the film. Moreover, we cannot exclude that at fluences as high as 600 mJ/cm^2, some structural degradation is likely to occur in the polysilicon film near the film–substrate region. Indeed, from simulations reported in fig. 2 it is observed that for a 600 mJ/cm^2 irradiation, the temperature of the interface is very close to 1920 K, which is the melting temperature of quartz. As a consequence, upon resolidification of the complex interface, the Si film could be strained thus explaining the slight increase of the Raman peak frequency and linewidth observed in fig. 7, around 600 mJ/cm^2 laser fluence.

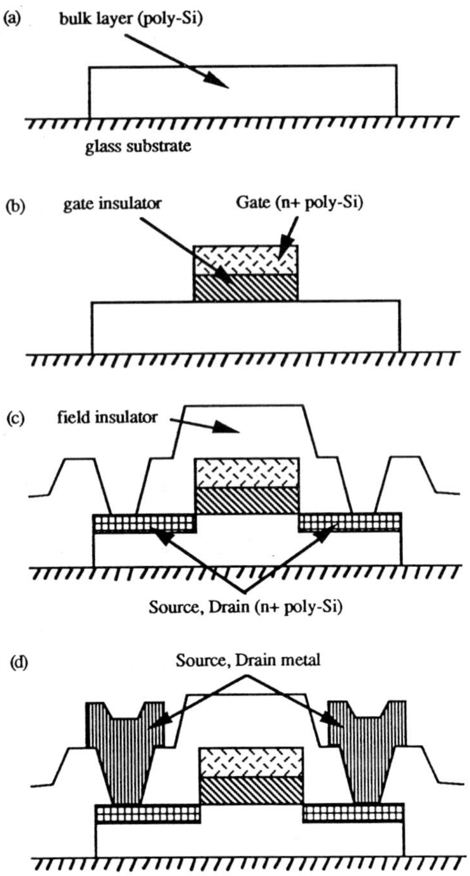

Fig. 9. Process flow for self-aligned poly-Si TFT.

TFT device characteristics

In order to check up on their device performances, n-channel poly-Si TFTs were fabricated using a simplified MOS structure (four photomasks). A scheme of the self-aligned process flow used in our experiments is shown in fig. 9.

A series of 200 nm thick a-Si layers, deposited by LPCVD at 550°C, onto quartz substrates were crystallized at various fluences using the pulsed ArF excimer laser. Subsequently, the substrates were annealed at 630°C in dry nitrogen during 10 h. Hence, poly-Si TFTs made by pulsed laser crystallization could be directly compared with those made by thermal crystallization. A 100 nm thick, high-quality SiO_2 gate insulator was used. This oxide was formed by LPCVD at 600°C using the organic precursor tetramethylcyclotetrasiloxane TOMCATS ($Si_4C_4O_4H_{16}$) source. An n^+ poly-Si gate was used. Self-alignment was obtained by phosphorus (70 keV, 5×10^{15} cm^{-2}) ion implantation to form the doped source and drain regions. An enhanced H-passivation step was performed by a modified sintering [27].

As shown on the drain current-versus-gate voltage characteristics (I_D–V_G) of fig. 10, high-quality polysilicon TFTs can be prepared by pulsed excimer laser crystallization.

Summarized in table 2 are the typical characteristics in terms of field-effect mobility (μ_{fe}), threshold voltage (V_{th}), subthreshold slope (S), off current (I_{off} at -5 V) and on/off current ratio (I_{on}/I_{off}) of several series of n-channel poly-

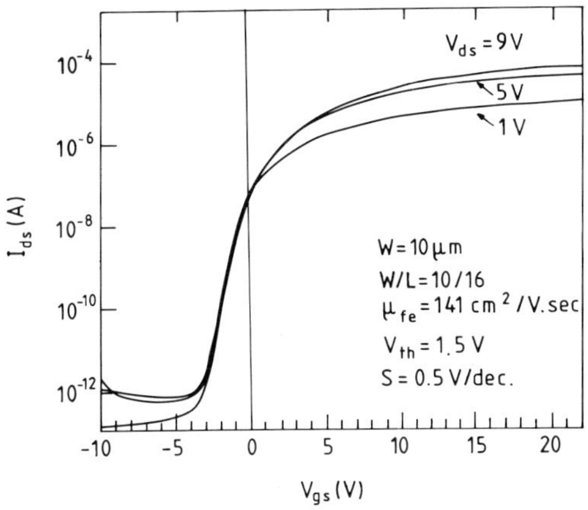

Fig. 10. Drain current–gate voltage characteristics of an n-channel poly-Si TFT made by pulsed ArF excimer laser crystallization ($E = 440$ mJ/cm^2).

Si TFTs prepared, on the same wafers, by thermal and laser crystallization, respectively. Also reported in table 2 are the performances of laser-processed TFTs published in the literature, which mainly originate from Japanese groups.

Our work shows a strong improvement of the electrical characteristics of the TFTs fabricated by laser processing, with a field effect mobility up to 140 cm^2/V · s (not more than 55 cm^2/V · s by thermal crystallization), an off current less than 1×10^{-12} A for a channel width of 10 μm and on/off current ratio exceeding 5×10^8. Furthermore, after laser processing, the threshold voltage and subthreshold slope are substantially improved. This could indicate that the concentration of defects at the oxide/polysilicon interface is lower after laser crystallization than by classical thermal treatment. Among the various possible explanations, we can suggest that some purification phenomenon occurs in the crystallized layer due to the simultaneous removal of surface contaminants and segregated impurities (for example metals) through the laser melting process.

Our results, in table 2, also show the strong influence of the laser energy density on the final device quality. The best performances are obtained above 400 mJ/cm^2. In this range of values, the whole 200 nm thick a-Si film deposited on quartz substrate is molten, as deduced from the simulations in fig. 1. In addition, these fluences give rise to large-size ($\geq$ 100 nm) grains of polysilicon as revealed by TEM experiments (fig. 8). An interesting explanation of this behaviour was given by Bachrach et al. [28]: once the melted zone reaches the substrate, superheating of liquid silicon above the melting temperature is possible and the thermal model further shows that the film reaches a rather uniform temperature due to the large thermal conductivity of the melt relative to the quartz substrate (table 1) which stays in the solid phase in a large range of laser fluences. With the excimer laser the melting threshold of quartz, at the silicon–substrate interface, is reached only at 600 mJ/cm^2 as deduced from the simulations of fig. 2. During the cooling stage of the process, the crystallization of molten Si is directly initiated from the amorphous substrate where no (or very little) nucleation sites exist, so that fewer grains are initiated at the same time therefore promoting crystallites of larger dimension since competition between them is reduced.

We must, however, notice that in spite of the fact that grains achieved by pulsed laser processing ($\sim$ 100 nm with the excimer) are smaller than those obtained by thermal crystallization ($\sim$ 500 nm [3]), higher mobility values are measured in the first case. This could suggest that the structural quality of grains and especially grain boundaries has a strong influence on the final value of μ_{fe}. The limitation of the grain size obtained by pulsed laser annealing is mainly related to the high solidification velocity of molten Si which reaches several meters per second with the ex-

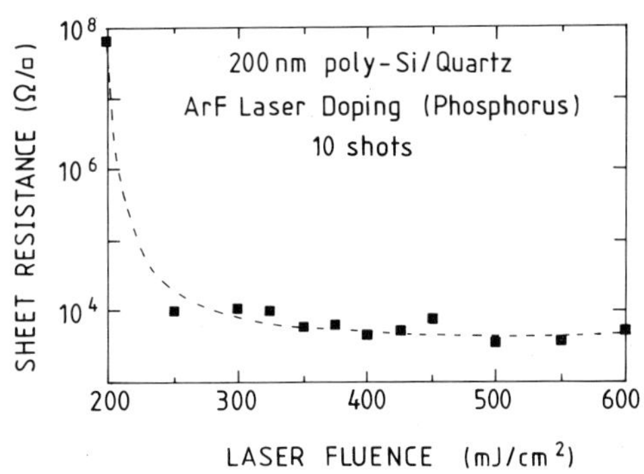

Fig. 11. Sheet resistance of P-doped poly-Si films after 10-shot ArF laser irradiation at various fluences.

Table 2
Comparison of the characteristics of the poly-Si TFTs made by thermal and laser crystallization

Cystallization process	Thermal anneal (630°C) This work	ArF (20 ns) 220 mJ/cm² This work	ArF (20 ns) 440 mJ/cm² This work	XeCl (30 ns) 240 mJ/cm² [8]	XeCl (35 ns) 160 mJ/cm² [9]	XeCl (35 ns) 300 mJ/cm² [29]	ArF 360 mJ/cm² [30]	ArF (17 ns) 300 mJ/cm² [31]		KrF 350 mJ/cm² [32]	
Sub. T (°C) during laser cryst.		Ambient	Ambient	Ambient	Ambient	Ambient	Ambient	Ambient	400	400	400
TFT fabrication max. T (°C)	630	630	630	250	260	600	350	400	400	400	400
Poly-Si film thickness (nm)	200	200	200	30	150	50	50	100	100	100	50
Grain size (nm)	500		60–100	10–60	30–200			30–50	200–300		200–300
μ (cm²/V·s)	55	90	141	54	102	120	110	100	230	285	329
V_{th} (V)	5	2.5	1.5			4					6.7
S (V/dec)	1.05	0.9	0.5	0.38		0.55					0.37
I_{off} (A) for $V_{ds}=5$ V	2.5×10^{-11}	1×10^{-11}	$<10^{-12}$			10^{-11}					
I_{on}/I_{off}	4×10^{6}	3×10^{7}	$>5\times10^{8}$	1.5×10^{7}	$>10^{6}$	2×10^{7}	$>10^{6}$	$>10^{6}$			7.6×10^{7}

This work: $W = 10$ μm, $W/L = 10/16$, $V_{gs} = 20$ V, $V_{ds} = 5$ V.

cimer laser. As shown by Kuriyama et al. [31], a reduction of this solid–liquid interface velocity by heating the substrate (~ 400°C) during laser processing, leads to an enlargement of the polysilicon grain size (up to 300 nm with the excimer laser), and subsequent enhancement of the field-effect mobility (up to 230 cm²/V · s for 100 nm thick poly-Si, table 2).

Using the same procedure as before (substrate heating at 400°C during laser processing), Zhang et al. [32], reported this year, for the first time, field-effect mobilities in excess of 300 cm²/V · s (table 2), by working with very thin poly-Si films of 50 nm in thickness, which demonstrates the strong interest of pulsed laser crystallization to fabricate high-performance TFTs, at low temperature.

4. Pulsed excimer laser doping

Poly-Si films were prepared by thermal heating (650°C, 13 h) of 200 nm a-Si layers deposited onto 3 inch diameter quartz wafers by LPCVD at 550°C. After a standard cleaning procedure, the polysilicon surface was coated with a spin-on silicate glass (SOG) film containing phosphorus dopant (phosphorofilm from Emulsitone). After deposition, the samples were prebaked at 100°C for 30 min using a hot plate. The thickness of the

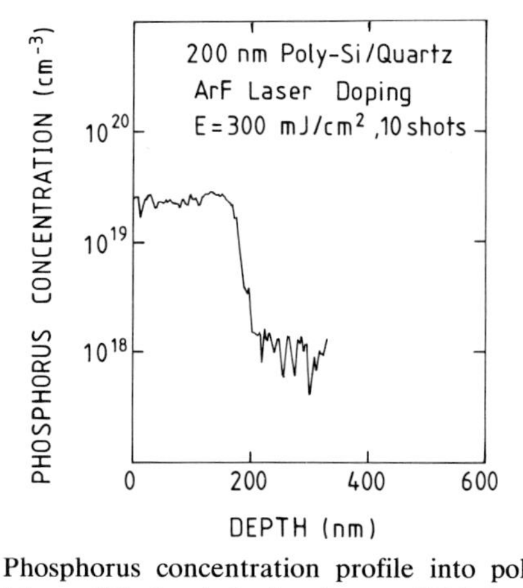

Fig. 12. Phosphorus concentration profile into poly-Si film after ArF laser (10 shots) doping from a SOG layer (SIMS measurements).

Table 3
Comparison of the characteristics of poly-Si TFTs made by thermal and laser doping

Starting material	Fabrication	I_{on}/I_{off}	μ ($cm^2/V \cdot s$)	S (V/dec)	V_{th} (V)	References
Poly-Si	P-doped SOG ArF laser diff.	1.5×10^5	51	1.09	4	This work
Poly-Si	P ion impl. Thermal ann. (630°C)	5×10^5	56	1.1	5	This work
Poly-Si	P-doped μc-Si Thermal diff. (630°C)	10^6	17	–	8.3	[33]
Poly-Si	PCl_3 source ArF laser diff.	10^4	1	–	22	[10]
Poly-Si	PH_3 ion impl. XeCl laser ann.	2×10^6	28	–	6.5	[11]
a-Si (250°C)	P-doped a-Si XeCl laser cryst.	1.5×10^7	54	0.38	5	[8]

dopant layers we used ranged typically between 100 and 300 nm. The pulsed ArF excimer laser beam passed through the transparent SOG layer to irradiate the Si surface. As revealed by the sheet-resistance measurements of fig. 11, above a laser fluence of 250 mJ/cm^2, polysilicon is locally melted and P dopant entered the molten region from the SOG layer by liquid-phase diffusion. It results in a sharp decrease of the sheet resistance from the substrate value ($\sim 10^{10}$ $\Omega/\square$) down to 5 $k\Omega/\square$, following a 10-shot irradiation at a fluence of 400 mJ/cm^2. Due to the fast diffusion of the dopant into the molten layer ($\sim 10^{-4}$ cm^2/s) [22], the application of multipulse laser treatment allows an optimization of the dopant distribution profiles (P surface concentration and junction depth) which present a box-like shape as revealed by secondary ion mass spectroscopy (SIMS) measurements of fig. 12.

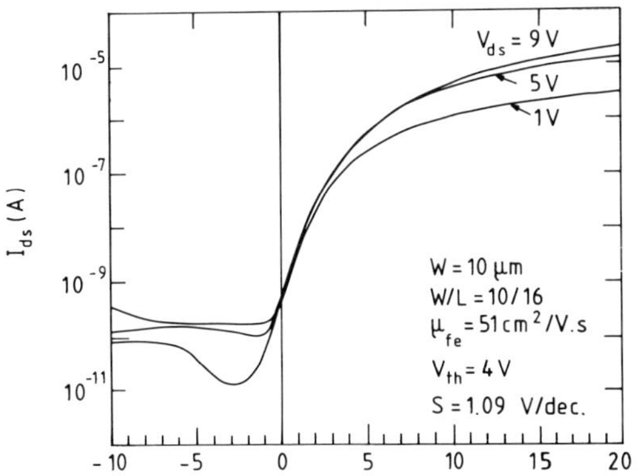

Fig. 13. Drain current–gate voltage characteristics of an n-channel poly-Si TFT made by ArF laser (10 shots) doping from a SOG layer.

TFT device characteristics

This new laser doping technique has been applied to form the source and drain regions of n-channel poly-Si TFTs fabricated using the self-aligned process flow schematically presented in fig. 9. In this case, the 200 nm thick active poly-Si layer was prepared by thermal crystallization of undoped LPCVD a-Si films.

The phosphorus-ion implantation step, for source and drain formation, was replaced by laser doping from P-doped SOG layer (100 nm thick), following the process described above.

Fig. 13 shows the drain current-versus-gate voltage (I_D–V_G) characteristics of a laser-doped poly-Si TFT with a gate width of 10 μm and a gate length of 16 μm. For comparison, TFT devices were formed on other wafers by ion implantation and thermal annealing [22]. Their characteristics are summarized in table 3. Almost identical results are obtained using the two doping techniques ($\mu_{fe} \approx 50$ to 60 $cm^2/V \cdot s$). We must, however, notice that some technological problems have to be solved when using laser doping from a solid source which requires the deposition of uniform SOG layers to have reproducible results on a large area. In addition, the device performances could be influenced by the generation of electrical defects at the SOG/poly-Si interface by laser irradiation.

Table 3 also lists a number of published TFT results obtained by using various thermal (low temperature) or laser-doping processes. These results show that pulsed laser doping from a spin-on source appears very competitive, against another techniques, to form source and drain regions of poly-Si TFTs.

5. Conclusion

This work has demonstrated the possibility to fabricate high-mobility ($\mu_{fe} > 140$ $cm^2/V \cdot s$) poly-Si TFTs, at low substrate temperature ($\leq$ 600°C), by pulsed excimer laser processing. Both laser crystallization and doping from spin-on glass source appear very promising to replace standard technologies, i.e. thermal crystallization and ion implantation, in the self-aligned process flow of the TFT.

References

[1] H.W. Lam and M.J. Thompson, Eds., Mater. Res. Soc. Symp. Proc., Vol. 33 (North-Holland, New York, 1984).

[2] A. Chiang, M.W. Geis and L. Pfeiffer, Eds., Mater. Res. Soc. Symp. Proc., Vol. 53 (Materials Research Society, Pittsburgh, PA, 1986).

[3] K. Kobayashi, J. Nijs and R. Mertens, J. Appl. Phys. 65 (1989) 2541.

[4] A. Mimura, N. Konishi, K. Ono, J. Ohwada, Y. Hosokawa, Y. Ono, T. Suzuki, K. Miyata and H. Kawakami, IEEE Trans. Electron Devices ED-36 (1989) 351.

[5] T. Nishimura, A. Ishizu, T. Matsumoto and Y. Akasaka, ref. [1], p. 221.

[6] T. Unagami and O. Kogure, IEEE Trans. Electron Devices ED-35 (1988) 314.

[7] T. Serikawa, S. Shirai, A. Okamoto and S. Suyama, IEEE Trans. Electron Devices ED-36 (1989) 1929.

[8] T. Sameshima, M. Hara and S. Usui, Jpn. J. Appl. Phys. 28 (1989) 1789.

[9] K. Sera, F. Okumura, H. Uchida, S. Itoh, S. Kaneko and K. Hotta, IEEE Trans. Electron Devices ED-36 (1989) 2868.

[10] P. Coxon, M. Lloyd and P. Migliorato, Appl. Phys. Lett. 48 (1986) 1785.

[11] G. Kawachi, T. Aoyama, T. Suzuki, A. Mimura, Y. Ohno, N. Konishi and Y. Mochizuki, Ext. Abstr. 22nd Int. Conf. on Solid State Devices and Materials, Sendai, 1990, p. 971.

[12] T.F. Deutsch, in: Laser Diagnostics and Photochemical Processing for Semiconductor Devices, Eds. R.M. Osgood, S.R.J. Brueck and H.R. Schlossberg, Mater. Res. Soc. Symp. Proc., Vol. 17 (North-Holland, New York, 1983) p. 225.

[13] E. Fogarassy, D.H. Lowndes, J. Narayan and C.W. White, J. Appl. Phys. 58 (1985) 2167.

[14] T.W. Sigmon, in: Photon, Beam and Plasma Stimulated Chemical Processes at Surfaces, Eds. V.M. Donnelly, I.P. Herman and M. Hirose, Mater. Res. Soc. Symp. Proc., Vol. 75 (Materials Research Society, Pittsburgh, PA, 1987) p. 619.

[15] R.Z. Bachrach, K. Winer, J.B. Boyce, S.E. Ready, R.I. Johnson and G.B. Anderson, J. Electron Mater. 19 (1990) 241.

[16] K. Winer, R.Z. Bachrach, R.I. Johnson, S.E. Ready, G.B. Anderson and J.B. Boyce, in: Materials Issues in Microcrystalline Semiconductors, Eds. P.M. Fauchet, K. Tanaka and C.C. Tsai, Mater. Res. Soc. Symp. Proc., Vol. 164 (Materials Research Society, Pittsburgh, PA, 1990) p. 183.

[17] M. Elliq, E. Fogarassy, J.P. Stoquert, C. Fuchs, S. de Unamuno, B. Prevot and H. Pattyn, Appl. Surf. Sci. 46 (1990) 378.

[18] E.L. Mathé, A. Naudon, M. Elliq, E. Fogarassy and S. de Unamuno, Appl. Surf. Sci. 54 (1992) 392.

[19] E. Fogarassy, B. Prévot, S. de Unamuno, M. Elliq, H. Pattyn, E.L. Mathé and A. Naudon, Appl. Phys. A, to be published.

[20] E. Fogarassy, R. Stuck, J.C. Muller, M. Hodeau, A. Wattiaux, M. Toulemonde and P. Siffert, in: Proc. 3rd EC Photovoltaic Energy Conf., Ed. W. Palz (Reidel, Dordrecht, 1980) p. 639.

[21] K. Sera, F. Okumura, S. Kaneko, S. Itoh, K. Hotta and H. Hoshino, J. Appl. Phys. 67 (1990) 2359.

[22] A. Slaoui, M. Elliq, H. Pattyn, E. Fogarassy, S. de Unamuno and R. Stuck, in: Photons and Low Energy Particles in Surface Processing, Eds. C. Ashby, J.H. Brannon and S. Pang, Mater. Res. Soc. Symp. Proc., Vol. 236 (Materials Research Society, Pittsburgh, PA, 1992).

[23] S. de Unamuno and E. Fogarassy, Appl. Surf. Sci. 36 (1989) 1.

[24] B. Prévot and J. Wagner, Prog. Cryst. Growth Charact. 22 (1991) 245.

[25] P.M. Fauchet and I.H. Campbell, CRC Crit. Rev. Solid State Mater. Sci. 14, Suppl. 1 (1988) S79.

[26] J. Gonzalez-Hernandez, G.H. Azabayejani, R. Tsu and F.H. Pollak, Appl. Phys. Lett. 47 (1985) 1350.

[27] H. Pattyn, K. Baert, P. Debenest, M. Heyns, M. Schaikers, J. Nijs and R. Mertens, ref. [11], p. 959.

[28] R.Z. Bachrach, J.B. Boyce, S.E. Ready and G.B. Ander-

son, in: Polycrystalline Semiconductors II, Eds. J.H. Werner and H.P. Strunk, Vol. 54 of Springer Proceedings in Physics (Springer, Berlin, 1991) p. 330.

[29] K. Sera, H. Asada, F. Okumura, H. Tanabe, K. Nakamura, A. Tada and S. Kaneko, Ext. Abstr. 23rd Int. Conf. on Solid State Devices and Materials, Yokohama, 1991, p. 590.

[30] K. Shimizu, H. Hosoya, O. Sugiura and M. Matsumura, Jpn. J. Appl. Phys. 30 (1991) 3704.

[31] H. Kuriyama, S. Kiyama, S. Noguchi, T. Kuwahara, S. Ishida, T. Nohda, K. Sano, H. Iwata, H. Kawata, M. Osumi, S. Tsuda, S. Nakano and Y. Kuwano, ref. [30], p. 3700.

[32] H. Zhang, N. Kusumoto, T. Inushima and S. Yamazaki, IEEE Electron Device Lett. EDL-13 (1992) 297.

[33] K. Kobayashi, J. Nijs and R. Mertens, J. Appl. Phys. 65 (1989) 2541.

Applied Surface Science 69 (1993) 242–248
North-Holland

applied
surface science

Laser-induced modifications in a-C:H thin films

J. Spousta, J. Perrière, A. Laurent

Groupe de Physique des Solides, URA 17, Universités Paris VII et Paris VI, Tour 23, 2 Place Jussieu, 75251 Paris Cedex 05, France

E. Fogarassy, B. Prevot and S. de Unamuno

Laboratoire PHASE, CRN, UPR 292, 23 Rue du Loess, 67037 Strasbourg Cedex, France

Received 2 June 1992; accepted for publication 30 October 1992

We have studied the transformations of amorphous hydrogenated carbon films (a-C:H) induced by pulsed laser irradiations. By the complementary use of isotopic (hydrogen/deuterium) tracing experiments and nuclear microanalysis, the changes induced by the laser treatments in the elemental composition of the films were evidenced. The loss of deuterium atoms was accompanied by a graphitization of the films. In addition, even at laser intensities much lower than the ablation threshold, a large exchange phenomenon was evidenced between the deuterium atoms initially incorporated in the amorphous carbon layers, and hydrogen atoms coming from the ambient water vapour. These findings are certainly related to the thermal effects induced in the a-C:H films by the laser pulses.

1. Introduction

The amorphous hydrogenated carbon films (a-C:H) represent a metastable form of carbonaceous material, between graphite and diamond. The precise properties of such films (optical band gap, refractive index, optical absorption, electrical resistivity, microhardness, etc.) are known to be strongly related to their hydrogen content, and thus to their growth process [1]. Such metastable materials can be easily changed by various ways in which energy is deposited in the films leading mainly to the modification of the hydrogen and carbon microstructure, and to the loss of hydrogen. For example, annealing at moderate temperature ($T < 600°C$), or irradiation by electron or ion beams leads to the release of hydrogen atoms, and to a decrease of resistivity by orders of magnitude, suggesting therefore some process of graphitization of the a-C:H films [2].

Laser irradiation of a-C:H films also leads to material transformations. Depending upon the precise laser intensity, various phenomena can occur: at sufficiently high intensity, laser ablation is observed, leading to material removal, while a decreasing intensity is accompanied by the graphitization of the films (hydrogen loss) [3], and at very low intensity there is no observable interaction between the laser and the film.

In this paper, we describe the effects of laser irradiation at intensities lower than the ablation threshold, on two different kinds of a-C:H films, i.e. "polymer like" films with a high hydrogen concentration (H/C ≈ 1) and "diamond like" films with a reduced hydrogen concentration (H/C ≈ 0.5). We will emphasize the changes in elemental composition of the films and correlate them to the modifications in the physico-chemical nature of the films. Isotopic tracing experiments were carried out, using a deuterium-enriched amorphous carbon film. An exchange phenomenon was thus evidenced between the atoms initially present in the films and atoms coming from the ambient during laser irradiation.

2. Experimental

The a-C:H films were deposited onto n-type Si single-crystal substrates in a DC multipolar plasma system as described elsewhere [4]. In this plasma set-up, the most important deposition parameter is the substrate bias which determines the impact energy of the film-forming particles. In fact, the a-C:H films grown in this system incorporate significant amounts of hydrogen, which decrease with increasing substrate bias (i.e. increasing values of the energy of the impinging ions). Pure methane gas was used, either with the natural hydrogen isotopic composition (i.e. CH_4), or highly enriched in deuterium (up to 90%). This deuterated methane (CD_4) was used for the isotopic tracing experiments [5] which were carried out in order to determine the effects of the laser treatments on the amorphous carbon layers.

The a-C:H (or a-C:D) films were irradiated under atmospheric pressure using a pulsed ArF excimer laser (Lambda Physik) providing at 193 nm wavelength, pulses of 20 ns duration with energy densities between 0.05 to 1 J/cm^2. A frequency-tripled Nd:YAG laser (BMI) delivering at 354 nm, energy densities ranging between 0.1 and 0.5 J/cm^2 per pulse of 5 ns duration was also used.

The in-depth compositions of the a-C:H films as deposited and after the subsequent laser irradiations, were routinely studied by nuclear microanalysis. The overall carbon and oxygen contents of the films were determined by nuclear reaction analysis (NRA), through the following reactions ^{12}C (d, p) ^{13}C and ^{16}O (d, p) ^{17}O, induced respectively by 965 and 850 keV deuteron beams. Rutherford backscattering spectrometry (RBS) was used to determine the film thickness and to study the possible film–substrate reactions after laser irradiations. The hydrogen and deuterium contents, and their in-depth locations were determined by elastic recoil detection analysis (ERDA). The precise depth profiles were obtained by the SENRAS simulation program [6].

The a-C:H films were also investigated by Raman spectroscopy in order to study their structural properties as well as their possible modification upon the UV laser irradiations. Spectra were obtained using photons ($E = 2.7$ eV) incident at 76° with respect to the normal of the film. The

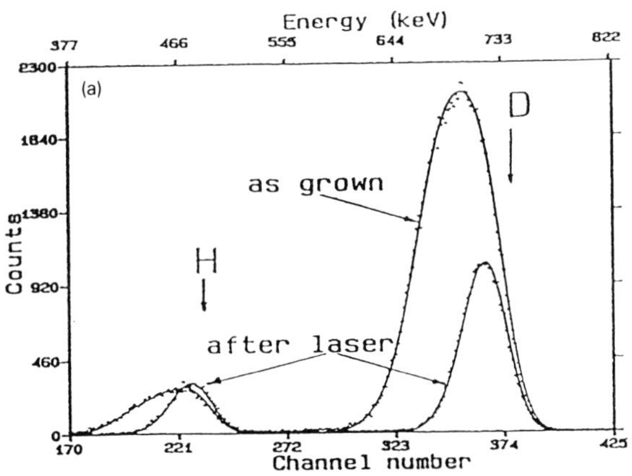

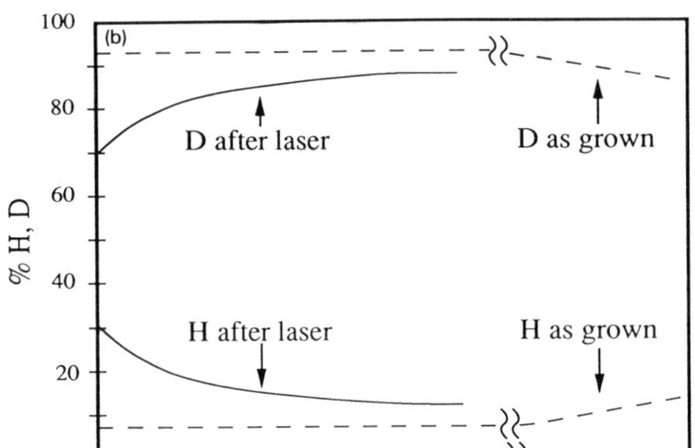

Fig. 1. (a) ERDA spectra (dots) and SENRAS computer simulations (full lines) of "polymer like" a-C:D thin films, before and after laser irradiation by a fluence of 100 mJ/cm^2. (b) H and D in-depth profiles in "polymer like" films, before and after laser irradiation by a fluence of 100 mJ/cm^2, deduced from the ERDA spectra by the SENRAS computer simulation.

laser power was limited to values not exceeding 250 mW and the optical resolution of the double monochromator was set to 6 cm^{-1}.

3. Results

During plasma deposition, the energetic ionic species act as structural modifiers making the films harder and stronger, while radicals act primarily as the source for the deposition of the polymeric species. Thus, "polymer like" films (with the ratio (D + H)/C equal to 1) and "diamond like" films (with (D + H)/C equal to 0.6) have been grown in this work. The thicknesses of the films were respectively 490×10^{15} C atoms/cm^2 (~ 50 nm) for "polymer like" films and 1750×10^{15} C atoms/cm^2 (~ 200 nm) for "diamond like" films. Whatever the precise nature of the films, and whatever the photon wavelength and laser intensity, some general changes in the elemental composition were evidenced. As a matter of fact, figs. 1a and 2a present typical ERDA spectra recorded in this work on as-deposited films ("polymer like" and "diamond like", respectively), and after UV laser irradiations with one single pulse.

These spectra present the main effects associated with laser irradiations. First a decrease in the deuterium content can be deduced in these spectra, characterized by the decrease in the deuterium yield after the laser treatment. Moreover, a large change is also observed in the D and H depth distributions in the films, as shown in figs. 1b and 2b presenting the H and D profiles deduced from the ERDA spectra by the computer simulation. In fact, the near-surface region of the a-C:D films shows a deuterium depletion and a hydrogen enrichment after the laser treatment. This behaviour is less pronounced for the "polymer like" film, as shown in fig. 1b, while a practically pure hydrogen layer is observed in the near-surface region in fig. 2b. Therefore, an exchange phenomenon occurs during the laser irradiations between the deuterium initially fixed in the films and hydrogen atoms coming from the ambient.

A decrease of the a-C:D film thickness can be also deduced from these ERDA spectra. This phenomenon was checked by RBS analysis, while the nuclear reaction analysis indicates a decrease of the overall carbon content for both kinds of films, and in the case of "diamond like" films an oxygen incorporation occurs during laser irradiations. The decrease in film thickness corresponds to an etching phenomenon, i.e. during the laser pulse, the temperature of the film increases, and the oxygen species of the ambient can react with

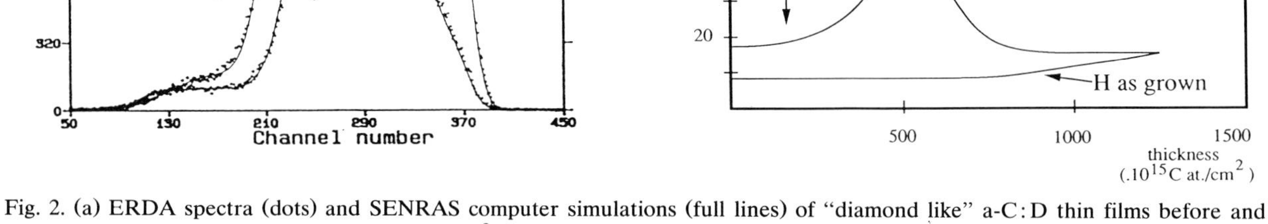

Fig. 2. (a) ERDA spectra (dots) and SENRAS computer simulations (full lines) of "diamond like" a-C:D thin films before and after laser irradiation by a fluence of 200 mJ/cm^2. (b) H and D in-depth profiles in "diamond like" a-C:D thin films before and after laser irradiation by a fluence of 200 mJ/cm^2, deduced from the ERDA spectra by the SENRAS computer simulation.

Table 1
"Polymer like" a-C:D thin films

Laser fluence (mJ/cm^2)	$\frac{D+H}{C}$	Δ thickness (×10^{15} C atoms/cm^2)	Δ O gain (×10^{15} O atoms/cm^2)
100	0.98	240	−25
200	0.76	365	−50
350	0.68	430	−56
500	0.56	450	−56
As grown	1.10	510	83

the carbon of the films, leading to the formation of a volatile compound (CO_2). This thermally assisted chemical etching was found to increase with the laser intensity, while no true differences were found with a change of the laser wavelength (193 and 354 nm).

Tables 1 and 2 summarize all these changes in the elemental composition for a-C:D films, irradiated at 193 nm photon wavelength at various intensities. Table 1 shows that for "polymer like" films, the number of oxygen atoms decreases with increasing laser intensities, in contrast to the "diamond like" films, in which the content of the oxygen atoms increases (see table 2). In fact, during the laser irradiation, the oxygen atoms (initially incorporated in the films) are lost due to the etching, while oxygen atoms coming from the ambient can be fixed in the surface region. For "diamond like" films, this last process seems to play a major role (i.e. increase in the oxygen content), while the reverse is observed for "polymer like" films. This difference can be related to

Table 2
"Diamond like" a-C:D thin films

Laser fluence (mJ/cm^2)	$\frac{D+H}{C}$	Δ thickness (×10^{15} C atoms/cm^2)	Δ O gain (×10^{15} O atoms/cm^2)
50	0.53	225	+65
200	0.50	350	+100
350	0.48	400	+107
600	0.48	470	+110
As grown	0.61	1750	95

Δ O = after irradiation − as deposited.

differences in initial thickness [7] and initial deuterium concentration of these two kinds of films, leading therefore to differences in the thermal effects associated with the laser irradiations.

These two tables clearly show a decrease of the overall content in the light elements (D and H). The Raman investigations confirm some of the above observations and give also complementary information. As a function of the nature of the a-C:H films, quite different Raman responses are observed. As-deposited "polymer like" films are characterized by flat and featureless spectra in the energy range extending from 1000 to 1900 cm^{-1}. This suggests that no structural order exists within the depth probed by the visible photons, or that the concentration of graphitic carbon bonds is too weak to give a significant Raman signal. This is probably due to the high concentration of hydrogen present in these films, which inhibits the formation of the three fold-coordinated carbon structure. This situation does not change upon laser irradiation, with the exception of the observation of a new but weak band at about 970 cm^{-1}, which originates undoubtedly from the Si substrate, confirming that some etching processes occur during the laser irradiation.

In contrast, the Raman spectrum of an as-deposited "diamond like" film shows typical features which consist in two broad bands located around 1570 and 1365 cm^{-1}. These bands are known to correspond to the G (for graphitic) and D (for disorder-induced) peaks respectively, which are usually observed for glow-discharge deposited, ion-beam sputtered or simply evaporated a-C:H films [8,9]. After laser irradiation the previous Raman features are almost preserved. However, closer inspection of the spectra reveals some subtle modifications. For irradiations lower than 300 mJ/cm^2, the overall scattering level is found to decrease by about 40%, whereas light elastically scattered at the film surface increases markedly. This behaviour can be explained by two independent, but possibly additive, phenomena: either a reduction of the layer thickness in which the Raman signal is generated, and/or an increase of the surface roughness of the film, which in turn weakens the Raman yield. The first

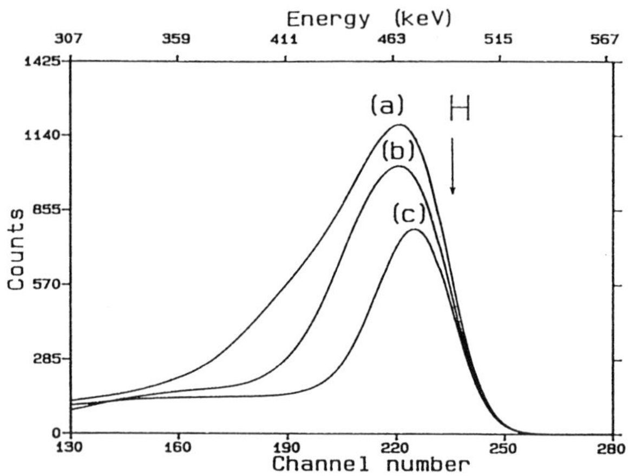

Fig. 3. Hydrogen contribution to the ERDA spectra of the "diamond like" films after different laser treatments: (a) corresponds to a 600 mJ/cm^2 laser fluence, (b) 200 mJ/cm^2 and (c) 50 mJ/cm^2.

process corresponds well to the ERDA and RBS analyses, the second was not studied in the present work, but similar observations have been reported for laser-irradiated a-Si:H films [10]. For irradiations performed at higher energy fluences, a reverse evolution is observed, suggesting that more complex mechanisms occur, including melting/resolidification processes, which could improve the structural quality of the film as well as of its free surface. This is particularly visible on film irradiated at the highest (600 mJ/cm^2) energy, where both the D/G peak intensity ratio and the G peak energy increase.

The ERDA spectra in fig. 2 clearly demonstrate an exchange phenomenon which is not only a surface process. In fact, deuterium atoms initially present in the bulk of the films have been replaced by hydrogen atoms during the laser treatment. The importance of this exchange phenomenon, and the in-depth distribution of the exchanged atoms vary with the laser intensity. This can be seen in fig. 3, which represents the hydrogen contribution to the ERDA spectra (determined by the SENRAS program) of the "diamond like" serie. This figure shows an increase and a broadening of the H yield with the laser intensity. These points characterize an increasing number of incorporated hydrogen atoms, and an increase in their depth distribution. Similar conclusions can be drawn, for the effects of the number of pulses, from the ERDA spectra of "diamond like" films presented in fig. 4. Moreover, for the high number of pulses, a saturation in the hydrogen incorporation can be deduced from fig. 4. This effect corresponds to the fact that for these samples practically all the deuterium atoms have been expelled from the films and replaced by hydrogen atoms.

The origin of the incorporated hydrogen atoms would appear to be the water vapour present in the ambient during the laser treatment. In fact, the OH present, near or at the surface of the sample, can be dissociated by the incident photon flux, and the resulting atomic hydrogen and oxygen could react with the films. The hydrogen atoms can be injected in the film and diffused inward, probably through a laser melting process, while the very active atomic oxygen can chemically react with carbon atoms. Thus they can be fixed in the near-surface region of the film as it is observed in the case of "diamond like" films, and can contribute to the etching phenomenon.

The major difficulty in the interpretation of the pulsed laser processing is the accurate evaluation of the thermal contribution resulting from the interaction between the incoming light and the solid, which cannot be measured experimentally. For these reasons, surface melt dynamics

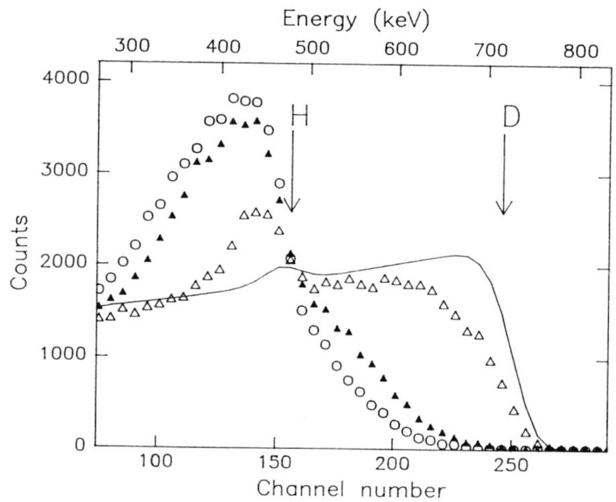

Fig. 4. ERDA spectra of "diamond like" films as grown (———) and after: 500 laser pulses (○), 100 laser pulses (▲) and 10 laser pulses (△).

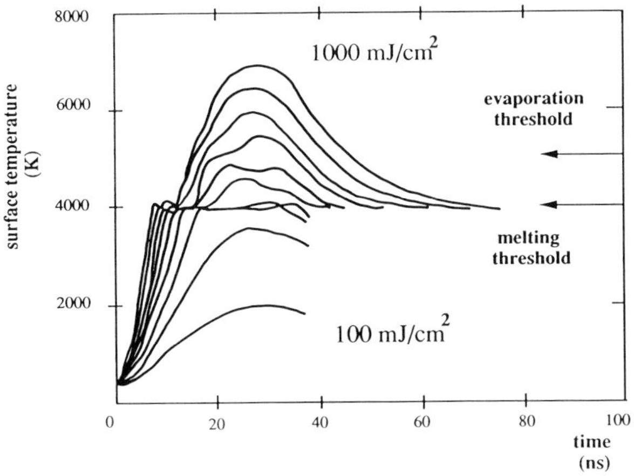

Fig. 5. Time dependence of the surface temperature of a 200 nm thick carbon layer on Si substrate, irradiated with an excimer laser (193 nm, 20 ns pulse) for different laser fluences (from 100 to 1000 mJ/cm^2) obtained by solving the heat flow equation.

under laser irradiations must be simulated by solving the heat flow equation. The melting threshold, melt duration and depth of fusion could be calculated in different irradiation conditions.

Because the optical and thermal properties of hydrogenated amorphous carbon, at 193 nm, are poorly known, especially when the H concentration is very high, the calculations were only performed for hydrogen-free a-C layers. This means that they can just give a rough estimate of the thermal effect induced by the excimer laser in hydrogenated films. In fig. 5 is reported the time-dependent surface temperature rise of a 200 nm thick carbon layer on Si, irradiated at various laser fluences (from 100 to 1000 mJ/cm^2). At around 250 mJ/cm^2, the temperature of an a-C film reaches a surface value of about 4000 K, corresponding to its melting threshold. It is interesting to notice that above 200 mJ/cm^2, ERDA experiments have shown strong in-depth exchange phenomena between D and H in "diamond like" films (fig. 3), which could be correlated with the formation of a molten layer. At laser fluences higher than 600 mJ/cm^2, the surface temperature reaches the boiling point of carbon (about 5000 K). This could partly explain the strong diminution of the film thickness after high fluence laser irradiation, as reported in the tables 1 and 2.

4. Conclusions

In this work, it was found that the most pronounced effects in a-C:D films resulting from pulsed laser irradiations are the deuterium depletion, and the exchange phenomenon between deuterium atoms fixed in the films during the plasma deposition and hydrogen atoms coming from the ambient. It has been previously shown [11] that, during thermal annealing, hydrogen leaves the a-C:H films in molecular form, i.e. molecular recombination occurs in the bulk, followed by diffusion as molecules through the layer until they are lost into the vacuum. A similar description can be proposed for the loss of deuterium under laser irradiation: once a D molecule has been formed in the bulk, it diffuses out of the material, in a purely interstitial way, without further interactions.

In addition to this outward diffusion, the exchange phenomenon implies the existence of an inward diffusion of hydrogen in atomic form. These atoms injected in the films at the surface, migrate into the material until they are trapped on a dangling bond. Following our results, the characteristic penetration length of these atoms (before their fixation) is an increasing function of the laser intensity or number of pulses. Taking into account the estimation of the temperature rise during the laser pulses, the driving force for all the atomic movements in the films is believed to be the temperature.

Acknowledgement

This work was supported by the CNRS (GDR 86 and GRECO 57).

References

[1] J. Robertson, Adv. Phys. 35 (1986) 317.

[2] M.E. Adel, O. Amir, R. Kalish and L.C. Feldman, J. Appl. Phys. 66 (1989) 3248.

[3] S. Prawer, R. Kalish and M. Adel, Appl. Phys. Lett. 48 (1986) 1585.
[4] J. Perrière, J. Siejka, J.P. Enard, A. Laurent and F.M. d'Heurle, Mater. Res. Soc. Symp. Proc. 38 (1984) 443.
[5] J. Perrière, A. Laurent and J.P. Enard, Mater. Sci. Eng. B 11 (1992) 347.
[6] G. Vizkelethy, Nucl. Instr. Meth. B 45 (1990) 1.
[7] A. Jadin, I.V. Filiouguine, M. Wautelet and L.D. Laude, Appl. Surf. Sci. 46 (1990) 375.
[8] R.O. Dillon, J.A. Woollam and V. Katkanant, Phys. Rev. B 29 (1984) 3482.
[9] N. Wada, P.J. Gaczi and S.A. Solin, J. Non-Cryst. Solids 35 (1980) 543.
[10] M. Elliq, E. Fogarassy, C. Fuchs, J.P. Stoquert, S. de Unamuno, B.Prevot and E.L. Mathe, Appl. Surf. Sci. 54 (1992) 35.
[11] C. Wild and P. Koidl, Appl. Phys. Lett. 51 (1987) 1506.

Applied Surface Science 69 (1993) 249–252
North-Holland

applied
surface science

Shallow p–n junctions produced by laser doping with boron silicate glass

D. Bollmann, G. Neumayer, R. Buchner and K. Haberger
Fraunhofer Institute for Solid State Technology, Hansastrasse 27d, D-8000 München 21, Germany

Received 6 June 1992; accepted for publication 30 October 1992

Shallow junctions have been fabricated using an excimer and a CO_2 laser as heating source. The diffusion was made out of a silicate glass layer. The produced diodes were electrically characterized. The concentration profiles measured by SIMS agree with simulations by ICECREM and have a depth of about 100 nm.

1. Introduction

In the development of sub-micron CMOS devices the formation of ultra-shallow junctions is becoming a challenging task. The usual ion implantation has a limited depth control because of channeling and the necessity of an annealing step that broadens the profiles [1]. In the now favoured processes using silicides [2], heavy metals are introduced that are not always of advantage. But the ability of a solid source process to control the doping profile in small scales is clearly shown [3].

Laser driven out-diffusion from solid source layers could be a promising way for the production of very flat p–n junctions. It uses short laser pulses for transient heating of the region to be doped. Thus the temperature–time budget can be optimized. The dopant is contained in a silicate glass which may be removed after the treatment.

The aim of this work is to find a set of optimum process parameters and to determine a relation between them and the resulting doping profile.

2. Experimental

2.1. Process with an excimer laser

We used an excimer laser of the type LPX 100. Used with krypton fluoride it emitted an ultraviolet radiation with a wavelength of 248 nm. This is transmitted by the oxide and absorbed at the surface of the silicon. The pulse duration here is only 12 ns, which means nearly instant heating. For a numerical simulation of the diffusion process, this time is too short.

A quartz lense concentrated the laser beam, see fig. 1. Thus a focus area with a nearly rectangular shape was formed with an energy density of 2.7 J/cm^2, which is below the threshold of surface damage. In order to increase the dopant concentration, the pulses were repeated on the same spot. In this setup the exposure was static without scan.

The solid source layer was boron silicate glass (BSG) produced by chemical vapour deposition. The boron concentration in the source layer was about 4%, which corresponds to 2×10^{21} cm^{-3}. The samples were analysed by SIMS.

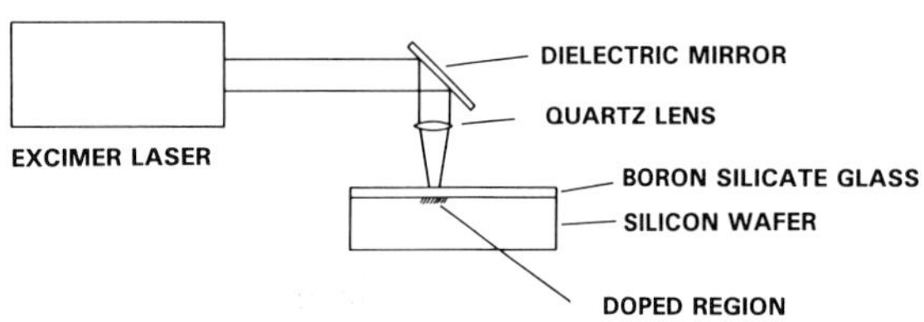

Fig. 1. The wafer is irradiated by the partly focussed beam of an excimer laser. During transient heating, the dopant is diffusing into the silicon.

2.2. Process with a CO_2 laser

As a second energy source we used a high-power CO_2 laser that emits infrared radiation with a wavelength of 10.6 μm. This is well-absorbed in the oxide layer that contains the dopant, but is transmitted by the bulk silicon. Therefore, the thermal stress is concentrated at the region where the diffusion takes place.

The laser was pulsed with an energy of 3 J and a pulse duration in the range from 8–25 ms. This time was short enough for rapid cooling after the end of the pulse. The bulk of the wafer is held at a preheating temperature of 450°C in order to reduce thermal stress. The output power of the laser was adjusted in a way that the pulse energy was constant.

The wafer was moved by an *x–y*-table which results in a scanning of the laser beam with partial overlapping. In the pulse mode it was possible to reduce the scanning velocity without a change in the spatial temperature distribution. The pulses were separated by sufficiently long pauses for thermal equilibrium. Thus we achieved an effective pulse number of 160 over the area of several chips with a good homogeneity. When the method is more established the pauses may be shortened in order to accelerate the process.

2.3. Sample preparation

In order to show the applicability of the p–n junctions formed by laser doping we developed a process for diodes with metallization. Therefore we were able to determine the electrical parameters of the doped samples. The process consists of the following steps (see fig. 2):

(1) thermal oxidation of the field oxide as a diffusion barrier,
(2) the windows for doping are opened by photolithography,
(3) the phosphorus silicate glass (PSG) as source layer is deposited by CVD. For better absorption of the laser beam, a second oxide (TEOS) is deposited,
(4) the dopant diffuses into the silicon when irradiated with the laser,
(5) the second mask opens the contact holes,

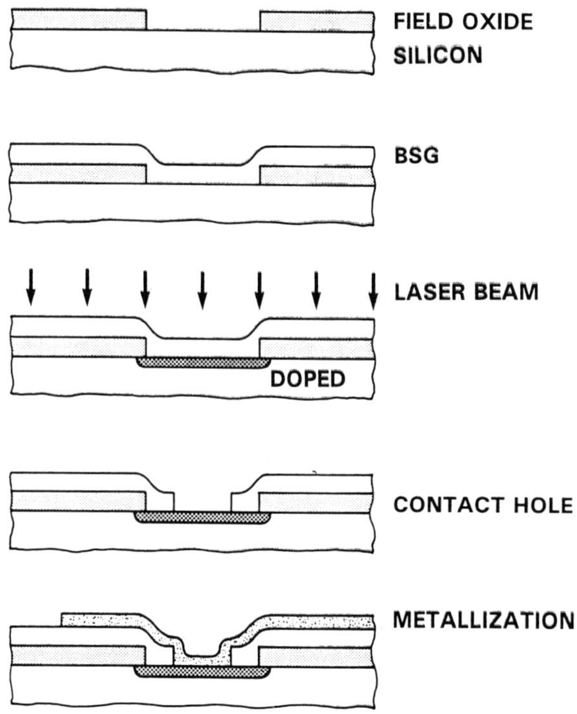

Fig. 2. The process used for the production of diodes by laser doping.

(6) aluminium is sputtered as metallization, and
(7) the third mask defines the interconnections and the bond pads.

The set of masks was made for test purposes and mainly consists of diodes of different sizes. The size of the square diodes varied in steps of factor 2 from 20 μm up to 2.5 mm. They were arranged in groups with and without bond pads.

In addition, a van der Pauw structure and a meander rendered the determination of the sheet resistance possible. The meander had a length of 4800 and a width of 3.0 μm, which means a relation of 1600.

3. Simulation of the doping process

In comparison with the pulse mode the laser was also driven in the CW mode. But here the lateral temperature distribution was not symmetric in the scan direction. A hot tail formed on the back, as can be seen in fig. 3 (bottom), which is the result of a simulation by the finite differences

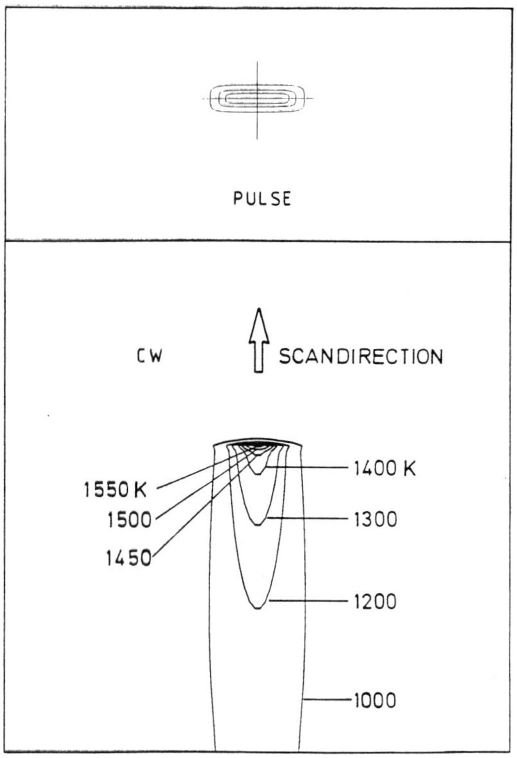

Fig. 3. The temperature distribution at the wafer surface was simulated (bottom) for a scanning CW-beam, and (top) for a pulsed beam.

method. The shape also depends on the scan velocity. In the case of the pulse mode, the temperature distribution is symmetric and more homogeneous, see fig. 3 (top). Therefore, the numerical treatment of the diffusion process can be reduced from 3 to 2 dimensions.

The diffusion of the dopant was simulated by the conventional program ICECREM [5]. The transient heating and cooling cycle was replaced by the assumption of a constant temperature of 1350°C for a time of 30 ms. The dopant source was taken as reservoir at the surface of the silicon with a concentration of 2×10^{21} cm^{-3}. The calculation is in agreement with points measured by spreading resistance.

4. Results

The resulting dopant profiles of the samples treated with the excimer laser were determined by SIMS. The measured maximum concentrations in the doped layer range from 2×10^{19}–5×10^{18} cm^{-3}. We varied the number of pulses in order to increase the concentraton. The experiments showed a normal diffusion profile in the case of one pulse, see fig. 4, and a plateau in the case of 10 pulses. The resulting depth of the p–n junction formed by 1 excimer pulse was about 150 nm. It increased to 300 nm in the case of 50 pulses and remained constant at a higher number of pulses. Then a saturation effect seemed to limit further diffusion. The samples treated with the CO_2 laser were examined by SIMS, too. The depth of the p–n junction was in the range of 100 nm, as reported earlier [4].

The sheet resistance of the doped layer was measured with a van der Pauw structure and a meander. This showed that it was necessary to apply several overlapping pulses to form a well defined depletion region. In the case of less than 10 pulses the meander was short-circuited.

Crystal defects might be created by the high temperature gradients during laser irradiation. A measure of the crystal quality in the space charge region is the leakage current of a diode. In our experiments it was below 1 nA at a voltage of -5 V for diodes a junction area of 40×40 μm^2, see fig. 5. Best results were achieved with multiple overlapping pulses. This shows that it is possible

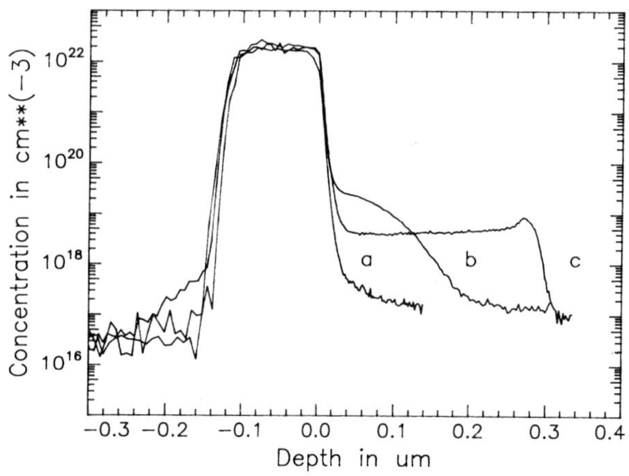

Fig. 4. The SIMS analysis shows the doping profile in the depth range of 0–0.2 μm and the BSG layer on the left. The profiles were obtained (a) without exposure, (b) with one pulse, and (c) with 50 pulses.

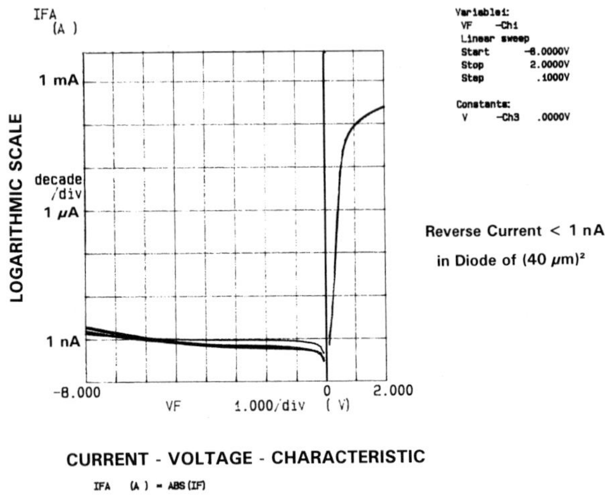

Fig. 5. The electrical measurements show the current–voltage characteristic with reverse current.

to produce microelectronic devices by laser doping with sufficiently low thermal stress.

The measurement of the current–voltage characteristic was repeated with a set of 200 identical treated diodes in order to test the reliability of the method. The leakage current is entered in a histogram, fig. 6. It shows a typical Poisson distribution with few runaways. They are cumulated at the right end of the scale.

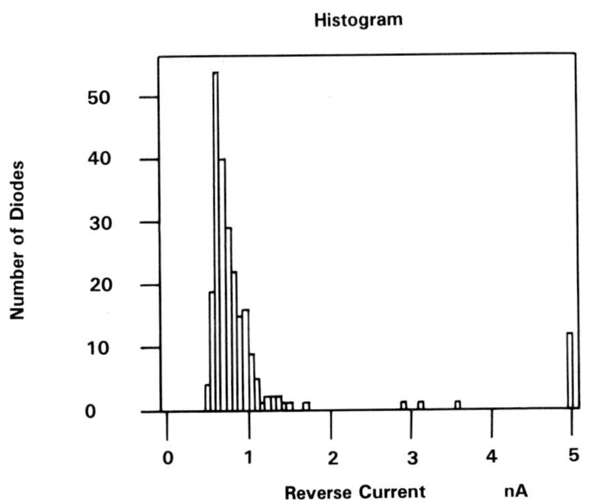

Fig. 6. A histogram shows the distribution of the measured reverse current.

5. Conclusion

Diodes and some test devices were produced with the laser doing process. The SIMS analysis showed the p–n junction to be in a depth of 150 nm. Electrical measurements showed that it is necessary to apply a large number of overlapping pulses in order to produce diodes with well defined p–n junctions and with limited thermal stress and low leakage current.

Acknowledgements

The authors thank Mrs. B. Schmiedt for fabricating the samples in the CMOS-line and W. Muth for his help with the electrical measurements and the design of the photolithographic masks. This work was partly funded by the JESSI project.

References

[1] H. Ryssel and I. Ruge, Ion Implantation (Teubner, Stuttgart, 1975).
[2] H. Jiang et al., J. Electrochem. Soc. 139 (1992) 196.
[3] C. Hill, J. Vac. Sci. Technol. B 10 (1992) 289.
[4] G. Stock, D. Bollmann, R. Buchner, G. Neumayer and K. Haberger, Appl. Surf. Sci. 46 (1990) 348.
[5] P. Pichler et al., ICECREM version 4.1, FhG-AIS Erlangen.

Applied Surface Science 69 (1993) 253–257
North-Holland

applied
surface science

Single-shot micro-patterning of polymer surfaces by UV incubation/dye laser ablation using photochromism

S. Preuss and M. Stuke
Max-Planck-Institut für biophysikalische Chemie, P.O. Box 2841, 3400 Göttingen, Germany

Received 2 June 1992; accepted for publication 30 October 1992

The optical absorption of subsurface polymer material can be increased by low intensity UV irradiation (incubation) with spatial control using a suitable contact mask. Spatially selective ablation can then be achieved with large area pulses at a wavelength at which only the modified but not the virgin material absorbs the laser light. With this two-step process, micron or even submicron patterns can be obtained in a simple way. By doping polymers with photochromic compounds the efficiency of this process is improved drastically. The number of required pulses is reduced to the minimum, that is, micron sized structures can be generated on various polymer surfaces with *one single* pulse for incubation and ablation, respectively.

1. Introduction

Surfaces of various materials can be patterned by exposure to a spatially defined laser beam of suitable fluence [1,2]. The easiest way for large area patterning would be the use of free standing contact masks. However, often fluences are required that affect the mask material itself and/or the mask is contaminated by the ablation plume. For micron or even submicron size patterning stable free standing contact masks are not easily available [3] due to the lack of mechanical stability. Metal masks on transparent carriers such as chromium on quartz can be used in this case. Obviously, the ablation products cannot escape through such a mask and hence, the mask would become immediately contaminated. The suitability of other possible arrangements for direct ablation patterning such as imaging and fringe patterning have been discussed in detail elsewhere [4].

We have recently [4,5] introduced a two step process, that offers certain advantages over direct laser ablation and is suitable for the generation of micron or even submicron patterning: in the first step the substrate is exposed spatially defined to low fluence UV laser or UV lamp irradiation using a contact mask. These fluences are not sufficient to cause ablation, but can modify the substrate properties such as the optical density (OD) in the irradiated regions by photochemical reactions (incubation). E.g., the OD of polymethylmethacrylate (PMMA) in the near UV region is increased considerably when exposed to KrF excimer laser pulses ($\lambda = 248$ nm) [5]. After the first step, the mask is removed. In the second step, the laser is set to a wavelength, at which it is only absorbed by the incubated material but not by the virgin material. Thus, the incubated areas can be removed (ablation) selectively by exposing the surface to large area laser pulses of suitable fluence without damage to the remaining surface.

By separating the information transfer to the polymer surface (step I, incubation) from the removal of material (step II, ablation) the following advantages are achieved: A contact mask can be used for the first step, because only low fluences are needed and no ablation products, which would contaminate the contact mask in a conventional laser ablation process, are produced. Information transfer can be performed through a transparent reference block (e.g. quartz or calcium fluoride), because no ablation products have to escape from the surface. Technical polymer surfaces can be planarized by pressing them against the reference block. Therefore, technical polymer surfaces can be patterned over macroscopic areas (e.g. by using projection devices) with high resolution, which is virtually impossible with conventional direct laser ablation.

The use of a contact mask offers an additional advantage: by irradiating the surface of a closely spaced object through a small aperture of a contact mask, rather than imaging a beam of radiation onto the object, resolution of the order of the aperture size rather than of the order of the wavelength can be achieved [6], so that subwavelength sized patterning becomes possible.

We have shown that micron and submicron structures can be produced reliably and with good quality on undoped PMMA with this process. However, several hundred pulses were required for incubation and ablation, respectively. The large number of pulses is due to the weak increase of optical density (OD) of undoped PMMA in the incubation step (see fig. 1 in ref. [4]). To decrease the number of pulses we used PMMA samples doped [7] with 6-nitro-BIPS (= 6-nitro-1′,3′,3′-trimethyl-spiro-[2H-1-benzopyran-2,2′-indoline]). This photochromic [8] molecule (fig. 1) belongs to a group of spyropyrans that undergoes a ring opening reaction when irradiated with UV light. In the closed form, the π-electron systems of both parts of the molecule do not interact significantly whereas in the open form the π-electrons conjugate over the entire molecule leading to an intense absorption in the visible region (fig. 2). We have chosen this molecule, because it shows a strong photochromic effect with a quantum yield of about 1 [9] in a PMMA matrix, its thermal fading reaction is comparatively slow and it is easily synthesised by condensation from 1,3,3-trimethyl-2-methyleneindoline with 2-hydroxy-5-nitro-benzenecarbaldehyde [10].

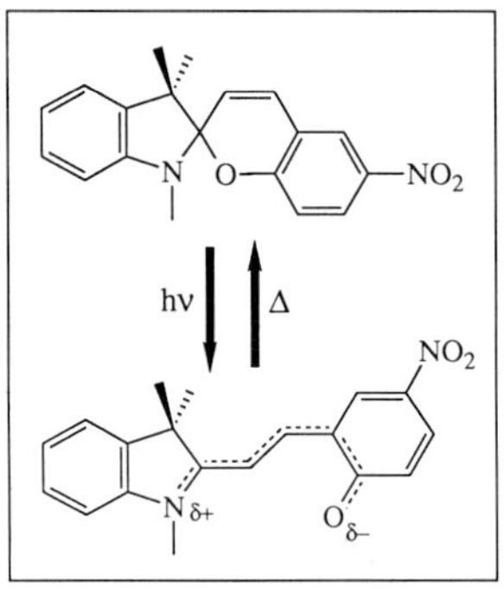

Fig. 1. Structure of 6-nitro-BIPS, its photochemical ring-opening reaction (coloring) and thermal back reaction (bleaching).

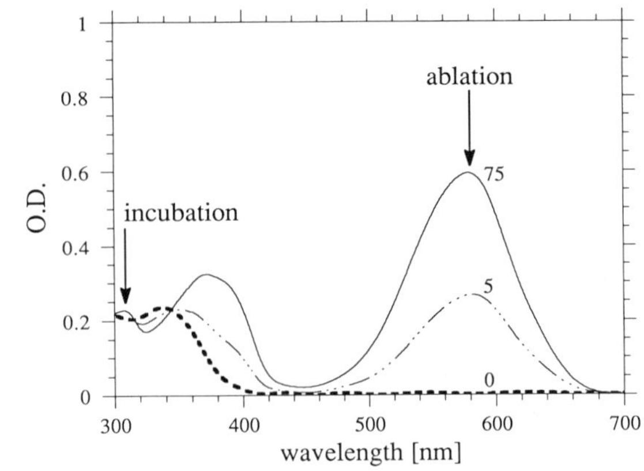

Fig. 2. Absorption spectra of a spin-on film (1.3 μm thick, on quartz) of PMMA doped with 5 wt% 6-nitro-BIPS after irradiation with various numbers of XeCl excimer laser pulses (308 nm, 10 mJ/cm^2).

2. Experimental

Films of doped PMMA were produced by spin-coating from a solution of PMMA and 6-nitro-BIPS in trichloroethylene on PMMA plates for incubation/ablation processing or on fused silica (Heraeus, Suprasil I) for UV and visible absorption spectroscopy. The experimental procedure is shown in fig. 3. For incubation, a XeCl excimer laser (308 nm, Lambda Physik EMG 203) was applied to irradiate the samples through contact masks (chromium on quartz). To obtain a homogeneous beam profile on the substrate surface a beam homogenizer based on a cylindrical lens array was used. For ablation, an excimer pumped dye laser (Lambda Physik FL 3002, Rhodamin 6G) was set to the maximum of the induced absorption at $\lambda = 580$ nm (see fig. 2). Weak focusing with a quartz lens was necessary to obtain appropriate fluences for selective ablation.

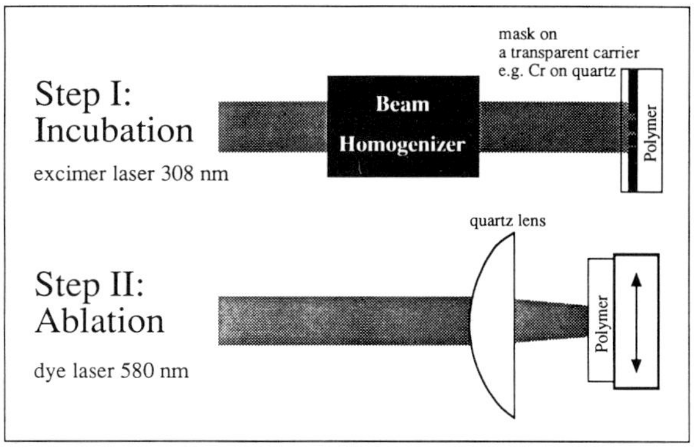

Fig. 3. Experimental procedure: in the first step, the sample is exposed spatially defined to UV irradiation by using a contact mask. In the second step, the UV exposed areas are selectively removed by a weakly focused dye laser pulse. To process the whole surface, the sample is moved by a translation stage during the ablation process to scan the surface.

To expose the whole surface to the ablation pulses, the substrate was moved perpendicular to the laser beam during the ablation process. Absorption measurements were carried out with a commercial spectrophotometer.

3. Results and discussion

The absorption spectra of a PMMA film doped with 6-nitro-BIPS is shown in fig. 2. The unirradiated film only absorbs below 400 whereas the irradiated film strongly absorbs in the visible with the maximum at 580 nm. The increase of OD at $\lambda = 580$ nm (fig. 4) depending on the number of UV laser pulses (308 nm, 10 mJ/cm^2) is almost linear for the first 10 pulses and saturates after about 30 pulses. Between 100 and 200 pulses the OD stays almost constant at its maximum ($OD_{max} = 0.61$, $\alpha = 4500$ cm^{-1}). At room temperature, bleaching by the thermal back reaction is slow enough (the OD decreases within one hour just to about 85% of its initial value) to allow subsequent processing. At larger pulse numbers (> 300), the OD decreases (see insert fig. 4) due to thermophotodegradation of the dopant [11]. The photochemical coloring reaction can also be induced by other UV sources such as low intensity mercury UV lamps. In this case, no degradation occurred due to negligible heating by the lamps. Thus, the application of low intensity UV irradiation (UV lamps or excimer lasers at very low fluence) in the incubation step can be advantageous, when longer irradiation times (ca. 1 min) can be accepted.

Using thermophotodegradation, the generation of the reverse image of a mask is also possible: first, the dopant is degraded with spatial control via the contact mask by a large number of excimer laser pulses or UV lamp irradiation plus additional substrate heating. After removal of the mask, the surface is exposed to low intensity UV light at room temperature to incubate the whole surface except the regions where the dopant has been destroyed. After ablation, the reverse pattern on the polymer surface could then be obtained.

The SEM photograph of a doped PMMA sample (5 wt% 6-nitro-BIPS) after incubation/ablation processing is shown in fig. 5. Only 5 incubation pulses (308 nm, 10 mJ/cm^2) and *one single* ablation pulse (580 nm, 1.9 J/cm^2) have been applied, compared to about 1000 pulses required for processing in the case of undoped PMMA. The structure of the mask (circular apertures of 2.5 μm diameter spaced by 15 μm) was transferred precisely to the polymer surface. The

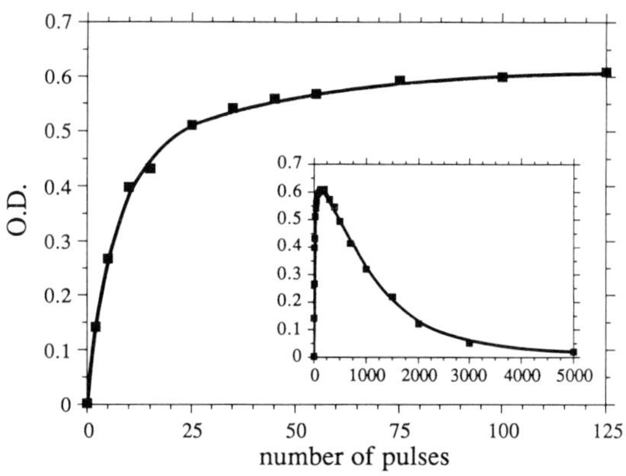

Fig. 4. Rise of optical density at $\lambda = 580$ nm of a spin-on film (1.3 μm thick, on quartz) of PMMA doped with 5 wt% 6-nitro-BIPS after irradiation with various numbers of XeCl excimer laser pulses (308 nm, 10 mJ/cm^2). Insert: at larger pulse numbers, the OD decreases due to thermophotodegradation of the dopant.

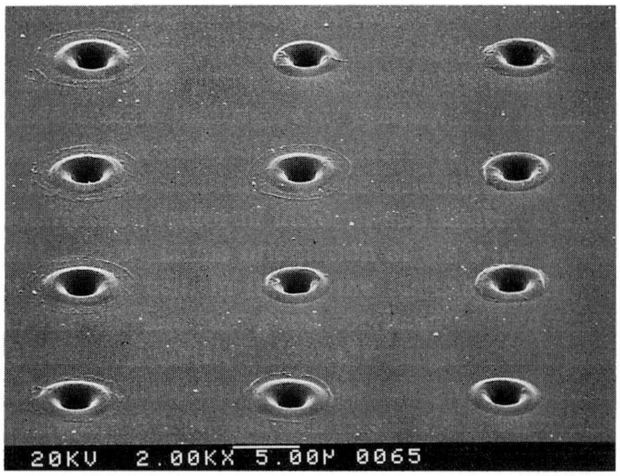

Fig. 5. SEM-photograph (tilt angle 55°) of a doped PMMA (5 wt% 6-nitro-BIPS) surface after processing, incubation with 5 XeCl excimer pulses (10 mJ/cm^2) through circular apertures of 2.5 μm diameter, ablation with one dye laser pulse (580 nm, 1.9 J/cm^2).

structures (see detailed view in fig. 6) are about 2 μm deep (ablation rate ~ 2 μm/pulse) and have a bulge on the top. This is probably due to an increased thermal component of the visible dye laser ablation at $\lambda = 580$ nm compared to the UV excimer laser ablation at $\lambda = 308$ nm, which was used for the selective ablation of undoped PMMA films, but can also be attributed to an insufficient adhesion of the doped film to the bulk material.

Thermal bleaching during the first ablation pulse is obviously slow enough to allow sufficient energy transfer to the substrate to ablate the incubation sites. The application of a second dye laser pulse is useless due to complete bleaching after the first pulse. If more than one ablation pulse is needed, e.g. for the generation of deep structures, photochemical compounds without fading reactions have to be used instead of photochromic compounds as dopants.

Fig. 7 shows, that it is possible to generate micro-patterns even by *one single shot for incubation and ablation, respectively*. In this experiment, *one* excimer laser pulse at 308 nm (80 mJ/cm^2) was used to incubate PMMA doped with 20 wt% 6-nitro-BIPS through a contact mask with square apertures of 7 μm. For selective ablation, *one* dye laser pulse at 580 nm (1 J/cm^2) was applied per area. Even after single shot processing, the structure of the mask is reproduced reliably on the surface. Due to diffraction, Fresnel patterns are generated on the bottom of the ablated structures.

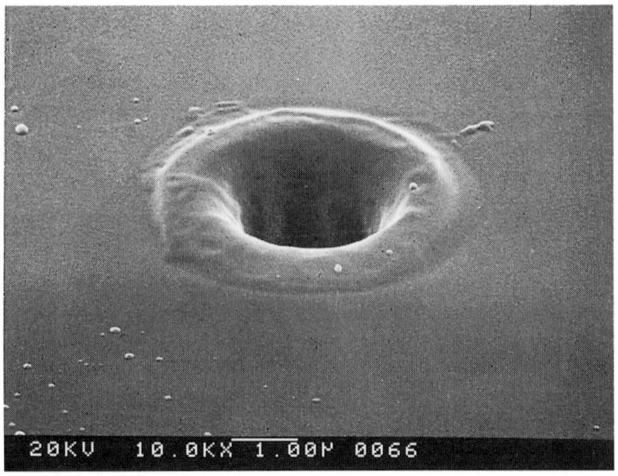

Fig. 6. Detailed view (tilt angle 55°) of one of the structures in fig. 5.

Considering the feasibility of our process, it should be mentioned that the process window is very large and appropriate parameters for surface patterning can be found easily: In the incubation step, a significant modification of the optical absorption between modified and virgin material is

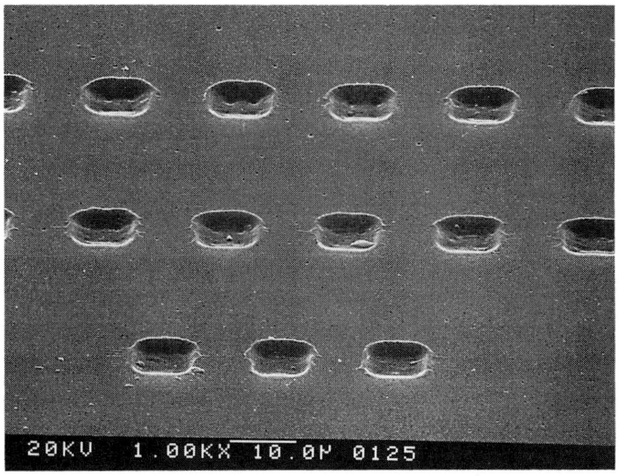

Fig. 7. SEM photograph (tilt angle 55°) of a doped PMMA (20 wt% 6-nitro-BIPS) surface after processing, incubation with *one* XeCl excimer pulse (80 mJ/cm^2) through square apertures of 7 μm, ablation with *one* dye laser pulse (580 nm, 1 J/cm^2).

required. The dose needed for this purpose is not critical, since this can be achieved with pulse numbers ranging from one to several hundreds (compare fig. 2 and fig. 4). The fluences for ablation can be varied over a wide range due to the large difference of the ablation thresholds Φ_T for incubated and virgin material ($\Phi_T \sim 1/\alpha$) [12]. In fact, there is rather thermal damage than well controlled ablation of the unincubated sample at $\lambda > 400$ nm due to the lack of absorption. Suitable fluences for selective ablation at 580 nm are therefore ranging from the ablation threshold of the incubation sites (< 1 J/cm^2, depending on the incubation dose and the dopant concentration) to the damage threshold of the unincubated sample (> 10 J/cm^2). Best results are obtained at 1–3 J/cm^2.

4. Conclusion

We have shown, that polymer surfaces can be patterned by means of a simple and reliable two step process. This process offers several advantages over conventional direct laser ablation and produces patterns in the micron or even submicron range. In the first step, absorption in the subsurface material is induced with spatial control by photochemical reactions (incubation step). In the second step, selective removal of the incubated areas is achieved by laser ablation at a wavelength at which only the incubated areas absorb (ablation step). Compared to recent results, we have improved this process, by doping PMMA films with a photochromic compound. Thereby, we could reduce the number of required pulses from several hundreds for undoped PMMA to the minimum of *one single* pulse for incubation and ablation, respectively. Furthermore, the process is not limited to PMMA (or any material that rises itself the OD) anymore. By doping with 6-nitro-BIPS, various polymers such as polycarbonate, polyethylene terephtalate etc., can be processed in the same way. Micron-sized structures of various shapes have been produced.

Acknowledgements

We would like to thank B. Albert (BASF AG) for providing 6-nitro-BIPS, C. David for SEM photographs and K. Müller for technical assistance. Support by F.P. Schäfer and financial support by BMFT 13N 5398/7 is gratefully acknowledged. S. Preuss thanks the Fonds der Chemischen Industrie for a scholarship.

References

[1] R. Srinivasan and V. Mayne-Banton, Appl. Phys. Lett. 41 (1982) 576.
[2] Y. Kawamura, K. Toyoda and S. Namba, Appl. Phys. Lett. 40 (1982) 374.
[3] F. Paschke, W. Fallmann and H. Löschner, Eds., Microcircuit Engeneering (North-Holland, Amsterdam, 1989).
[4] S. Preuss, H.-C. Langowski, T. Damm and M. Stuke, Appl. Phys. A 54 (1992) 360.
[5] S. Küper and M. Stuke, Appl. Phys. A 49 (1989) 211.
[6] Y. Leviatan, J. Appl. Phys. 60 (1986) 1577.
[7] M. Bolle, K. Luther, J. Troe, J. Ihlemann and H. Gerhardt, Appl. Surf. Sci. 46 (1990) 279.
[8] For recent reviews, see: H. Dürr and H. Bouas-Laurent, Eds., Photochromism (Elsevier, Amsterdam, 1990); H. Dürr, Angew. Chem. Int. Ed. Engl. 28 (1989) 413.
[9] A.V. Shabya, K.B. Demidov and Y.N. Polyakov, Opt. Spektrosk. 20 (1966) 738.
[10] C.F. Koelsch and W.R. Workman, J. Am. Chem. Soc. 74 (1952) 6288.
[11] R. Guglielmetti, in: H. Dürr and H. Bouas-Laurent, Eds., Photochromism (Elsevier, Amsterdam, 1990) p. 855.
[12] H.H.G. Jellinek and R. Srinivasan, J. Phys. Chem. 88 (1984) 3048.

Applied Surface Science 69 (1993) 258–261
North-Holland

applied surface science

SF_6 sensitized CO_2 laser CVD of amorphous silicon

E. Golusda, P. Hessenthaler, G. Mollekopf and H. Stafast *

Battelle Europe, Am Römerhof 35, D-6000 Frankfurt am Main 90, Germany

Received 2 June 1992; accepted for publication 30 October 1992

Sensitized CW CO_2 laser CVD of a-Si:H from SiH_4 was performed with SF_6 as a strong but chemically inert absorber. Relative to CO_2 laser CVD without SF_6, the CO_2 laser power, the total and the silicon hydride partial pressures could be substantially reduced. The processing gas was investigated by in situ mass analysis. Polysilanes (Si_nH_{2n+2}, $n \geq 3$) were not detected. The a-Si:H thin films obtained with SF_6 sensitized CO_2 laser CVD at a substrate temperature of 300°C contained about 10 at% of hydrogen compared to 20–25 at% with conventional CO_2 laser CVD. Furthermore, the maximum photoconductivity of the a-Si:H thin films could be raised from below $10^{-5}\ \Omega^{-1}\cdot cm^{-1}$ to $10^{-4}\ \Omega^{-1}\cdot cm^{-1}$ using SF_6 sensitization.

1. Introduction

So far, continuous wave (CW) CO_2 laser CVD of amorphous hydrogenated silicon (a-Si:H) from SiH_4 or Si_2H_6 was performed with pure silanes or silanes diluted with inert gases like He and Ar [1–5]. Under these conditions the applied silane gas simultaneously serves as absorbing and as chemically reactive species. On the one hand, a high silane concentration is a prerequisite for efficient CO_2 laser absorption but, on the other hand, favours unwanted gas-phase nucleation (powder formation). There remains a narrow window of acceptable processing parameters between low absorption (slow film formation) and powder formation. The a-Si:H thin films obtained by CO_2 laser CVD so far suffer from their high hydrogen content and show relatively poor electronic quality, especially with respect to their photoconductivity, defect density, and Urbach energy [6]. In this paper sensitized CO_2 laser CVD is applied using sulfur hexafluoride (SF_6), a strong absorber of 10.6 μm CO_2 laser light and a chemically inert compound. Thus CO_2 laser absorption is separate from the concentration of the chemically reactive silicon hydride species in the gas phase which can now be optimized independently.

2. Experimental

The two apparatuses for a-Si:H thin film deposition by CW CO_2 laser CVD were described previously [5,7]. Briefly, the horizontal CO_2 laser beam entered and left the deposition reactors through KCl windows. It passed the substrates (Corning glass 7059 or Si⟨111⟩) in a parallel configuration at a distance of 8 mm between the laser beam axis and the vertical substrate surface. The inlet for the SF_6/SiH_4 processing gas mixture was located in the centre of the CVD reactor while each window was flushed during deposition with He or Ar. The laser power transmitted through the reactor was measured and stabilised to ±1% using a home-built power control unit. All gas fluxes were regulated by mass flow controllers. The total processing gas pressure was measured by a capacitance manometer and kept constant by means of a regulated throttle valve.

One deposition system was equipped with an in situ mass analysis: the processing gas was probed directly in the substrate plane through a

* To whom correspondence should be addressed.

pinhole in the heating block and passed via an effusive molecular beam into the ion source of a quadrupole mass spectrometer (QMS). Its sensitivities for Ar, SF_6, SiH_4, and Si_2H_6 were calibrated. Practically perfect mixing of the processing and flushing gases was observed in front of the nozzle.

The thickness of the a-Si:H films was measured using a profilometer (Dektak 3030 ST). Thin films on silicon substrates were further investigated by Fourier transform infrared spectrometry (FTIR, Bruker IFF 88). Their integral absorption between 530 and 750 cm^{-1} normalized to the film thickness served as a measure of their total hydrogen content. The related conversion factor was derived from absolute hydrogen depth profiles obtained by nuclear p-γ reaction with some selected samples of the same set. With a-Si:H thin films on Corning glass the electrical conductivities at room temperature were determined in the dark and under approximate AM 1 illumination (100 mW/cm^2) using a highly sensitive electrometer (Keithley model 602).

3. Results

The absorption of 10.6 μm CO_2 laser light (P(20) line of the 00°1–10°0 transition) by SF_6 is much more efficient than that by SiH_4 (fig. 1). The data were recorded at two laser power (intensity) levels of $I_0 = 100$ and 160 W measured behind the empty cell ($l = 38$ cm), respectively. The laser beam had a diameter of 10 mm and a close to rectangular intensity profile. The laser beam portion I_A absorbed by the gas is normalized relative to I_0 (ordinate scale). With both gases the absorption behaviour represented by the I_A/I_0 values as a function of pressure deviates from Beer's law (fig. 1).

A gas mixture of 0.49 mbar SiH_4, 0.49 mbar SF_6, and 1.2 mbar Ar was irradiated at various CO_2 laser power levels up to 190 W while keeping the substrate block temperature constant at 300°C. At every level of laser power the stationary partial pressure of SiH_4 was determined using the QMS (fig. 2a). Its value decreases monotonically with laser power above a threshold of about

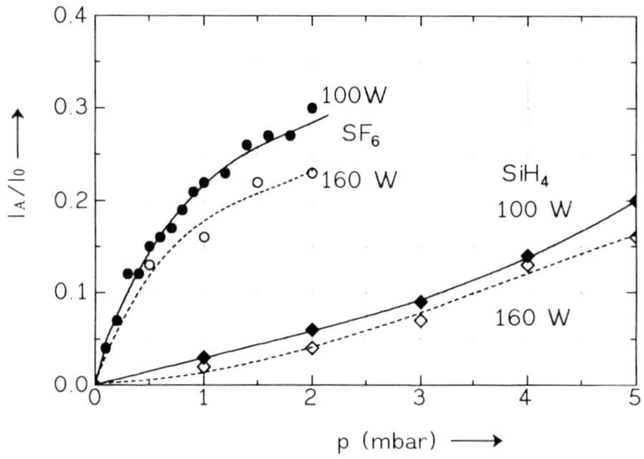

Fig. 1. Relative CO_2 laser absorption I_A/I_0 by SF_6 and SiH_4 measured at two levels of incident laser power as a function of gas pressure p (cf. text).

50 W. The same power threshold is observed with the formation of disilane (fig. 2b). In contrast to the monotonic SiH_4 decrease the stationary concentration of Si_2H_6 passes a maximum at about

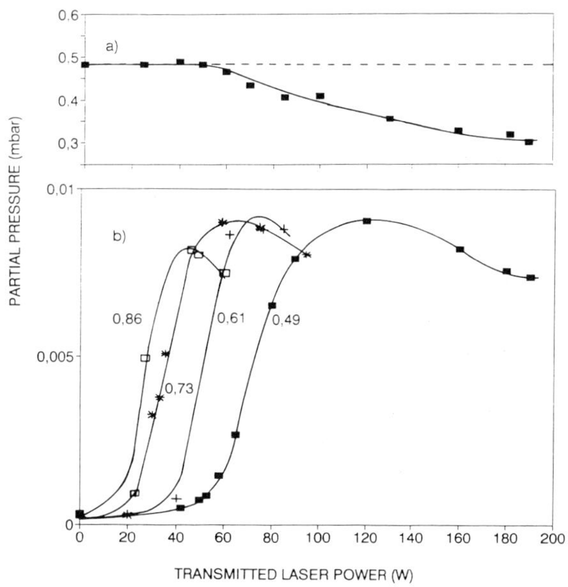

Fig. 2. Stationary partial pressures of (a) SiH_4 and (b) Si_2H_6 as a function of CO_2 laser power measured behind the reactor after passing the processing gas mixture of 0.49 mbar SiH_4, 1.2 mbar Ar, and 0.49 mbar SF_6; in addition similar mixtures in (b) contained 0.61, 0.73, and 0.86 mbar SF_6, respectively.

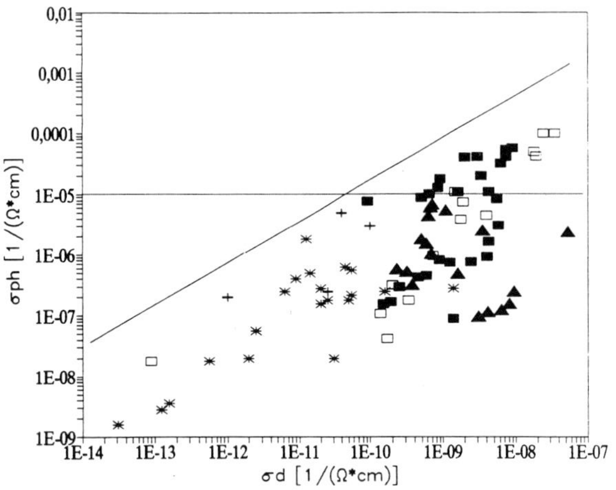

Fig. 3. Electrical photoconductivity σ_{ph} of a-Si:H thin films plotted versus their dark conductivity σ_d for samples obtained via conventional CW CO_2 laser CVD by Meunier et al. (+, [2]), Hess et al. (∗, [3]), Stafast et al. (▲, unpublished results) and via SF_6 sensitized CO_2 laser CVD (□: small reactor with gas analysis, ■: large reactor, this work) in comparison with values of a-Si:H films from plasma CVD (straight diagonal line, [8]).

120 and at 190 W amounts to about 80% of its maximum value.

Fig. 2b displays the stationary Si_2H_6 pressure as a function of CO_2 laser power when applying various SF_6 pressures under otherwise identical reaction conditions. Obviously the onset of Si_2H_6 formation and its maximum pressure are shifted towards lower laser power upon increasing the SF_6 pressure. The behaviour of the SiH_4 partial pressure in these experiments was found equivalent to that with 0.49 mbar SF_6 shown in fig. 2a. Furthermore, Si_3H_8 or higher silanes could not be detected during any of these experiments. Without SF_6 under otherwise identical reaction conditions no decomposition of SiH_4 and no formation of Si_2H_6 could be detected even at the highest laser power.

The electrical conductivities of the a-Si:H thin films obtained by SF_6 sensitized CO_2 laser CVD are summarized in fig. 3. Their photo conductivity σ_{ph} clearly exceeds 10^{-5} and some data are found in the 10^{-4} $\Omega^{-1} \cdot cm^{-1}$ region. In these experiments 0.01 mbar SiH_4, 2.16 mbar SF_6 and 0.83 mbar He were applied. For one of these a-Si:H thin films the optical bandgap was determined and it amounted to 1.75 eV. The hydrogen content was found to be about 10 at% or slightly below.

4. Discussion

One promising technical application of a-Si:H thin films is solar cell fabrication. For this purpose a-Si:H with a high σ_{ph}/σ_d ratio and a high σ_{ph} value is required. Furthermore, this film quality should be achieved at a low film deposition temperature to avoid ion diffusion during solar cell fabrication. Fulfillment of these requirements is a main goal of the present work on CW CO_2 laser CVD, which appears to be a promising method for "soft" deposition of high quality interface layers.

Looking at the substrate temperature dependence of the a-Si:H conductivities obtained via pulsed and CW CO_2 laser CVD from SiH_4 (and buffer gas), the maximum σ_{ph} values are found at 300 and 400°C, respectively [6]. This 100°C temperature difference appears astonishing in view of identical primary reaction channels following both routes of SiH_4 activation [9]. It may, however, be due to a higher maximum gas temperature achieved in pulsed CO_2 laser CVD compared to that in the CW mode. While heat conduction is negligible on the short microsecond time scale of pulsed activation, CW laser induced gas heating has to compete with heat flow to the cold surroundings. Raising the processing gas temperature by increasing the laser power and/or the SiH_4 pressure leads, however, to unwanted powder formation. SF_6 sensitization, on the other hand, can enhance the laser absorption (fig. 1) and at the same time allows to reduce the SiH_4 pressure. Thus it has affected the conditions and results of cw CO_2 laser CVD with respect to the total pressure, the partial silicon hydride pressures, and the hydrogen content as well as the electrical conductivities of the a-Si:H thin films.

With regard to the processing gas, the total pressure could be reduced from previously 5 [5–7] to 1–3 mbar. The SiH_4 partial pressure was lowered down to 0.01 compared to 1–2 mbar in conventional cw CO_2 laser CVD in the same setup. With SF_6 sensitization the conversion of the SiH_4 educt compound was pushed to a high degree. Trisilane and higher polysilanes are no longer present even at 300°C substrate temperature whereas polysilanes up to pentasilane were found in previous non-sensitized experiments even at 400°C [7]. The absence of polysilane as well as the decrease in Si_2H_6 at high laser power (fig. 2) is consistent with decomposition of these silicon hydrides at a high processing gas temperature.

Large changes due to SF_6 sensitization are also found in the a-Si:H thin films deposited at 300°C. Their hydrogen content now is of the order of 10 compared to previous 20–25 at%. The photo conductivity of the order of 10^{-4} (previously below 10^{-5} $\Omega^{-1} \cdot cm^{-1}$, fig. 3) was obtained with the lowest SiH_4 partial pressure applied in these experiments. A plot of σ_{ph} versus σ_d (fig. 3) appears to be a convenient data presentation [8] with respect to the above requirements for solar cell material because σ_{ph} and σ_{ph}/σ_d are documented simultaneously. The data of a-Si:H from plasma CVD under optimized conditions [8] may serve as a reference for high electronic film quality. Following this guideline the need for further improvements in sensitized CO_2 laser CVD is evident. Additional thin film diagnostics like determination of the IR absorption, defect density, Urbach energy, and possible SF_6 contaminations are presently under investigation.

Summarizing, the high hydrogen content of a-Si:H thin films obtained by CW CO_2 laser CVD below 400°C so far appeared as a main drawback to high film quality. Applying SF_6 sensitization to CW CO_2 laser CVD allowed to reduce the total pressure and the partial silicon hydride pressures. As a result, a-Si:H thin films with considerably less hydrogen and relatively high photoconductivity values about 10^{-4} $\Omega^{-1} \cdot cm^{-1}$ were obtained. In addition, sensitization opens up new possibilities of thin film deposition because laser absorption by the film forming educt compound(s) is no longer required.

Acknowledgements

Funding of this work by the German Ministry of Research and Technology (BMFT) under contract number 0328965 A is gratefully acknowledged. Thanks are due to Professor Bethge and Dr. Baumann (IKF, University of Frankfurt) for their nuclear p-γ determination of the hydrogen depth profiles. The authors thank Dr. Richter and Mr. Krüger for measuring electronic film properties and FTIR spectra, respectively. Technical assistance by K.-D. Lühmann and W. Klück is also gratefully acknowledged.

References

[1] I. Gianinoni and M. Musci, J. Non-Cryst. Solids 77/78 (1985) 743.

[2] M. Meunier, J.H. Flint, J.S. Haggerty and D. Adler, J. Appl. Phys. 62 (1987) 2812; 2822.

[3] K. Hesch, H. Karstens and P. Hess, Thin Solid Films 218 (1992) 29.

[4] P. Gonzalez, M.D. Fernandez, B. Leon and M. Perez-Amor, Proc. SPIE 1022 (1988) 97.

[5] E. Golusda, R. Lange, K.-D. Lühmann, G. Mollekopf, M. Wacker and H. Stafast, Appl. Surf. Sci. 54 (1992) 30.

[6] E. Golusda, K.-D. Lühmann, G. Mollekopf, H. Stafast and M. Wacker, Ber. Bunsenges. Phys. Chem. 95 (1991) 1414.

[7] E. Golusda, K.-D. Lühmann, G. Mollekopf, M. Wacker and H. Stafast, Proc. 10th Eur. Photovolt. Solar Energy Conf., 8–12 April 1991, Lisbon, Portugal, p. 143.

[8] W. Beyer and B. Hoheisel, Solid State Commun. 47 (1983) 573.

[9] H. Stafast, Appl. Phys. A 45 (1988) 93.

Applied Surface Science 69 (1993) 262–266
North-Holland

applied
surface science

Effects of UV light on the deposition kinetics and optoelectronic properties of a-Si:H films deposited by RF glow discharge

N. Layadi, P. Roca i Cabarrocas, V. Yakovlev and B. Drévillon

Laboratoire de Physique des Interfaces et des Couches Minces, Ecole Polytechnique, F-91128 Palaiseau Cedex, France

Received 2 June 1992; accepted for publication 30 October 1992

We here present the results of a study of the growth kinetics and optoelectronic properties of a-Si:H films deposited on glass substrates by RF glow discharge and submitted to the UV light emitted by an XeCl excimer laser during their growth. The comparison to reference samples deposited under the same plasma conditions but without the UV irradiation shows that the UV light produces an acceleration of the initial stage of the deposition and a small increase of the average deposition rate. The UV light induces an increase of the film porosity and small changes on the optical properties and hydrogen content of the films. The activation of the initial stage of deposition and the increase of the film porosity were not observed either on crystalline silicon or on a-Si:H substrates. Whereas the defect density and the Urbach energy of the films are slightly reduced by the exposure to the UV light, the dark conductivity is strongly increased for the films deposited at low substrate temperature (50°C).

1. Introduction

Hydrogenated amorphous silicon (a-Si:H) is a thin-film semiconductor which has received much attention during the last decade. Indeed, its optoelectronic properties make it a good candidate for many thin-film semiconductor applications (solar cells, thin-film transistors, etc.). As a consequence, a lot of research efforts have been focused on the plasma processes involved in the plasma-enhanced chemical vapor deposition (PECVD) of a-Si:H and on the effects of different external parameters (RF power, flow rate, pressure, and substrate temperature) and intrinsic plasma parameters (ion bombardment and geometry) which determine the optoelectronic properties of the a-Si:H films. However, few studies have addressed the effects of the UV radiation on the film growth processes and optoelectronic properties even though it is obvious that the UV light emitted by the plasma can modify the chemistry of the discharge as well as the chemistry of the a-Si:H surface. In particular, in the a-Si:H deposition from a silane plasma (SiH_x radicals) the growing surface is saturated by hydrogen and the UV photons with energies higher than the Si–H bonding energy (3.34 eV) will be able to break this bond and thus change the hydrogen coverage of the surface. Moreover, the results published so far indicate rather different effects of the UV light when compared to the samples deposited under the same conditions without UV light: (i) enhancement [1] or decrease [2] of the oxygen content; (ii) reduction [3,4] or no effect [5] on the bonded hydrogen content; (iii) increase [4] or no effect [5] on the defect density. These apparent discrepancies could reveal the complexity of the processes involved in the PECVD of a-Si:H assisted by UV light; and thus very different and even opposite effects can be obtained depending on the kind of UV light source used (laser or lamp); the nature of the substrate and its thermal conductivity; and the plasma conditions used for a-Si:H deposition as the UV effects may or may not be masked by the glow discharge.

In this work, reference samples were deposited on various substrates at 50, 100 and 200°C (temperature of the sample surface calibrated with a flat Pt-100 resistor attached to the glass surface)

0169-4332/93/$06.00

and their optoelectronic properties were compared to those of films deposited under the same RF glow-discharge conditions but exposed to the laser radiation during their growth.

2. Experimental

The a-Si : H films were deposited by the decomposition of pure silane in a capacitively coupled RF (13.56 MHz) glow-discharge reactor consisting of two parallel plates of 12 cm diameter and 3.7 cm apart. The plasma was confined between the two electrodes by a grounded stainless steel cylinder equipped with apertures for the UV laser irradiation and the ellipsometry measurements. Prior to deposition the system was pumped down to 10^{-6} Torr by a turbomolecular pump. The ellipsometric angles Ψ and Δ were measured in situ with a UV–visible spectroscopic phase-modulated ellipsometer described elsewhere [6]. For kinetic studies, the photon energy was fixed at 3.5 eV which corresponds to the maximum of the imaginary part of the dielectric function ϵ_2 of a-Si : H and thus gives a high sensitivity to any surface modification. A XeCl excimer laser delivering pulses of 40 ns duration with a maximum energy of 300 mJ/pulse at a repetition rate of 1 Hz was used as a UV source. The pulse energy of the laser at the location of the substrate inside the reactor was measured with a Joule meter. The following deposition conditions were used: silane flow rate: 12 sccm, total pressure 70 mTorr, RF power 2 W and an average laser intensity of 30 mW/cm^2.

The thickness of the films was accurately measured by profilometer measurements and computed from the fit to the spectroscopic measurements. The absorption coefficient down to 0.8 eV was determined by photothermal deflection spectroscopy (PDS) measurements on a-Si : H films deposited on Corning 7059 glass substrates [7]. The Urbach energy, related to the disorder of the film, and the defect density were respectively deduced from the exponential part of the absorption coefficient and the excess subgap absorption according to Amer and Jackson [7]. After PDS measurements, 2 mm apart chromium contacts were thermally evaporated on the a-Si : H films to measure their dark conductivity and activation energies. Crystalline silicon substrates were used for infrared transmission measurements to check the effect of the UV light on the hydrogen bonding and hydrogen content.

3. Results and discussion

In fig. 1 we compare the evolution of the ellipsometric Ψ and Δ angles during the deposition of a-Si : H on Corning 7059 glass substrates at 50°C. We observe that the initial stage of the deposition follows the same trajectory, independently of the UV light. The initial trajectory can be described by the formation of a chemical interface between the glass substrate and the a-Si : H film as previously reported [8,9]. However, there is an increase by a factor of 1.4 of the initial stage of the deposition in the UV-assisted PECVD. This increase is also observed for films deposited at 200°C. In fact, if we compare the deposition at 50 and 200°C without UV we observe that the chemical interaction is faster at 200°C. Thus from the kinetic analysis of the initial deposition process we conclude that the ini-

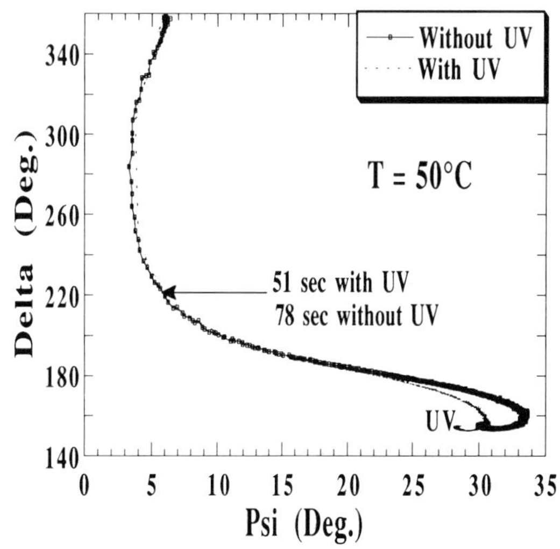

Fig. 1. Real-time ellipsometry examination at 3.5 eV of the deposition at 50°C of a-Si : H on Corning glass substrates with and without UV-assisted PECVD.

tial stage of the deposition of a-Si:H on glass substrate corresponds to the formation of a chemical interface and that this process is modified by the UV light. The UV light speeds up the formation of the interface but does not modify the growth mechanism. Moreover, we find that the deposition rate r_d, deduced from the ratio between the film thickness and the deposition time, is higher for the films deposited under UV light and increases from 0.27 to 0.31 Å/s at 50°C and from 0.36 to 0.45 Å/s at 200°C. We note that as previously reported [10] r_d increases with T_s and that the enhancement of r_d due to the UV light is smaller than the one due to the increase of T_s from 50 to 200°C. Thus, at the initial stage of deposition we observe an increase of the interaction of a-Si:H with the glass substrate for the UV-assisted PECVD.

After the initial stage of deposition, Ψ and Δ converge towards the values of a-Si:H. However, the point of convergence is different for the films deposited with and without UV, which indicates that UV irradiation modifies the bulk properties of a-Si:H. Indeed, the imaginary part of the pseudodielectric function (ϵ_2) determined by spectroscopic ellipsometry measurements (figs. 2 and 3) shows that the UV light produces a decrease of the maximum of ϵ_2 together with a slight shift of the spectra towards higher energy, in particular for the films deposited at 200°C. The effect of the UV light can be described at both temperatures as an increase of the void volume fraction f_v without modification of the surface roughness. At 50°C f_v increases from 0.13 to 0.23 (fig. 2) whereas at 200°C f_v increases from 0.04 to 0.17 (fig. 3). We speculate that this increase of f_v is produced by the sharp increase of the temperature at the surface of the sample which will produce a diffusion of hydrogen which either accumulates inside the film and leads to the formation of voids or diffuses out of the film [11]. One could then expect a large reduction of the hydrogen content of the films deposited by PECVD assisted by UV light. However, from IR measurements we only observe a slight decrease of the hydrogen content and this can be explained by the hydrogenation of the growing film by the plasma. As a consequence, the hydrogen content is determined by the plasma process and the film microstructure by the UV light. We note that the increase of the initial stage of the interaction with the substrate and the modification of the bulk a-Si:H properties are only observed on Corning glass substrates but not on crystalline or a-Si:H substrates (≥ 1000 Å thick a-Si:H films deposited on glass substrates). This reveals the importance of the substrate in the UV light-assisted PECVD.

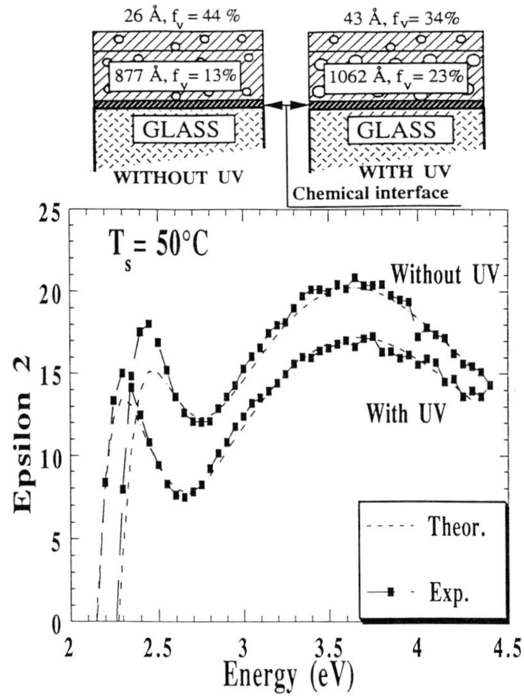

Fig. 2. Energy dependence of the imaginary part of the dielectric function for films deposited with and without UV treatment at 50°C.

Fig. 4 shows the absorption coefficient of a-Si:H films deposited at 200°C with and without UV-assisted deposition. We note that at high energy and in the exponential region the absorption coefficient is the same for both films. This indicates that neither the optical gap nor the disorder of the film are strongly affected by the UV irradiation. The only difference appears in the low-energy range, which corresponds to the subgap absorption and indicates that the defect density is slightly smaller in the films deposited

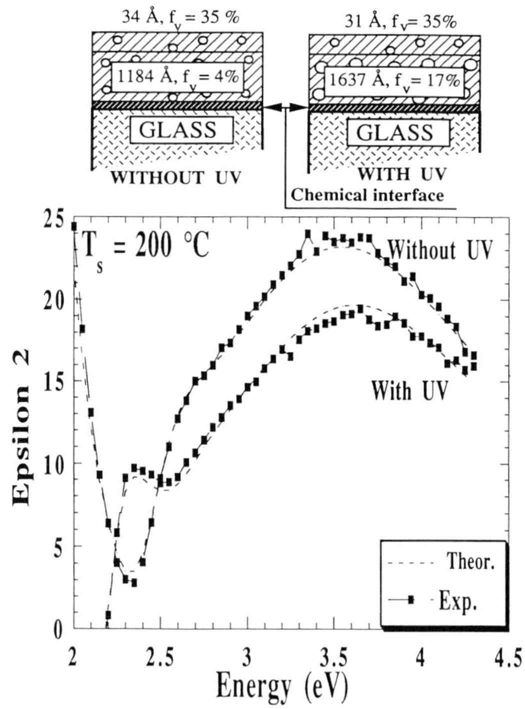

Fig. 3. Energy dependence of the imaginary part of the dielectric function for films deposited with and without UV treatment at 200°C.

with UV. The inverse slope of the exponential part gives a characteristic energy $E_0 = 54$ meV and the integration of the excess absorption at low energy gives a defect density of $\sim(2\text{–}3) \times 10^{16}$ cm^{-3}. The same effect of the UV light was observed for a-Si:H films deposited at 50, 100 and 200°C.

The slight effect of the UV light on the defect density is contrasted by the strong changes in the temperature dependence of the dark conductivity of a-Si:H films deposited at 50°C. Indeed, there is an increase of the conductivity by four orders of magnitude and a decrease of its activation energy (fig. 5). Such strong increase of the dark conductivity can be attributed either to the enhancement of the incorporation of impurities by the UV light or to the change of the microstructure of the film. Now, even though it is well known that the increase of the impurity content, nitrogen and oxygen in particular, results in a decrease of the film conductivity, we point out that in order to obtain such a high value of the dark conductivity, it would be necessary to incorporate of the order of 1% of oxygen or nitrogen [12], i.e. to increase the impurity content by at least two orders of magnitude. Moreover, this effect appears for the films deposited at 50°C but

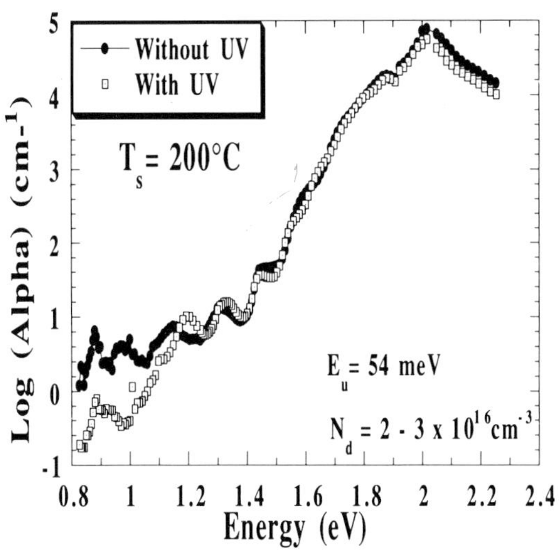

Fig. 4. Energy dependence of the absorption coefficient of a-Si:H films deposited with and without UV treatment at 200°C.

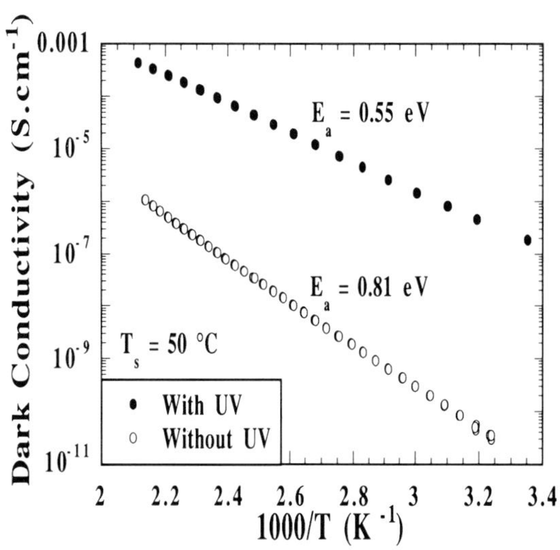

Fig. 5. Temperature dependence of the dark conductivity of a-Si:H films deposited with and without UV treatment at 50°C.

not at 100 and 200°C. We propose that the increase of the conductivity is correlated to the increase of the void fraction of the films and to the development of some kind of microstructure which can be deduced as a non-microcrystalline phase from the spectroscopic ellipsometry measurements. However, at 200°C there is an increase of f_v but no drastic effect on the dark conductivity; further measurements are necessary to explain these differences.

4. Conclusion

We have compared the growth kinetics and optoelectronic properties of a-Si:H films deposited on Corning 7059 glass substrates by PECVD of pure silane to those of a-Si:H films deposited by PECVD of pure silane assisted by the UV light emitted by an excimer XeCl laser. The in-situ ellipsometry study of the deposition kinetics has revealed that the UV light accelerates the initial interaction of the a-Si:H film with the glass substrate and increases the void volume fraction of the films. These effects are not observed on crystalline and a-Si:H substrates which clearly indicates the strong influence of the substrate. The density of states and the dark conductivity of the films deposited at 100 and 200°C are slightly affected by the UV light. However, the dark conductivity of the films deposited at 50°C increases by four orders of magnitude, which is correlated to the large increase of the void volume fraction of the film (23%). As a conclusion, the PECVD of a-Si:H assisted by UV light is a complex process as the UV light affects both the plasma and surface chemistry, and the effects of the UV light may be masked by those of the plasma.

Acknowledgements

We thank M. Stchakovsky, B. Schlayen, J.C. Vitu and Y. Josserand from ISA Jobin-Yvon for their help in the in-situ ellipsometry set-up.

References

[1] R. Martins, N. Carvalho, E. Fortunato, A. Maçarico, M. Santos, I. Baia, M. Viera and L. Guimaraes, J. Non-Cryst. Solids 97/98 (1987) 1399.
[2] Yao-Hua Yang, Thèse de doctorat de l'Institut National Polytechnique de Grenoble, 1987 (in French).
[3] M. Yamanaka, Y. Hayashi and I. Sakata, Jpn. J. Appl. Phys. 29 (1990) L217.
[4] S. Koynov, A. Toneva and K. Kougia, J. Non-Cryst. Solids 137/138 (1991) 685.
[5] A. Suzuki, Y. Toyoshima, P. McElney and A. Matsuda, Jpn. J. Appl. Phys. 30 (1991) L790.
[6] B. Drévillon, J.Y. Parey, M. Stchakovsky, R. Benferhat, Y. Josserand and B. Schlayen, Proc. SPIE 1888 (1989) 174.
[7] N.M. Amer and W.B. Jackson, Semicond. Semimet. 21B (1984) 83.
[8] A.M. Antoine and B. Drévillon, J. Appl. Phys. 63 (1988) 360.
[9] N. Blayo and B. Drévillon, Appl. Phys. Lett. 57 (1990) 786.
[10] P. Roca i Cabarrocas, J.Z. Liu, H.R. Park, A. Maruyama and S. Wagner, J. Non-Cryst. Solids 114 (1989) 190.
[11] E.L. Mathé, A. Naudon, M. Elliq, E. Fogarassy and S. de Unamuno, Appl. Surf. Sci. 54 (1992) 392.
[12] A. Morimoto, M. Matsumoto, M. Yoshita, M. Kumeda and T. Shimizu, Appl. Phys. Lett. 59 (1991) 2130.

Applied Surface Science 69 (1993) 267–271
North-Holland

applied
surface science

Laser-induced CVD of titanium diboride and the influence of atomic hydrogen

J. Elders and J.D.W. van Voorst

Laboratory for Physical Chemistry, University of Amsterdam, Nieuwe Achtergracht 127, 1018 ZW Amsterdam, Netherlands

Received 2 June 1992, accepted for publication 30 Oktober 1992

CO_2 laser-induced chemical vapor deposition of titanium diboride (TiB_2) has been performed on mullite substrates by the hydrogen reduction of titanium tetrachloride and boron trichloride. The influence of the hydrogen concentration on the morphology, growth rate and nucleation rate will be discussed. These results suggest that atomic hydrogen may play an important role in the deposition process. An atomic hydrogen source, a hot filament, has therefore been constructed. The results of the hot filament-enhanced depositions on thermally oxidized silicon will be presented.

1. Introduction

Titanium diboride (TiB_2) has several promising properties, such as the high hardness, the excellent resistance against corrosion and chemical attack, the resistance against wear and erosion, the high melting point and the low bulk resistivity. These properties make titanium diboride a promising candidate for applications in rocket nozzles, combustion engines, jet engine turbines, electrical contacts, thermocouple protection tubes, cathodes in aluminum reduction cells, fusion devices requiring low Z refractory materials [1], integrated circuit metallization [2] and as a diffusion barrier between silicon and aluminum [3].

In this paper the CO_2 laser-induced chemical vapor deposition process of titanium diboride according to eq. (1) will be described,

$$TiCl_4 + 2\ BCl_3 + 5H_2 \rightarrow TiB_2 + 10\ HCl. \qquad (1)$$

The CO_2 laser heats the substrate locally which induces a surface reaction [4]. The CO_2 laser-induced deposition process is performed on mullite substrates ($3Al_2O_3 \cdot SiO_2$) at 20 mbar.

Results of the growth rate as a function of the temperature and the hydrogen–chlorine ratio will be presented. As will be shown, the growth rates and the morphologies depend to a large extent on the hydrogen concentration. Photolytically enhanced deposition of titanium diboride [4], and the influence of the hydrogen concentration suggest that atomic hydrogen may play an important role in the deposition process. An atomic hydrogen source, a hot filament, has therefore been constructed in order to study the influence of atomic hydrogen. The hot filament-enhanced deposition process is performed on thermally oxidized silicon substrates at 1 mbar. Growth rates, morphologies, wavelength dispersive X-ray (WDX) analyses and X-ray diffraction (XRD) results will be presented.

2. Experimental

The experimental setup has been described elsewhere [4–7]. In short, it comprises a reaction chamber (29 cm high and 26 cm in diameter), a pumping unit, a CW CO_2 laser and a reactant gas supply system. The reactant gas supplies are regulated by mass flow controllers (ASM). For further information on the CO_2 laser, pumping unit and gas flow system see e.g. ref. [4]. Typical deposition conditions are 20 sccm $TiCl_4$, 100 sccm BCl_3, 1000 sccm H_2, 600 sccm Ar, 20 mbar and temperatures varying from 700 to 1050°C.

The temperature of the heated substrate is measured by a two-color infrared temperature meter (Ircon, SRL-14C05) and a disappearing filament pyrometer (Leeds & Northrup, 8633-c). The two-color infrared temperature meter is used in a feed-back loop in order to stabilize the substrate temperature.

The thickness of the mullite substrates varies from 2–4 mm, the coating thickness has therefore been determined by breaking the samples and studying the fracture surface by the scanning electron microscope (SEM). The thickness is defined as the maximum height of the deposited film above the substrate surface. An error is introduced by this method when comparing coatings of which the roughness and the density differ. Unlike the mullite substrates, the thickness of the thermally oxidized silicon substrates is uniform (i.e. 700 μm thick with an oxide layer of 400 nm). For this reason, the coating thickness of the depositions on this substrate can be measured by a micrometer. However, since on these substrates deposition at the back of the substrate often occurred, the accuracy of the thickness measurement is decreased. Since the initial goal is the qualitative study of the role of atomic hydrogen, the order of magnitude will be given as a rough indication of the growth rate differences.

Hot tungsten filaments varying from 1–2 cm in length and a diameter of 0.1 mm have been used to dissociate molecular hydrogen. The recombination of atomic hydrogen into molecular hydrogen in the gas phase occurs via a three-body collision. It is estimated that the atomic hydrogen atoms can diffuse several centimeters before recombining [8]. Nevertheless, the distance between the source and the substrate has been minimized and was never larger than 2 cm. A cap has been constructed to prevent the detection of the radiation of the hot filament by the infrared temperature meter. Above a filament temperature of 1800°C the atomic hydrogen production increases strongly with increasing temperature [9]. The temperature of the tungsten filament has been measured with a disappearing filament pyrometer. For these experiments the pressure has been reduced to 1 mbar to minimize recombination of the atomic hydrogen. At this pressure, the growth rate is much lower than at 20 mbar [6]. The gas flows have been reduced because of the limited pumping capacity in order to obtain this pressure. In all the depositions on thermally oxidized silicon, the reactant gas flows were 8 sccm $TiCl_4$, 30 sccm BCl_3 and 500 sccm H_2 and the total argon flow along the windows was 110 ± 20 sccm. Before mounting the substrates, the samples were etched in 1% HF, rinsed in demineralized water and finally rinsed in isopropanol.

The hot filament supplied additional heating of the substrate: in these experiments the average CO_2 laser power was less than without the hot filament. However, the hot filament caused large temperature oscillations as well.

3. Results

3.1. CO_2 laser-induced CVD

The activation energy of the CO_2 laser-induced deposition process on mullite is in the reaction-limited regime (i.e. up to 1080°C) 230 ± 37 kJ/mol and in the mass-diffusion-limited regime 36 ± 23 kJ/mol. These regimes are clearly distinguished by their morphologies as well. The growth rates vary from 0.4 μm/min at 870°C up to 8.1 μm/min at 1050°C and 31.8 μm/min at 1150°C. These results have been obtained at 20 mbar with 20 sccm $TiCl_4$, 100 sccm BCl_3, 1000 sccm H_2 and 600 sccm argon (further named center point deposition conditions). However, when varying the hydrogen concentration, the growth rates as well as the morphology change drastically. In table 1 the growth rate as a function of the hydrogen concentration is given and is compared to the growth rate obtained at center point deposition conditions. Table 1 shows that the growth rate is highly sensitive to the hydrogen concentration and that this dependence changes with temperature. When determining the apparent activation energy, it will change with changing hydrogen concentration. As has been reported [4], the estimated apparent activation energy is over 600 kJ/mol at 20 sccm $TiCl_4$, 100 sccm BCl_3 and 500 sccm H_2, 413 kJ/mol at 775 sccm H_2, 230 ± 37 kJ/mol at 1000 sccm H_2 and 93 kJ/mol at 3400 sccm H_2. Since the density of the coatings varies, the observed trend is of main interest, not the absolute values. The morphologies vary from dendritic at low hydrogen concentrations, to smooth at high hydrogen concentrations (see e.g. ref. [4]). At high hydrogen concentrations (i.e. 3400 sccm H_2, other parameters at center point deposition conditions) the deposits are amorphous. In fact, the wavelength-dispersive X-ray analysis of an amorphous coating indicated that the coating is overstoichiometric in boron and that the boron content in the coating varied as a function of the location. At center point deposition conditions the deposits did generate an X-ray diffraction signal.

Table 1
The growth rate (μm/min) as a function of the reactant gas ratio

No.	Gas flow (sccm) ($TiCl_4:BCl_3:H_2$)	T (°C)	Growth rate (μm/min)	Growth rate (μm/min) at center point
1	20:100:500	1250	34	–
2	20:100:500	1135	0	26.2
3	20:100:730	980	0	2.7
4	20:100:750	1025	0	5.6
5	20:100:750	1040	0.1	7.0
6	20:100:750	1050	0	8.1
7	20:100:750	1150	14.2	31.8
8	20:100:750	1000	0.1	3.8
9	20:100:800	1100	12.7	16.5
10	20:100:800	1100	0.3	16.5
11	20:100:1530	870	0.1	0.4
12	20:100:1500	1020	27	5.2
13	20:100:2000	1090 ± 60	26.3	14.4
14	10:50:1010	1005	24.1	4.1
15	10:50:1010	1030 ± 25	25.8	6.0
16	10:50:1520	1025 ± 25	3.7	5.6
17	10:50:1520	1010	4.1	4.4
18	12:50:1700	925	0.7	1.0
19	12:50:1700	1060	1.5	10.1

The argon flow was constant (600 sccm), the pressure was 20 mbar and the substrate was mullite. The table at the right side shows the growth rate at the center point deposition conditions as determined from the Arrhenius plot. The growth rate at 1250°C (no. 1) at the center point deposition conditions is not given because it has not been determined experimentally.

Table 2
An overview of depositions on thermally oxidized silicon

No.	$T_{substrate}$ (°C)	Deposition time (min)	$T_{filament}$ (°C)	Growth rate (μm/min)
1	1185 ± 65	20	RT	4.8
2	1000 ± 30	34	RT	1.5 (fig. 2)
3	1040	10	RT	–
4	1005	10	RT	–
5	975 ± 15	30	RT	0.7
6	975 ± 15	60	RT	0.7
7	890 ± 40	10	1900	0.6
8	985 ± 85	10	1945	3
9	1035 ± 85	4	1945	1.6 (fig. 1)

CO_2 laser-induced and hot filament-enhanced results are shown. The deposition conditions are 1 mbar, 8 sccm $TiCl_4$, 30 sccm BCl_3, 500 sccm H_2 and 110 ± 20 sccm argon. A rough indication of the growth rate is given, as determined with a micrometer. Because of a large degree of uncertainties, e.g. originating from the fact that at higher temperatures deposition at the back of the substrate occurred, the numbers should be seen as estimated growth rates. RT denotes room temperature.

At high hydrogen concentration (i.e. 3400 sccm H_2), a linear extrapolation of the coating thickness versus time plot shows that the line crosses the origin, in contrast to the results obtained at center point deposition conditions [7].

3.2. Hot filament-induced CVD

As has been mentioned, in the hot filament-enhanced depositions the pressure has been re-

Fig. 1. SEM photograph of a hot filament-induced CVD process. The deposition time was four min, the substrate temperature was 1035 and the filament temperature was 1945°C (no. 9 of table 2). The bar indicates 9.35 μm.

Fig. 2. SEM photograph of CO_2 laser-induced CVD. The deposition time was 34 min and the substrate temperature was $1000 \pm 30°C$ (no. 2 of table 2). The bar indicates 9.01 μm.

duced to 1 mbar. Since the growth rate at 1 is lower than at 20 mbar [6] and since the reactant gas flow is different, the growth rates cannot be compared with those obtained in the CO_2 laser-induced deposition process. CO_2 laser-induced depositions have therefore been performed on the same substrate at the same deposition parameters. The results of these depositions are presented in table 2. As can be observed, the depositions performed with a hot filament generate higher growth rates compared to those without filament. Wavelength dispersive X-ray analysis of a hot filament-enhanced deposit resulted in stoichiometric titanium diboride throughout the whole deposit, unlike the CO_2 laser-induced deposit. In the CO_2 laser-induced deposit the coating was overstoichiometric in boron at the edge of the deposit, just like the deposits on mullite. The morphologies of the hot filament-enhanced depositions is different as well. As is shown in the figures, the morphology of the a hot filament-enhanced deposit (fig. 1) is smooth unlike the morphology of a deposit without hot filament (fig. 2) while the growth rates were similar.

4. Discussion

In order to grow a coating on a substrate of a different material, allonucleation should first occur, i.e. the deposition of titanium diboride onto a different material. Before the whole substrate surface is filled in with the titanium diboride, allonucleation may occur on an exposed substrate surface and isonucleation may occur on the deposited material itself. At the same time the crystal lattice may be filled in by boron and titanium atoms, resulting in crystal structures which may become macroscopically observable. When the titanium and boron chlorides arrive on the substrate surface, the chlorides will be stripped by the hydrogen and, depending on the deposition parameters, the reactants will migrate over the substrate surface until an active site is reached. However, the extent to which allonucleation, isonucleation and crystal growth appear, depends on the deposition parameters.

In the high temperature regime, the growth rates as a function of the hydrogen concentration are of the same order. However, in the low temperature regime the growth rate as a function of the hydrogen concentration varies drastically: at a low hydrogen concentration there is no titanium diboride growth observed, indicating that the nucleation is hampered. At low hydrogen concentration, a linear extrapolation of the coating thickness versus time plot towards the origin indicates that the line does not cross the origin [7]. This also supports the view that nuclei formation is hampered.

At high hydrogen concentrations a different behavior is observed in the low temperature regime: smooth coatings are observed and at sufficiently high hydrogen concentration, these coatings are amorphous, indicating that the nucleation rate is very high. If, in the high hydrogen concentration limit, the coating thickness versus time plot is extrapolated, the line will cross the origin [7], implying that the growth starts immediately after the beginning of the deposition process.

The above-mentioned trend in the activation energy as a function of the hydrogen concentration can thus be understood in terms of a large influence of the hydrogen concentration on the nucleation rate.

The growth rates obtained with the hot filament-enhanced CVD process are higher than obtained with the CO_2 laser-induced CVD process and the growth rate seems to increase with increasing filament temperature (cf. table 2) just like the atomic hydrogen concentration [2]. The atomic hydrogen is thus responsible for the increased growth rate. Besides an increase of the growth rate, the hot filament-enhanced deposition process resulted in stoichiometric titanium diboride throughout the whole deposit, as WDX analysis indicated.

As has been mentioned before, the morphologies of the hot filament-enhanced CVD do not show dendritic structures, (cf. fig. 2) while the growth rates are high. At these growth rates, morphologies characteristic for a diffusion-limited mechanism are observed for the laser-induced deposition of titanium diboride, as is illustrated in fig. 1. The creation of atomic hydrogen is thus of utmost importance for the nucleation of the titanium diboride deposition process.

5. Conclusion

The growth rates as well as the morphologies of the CO_2 laser-induced CVD experiments indicated that the hydrogen concentration plays an important role in the nucleation process. In the low temperature regime the growth rate appeared to be more sensitive to the hydrogen concentration than in the high temperature regime. At sufficiently high hydrogen concentrations the deposits were smooth and amorphous, indicating the importance of the hydrogen concentration for the nucleation rate of the deposition process. In order to study the influence of atomic hydrogen, an atomic hydrogen source, a hot filament, has been introduced. The hot filament-enhanced depositions showed that the introduction of atomic hydrogen generates higher growth rates, smooth morphologies and stoichiometric titanium diboride throughout the whole deposit. Atomic hydrogen is therefore important for the nucleation of the titanium diboride deposition process.

References

[1] C.I.H. Ashby, J. Nucl. Mater. 122/123 (1984) 1406.
[2] J.R. Shappirio and J.J. Finnegan, Thin Solid Films 107 (1983) 81.
[3] T. Larsson, H.O. Blom, S. Berg and M. Östling, Thin Solid Films 172 (1989) 133.
[4] J. Elders and J.D.W. van Voorst, Appl. Surf. Sci. 52 (1992) 135.
[5] J. Elders, D. Bebelaar and J.D.W. van Voorst, Appl. Surf. Sci. 46 (1990) 215.
[6] J. Elders, P.A. Quist, B. Rooswijk, J.D.W. van Voorst and J. van Nieuwkoop, Surf. Coating Technol. 45 (1991) 105.
[7] J. Elders, Laser-Induced Chemical Vapor Deposition of Titanium Diboride, Doctoral Thesis, University of Amsterdam, Amsterdam, 1992.
[8] F.G. Celii and J.E. Butle, Appl. Phys. Lett. 54 (1989) 1031.
[9] F. Jansen, I. Chen and M.A. Machonkin, J. Appl. Phys. 66 (1989) 5749.

Applied Surface Science 69 (1993) 272–276
North-Holland

applied surface science

Laser induced fluorescence spectroscopy of WF_6/M ($M = H_2$, Ar) gas mixtures during LCVD of W at a high laser energy density

P. Heszler *[,1], P. Mogyorósi [2] and J.O. Carlsson

Thin Film and Surface Chemistry Group, Department of Chemistry, Uppsala University, P.O. Box 531, S-751 21 Uppsala, Sweden

Received 2 June 1992; accepted for publication 30 October 1992

ArF excimer laser induced fluorescence spectroscopy has been performed for WF_6, and for gas mixtures of WF_6/H_2 and WF_6/Ar at a high (~ 1 J/cm^2) laser fluence. The recorded spectra have line and band structures, which originate from both W atoms and W ions produced after photolysis of WF_6. Light was also emitted for a wavelength shorter than the laser excitation wavelength. This was due to the excitation of and the emission from the W ions. There were no differences between the spectra obtained from pure WF_6 and those obtained from the WF_6/Ar mixture. However, using a WF_6/H_2 gas mixture, the intensity of the scattered laser light increased significantly. This is explained by W cluster formation in the gas phase which was facilitated by the presence of H_2. The intensity of a characteristic W line was measured for different total pressures and different WF_6 partial pressures for the WF_6/Ar mixture. Two linear regimes could be observed in both the pressure dependence curves. For lower pressures (first part of the curve) the slope of the line was steep, while the second part (higher pressures) was less steep. This observation might be explained by the formation of W atoms in the vapour at low pressures, while higher pressures yielded such a high W concentration in the vapour that W clusters were nucleated. The intensity of a characteristic W line had a strong nonlinear dependence on the laser fluence. Model calculation was performed to explain this nonlinearity and the agreement with the experimental results was good.

1. Introduction

Among the promising applications of the pulsed ArF excimer laser is one assisting the CVD of tungsten from WF_6 (LCVD). The main advantages of applying the laser technique are to increase the W deposition rate and to decrease the deposition temperature of CVD [1,2].

Light emission occurs during ArF LCVD of W [3]. Using a WF_6/H_2/Ar gas mixture, ~ 100 mJ/cm^2 laser fluence and a repetition rate of at least 5 Hz, a broad band spectrum (FWHM ~ 300 nm, white colour, by eyes), which originates from excited W clusters with a size of ~ 15 nm, can be observed [4]. By focussing the laser beam to ~ 1 J/cm^2 in laser fluence bluish light was generated and in addition to the broad band spectrum discrete lines, which mostly could be attributed to W, was obtained. However, several lines in the spectrum could not be identified. The bluish light was also observed for the WF_6/Ar gas mixture. Production of tungsten as well as W^+ during ArF excimer laser irradiation of WF_6 has been detected by using mass spectrometry [5].

Spectroscopic investigations can give useful information on reaction pathways in the vapour during the deposition processes. In this paper the origin of the line spectrum was studied at a high laser fluence (~ 1 J/cm^2) for different gas mixtures (pure WF_6, WF_6/Ar and WF_6/H_2). We used only gas mixtures of two components to avoid the broad band radiation [3]. In addition the influence of pressure and laser fluence, respectively on the emitted light intensity was examined. Finally model calculation for the emitted light intensity was performed on the basis of rate equations.

* To whom all correspondence should be addressed.
[1] On leave from the Research Group of Laser Physics of the Hungarian Academy of Sciences.
[2] Present address: Laser and Vacuum Sci., P.O. Box 1191 H-6701 Szeged, Hungary.

2. Experimental

The experimental set-up is described below. A stainless-steel vacuum chamber was used as a reaction chamber with quartz top, front and rear windows. The gases were of high purity (WF_6 99.998%; Ar 99.9997%; H_2 99.9999%). Flow rates were controlled by calibrated mass flow meters and the total pressure in the chamber was set by a gate valve at the exit of the chamber. Ar was used to purge the windows to avoid deposition. The air leakage rate into the deposition system corresponded to an air impurity level of about 2 ppm in the reaction gas. Before starting the experiments the reactor was purged with the reaction gas mixture. The gas composition was $WF_6/M = 1:200$ (M = Ar, H_2) in most of the experiments and the total pressure was 4 mbar. The ArF excimer laser ($\lambda = 193$ nm, FWHM 17 ns) beam was focussed into the gas mixture by a cylindrical lens with 380 mm focal length and the beam size was $\sim 20 \times 0.8$ mm^2 in the examined region. The laser fluence and the repetition rate were ~ 1 J/cm^2 and 20 Hz, respectively. The laser induced fluorescent light was observed perpendicular to the laser beam. The light was coupled into a quartz optical fibre which was connected to a Czerny–Turner type of spectrograph. The three gratings used in the spectrograph allowed for 1.8 and 0.45 nm, and 1.2 Å resolution, respectively. The spectra were recorded by an optical multichanel analyzer (OMA III of EG& G). The exposure intervals were 1 s. The intensity measurements were performed for the 429.5 nm W line at a fixed channel of the OMA. The spectra were not corrected for the sensitivity curve of the detector. The experiments were performed at room temperature.

3. Results and discussion

3.1. Origin of the spectra

The spectrum obtained from pure WF_6 gas is shown in fig. 1a. The spectral lines (except the scattered ArF lines) may originate from WF_6, W subfluorides or from elemental W produced by

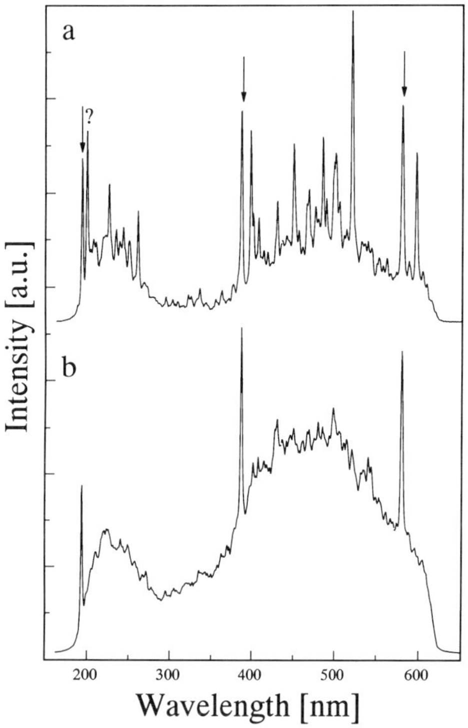

Fig. 1. Spectra recorded from (a) pure WF_6 (30 μbar pressure, laser fluence ~ 1 J/cm^2), and (b) from a W plasma in air (W plate irradiated by ~ 50 J/cm^2 laser fluence). The lines indicated by arrows are the first, second and third order of the scattered ArF laser line.

the photolysis. For line identification a reference experiment was performed by focussing the laser beam onto a pure W plate at a laser fluence of $\sim$ 50 J/cm^2. The spectrum recorded from the generated plasma is presented in fig. 1b. As can be seen, the two spectra have common peaks. A careful analysis of these two spectra using higher (0.45 nm and 1.2 Å) resolutions showed that all the lines obtained from pure WF_6 appeared also in the spectrum of the W plasma. This indicates that the spectral lines in the WF_6 spectrum originate from excited elemental W and from W ions. However, we found an interesting exception, namely the line at 199.0 nm in the spectrum of pure WF_6 (and WF_6/Ar mixture, see fig. 2a, indicated by question marks in fig. 1a) had a $\sim$ 10 times higher intensity than in the W plasma spectrum. The detailed part of this region is shown in fig. 2. However, this intensity difference cannot be explained. Furthermore, there is no spectral line emitted from W atoms below 209 [6] but light emission was obtained below the 193 nm excitation wavelength from both WF_6 and from W, see fig. 2. Since the lines must originate from W and since only the W ions can emit spectral lines below 209 nm, we can assume that these lines are emitted from W ions.

To produce tungsten from pure WF_6 in the vapour the following photolytic reaction is required:

$$WF_i + h\nu_l \rightarrow WF_{i-1} + F, \quad i = 1\text{–}6, \tag{1}$$

and for the light emission

$$W + h\nu_l \rightarrow W^* \rightarrow W + h\nu_e, \tag{2}$$

where ν_l is the laser photon and ν_e is the emitted photon frequency, respectively. Since the energy of the laser photon is 6.43 eV and the first ionization potential of W is 7.98 V [7], we have to

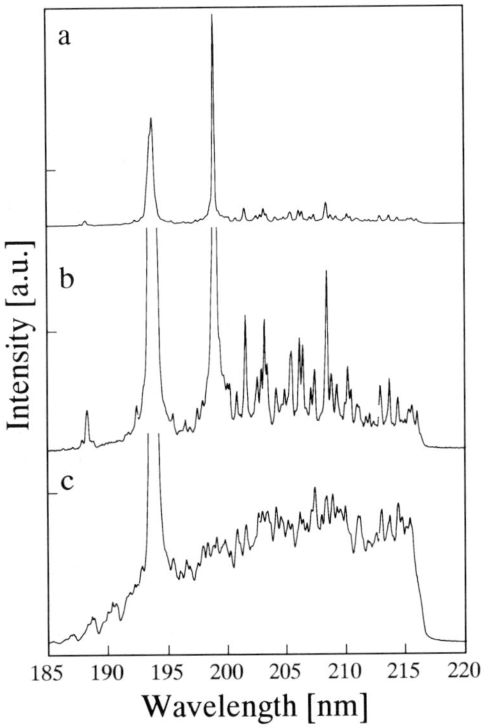

Fig. 2. High resolution spectra from (a), (b) WF_6/Ar mixture (0.5% WF_6 in Ar, total pressure 4 mbar, ~1 J/cm^2 laser fluence), and (c) from a W plasma in air (W plate irradiated by ~ 50 J/cm^2 laser fluence).

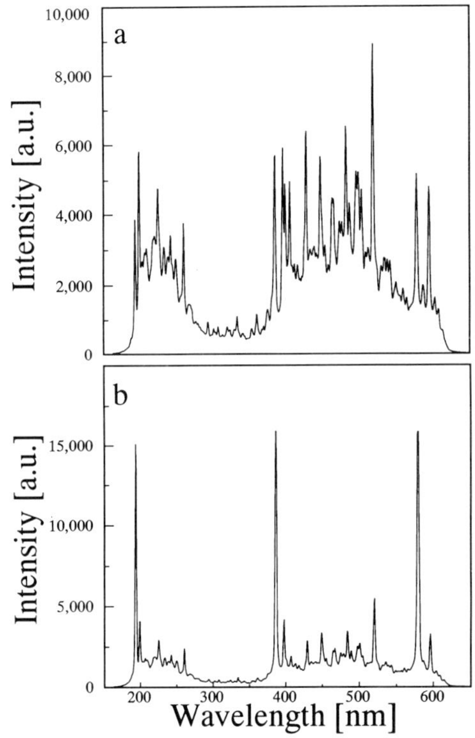

Fig. 3. Spectra obtained from (a) WF_6/Ar and from (b) WF_6/H_2 gas mixtures. Gas composition 0.5% WF_6, total pressure 4 mbar, laser fluence ~ 1 J/cm^2.

assume either a two-photon absorption process for W atoms or that the subsequent steps of the photolysis produce such highly excited W atoms which can be ionized by one-photon absorption processes. Every step of the photolysis can leave a considerable energy in the remaining molecule since the average W–F bond energy is about 5.3 eV [8].

3.2. Comparison of different gas mixtures

A spectrum recorded from the WF_6/Ar gas mixture is shown in fig. 3a. The only difference between the spectra in figs. 3a and 1a (pure WF_6) is that in the presence of Ar the line splitting is more pronounced. However, when a WF_6/H_2 gas mixture was used, the intensity of the scattered ArF laser lines increased by a factor of 4–5 compared to the WF_6/Ar gas mixture, which is explained by the fact that H_2 favours W cluster formation in the gas phase. Formation of W clusters indicated by broad band light emission upon irradiation of $WF_6/H_2/Ar$ gas composition has been reported in a previous paper [4]. According to our calculations the cluster formation occurs via H_2 reduction of W subfluorides [9]. Using only a WF_6/H_2 gas mixture, the broad band light emission cannot be seen anymore. A possible explanation might be that light emission from the excited W clusters requires collisions with Ar. However, it cannot be excluded that the Ar has other roles in the W cluster formation as well. Finally, it was also observed that using the WF_6/H_2 gas mixture and a laser fluence of 100 mJ/cm^2, no scattered light could be observed [10]. This may be explained by the fact that at lower laser fluences the concentration of the W atoms in the vapour is too small to produce W clusters.

3.3. Pressure dependences

The influence of the WF_6 partial pressure and total pressure, respectively on the intensity of a characteristic W line at 429.6 nm (we note that this line in our spectra corresponds to the superposition of the 429.46 and the 430.21 nm W lines which were not resolved with the 150/mm grating) are shown in fig. 4. As can be seen, the two curves are very similar and consist of two linear parts. For the lower pressures (first part of the

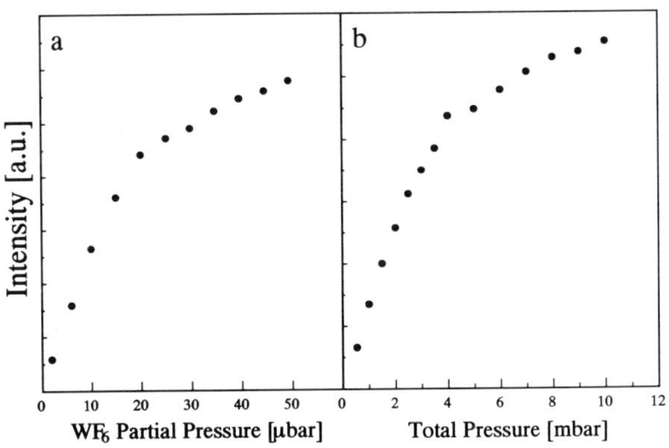

Fig. 4. Intensity of the emitted 429.6 nm W line (a) as a function of WF_6 partial pressure (total pressure 4 mbar), and (b) as a function of total pressure 0.5% WF_6 in Ar. Laser fluence ~ 1 J/cm^2.

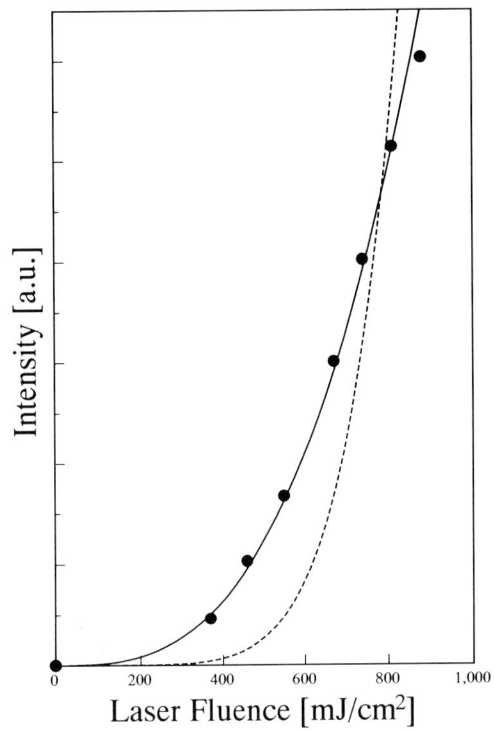

Fig. 5. Intensity of the emitted 429.6 nm W line as a function of laser fluence (solid line measured, dashed line calculated). 0.5% WF_6 in Ar, total pressure 4 mbar.

curves) the slopes of the lines are steeper than at higher pressures. Furthermore, the discontinuity in the curves occurs at the same WF_6 partial pressure which means that the observed effect depends only on the WF_6 partial pressure. However, one should expect that the intensity of the emitted light should be linear with the WF_6 partial pressure, since the laser contributing part to the W deposition rate is linear with the WF_6 partial pressure [11]. The discontinuities in the curves are attributed to a change in deposition mechanism. For a low WF_6 partial pressure, tungsten is deposited atom by atom. However, for a higher WF_6 partial pressure homogeneous tungsten nucleation in the vapour takes place and tungsten clusters are formed.

3.4. Laser fluence dependence

The intensity of the 429.6 nm W spectral line is shown as a function of the laser fluence in fig. 5. As can be seen, the curve is not linear which is explained by the nonlinearity in the subsequent steps of the photolysis of WF_6. The dashed curve is calculated from rate equations describing the photolysis, see details in ref. [9]. (Since we do not know the exact transmission functions of the optical and electronic systems used, the calculated and measured curves were occasionally matched at 800 mJ/cm^2 laser fluence.) The difference between the measured and calculated curves is explained by the fact that the absorption cross sections of the W subfluorides, produced by the photolysis, are not known.

4. Conclusions

Both tungsten atoms and ions are formed upon irradiation of WF_6, WF_6/Ar and WF_6/H_2 gas mixtures. The tungsten atoms are produced in subsequent photolytic steps combined with hydrogen reduction when a WF_6/H_2 gas mixture is used. The tungsten ions may either be formed after a two-photon absorption of tungsten atoms or after a single-photon absorption of a highly excited tungsten atom obtained in the last reaction step. This is also consistent with a mass spectrometric study [5].

There is no significant difference between the spectra recorded from pure WF_6 and WF_6/Ar gas mixtures, respectively. However, when using a WF_6/H_2 gas mixture the intensity of the scattered ArF laser lines increased considerably, since H_2 favours W cluster formation in the gas phase.

The intensity of a characteristic W line increased with the WF_6 partial and the total pressures for the WF_6/Ar gas mixture, respectively. For lower pressures the slope of the pressure dependence curve was steep, while a less steep slope was obtained at higher pressures. The change in the slope of the curves was attributed to a change in deposition mechanism from atom to cluster deposition.

The influence of laser fluence on the intensity of the characteristic W line was not linear, which is due to the nonlinear behaviour of the subsequent steps of the photolysis. Relatively good agreement between measured and calculated curves was obtained.

Acknowledgement

Financial support by the Swedish Natural Science Research Council is gratefully acknowledged.

References

[1] R.L. Krans, A. Bernstein and W.C. Sinke, Mater. Res. Soc. Symp. Proc. 181 (1990) 597.
[2] A. Shintani, S. Tsuzuku, E. Nishitani and M. Nakatani, J. Appl. Phys. 61 (1987) 2365.
[3] P. Mogyorósi, J.O. Carlsson and M. Moradi, Appl. Surf. Sci. 54 (1992) 46.
[4] P. Heszler, P. Mogyorósi and J.O. Carlsson, Mater. Res. Soc. Symp. Proc. 236 (1992) 123.
[5] M. Stuke, Ber. Bunsenges. Phys. Chem. 86 (1982) 837.
[6] Wavelength and Transition Probabilities for Atoms and Atomic Ions, NSRDS-NBS 68 (1980).
[7] R.C. Weast, Ed., CRC Handbook of Chemistry and Physics, 59th ed. (CRC Press, Boca Raton, FL, 1978).
[8] D.E. Hildebrand, J. Chem. Phys. 62 (1975) 3074.
[9] P. Heszler, P. Mogyórosi and J.O. Carlsson, to be published.
[10] P. Heszler and J.O. Carlsson, to be published.
[11] H. Matsuhashi, S. Nishikawa and S. Ohno, Mater. Res. Soc. Symp. Proc. 129 (1989) 63.

Applied Surface Science 69 (1993) 277–280
North-Holland

applied
surface science

Low temperature deposition of SiO_2 and PSG using SiH_4, N_2O and phosphorous vapour for damage-free passivation of InP-based PIN diodes by plasma- and photo-assisted LPCVD

R. Riemenschneider, N. DasGupta, R. Schütz, H.L. Hartnagel
Institute for High Frequency Electronics, TU Darmstadt, Merckstrasse 25, D-6100 Darmstadt, Germany

and

H. Kräutle
Research Center of the German TELEKOM, Am Kavaleriesand 3, D-6100 Darmstadt, Germany

Received 2 June 1992; accepted for publication 30 October 1992

An improved deposition technique of phospho-silicate glass (PSG) for the passivation of InP has been developed using a low-temperature plasma-enhanced chemical vapour deposition (PECVD). Leakage-current measurements, capacitance–voltage analyses (*C–V* curves) and deep-level transient spectroscopy (DLTS) have been employed to compare this passivation technique with a commonly applied plasma CVD and a photo-assisted CVD of silicon oxide. Improvements regarding interface states and deep-level traps have been achieved with a phosphorous vapour present during the insulator growth. The measurement of leakage on a semi-insulating InP surface, which is very relevant for the reliability and performance of InP-based PIN diodes, provides reliable results to detect surface damage and material degradation. The optimised plasma deposition of phospho-silicate glass at a temperature of 300°C with a low phosphorous concentration ($<$ 10 wt% P) prevents surface leakage and does not show plasma damage.

1. Introduction

A real interest in the passivation of InP-based PIN diodes, which are commonly used as receiver elements in optoelectronic devices, lies in the investigation of surface-leakage currents. Plasma-enhanced CVD provides a good quality of the deposited films (mechanically strong, electrically stable), but PECVD will cause surface damage due to ion bombardment on the semiconductor surface. As reported in ref. [1], large instabilities in the $I(V)$ characteristics of these PIN diodes due to interface traps and surface leakage have been observed after plasma passivation.

Photo-assisted CVD does not affect the semiconductor surface. However, these films suffer from a lower density and an unstable interface region depending on the incorporation of intermediate reaction products, which are created during the film formation. Thus, these samples have to be accurately prepared in order to remove the native oxide and avoid contamination. In general, the deposition of heteromorphic insulators on III–V semiconductors is a compromise according to the substrate temperature, since a deposition at a high temperature will improve insulator and interface properties, but in contrast to that will cause degradation of the semiconductor surface due to the out-diffusion of the Group V element, e.g. phosphorous.

Hence, problems which are arising by the insulator deposition on InP are a mechanical or electrical instability of the interface, plasma damage and phosphorous deficiency at the semiconductor surface. Improvements of InP-device characteristics have been achieved when the insulating layer is deposited with an overpressure of phosphorous present in the growth ([2–6]). Plasma damage will, as shown in this paper, increase surface leakage currents and cause surface charge by ion bombardment, whereas interface instabilities will lead to a high surface-state density. Both effects have been studied by $C(V)$ measurements.

2. Experimental

In our experiments, a low-pressure CVD system (LPCVD, see fig. 1) has been used for the deposition of SiO_2 and PSG. We are using SiH_4 and N_2O as gas sources which are excited inside the reactor either by plasma or UV light. The RF plasma deposition (at 27 MHz and 50 W) has been performed at a temperature of 300°C and at a pressure of 0.1 mbar. The photo-assisted CVD for SiO_2 deposition uses UV light (254 nm spectral line) for gas excitation at a low temperature (100–150°C) and at 10 mbar. Atomic oxygen is formed in this photo process by a mercury-catalysed dissociation of N_2O followed by the reaction with silane. Furthermore, the same system is used for the deposition of phospho-silicate glass just by evaporating solid phosphorous (P_4) into the gas stream which is then oxidised together with silicon to $SiO_2:P_2O_5$. The evaporation cell consists of a Knudsen cell which is well known in MBE technology.

We did not achieve promising results using the UV light-assisted technique for the deposition of PSG. Therefore, we have concentrated on the optimisation of the plasma process. The new idea was to use an electrical shield which will protect the samples and prevent a loading effect on the semiconductor surface. This principle is similar to a barrel reactor where the electrical-field strength is negligible inside the cavity and ion bombardment is avoided.

The phosphorous concentration in the films varies with the temperature of the evaporation cell, it has been determined by X-ray photoelectron spectroscopy. Leakage currents on the InP surface have been studied on semi-insulating InP by taking $I(V)$ characteristics. We have used MIS structures of undoped InP to carry out $C(V)$ measurements at a frequency of 1 MHz and we have taken n^--bulk GaAs ($N_D = (3–4) \times 10^{17}$ cm^{-3}) to build Schottky diodes for DLTS. Using these results we intend to compare the commonly known PECVD and the photo-CVD of SiO_2 with an improved deposition of phospho-silicate glass.

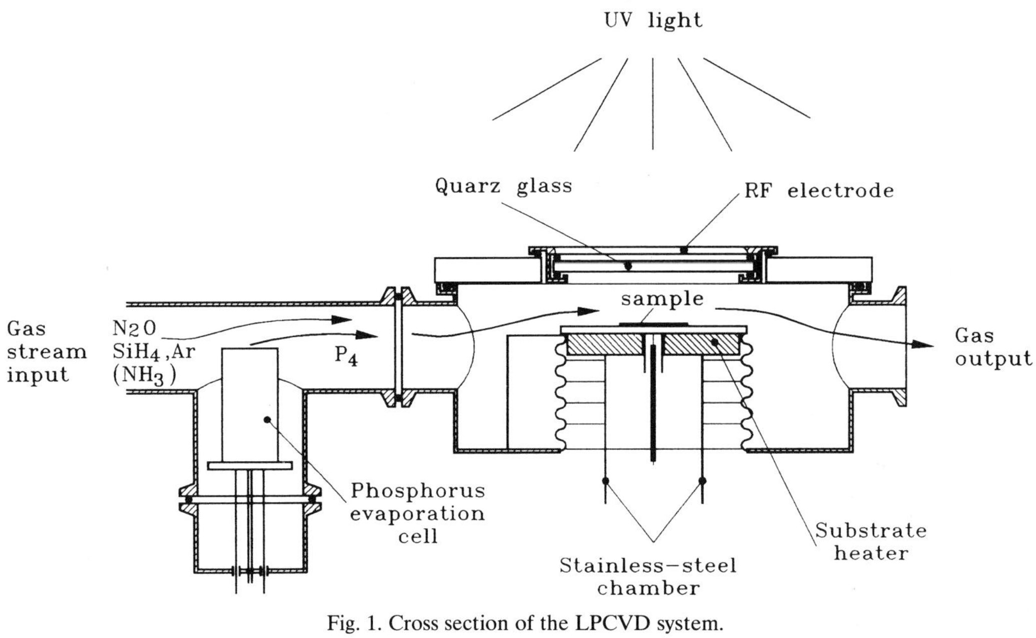

Fig. 1. Cross section of the LPCVD system.

3. Results

The phosphorous concentration in the PSG films strongly depends on the phosphorous vapour present in the film growth. A high amount of phosphorous in the insulating layer will deteriorate its physical and electrical properties, e.g. etch rate, dielectric constant, electrical breakdown field strength. Surface leakage measurements have been performed in order to compare the SiO_2 deposition of PECVD and photo-CVD to a plasma-deposited PSG with a low phosphorous concentration (less than 10 wt% P). For all analysed samples the resistance is almost constant when rather low voltages are applied. The leakage current increases abruptly at the voltage, where the structure is getting conductive. This point may be assumed as a breakdown, which is sensitive to any surface effects. Photo-CVD shows a breakdown at 30 V (fig. 2), while plasma damage strongly decreases the breakdown voltage down to 5–10 V. The PSG passivation with a very low P content exhibits an improved $I(V)$ characteristic with an electrical breakdown at 60 V, as shown in fig. 3.

C–V measurements on InP-MIS diodes give further information about the interface properties. Taking into account the flatband voltage and the hysteresis of the obtained curves we can conclude that plasma damage leads to a high density of fixed interface and oxide charges. We observe a significant shift in the reverse-biased direction of the plasma-passivated diode (– 10 V). Regarding the C–V curve of photo-CVD passivated samples (see fig. 4) an increased hysteresis is related to a high amount of interface states. The results of the PSG deposition (in fig. 5) shows neither a large shifting due to plasma damage nor a large hysteresis. As found by DLTS, the common PECVD and photo-CVD create a significant peak in the DLTS spectrum (see fig. 6). The DLTS

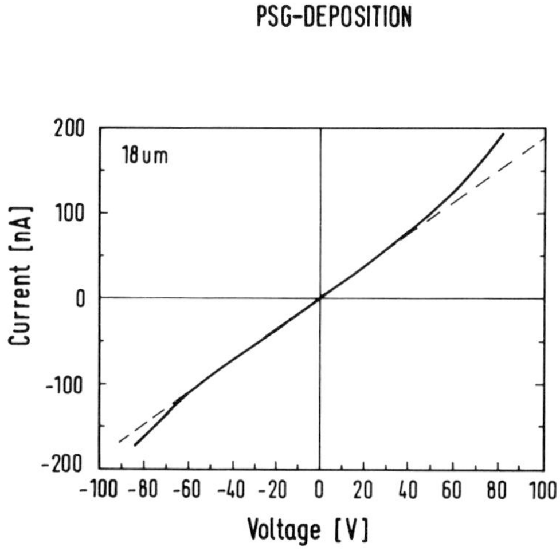

Fig. 3. $I(V)$ characteristic of a PSG sample (2 wt% P) on s.i. InP, distance 18 μm. The breakdown voltage has been increased up to 60 V.

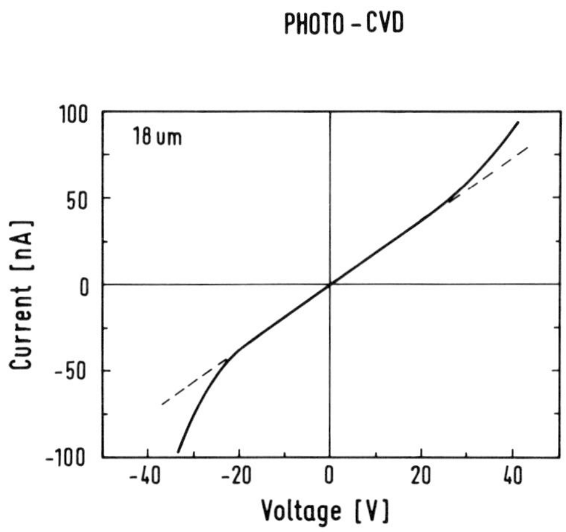

Fig. 2. $I(V)$ characteristic of photo-CVD SiO_2 on s.i. InP, distance 18 μm. We observe a breakdown voltage at 30 V.

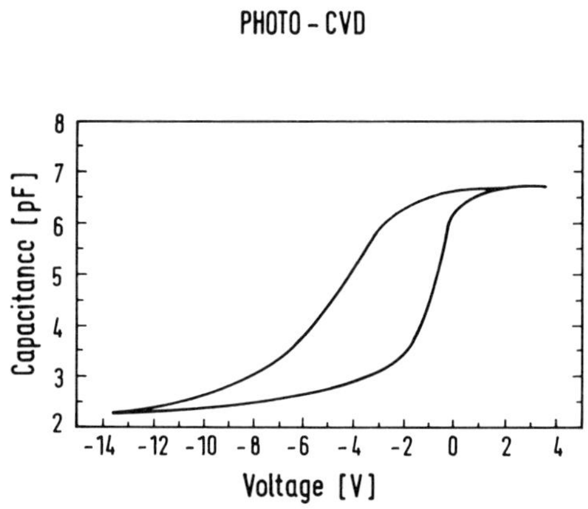

Fig. 4. C–V curve at 1 MHz of a MIS structure on InP using SiO_2, 100 nm thickness. The insulator which was deposited at 130°C by photo-CVD exhibits a large hysteresis.

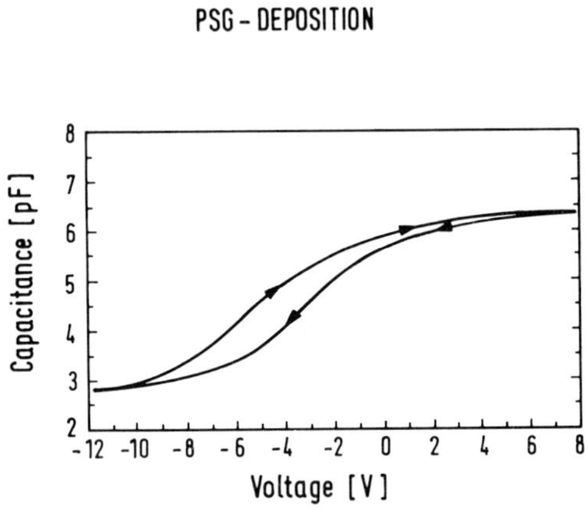

Fig. 5. $C-V$ curve of a PSG sample (2 wt% P), thickness 100 nm. The PSG layer which was grown at 300°C by plasma CVD shows low hysteresis and low voltage shifting.

spectrum of the examined PSG sample exhibits a rather flat line.

Thus, DLTS and $C(V)$ measurements strongly correlate with the performed leakage-current analysis. CV results and DLTS spectra of PECVD and photo-CVD differ significantly due to plasma damage and a photo-inherent failure. While the observed defects in plasma CVD may easily be related to ion bombardment and a degradation of

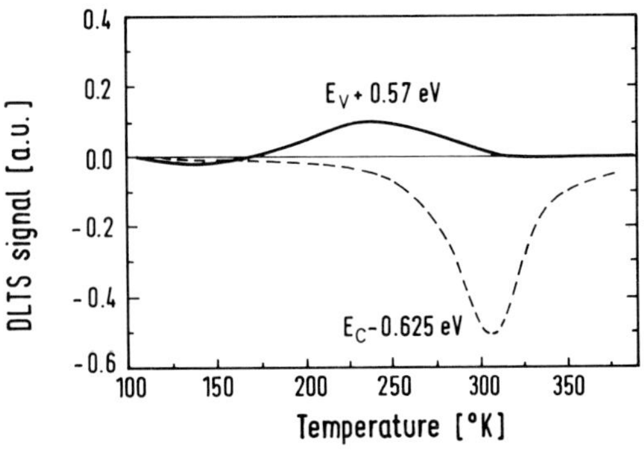

Fig. 6. DLTS spectra of GaAs-Schottky diodes. The structure has been passivated by SiO_2 using photo-CVD and PECVD. The positive peak is related to plasma damage; we observe a negative peak after photo-CVD deposition.

the semiconductor surface, the explanation of the results obtained from photo-CVD is more complicated. We suspect a photo-inherent failure which is possibly caused by an unstable, nonuniform interface, by a semiconductor surface damage (dangling bonds, P vacancies), or by surface contamination (Hg incorporation, rest contaminations in the reactor).

4. Conclusions

We have demonstrated an improved plasma deposition of phospho-silicate glass at a temperature of 300°C with a low phosphorous concentration (< 10 wt% P) for the passivation of InP-based PIN diodes. We obtained a high quality of the films as expected from the commonly applied plasma CVD and we could prevent any plasma damage. The described PSG samples obviously have good interface properties and a high resistivity. The P content should be less than 10 wt% P, because a higher amount of phosphorous leads again to a degradation of the interfacial properties.

We have used leakage-current measurements on semi-insulating InP as a promising method to detect surface damage and material degradation of the InP surface. We have shown a strong correlation between the leakage-current measurements and the obtained $C-V$ curves and DLTS spectra.

References

[1] A. Scavennec et al., Solid State Electron. 33 (1990) 429.
[2] S.K. Krawczyk et al., Conference on Insulating Films on Semiconductors, April 1991.
[3] J.W. Osenbach et al., J. Appl. Phys. 67 (1990) 6830.
[4] W. Kern and G.L. Schnable, RCA Rev. 43 (1982) 423.
[5] R.P.H. Chang et al., Appl. Phys. Lett. 30 (1977) 657.
[6] R.A. Bowling and G.B. Larrabee, J. Electrochem. Soc. 132 (1985) 141.

Applied Surface Science 69 (1993) 281–284
North-Holland

applied
surface science

The role of silane and N_2O in the CO_2 laser-CVD of silicon oxide films

D. Fernández, P. González, J. Pou, E. García, J. Serra, B. León and M. Pérez-Amor
Departamento de Física Aplicada, Universidad de Vigo, P.O. Box 62, 36280 Vigo, Spain

Received 2 June 1992; accepted for publication 30 October 1992

The laser-CVD of silicon oxide films on silicon substrates has been carried out using a CW-CO_2 laser in a parallel configuration, and with SiH_4 and N_2O as precursor gases. In this paper we report the influence of the reactant partial pressures on the film growth rate. The individual apparent orders of the chemical reaction for SiH_4 and N_2O have been determined.

1. Introduction

The laser-induced CVD is a low-temperature technique, that offers potential advantages for producing silicon oxide films. In a parallel configuration substrate temperature and gas-phase parameters may be controlled independently. Using a CO_2 laser may bring about benefits because of its economy, reliability, and easy handling compared to other UV sources. Moreover, it is already widely used in the industrial world.

With SiH_4 and N_2O as precursor gases, the 10.6 μm line of the CW-CO_2 laser is resonantly absorbed by the SiH_4 molecules. Energy redistribution by intermolecular collisions leads to gas heating [1], which controls the gas-phase reaction [2]. For mixture ratios $\varphi = P_{N_2O}/P_{SiH_4} > 25$ the deposited films are stoichiometric [3]. In this paper we study the influence of the SiH_4 and N_2O partial pressures on the film growth rate.

2. Experimental

The experimental set-up was described in detail in a previous paper [4]. Basically, it consists of a CW-CO_2 laser beam tuned at 10.6 μm passing through the reactant gases, SiH_4 and N_2O, parallel to the substrate surface. The power density of the focused beam was fixed at 700 W/cm^2. The transmitted power is monitored at the exit window. An Ar stream is used to purge the entrance window. Substrates are Si(100) wafers placed on a temperature-controlled heater. The composition and flow rate of the gas mixture are adjusted using electronic mass flow apparatus. The pressure is measured by a capacitance gauge.

3. Results

3.1. Influence of silane concentration

The influence of silane on the growth rate and the properties of the deposited films has been studied in the stoichiometric regime ($\varphi > 25$). Growth rates versus P_{SiH_4} are depicted in fig. 1. As the absorption coefficient of the gas mixture linearly increases with P_{SiH_4} [3], the laser energy coupled to the gas mass is enhanced. Since silane is the minor component in the gas mixture, it can be assumed that heat loss phenomena are not affected by the [SiH_4] change under our experimental conditions. The peak gas temperature in these experiments has been calculated using the model developed by our group [2]. The calculations reported in fig. 2 show that the gas temperature increases with silane pressure and leads to

0169-4332/93/$06.00

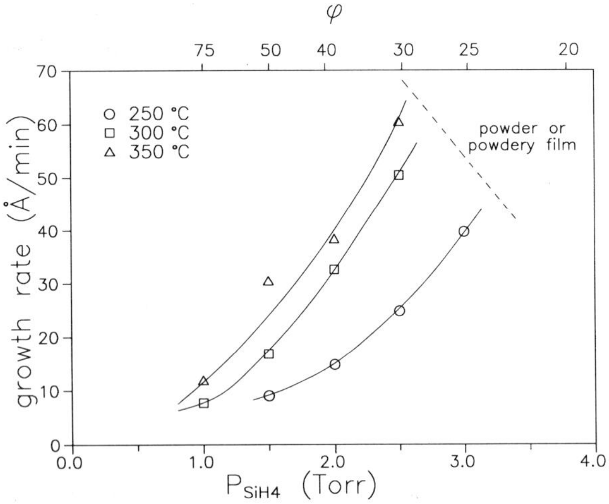

Fig. 1. Growth rate versus P_{SiH_4} at three different substrate temperatures of 250, 300, and 350°C. $P_{N_2O} = 76$ Torr, $P_{Ar} = 69$ Torr, beam–substrate distance 3.0 mm, $\phi_{N_2O} + \phi_{SiH_4} = 8.5$ sccm, $\phi_{Ar} = 7.5$ sccm.

higher film growth rate. On the other hand, the enhancement of growth rates at higher P_{SiH_4} can also be due to the effect of the increasing reactant concentration on the chemical reaction rate. Furthermore, for a constant value of P_{SiH_4}, higher growth rates are obtained the higher the substrate temperature. This is due to the contribution of the substrate heater to the gas heating.

For each substrate temperature, there is a value of P_{SiH_4} above which dense films do not grow, but powder or powdery film instead. High

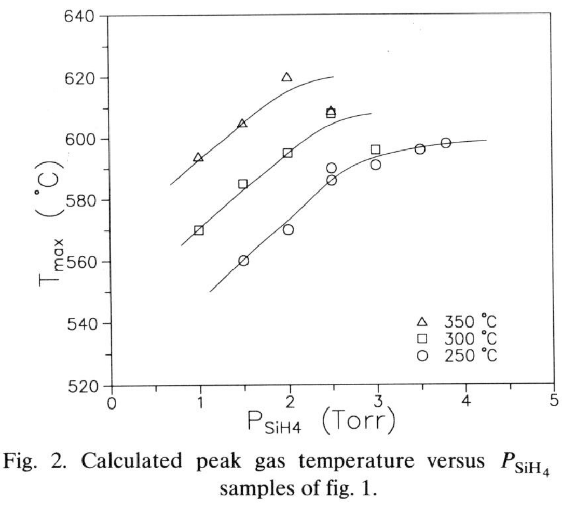

Fig. 2. Calculated peak gas temperature versus P_{SiH_4} for samples of fig. 1.

Table 1
Refractive indices and FT-IR spectra characteristics of samples of fig. 1

P_{SiH_4} (Torr)	T_s (°C)	ν (cm^{-1})	$\Delta\nu$ (cm^{-1})	Refractive index
1.5	250	1058	71	1.459
2.0	250	1058	74	1.456
2.5	250	1060	75	1.450
3.0	250	1062	64	1.447
1.0	300	1060	72	1.453
1.5	300	1060	72	1.453
2.0	300	1060	70	1.448
2.5	300	1061	78	1.447
1.0	350	1060	72	1.450
1.5	350	1062	72	1.450
2.0	350	1060	70	1.449
2.5	350	1062	71	1.444

silane partial pressures cause the increase both of the gas temperature and reactant concentration above a limit at which homogeneous nucleation and powder formation are favoured against film growth.

The refractive indices reported in table 1 are close to the values corresponding to a SiO_2 film, but present a small tendency to decrease at higher silane pressure and substrate temperatures.

Film composition has been studied by FT-IR spectrophotometry. IR spectra show the absorption bands of the Si–O–Si bonds at approximately 1060, 810 and 450 cm^{-1}, respectively. In addition, small bands at 2265 and 880 cm^{-1} indicate the presence of Si–H bonds in a nearly SiO_2 environment [5]. No evidence of either carbon or nitrogen-related peaks has been detected. The position ν and the full-width at half-maximum $\Delta\nu$ of the Si–O–Si stretching band are constant (table 1) at approximately 1060 and 75 cm^{-1}, respectively. These data show that all films have a good stoichiometry, and for increasing P_{SiH_4} the density slightly decreases [6].

3.2. Influence of N_2O concentration

Previously [3], we had studied the evolution of the deposition rate as a function of N_2O pressure (P_{N_2O}) in the limited range of $\varphi < 10$ and with an

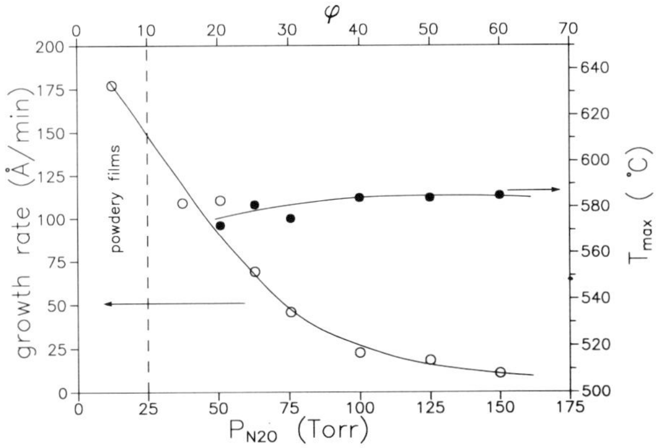

Fig. 3. Growth rate and peak gas temperature versus P_{N_2O}. $T_s = 250°C$, $P_{SiH_4} = 2.5$ Torr, $P_{Ar} = 69$ Torr, beam–substrate distance 3.0 mm, $\phi_{SiH_4} = 0.27$ sccm, $\phi_{Ar} = 7.5$ sccm, $\phi_{N_2O} =$ 1.3–16.2 sccm.

unfocused beam. The gas mixture absorption coefficient was independent of P_{N_2O}.

In the present paper we report the influence of P_{N_2O} on the film deposition rate for $10 \leq \varphi \leq 60$ and other processing conditions. Unlike reported in ref. [3], when the gas mole ratio is changed in a much wider range, the gas transmission diminishes as P_{N_2O} raises. N_2O by itself does not absorb the laser radiation [3], but it enhances the mixture absorption by collisional broadening of the silane absorption band [7]. As the thermal conductivity of the mixture also increases with P_{N_2O}, the calculated peak gas temperature remains practically constant, as depicted in fig. 3. In this case, for $\varphi \leq 10$ powder or powdery films are obtained, since with a focused beam much higher precursor generation rates are reached. This behaviour is similar to that mentioned at high P_{SiH_4}. For higher φ values the growth rate rapidly diminishes as P_{N_2O} increases.

4. Discussion

Previous studies of the reaction kinetics in our system [2] have revealed that the deposition process is controlled by a homogeneous chemical reaction. The film growth rate is driven by the peak gas temperature. Such studies, carried out at a fixed gas composition ($\varphi = 30$), have given a value for the apparent activation energy of the process of $E_a = 45.5$ kcal/mol, as well as an apparent overall order of the reaction of -1.5. Therefore, the rate v of the homogeneous reaction between SiH_4 and N_2O yielding the film precursor species can be expressed as

$$v = k_0 \, e^{-E_a/RT_{max}} P_{SiH_4}^a P_{N_2O}^b, \quad (1)$$

where k_0 is the frequency factor. The constants a and b are the individual orders of the reaction ($a + b = -1.5$). T_{max} is the peak temperature reached by the gas at the axis of the laser beam.

As the calculated peak gas temperature at increasing P_{N_2O} is practically constant (fig. 3), from eq. (1) it follows that b may be determined from the growth rate data. When plotting the growth rate versus $1/P_{N_2O}^2$ a straight line is obtained with a very good correlation coefficient (0.994). Therefore, the order of the reaction for N_2O is $b = -2$. The order of the reaction for silane $a = 0.5$ is immediately obtained.

To confirm the good agreement between the theoretical and experimental data it is necessary to consider the variations of the calculated peak gas temperature with P_{SiH_4}. As in these experiments P_{N_2O} was kept constant, it is possible to obtain from eq. (1) a P_{SiH_4}-independent Arrhenius equation as follows:

$$\frac{v}{P_{SiH_4}^{0.5}} = \frac{k_0}{P_{N_2O}^2} e^{-E_a/RT_{max}}. \quad (2)$$

Fig. 4 shows an Arrhenius plot of eq. (2) demonstrating that the order of the reaction for silane is $a = 0.5$. The slope of the straight line gives an apparent activation energy of $E_a = 46.6$ kcal/mol, very close to the previously obtained value [2].

The a and b values should be explained by a kinetic mechanism for the chemical reaction. Based upong experimental results, Chapple-Sokol et al. [8] have proposed a kinetic model for the atmospheric CVD of SiO_2 films from SiH_4–N_2O for $\varphi < 80$. Their apparent activation energy is 45 kcal/mol and, for our low-temperature process, the growth rate curve as a function of P_{N_2O} and P_{SiH_4} has the same behaviour. In addition, we have been able to specify the apparent order of the reactions for both precursors. Moreover, the

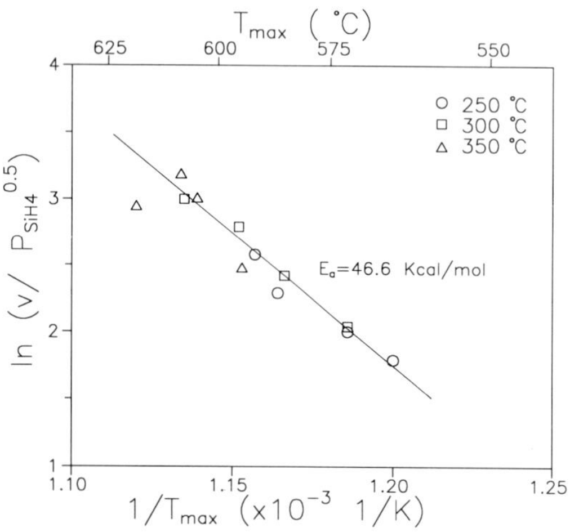

Fig. 4. P_{SiH_4}-independent Arrhenius plot for samples of fig. 1.

properties of these films show the same dependence for both experiments.

Taking into account these similarities, our LCVD process may probably proceed by the mechanism proposed by Chapple-Sokol, initiated by the homogeneous decompostion of silane:

$SiH_4 \rightleftarrows SiH_2 + H_2$

Then, SiH_2 follows two possible main reaction paths:

1) $SiH_2 + SiH_4 \rightarrow$ higher silanes $\rightarrow \ldots$
$\rightarrow$ a-Si:H film

2) $SiH_2 + N_2O \rightarrow H_2SiO \rightarrow \ldots \rightarrow SiO_2$ film

Path 1) can take place in the absence of N_2O. At higher φ, path 2) becomes more favoured at the expense of path 1). Thus, the deposited film oxidizes progressively, in agreement with our experimental results [3]. Moreover, the silanone (H_2SiO) chemistry is still not well-known, and it is impossible to elucidate all the steps following silanone generation.

In spite of the growth rate of SiO_2 films being low for higher φ's, higher growth rates can be achieved by adjusting other processing parameters, such as laser power, which is limited in our experimental system. In fact, by raising the laser power from 21 to 26 W we have obtained SiO_2 growth rates four times higher. In this way our process could be optimized for industrial applications. Potential applications for our low-temperature film deposition technique are protective or antireflecting coatings on several types of substrates (metals, plastics, glass, etc.), which are presently tested.

5. Conclusions

Our investigations dealt with the influence of the SiH_4 and N_2O partial pressures on the conditions of deposition of SiO_2 by CW-CO_2 laser-CVD. We were able to deduce the individual apparent orders of the reactions for both precursors, which can be studied on the basis of a kinetic mechanism proposed by Chapple-Sokol et al. [8]. Moreover, once more we have found that the peak gas temperature drives the film growth rate, in agreement with our previous results.

Acknowledgments

This work has been partially supported under CEE-DGXII contracts MA1E 0029 UK and SCI*.0201.C (JR), and CICYT project MAT 88-0733.

References

[1] H. Stafast, Appl. Phys. A 45 (1988) 93.
[2] D. Fernández, C. Garrido, P. González, J. Pou, E. García, J. Serra, B. León and M. Pérez-Amor, submitted.
[3] D. Fernández, P. González, J. Pou, B. León and M. Pérez-Amor, Appl. Surf. Sci. 54 (1992) 112.
[4] D. Fernández, P. González, J. Pou, E. García, B. León and M. Pérez-Amor, Thin Solid Films 220 (1993) 100.
[5] D.V. Tsu, G. Lucovsky and B.N. Davidson, Phys. Rev. B 40 (1989) 1975.
[6] W.A. Pliskin, J. Vac. Sci. Technol. 14 (1977) 1064.
[7] E.D. Hinkley, K.W. Nill and F.A. Blum, in: Laser Spectroscopy of Atoms and Molecules, Ed. H. Walther (Springer, Berlin, 1976) p. 162.
[8] J.D. Chapple-Sokol, C.J. Giunta and R.G. Gordon, J. Electrochem. Soc. 136 (1989) 2993.

Applied Surface Science 69 (1993) 285–289
North-Holland

applied surface science

"In situ" XPS study of the photoassisted reduction of noble-metal cations on TiO_2

A. Fernández and A.R. González-Elipe

Instituto de Ciencia de Materiales de Sevilla and Departamento de Química Inorgánica, Universidad de Sevilla, P.O. Box 1.115, 41080 Sevilla, Spain

Received 2 June 1992; accepted for publication 30 October 1992

In this work we have carried out an "in situ" XPS analysis of the photoassisted reduction of noble-metal cations supported on titanium oxide by irradiation of Rh^{3+}–, Pt^{4+}–, Pd^{2+}– and Ag^{+}–TiO_2 precursors in the presence of an i-PrOH/H_2O vapour pressure. This analysis has enabled a comparison of the photoreduction rate of the different metals. It has also shown that the metal dispersion that can be obtained working at the gas/solid interface is higher than that obtained during the photoreduction of the same metals on TiO_2 in an i-PrOH/H_2O suspension at the liquid/solid interface. In addition, it has been observed that the reduction of a polyvalent metal cation occurs step by step, so that while the reduction of Ag^{+} and Pd^{2+} proceeds rapidly up to the metallic state, other metals like Pt^{4+} or Rh^{3+} are only partially reduced up to Pt^{2+} or Rh^{+}. The role of the small metallic clusters on the surface of TiO_2 as traps for the photogenerated electrons has also been highlighted.

1. Introduction

Metal cations with the appropriate redox potentials can be reduced at room temperature by photoelectrons created by band-gap illumination of common semiconductors. The metals normally used have been Pt, Pd, Rh, Au, Ir and Ag and they were photoreduced on various semiconductors mainly oxides such as ZnO, WO_3 and predominantly TiO_2 (refs. [1–10], and references therein). Most of these studies have been carried out by illumination of a suspension of the semiconductor in an aqueous solution of the corresponding metal salt. However, in order to prepare metal/semiconductor devices with metallic particles of ultrafine dimensions we have studied the photoreduction process by illumination at the gas/solid interface of a TiO_2 powder impregnated with a metallic salt. In this way we can avoid the mass transport phenomena which are very important at the liquid/solid interface, thus we expect the achievement of a high dispersion of the metallic phase. It should be emphasized that this methodology could be applied to the preparation of thin layers while the fabrication of ultrafine metallic particles in porous and non-porous matrices may lead to many possible device applications including heterogeneous catalysts, non-linear optic devices and high-voltage switching devices based on interparticle tunneling [11]. Finally, the "in situ" XPS analysis of the photoreduction process has enabled important conclusions to be obtained for the mechanism of the reaction.

2. Experimental

TiO_2 was Degussa P-25 (50 m^2 g^{-1}, non-porous, ~ 70% anatase). The Rh^{3+}–, Pt^{4+}–, Pd^{2+}– and Ag^{+}–TiO_2 precursor materials were prepared by incipient wetness impregnation of the powdered oxide with aqueous solutions of the $RhCl_3$, H_2PtCl_6, $Pd(NO_3)_2$ and $Ag(NO_3)$ metal salts, respectively. Metal loadings of 2% by weight were used in this study. Pure grade reagents were purchased either from Merck ($RhCl_3$, H_2PtCl_6 and $AgNO_3$) or from Ventrom ($Pd(NO_3)_2$.

The illumination experiments at the liquid/solid interface were carried out by irradiation with a high-pressure mercury lamp (Osram HBO 200 W) of 100 mg of the TiO_2 material suspended in 25 ml of an i-PrOH/H_2O (v/v, 1:1) solution containing the metallic salt. A Pyrex cell in which the suspension can be purged with argon prior to irradiation was used for these experiments.

The illumination at the gas/solid interface was carried out in the pretreatment chamber of the XPS spectrometer in which the samples in the form of pressed pellets (5×5 mm^2) were irradiated in the presence of 3×10^{-1} Torr of the vapour pressure of an i-PrOH/H_2O (v/v, 1:1) mixture. Illumination was carried out through a Pyrex window using the light of a high-pressure Hg lamp (Philips HPK 125 W) focussed with a lens on top of the sample holder so that the complete sample area was irradiated.

A water filter for IR radiation was used in all the experiments to avoid additional heating of the sample. Energy outputs of about 400 mW/cm^2 have been measured at the sample holder in which a thermocouple attached to the rear allowed control of the temperature during the experiments, which was not higher than 323 K.

XPS spectra were recorded with a Leybold-Heraeus LHS-10 spectrometer using Mg Kα radiation as the excitation source and a pass energy of 50 eV. Binding energy (BE) reference was taken at the Ti $2p_{3/2}$ level of Ti^{4+} at 458.5 eV. The spectra were submitted to deconvolution by use of a HP-1000E computer on line with the spectrometer assuming Gaussian/Lorentzian curves. Sensitivity factors supplied with the instrument were employed to calculate the atomic percentages.

3. Results and discussion

Previous studies on the photocatalytic reduction of noble metals on TiO_2 water suspensions have enabled different metals to be ordered according to their photoreduction rates [2,6]:

$$Ag > Pd > Au > Pt \gg Rh \gg Ir \gg Cu = Ni \approx 0.$$

On the basis of these results we have selected four metals (Ag, Pd, Pt and Rh) with different photoreduction rates. In addition, we have carried out our experiments in the presence of an i-PrOH/H_2O mixture, either in the gas or in the liquid phase, because alcohols act as very efficient hole scavengers, thus preventing the recombination of the electron–hole pairs generated during irradiation [2].

Figs. 1 and 2 show the Ag(MVV) Auger peak and the Pd(3d), Pt(4f) and Rh(3d) photoelectron peaks of the original M^{n+}–TiO_2 precursor samples and their evolution in the presence of an i-PrOH/H_2O vapour pressure in the dark and under irradiation in the pretreatment chamber of the XPS spectrometer. We have registered the Ag(MVV) Auger peak because the Ag(3d) photoelectron peak shifts only slightly by changes in the oxidation state of silver. From these figures we observe that there is no change in the total intensity of the peaks so that no agglomeration or formation of big particles occurs during the photoreduction process under these experimental conditions. In fact, table 1 shows the intensities of the XPS peaks of Ag(3d), Pd(3d), Pt(4f) and Rh(3d) normalized to the intensity of the Ti(2p) peak for the M–TiO_2 (M = Ag, Pd, Pt, Rh) samples obtained by the photochemical reduction of the M^{n+}–TiO_2 precursors in the presence of i-PrOH/H_2O either at the liquid/solid interface (working in a suspension) or at the gas/solid interface (working in the presence of a vapour pressure). The higher normalized intensities obtained in this latter case for the same metal loadings indicate a higher dispersion (or a smaller particle size) for the metallic phase in these samples. In fact, at the gas/solid interface the M^{n+} species adsorbed at the TiO_2 surface will have only a low mobility. However, at the liquid/solid interface the oxidized M^{n+} species adsorbed on the TiO_2 are in equilibrium with those present in solution and the reduction can in this case progress by the growth of metallic nuclei formed during the first stages of the process. This mechanism of nucleation and growth is well established in the photocatalytic reduction of noble-metal cations by irradiation of suspensions of powdered semiconductors in aqueous solutions and leads to the agglomeration of the metallic phase [6]. As a reference on the particle sizes we have measured by transmission electron microscopy the size of the silver particles in the Ag–TiO_2 sample obtained by irradiation in a suspension given diameters ranging from 50 to 150 Å. This is the upper limit of the size for the metallic phase in this study since silver is the metal that gives rise to the biggest particles [7].

In fig. 1 we have depicted the changes in the position of the Ag(MVV) Auger peak during the $Ag^+ \rightarrow Ag^0$ reduction process in the Ag–TiO_2 sample together with the deconvolution of the Pd(3d) photoelectron peak for the Pd–TiO_2 sample showing the evolution from a mixture of Pd^{2+} (BE 337.0 eV) and Pd^0 (BE 335.5 eV) to pure Pd^0. This figure shows that the reduction processes Ag^+/Ag^0 and Pd^{2+}/Pd^0 are very quick and efficient. Even by allowing the samples to evacuate at UHV conditions or to contact in the dark during several hours with the i-PrOH/H_2O vapour it is possible to reach some reduction which is highly enhanced by irradiation. The photoassisted reduction due to the irradiation of the samples with the X-ray source of the spectrometer may not be discarded at that point. Slower photoreduction rates are however observed in fig. 2 where we have depicted the evolution of the photoelectron peaks for the platinum and rhodium samples in agreement with previous results at the liquid/solid interface. Worthy of mentioning in this figure is the observation of the

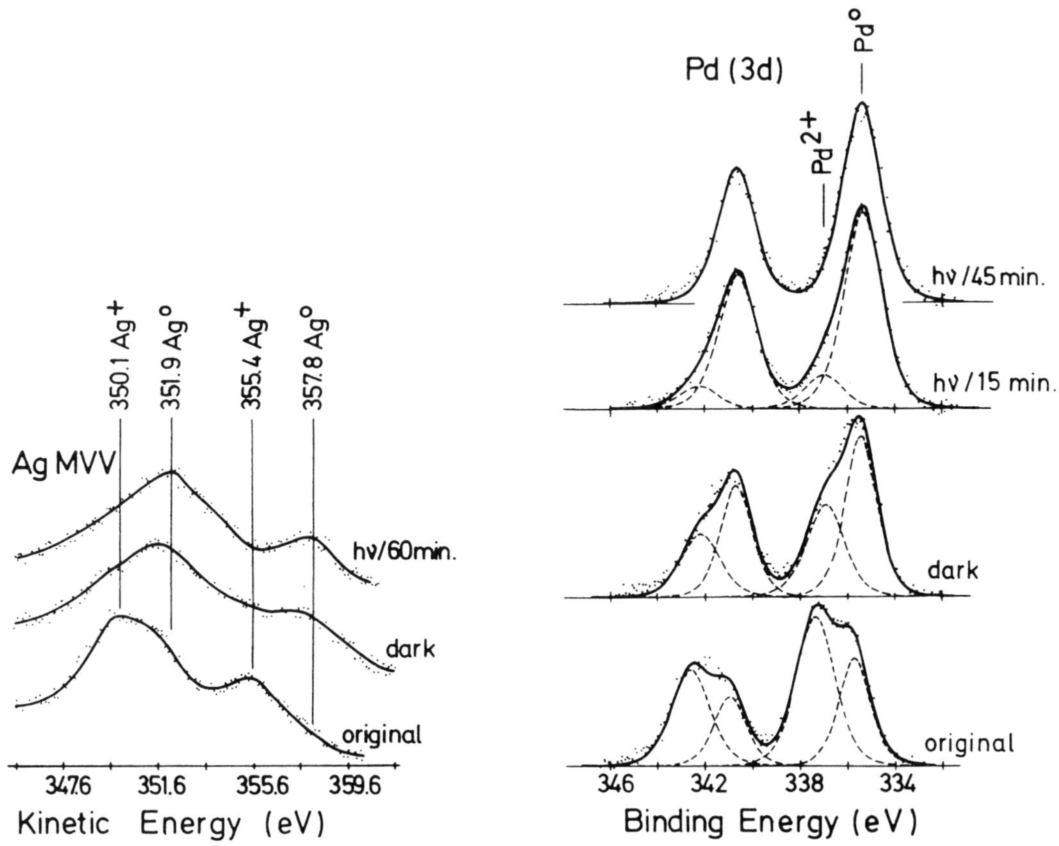

Fig. 1. Evolution in the dark and under illumination of the Ag(MVV) Auger peak and the Pd(3d) photoelectron peak of the Ag–TiO_2 and Pd–TiO_2 precursor samples, respectively.

Table 1
Intensities of the Ag(3d), Pd(3d), Pt(4f) and Rh(3d) photoelectron peaks for the different $M-TiO_2$ samples referenced to the intensity of the Ti(2p) peak

Sample	I_M/I_{Ti}	
	Liquid/solid	Gas/solid
$Ag-TiO_2$	0.003	0.089
$Pd-TiO_2$	0.030	0.045
$Pt-TiO_2$	0.017	0.029
$Rh-TiO_2$	0.019	0.045

step-by-step reduction of a polyvalent metal cation by detecting the evolution of the different oxidation states of rhodium according to: Rh^{3+} (BE 309.1 eV) → Rh^+ (BE 308.0 eV) → Rh^0 (BE 307.0 eV). In addition, we observe also that the Rh^{3+} species are mainly photoreduced up to Rh^+ leading via the formation of small amounts of Rh^0 to a steady-state situation as shown in fig. 3 where we have represented the evolution of the different oxidation states as a function of the irradiation time. This behaviour contrasts with that observed at the liquid/solid interface [2,6] where it is possible to reach an important reduction of the Rh^{3+} species to Rh^0. We interpret these effects by assuming a more favourable kinetics for the $Rh^{3+} \rightarrow Rh^+$ process than that for the $Rh^+ \rightarrow Rh^0$ on the TiO_2 support. However, on already formed Rh^0 metallic nuclei, the necessary overpotential

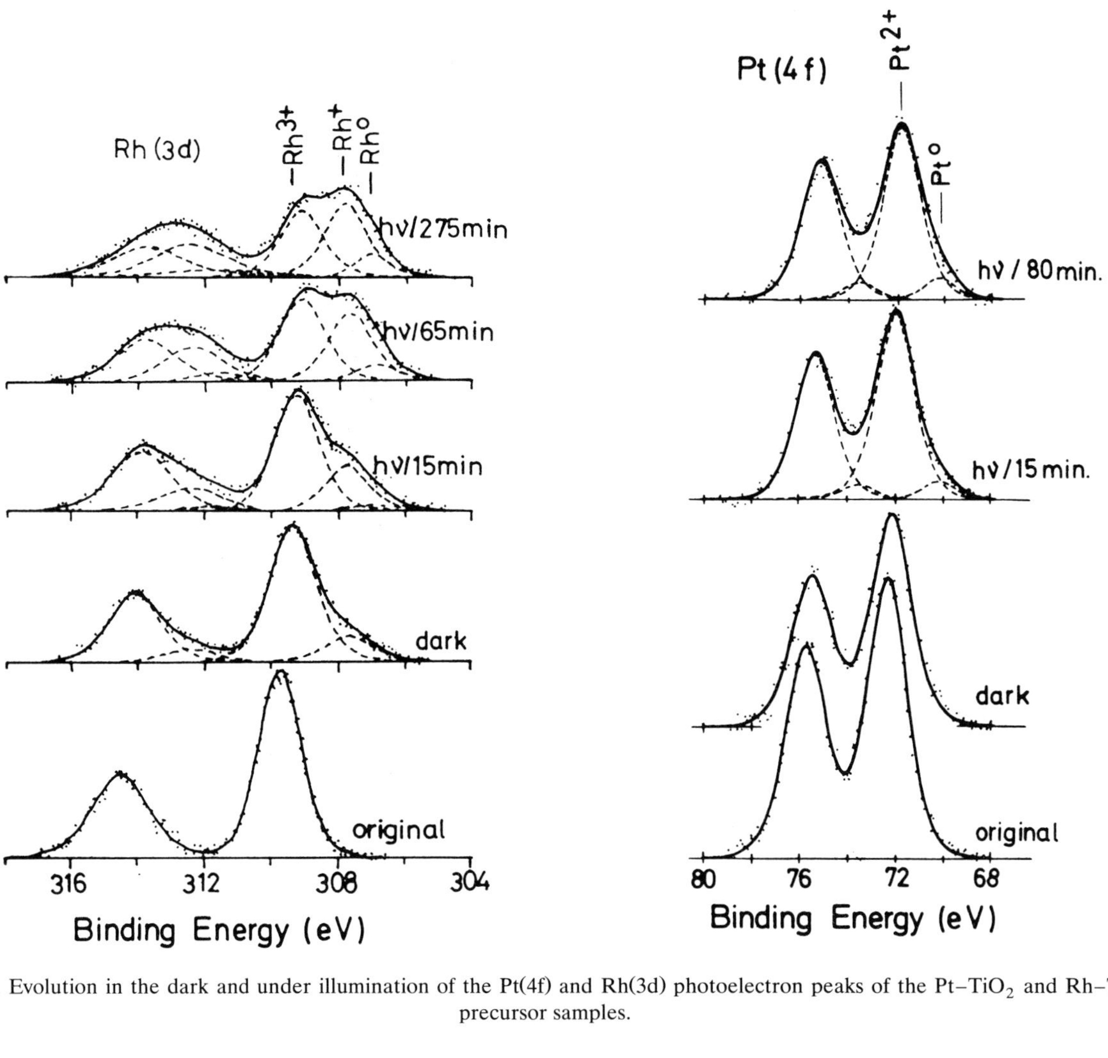

Fig. 2. Evolution in the dark and under illumination of the Pt(4f) and Rh(3d) photoelectron peaks of the $Pt-TiO_2$ and $Rh-TiO_2$ precursor samples.

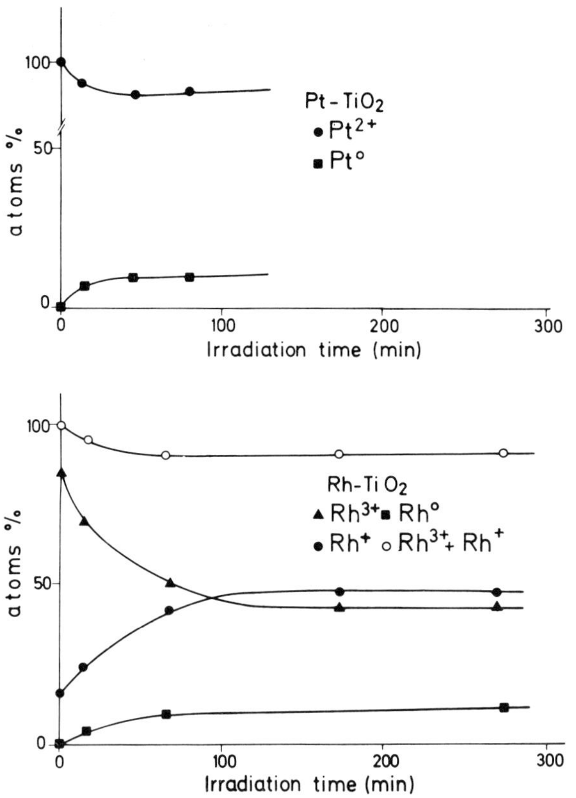

Fig. 3. Evolution of the different oxidation states during the illumination of the $Pt-TiO_2$ and $Rh-TiO_2$ samples.

for the $Rh^+ \rightarrow Rh^0$ process should be much smaller than on the TiO_2 surface, and this would explain the fact that working at the liquid/solid interface it is possible to reduce all the Rh^{3+} species to Rh^0. In this case the Rh^{n+} oxidized species in the solution can be fully reduced on previously formed Rh^0 nuclei. At the same time these Rh^0 clusters behave as traps for the photogenerated electrons so that working at the gas/solid interface the Rh^{3+} species anchored on the TiO_2 support cannot be totally reduced up to Rh^+ (or Rh^0) due to the low-phase mobility under these experimental conditions. In the case of platinum (fig. 2) we observe that the Pt^{4+} species are very unstable and are reduced up to Pt^{2+} (BE 72.2 eV) at the ultrahigh vacuum conditions of the XPS spectrometer ($P = 10^{-8}$ Torr). The formation of metallic Pt^0 (BE 70.2 eV) small clusters which act as traps for the photogenerated electrons leads, as observed in the case of rhodium, to a steady-state situation (fig 3).

From the above-described results we can conclude that it is not easy to extend the numerous results of the liquid/solid interface to the gas/solid interface. On the other hand, silver and palladium appear as the most appropriate metals to achieve a full reduction under the experimental conditions presented in this work by using conventional high-pressure Hg lamps.

Finally, the application of this method to the synthesis of $M-TiO_2$ devices containing small metallic particles should be emphasized.

Acknowledgements

The authors thank the DGICYT (Proj. No.: PB91-0835) and Junta de Andalucia for financial support.

References

[1] B. Krauetler and A.J. Bard, J. Am. Chem. Soc. 100 (1978) 4317.

[2] K.H. Stadler and H.P Boehm, in: Proc. 8th Int. Congr. on Catalysis, Berlin, Vol. 4 (Verlag Chemie, Weinheim, 1984) p. 803.

[3] S. Sato, J. Catal. 92 (1985) 11.

[4] H. Nakamatsu, T. Kawai, A. Koreeda and S. Kawai, J. Chem. Soc. Faraday Trans. I, 82 (1985) 527.

[5] E. Borgarello, N. Serpore, G. Emo, R. Harris, E. Pelizzetti and C. Minero, Inorg. Chem. 25 (1986) 4499.

[6] J.M. Herrmann, J. Disdier, P. Pichat and C. Leclercq, in: Preparation of Catalysts IV, Eds. B. Delmon, P. Grange, P.A. Gacobs and G. Poncelet (Elsevier, Amsterdam, 1987) p. 285.

[7] J.M. Herrmann, J. Disdier and P. Pichat, J. Catal. 113 (1988) 72.

[8] A. Fernández, G. Munuera, A.R. González-Elipe, J.P. Espinós, J.M. Herrmann, P. Pichat and C. Leclercq, Appl. Catal. 57 (1990) 191;
J.M. Herrmann, J. Disdier, P. Pichat, A. Fernández, A.R. González-Elipe, G. Munuera and C. Leclereq, J. Catal. 132 (1991) 490.

[9] J.W.M. Jacobs and D. Schryvers, J. Catal. 103 (1987) 436.

[10] H. Hada, Y. Yorezawa, M. Ishino and H. Tanemura, J. Chem. Soc. Faraday Trans. I, 78 (1982) 2677.

[11] E.J.A. Pope and J.D. Mackenzi, MRS Bull. (March 1988) 20.

Applied Surface Science 69 (1993) 290–298
North-Holland

applied
surface science

Laser induced plasma formation by picosecond pulse irradiation

W. Marine [a], M. Gerri [a], P. Thomsen-Schmidt [b] and J.M. Scotto d'Aniello [a]

[a] *URA CNRS 783, Faculté des Sciences de Luminy, Département de Physique, Case 901, 13288 Marseille Cedex 9, France*
[b] *Optisches Institut, TU Berlin, Berlin, Germany*

Received 2 June 1992; accepted for publication 30 October 1992

The time of flight measurements by optical and ion-probe methods have been performed during plasma expansion induced by picosecond laser pulse. The time and angle resolved measurements show two distinct ion and electron populations. Faster species show a broad angular distribution with shift of this distribution towards the laser beam. Slow species show a forward direction in space. The overall positive current was observed in the forward direction relatively to the target. At the higher angles, plasma expansion is characterized by time/space charge separation. The spatial edge and the time trailing edge of the plume exhibit overall neutrality.

1. Introduction

Pulsed laser ablation is one of the most promising laboratory methods for thin film deposition. The development of this technique and the processing control depend on an understanding of the basic processes associated with laser beam–target and collisional particle interaction during laser induced plasma (LIP) expansion. The manipulation of the nucleation/growth process during film deposition [1,2] is a very attractive branch in the field of laser ablation. So the effect of an applied electrical field on the LIP plume expansion must be studied more carefully. The electrical measurement during the process control is also one of the suitable methods.

In this paper, we report the results of the optical and ion-probe time-of-flight (TOF) measurements during LIP expansion. For ablation, we used the picosecond laser pulse. The picosecond pulse laser ablation allows to reduce the temperature and velocity dispersion of the ablated species, so that expansion dynamics can be properly observed in the nanosecond time scale [3]. We used copper and Si targets because there are many results on this model system [4–7].

2. Experimental set-up

From an experimental point of view, a primary feature of the present work was the use of picosecond laser pulses to induce rapid excitation and heating of the target surface with following time resolved observation of the expansion dynamics. For excitation, we used 35 ps ($\lambda = 1.06$ μm) pulses selected from the pulse train of the mode-locked YAG laser. The targets were irradiated in a cylindrical stainless-steel vacuum chamber with residual pressure smaller than 10^{-6} Torr. The laser beam was introduced at an angle of incidence of 40° from the target normal. For the detection, we used two techniques: the time resolved emission spectroscopy and the time resolved ion probe measurements with simultaneous monitoring of the transient target voltage induced by irradiation.

During the ablation, optical emission from an LIP plume was imaged into the 150 μm slit of the grating spectrometer. The optical image could be translated to monitor different zones of the expanding plume. The time-of-flight spectra from the LIP plume was studied by using a rapid photomultiplier tube (2 ns rise time) and a 400

MHz storage oscilloscope. The spatial and spectral resolution were 250 μm and 0.5 Å, respectively.

An ion probe, consisting of a 7×7 mm^2 copper plate, was mounted on an $XYZ\Theta$ manipulator. The target–probe distance could be varied between 10 and 63 mm. The ion probe was floated with respect to the vacuum chamber by terminating the coaxial line at a digitizing 350 MHz oscilloscope. A bias 0–$\pm$100 V was used to reject electrons or ions by reversing the voltage across the collecting electrodes. Another similar probe was installed on a $Z\Theta$ manipulator so that angle resolved charge variation could be monitored. In this case, the first probe was used as a polarizer. The simplified experimental set-up is shown in fig. 1.

The target was placed on an insulating rotating holder and the voltage variation was measured between the target and ground via a 50 Ω resistance. The targets were a polycrystalline copper foil and α-Si. The α-Si was obtained by As^+ ion implantation (70 keV) of Si wafers at a dose of 5×10^{15} cm^2. Before irradiation, targets were chemically cleaned, rinsed and immediately placed in the vacuum chamber. Every fresh area of the target was irradiated with a minimum of 10 shots before collecting the signals.

The whole detection system was triggered by laser pulses via a p-i-n diode and a small portion of the beam was used to monitor the irradiation energy.

3. Results and discussion

In this work, we used two kinds of targets, α-Si and Cu. Despite the different nature of these targets, semiconductor and metal, similar results concerning the time evolution of LIP expansion were observed. The difference was found only in the absolute values of the expansion velocities and electron and ion concentrations. First we present the results of the ion probe measurements from a Cu target and then we compare these ion probe results with the optical TOF measurements from the both targets. The ablation threshold, W_{th}, of Si was determined by an optical pump and probe experiment in air [8], and it was about $2W_m$, where $W_m = 0.25$ J/cm^2 is the melting threshold. During ablation in vacuum, a ion emission occurs at fluence $W_{th} = 2W_m$ and at slightly higher fluence (about 10–15%) a photon emission was observed. We define this value as the ablation threshold of Si. In a similar way, we determined the ablation threshold of Cu. W_{th} of Cu was ~ 0.5 J/cm^2.

The majority of measurements was performed at laser fluences of about 2 to $10W_{th}$. These conditions corresponded to the formation of a

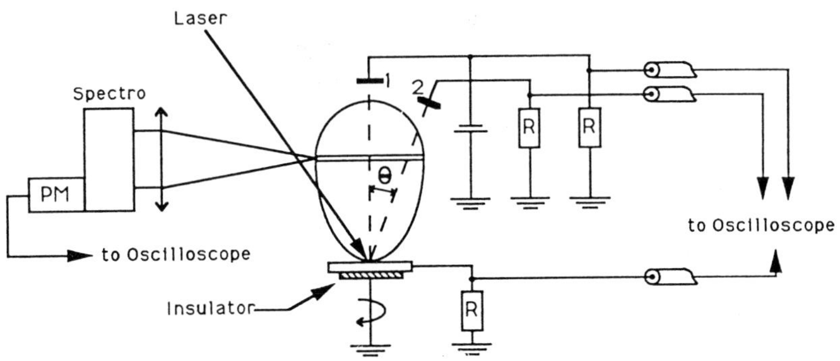

Fig. 1. Experimental set-up for optical and ion-probe time-of-flight measurements. Detector (1) was used for ion-probe measurements at $\theta = 0$. For angle-resolved measurements, detector (2) was used. In this case, bias voltage was applied between detector/polarizer (1) and ground. R is a 50 Ω resistor. The target was placed into an insulator holder. PM is a photomultiplier.

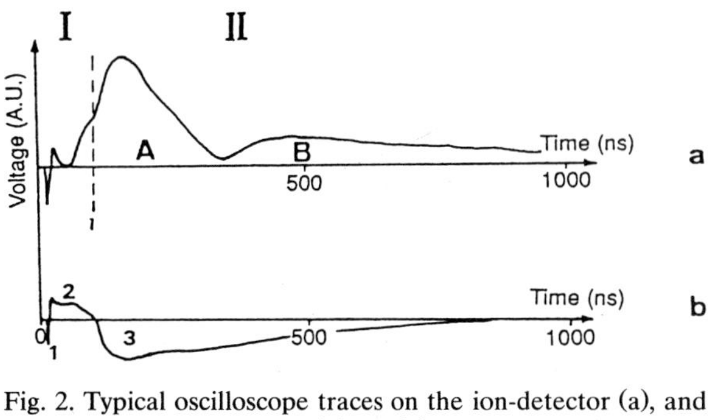

Fig. 2. Typical oscilloscope traces on the ion-detector (a), and the transient target signal (b). In (a), zones I and II delimit two time zones (see text). I is the zone where the fastest particles were observed. II corresponds to the arrival time of the majority particles. A is the first positive peak corresponding to the high energy particles. B is the signal from Cu^{2+} population. In (b) peaks (1) and (2) are the peaks given by ion and electron emission from the target. (3) is the negative signal induced by the secondary electron emission from the detector.

bright LIP plume, and signals from positive ions and negative species could be easily recorded.

Fig. 2 shows a typical oscilloscope trace obtained with zero bias on the ion probe (fig. 2a) and the corresponding transient target voltage (fig. 2b). The signal from the collector exhibits a net positive voltage except a rapid negative part at the beginning of the pulse. Because of the relatively complex structure of the collector response, to simplify the discussion, we delimit two time zones. They are labelled by Roman numerals in fig. 2a. In zone I, between zero time corresponding to the laser pulse and the leading edge of the signal A, we observed very rapid electron and ion signals. The time interval II corresponds to particle arrival times greater than 100 ns. In this time interval most of the ions and electrons arrived on the collector. Interesting is the fact that the collector signal, at zero bias, displays a net positive current. This shows that the ion concentration is higher at low collecting angles and the LIP plume does not exhibit overall electrical neutrality.

The target transient voltage shows a complex bipolar signal (fig. 2b). At low laser fluence, it consists of a rapid negative peak at zero delay time (labelled 1 in fig. 2b) and a positive broad signal (2). The duration of the latter one is about 100–200 ns depending on the laser energy. When the laser energy increases, the new negative peak [peak (3)] appears at a longer delay time than peak (2). The transient target voltage obviously indicates the higher ejection rate of the positive particles just after laser pulse [peak (1)] and an emission of the negative ones at longer time [peak (2)]. From the studies of the intensity and of the delay time variation of the peak (3) versus laser fluence, bias voltage and target–detector separation, we conclude that this peak corresponds to the arrival of the negative charges on the target. This point will be discussed later.

The intensities of both, target and ion probe signals, depend strongly on the target surface state. The amplitude increasing and the ion distribution shifting to the higher translation energy were observed when fresh unirradiated areas of the target were exposed to the laser pulse. Thus all experiments were carried out on surface treated by multiple shots. After 7–10 shots, the signals stabilized. However, when the time between two laser shots was more than 2–3 min, we also observed the ion signal variation. Characteristic of this delayed irradiation was an intensity increase of the leading edge of the ion signal. In this case, the peak A (fig. 2a) exhibited a double structure. These effects can be explained by surface contamination or by rapid partial surface oxidation due to relatively low vacuum (10^{-7} Torr). Another possibility is a reconstruction of the surface states as was discussed previously [9]. These qualitative observations show that the mechanism responsible for high energy ion production is strongly dependent on the surface states. We believe that this kind of ions was produced by laser induced desorption/ablation just before the melting, because if the surface melting occurs, all information on the surface state must have been lost.

We cannot give full description of the signal evolution from the time zone I, because of a relatively low time resolution (2 ns) of our system. Moreover, the concentration of particles in this zone is only about 0.01–0.1% of the total number of the ejected ions. Thus these particles do not have strong influence on the laser deposition process. However, from the measurements of the arrival delay time versus target–collector distance, the translation velocity of the positive ions (first positive peak in fig. 2a) is about $(2–3) \times 10^8$ cm/s. The translation energy of the corresponding electron population is about 75 eV.

In the time zone II, two different ion populations can be clearly seen from fig. 2a. The first population (A) has a translation velocity of about $(1.2–2) \times 10^7$ cm/s. The second one (B) expands with lower velocity $(3–4) \times 10^6$ cm/s. It is evident that at zero bias all particles, negative and positive, will be collected by the detector. The detector response will be proportional to the space charge versus time. In fig. 3, we show the evolution of the detector signal for different angles θ relative to the target normal.

This figure shows that the preponderant portion of the positive charges is located at the angle of about 15°. At higher angles, the space charge is mostly negative. When we applied a negative voltage between the polarizer and the ground, the negative species were repelled to the higher angles and the positive accelerated to the polarizer. Fig. 4 shows the maximum voltage on the detector (2) (see experimental set-up) as the function of the detection angle for 0, −10 and −100 V bias. The relative variation of the maxima of the negative and positive signal shows a non-linear process. In fact, we observed more significant increase of the signal intensity from the positive species than from the negative ones. However, the time integrated intensity of both signals shows much lower variation of the relative charge versus negative bias. The non-linear increase of the positive signal can be explained by secondary electron emission on the leading edge of the ion population A.

Another peculiar feature can be seen from fig. 3 if we analyze the intensity variation of the electron and ion signals within the time zone I. These signals have a broad angular distribution with intensity increase for higher angles of detection. Measurements for positive and negative angles relative to the target normal show that this

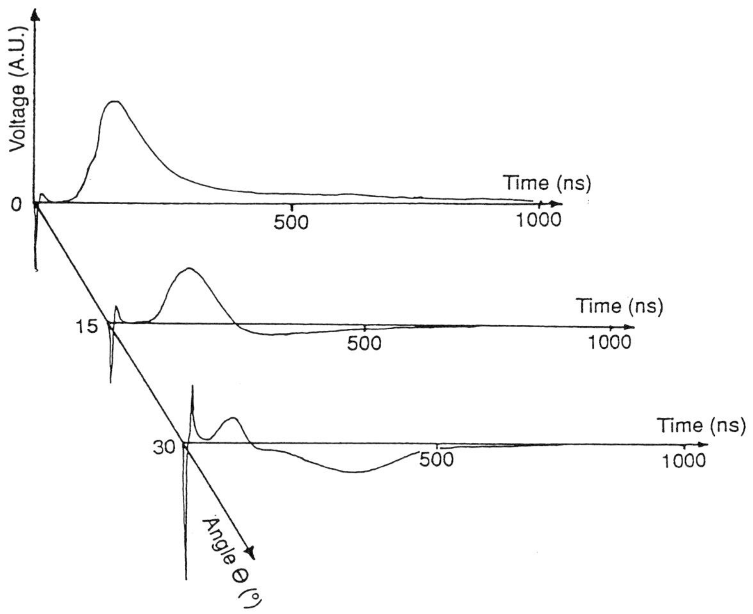

Fig. 3. Typical oscilloscope traces of the ion-probe signal at zero bias for different angles θ. Where θ is an angle relative to the target normal, the laser energy was 3 J/cm^2.

distribution is inclined toward the laser incidence.

The results for the time/angle resolved expansion suggest that the particles from the second time zone expand with more forward direction in space than the faster particles from the zone I. This clearly shows two different mechanisms of the laser beam–target interaction during laser ablation.

We do not discuss the mechanisms of the particle expansion from the time zone I because it is necessary to determine more carefully the influence of the surface states and the surface preparation on the desorption mechanism. Generation of the high-energy electrons, due to laser ablation, is not a new phenomenon [10]. Many results have been explained by a multiphoton process for electron emission and by a thermally

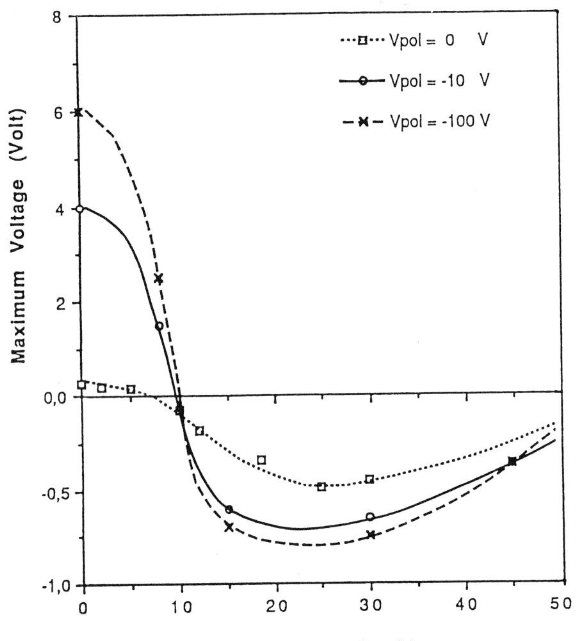

Fig. 4. Maximum voltage on the detector (2) as a function of the detection angles. In this case, ion-probe 1 was used as a polarizer. The bias voltages were $V=0$ (fine dashed line), -10 (solid line) and -100 V (large dashed line), respectively. The laser energy was 4 J/cm^2. The line is given to guide the eyes.

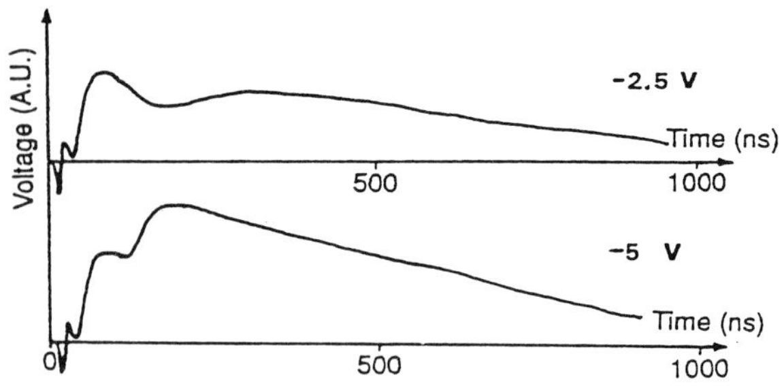

Fig. 5. Ion-probe signal at $\theta=0$ for different negative bias.

initiated process for ion emission [11–13]. One of the possible explanations is the ion acceleration by a self-consistent electric field arising after particle generation. This field appears due to the separation of the charges [14] because the lighter particles (electrons) expand more rapidly. However, an unusual spatial distribution of the fast electrons and ions can indicate the influence of the laser field on the desorption mechanism.

The oscilloscope traces of a positive detector signal at $\theta=0$ for different negative bias are shown in fig. 5. When the bias increases, the amplitude of the peak B increases and its maxima shifts to the shorter time. At the bias voltage about -6 to -7 V, the detector signal, from zone I, exhibits an apparent single peak profile with much broader distribution than at zero bias.

The angle-resolved analysis of the plume charge variation versus bias voltage shows that the intensity increase of the peak B is mostly due to attraction and acceleration of the ions localized at the back time/space edge of the LIP plume. This observation also implicates a quasi-neutrality of the particle beam within a trailing part of this one. We note a low influence of the bias voltage on the intensity and maxima position of the peak A. This can be explained by low spatial divergence of the ion population A.

Increasing the bias voltage above -10 V results in a progressive shift of the peak maxima to the greater time. A similar effect has been observed earlier and was attributed to screening of the electrical fields by the neutral species [15] or to generation of the positive signal on the detector by secondary electrons upon arrival of energetic neutral species [16].

From optical time-of-flight (TOF) measurements we found the expansion velocity of the neutral copper ($v = 6 \times 10^5$ cm/s) and the copper ions of the Cu^+ and Cu^{2+}. Corresponding velocities are 1.4×10^6 and 2.9×10^6 cm/s. Similar measurements were performed on the Si target [8]. The main conclusions of these experiments are: the Coulomb force strongly influences the velocity distribution of the ejected species; the velocity of the neutral atoms is much lower than the velocity of the ions; the velocities of the charged particles are proportional to their charge and inversely proportional to the square root of the mass.

After picosecond laser ablation, the optical TOF spectra allow to observe the expansion dynamics at distances up to 15 mm. At distances of about 10 mm, the leading edge of the TOF distribution of the neutral particles shifts to the higher energy. We explained this effect [3] by three-body recombination process via the reaction

$$A^n + e + e \rightarrow A^{(n-1)*} + e,$$

where A^n is a ion with corresponding positive charge n. (*) indicates an excited state.

However, if the three-body recombination has an important influence on the energy distribution of the ejected atoms, there is no significant influence on the concentration of the neutral species. Ion fraction in the LIP plume was estimated less than 10% [5,6].

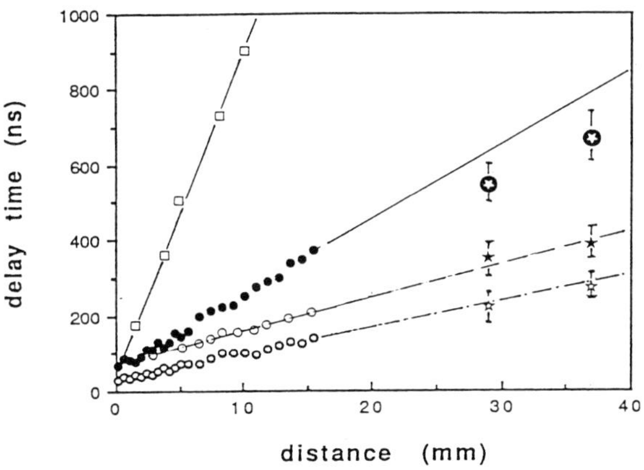

Fig. 6. Delay time of optical TOF spectra from Si^+ (full dots), Si^{2+} (open dots), Si^{3+} (open thick dots) and Si^0 (open squares) versus target–detector distance. The delay time of the signal from ion probe measurements is shown by stars.

Fig. 6 shows the delay time of TOF spectra as a function of distance. This experiment was carried out on Si. In this case we could identify ions with a charge up to 3. The circles and squares on fig. 5 correspond to the arrival time of Si^n found from optical TOF spectra. The stars indicate the position of the ion probe signal. The optical and ion probe measurements show a free-expansion regime of the highly ionized atoms and velocity increase of the Si^+. This velocity increase indicates an energy transfer from highly to low ionized particles via three-body recombination. Moreover, we do not find any electrical signal which can be induced by neutral particles via secondary electron emission. This fact shows that the energy of the neutral is too low to induce an emission of secondary electrons.

In the copper case, the translation velocity of the Cu^{2+} is very close to the translation velocity of the peak B [$(3.5\text{–}4) \times 10^6$ cm/s] of the ion signal. This velocity shift can be also explained, as in the Si case, by three-body recombination. We could not determine the particle charge corresponding to the population A because the emission lines from highly ionized Cu are localized in the UV domain.

The translation energy of the population A is about several thousands eV. This highly energetical ions can induce the secondary electron emission. In order to test this possibility, we measured simultaneously the transient target voltage and the detector signal. From response of the target on the applied bias voltage, positive or negative, we conclude that the negative peak (3) (fig. 2b) corresponds to the arrival of negative species on the target.

Fig. 7 shows the intensity of the target peak (3) at −50 V, versus distance (d) between detector and target. This peak was normalized on the intensity of the positive detector signal. When d decreases, the negative target signals increase. This implies that a greater number of the negative particles is collected by the target. The time position of the peak (3) (fig. 2) shows that the negative particles collected by the target were

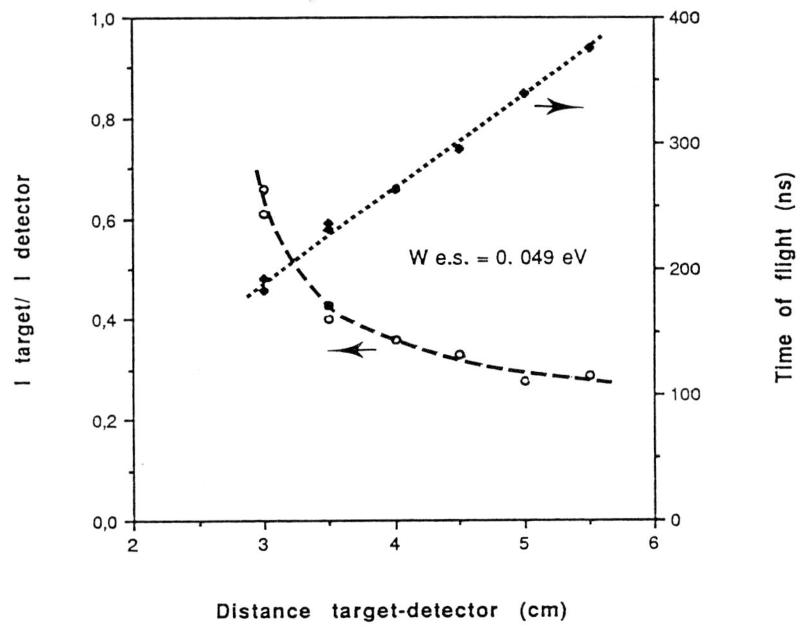

Fig. 7. (a) Relative intensity variation of the negative target and positive detector signal versus distance between detector and target. (b) Delay time of the target peak (3) versus distance. The bias voltage and laser energy were −50 V and 3 J/cm^2, respectively.

generated before the maxima of the ion signal on the detector. Thus, only the particles from the leading edge of the ion population can efficiently produce the secondary electrons by impact on the detector surface. In this case, the secondary electron emission cannot be responsible of the shift to greater time of the ion probe signal when the bias voltage increases. One of the possible explanation of this effect is the following. At a low bias voltage, only Cu^{2+} ions are efficiently extracted from the LIP plume. Due to screening effects, there is no significant influence of the electrical field on the acceleration of the Cu^+ ions located within the neutral back edge of the plume. Under higher bias voltage, Cu^+ ions can be extracted from the neutral part of the plume. Assuming a higher concentration of the Cu^+ than Cu^{2+}, the resulting ion signal will be shifted to greater time because of the lower acceleration and the lower initial velocity of the Cu^+.

In fig. 8, we show the "I–V" characteristics of the ion probe. The negative current I on this figure corresponds to the positive voltage of the detector and vice-versa. TOF signals from electron populations were obtained by reversing the collector bias to repel positive species. At positive bias, up to +2.5–3 V, we observed a positive

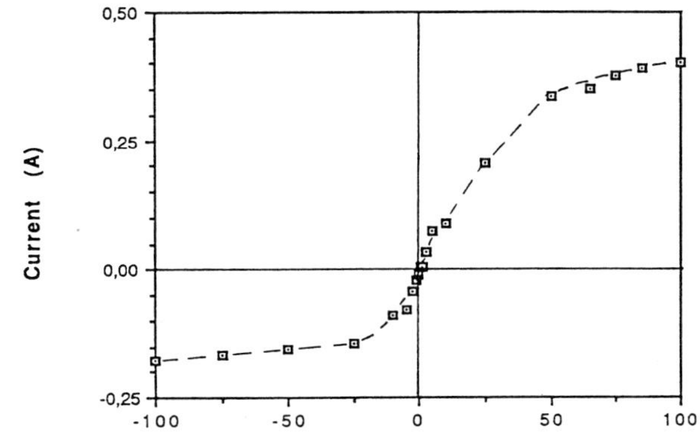

Fig. 8. "I–V" characteristics of the ion probe. The target–detector distance was 4.5 cm and the laser energy was 3 J/cm^2.

signal; at higher positive bias, mixed ion-electron signals were observed. At a bias more than +5 V, only an electron signal was detected. At relatively high bias of about +20 V we observed a double negative peak. The time position of the first negative peak is slightly delayed relative to the time position of ion signal A from the time zone II. The second negative peak is located at the same time zone as the positive peak B (fig. 2a). This negative peak shows a broad distribution similar to the corresponding ion distribution. The time integrated signal of the first negative peak is about 5–10% of the intensity of the second one. Thus assuming a Maxwellian distribution of the electron energy and neglecting the first peak, we calculated the electron temperature of the second electron distribution. The electron current is given by

$$I_p = \tfrac{1}{4} n_e A \sqrt{T_e} \exp\left[-\frac{e(V_s - V_p)}{kT_e} \right],$$

where V_s and V_p are plasma space potential (+40 V in our case) and bias potential. n_e and T_e are the electron concentration and electron temperature. A is a constant.

A plot of the natural logarithm of the probe current as a function of the probe voltage V_p should yield a straight line with a slope equal to the inverse of the average electron temperature e/kT_e.The fit of the "I–V" curve gives $T_e = 7.15$ eV at a laser fluence about 3 J/cm^2. This temperature is in good agreement with $T_e = 7$ eV which was found from the measurements of the transient target voltage after excimer laser ablation [5].

Finally, from the time integrated signals of the ions and the negative peak of the target due to secondary electron emission, we estimated the number of ions arrived on the detector at the ion current saturation regime. This number is 10^{11} $[e]$/pulse, where e is the electron charge. In this estimation, we used a cos θ distribution of the secondary electron and the assumption that most of the ions are Cu^+. Typically, during picosecond ablation, the number of ejected species is about 10^{13} atoms/pulse. Thus the ion flux is about 1%.

4. Conclusion

The present results have shown that ions and electrons ejected from metallic or semiconducting targets have two different populations with specific space distribution. A "low energy" population expands with a very forward direction in space. Faster population has a broad angular distribution inclined towards the laser beam direction. Probably these two populations are induced by different laser–target interaction processes. The angle-resolved measurements show an overall positive space charge in the forward direction relative to the target and the positive –negative charge separation at higher angles. The spatial edge and the trailing edge of the LIP plume exhibit over neutrality.

In agreement with the work of Brongersma et al. [16], it has been shown that the secondary electron generated from the ion probe surface can induce a complementary positive signal during ion probe measurements. However, this signal is mostly given by high energy ions.

References

[1] H. Izumi, K. Ohata, T. Hase, K. Suzuki, T. Morishita and S. Tanaka, J. Appl. Phys. 68 (1990) 6331.
[2] H.S. Kwok, H.S. Kim, S. Witanachchi, E. Petrou, J.P. Zheng, S. Patel, E. Narumi and D.T. Shaw, Appl. Phys. Lett. 59 (1991) 3643.
[3] W. Marine, J.M. Scotto d'Aniello, M. Gerri and P. Thomson-Schmidt, in: Laser Ablation of Electronic Materials: Basic Mechanism and Application, Eds. E. Fogarassy and S. Lazare (North-Holland, Amsterdam, 1992) p. 89.
[4] R.J. von Gutfeld and R.W. Dreyfus, Appl. Phys. Lett. 54 (1989) 1212.
[5] P.E. Dyer, Appl. Phys. Lett. 55 (1989) 1630.
[6] J.C.S. Kools, S.H. Brongersma, E. van de Riet and J. Dieleman, Appl. Phys. B 53 (1991) 125.
[7] A.M. Malvezzi, J.M. Liu and N. Bloombergen, Mater. Res. Symp. Proc. 23 (1984) 195.
[8] W. Marine, J.M. Scotto d'Aniello and J. Marfaing, Appl. Surf. Sci. 46 (1990) 239.
[9] P.G. Strupp, P.C. Stair and E. Weitz, J. Appl. Phys. 69 (1991) 3472.
[10] Yu.A. Bykovskii, S.M. Sil'nov, B.Yu. Sharkov and S.M. Shuvalov, Sov. J. Quantum Electron. 5 (1975) 538.
[11] P. Agostini and G. Petite, Contemp. Phys. 29 (1988) 57.

[12] C.O. Park, H.W. Lee, T.D. Lee and J.K. Kim, Appl. Phys. Lett. 52 (1988) 368.
[13] Ph. Martin, R. Trainham, P. Agostini and G. Petite, Phys. Rev. B 45 (1992) 69.
[14] Yu.A. Bykovskii, N.N. Degtyarenko, V.F. Elesin, V.E. Kondrashov and E.E. Lovetskii, Sov. Phys.-Tech. Phys. 18 (1974) 1597.
[15] D. Lubben, S. Barnett, K. Suzuki, S. Gorbatkin and J.E. Green, J. Vac. Sci. Technol. B 3 (1985) 969.
[16] S.H. Brongersma, J.C.S. Kools, T.S. Baller, H.C.W. Beijerinck and J. Dieleman, Appl. Phys. Lett. 59 (1991) 1311.

Applied Surface Science 69 (1993) 299–304
North-Holland

applied
surface science

Large area photochemical dry etching of polyimide with excimer UV lamps

Jun Ying Zhang, Hilmar Esrom

Corporate Research, Asea Brown Boveri AG, D-6900 Heidelberg, Germany

Ulrich Kogelschatz

Corporate Research, Asea Brown Boveri A.G., CH-5405 Baden, Switzerland

and

Gerhard Emig

Institut für Chemische Technik, Universität Karlsruhe, D-7500 Karlsruhe, Germany

Received 2 June 1992; accepted for publication 30 October 1992

Large area dry etching and surface modification of polyimide (PI) films was achieved with an incoherent excimer UV source. The transmittance change of thin PI films on fused quartz substrates was determined with an UV spectrophotometer after different exposure times at three different wavelengths: $\lambda = 172$ (Xe_2^*), 222 (KrCl*) and 308 nm (XeCl*), respectively. Etch rates of PI were measured at the different wavelengths as a function of UV intensity, exposure time and background pressure in the exposure reactor. The morphology of the exposed areas of the PI films was investigated with SEM. The results are compared with data obtained with mercury UV lamps and excimer laser experiments published by several groups.

1. Introduction

The utility of photochemical dry etching of polymers with UV radiation in microelectronics, optics, multilayer printed circuit boards, packaging techniques and surgery is well recognized [1,2]. In packaging and microcircuit fabrication, polyimides have drawn particular interest as potential interlevel dielectrics because of their favourable chemical, mechanical and electrical properties [3,4]. The use of PI films as an interlayer dielectric material and the need to etch precise vias in these films require better control of the local polymer etching processes and more understanding of the etching mechanism [5]. A number of studies of PI have been conducted with respect to aspects such as surface modification [6], etch rates, laser wavelengths, fluence [7,8], etc. The etching of PI by an ultraviolet excimer laser pulse is thought to be mainly photochemical decomposition brought about by a multiphoton excitation process to a high electronic state [8].

However, it is difficult to expose large areas (several m^2) of polymers in an efficient way with excimer lasers because of their relatively small pulse energy and low repetition rate and their characteristic threshold pulse energy below which no significant removal of polymeric layers is observed. On the other hand, as far as lamps are concerned, due to UV intensity limitations of e.g. low pressure mercury lamps, the achieved etch rates of polymers like PI were very small [9]. However, an incoherent excimer UV source

[10,11] was demonstrated to be very useful for large area photochemical dry etching of PMMA [12].

In the present paper, we report on dry etching of PI by using incoherent excimer UV radiation at different wavelengths $\lambda = 172$ (Xe_2^*), 222 ($KrCl^*$) and 308 nm ($XeCl^*$), respectively. The etch rates of PI were determined as a function of UV intensity and exposure time at different background pressures. Differences to UV induced etching of PMMA will be discussed.

2. Experimental

In our experiments we used incoherent excimer radiation from a dielectric barrier discharge (silent discharge) operating in pure xenon, gas mixtures of krypton/chlorine and xenon/chlorine providing intense narrow band radiation at $\lambda = 172$ (Xe_2^*), 222 ($KrCl^*$) and 308 nm ($XeCl^*$), respectively. The discharge was initiated in an annular gap between coaxial quartz tubes [11]. The outer quartz tube had a diameter of 3 and an active length of 15 cm defined by the length of the grounded wire mesh acting as transparent outer electrode. The radial width of the annular discharge gap was about 0.8 cm. The excimer UV sources were operated at frequencies of about 20 kHz and voltage amplitudes up to 20 kV. More details about this special excimer UV source can be found in the literature [10,11]. Irradiation of the polyimides was carried out in a small exposure reactor [13]. The distance between the samples and the excimer source was 1 cm.

A stock solution of PI (Merck) was diluted in appropriate solvents (Merck, selectiplast HTR D-2). The PI solution was spin-coated on quartz substrates (Suprasil I, Heraeus) for transmission measurements. Spin coatings of PMMA (Röhm, Darmstadt) on quartz substrates (Suprasil I) were prepared by using a PMMA/trichloroethylene solution. Commercial samples of PI (DuPont; Kapton™, thickness = 75 μm) and PMMA (Röhm) used for the etching experiments were cleaned with alcohol before the experiments. All irradiations were performed at room temperature.

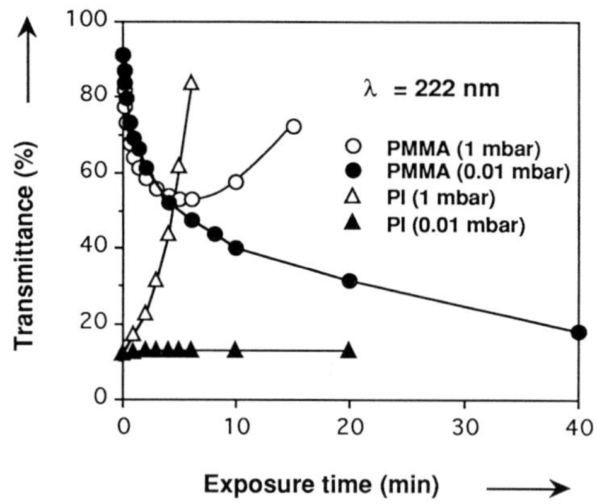

Fig. 1. Transmittance of thin PI films (0.2 μm, $\lambda_{tr} = 250$ nm) and thin PMMA films (0.6 μm, $\lambda_{tr} = 280$ nm) on quartz after different exposure times (min) with incoherent excimer UV radiation at $\lambda = 222$ nm (20 mW cm^{-2}).

The etch rate of the polymers was obtained by measurement of the depth of a hole which resulted from irradiation through a contact mask under a variety of experimental conditions. The depth measurements were performed on a Sloan Dektak stylus profilometer.

The morphology of the irradiated PI films was investigated by scanning electron microscopy (SEM) (JEOL JXA-733).

Transmission changes of thin polymer films were measured with an UV spectrophotometer (Beckman) after different exposure times at the three wavelengths used.

3. Results and discussion

3.1. Transmission changes during UV irradiation

Fig. 1 shows the UV transmission of thin PI films (thickness: 0.2 μm) and thin PMMA films (thickness: 0.6 μm) during UV irradiation (20 mW cm^{-2}) at $\lambda = 222$ nm ($KrCl^*$) at different background pressures of 1 and 0.01 mbar. At the background pressure of 1 mbar, the UV irradiation rapidly leads to a decrease of UV transmission of PMMA due to the formation of aldehyde groups or conjugated double bonds [14]. After 5 min exposure time UV transmission of PMMA increases due to photo-oxidative etching of thin

PMMA films. At a low background pressure of 0.01 mbar, UV transmission of thin PMMA films decreases still after very large exposure times (> 3 h). In this case only the chemical structure of PMMA changes and no etching of PMMA occurred. As opposed to PMMA, UV transmission of thin PI films increases immediately after irradiation due to etching of thin PI films at 1 mbar. At a low background pressure of 0.01 mbar, no change of UV transmission was observed by the irradiation of PI. Therefore, no etching and decomposition of PI during the UV irradiation occurs at the latter pressure.

These experimental results can be explained as follows (the observed pressure effects will be discussed in the next section).

After short exposure times the chemical constitution of PMMA has changed, while no etching is observable. These changes in the chemical structure of PMMA have been observed by Küper and Stuke in the infrared spectrum after subjecting the samples to excimer laser pulses at 248 nm [15]. They observed that after laser exposure unsaturated aliphatic compounds exist, which can be clearly identified by their absorption at 3078 and 1643 cm^{-1}. The repeating structural unit of the PMMA is as follows:

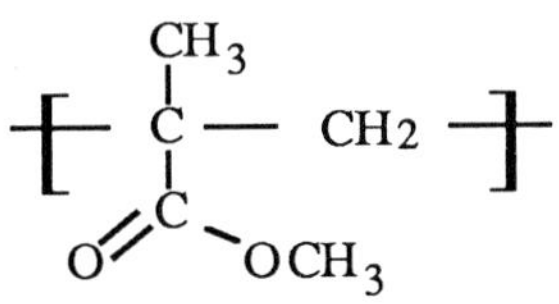
PMMA

These units become saturated aliphatic ester by irradiation-induced removal of an H atom. Unsaturated compounds are known to absorb at longer UV wavelengths than their saturated derivatives. Therefore, a broad UV absorption between 260 and 400 nm is observed after a short exposure time in our excimer lamp experiments. In comparison with the chemical structure of PMMA, PI is an aromatic derivative with repeating structural units as follows:

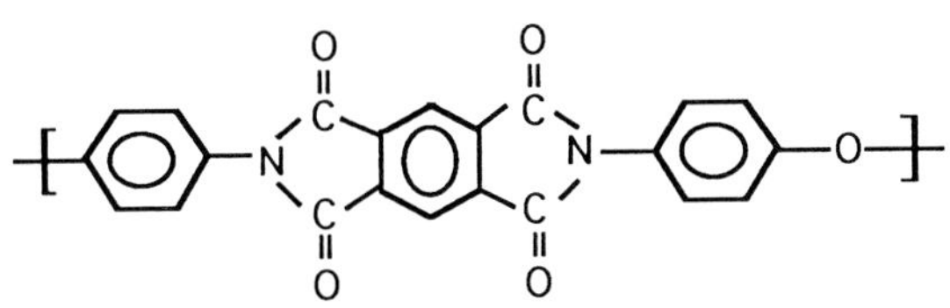
Polyimide

The absorption of photons in PI would cause a photoreaction first in the weak bonds of the PI molecule which are in the imide groups. It will most probably result in initial loss of carbon monoxide and followed by the loss of other smaller products such as CO_2, H_2O, HCN, and benzene [16]. Therefore, UV exposure leads to the increase of transmission of thin PI film immediately.

Further decomposition experiments of PI were performed under different background pressures. In fig. 2 the transmittance of PI (at $\lambda = 250$ nm) is shown as a function of the exposure time [20 mW cm^{-2}; $\lambda = 222$ nm (KrCl*)]. The increase of the etch rate by lowering the pressure in the exposure reactor is apparent.

Furthermore, a wavelength dependence of the etch rate of PI was observed. The results given in fig. 3 were obtained with irradiation at $\lambda = 172$ (Xe_2^*), 222 (KrCl*) and 308 nm (XeCl*) at a constant background pressure ($p = 10$ mbar) in the exposure reactor. The different etch rates correlate with the different absorption coefficients at the wavelengths of excimer UV radiation. The highest etch rate was achieved in case of radiation at $\lambda = 172$ nm due to the higher absorption of PI in the vacuum ultraviolet $\lambda < 200$ nm [17].

3.2. Etching of PI with incoherent excimer UV radiation

Excimer laser etching of polymers is characterized by the wavelength dependence of the etch

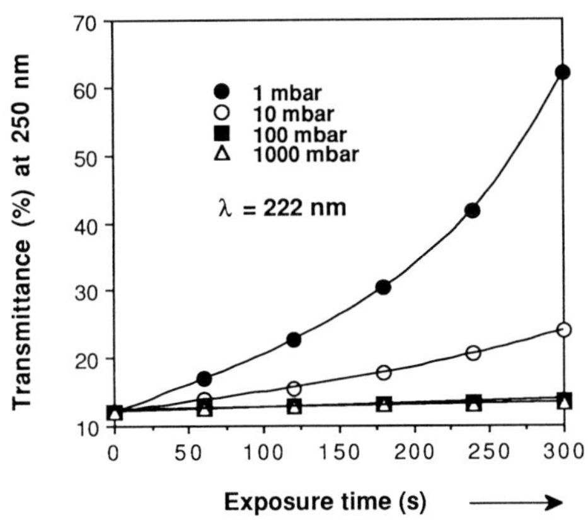

Fig. 2. Transmittance at $\lambda_{tr} = 250$ nm of thin PI films (0.2 μm) on quartz after different exposure time (s) with incoherent excimer UV radiation ($\lambda = 222$ nm, 20 mW cm^{-2}) at different background pressures.

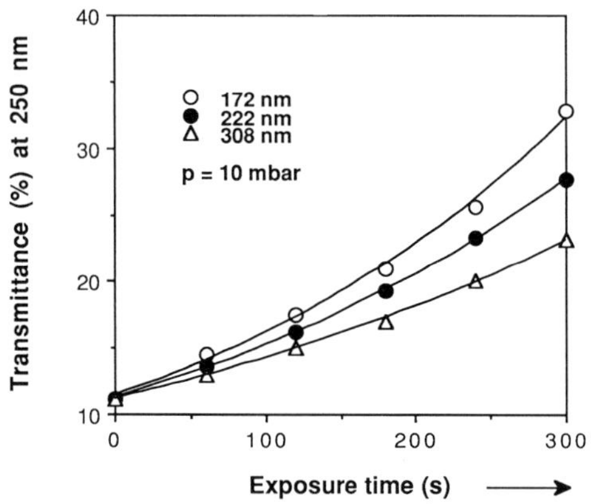

Fig. 3. Transmittance at $\lambda_{tr} = 250$ nm of thin PI films (0.2 μm) on quartz after different exposure times (s) with incoherent excimer UV radiation at different wavelengths: $\lambda = 172$ nm (20 mW cm^{-2}), 222 nm (20 mW cm^{-2}) and 308 nm (20 mW cm^{-2}), respectively.

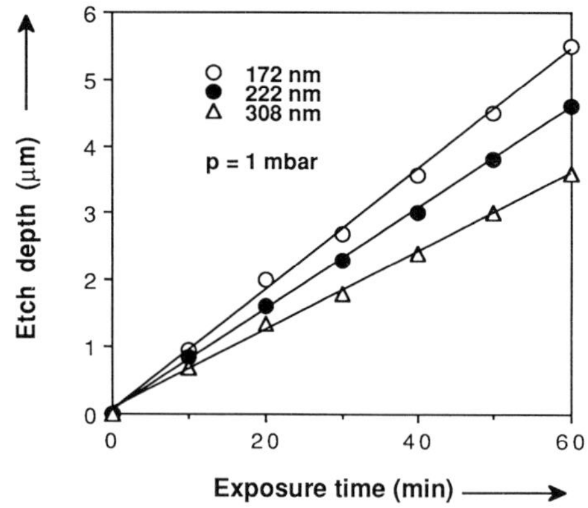

Fig. 4. Etch depth (μm) of PI as a function of the exposure time (min) at different wavelengths: $\lambda = 172$ nm (30 mW cm^{-2}), 222 nm (30 mW cm^{-2}) and 308 nm (40 mW cm^{-2}).

rate [4] attributed to the absorption properties of the material. This behaviour is also found in our case of photoetching with incoherent excimer radiation as illustrated in fig. 4. Due to the stronger absorption of PI in the VUV region the highest etch rate of 0.1 μm min^{-1} was determined for radiation at $\lambda = 172$ nm under the chosen parameters (30 mW cm^{-2}; $p = 1$ mbar). Note that the 308 nm radiation with the highest UV intensity of 40 mW cm^{-2} leads to the lowest etch rate according to the low absorption of PI at $\lambda = 308$ nm.

In fig. 5 the etch rate of PI is shown as a function of the UV intensity at the three wavelengths used. Typical for UV excimer laser etching is the flattening of the etch curves at high UV intensity [4]. Apparently we are still in the linear range of the etch curves at the investigated UV intensities. An enhancement of the etch rate by using higher photon fluxes of more powerful excimer lamps seems possible.

Etching of PI by irradiation with UV from a mercury lamp at $\lambda = 185$ nm (UV intensity: 2.5 mW cm^{-2}) was investigated by Srinivasan and Lazare [9]. The 254 nm radiation was not filtered out. The etching experiments were performed in air. The etch rate of PI was determined to be about 0.001 μm min^{-1}. These results showed the effect of the 185 nm continuous radiation. The etching mechanism was interpreted as a photo-oxidative etching of the polymer surface which was supported by ESCA investigations. However, with incoherent radiation at $\lambda = 172$ nm, we estimated etch rates 0.1 μm min^{-1} in case of PI at a low background pressure of 1 mbar.

At lower background pressures of 0.1 mbar the etch rates of PMMA and PI decrease to about 0.01 μm min^{-1}. The pronounced maximum of the etch rates at a background pressure of 1 mbar (fig. 6) is believed to be caused by two opposing effects. At very low pressures photo-oxidation is reduced because of the lack of oxygen in the background gas. If the pressure is too high the

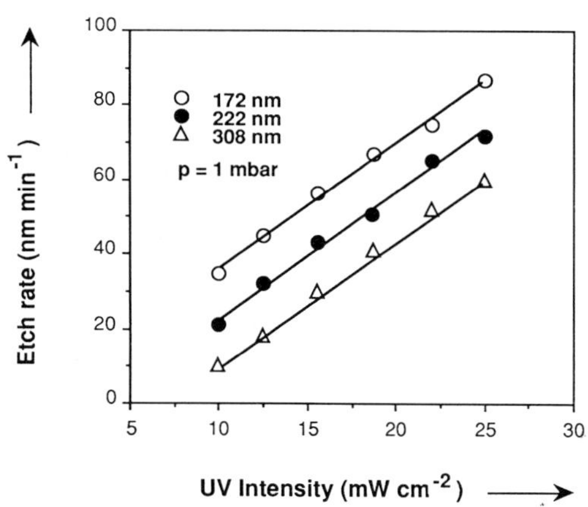

Fig. 5. Etch rate of PI as a function of the UV intensity at different wavelengths: $\lambda = 172$ nm, 222 nm and 308 nm, respectively.

short wavelength radiation is attenuated and the removal of volatile decomposition products is hampered. In between, at a pressure of about 1 mbar, apparently optimum conditions exist for the oxidation of the polymer surface and its decomposition resulting from photolytic chain scission. The most active species are most likely the free radicals $O(^1D)$ and $O(^3P)$ as well as excited molecular oxygen species [18].

In comparing the results reported here with those of other workers obtained with conventional mercury lamps [9,17,19], using the incoherent excimer UV source, the etch rate of PI can be enhanced by a factor of 100. Perhaps, a remark about the temporal behaviour of the light emission of the different UV sources should be made. While the intensity of low-pressure mercury lamps is modulated with twice the line frequency, the excimer UV sources emit bursts of extremely short UV pulses [11]. The duration of the UV pulses is of the order of ns, the repetition frequency of the bursts was 40 kHz.

3.3. The morphology of the exposed areas of PI film

Polymer surfaces are often smooth, but excimer laser induced etching can lead to alteration of the surface morphology. The roughness of the UV irradiated area can be changed in a wide range of the laser parameters (wavelength, pulse energy, pulse rate, number of pulses, etc.) and depends also on the properties of the material to be etched. Taylor et al. [20] showed that in case of PI the characteristic scale of the conical structures of the etched surface depends on the laser fluence of the excimer laser radiation at 248 and 308 nm. The width of the cones is increased by raising the fluence.

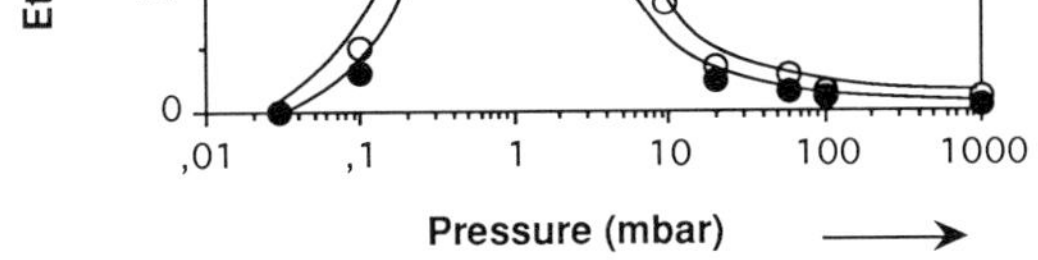

Fig. 6. Plot of the etch rate as a function of background pressure with incoherent excimer UV radiation for PI and PMMA ($\lambda = 222$ nm, 20 mW cm^{-2}).

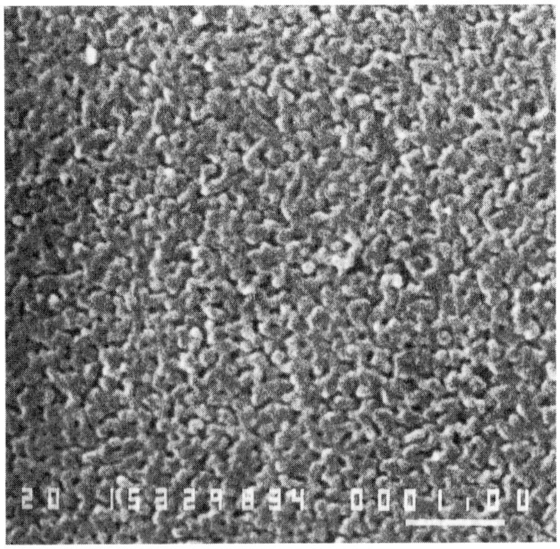

Fig. 7. SEM photograph of the typical surface modification of PI after irradiation with incoherent excimer UV radiation at $\lambda = 222$ nm (scale shown = 1 μm).

In fig. 7 the typical surface roughness observed in the etched region of PI after irradiation with the excimer lamp at $\lambda = 222$ nm is shown. Compared with the excimer laser experiments [20] the characteristic size of the roughness features is smaller. No cone formation was observed in the etched region. Therefore, the surface modification with incoherent excimer radiation appears to be different to surface alteration induced by excimer laser irradiation.

4. Conclusions

Dry etching and surface modification of PI with incoherent radiation from a silent discharge excimer UV source was demonstrated. The efficiency of the surface alteration depends mainly on the wavelength of the excimer UV radiation and is related to the absorption coefficient of PI. Oxygen plays a major role in the photolytic etching process as we also found in case of PMMA. Compared with conventional UV mercury lamps etch rates higher by a factor of 100 can be achieved with the incoherent excimer UV source.

Acknowledgements

The authors would like to thank J. Demny and U. Feller, Asea Brown Boveri, Heidelberg, for SEM investigations.

References

[1] F. Kokai, H. Saito and T. Fujioka, Macromolecules 23 (1990) 674.
[2] R. Srinivasan and B. Braren, J. Polym. Sci. 22 (1984) 2601.
[3] A.T. Barfknecht, J.P. Partridge, C.J. Chen and C.-Y. Li, Eds., Mater. Res. Soc. Proc. 167 (1990).
[4] E.D. Feit and C.W. Williams, Eds., Polymer Materials for Electronic Applications, Vol. 184 (American Chemical Society, Washington, DC, 1982).
[5] R. Srinivasan and B. Braren, in: Lasers in Polymer Science and Technology: Applications III, Eds. J.-P. Fouassier and J.F. Rabek (CRC Press, Boca Raton, FL, 1989).
[6] P.E. Dyer, S.D. Jenkins and J. Sidhu, Appl. Phys. Lett. 49 (1986) 453.
[7] P.E. Dyer and J. Sidhu, J. Appl. Phys. 57 (1985) 1420.
[8] R. Srinivasan, B. Braren and R.W. Dreyfus, J. Appl. Phys. 61 (1987) 372.
[9] R. Srinivasan and S. Lazare, Polymer 26 (1985) 1297.
[10] B. Eliasson and U. Kogelschatz, Appl. Phys. B 46 (1988) 299.
[11] U. Kogelschatz, Pure Appl. Chem. 62 (1990) 1667.
[12] H. Esrom, J.Y. Zhang and U. Kogelschatz, Mater. Res. Soc. Proc. 236 (1992) 39.
[13] H. Esrom, J. Demny and U. Kogelschatz, Chemtronics 4 (1989) 202.
[14] A.R. Schultz, J. Phys. Chem. 65 (1961) 967.
[15] S. Küper and M. Stuke, Appl. Phys. A 49 (1989) 211.
[16] J.H. Brannon, J.R. Lankard, A.I. Baise, F. Burns and J. Kaufman, J. Appl. Phys. 58 (1985) 2036.
[17] H.R. Philipp, H.S. Cole, Y.S. Liu and T.A. Sitnik, Appl. Phys. Lett. 48 (1986) 192.
[18] B. Eliasson and U. Kogelschatz, OZONE Sci. Eng. 13 (1991) 365.
[19] M. Bolle and S. Lazare, Appl. Surf. Sci. 54 (1992) 471.
[20] R.S. Taylor, K.E. Leopold, D.L. Singleton, G. Paraskevopoulos and R.S. Irwin, J. Appl. Phys. 64 (1988) 2815.

Applied Surface Science 69 (1993) 305–309
North-Holland

applied
surface science

Laser-induced deposition of aluminium on gallium arsenide and silicon nitride from trimethylamine alane

T. Frugier [a], A. Boulahia [a], A. Sayah [a], D. Tonneau [a], J.E. Bourée [a], J.M. Siffre [b] and D. Mencaraglia [c]

[a] *Laboratoire de Physique des Solides, CNRS, 92195 Meudon, France*
[b] *Laboratoire de Physico-Chimie des Surfaces, ENSCP, 75231 Paris, France*
[c] *Laboratoire de Génie Electrique de Paris, ESE, 91192 Gif-sur-Yvette, France*

Received 2 June 1992; accepted for publication 30 October 1992

Aluminium is a highly conductive metal, which is extensively used in the microelectronics industry for semiconductor and interconnect technology. Laser-induced chemical vapour deposition is a very attractive technique for the direct writing of micron-sized aluminium features of high quality with high deposition rates. The use of a solid precursor, an adduct called trimethylamine alane (TMAAH) having interesting physical properties, is reported as a good source for laser-induced deposition of high-purity aluminium films on GaAs- and Si_3N_4-coated GaAs substrates as well as on Al predeposited lines. Thus 30 μm wide, 0.5 μm thick lines, exempt from C, have been deposited using a focused CW visible laser (λ = 514 nm), at scanning speeds in the range 2–30 μm/s. The laser-induced temperatures, calculated at the centre of the spot, are lower than 300°C. Resistivity measurements performed on Al lines deposited on Si_3N_4-coated GaAs substrates give typically 10 $\mu\Omega \cdot$ cm. The very different deposition rates on GaAs or Si_3N_4-coated GaAs substrates and on Al predeposited lines, which are observed for the TMAAH complex, are consistent with surface physical chemistry studies, which show that both the surface nature and the chemisorbed species play a leading role in the thermal decomposition of the precursor. Thus, it is demonstrated that a laser-writing technique, based on a surface-controlled pyrolytic process of a well defined precursor can be used for in situ reparation or customization of integrated circuits.

1. Introduction

Laser direct writing of metals has attracted increasing interest in recent years as a non-photolithographic process for localized metallization patterns on microchips [1]. This technique has recently been used for integrated circuit customization [2] and to repair prototype circuits in a large-scale-integration (LSI) development unit [3].

Aluminium, a highly conductive metal, is extensively used in the microelectronics industry for semiconductor and interconnect technology. However, up to now, it has been difficult to write directly high-purity aluminium lines on silicon or on silica, due to the use of available trialkylaluminium precursors. The dissociation chemistry of trimethylaluminium (TMA) leads to carbon incorporation coming from organic ligands [4], whereas deposition from tri-isobutylaluminium (TIBA) is extremely slow because of its very low vapour pressure [5]. Thus new precursors are being used, like dimethylaluminium hydride (DMAH) [6] or trimethylamine aluminium hydride (TMAAH) [7,8], which provide deposits containing little or no carbon inclusion. TMAAH, also known as trimethylamine alane, is the 2:1 adduct of trimethylamine to aluminium hydride, which has interesting physical properties such as significant vapour pressure (2 mbar at room temperature), a low pyrophoricity, a weak Al–N bond and no direct Al–C bond [9,10]. TMAAH is thermally decomposed to aluminium metal below 300°C [7].

We report in this work the use of TMAAH as a good source for laser-induced deposition of high-purity aluminium lines on gallium arsenide

0169-4332/93/$06.00

and on silicon nitride coated GaAs substrates, through a photothermal process, using a CW visible laser ($\lambda = 514$ nm). Some laser deposition experiments have also been performed on Al predeposited lines, obtained by electron-gun evaporation under high vacuum.

2. Experimental

The experimental technique and apparatus used in this study have been described in detail elsewhere [11].

Samples used were (100)-oriented semi-insulating GaAs and silicon nitride film (0.2 μm thick) deposited on GaAs by CNET-Laboratoire de Bagneux. The GaAs samples were degreased with a conventional method, then etched in 2:1:8 ammonia solution. The Si_3N_4-coated GaAs substrates were used as processed. Special care has been taken just after the introduction of the samples into the chemical cell. This high-vacuum cell is evacuated to 10^{-8} Torr and degassed at 120°C during 20 h in order to desorb the water molecules. Then the pipes are degassed while flowing the purified hydrogen (carrier gas) until the detected H_2O concentration attains 0.7 ppm vol. TMAAH was purchased from SMI Society, Organometallic Division. The temperature of the bath is maintained at 35 or 45°C, depending on the needed vapour pressure. TMAAH is carried by H_2 (15 sccm) and further diluted in H_2 with a dilution ratio in the range 0.09–0.17.

The substrate temperature was typically maintained at 50°C.

The laser used was a CW multiline (488–514 nm) Ar ion laser with an output power in the range 0.2 to 0.8 W. The spot diameter, measured with a CdTe Schottky diode [11] was about 22 μm.

3. Results

For most runs, a GaAs substrate and a GaAs one coated with a 0.2 μm thick silicon nitride layer were placed in the cell, in order to study the different behaviour in the nucleation and subsequent growth steps. In some cases, aluminium lines were deposited across 0.2 μm thick aluminium pads, predeposited through a mask on the Si_3N_4-coated GaAs wafers, in order to measure the resistance of the laser-deposited lines. A four-point electrical probe or an AC two-point bridge were used to measure the resistance. The resistivity was calculated from line thickness and width measurements performed using a stylus profilometer. The lines were usually examined by scanning electron microscopy (SEM). Several lines were examined by SEM equipped with energy-dispersive X-ray analysis (SEM + EDX manufactured by LINK) dedicated to light elements or, for further investigation, by Auger electron spectroscopy (AES).

For each experimental condition, dots or lines were deposited either by fixing or usually by translating the substrate, inside the cell, in the focal plane of the lens.

3.1. Deposition on Si_3N_4-coated GaAs substrates

No deposit was observed for laser powers lower than 0.25 W. Thus most of the lines were written with a power in the range of 0.3 to 0.5 W and a scanning velocity between 2 and 32 μm/s. Fig. 1a shows a part of a well defined 3 mm long, 25 μm wide and 0.2 μm thick aluminium line. This line was deposited under 0.5 W laser power, 4 μm/s scanning speed and 0.11 dilution ratio. The roughness of the line, characterized by a distribution of grain sizes in the range 0.1–1 μm (see fig. 1b), indicates that the initial nucleation is not uniform. The measured resistance between the two pads shown in fig. 1a is 19 Ω, which leads to a calculated resistivity of 10 $\mu\Omega \cdot$cm. The calculated resistivity for several lines was about three to four times the bulk resistivity of aluminium and was slightly dependent of the writing speed. The vertical growth rate was estimated at 400 Å/s, by plotting the line thickness versus the dwell time.

Fig. 2 shows a SEM + EDX spectrum, performed at low electron voltage (3.6 kV), of a line drawn on Si_3N_4-coated GaAs at 0.7 W laser power, 2 μm/s scanning speed and 0.09 dilution ratio. This spectrum, which shows a tiny C peak and a small O peak, is similar to that obtained

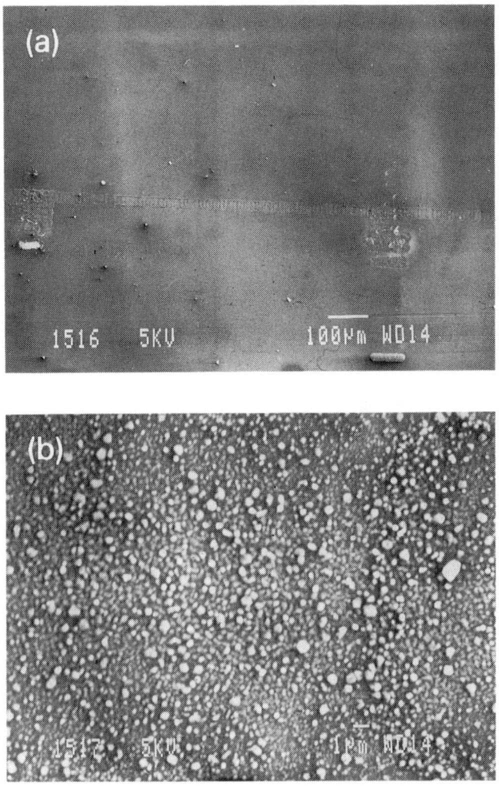

Fig. 1. SEM micrograph of a line drawn on Si_3N_4-coated GaAs at 0.5 W laser power, 4 μm/s scanning speed and 0.11 dilution ratio: (a) general view; (b) detailed view showing the line microstructure.

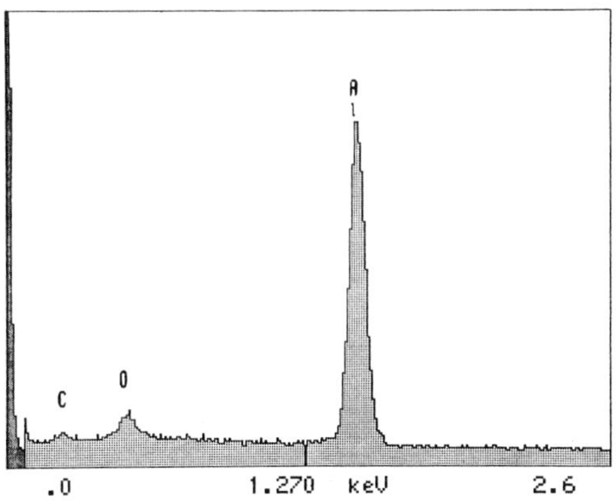

Fig. 2. SEM + EDX spectrum of a line drawn on Si_3N_4-coated GaAs at 0.7 W laser power, 2 μm/s scanning speed and 0.09 dilution ratio.

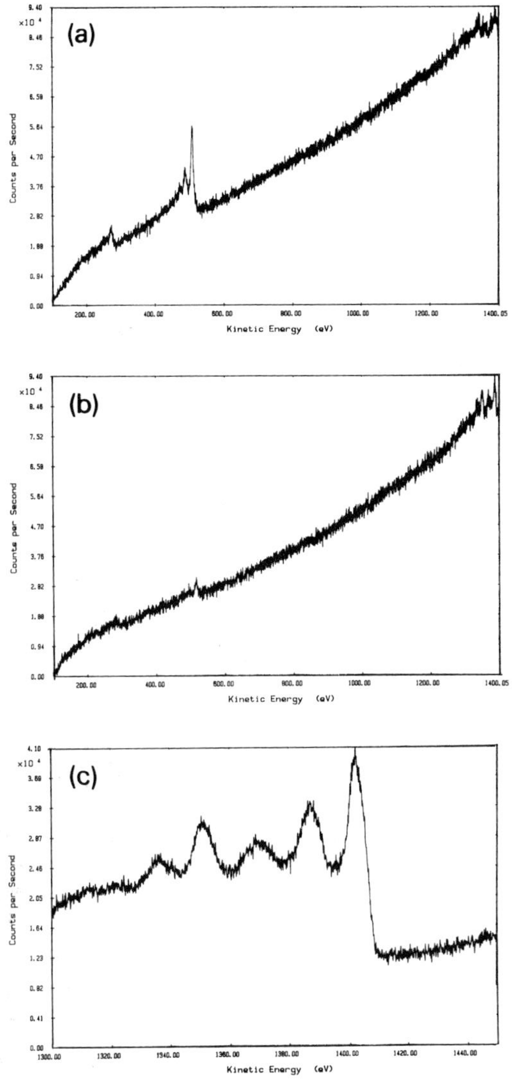

Fig. 3. AES spectra of the line defined in fig. 2: (a) spectrum obtained after a rapid etching with an argon-ion gun (5 kV, 10 $\mu A/cm^2$, 2 min); (b) spectrum obtained after a longer etching (5 kV, 40 $\mu A/cm^2$, 8 min); (c) closed view of the (b) spectrum in the range 1.3–1.45 keV.

from a commercial aluminium sheet taken as a reference. At higher voltage (7.2 kV), the C and O peaks disappear, which confirms that these elements are only observed on the surface. To go further, the same line was analyzed by AES: after a rapid etching (argon-ion sputtering at 5 kV, 10

$\mu A/cm^2$ during 2 min), a small carbon peak and a large oxygen peak indicate (fig. 3a) that C is observed in the first monolayers while Al is heavily oxidized up to 15–20 Å. A longer etching (5 kV, 40 $\mu A/cm^2$, 8 min) reveals a relatively pure aluminium (95–97 at%) with trace amounts of oxygen (less than 3–5 at%) as the only impurity (see figs. 3b and 3c). These Auger data confirm that the TMAAH precursor does not introduce any C in the deposited film [7,8]. Moreover, they are consistent with the resistivity measurements mentioned above.

3.2. Deposition on GaAs substrates

For the same experimental conditions as in the preceding section, it was more difficult to deposit dots and lines. These deposits were very rough. They were composed essentially of very thin layers and oxidized clusters. SEM + EDX as well as AES data confirmed that the films were highly oxygen-contaminated. In an earlier study [12], it was observed that a high-purity aluminium film was deposited on GaAs in a two-step process. First, the substrate was maintained at 50°C in a flow of TMAAH–H_2 mixture, then the substrate was heated up to a surface temperature of 120°C during 3 h in the same gas flow. The growth rate was estimated as 100 Å/min.

These two experiments show clearly the importance of the nucleation step before the growth and provide evidence that both surface and gas-phase reactions contribute significantly to the growth [11].

3.3. Deposition on predeposited aluminium lines

Within our available experimental conditions (the laser power being limited to 1 W), it has not been possible to deposit dots or lines across the predeposited Al lines, even though it is well known that film nucleation and growth is easier on the pure Al metal surface than on GaAs [9,10]. This observation confirms the leading role played by the surface in the decomposition of the precursor.

4. Discussion

The temperature induced by the laser beam at the centre of the impact has been calculated for the three kinds of substrates maintained at a fixed temperature of 50°C, taking into account the surface reflectivity of GaAs, Si_3N_4-coated GaAs and aluminium (42%, 34% and 92%, respectively). From fig. 4, it is clear that, for a 0.6 W laser power, the local temperature never exceeds 300°C. This means that the surface stoichiometry of the GaAs substrate remains unchanged during the experiments.

The discussion below is based on the studies relative to the surface reactivity of TMAAH regarding GaAs and Al [9,10], using temperature-programmed desorption (TPD) data.

4.1. Deposition on Si_3N_4-coated GaAs substrates

The fact that no deposition is observed at $P = 0.2$ W (induced temperature of about 110°C) is consistent with the fact that TMAAH is not thermally decomposed below 380 K [9,10]. For $P = 0.3$ to 0.5 W, the induced temperature on the surface ranges from 150 to 250°C and TMAAH provides Al deposits, only if the surface is prenucleated. The nucleation takes place if the substrate temperature is not too high in order to

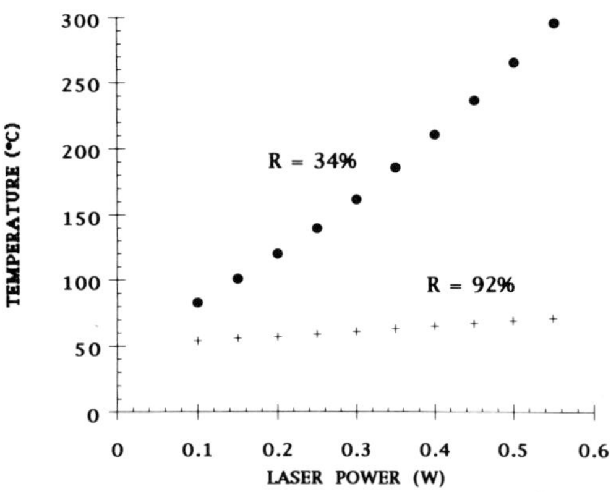

Fig. 4. Laser power dependence of the temperature induced at the centre of the laser spot for a substrate reflectivity of 34% (Si_3N_4-coated GaAs) and 92% (Al). The temperature of the whole substrate was maintained at 50°C and the laser spot diameter was 22 μm.

allow chemical adsorption (molecular desorption of TMAAH has been observed near 200 K [9,10]). High purity deposits analyzed by AES (fig. 3) come from the high stability of silicon nitride surface regarding oxygen.

4.2. Deposition on GaAs substrates

It has been shown using quadrupole mass spectrometry [9] that the TMAAH complex reacts on the clean GaAs surface at temperatures higher than 450 K. On the other hand oxygen, as determined by AES, is the only impurity observed at the GaAs surface and results strictly from environment exposure. However, O poisons the surface sites and forms stable alumina when TMAAH is decomposed: oxygen inhibits further nucleation and growth [10,11]. This explains the formation of alumina clusters, for laser power ranging between 0.4 and 0.6 W.

4.3. Deposition on predeposited Al lines

The temperature induced by the laser on the Al surface reaches 80°C for a laser power of 1 W (see fig. 4). This means that TMAAH cannot be thermally decomposed at this laser power. If the surface of Al is transformed into a very thin (20 Å) Al_2O_3 film, the decomposition of TMAAH becomes more difficult, the threshold temperature being around 470 K [10].

5. Conclusion

Aluminium lines have been deposited on Si_3N_4-coated GaAs substrates by laser thermal decomposition of trimethylamine alane at moderate temperatures ranging from 420 to 570 K. The purity of the aluminium deposits is quite high and the oxygen contamination is lower than 5 at%. The corresponding resistivity is about 10 $\mu\Omega \cdot$cm for a 2000 Å thick line. On GaAs substrate, the line thicknesses are smaller (15% less) due to the enhanced nucleation barrier. The lines are conductor only if the GaAs/Al interface is exempt from oxygen.

Due to the high reflectivity of Al compared to that of GaAs or Si_3N_4, it has not been possible by laser processing to deposit aluminium dots or lines on predeposited aluminium patterns. These apparent contradictory results confirm that laser-induced deposition of aluminium from TMAAH is mainly controlled by surface reactions.

Direct writing of aluminium tracks by laser-induced thermal decomposition of TMAAH seems very attractive for IC rewiring applications on GaAs.

Acknowledgements

This work was supported by an EEC SCIENCE programme (contract No. SCI*. 0 201.C). The authors would like to thank Dr. J. Flicstein, CNET Bagneux, for supplying the Si_3N_4 coatings on GaAs substrates.

References

[1] For a general review, see:
Laser Microfabrication: Thin Film Processes and Lithography, Eds. D.J. Ehrlich and J.Y. Tsao (Academic Press, New York, 1989);
Photochemical Processing of Electronic Materials, Eds. I.W. Boyd and R.B. Jackman (Academic Press, London, 1992).

[2] J.G. Black, S.P. Doran, M. Rothschild and D.J. Ehrlich, Appl. Phys. Lett. 50 (1987) 1016.

[3] S. Kishida, Y. Morishige, F. Uesugi and H. Yokoyama, in: Technical Digest (CLEO, Baltimore, MD, 1987) p. 282.

[4] D.W. Squire, C.S. Dulcey and M.C. Lin, J. Vac. Sci. Technol. B 3 (1985) 1513.

[5] B.E. Bent, R.G. Nuzzo and L.H. Dubois, J. Vac. Sci. Technol. A 6 (1988) 1920.

[6] T. Cacouris, G. Scelsi, P. Shaw, R. Scarmozzino, R.M. Osgood and R.R. Krchnavek, Appl. Phys. Lett. 52 (1988) 1865.

[7] T.H. Baum, C.E. Larson and R.L. Jackson, Appl. Phys. Lett. 55 (1989) 1264.

[8] D.B. Beach, S.E. Blum and F.K. Le Goues, J. Vac. Sci. Technol. A 7 (1989) 3117.

[9] J.S. Foord, A.J. Murrell, D. O'Hare, N.K. Singh, A.T.S. Wee and T.J. Whitaker, Chemtronics 4 (1989) 262.

[10] L.H. Dubois, B.R. Zegarski, C.-T. Kao and R.G. Nuzzo, Surf. Sci. 236 (1990) 77.

[11] J.E. Bourée and J. Flicstein, NATO ASI Ser. E 139 (1988) 121; B 198 (1989) 33;
J.E. Bourée et al., Mater. Res. Soc. Symp. Proc. 129 (1989) 251.

[12] A. Sayah, J. Fernandez, J.E. Bourée, D. Tonneau, H. Orain and J.F. Rommeluère, in: Proc. 6th Int. Workshop on Physics of Semiconductor Devices (1991) p. 617.

Applied Surface Science 69 (1993) 310–316
North-Holland

applied surface science

A comparison of SiO_2 thin-film UVCVD rates and mechanisms from SiH_4/O_2 and SiH_4/N_2O precursor gases by surface-sensitive infrared spectroscopy

C. Debauche, C. Licoppe and J. Flicstein

Laboratoire de Bagneux, Centre National d'Etudes des Telecommunications, France Telecom, Paris B, France

Received 2 June 1992; accepted for publication 30 October 1992

With the help of a surface-sensitive multiple internal reflection technique, we have been able to measure photodeposition rates of silica with a low-pressure mercury lamp down to room temperature. At such a temperature, the deposition rates with O_2 or N_2O as the oxidizing precursor are equal. While the oxygen-induced photodeposition rate increases rapidly with temperature in the O_2 system, it saturates at a low value, for a temperature higher than 100°C in the N_2O system. In the first case the surface is covered with hydroxyl groups, adsorbed water, mono- and dihydride species. In the second case we reveal an almost water- and hydroxyl-free bulk and surface material, and the presence of few nitrogen-containing species. These features make the silica films prepared in this latter way an interesting material for various technological processes.

1. Introduction

UV-induced chemical vapour deposition (UVCVD) has recently received much attention in fragile semiconductor processing because of its low process temperature and the small damage it creates on semiconductor surfaces. Device-quality dielectric materials such as SiO_2 are increasingly being deposited by this process. SiO_2 is currently photodeposited mostly from SiH_4/O_2 and from SiH_4/N_2O mixture [1,2]. Though this last system has been less studied, it offers some advantages. Numerous reports have indicated recently that the presence of nitrogen in SiO_2 may beneficially affect its properties [3,4]. It has been shown that re-oxidized/nitrided SiO_2 has superior electrical properties as compared with pure SiO_2 due to nitrogen pile-up at the Si/SiO_2 interface [5]. The development of in-situ analysis techniques such as multiple internal reflection infrared spectrometry [6] has allowed a new understanding of UVCVD surface mechanisms in the recent years. In this study, this technique is used to investigate the deposition of SiO_2 ultra-thin films from direct UV photolysis of SiH_4/O_2 and SiH_4/N_2O mixtures at substrate temperatures between 20 and 200°C. Deposition kinetics, film structure and surface mechanisms for deposition in the SiH_4/N_2O system are studied and systematically compared to UVCVD SiO_2 from the SiH_4/O_2 mixture. The specific interests of SiO_2 films prepared from the SiH_4/N_2O system will be discussed in conclusion.

2. Experimental

The experimental apparatus is composed of a rapid thermal chemical vapour deposition (CVD) vessel [7] equipped with a low-pressure mercury lamp providing UV lines at 185 and 254 nm and coupled to a modified FTIR Nicolet 740 SX spectrometer through ZnSe windows. UVCVD occurred on one side of a semiconductor prism cut into a rectangular shape, 50 mm long, 5 mm wide and 1 mm thick with both edges bevelled at

0169-4332/93/$06.00

45°. This experimental geometry allows for about 40 total internal reflections on each face. The sensitivity of the system is $\Delta R/R = 5 \times 10^{-5}$ at 2000 cm^{-1} with 1000 scans of the interferometer, so that 0.02 layers of SiO_2 could be detected on InP. Data are taken in unpolarized light with a resolution of 4 cm^{-1}. We have performed many experiments with similar results on wafers made of InP, for which low-temperature UVCVD of SiO_2 is a very important technological process. But undoped germanium prisms were used in this study, because they are transparent down to a wavelength of 850 cm^{-1}, while InP wafers absorb infrared light below 1000 cm^{-1}. Germanium wafers allow the observation of the absorption bands assigned to Si–H bending modes, which lie in the region extending between 850 and 1000 cm^{-1}. After degreasing with the usual trichoroethylene–acetone–methanol sequence to reduce the hydrocarbon groups on the substrate surface, the germanium wafers were treated with the following de-oxidizing procedure [8]: 1 : 1 : 4 HCl(38%) : H_2O_2(30%) : DI water for 5 min, DI water rinse, then H_2O_2(30%) for 5 min, DI water rinse, 1 : 10 HF(48%) : DI water until hydrophobic (about 2 min), and a final DI water rinse.

3. Results and discussion

3.1. Comparison of UVCVD SiO_2 deposited films using SiH_4/O_2 and SiH_4/N_2O precursor gas systems

To determine the behaviour of deposition mechanisms of both the oxides, thickness versus time data as a function of process temperature were taken. Using the SiH_4/O_2 and SiH_4/N_2O precursor gas mixtures diluted in nitrogen atmosphere alternately, photochemical deposition of SiO_2 was performed at a given temperature, between room temperature (RT) and 300°C. All growth runs were achieved at a total pressure of 10 Torr with flow rate ratios 5/100 for the two mixtures. The dilution of SiH_4 (SiH_4/N_2) was fixed at 1/100 unless specified otherwise. After each growth run, an in-situ absorption spectrum was taken. For both substrate temperatures, RT and 100°C, Si–O stretching intensity data (around 1060 cm^{-1}) are reported versus UV irradiation time in fig. 1a. The film thicknesses were derived from the intensity of the Si–O stretching band localized around 1060 cm^{-1}, with the empirical correspondence rule that an absorbance of 60 ×

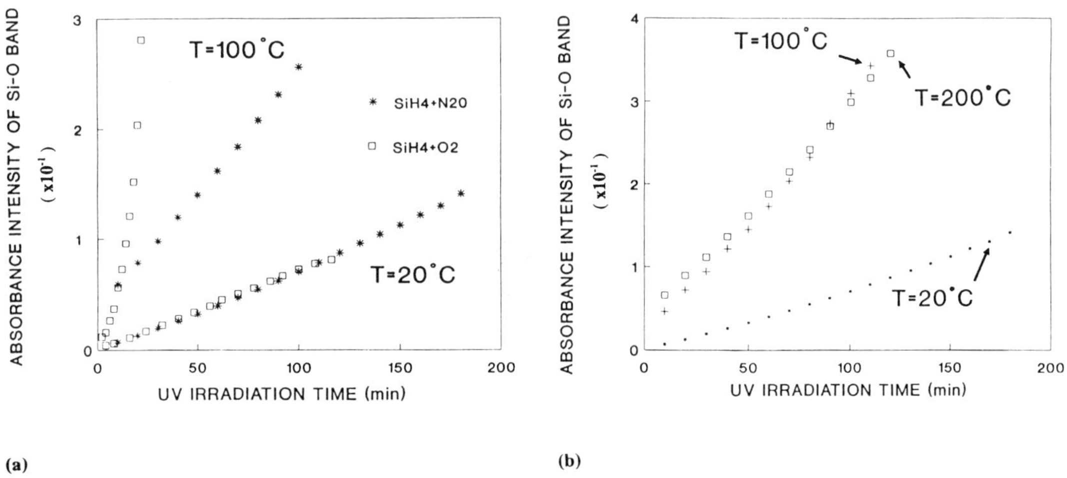

Fig. 1. (a) Intensity of the Si–O band around 1060 cm^{-1} versus UV irradiation time for silicon dioxide deposited at different temperatures, using the $SiH_4 + O_2$ (square) and $SiH_4 + N_2O$ (star) systems. (b) Intensity of the Si–O band around 1060 cm^{-1} versus UV irradiation time for a silicon dioxide deposited at three temperatures 20°C (point), 100°C (cross) and 200°C (square), using the $SiH_4 + N_2O$ system.

10^{-4} corresponds to the equivalent of one monolayer of deposited SiO_2. The thickness of this monolayer is estimated as 2.3 Å [9]. Therefore, the slopes measure the deposition rates. In sharp contrast to what happens at higher temperatures, one observes the same deposition rate at room temperature. Considering the SiH_4/N_2O system alone, successive silica depositions were performed at three temperatures, RT, 100 and 200°C. The corresponding Si–O stretching band intensity is plotted in fig. 1b, as a function of UV irradiation time in minutes. If the room-temperature deposition rates are lower than the deposition rates obtained at 100 and 200°C, photodeposition kinetics vary very little for substrate temperatures between 100 and 300°C. In the case of SiO_2 from SiH_4/N_2O, it appears that the temperature does not play the same activation role as in the SiO_2 deposition from SiH_4/O_2. In summary, the factor which controls the deposition rate in the nitrided SiO_2 production is not the temperature.

3.2. Chemical composition of UVCVD SiO_2 films using the SiH_4/N_2O system at various deposition temperatures

Photochemical deposition of SiO_2 using a SiH_4/N_2O mixture was achieved at temperatures between room temperature (RT) and 200°C. At these temperatures, purely thermal CVD was observed to be negligible, so that the deposition parameters are controlled by photochemical processes. Growth runs were performed in a SiH_4/N_2O mixture under UV irradiation with flow rates 5/100 sccm and a total pressure of 10 Torr. Fig. 2 shows the differential infrared spectra observed after 8 or 9 successive runs at a substrate temperature of 20°C (fig. 2a), 100°C (fig. 2b) and 200°C (fig. 2c) in the region of the Si–O stretching resonances. Each spectrum n in fig. 2 must be interpreted as adding the contribution of the n previous deposition runs. At these three temperatures, each run duration was chosen so as to deposit around 8 or 9 layers of thickness 10 Å, in

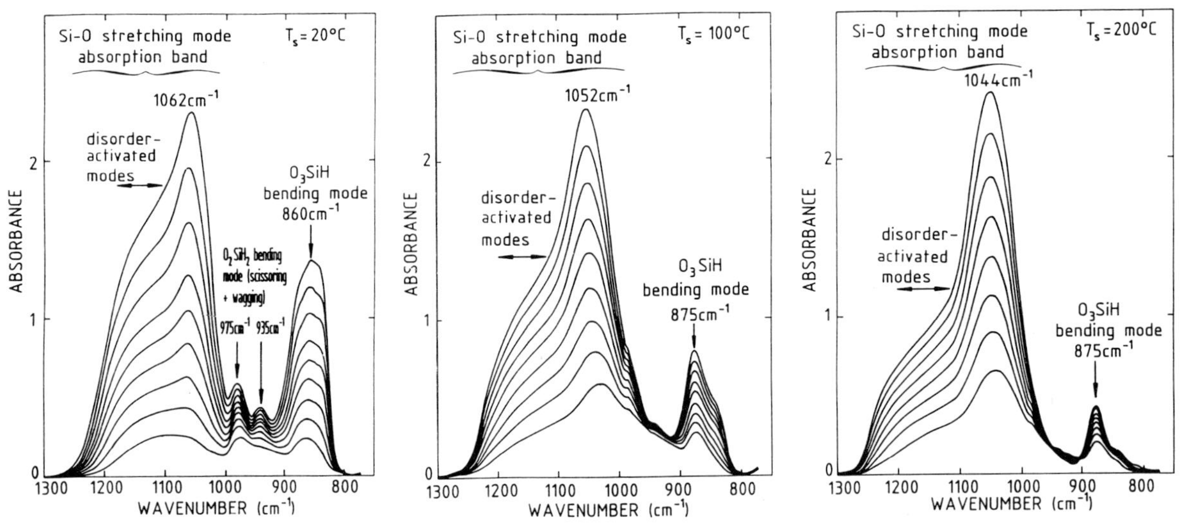

(a) (b) (c)

Fig. 2. Multiple infrared reflection–absorption spectra in the Si–O stretching region for consecutive UVCVD growth cycles of SiO_2 deposited from the $SiH_4 + N_2O$ mixture, performed at room temperature (a), 100°C (b) and 200°C (c). Spectrum n shows the difference in the absorbance spectrum between the sample after n growth cycles and the substrate reference.

order to obtain a final layer of 80–90 Å. Fig. 2a shows the spectra obtained in the room-temperature runs. These films exhibited a very high amount of porosity and surface roughness. This is seen from the high level of absorption in the 1100–1200 cm^{-1} range, which has been assigned to disorder-activated modes [10,11], and which systematically correlated in our experiments with a high density of Si–H bonds which locally terminates the amorphous network. In this wavenumber range (800–1300 cm^{-1}), infrared spectra also show a decrease of the Si–H and Si–H_2 bond densities with increasing deposition temperature, directly correlated with the absorbance intensity data of the Si–H and Si–H_2 stretching mode absorptions in the 2000–2300 cm^{-1} range. Therefore, infrared spectra show a shift of the Si–O stretching mode absorption towards the low energies at increasing temperatures. This shift is equal to 18 cm^{-1}. By contrast, in the case of deposited SiO_2 films using SiH_4/O_2 precursor gases, infrared spectra show the opposite behaviour, that is a shift of this Si–O peak towards the high energies when the deposited film temperature increases. In the case described here, this shift towards the low energies is due probably to the nitrogen incorporation. As a matter of fact, the bond stretching frequencies of the various molecular groups and in particular that of the Si–O bond vary systematically with the chemical nature of the matrix in which this bond is imbedded. In oxynitride film the most intense line in infrared spectra shifts towards low energies with decreasing oxygen concentration and increasing nitrogen concentration [12]. In the infrared spectrum of oxynitride material, very large peaks located between 835 and 1000 cm^{-1} indicate simultaneous, nitrogen and oxygen bonding in the alloy material. The small shift in Si–O bond stretching frequencies in SiO_2 deposited from SiH_4/N_2O mixture does not give precisely the nitrogen concentration but provides an order of magnitude of a few percents. The presence of a narrow peak localized at 3380 cm^{-1}, assigned to the stretching mode of N–H bonds corroborates the hypothesis of nitrogen incorporation (fig. 3). The reaction paths which allow the incorporation of nitrogen are not completely understood. Two reaction

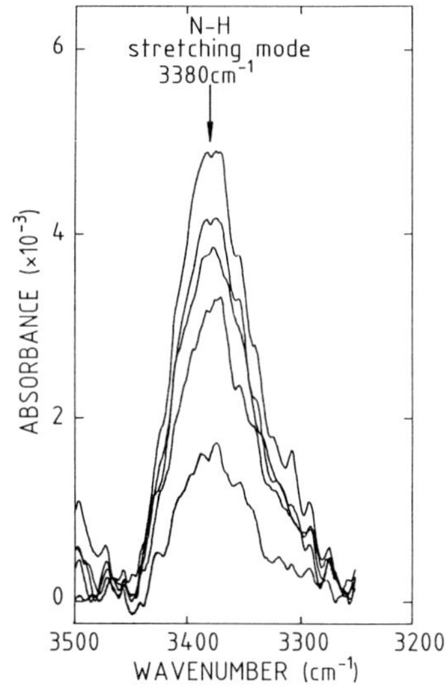

Fig. 3. Multiple infrared reflection–absorption spectra in the 3200–3500 cm^{-1} wavenumber region for consecutive UVCVD growth cycles of SiO_2 deposited from the SiH_4+N_2O mixture, showing the absorption due to the N–H stretching vibration mode at 3380 cm^{-1}.

paths appear possible, a direct adsorption of N_2O gas on the surface, which is observed in the IR spectrum from two narrow lines localized at 2234 and 2211 cm^{-1} and which can be directly photolyzed on the surface in the adsorbed phase. The nitrogen incorporation can equally come from the reaction between NO radicals produced in gas-phase reactions and the deposited layer surface. As a matter of fact, the photodissociation of N_2O gas is known to produce atomic oxygen $O(^1D)$, which reacts mostly with N_2O gas to produce O_2 and NO radicals [13]. Otherwise, the silicon hydride and dihydride densities were observed to be more important in nitrided UVCVD produced SiO_2 from SiH_4/N_2O mixture than from SiH_4/O_2, by about one order of magnitude for Si–H_2. This point will be discussed in the following section.

3.3. *Behaviour of the UVCVD SiO_2 surface under different gaseous treatments*

Since the UV photodissociation of N_2O produces atomic oxygen $O(^1D)$, which is known to be able to oxidize X–H species into OH groups and since the behaviour of silane exposure under UV irradiation on hydrated silica film surface was studied recently [14], we have used surface-sensitive infrared spectroscopy to try to give some new data in surface mechanisms in this deposition system. This way, we have deposited an ultra-thin film of silica with a SiH_4/N_2O mixture under UV irradiation by a low-pressure mercury lamp using the experimental conditions described above. The temperature and the duration of the deposition were fixed at 100°C and 30 min, respectively. A reference spectrum was taken after deposition, showing no absorption due to hydroxyl and adsorbed water groups in the spectra range 3300–3700 cm^{-1}, and the presence of a narrow line at 3385 cm^{-1} characterizing N–H bonds. The film thickness could be estimated from the absorbance of the Si–O stretching mode absorption band to be about 45 Å. The film was then first exposed to N_2O and then to SiH_4. Both gaseous treatments were performed during 10 min each, at a pressure of 2 Torr and a temperature of 100°C with the UV lamp on. Figs. 4a and 4b show the changes in film structure due to the gaseous treatments, respectively in the 3400–4000 and 2000–2300 cm^{-1} spectral regions. Curves labelled (a) in figs. 4a and 4b show the effect of the N_2O treatment on the film surface. A single narrow line appears at 3740 cm^{-1}, characteristic of *isolated* SiOH groups. These hydroxyl groups are separated by more than 3 Å [15]. At the same time, fig. 4b exhibits a consumption of Si–H and Si–H_2 localized at 2255 and 2202 cm^{-1}, respectively [14]. The 2165 cm^{-1} line corresponds probably to Si–H or Si–H_2 in the silicon dioxide matrix where one N atom (O_{3-x}NSi–H_x; $x = 1, 2$) is incorporated. O atoms being more electronegative than N atoms, the line position is downshifted from its position in the silicon dioxide matrix (O_{4-x}Si–H_x; $x = 1, 2$). This peak appears when the deposition temperature was 100 or 200°C in agreement with previously described observations, that the incorporation of N atoms increases with the deposition temperature. The data show that even if no hydroxyl group was present on the films, the oxidation of Si–H and Si–H_2 species by photodissociated N_2O into Si–OH groups is a possible reaction path. Curves labelled (b) in figs. 4a and 4b show the changes occurring during a subsequent exposure to silane and UV at 100°C. This treat-

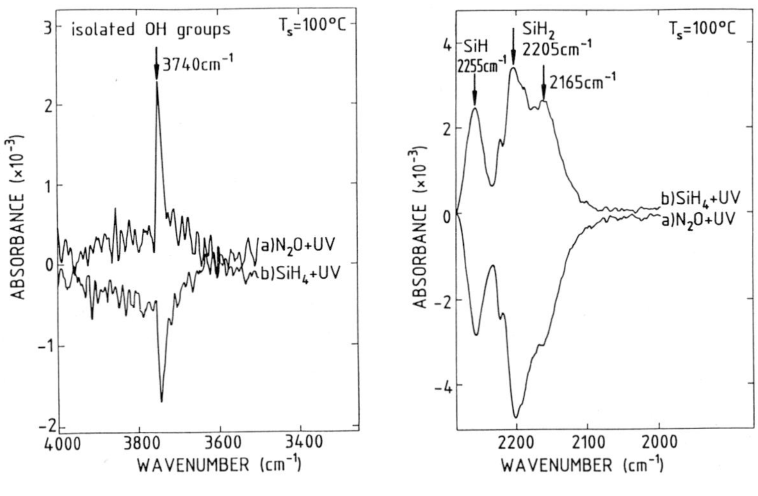

(a) (b)

Fig. 4. (a) Differential spectra in the OH region of an as-deposited ultra-thin silica film following N_2O + UV exposure (curve a) and then SiH_4 + UV exposure (curve b). These two specific gaseous treatments were performed at a substrate temperature of 100°C, a pressure of 2 Torr and an irradiation time of 15 min. A positive absorbance peak means addition of isolated Si–OH bonds while a negative peak indicates removal of such a bond during the treatment. (b) Modifications in the Si–H region of the as-deposited ultra-thin silica film following gaseous treatments described in fig. 3a.

ment leads to the suppression of Si–OH groups and the formation of Si–O–Si, Si–H and $Si–H_2$ bonds in agreement with the observation that silane reacts with photo-excited hydrated surfaces [14,16]. The reaction for surface oxidation of hydride and dihydride species appears very slow compared to deposition kinetics, so that incorporation of OH is very weak, explaining the appearance of the specific line attributed to *isolated* Si–OH. The opposite occurs in the case of the photolyzed O_2 treatment on pure SiO_2 surface [14]. This treatment produces an important amount of H_2O adsorbed and Si–OH groups, characterized by a large absorption band between 3300 and 3700 cm^{-1}. If the silanization step appears fast, the photo-induced oxidation step appears very slow. With molecular oxygen as a precursor the rate of oxidation is faster. Hence the surface of films deposited by UVCVD from SiH_4/O_2 mixture should be covered with a significant amount of hydroxyl and water groups. In this scheme, the surface of films deposited by UVCVD from SiH_4/N_2O mixture should on the contrary be covered with a significant amount of hydride and dihydride species in agreement with the observations. The low atomic oxygen production and the presence of N_2O gas and NO radicals in the gas explain respectively: (i) the low surface reaction kinetics with hydrided species, hence the high density of hydride and dihydride species; (ii) the reaction of N_2O gas and NO radicals on the surface to form N–H bonds. A second surface mechanism was apparent during these treatments, namely direct adsorption of N_2O gas to the surface. The importance of this reaction in the overall deposition mechanisms can presently be quantified. When N_2O gas exposure alone is performed on the layer surface, one observes in the IR spectrum two narrow lines localized at 2234 and 2211 cm^{-1} with a full-width at medium-height (FWMH) of 25 and 33 cm^{-1}, respectively. In fig. 3a, the FWMH of two lines localized at 2255 and 2205 cm^{-1} are greater because of the broadening due to the bond-angle variation in amorphous networks. As a matter of fact, the spectrum is the superposition of the two lines due to the N_2O gas adsorption (in positive) and of the Si–H stretching lines (in negative). The reverse spectrum is observed under exposures to SiH_4 and UV irradiation. It is therefore observed that the silane can react with the N_2O adsorbed species under UV irradiation, so that direct photolysis of the adsorbed layer is a parallel reaction path also accounting for SiO_2 deposition.

4. Conclusions

In summary, we have shown that the deposition rate of SiO_2 from SiH_4/N_2O mixture was independent of the temperature in the range 100–300°C. With molecular oxygen as a precursor, the deposition rate increases with the temperature. After multiple reactions in the gas phase described in the previous section, the low production of atomic oxygen does not allow a fast oxidation reaction with hydride and dihydride species present at the surface layer. Therefore, it induces a vanishing amount of silanol species characterized by the presence of a narrow single line assigned to *isolated* Si–OH. This observation extends to layers with thickness up to 2500 Å, where infrared transmission spectra do not show any measurable absorption due to Si–OH groups and H_2O. In the overall deposition mechanisms, this reduced photo-oxidation reaction contributes probably to the low deposition rate of SiO_2 from SiH_4/N_2O mixture.

References

[1] P.K. Boyer, G.A. Roche, W.H. Ritchie and G.J. Collins, Appl. Phys. Lett. 40 (1982) 716.
[2] M. Petitjean, N. Proust, J.F. Chapeaublanc and J. Perrin, Appl. Surf. Sci. 46 (1990) 189.
[3] T. Ito, T. Nakamura and H. Ishikawa, IEEE Trans. Electron Devices ED-29 (1982) 498.
[4] M.M. Moslehi, K.C. Saraswat and S.C. Shatas, Appl. Phys. Lett. 47 (1989) 1113.
[5] T. Hori, H. Iwasaki and K. Tsuji, IEEE Trans. Electron Devices 36 (1989) 340.
[6] Y.J. Chabal, Surf. Sci. 46 (1990) 175.
[7] Y.I. Nissim, J.M. Moison, F. Houzay, F. Lebland, C. Licoppe and M. Bensoussan, Appl. Surf. Sci. 46 (1990) 175.
[8] D.C. Paine, J.J. Rosenberg, S.C. Martin, D. Luo and Kawazaki, Appl. Phys. Lett. 57 (1990) 1443.

[9] C. Licoppe, C. Meriadec, Y.I. Nissim and J.M. Moison, Appl. Surf. Sci. 54 (1992) 445.
[10] C.T. Kirk, Phys. Rev. B 38 (1988) 1255.
[11] P. Lange, J. Appl. Phys. 66 (1989) 201.
[12] F. Lebland, C. Licoppe, Y. Gao and Y.I. Nissim, Appl. Surf. Sci. 24 (1991) 125.
[13] D.L. Baulch, R.A. Cox, P.J. Crutzen, R.F. Hampson, Jr., J.A. Kerr, J. Troe and R.T. Watson, J. Phys. Chem. Ref. Data 11 (1982) 327.
[14] C. Licoppe and C. Debauche, Phys. Rev. B, in press.
[15] V.F. Kiselev and O.V. Krylov, Adsorption Processes on Semiconductor and Dielectric Surfaces (Springer, Berlin, 1985) p. 116.
[16] C. Licoppe, Y.I. Nissim and J.M. Moison, Phys. Rev. B 45 (1992) 6275.
[17] C. Debauche, C. Licoppe, J. Flicstein and R.A.B. Devine, Appl. Phys. Lett. 61 (1992) 306.

Applied Surface Science 69 (1993) 317–320
North-Holland

applied
surface science

Ar^+ laser-induced forward transfer (LIFT): a novel method for micrometer-size surface patterning

Z. Tóth [a], T. Szörényi [a] and A.L. Tóth [b]

[a] *Research Group on Laser Physics of the Hungarian Academy of Sciences, Dóm tér 9, H-6720 Szeged, Hungary*
[b] *Institute for Technical Physics, P.O. Box 76, H-1325 Budapest, Hungary*

Received 2 June 1992; accepted for publication 30 October 1992

As a result of a systematic study on laser-induced forward transfer of tungsten (W) thin films by a focused Ar^+ laser ($\lambda = 514$ nm, $2w_0 = 3$ μm), it is shown that deposition of 2–3 μm diameter W spots of 100 nm thickness and 100% coverage is attainable at low processing powers (50–100 mW).

1. Introduction

The transfer of thin absorbing films from a transparent support, onto a target substrate in close proximity to the film, with single laser pulses is a challenging novel approach to surface patterning. The feasibility of this technique, the laser-induced forward transfer (LIFT), has already been reported for metal layers [1–8], Y–Ba–Cu–O and Bi–Sr–Ca–Cu–O superconducting thin films [9], metal oxides [10] and compound semiconductors [11,12], using pulsed lasers (5 ns < FWHM < 100 ns).

The generally applied pulsed laser processing of thin films in the high fluence regimes (up to some J/cm^2) leads to excessive heating rates causing a fast vaporization of the illuminated area. The flight of the ablated material to the surface to be patterned is turbulent, resulting in deposits that spread beyond the print area and usually are of poor morphology. This effect severely restricts the spatial resolution of this technique: the smallest pattern sizes reported are in the regime of some tens of μm [1–4,9].

To avoid the spreading of the material during transport between the support and the target an obvious solution is to ablate and transfer the material (at least partially) in solid phase as described in refs. [6–9]. In this contribution we demonstrate that a focused beam of TEM_{00} mode of an Ar^+ laser allows deposition of metal patterns of μm-size lateral dimension and of improved quality. Relatively low processing powers (hundreds of mW) and pulse durations up to milliseconds are suitable to transfer spots in solid phase in a single piece.

2. Experimental

Tungsten films of 100 nm thickness vacuum evaporated onto glass substrates are used as source material. Because of its high melting and boiling points W is an ideal (model) material allowing processing in large power domains in both solid and molten phases. The measured reflectivity and transmission of the sample at $\lambda = 514.5$ nm are 30 and 1.5%, respectively. Glass and silicon plates are placed in contact with the W films and this sample is mounted on an *XY*-micrometer stage. The samples are illuminated with a focused beam of Gaussian spatial distribution of an Ar^+ laser. The beam diameter, $2w_0$ (w_0 is the radius of the $1/e^2$ decrease in intensity), measured by the scanning knife edge method, is 3 μm. The power impinging on the sample surface and the illumination time is set by an electro-optical modulator between 0 and 700 mW and from 1 μs to 1 ms, respectively. The ablated areas and the deposits are characterized by optical and scanning electron microscopy in secondary and backscattered electron imaging mode (SEM, SEI and BEI, respectively).

3. Results

The modification (i.e. rapid colouring) of the surface of the W film due to irradiation is observed at 18 mW absorbed power (all values quoted below are absorbed powers). The onset of ablation starts at 40–50 mW, practically independent of the length of the processing pulses above 10 μs. Processing with powers just above the ablation threshold results in the best quality of the prints. The reproducibility of the prints is relatively poor when using shorter pulses, and is increased with increasing pulse duration. Above 100 μs the process is stabilized, resulting in 2–3 μm diameter deposits. Fig. 1 shows a typical SEM micrograph of a transferred pattern on a Si surface. The edges of the spot on the substrate are of sharp contrast and scalloped, indicating that the material of the spot has been blown off in solid phase from the original film. The darker area in the center of the print (in the secondary electron image) suggests that this part has been thinned down. The most probable explanation is that this central part has melted as a result of the inhomogeneous heating by the Gaussian beam, and moved out due to surface tension forces [13]. In fig. 2 a secondary electron image of a spot is shown, the sample being tilted at 70° to make the off-surface edges visible. The deposits stick to the substrate in the inner parts only, the edges of the spots stand out from the surface of the substrate. This fact is also supported by (characteristic shadow-like) backscattered electron images using a detector in tangential position.

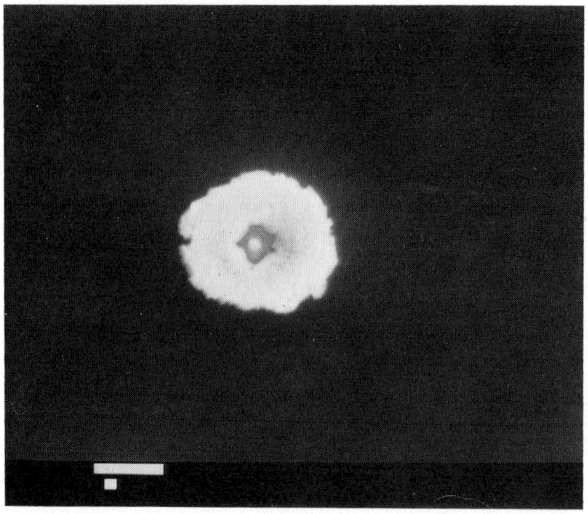

Fig. 1. Scanning electron micrograph of a pattern transferred onto a silicon surface by LIFT of a 100 nm thick W film by a pulse of 52 mW power and 50 μs duration. The white scaling bar indicates 1 μm (beam energy: 12.5 keV; sample current: 44 pA).

By increasing the irradiation power the effect of the inhomogeneous heating becomes dominating, even for shorter pulse durations (2–50 μs). Due to excessive heating the middle of the deposits is first partially, and subsequently com-

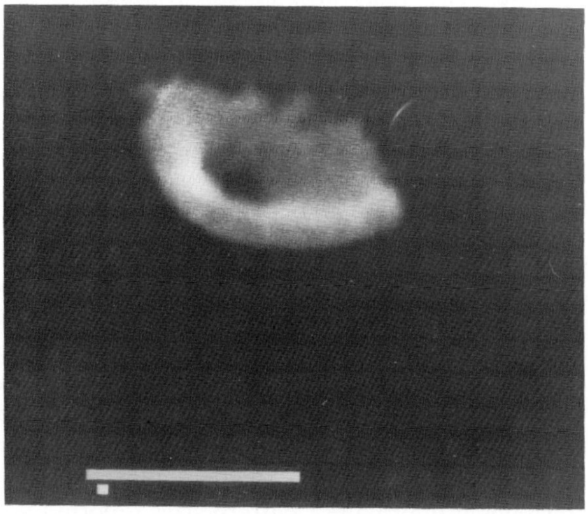

Fig. 2. Scaning electron micrograph of a pattern transferred by a pulse of 116 mW and 20 μs length. The sample is tilted with 70°; the scaling bar indicates 1 μm (beam energy: 25 keV; sample current: 20 pA).

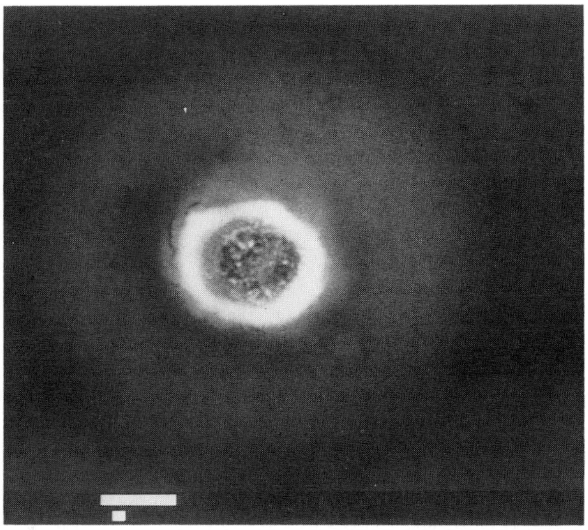

Fig. 3. Scanning electron micrograph of a pattern transferred by a 476 mW pulse of 200 μs duration. The scaling bar indicates 1 μm (beam energy: 12.5 keV; sample current: 67 pA).

pletely evaporated, leading to deteriorated patterns as shown in fig. 3. The diameter of the evaporated part is increasing with increasing power and illumination time.

4. Model calculations and discussion

To obtain a consistent explanation of the experimental results outlined above the evolution of spatial temperature profiles was calculated. The solution of the thermal conductivity equation for the layer system of cylindrical symmetry supposing no material loss was obtained numerically using the method of finite differences.

In fig. 4 the steady state lateral temperature distributions are shown for absorbed powers falling into the power regime applied. The main information from our calculation is that no melting can take place below ~ 130 mW absorbed power since even the possible maximum temperatures in the performed experiments are below the melting point, $T_m = 3653$ K. Therefore, the ablation starts in solid phase in the experimental power regime. Because no ablation is taken into account in the model, these results can only give an upper limit to the temperatures which can be reached.

As soon as the thermal contact between the support and the film breaks down and film transfer starts the temperature loss is due to radiation, while – depending on pulse duration – the processing beam still maintains an in-flight heating. The larger the distance between the supported film and the surface to be patterned, the longer the transfer time which can lead to excessive heating of the film material. Probably the melting of the inner parts of the deposits is due to this effect.

As the transferred pattern splashes to the substrate surface cooling or further heating of the pattern material takes place depending on the pulse duration and the thermal parameters of the substrate. In fig. 5 the maximum (steady state) temperatures are plotted for three characteristic cases. The middle curve gives the temperatures reached in the middle of the beam in the original W film on the glass support (fig. 5a). The upper and lower curves show the calculated maximum temperatures of 3 μm diameter W films of the same thickness transferred onto glass (fig. 5b) and silicon (fig. 5c) substrates, respectively. Since the thermal conductivity of the glass is small and the lateral heat flow from the localized spot is restricted, the temperatures become higher as compared to the case of the supported original

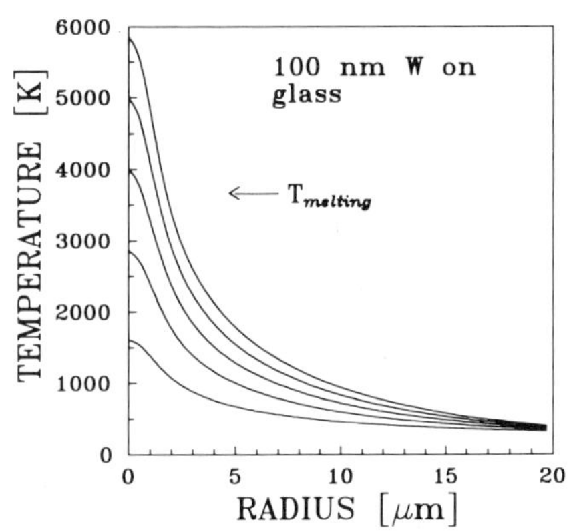

Fig. 4. Calculated steady state lateral temperature distributions in 100 nm thick W films, illuminated with a focused Ar$^+$ beam ($2w_0 = 3$ μm). The absorbed powers are 50, 100, 150, 200 and 250 mW, corresponding to curves of increasing temperatures.

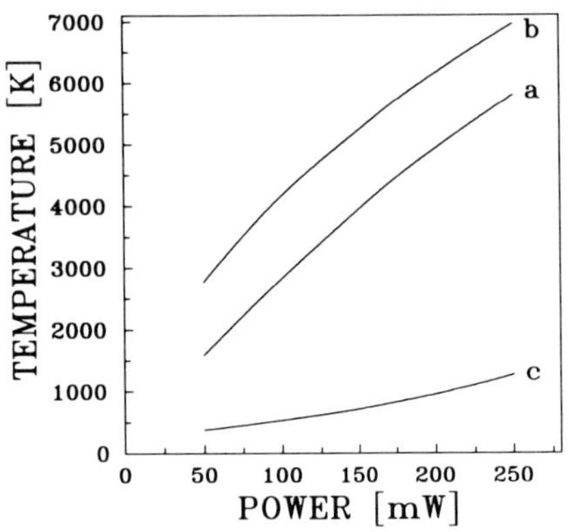

Fig. 5. Maximum temperatures as function of the absorbed power. (a) 100 nm thick W film on glass; (b) 100 nm thick 3 μm diameter W spot on glass; (c) 100 nm thick 3 μm diameter W spot on silicon.

(large area) thin film. This situation is relevant when transferring onto a glass substrate providing the pulse duration is considerabley longer than the time needed for ablation and transfer. The resulting annealing can modify the sticking and the morphology of the transferred pattern, so the exact setting of the illumination time is essential in determining its properties. When using silicon of high thermal conductivity as the substrate to be patterned, a fast cooling of the film material takes place after landing, so the morphology of this system will not depend rather sensitively on the duration of the annealing.

5. Conclusions

The possibility of LIFT by means of a focused Ar$^+$ laser beam is demonstrated. It is shown that best quality patterns can be attained by transferring thin films in solid phase. The deposited tungsten spots have a lateral dimension of 2–3 μm which is considerably lower than previously reported in the LIFT literature. A comparison of experimental data with temperature calculations reveals that the pulse duration plays a decisive role in determining the print properties. By using laser pulses of a duration ranging from 1 μs to 1 ms the possibility of in-flight heating and subsequent annealing is demonstrated. Further research is needed to optimize this annealing process and eliminate the damaging consequences of the inhomogeneous heating.

In summary, this procedure offers a simple, single step technology for surface patterning in the μm-regime.

Acknowledgements

This research was partially supported by the State Office for Technical Development, Hungary (OMFB 92-97-10-0004), and the Hungarian National Science Foundation (OTKA I/3 3060).

References

[1] J. Bohandy, B.F. Kim and F.J. Adrian, J. Appl. Phys. 60 (1986) 1538.

[2] F.J. Adrian, J. Bohandy, B.F. Kim and A.N. Jette, J. Vac. Sci. Technol. B 5 (1987) 1490.

[3] J. Bohandy, B.F. Kim, F.J. Adrian and A.N. Jette, J. Appl. Phys. 63 (1988) 1158.

[4] R.J. Baseman, A. Gupta, R.C. Sausa and C. Progler, in: Laser and Particle-Beam Chemical Processing for Microelectronics, Vol. 101, Eds. D.J. Ehrlich, G.S. Higashi and M.M. Oprysko (MRS, Pittsburg, PA, 1988) p. 237.

[5] D.E. Alexander, G.S. Was and F.J. Mayer, J. Mater. Sci. 23 (1988) 2181.

[6] P. Mogyorósi, T. Szörényi, K. Bali, Zs. Tóth and I. Hevesi, Appl. Surf. Sci. 36 (1989) 157.

[7] Zs. Tóth, Z. Kántor, P. Mogyorósi and T. Szörényi, in: Laser Assisted Processing II, Ed. L.D. Laude (Proc. SPIE 1279, 1990) p. 150.

[8] Z. Kántor, Z. Tóth and T. Szörényi, Appl. Phys. A 54 (1992) 170.

[9] E. Fogarassy, C. Fuchs, F. Kerherve, G. Hauchecorne and J. Perriere, J. Mater. Res. 4 (1989) 1082.

[10] J.A. Greer and T.E. Parker, in: Excimer Beam Applications, Eds. A.N. Pirri, B.P. Piwczyk (Proc. SPIE 998, 1988) p. 113.

[11] T. Szörényi and Z. Tóth, in: Laser Assisted Processing, Eds. L.D. Laude and G. Rauscher (Proc. SPIE 1022, 1989) p. 93.

[12] Z. Tóth and T. Szörényi, Appl. Phys. A 52 (1991) 273.

[13] D.J. Broer and L. Vriens, Appl. Phys. A 32 (1983) 107.

Applied Surface Science 69 (1993) 321–325
North-Holland

applied
surface science

Laser beam deflection monitoring of Nd:YAG laser ablation: pulse shape and repetition rate effects

Janez Diaci and Janez Možina
Faculty of Mechanical Engineering, University of Ljubljana, P.O. Box 394, 61000 Ljubljana, Slovenia

Received 2 June 1992; accepted for publication 30 October 1992

The laser beam deflection probe has been employed to study blast waves generated during ablation of metallic surfaces by sequences of 1.06 μm Nd:YAG laser pulses separated by less than 1 μs. A fluence threshold has been found, below which the effects of individual pulses can be resolved by the laser probe. Above that, the deflection signal has a similar form as if the surface were irradiated with a single pulse. Analysis of the signals in terms of the spherical blast wave theory shows that a pulse sequence generates a weaker blast wave than a single pulse of equal total energy. On the other hand, the sequence yields a higher etch depth than the single pulse.

1. Introduction

Laser beam deflection (LBD) technique has recently been introduced to the laser ablation studies as a quick and reliable means to determine the ablation thresholds [1]. Applicable to all kinds of surfaces and materials, it allows on-site and on-line monitoring of the etch depth [2] and provides valuable new information that complements the data obtained by other methods [3–5]. Analysis of the detected signal waveforms enables time resolved studies of the ablation plume features such as acoustic, shock and cooling waves [6–8].

In this contribution we examine ablation of metallic surfaces by sequences of 60 ns Nd:YAG laser pulses with repetition intervals shorter than 1 μs. The multipulse technique offers significant advantages, e.g. higher material removal rates or very fine machining quality, in some laser machining applications [9]. Though on a different temporal scale, the multipulse technique also appears to be a promising option for the ablation, especially for the systems employing fiberoptic power delivery, where the transmitted intensity should be kept low to prevent the damage of the fiber.

2. Experimental system and procedure

The schematic of our experimental setup is shown in fig. 1. A Q-switched laser of 1.06 μm wavelength was used as the ablation source. Various temporal profiles of the ablation pulse were achieved by variation of the optical pump energy. The ablation beam was focused to a spot of about 0.5 mm in diameter to achieve ablative fluence on the sample surface. Calibrated neutral density filters were inserted into the beam to vary the

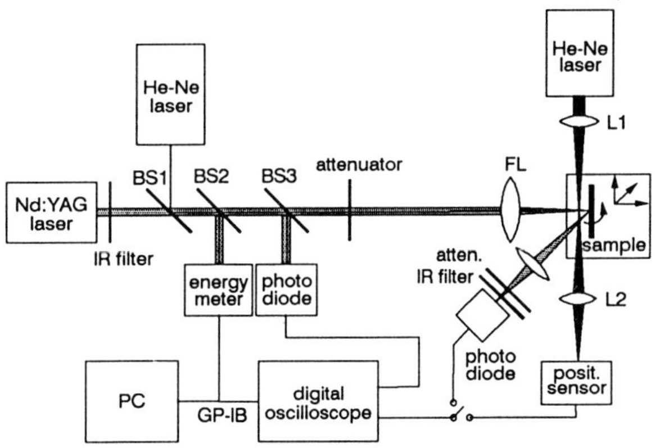

Fig. 1. Experimental setup.

incident energy on the sample surface. Laser pulse energy was monitored by an energy meter receiving a fraction of the pulse diverted by the beam splitter BS2.

A 5 mW He–Ne probe beam was aligned parallel to the irradiated surface and perpendicular to the ablation beam. Deflection of the probe beam due to the passage of the ablation plume was detected by a position sensitive detector with a 2.5 MHz bandwidth. Since the finite size of the probe beam limits the temporal resolution of the probe we focused the beam by a 100 mm focal length lens to a waist of 37 μm located above the irradiated spot. The probe thus allowed measurements with a bandwidth of 1.5 MHz, as estimated from the transit time of the acoustic wave across the waist of the probe beam. A 500 MHz photodiode, receiving a part of the ablation beam, was employed to detected the temporal profile of the ablation pulse and to provide the triggering signal for the oscilloscope. Both signals were amplified, digitized by a sampling oscilloscope and fed into a PC computer.

Steel and aluminum samples were ablated in air at atmospheric pressure. By changing the temporal profile, the total incident energy and the distance between the probe beam axis and the surface, a set of signals was acquired for each sample.

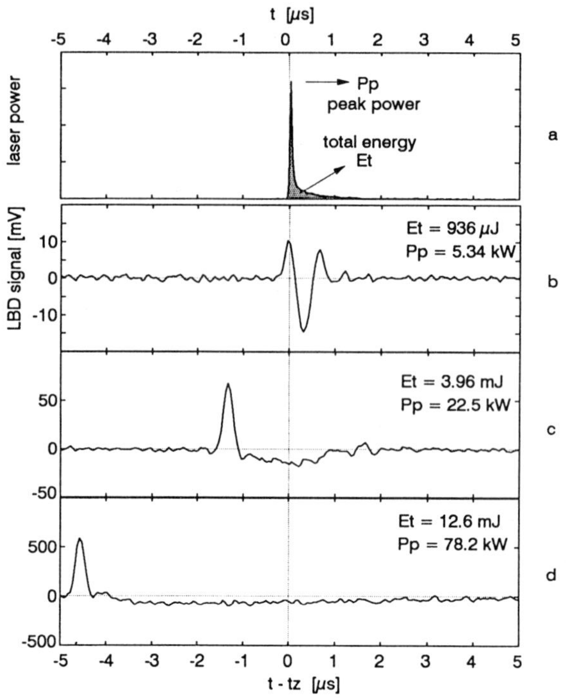

Fig. 2. Typical Nd:YAG laser single pulse (a) and the laser beam deflection (LBD) signals, (b)–(d), detected during irradiation of a mild steel surface by such pulses. With each LBD signal, the total incident energy E_t and the peak power P_p of the laser pulse are indicated. The deflection signals are shown retarded for the sound wave transit time $t_z = 8.60$ μs.

3. Results and discussion

To illustrate the ability of the LBD technique to resolve high repetition rate ablation events we present two sets of LBD signals acquired during irradiation of a mild steel surface. The set in figs. 2b–2d results from irradiation with single laser pulses at various total incident energies E_t and the other one, in figs. 3b–3e, from multiple laser pulses at similar energies E_t. To enable a comparison between the LBD signals and the corresponding temporal profiles of the ablation pulses, we depicted the latter ones in the upper (a) graphs of figs. 2 and 3 using the same time scale. Note that the LBD signals are retarded for the time t_z required by a sound wave to traverse the distance r_s between the irradiated spot and the probe beam axis. The initial positive part of the LBD signals corresponds to compression.

Let us discuss the signals of fig. 2 first. The distance r_s was 3.00 mm in this case, corresponding to the sound wave transit time $t_z = 8.60$ μs. By comparison of the LBD signals of various amplitudes (cf. figs. 2b–2d), we observe that below the ablation threshold, the wavefront transit time from the source to the probe equals the sound wave transit time $t_z = r_s/c_0$. The wavefront therefore propagates with constant (sonic) velocity c_0. Above the threshold, the wavefront transit time t_s is generally shorter than t_z and depends on the amplitude, indicating that a shock is formed on the wavefront. With the increase of energy the signals extend in both directions: the compression arrives sooner (as a result of the

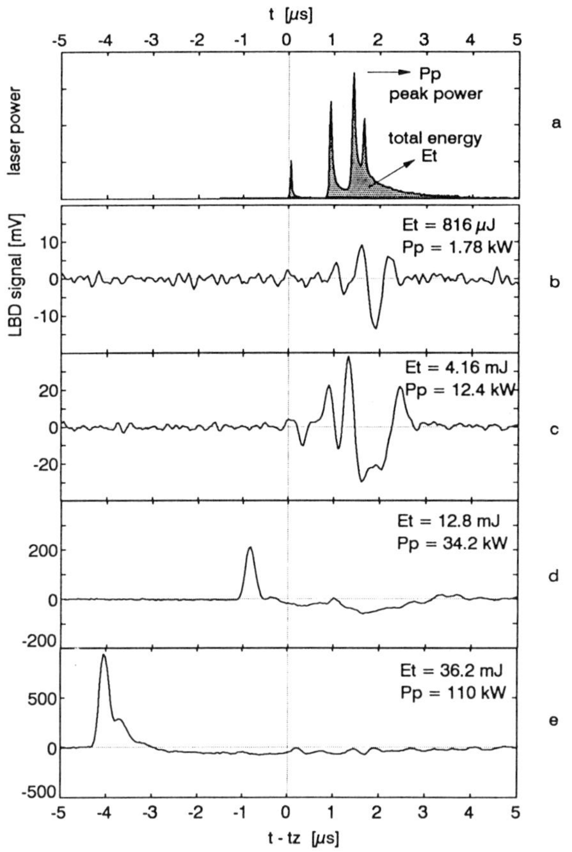

Fig. 3. Typical Nd : YAG laser multiple pulse (a) and the laser beam deflection (LBD) signals, (b)–(e), detected during irradiation of a mild steel surface by such pulses. With each LBD signal, the total incident energy E_t and the peak power P_p of the laser pulse are indicated. The deflection signals are shown retarded for the sound wave transit time $t_z = 11.43$ μs.

supersonic shock wavefront propagation) and the rarefaction lasts longer.

In the case of multiple pulse ablation (fig. 3), the distance r_s was 3.90 mm and the transit time $t_z = 11.43$ μs. At the lowest energy (fig. 3b) no plasma could be observed visually and no significant ablation occurred. The first three peaks of the laser pulse produced well separated bipolar deflection waveforms, which look quite similar to the waveforms observed with the single pulse. The waveform corresponding to the fourth peak cannot be separated from that of the third one presumably because of the limited bandwidth of the present configuration. With the increase of the incident laser energy above the ablation threshold (1.1 mJ), the second and the third peak in the LBD signal shift to the left (cf. figs. 3b and 3c) indicating that each of them corresponds to a separate shock. The time intervals between the peaks decrease with the increase of energy as a result of higher propagation speed of the shocks with higher amplitude. At about 6 mJ, the third shock catches up and merges with the second one, while the first wavelet, still propagating with acoustic velocity, indicates that no appreciable ablation takes place during the first peak of the laser pulse. Between 7 and 8 mJ, the merged shock overtakes the first wavelet and above that, we detect only single shocks: the signal (figs. 3d and 3e) consists of a single short positive pulse and a long negative tail resembling the waveforms observed with the single pulses of comparable energy. Above that threshold, the details of the interaction of individual pulses with the surface can not be resolved by the probe any more.

Recent studies [3,4,8] have shown that propagation of the spherical shock generated during the ablation process can be well described using the point explosion model [10]. Since there is no analytical solution in the intermediate shock region, which is our current interest, we used a published numerical solution [11]. The results of the theory are expressed in terms of dimensionless space $\xi = r/r_c$ and time $\tau = t/t_c$ coordinates with characteristic scaling parameters:

$$r_c = (\alpha E/p_0)^{1/3} \quad \text{and} \quad t_c = r_c/c_0, \tag{1}$$

that depend on blast energy E, pressure p_0 and sound velocity c_0 of the undisturbed gas. The constant α takes the value 1.175 for air [10].

In order to determine the blast energy E we used the theoretical shock trajectory $\tau_s = f(\xi_s)$ given in a tabular form [11]. Employing spline interpolation, we constructed an approximation of a function that would map τ_s/ξ_s into ξ_s. From the definition of the dimensionless coordinates it follows that:

$$\tau_s/\xi_s = t_s/t_z. \tag{2}$$

By inserting measured data into the right-hand side of eq. (2) we calculated the argument of the spline function and then found the corresponding

ξ_s as the spline function value. Using eqs. (1) we then determined E:

$$E = (p_0/\alpha)(r_s/\xi_s)^3. \quad (3)$$

The energy E_h contained in a hemispherical wave is assumed to be a half of the blast energy that would produce the same mechanical effect in the spherical case: $E_h = E/2$.

For single pulses, we took the moment of triggering the oscilloscope as the time origin, i.e. the instant of the release of the blast energy E. For multiple pulses producing merged shocks, however, we took the beginning of the third pulse as the time origin, because that pulse produced blast wave that overtook the preceding ones.

Comparing the results from the single and multiple pulses with equal total incident energy E_t (table 1), we find that the single pulses produce blast waves much more efficiently. For a 24.2 mJ single pulse we obtained a very high figure of 42% for the conversion efficiency. A possible explanation is that a laser supported absorption wave was created in the early phase of development of the ablation plume. In the pertinent intensity range (between 10^7 and 10^8 W/cm^2) the expanding plasma plume becomes highly absorbing for the incoming laser radiation and supplies the shock with energy which sustains its strength and high propagation velocity and postpones its decay. A preliminary investigation of the reflected laser light has shown that a significant decrease of the reflected power took place during the most intense part of the ablation pulse. The reflected power, however, did not diminish so drastically to allow the conclusion that a laser supported detonation wave was

Table 1
A comparison of the parameters of blast waves produced during ablation of a mild steel sample with single and multiple laser pulses

	E_t (mJ)	t_z (μs)	t_s (μs)	ξ_s	E_h (mJ)
Single pulse	12.6	8.60	3.72	0.71	3.2
	24.2	8.60	2.63	0.48	10.2
Multiple pulse	12.8	11.43	8.96	2.71	0.13
	25.4	11.43	6.57	1.16	1.7

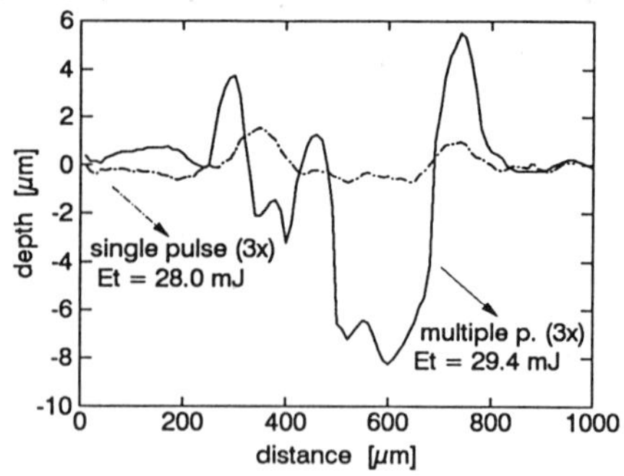

Fig. 4. Shape profiles of ablation craters produced by three consecutive irradiations of a stainless steel surface by single pulses (the dash-dot line) and multiple pulses (the solid line) with similar total energy.

formed, which would completely absorb the subsequent part of the ablation pulse [12,13].

Inspection of the irradiated surfaces shows in all cases a monotonous increase of the etch depth with pulse energy, confirming that no appreciable plasma shielding took place. The amplitude of the LBD signal, which also increases monotonically with pulse energy, could therefore be used for the on-line monitoring of the etch depth. In the case of multiple pulse irradiation, however, the amplitude changes so drastically when the consecutive shocks merge that an integral parameter, like mean square value of the signal, should be more appropriate for monitoring the total etch depth.

For the same total energy, the multiple pulse irradiation produces a higher etch depth than the single pulse (fig. 4), but it also yields a larger heat affected zone, more irregular shape of the crater with larger remnants of melted and resolidified material.

In conclusion, we find the laser probe a very suitable tool for monitoring the multipulse ablation. Because of the wide bandwidth and inherently differential behavior, it offers a possibility to resolve the effects of individual pulses in a high repetition rate pulse sequence.

References

[1] J.A. Sell, D.M. Heffelfinger, P. Ventzek and R.M. Gilgenbach, Appl. Phys. Lett. 55 (1989) 2435.

[2] J.A. Sell and D.M. Heffelfinger, Bull. Am. Phys. Soc. 35 (1990) 1528.
[3] P.E. Dyer and J. Sidhu, J. Appl. Phys. 64 (1988) 4657.
[4] T. Zyung, H. Kim, J.C. Postlewaite and D.D. Dlott, J. Appl. Phys. 65 (1989) 4548.
[5] R. Srinivasan, B. Braren, K.G. Casey and M. Yeh, Appl. Phys. Lett. 55 (1989) 2790.
[6] G. Koren, Appl. Phys. Lett. 51 (1987) 569.
[7] H. Guo and Q. Lou, Opt. Commun. 77 (1990) 381.
[8] J. Diaci and J. Možina, in: Physical Acoustics, Eds. O. Leroy and M.A. Breazeale (Plenum, New York, 1991) p. 291.
[9] N. Rykalin, A. Uglov and A. Kokora, Laser Machining and Welding (Mir, Moscow, 1978) [in Russian].
[10] L.I. Sedov, Similarity and Dimensional Methods in Mechanics (Nauka, Moscow, 1972) [in Russian].
[11] D.E. Okhotsimsky and Z.P. Vlasova, Zh. Vychisl. Mat. Mat. Fiz. 2 (1962) 107 [in Russian].
[12] E. Stürmer and M. von Allmen, J. Appl. Phys. 49 (1978) 5648.
[13] Y. Xia, L. Mei, C. Tan, X. Liu, Q. Wang and S. Yue, Appl. Phys. A 45 (1991) 425.

Applied Surface Science 69 (1993) 326–329
North-Holland

applied
surface science

Quest for high quality local electroless laser deposition from the liquid phase: decomposition of ammonium molybdate

K. Bali [a], Zs. Geretovszky [a], A.L. Tóth [b] and T. Szörényi [c]

[a] *Department of Experimental Physics, JATE University, Dóm tér 9, H-6720 Szeged, Hungary*
[b] *Institute for Technical Physics, P.O. Box 76, H-1325 Budapest, Hungary*
[c] *Research Group on Laser Physics, Dóm tér 9, H-6720 Szeged, Hungary*

Received 2 June 1992; accepted for publication 30 October 1992

A systematic study on concentration (0.001–0.1M), laser power (up to 200 mW) and scanning speed (from 10 to 200 μm/s) dependence of pyrolytic laser decomposition of ammonium molybdate is reported. It is shown that there exists an optimal concentration range in which best quality deposition is observed and all the disadvantageous features typical for slow writing can significantly be improved by the increase of scanning speed. At speeds exceeding 0.1 mm/s well defined lines without macrostructure and of excellent morphology can be written and all the important characteristics of the lines remain virtually unchanged in a wide range of processing parameters.

1. Introduction

It is apparent from a literature survey that laser deposition from solutions has attracted only limited attention as compared to LCVD and, more recently, decomposition of solid surface layers [1–6]. This is rather surprising since the use of liquid precursors offers obvious advantages: (*i*) a wide range of suitable precursors which are, in general, non-toxic; (*ii*) the high precursor concentration is conveniently adjustable over several orders of magnitude; (*iii*) the simplicity, since no complicated vacuum and gas handling systems or coating/removing techniques are needed unlike in the case of LCVD and deposition from spun-on layers, respectively, and (*iv*) the use of low-cost materials. Although the feasibility of metal deposition from liquid phase has already been demonstrated [1–3] the potentialities and limits of this technique are far from being thoroughly explored.

The choice of solutions of organic precursors has the advantage of (usually) readily accessible decomposition; however, the synthesis of such chemicals is more involved and special care is needed because of their toxicity. Contrary to this, solutions of inorganic compounds are, in general, inexpensive, more stable and their preparation is straightforward. With the exception of a few cases, when the solution contains the metal in elemental or in non-oxidized form, the most problematic step of deposition is the reduction of the metal containing ions. Since the use of external current sources [7,8] or the need for a substrate serving as an electron source driven by the laser irradiation [9] sets a severe limitation to the applicability of the process, a more general solution which can be universally applied is the production of electrons by the laser-initiated chemical reaction itself [10–13]. A promising class of precursors, containing reducing agents which are far too stable to react under "ordinary" conditions but can be decomposed under the extreme conditions produced at the irradiated surface, are ammonia-containing metal salt solutions [14].

In this contribution the first results of a systematic study, aiming at the determination of the fundamentals governing electroless pyrolytic decomposition of aqueous solutions of inorganic precursors, are reported using aqueous solutions of ammonium heptamolybdate as a model system.

2. Experimental

The Gaussian beam of the Ar^+ laser ($\lambda = 514.5$ nm) is focused on the surface of the substrate ($1/e^2$ diameter: $2w_0 = 4.4$ μm) covered by 0.001–0.1M $(NH_4)_6Mo_7O_{24}$ of 2 mm thickness. All solutions used are transparent in the visible range. The absorption coefficient measured at the wavelength of the Ar^+ laser using the limiting high concentration (0.1M) is 1.6×10^{-2} cm^{-1}. The substrates are glass microscope slides covered with 130 nm vacuum deposited W. Lines are deposited by moving the sample holder perpendicularly to the vertically aligned laser beam with scanning speeds up to 200 μm/s. The lines deposited are characterized by profilometry (TALYSTEP), optical and scanning electron microscopy.

3. Results and discussion

Contrary to the anticipation that higher precursor concentrations allow deposition at higher scanning speeds it has been found, to our surprise, that in the range of speeds applied there is a range of optimal concentrations between 5×10^{-3} and 10^{-2}M in which effective deposition of continuous lines of appropriate morphology is attainable. At low concentrations (10^{-3}M) continuous but still structured lines can only be obtained in a narrow power range between 20 and 50 mW. At high concentrations, although high deposition yield is observed, the quality of deposits is rather poor as verified by optical microscopy and SEM. In the case of 5×10^{-2}M roughly structured lines can be written above 70–90 mW and at lower powers a dot-like pattern is observed. In the case of 10^{-1}M solutions, on the other hand, no continuous lines can be deposited in the whole processing parameter range. This effect is not related to the practically negligible absorption of the precursor solution, but is an

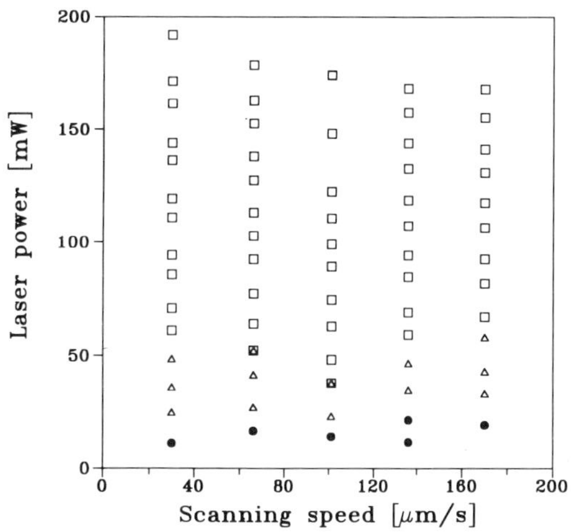

Fig. 1. Quality mapping of deposition from a 10^{-2}M aqueous solution of $(NH_4)_6Mo_7O_{24}$ in the laser power-scanning speed plane. Each symbol represents a different deposition grade: (●): no deposition; (△): dot-like deposition and (□): continuous line formation.

inherent feature of the deposition mechanism. Subsequently we shall focus our attention at the results measured in the optimal concentration range.

As shown in fig. 1 the threshold power for continuous line deposition is, nearly independently of the scanning speed, approximately 50 mW. This lower limit of line formation exactly coincides with the upper limit of bubble formation indicating that the quality of deposition is strongly correlated with the appearance of bubbles during deposition. This series of experiments sets a lower power limit; it can be concluded that the appropriate domain for well-adhering, continuous line deposition lies above 50 mW in the thoroughly examined 10–200 μm/s range.

A light microscopic examination of the stripes reveals that in a wide range of processing parameters material removal in the line center competes with deposition. As seen in fig. 2 this undesirable removal becomes dominant at higher powers and proceeds independently of deposition. Unlike the line formation threshold that of the removal shifts to higher power levels with increasing the scanning speed, and, as a consequence of increasing the speed, the useful domain of con-

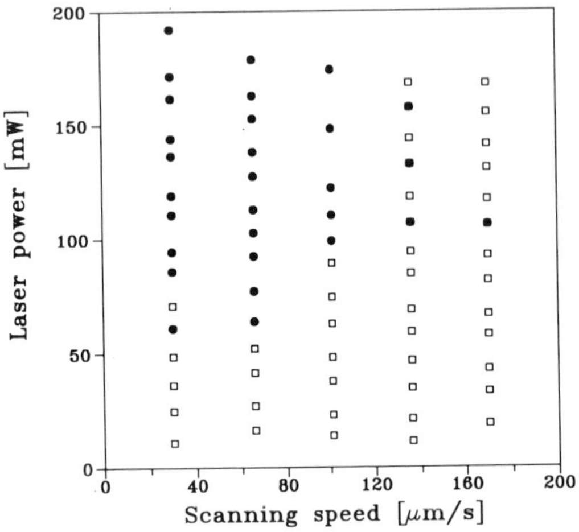

Fig. 2. Quality mapping of material removal from a 10^{-2}M aqueous solution of $(NH_4)_6Mo_7O_{24}$ in the laser power-scanning speed plane. (□): no removal and (●) removal.

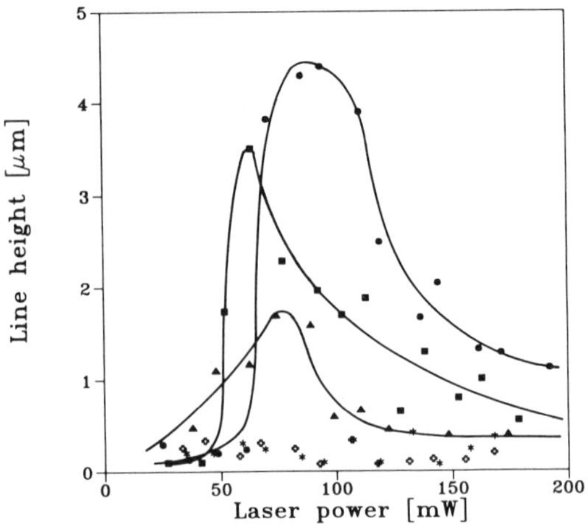

Fig. 3. Line height as a function of the laser power at five writing speeds: (●): 30 μm/s; (■): 66 μm/s; (▲): 101 μm/s; (∗): 136 μm/s and (✣): 170 μm/s. The curves are best fitting spline functions and are shown merely to highlight their biphasic character.

tinuous line deposition without removal can significantly be extended.

Throughout the range examined with increasing power the linewidth linearly increases from 5 to 25 μm, essentially independently of the scanning speed. The height versus power functions, on the other hand, display characteristic differences depending on the scanning speed (fig. 3). In the case of slow writing (< 136 μm/s) the height steeply increases on increasing the power and reaches a maximum just at the removal threshold (cf. fig. 2). The maximum values decrease with increasing speed. Beyond the removal threshold an abrupt decrease is observed. Applying higher speeds, the height remains constant at around 0.3 μm in the whole power range. The lack of the biphasic character is consistent with the observation that at speeds near and above 136 μm/s no material removal takes place.

The examination of the SEM micrographs further substantiates the notion that the writing speed plays an essential role in determining deposition characteristics. With increasing scanning speed there is a change in morphology: below 136 μm/s lines with periodic nodule structure and rough surface form (fig. 4), while at higher speeds practically structureless smooth lines with considerably smaller grain size (fig. 5) can be written.

As determined by energy and wavelenght dispersive X-ray analyses the composition of deposits is nearly stoichiometric MoO_3 in the whole domain of processing parameters. This composition is achieved via thermal decomposition of $(NH_4)_6Mo_7O_{24}$, however, the local temperature, even in the center of the Gaussian beam (as predicted also by parallel temperature distribu-

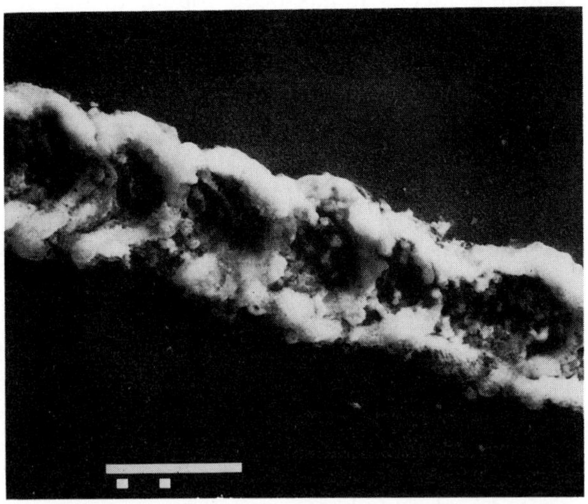

Fig. 4. Secondary electron image of a line written from a 10^{-2}M solution at a laser power of 77.2 mW and writing speed of 66 μm/s. The scaling bar denotes 10 μm.

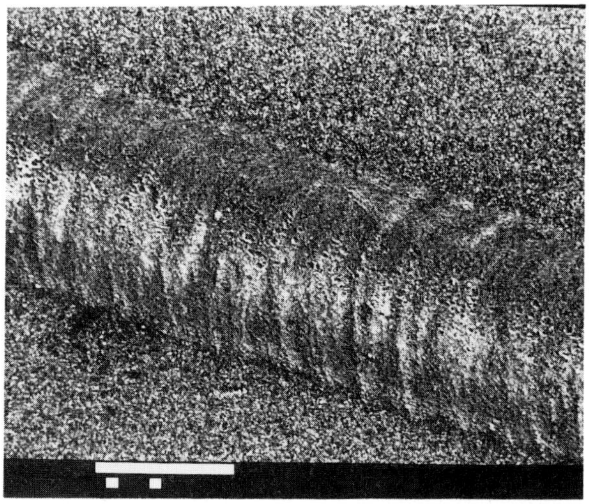

Fig. 5. Secondary electron image of a line written from a 10^{-2}M solution at a laser power of 94.4 mW and writing speed of 136 μm/s. The scaling bar denotes 10 μm.

tion calculations), is not sufficiently high to initiate a further decomposition of ammonia.

4. Conclusions

In summary, the potentiality of liquid phase deposition can best be utilized in a narrow concentration range (between 5×10^{-3} and 10^{-2}M) and the writing speed plays a crucial role in determining deposition characteristics. Above a laser power threshold of ~50 mW continuous lines can be obtained at any scanning speed while the material removal at high powers can efficiently be suppressed by increasing the speed. Deposition of well defined lines without macrostructure and of excellent morphology is possible at writing speeds > 0.1 mm/s. From technological point of view it is a definite advantage that, using high writing speeds, all the important characteristics of the lines remain virtually unchanged in a wide range of processing parameters. The main conclusion of this work is that under optimized conditions liquid phase deposition is a competitive technique with economic writing speeds and acceptable material properties.

Acknowledgement

This work was supported by the Hungarian National Science Foundation (OTKA I/3 3060).

References

[1] D. Bäuerle, Chemical Processing with Lasers, Volume 1 of Springer Series in Materials Science (Springer, Berlin, 1986).

[2] I.W. Boyd, Laser Processing of Thin Films and Microstructures – Oxidation, Deposition and Etching of Insulators, Volume 3 of Springer Series in Materials Science (Springer, Berlin, 1986).

[3] J. Haigh and M.R. Aylett, in: Laser Microfabrication, Eds. D.J. Ehrlich and J.Y. Tsao (Academic Press, San Diego, CA, 1989) p. 453.

[4] M. Widmer, P. Hoffmann, B. Lecohier, H. Solka, J. Philippoz and H. van den Bergh, in: Photochemical Processing of Electronic Materials (Academic Press, London, 1992) p. 433.

[5] P. Hoffmann, B. Lecohier, S. Goldoni and H. van den Bergh, Appl. Surf. Sci. 43 (1989) 54.

[6] S. Preuss and M. Stuke, Appl. Surf. Sci. 54 (1992) 308.

[7] R.J. von Gutfeld, R.E. Acosta and L.T. Romankiw, IBM J. Res. Dev. 26 (1982) 136.

[8] M.H. Gelchinski, L.T. Romankiw, D.R. Vigliotti and R.J. von Gutfeld, J. Electrochem. Soc. 132 (1985) 2575.

[9] J. Zahavi, S. Tamir, M. Rotel, G.J. Campisi, P.E. Pehrsson and M. Halliwell, in: Laser and Particle-Beam Chemical Processing for Microelectronics, Eds. D.J. Ehrlich, G.S. Higashi and M.M. Oprysko (Mater. Res. Soc. Proc. 101 (1988) 101).

[10] R.K. Montgomery and T.D. Mantei, Appl. Phys. Lett. 48 (1986) 493.

[11] J.W.M. Jacobs and C.J.C.M. Nillesen, J. Vac. Sci. Technol. B 8 (1990) 635.

[12] H. Yokoyama, S. Kishida and K. Washio, Appl. Phys. Lett. 44 (1984) 755.

[13] Y.F. Lu, M. Takai, T. Nakata, S. Nagatomo and S. Namba, Appl. Phys. A 52 (1991) 129.

[14] A.M.T.P. van der Putten, J.W.M. Jacobs, J.M.G. Rikken and K.G.C. de Kort, in: Laser Assisted Processing, Eds. L.D. Laude and G. Rauscher (SPIE Proc. 1022 (1989) 71).

Applied Surface Science 69 (1993) 330–334
North-Holland

applied
surface science

Time resolved reflectivity and transmission measurements during laser induced blow-off of thin metal films

Z. Tóth [a], J. Solis [b], C.N. Afonso [b], F. Vega [b] and T. Szörényi [a]

[a] Research Group on Laser Physics of the Hungarian Academy of Sciences, Dóm tér 9, 6720 Szeged, Hungary
[b] Instituto de Optica, CSIC, Serrano 121, 28006 Madrid, Spain

Received 2 June 1992; accepted for publication 30 October 1992

Time resolved reflectivity and transmission measurements with nanosecond time resolution are used to follow the evolution of ablation and material transport during Ar^+ and ArF excimer laser induced blow-off of W and Cr films deposited onto glass substrates. The analysis of the optical transients combined with heat flow calculations provides evidence that solid phase material removal is the dominant mechanism for the ablation of W films, while Cr is ablated in the form of vapor and small molten droplets.

1. Introduction

Laser ablation is the base of different methods in material processing of great technological importance such as the removal of opaque defects in optical masks [1] or the laser induced forward transfer (LIFT) of thin films for patterning applications [2–10]. The ablation of thin films is also one of the initial steps in the operation of trace element injectors [11]. The successful application of these techniques requires a clear comprehension of the mechanisms underlying the process of mass removal (and transfer) as a consequence of the laser irradiation. However, real time studies aiming to analyze the basic mechanisms of LIFT have been scarcely reported [12] and most of the works on ablation are performed in the fluence domain where evaporation is the dominant process for material removal [1–6,12–17]. Time resolved transmission measurements in thin gold films during laser induced blow-off, combined with rough estimations of the film temperature, have shown that blow-off occurs as the heated gold surface approaches boiling [12]. Contrary to this case, low temperature (solid state) laser induced material removal has been demonstrated [8–11] to occur in thin vanadium, titanium and chromium films.

In this paper we analyze the dynamics of material removal and transport during visible and UV laser induced blow-off of supported tungsten and chromium films. Since W has a high melting and boiling point, it is a suitable material for processing in a wide range of fluences. Chromium – with lower melting and boiling points – is studied for comparison. The dynamics of material removal is followed by time resolved reflectivity and transmission measurements performed during the laser induced blow-off. These measurements combined with detailed heat flow calculations will allow us to establish the existence of two different removal mechanisms: solid state removal for W and liquid–vapor phase removal for Cr.

2. Experimental

The samples are 100 and 60 nm thick W and Cr films, respectively, deposited by thermal evaporation onto glass substrates. The as-deposited samples are polycrystalline and the adherence of W and Cr films is fairly good and excellent, respectively. The experimental setup used to perform the irradiations and the real time optical measurements has been described in detail elsewhere [18]. The samples are irradiated with cavity-dumped Ar^+ (all lines) laser pulses (FWHM 10 ns) or ArF excimer laser ($\lambda = 193$ nm) pulses (FWHM 12 ns). The Ar^+ laser beam is focused onto the sample surface to a size of 10 μm ($1/e^2$ diameter) while the excimer beam is focused to a size of 0.8×3.1 mm^2 and 0.2×3.1 mm^2 for the W and Cr films, respectively. Maximum fluences of 330 mJ/cm^2 (Ar^+ laser) and 1.8 J/cm^2 (ArF laser) are used for processing. The reflectivity and the transmission of the samples is monitored in real time by means of a HeNe laser beam (632.8 nm) with a size at the irradiation site of 4.8 or 160 μm ($1/e^2$ diameter) for Ar^+ and ArF laser processing, respectively.

3. Results

Fig. 1 shows the evolution of the reflectivity and the transmission of a 100 nm thick W film upon irradiation with Ar^+ laser pulses. The blow-off pulse is followed by an abrupt reflectivity decrease which shifts to shorter times as the pulse energy density increases. It occurs within

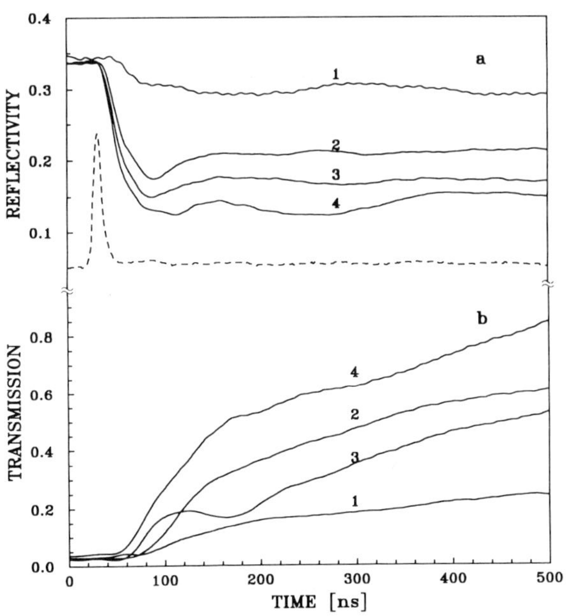

Fig. 1. Evolution of the reflectivity (a) and the transmission (b) of the W film processed by focused Ar^+ laser pulses. Incident fluences: (1) 92; (2) 185; (3) 233 and (4) 278 mJ/cm^2. The dashed curve indicates the processing pulse.

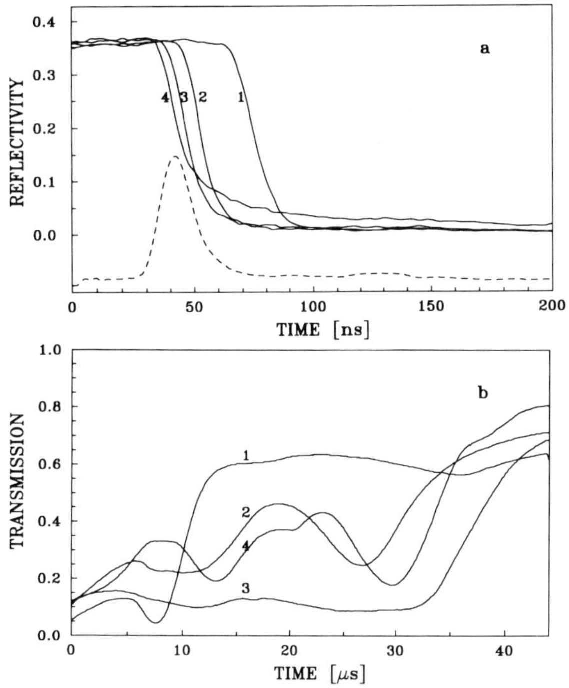

Fig. 2. Evolution of the reflectivity (a) of the W film processed by ArF excimer pulses. Incident fluences: (1) 321; (2) 581; (3) 1062 and (4) 1723 mJ/cm^2. The dashed curve indicates the processing laser pulse. Evolution of the transmission (b) at fluences: (1) 519; (2) 1180; (3) 1558 and (4) 1841 mJ/cm^2. The maximum of the processing pulse is taken as the origin of the time scale.

the first 50 ns after the blow-off pulse and small fluctuations of the reflectivity are observed at longer times. The final reflectivity level is always lower than the initial one being also a decreasing function of fluence. The changes expected in the reflectivity at 632.8 nm in the surface of a metal film upon melting is at the most a few percent of the initial value [19]. Since the final reflectivity level obtained in fig. 1 approaches the reflectivity level of the substrate as the fluence is increased, it can be easily concluded that the initial abrupt decrease in the reflectivity transients corresponds to the blow-off onset.

For comparison, fig. 2 shows similar results, but obtained by ArF laser processing. The main difference observed in the behavior of the reflectivity transients with respect to those obtained by Ar^+ processing is that now the reflectivity level achieved at times < 1 μs does not depend on fluence and it is surprisingly lower than the reflectivity of glass at the probing wavelength. In the case of Ar^+ laser processing the diameter of the ablated areas varies from 4 to 7.8 μm for increasing fluences and, therefore, the probe beam (4.8 μm, $1/e^2$ diameter, $d(1/e^2)$) samples areas in which both ablated and intact regions are present. The higher the ablation diameter, the lower the final reflectivity level measured. A minimum ablated region diameter of $1.84d(1/e^2)$ (~ 8.8 μm) would be required to observe the reflectivity value of the substrate [18,20]. For excimer laser processing the diameter of the probe beam (~ 160 μm) is always much smaller than the ablated region size and this effect is not observed.

The transmission transients (figs. 1b and 2b) provide complementary information both on the transformation of the material along its depth and on the transport of the removed material while crossing the path of the probe beam. Since most of the material removal occurs within the first 50 ns after the blow-off pulse according to the reflectivity transients, the evolution of the transmission transients for longer times has to be related to the interaction of the probe beam with the removed material on its way through the probe beam. In the case of Ar^+ laser pulses (fig. 1b) the final transmission level is reached within ~ 1 μs and fluctuations for shorter times are sometimes observed. When processing with ArF laser pulses, the transmission remains unaffected for times as long as 1 μs. On an extended time interval all the transients show a fluctuating-rising behavior, reaching the final transmission level > 50 μs after the blow-off pulse.

Processing 60 nm thick Cr films with Ar^+ and ArF pulses leads to similar behavior in reflectivity than the one observed in W films. Material removal occurs within a few tens of ns after the laser pulse, according to the reflectivity transients. The final transmission level is reached within ~ 200 ns (Ar^+ pulses) and ~ 500 ns (ArF pulses) after the processing pulse which shows that transport of the removed material occurs in a compressed time scale, compared to W. No fluctuations have ever been observed in the case of Cr.

4. Model calculations and discussion

The evolution of the temperature profiles has been calculated for all the studied configurations. For the case of Ar^+ laser processing, the solution of the heat flow equation with cylindrical symmetry was numerically obtained using the method of finite differences. In the case of ArF processing the calculation of the temperature profiles requires only the solution of the one-dimensional heat flow equation since the lateral extension of the heated material is orders of magnitude larger than the film thickness. In all calculations the temperature dependence of thermophysical data (for bulk material) [21,22] was taken into account whereas material partial evaporation was not considered.

The thermal conductivity of both metals is relatively high ($\kappa = 178$ and 89.5 W/mK at 300 and 3653 K for W and 90.3 and 64 W/mK at 300 and 1400 K for Cr, respectively), consequently the film temperature varies slightly in the direction perpendicular to the surface. Since the substrate has a much lower thermal conductivity, a great temperature gradient is produced near the film interface. The lateral distribution of the maximum temperature reached for Ar^+ laser processing of W and Cr films at several fluences is shown in figs. 3 and 4, respectively. The stars indicate the experimental values of the radii of the ablated regions for each fluence.

From fig. 3 it is clearly seen that in the fluence range of 90–180 mJ/cm^2 removal of W proceeds in solid phase. Furthermore, for the studied fluence range the temperature at the edge of the ablated holes never exceeds the melting temperature (3653 K). Complementary experiments in LIFT configuration (placing a glass plate in contact to the metal film in order to collect the removed material) were performed to analyze the transferred material. It consists of broken chips of metal for low fluences while for higher fluences traces of melting, as indicated by the rounded edges in the optical microscope, are observed in the deposited fragments. This gives further support for the ablation picture given above.

In the case of Cr films the temperature calcu-

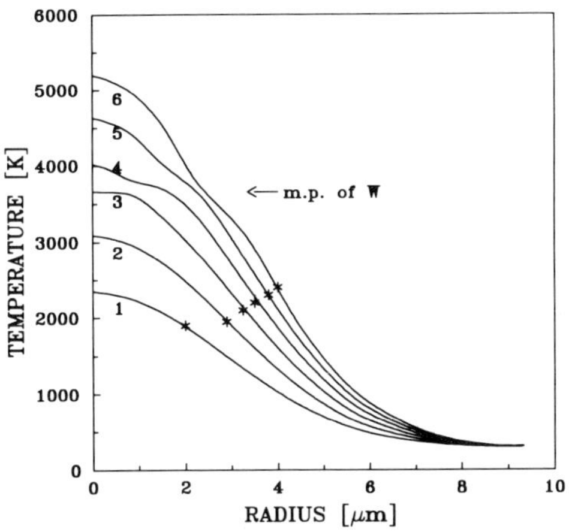

Fig. 3. Calculated distributions of maximum temperatures in the 100 nm thick W layer during Ar^+ laser processing with (1) 92; (2) 138; (3) 185; (4) 233 (5) 276 and (6) 325 mJ/cm^2 pulses. The stars indicate the radii of corresponding ablated areas.

lations (fig. 4) and complementary morphological analysis of the removed-material lead to the conclusion that ablation starts above the melting temperature (2148 K).

The differences observed in the transmission transients measured for the two materials are consistent with the different proposed ablation mechanisms for W and Cr. The presence of fragments of W removed in solid state and their random movement through the probe beam path cause the long time fluctuations observed in the transmission transients. The transport of large pieces compared to the size of the probe beam makes this effect more dominant in ArF than in Ar^+ laser processing. This effect is also responsible for the reflectivity level observed for times $<$ 1 μs in ArF processing, which is lower than that of the clean substrate. The HeNe beam reflected from the surface of these deformed pieces becomes out of the view angle of the detector, while the transmission remains constant in this time regime. The monotonous and faster increase observed in the transmission of Cr films is due to the ablation in the form of vapor and small molten droplets.

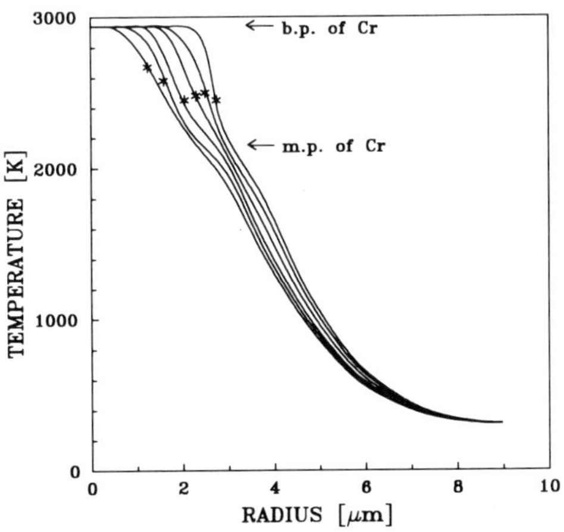

Fig. 4. Calculated distributions of maximum temperatures in the 60 nm thick Cr layer during Ar^+ laser processing with 208; 220; 233; 256; 279 and 298 mJ/cm^2 pulses for curves from left to the right, respectively. The stars indicate the radii of corresponding ablated areas.

5. Conclusion

Time resolved reflectivity and transmission measurements are used to follow the evolution of ablation and material transport during visible and UV laser induced blow-off of supported W and Cr films. The analysis of the optical transients combined with heat flow calculations evidence that solid phase material removal is the dominant mechanism for ablation of W films while Cr is mainly ablated in vapor phase and small molten droplets. In light of practical applications the use of LIFT of W in this low fluence domain allows the deposition of well-adhering micron-sized patterns of 100% coverage.

References

[1] D.J. Elliot, Integrated Circuit Mask Technology (McGraw-Hill, New York, 1985) p. 269.

[2] V.J. Zaleckas and H.C. Koo, Appl. Phys. Lett. 31 (1977) 615.

[3] J. Bohandy, B.F. Kim and F.J. Adrian, J. Appl. Phys. 60 (1986) 1538.

[4] F.J. Adrian, J. Bohandy, B.F. Kim and A.N. Jette, J. Vac. Sci. Technol. B 5 (1987) 1490.

[5] J. Bohandy, B.F. Kim, F.J. Adrian and A.N. Jette, J. Appl. Phys. 63 (1988) 1158.

[6] R.J. Baseman, A. Gupta, R.C. Sausa and C. Progler, in: Laser and Particle-Beam Chemical Processing for Micro-

electronics, Proc. MRS 101, Eds. D.J. Ehrlich, G.S. Higashi and M.M. Oprysko (MRS, Pittsburgh, PA, 1988) p. 237.

[7] E. Fogarassy, C. Fuchs, F. Kerherve, G. Hauchecorne and J. Perriere, J. Mater. Res. 4 (1989) 1082.

[8] P. Mogyorósi, T. Szörényi, K. Bali, Zs. Tóth and I. Hevesi, Appl. Surf. Sci. 36 (1989) 157.

[9] Z. Tóth, Z. Kántor, P. Mogyorósi and T. Szörényi, in: Laser Assisted Processing II, Ed. L.D. Laude, Proc. SPIE 1279 (1990) p. 150.

[10] Z. Kántor, Z. Tóth and T. Szörényi, Appl. Phys. A 54 (1992) 170.

[11] J.S. Bakos, P.N. Ignácz, J. Szigeti and J. Kovács, Appl. Phys. Lett. 51 (1987) 734.

[12] R.J. Baseman and N.M. Froberg, Appl. Phys. Lett. 55 (1989) 1841.

[13] J.E. Andrew, P.E. Dyer, R.D. Greenough and P.H. Key, Appl. Phys. Lett 43 (1983) 1076.

[14] V.P. Veiko, S.M. Metev, A.I. Kaidanov, M.N. Libenson and E.B. Jakovlev, J. Phys. D: Applied Physics 13 (1980) 1565.

[15] R.J. Baseman and J.C. Andreshak, in: Fundamentals of Beam–Solid Interactions and Transient Thermal Processing, Eds. M.J. Aziz, L.E. Rehn and B. Stritzker, Proc. MRS 100 (MRS, Pittsburgh, PA, 1988) p. 627.

[16] R.J. Baseman, N.M. Froberg, J.C. Andreshak and Z. Schlesinger, Appl. Phys. Lett. 56 (1990) 1412.

[17] R.M. Gilgenbach and P.L. Ventzek, Appl. Phys. Lett. 58 (1991) 1597.

[18] J. Solis and C.N. Afonso, J. Appl. Phys. 69 (1991) 2105.

[19] C.A. Macdonald and F. Spaepen, in: Laser Surface Treatment of Metals, Eds. P. Mazzoldi and C.N. Drapper, NATO ASI series, Vol. 116 (Nijhoff, Dordrecht, 1986).

[20] C.N. Afonso, J. Solis and F. Catalina, Appl. Surf. Sci. 43 (1989) 171.

[21] R.C. Weast, Ed., CRC Handbook of Chemistry and Physics, 59th ed. (CRC Press, Boca Raton, FL, 1978–79).

[22] Y.S. Touloukian, Ed., Thermophysical Properties of Matter, TPRC Data series (IFI/Plenum, New York, 1973).

Applied Surface Science 69 (1993) 335–339
North-Holland

applied
surface science

In situ growth of YBaCuO superconducting thin films by excimer laser ablation: influence of deposition and cooling parameters

C. Champeaux, P. Marchet, J. Aubreton, J.-P. Mercurio and A. Catherinot

URA 320 CNRS, Faculté des Sciences, 123 Avenue A. Thomas, 87060 Limoges Cedex, France

Received 2 June 1992; accepted for publication 30 October 1992

In situ YBaCuO thin films are deposited on MgO single-crystal by pulsed laser ablation using an excimer laser (KrF, 248 nm). Deposition and cooling conditions are investigated in order to obtain *c*-oriented samples. The influence of substrate temperature and oxygen pressure during both deposition and post-deposition processes on the structural (orientation, lattice parameters) and electrical (resistivity at 300 K, critical temperature, transition width) film properties is analyzed. These studies lead to reproducible high quality *c*-oriented thin films which exhibit a critical temperature of 90 K, a superconducting transition width less than 1 K and a good behavior in the normal state.

1. Introduction

Since the discovery of high-T_c superconductivity in 1987 [1], the elaboration of superconducting thin films has led to numerous investigations. Among the different deposition methods [2–5], the pulsed laser deposition (PLD) appears to be the most successful technique because of its ability to transfer the stoichiometry from the target to the substrate. However, to produce reproducible high quality YBaCuO thin films suitable for practical applications or basic physics studies, not only appropriate stoichiometry but also good crystallographic orientation is necessary. All these characteristics depend on simultaneous control of several physical parameters during both deposition and cooling steps. In this paper, we report the influence of the substrate temperature, the oxygen pressure during deposition and the environment (gas nature and pressure) during cooling on the structural and electrical properties of in situ laser-deposited YBaCuO thin films.

2. Experimental details

2.1. Target preparation

YBaCuO bulk targets are prepared from a corresponding mixture of 99.9% pure copper and yttrium oxides (CuO, Y_2O_3) and barium carbonate ($BaCO_3$). The powders are carefully mixed, ground and then calcined in an electrical furnace as follows: rise up to 850°C at 300°C/h, first dwell at 850°C for 20 h, second dwell at 950°C for 6 h and natural cooling. After crushing and sifting, the powders are pressed to a shape of a 2 cm diameter pellets, which are sintered at 975°C for 20 h under oxygen atmosphere. The superconductivity of the samples as-made is verified at 77 K by magnet levitation.

2.2. Deposition system

Deposition experiments are performed in an ultra-high vacuum cell equipped with an introduction chamber for targets and substrates manipulation. The beam of a pulsed excimer laser (248 nm, pulse duration 14 ns, repetition rate 10 Hz, energy per pulse 20–300 mJ) is focused by a tetrasyl quartz lens onto a rotating bulk superconducting YBaCuO target. According to the 35° incidence angle, the laser spot is elliptical and its size is typically 2.3 mm^2. The vaporized material, ejected normally to the target surface is deposited on a single-crystal (001) oriented MgO substrate placed opposite the target at 3.8 cm. The substrate can be heated up to about 820°C by a halogen lamp. Its temperature is estimated with a pyrometer in combination with a chromel–alumel thermocouple mounted at the edge of substrate holder. Several gas inlets allow the control of the deposition and cooling atmospheres.

2.3. Deposition parameters

The previous studies of the KrF laser-induced plasma plume created above a YBaCuO superconducting target by time-resolved optical spectroscopy [6] and mass spectrometry [7] advise us on deposition parameters, which lead to the stoichiometric transfer of cationic species (Y, Ba, Cu) from the target to the substrate, i.e., the laser fluence (3 J/cm^2) and the range of the oxygen pressure to be used during deposition (< 0.5 mbar). The deposition duration is 10 min, and leads to 300 nm thick films.

2.4. Thin films characterization

The structural properties of the deposited thin films are investigated by X-ray diffraction using a standard $\theta/2\theta$ diffractometer (Cu Kα radiation) in order to determine crystalline orientation. The detection of lattice plans parallel to the substrate surface allows a computation of lattice parameters. The electrical resistivity is measured by a standard DC four probes method in the 25–300 K range using a closed cycle helium cryostat. Temperature is monitored and given by a silicon diode.

3. Results

3.1. Substrate temperature

It is well recognized [8] that substrate temperature is one of the most important parameters for the growth of high quality superconducting thin films. In order to analyze its influence, several films are prepared at different substrate temperatures (T_s), as listed in table 1. All these films are mirror-like but their color evolves from brown at low temperatures (T_s < 620°C) to black at high temperature. The X-ray analysis (fig. 1) shows that the sample crystallinity changes with temperature. At low temperature (430–620°C), the films are amorphous. Then an increase of the deposition temperature improves the quality of the film crystallization and phase orientation. At 690°C, the strong diffraction peak at about 33° corresponds to the randomly oriented phase (103/013/110). At 725°C, this peak is no longer observed and the films are perfectly a-oriented. With increasing T_s, the intensity of the (h00) diffraction peaks decreases while that of (00l) peaks increases. The room temperature resistivity of these films decreases from a few k$\Omega \cdot$cm (T_s < 620°C) to a few m$\Omega \cdot$cm and their electrical characteristics versus temperature evolve from semiconducting (at 690°C) to superconducting behavior (at 790°C), as shown in fig. 2. These results are in agreement with the structural properties of the films and prove the important role of substrate temperature on the film structural properties.

Table 1
Deposition temperatures of the YBaCuO thin films on single crystal MgO(00l) samples (oxygen pressure 6×10^{-3} mbar, laser fluence 3 J/cm^2, natural cooling with 30 mbar of oxygen)

Sample	Temperature (°C)
D1	430
D2	620
D3	690
D4	725
D5	790

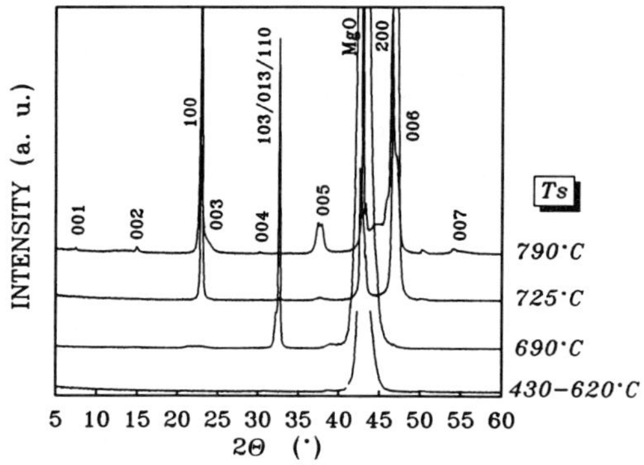

Fig. 1. X-ray diffraction patterns of YBaCuO films on MgO as functions of the substrate temperature.

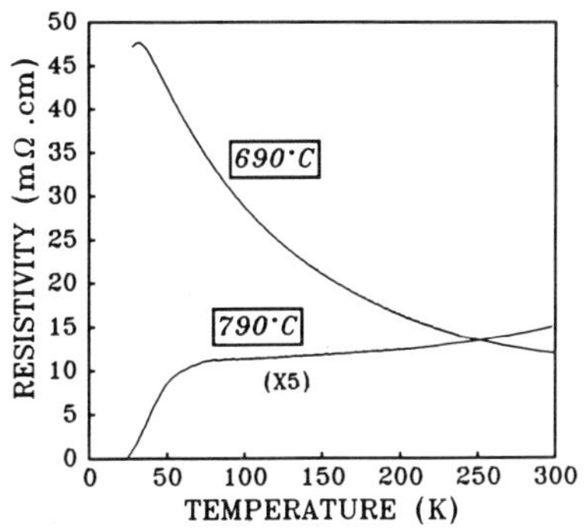

Fig. 2. Resistivity versus temperature curves for samples D3 (T_s = 690°C) and D5 (T_s = 790°C).

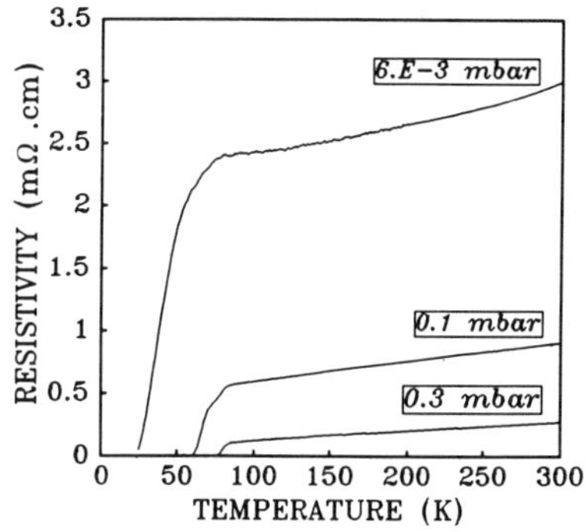

Fig. 3. Resistivity versus temperature curves for different deposition oxygen pressures.

3.2. Deposition oxygen pressure

The previous results show that substrate temperature during deposition should not the only parameter to be studied in order to improve high-T_c superconducting film quality. According to the previous experiments and our aim (*c*-oriented films), the selected substrate temperature is 790°C. The oxygen pressure will vary in the range of 6×10^{-3}–0.3 mbar. The other deposition parameters are identical as previously, that is to say: the laser fluence is 3 J/cm^2 and repetition rate, 10 Hz. After deposition, samples are cooled under oxygen at 30 mbar. Fig. 3 shows the change of the resistivity versus temperature curves with the deposition pressure. All the films are superconducting. The critical temperature (T_{onset}) increases (from 80 to 86 K) and the transition width decreases (from 32 to 6 K) when the deposition pressure increases. These electrical properties are in agreement with the structural ones. Indeed, the mixture of (h00) and (00l) orientations which is observed at low deposition pressure disappears on behalf of the single orientation (00l), at the highest pressures (fig. 4). The improvement of the critical temperature and transition width can be attributed to a more effective oxygen incorporation in the material at high pressures. This hypothesis is confirmed by the values of the lattice parameters *c*. In the present case, the *c* values, are respectively equal to 1.183 and 1.171 nm for 0.1 (sample D6) and 0.3 mbar (sample D8). Indeed, the lowering of the *c*-lattice parameter corresponds to an increase of the number of oxygen atoms in $YBa_2Cu_3O_x$ [9]. In fact, these two parameters *c* and *x* are linked by the relation

$$x = (1.2736 - c(\text{nm}))/0.01501$$

established for bulk samples by Ohkubo et al. [10].

Table 2
Deposition parameters, structural and electrical properties of YBaCuO thin films deposited on heated single crystal MgO(00l) substrates at 790°C (oxygen pressure 0.3 mbar and cooled down in different environments)

Sample	Cooling gas	Cooling pressure (mbar)	c (nm)	x (in YBa_2Cu_x)	ρ (300 K) (mΩ · cm)	T_{onset} (K)	T_{offset} (K)	ΔT (K)
D12	Vacuum	–	1.185	5.9	1200	–	–	–
D13	Argon	330	1.175	6.59	0.55	85	56.6	20
D7	Oxygen	2	1.174	6.65	1.3	88	58	18
D8	Oxygen	30	1.171	6.81	0.28	86	75.5	6
D9	Oxygen	160	1.169	6.95	0.25	89	87	1
D10	Oxygen	330	1.170	6.91	0.35	89	86.4	1.3
D11	Oxygen	930	1.169	6.98	0.2	89.2	86.2	2

3.3. Choice of gas and pressure during cooling step

The former studies emphasize the influence of the substrate temperature and the deposition oxygen pressure on the crystalline orientation. when the temperature is 790°C and the pressure 0.3 mbar, *c*-oriented superconducting thin films are obtained, but their electrical properties are not optimal (transition width 6 K). That is why the influence of the nature (vacuum, argon and oxygen atmosphere) and the pressure of the gas present during cooling is investigated. The deposition conditions are reported in table 2.

The films are found highly-*c* oriented independently on the nature and the pressure of the gas present during cooling. This observation demonstrates that the deposition conditions themselves (substrate temperature, oxygen pressure, laser fluence, target–substrate distance, etc.) determine the crystalline orientation. Fig. 5 shows the temperature dependence of resistivity for the films of this experiment. We notice that the film cooled down in vacuum is semiconductive while those, cooled down in argon or oxygen atmospheres, are superconducting. This is attributed to the difference in *c* lattice parameter and, therefore, in the oxygen content x (in $YBa_2Cu_3O_x$). Indeed, the sample D12, cooled in vacuum condition, is oxygen deficient ($x = 5.9$). This result, in agreement to Ohkubo's observations [10], leads to the two following hypotheses:
- the oxygen incorporation during the deposition process is insufficient,

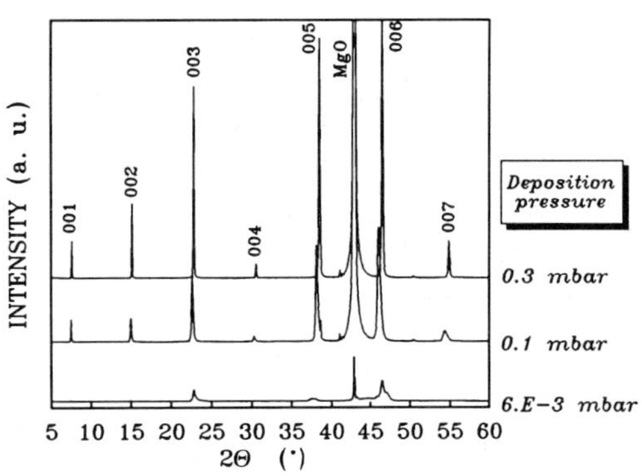

Fig. 4. X-ray diffraction patterns of YBaCuO films deposited on MgO at $T_s = 790°C$ as functions of the deposition pressure.

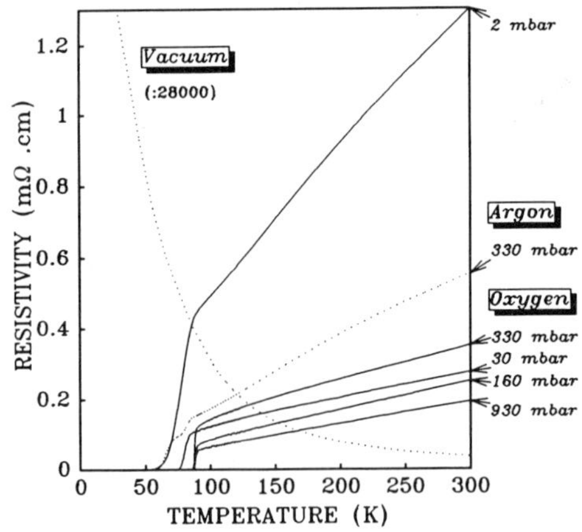

Fig. 5. Resistivity as a function of temperature for films cooled at different residual ambient gas.

- the oxygen contained in the film at the end of the deposition process has outdiffused from the film during the cooling process in vacuum.

In order to try to conclude, we can examine the sample D13, which is cooled down in argon pressure of 330 mbar. Its oxygen content, greater than in the case of cooling in vacuum, yields to a superconducting behavior, as shown in fig. 5. The transition width, of about 20 K, indicates that the incorporation of oxygen into the film during deposition is certainly not sufficient, according to the first hypothesis. But, it is not possible now to confirm or cancel the second hypothesis. Indeed, oxygen can outdiffuse, like in the case of cooling in vacuum, from the film to the gaseous phase (argon).

However, this experiment shows that the superconducting YBaCuO phase grows epitaxially on MgO during the deposition. This result, evident from our experiment, disagree with the conclusions of Ohkubo et al. [10] and Ying et al. [11].

These results show that cooling in oxygen is primordial to obtain high quality superconducting thin films. So, we have grown several films samples which were cooled down in different oxygen pressures (table 2). All these films are superconducting (fig. 5) and have a good metallic behavior above the transition (ratio of the resistivities at 300 and 100 K of about 3). The transition width varies from 18 to 1 K and presents an optimum for the oxygen pressures during cooling in the range of 160–330 mbar.

4. Conclusion

In situ growth of high quality superconducting thin films depends not only on the deposition parameters but on the cooling process. The determinant role of the substrate temperature and the oxygen pressure during deposition on the crystalline orientation, has been shown. The optimization of the electrical properties needs adequate oxygen pressure during the cooling process.

Acknowledgement

This work is partly supported by DRET under contract No. 88-084.

References

[1] J.G. Bednorz and K.A. Muller, Z. Phys. B 64 (1986) 189.
[2] D.K. Lathrop, S.E. Russek and R.A. Buhrma, Appl. Phys. Lett. 51 (1987) 1462.
[3] K. Shinohara, V. Matijasevic, P.A. Rosenthal, A. Marshall, R.H. Hammond and M.R. Beasley, Appl. Phys. Lett. 58 (1991) 756.
[4] I.D. Raistrick, M. Hawley, J.G. Beery, F.H. Garzon and R.J. Houlton, Appl. Phys. Lett. 59 (1991) 3177.
[5] S. Matsuno, F. Uchikawa and K. Yoshizaki, Jpn. J. Appl. Phys. 29 (1990) L947.
[6] C. Girault, D. Damiani, C. Champeaux, P. Marchet, C. Germain, J.P. Mercurio, J. Aubreton and A. Catherinot, Appl. Surf. Sci. 46 (1990) 78; Appl. Phys. Lett. 54 (1989) 2035.
[7] C. Champeaux, D. Damiani, C. Girault, P. Marchet, J. Aubreton, J.P. Mercurio and A. Catherinot, Laser Ablation of Electronic Materials: Basic mechanisms and Applications, Eds. E. Fogarassy and S. Lazare (Elsevier, Amsterdam, 1992) p. 141.
[8] See, for example: W. Shi, J. Shi, J. Sun, W. Yao and Zh. Qi, Appl. Phys. Lett. 57 (1990) 822.
[9] R.J. Cava, B. Batlogg, C.H. Chen, E.A. Rietman, S.M. Zahurak and D. Werder, Phys. Rev. B 36 (1987) 5719.
[10] M. Ohkubo, T. Kachi and T. Hioki, J. Appl. Phys. 68 (1990) 1782.
[11] Q.Y. Ying, H.S. Kim, D.T. Shaw and H.S. Kwok, Appl. Phys. Lett. 55 (1989) 1041.

Applied Surface Science 69 (1993) 340–344
North-Holland

applied
surface science

Laser studies of polystyrene precursors performed through resonant two photon ionization processes in a supersonic molecular beam

A. Giardini Guidoni *, S. Piccirillo, M. Coreno
Dipartimento di Chimica, Università La "Sapienza", P.le A. Moro 5, I-00185 Roma, Italy

M. Snels, A. Morone
Istituto Materiali Speciali, CNR, Area Industriale di Tito, I-85100 Tito (Pz), Italy

and

R. Teghil
Dipartimento di Chimica, Università della Basilicata, Via N. Sauro 85, I-85100 Potenza, Italy

Received 2 June 1992; accepted for publication 30 October 1992

Laser-induced resonant two photon ionization (R2PI) of styrene and 4-fluorostyrene, pure or mixed with various molecules, has been studied in a supersonic molecular beam and ionic clusters have been detected by time-of-flight (TOF) mass spectrometry. The wavelength spectra of the heterogeneous clusters show specific shifts of the 0_0^0 band due to van der Waals intracluster interaction.

1. Introduction

The study of the energetics and dynamics of large van der Waals (vdW) complexes, such as organic aromatic systems bounded to solvent molecules, has been extensively reported in the literature [1–4]. Recently, much attention has been paid to these cluster species with the aim to clarify nucleation phenomena. Photophysical and photochemical studies of these clusters give information on solvent effects, intracluster reactions, charge transfer processes and on photopolymerization phenomena. Model calculations of potential surfaces of vdW complexes have provided a knowledge of the intracluster binding energies and vdW vibrational modes in many systems [1]. The investigation of the displacement of the frequency of an electronic absorption band in a complex formed by chromophore and solvent gives information on the intracluster interaction. R2PI spectra of many clusters have been measured using a TOF mass spectrometer; in these experiments intracluster reactions can take place and, due to the low binding energy of vdW clusters, dissociative ionization can occur. Interesting results have been obtained recently on the ion chemistry in clusters with well-defined microsolvation environment. Typical examples are nucleophilic substitution taking place in aromatic radical cations surrounded by ammonia molecules [2].

The present work presents data on the spectroscopic shift of styrene (STY) and 4-fluorostyrene (FSTY) molecules clustered to various atoms and compounds such as Ar, N_2, NH_3 and a few ammines. The investigation of this chro-

* To whom correspondence should be addressed.

mophore-solvent laser-induced chemistry can shed some light on the processes leading to polymerization.

2. Experimental

The experimental apparatus for the production and detection of clusters consists of two independently pumped vacuum chambers separated by a skimmer of 1 mm inner diameter as shown in the scheme of fig. 1. The clusters are formed in a supersonic expansion of He, seeded with the aromatic systems, pure or mixed with solvent molecules in a suitable ratio. The stagnation pressure can be varied between 0.5 and 4 bar. The expansion in the first chamber is obtained through a pulsed valve (General Valve) with a nozzle at 400 μm inner diameter and aperture time up to 600 μs. The repetition rate is 10 Hz. The skimmed beam passes after 40 cm path between the plates of a homemade TOF mass spectrometer. Resolving power of the TOF is 1 mass unit at $m/e = 200$.

The clusters are ionized in a R2PI process by using a frequency doubled dye laser pumped by a Nd:YAG (Quantel model 710). The relative intensity of the ion signal depends on the expansion conditions: stagnation pressure, time lag between the dye laser pulse relative to the gas pulse, laser wavelength. In what follows R2PI experiments as a function of laser frequency on a given mass and TOF spectra on various mixtures will be reported and discussed. The compounds studied are STY and FSTY, pure or mixed with Ar, N_2 or with bases B as NH_3, $C_2H_5NH_2$ (MEA), $(C_2H_5)_2NH$ (DEA) and $(C_2H_5)_3N$ (TEA). All the systems are used as supplied by Aldrich Chemical without any further purification.

3. Results

In this section we present the mass spectra detected following one color R2PI at a given wavelength. A few examples will be reported for both STY and FSTY. The spectral dependence for some selected monomer and cluster masses will be shown and spectral shifts identified.

TOF mass spectra. The R2PI TOF mass spectrum of pure STY measured at an energy value corresponding to one photon energy of 34 602 cm^{-1} is displayed in fig. 2. Beside the STY^+ peak

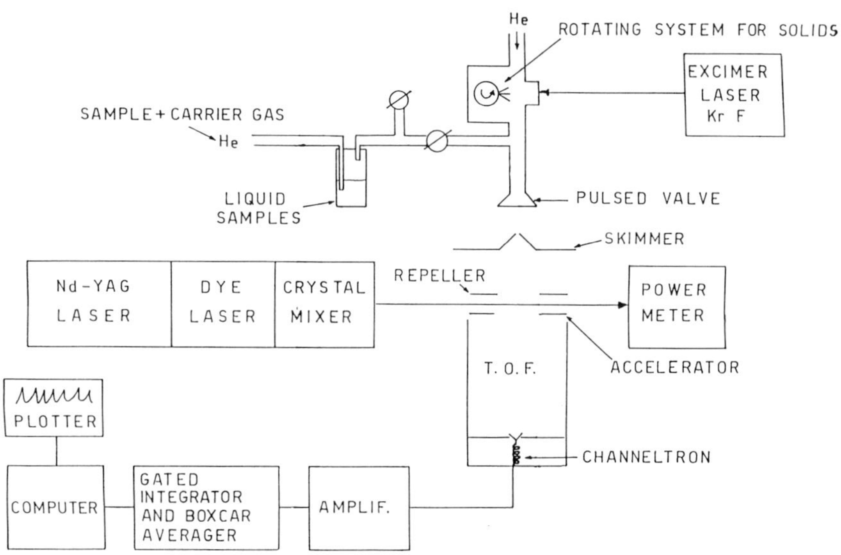

Fig. 1. Scheme of the supersonic beam apparatus.

other peaks belonging to $(STY)_n^+$ clusters can be identified with n up to 3. A mass spectrum with a larger number of peaks, corresponding to ions identified as $[(STY)_n(NH_3)_m]^+$, with n and m depending on the NH_3/STY ratio, is detected in an expansion experiment of a STY–NH_3 mixture. Similar results are observed by expanding FSTY pure or mixed with other molecules. A typical mass spectrum of FSTY is reported in fig. 3a. $(FSTY)_n^+$ clusters are displayed up to $n = 7$. In fig. 3b several sequences $(FSTY)_n(MEA)_m^+$ are indicated with n up to 2 and m up to 4, together with ions due to cluster fragmentation upon ionization. The fragments can be identified as the protonated clusters of the base B_iH^+ ($i = 1,2,3,4$). Analogous series are observed from FSTY–DEA and FSTY–TEA mixtures. The expansion of FSTY–NH_3 mixtures of various compositions ranging from 5 to 50% in NH_3 generates $(FSTY)_n(NH_3)_m^+$ clusters which appear to dissociate into non-protonated ammonia clusters.

Spectroscopy. Figs. 4 and 5 show the mass selected R2PI spectrum of STY and FSTY as a function of the wavelength in the range 34364–34875 cm^{-1}. Spectral frequencies and assignment of vibronic excitation of the $\tilde{A}^1A' \leftarrow \tilde{X}^1A'$ electronic transition are shown. The data, in agreement with previous work [6,7], are summarized in tables 1 and 2. Fig. 6 depicts the wavelength spectrum of FSTY, FSTY–N_2, FSTY–Ar, FSTY–MEA ions. The spectral shifts $\Delta\nu$ of the

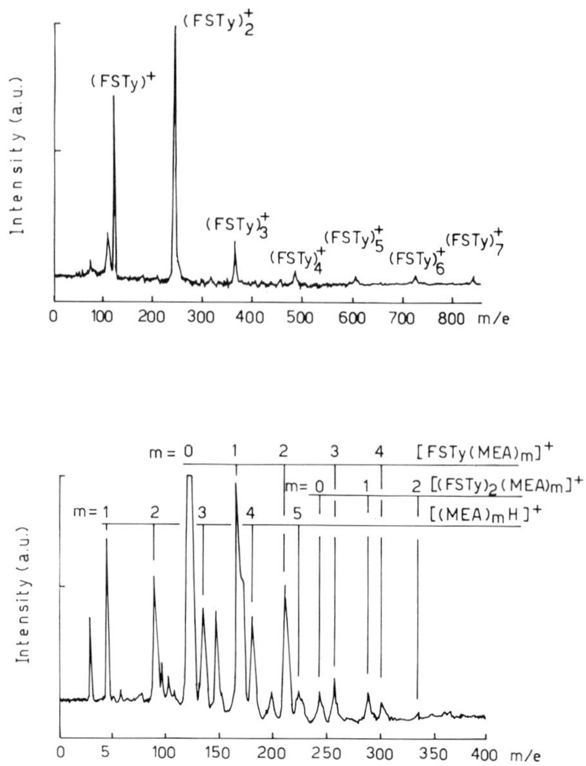

Fig. 3. R2PI TOF mass spectra of (a) FSTY, (b) FSTY–MEA mixture at $\lambda = 292.0$ nm.

electronic origin relative to the bare FSTY are −49, −39 and −31 cm^{-1} for N_2, Ar and MEA, respectively. In all spectra several structures are found towards the blue region of the main resonance peak as also reported in table 2. The shifts correspond to low frequency excitations of the adducts and the values are comparable with the model calculation of vibrational vdW models for rigid benzene–N_2 clusters [1]. The R2PI wavelength spectra of $(STY)_n^+$, $(FSTY)_n^+$, $STY_nB_m^+$, $FSTY_nB_m^+$ and B_nH^+ clusters yield a structureless broad peak whose maximum is redshifted with respect to the electronic origin of the monomer. In few cases an outline of the vibrational structure is still discernible as shown in fig. 6d.

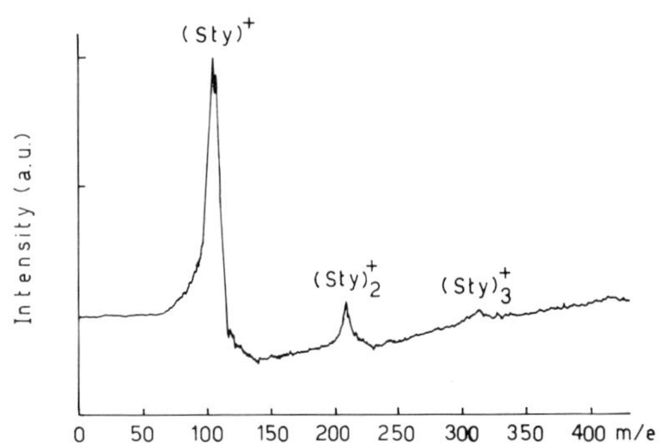

Fig. 2. R2PI TOF mass spectra of STY at $\lambda = 289.0$ nm.

Reactivity. This section concerns the intracluster reactivity leading to intracluster charge exchange and proton transfer.

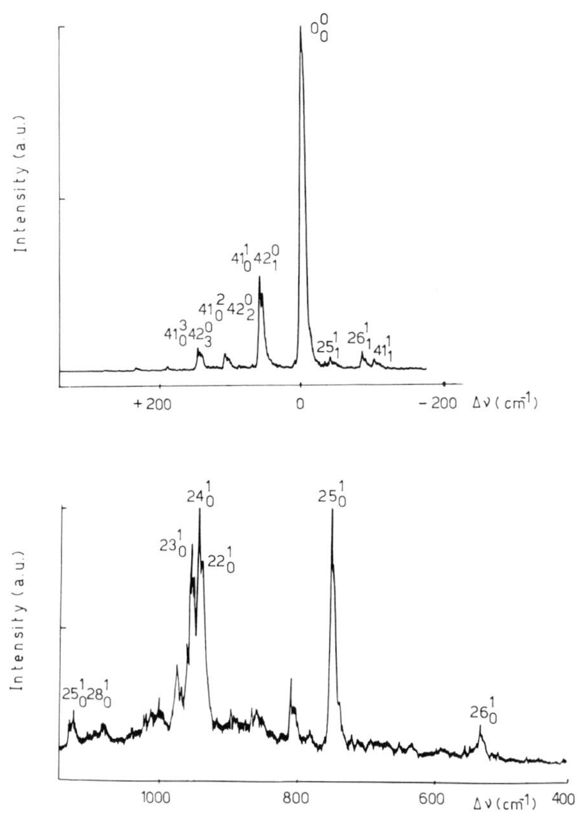

Fig. 4. One color R2PI spectrum of STY^+ relative to the electronic origin (34761 cm^{-1}).

STY. The energy used in these experiments to ionize the STY molecule is 8.6 eV, about 0.4 eV larger than the measured IP. STY proton affinity (PA) is 201 kcal/mol [8]. In the R2PI mass spectrum of STY also the $STYH^+$ could be present. We propose the following reaction scheme: $(STY)_2^+ \rightarrow STYH^+ + (C_8H_7)^{\cdot}$; $(STY)_3^+ \rightarrow STYH^+ + STY + (C_8H_7)^{\cdot}$. The wavelength spectra of $(STY)_2^+$ and $(STY)_3^+$ appear to confirm this hypothesis.

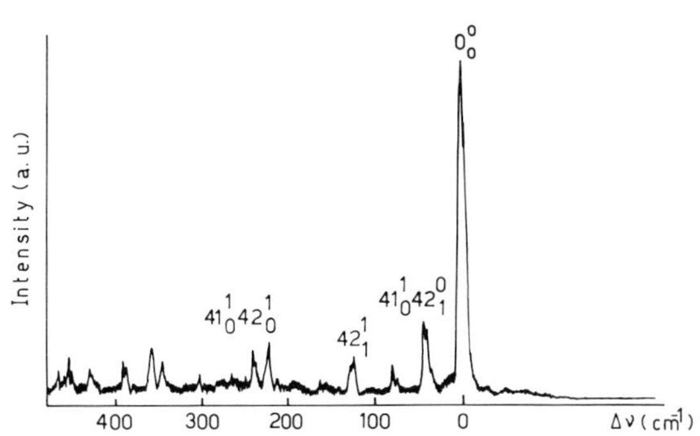

Fig. 5. One color R2PI spectrum of $FSTY^+$ ion relative to the electronic origin (34314 cm^{-1}).

Table 1
Analysis of the R2PI styrene spectrum

$\Delta\nu$ (cm^{-1}) [a]	Assignment	$\Delta\nu$ (cm^{-1}) [a]	Assignment
−100	41^1_1	196	$41^1_0 42^1_2$
−87		527	26^1_0
−80	26^1_1	748	25^1_0
−36		779	
−29	25^1_1	807	
−23		858	
−11	$41^0_1 42^1_0$	865	
−6		896	
0	0^0_0	939	22^1_0
8.5	29^1_1	943	24^1_0
27		956	23^1_0
47		964	20^1_0
54		978	19^1_0 (?)
57	$41^1_0 42^0_1$	1006	
103		1018	
108	$41^2_0 42^0_2$	1090	$25^1_0 40^1_0 41^1_0$
147	42^1_1	1135	$25^1_0 28^1_0$
150	$41^3_0 42^0_3$		

[a] Separation from STY origin band (34761 cm^{-1}).

Table 2

FSTY [a] $\Delta\nu$ (cm^{-1})	Assign-ment	FSTY· Ar [b] $\Delta\nu$ (cm^{-1})	FSTY· N_2 [c] $\Delta\nu$ (cm^{-1})	FSTY· MEA [d] $\Delta\nu$ (cm^{-1})
0	0^0_0	0	0	0
43	$41^1_0 42^0_1$	13	7	20
73		28	13	36
120	42^1_1	40	31	
219		46	41	
238	$41^1_0 42^1_0$	51	51	
302		55	66	
344		62	81	
362		70		
393		78		
431				
458				

[a] Separation from FSTY origin band (34314 cm^{-1}).
[b] Separation from electronic origin (34275 cm^{-1}).
[c] Separation from electronic origin (34265 cm^{-1}).
[d] Separation from electronic origin (34283 cm^{-1}).

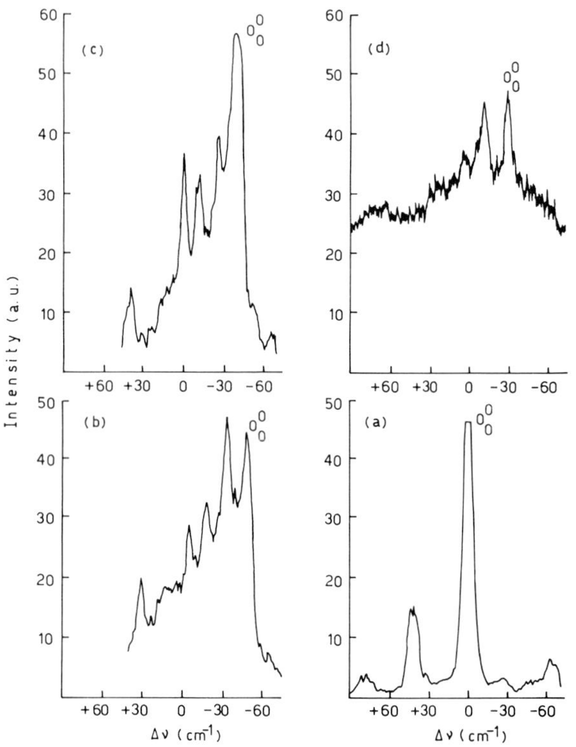

Fig. 6. One color R2PI spectrum of (a) $FSTY^+$, (b) $(FSTY \cdot N_2)^+$, (c) $(FSTY \cdot Ar)^+$, (d) $(FSTY \cdot MEA)^+$; the frequency shift is relative to the electronic origin of $FSTY^+$.

FSTY. In FSTY the first IP is 8.15 eV [9] and the energy of two photon at the electronic origin is 8.50 eV. This energy is considerably lower than the adiabatic first IP of NH_3 (10.17 eV) and MEA (9.10 eV), and slightly higher than that of DEA (8.51 eV) and TEA (7.50 eV) [8]. The PA's of all the bases studied range from 204 kcal/mol in case of NH_3 to 232.3 kcal/mol in case of TEA and the PA of FSTY is not known. The presence of protonated bases in the mass spectra, except for NH_3, indicates that proton transfer reactions occur. Here only the mixture FSTY–MEA will be discussed. The R2PI wavelength spectrum of $MEAH^+$ appears to have a maximum shifted at 60 cm^{-1} of the 0^0_0 electronic origin of FSTY and a width (FWHM) of about 230 cm^{-1}. This could be related to intracluster proton transfer from the dimer or medium size clusters $[FSTY(MEA)_m]^+$, as already observed for similar systems [3–5].

Further work is in progress along this line in order to ascertain the laser-induced protonation and polymerization processes in all the clusters studied.

Acknowledgements

We are grateful to Professor A. Mele for valuable discussions and critically reading the manuscript. This work was partially supported by CNR, P.F. Chimica Fine and Consorzio INFM.

References

[1] E.R. Bernstein, in: Atomic and Molecular Clusters, Ed. E.R. Bernstein (Elsevier, Amsterdam, 1990) p. 551.
[2] C. Riehn, C. Lahmann and B. Brutschy, J. Phys. Chem. 96 (1992) 3696.
[3] J.J. Breen, W.B. Tzeng, K. Kilgore, R.G. Keese and A.W. Castleman, Jr., J. Chem. Phys. 90 (1989) 19.
[4] O. Chesnovsky and S. Leutwyler, J. Chem. Phys. 88 (1988) 4127.
[5] C. Jouvet, C. Lardeux-Dedonder, M. Richard-Viard, D. Solgadi and A. Tramer, J. Phys. Chem. 94 (1990) 5041.
[6] J.A. Syage, F. Al Adel and A.H. Zewail, Chem. Phys. Lett. 103 (1983) 15;
L. Chia, L. Goodman and J.G. Phillis, J. Chem. Phys. 79 (1983) 593.
[7] J.M. Hollas and M. Zobir bin Hussein, J. Mol. Spectrosc. 145 (1991) 89.
[8] S.G. Lias, J.F. Liebman and R.D. Levin, J. Phys. Chem. Ref. Data 13 (1984) 695.
[9] C. Cauletti, M. De Simoni and P. Andreocci, unpublished results.

Applied Surface Science 69 (1993) 345–349
North-Holland

applied
surface science

Tungsten silicide formation by multipulse excimer laser irradiation

S. Luby, E. Majkova
Institute of Physics, Slovak Academy of Sciences, Bratislava, Slovak Republic

E. D'Anna, A. Luches, M. Martino, A. Tufano
Department of Physics, University of Lecce, Lecce, Italy

and

G. Majni
Department of Materials and Earth Sciences, University of Ancona, Ancona, Italy

Received 2 June 1992; accepted for publication 30 October 1992

The formation of tungsten silicide induced by XeCl laser irradiation of W(150 nm)/Si and W(500 nm)/Si samples was studied at laser fluences ranging from 0.6 to 1.8 J/cm^2. Up to 300 subsequent laser pulses were directed to the same irradiation site. After irradiation the samples were examined by different diagnostic techniques: Rutherford backscattering spectroscopy, X-ray scattering, resistometry and surface profilometry. Complete reaction of the 150 nm W film with silicon was obtained between 30 and 100 laser pulses at the fluence of 1.2 J/cm^2, and after 30 laser pulses and at the fluence of 1.5 J/cm^2. The sheet resistance of these silicides was 5–10 Ω. At the used fluences, only the onset of silicide synthesis at the W–Si interface was observed for the 500 nm W film. Numerical computations of the evolution and depth profiles of the temperature in the samples as a consequence of a laser pulse were performed and compared to the experimental results.

1. Introduction

Tungsten silicides have raised growing interest in the past years due to their use for ohmic contacts, diffusion barriers, interconnects, gates, adhesion promoters, diffusion sources and Schottky contacts in silicon and GaAs integrated circuit technology (see, e.g., refs. [1,2]).

The interest in laser synthesis of tungsten silicides arises from the fact that the furnace formation of tungsten disilicide in metal/silicon structures requires annealing at very high temperatures and WSi_2 formation is retarded when interfacial oxygen is present [3]. On the other hand, laser pulsed synthesis is a low thermal budget process and silicide formation via interfacial melting substantially eliminates the contamination problems. Moreover, rapid solidification takes place far from thermodynamic equilibrium and formation of metastable phases – which cannot be obtained in conventional furnace annealing – could be expected. Laser radiation is strongly coupled with tungsten, specially in the near ultraviolet region. For a XeCl laser (308 nm) the normal incidence reflectivity is 0.45 only. Hence, the problems with laser energy efficiency are partly alleviated.

With regard to the above-mentioned facts and to the limited knowledge of tungsten silicide formation by excimer laser irradiation [4], the topic is studied in this paper more thoroughly.

2. Experimental details

Tungsten films with thicknesses $d_W = 150$ and 500 nm were sputtered onto ⟨100⟩Si wafers 300 μm thick with a resistivity of 2–5 Ω · cm, at the substrate temperature of 150°C. The deposition chamber of our Z550 LH equipment was first evacuated down to 10^{-4} Pa, then the deposition was performed in high purity Ar. The thickness of the W layers was measured using a quartz crystal monitor and it was confirmed by Rutherford backscattering spectrometry (RBS).

The sheet resistance of the as-deposited samples was 1.2–1.4 Ω and 0.36–0.45 Ω for W films 150 and 500 nm thick, respectively. The resistivity of as-deposited tungsten was 18 $\mu\Omega \cdot$cm. The resistance was measured in linear 4-point configuration.

The samples were irradiated by using a Lambda Physik 315i XeCl excimer laser in a chamber evacuated down to 4×10^{-3} Pa. Some samples were also processed in very pure nitrogen (< 2 ppm O_2 contamination) at superatmospheric pressure (1100 mbar). The laser fluences were $F = 0.6$, 0.9, 1.2, 1.5 and 1.8 J/cm^2, the superimposed number of shots was $n = 1$–300, the pulse duration was $\tau = 30$ ns and the repetition rates of pulses were $f = 3$, 5 and 50 Hz.

The irradiated samples were analyzed by RBS using $^4He^+$ ions at 2 MeV, by X-ray diffraction (XRD) using Cu Kα radiation and by surface profilometry. The results were correlated with the sheet resistance data.

3. Temperature and heat conduction calculations

The temperature of the metal–silicon interface is the crucial parameter for both intermixing and silicide formation [4]. Since the direct measurement of temperature is complicated, due to the very fast heating and cooling rates, numerical computations were performed. The details can be found in authors' previous papers [4,5].

Typical results of numerical simulations are shown in fig. 1. A summary of the results for various fluences and both thicknesses of the W film are reported in table 1. According to the data of table 1, in our experimental conditions only silicon should melt ($T_{mSi} = 1685$ K), while tungsten ($T_{mW} = 3650$ K) should remain solid even for $F = 1.8$ J/cm^2.

It is also important to know the conditions of heat conduction in the multipulse regime. It is well known that silicon is an excellent heat conductor. Its thermal diffusion coefficient is $D_{Si} =$

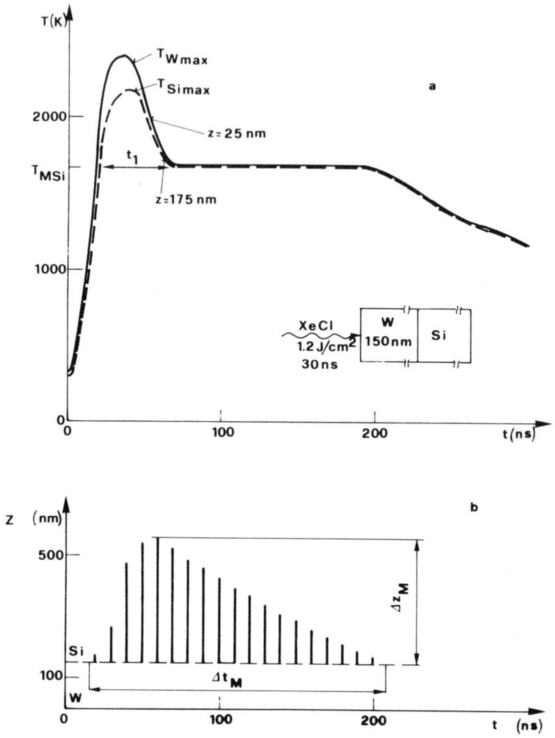

Fig. 1. Temperature and melting simulation for W(150 nm)/Si sample irradiated at laser fluence $F = 1.2$ J/cm^2. (a) (———): Temperature for depth $z = 25$ nm (near to the W surface), (— — —): temperature for $z = 175$ nm (near to the Si surface); (b) maximum depth Δz_m and duration Δt_m of melting.

Table 1
Summarization of temperature and melting computations in W(150–500 nm)/Si, as a consequence of a single XeCl laser pulse (30 ns); $T_{W_{max}}$: maximum temperature for depth $z = 25$ nm (near to W surface); $T_{Si_{max}}$: maximum temperature for $z = 175$ nm (near to Si surface); Δz_m: depth of melt; Δt_m: time of melting

Thickness of W (nm)	Fluence (J/cm^2)	$T_{W_{max}}$ (K)	$T_{Si_{max}}$ (K)	Δz_m (nm)	Δt_m (ns)
150	0.9	2030	1870	200	88
	1.2	2405	2200	400	193
	1.5	3025	2715	630	285
	1.8	3330	3000	805	375
500	0.9	2005	1705	50	20
	1.2	2360	1850	250	116
	1.5	2940	2165	450	210
	1.8	3150	2380	660	280

10^{-4} m^2/s at 300 K and it preserves a value $> 10^{-5}$ m^2/s up to 1500 K. So, the heat delivered by the laser pulse diffuses across the Si substrate (300 μm) in a very short time: $t = d^2/4D_{Si} \approx 0.2$–2 ms. This time is one or two orders of magnitude lower than the time interval between two successive shots at repetition rates of 50 Hz and 5 Hz, respectively. Thus, laser pulses on the samples are thermally independent and the accumulation of heat in the samples submitted to multipulse irradiation is not expected.

4. Experimental results

We observed by XRD that in our samples only tetragonal WSi_2 was formed. Metal-rich W_5Si_3 and metastable hexagonal WSi_2 were not formed. WSi_2 was identified by (002), (110), (101), (103) and (112) peaks. With increasing fluence and number of pulses they appeared in the above-mentioned sequence, as a rule. As-deposited bcc tungsten films were (110) textured ($2\theta_{exp} =$ 40.25°). Formation of WSi_2 was preceded by appearance of a $2\theta_{exp} = 40.5$–40.6° peak which could be caused by the shrinkage of the W lattice due to the dissolution of Si in it. Peaks corresponding to tungsten nitride in the samples irradiated in N_2 atmosphere were not observed.

At the fluence of 0.6 J/cm^2 no change of resistance, XRD and RBS spectra were observed. XRD spectra show that for vacuum processing with $F = 0.9$ J/cm^2 pulses a mixed structure of W and WSi_2 appears. With an increasing number of shots at the fluence of 1.2 J/cm^2 pure WSi_2 phase develops. For 500 nm W films the transformation is shifted towards higher laser fluences. After irradiation in N_2 atmosphere the same phases were found by XRD as for vacuum processing. However, some differences were observed in resistance and RBS data. They will be further discussed.

From RBS spectra a picture of the $W \rightarrow WSi_2$ transformation congruent with XRD results follows. With increasing number of pulses the mixing of W and Si is more distinct (see, e.g., fig. 2 for 150 nm W films). In the structure there are no sharp interfaces, which is typical for laser irradiation with nanosecond pulses [6]. At the thickness

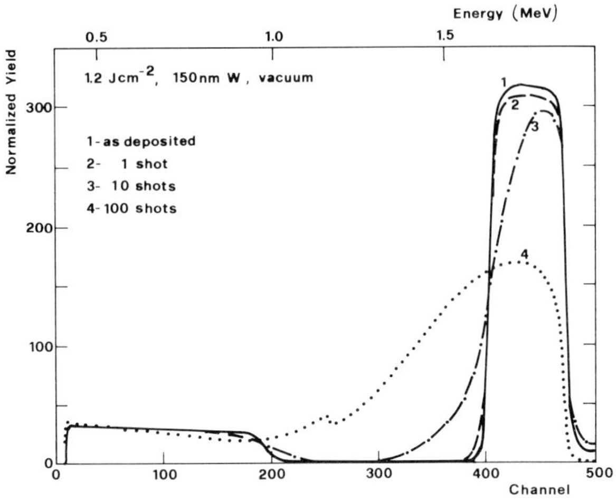

Fig. 2. RBS spectra for a W(150 nm)/Si sample processed at $F = 1.2$ J/cm^2 in vacuum.

of intermixed phases $d_{mix} \approx 2d_W$, formation of silicide was completed according to XRD measurements. This situation is set up somewhere between profiles 3 and 4 in fig. 2, i.e. between 30 and 100 laser pulses for $F = 1.2$ J/cm^2 and for 30 pulses for $F = 1.5$ J/cm^2. Then the resistance of the silicide layer remains the lowest. With increasing number of pulses W penetrates deeper into the sample and the resistance increases. Thus, the best conditions for silicide synthesis correspond to the threshold values of fluence and number of shots under which $W \rightarrow WSi_2$ transformation is completed. The corresponding resistance of the silicide layer obtained by irradiation of W(150 nm)/Si samples in vacuum is 5–10 Ω. The depth-averaged resistivity could be estimated as $\rho = (5\text{–}10)$ $\Omega \times 300$ nm $= 150$–300 $\mu\Omega \cdot$ cm. The surface of these samples after irradiation is flat.

In N_2 ambient atmosphere the formation of silicides is accelerated, but at a higher number of shots ablation takes place. These effects were evident in the first place in W(150 nm)/Si samples at $F = 1.5$ and 1.8 J/cm^2. The resistance increase is striking in this case. For ablated samples the resistance approaches to the value of Si substrate (95 Ω), as measured with the 4-point probe.

With W(500 nm)/Si samples a similar resistance increase was observed for both vacuum and

Table 2
Apparent interdiffusion coefficients at the W/Si interface related to one laser pulse at a given fluence F; they are calculated at $C/C_0 = 0.25$ concentration ratio for two times of melting: D_1 is for the time t_1 of the temperature overshot (see fig. 1), D_2 is for the time of melting Δt_m (see table 1)

Thickness of W (nm)	Ambient	F (J/cm^2)	D_1 (cm^2/s)	D_2 (cm^2/s)
150	Vacuum	1.2	2.38×10^{-5}	6.61×10^{-6}
150	Vacuum	1.5	1.37×10^{-5}	3.02×10^{-6}
150	Vacuum	1.8	3.27×10^{-5}	7.21×10^{-6}
150	Nitrogen	1.5	3.94×10^{-5}	9.09×10^{-6}
150	Nitrogen	1.8	1.67×10^{-4}	3.85×10^{-5}
500	Nitrogen	1.5	2.00×10^{-4}	7.64×10^{-5}
500	Nitrogen	1.8	2.85×10^{-4}	8.95×10^{-5}

N_2 processing. It was caused by cracking of the metal film and even its ablation for $n > 100$ at $F = 1.5–1.8$ J/cm^2. For these samples the influence of N_2 upon reactions was overlapped by other phenomena.

From the RBS spectra, concentration profiles of Si and W were calculated. The program simulates theoretical spectra which reflect the assumed concentration profile and compares them with the experimental ones. From these profiles and temperature simulations the interdiffusion coefficients at the W/Si solid/liquid interface were estimated. The calculations were performed only for the single-pulse regime. The results are shown in table 2.

5. Discussion

The silicide formation in our samples proceeds via interdiffusion and mutual dissolution at the W(solid)–Si(liquid) interface. Assuming that the dissolution of tungsten in molten silicon is a rate-limiting process, the formation of the silicon-rich WSi_2 phase is expected, according to the Si–W phase diagram, for short (ns) irradiation times. Our experimental results confirm this hypothesis. The kinetics of the transformation $W \rightarrow WSi_2$ corresponds quite well to temperature simulations (table 1). It shows that $d_W = 500$ nm should be considered as an upper limit of tungsten thickness in the XeCl laser synthesis of tungsten silicide because various complications appeared in this case, as mentioned above.

Usually, in laser processing of solid structures, melting proceeds from the surface into the depth. In our case melting starts at the interface. Hence, problems arise due to internal stress in tungsten. The stress is compressive or tensile, depending on the film deposition method and the gas (Ar) content in the sputtered film [7]. During laser heating of the W/Si sample the compressive component of the stress increases only slightly because the thermal expansion coefficient $\alpha_W = 4.5 \times 10^{-6}$ K^{-1} is not much higher than $\alpha_{Si} = 2.6 \times 10^{-6}$ K^{-1}. But after melting of silicon at the interface, the stress in the tungsten is not longer compensated by adhesion to the substrate. Thus, tungsten becomes rougher by wrinkling and, in contrast, it cracks during resolidification. Such degradation is more dangerous with thicker films, where the chance to relieve the stress is smaller. This picture corresponds to our observations, where cracking was typical for 500 nm W films. The higher efficiency of ablation of these samples could be due to limited thermal conductivity of cracked surfaces. The acceleration of silicide formation and ablation of tungsten when laser irradiation is performed in a nitrogen atmosphere is not understood up to now. One can speculate about the stronger laser coupling of radiation with a nitrided tungsten surface and ablation of thermodynamically less stable WN. These complex phenomena could be responsible also for the efficiency of the multipulse regime of laser processing of our samples, where the subsequent pulses find the surface less reflective.

6. Summary

Tungsten disilicide with a tetragonal structure was prepared in a multipulse regime by XeCl laser ns pulse irradiation of tungsten films, 150 and 500 nm thick, sputtered on monocrystalline silicon. The lowest resistance and resistivity of disilicide synthesized in vacuum were 5–10 Ω and 150–300 $\mu\Omega \cdot$ cm, respectively. Irradiation in a nitrogen atmosphere accelerated both the silicide formation and the surface ablation of the processed samples with thinner tungsten films. With thicker tungsten films these phenomena were overlapped by ablation that was independent from the working ambient and that could be enhanced due to cracking of the sample surface. From temperature calculations it follows that in our experiments silicides formed by interdiffusion and intermixing at the solid-tungsten–liquid-silicon interface.

Acknowledgement

This work was supported in part by the Consiglio Nazionale delle Ricerche.

References

[1] M. Hammar, S.L. Zhang, R. Buchta and T. Johanson, Thin Solid Films 185 (1990) 9.

[2] J. Willer, M. Heinzle, N. Arnold and D. Ristow, Solid State Electron. 33 (1990) 571.

[3] S.L. Zhang, U. Smith, R. Buchta and M. Ostling, J. Appl. Phys. 69 (1991) 213.

[4] E. D'Anna, G. Leggieri and A. Luches, Appl. Phys. A 45 (1988) 325.

[5] E. D'Anna, G. Leggieri, A. Luches, G. Majni and G. Ottaviani, Thin Solid Films 136 (1986) 93.

[6] E. D'Anna, G. Leggieri, A. Luches, S. Luby and J. Zemek, Appl. Phys. A 48 (1989) 503.

[7] M. Itoh, M. Hari and S. Nadahara, J. Vac. Sci. Technol. B 9 (1991) 149.

Applied Surface Science 69 (1993) 350–354
North-Holland

applied
surface science

An optical in-situ method for layer growth characterization

N. Dietz and H.J. Lewerenz

Hahn-Meitner-Institut Berlin, Bereich Photochemische Energieumwandlung, Postfach 390128, Glienicker Strasse 100, D-1000 Berlin 39, Germany

Received 2 June 1992; accepted for publication 30 October 1992

An optical method is presented which allows the in-situ characterization of growing films on solid substrates. The technique is based on the changes in reflectivity of p-polarized light at the substrate Brewster angle. The changes in the reflectivity are shown to be large enough to monitor layer growth as well as to reveal the film thickness and the optical constants of the film.

1. Introduction

At present, the characterization of the growth of thin films is being intensively pursued [1,2]. In-situ techniques such as combined XPS (X-ray photoelectron spectroscopy) and LEISS (low energy ion scattering spectroscopy) [3], AES (Auger electron spectroscopy), RHEED (reflection high energy electron diffraction) and ellipsometry are frequently employed. These methods are rather complex and need considerable technical equipment. In the present work we present a simple method which makes use of the relaxation of the Brewster angle condition for p-polarized light [4,5] of a substrate material upon film growth. The applicability of the method will be demonstrated for a film growing on top of a silicon substrate taking into account the experimental limitations through angle divergence and depolarization effects.

For p-polarized light, Si exhibits a reflectivity minimum (pseudo-Brewster angle) at $\varphi_B = 75.637°$ ($\epsilon_{s1} = 15.25$, $\epsilon_{s2} = 0.17$) for illumination with a He/Ne laser ($\lambda = 632.8$ nm). We consider the change in reflectivity R_p at this fixed angle due to the growth of a layer on silicon. To simplify matters, it has been assumed that this layer is homogeneous and grows layer by layer (Frank–van der Merwe growth mechanism) [6].

2. Measurement procedure

The measurement principle is shown in fig. 1. A He/Ne laser (1–3 mW) is used as light source. The output power has to be stabilized and a beam-splitting unit is needed to measure the light intensity accurately enough. The polarizer needs an extinction ratio better than 10^6 for the p-polarized component in order to achieve a high resolution of the reflectivity minimum of the Si substrate. This condition is fulfilled by a Glan–Thompson prism for instance. The detection of the beam reflected at the sample is done by a cooled Si photodiode with an additional low noise amplifier. The configuration allows detection of changes in R_p of the order of 1×10^{-5}. The investigation of the initial stages of deposition is

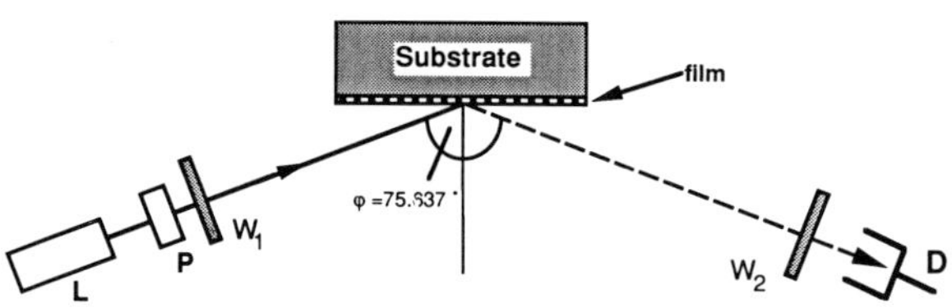

Fig. 1. Schematical diagram of the experimental setup; L: He/Ne-laser; P: polarizer; W1, W2 are the windows into the vacuum chamber; D: detector. Angle of incidence $\varphi_B = 75.637°$ (Brewster angle for the silicon substrate at 632.8 nm with $\epsilon_{s1} = 15.25$ and $\epsilon_{s2} = 0.17$).

usually made with a lock-in technique for enhanced sensitivity.

The laser beam divergence is about 1 mrad ($\approx 0.06°$) and the polarizer extinction ratio $(R_s/R_p) \approx 10^{-6}$. While passing the polarized light through tempered window glasses into the vacuum chamber a reduction the polarization extinction ratio (stress induced birefringence) to about 10^{-5} has to be assumed.

3. Model consideration

The change in reflectivity R_p of a homogeneous film growing layer by layer can be calculated using Fresnels equations for a three layer system [7] (ambient/film/substrate):

$$rr_p = \frac{r_{01_p} + r_{12_p}\, e^{-i2\phi_1}}{1 + r_{01_p} r_{12_p}\, e^{-i2\phi_1}}, \tag{1}$$

where r_{01_p} and r_{12_p} are the reflectivity coefficients

$$r_{01_p} = \frac{\epsilon_f \cos\varphi_0 + \sqrt{\epsilon_a}\sqrt{\epsilon_f - \epsilon_a \sin^2\varphi_0}}{\epsilon_f \cos\varphi_0 + \sqrt{\epsilon_a}\sqrt{\epsilon_f + \epsilon_a \sin^2\varphi_0}} \tag{2a}$$

and

$$r_{12_p} = \frac{\epsilon_s\sqrt{\epsilon_f - \epsilon_a \sin^2\varphi_0} - \epsilon_f\sqrt{\epsilon_s - \epsilon_a \sin^2\varphi_0}}{\epsilon_s\sqrt{\epsilon_f - \epsilon_a \sin^2\varphi_0} + \epsilon_f\sqrt{\epsilon_s - \epsilon_a \sin^2\varphi_0}} \tag{2b}$$

for p-polarized light at the ambient/film and film/substrate interfaces, respectively. The phase factor ϕ_1 in eq. (1) is given by

$$\phi_1 = \frac{2\pi d_{f1}}{\lambda}\sqrt{\epsilon_f - \epsilon_a \sin^2\varphi_0}\,. \tag{3}$$

The film's thickness is d_{f1}, φ_0 is the angle of incidence and ϵ_s, ϵ_f, ϵ_a are the complex dielectric functions of the substrate, film and the ambient, respectively.

In fig. 2a, $R_p = rr_p \times rr_p^*$ is plotted for transparent films ϵ_f with $\epsilon_{f1} = 8$, 6, 4 and 2 for a thickness range between 0 and 5000 Å. A pronounced modulation of R_p is calculated which has the highest modulation depth for the smallest ϵ_{f1} value. Fig. 2b shows changes in reflectivity during the first 100 Å film growth.

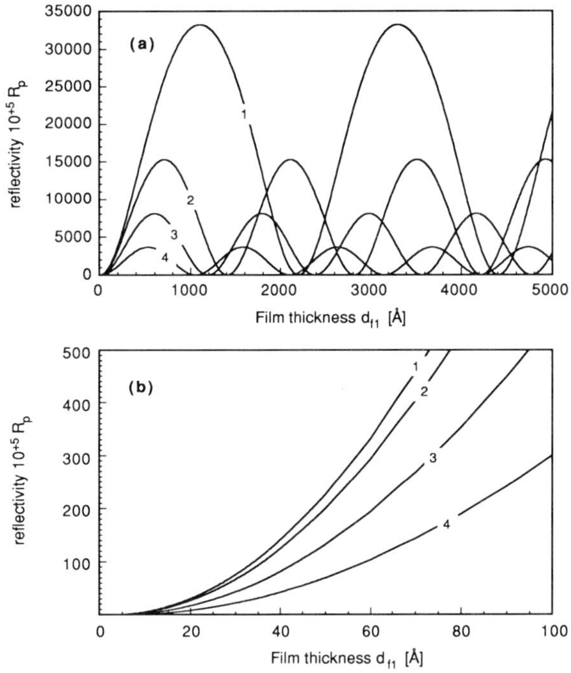

Fig. 2. (a) Calculated changes in reflectivity R_p by growing films (optical constant ϵ_f) on Si-substrate (transparent films ϵ_f with 1: $\epsilon_{f1} = 3$; 2: $\epsilon_{f1} = 6$; 3: $\epsilon_{f1} = 8$ and 4: $\epsilon_{f1} = 10$). Angle of incidence $\varphi_B = 75.637°$. (b) Changes in reflectivity for film growth up to 100 Å.

The occurring minima and maxima as well as the modulation height in the reflectivity are related to the optical constant of the grown film. The sensitivity of the technique is highly dependent on the magnitude between the occurring minima and maxima during the growth. At the beginning of the growth process, after having chosen the angle of incidence at the Brewster angle of silicon, the determined reflectivity is close to the first reflection minimum, with a maximal change in reflectivity. Therefore, a high sensitivity range during the layer growth (seven orders of magnitude) should be available to determine the conditions of growth.

Fig. 3 shows changes in reflectivity between the occurring maxima and minima, as a function of the optical constant of the grown transparent film ϵ_f. For optical constants of the film close to the optical constant of the substrate ($\epsilon_{s1} = 15.25$), the reflectivity range between the occurring minima and maxima decreases and is close to zero if

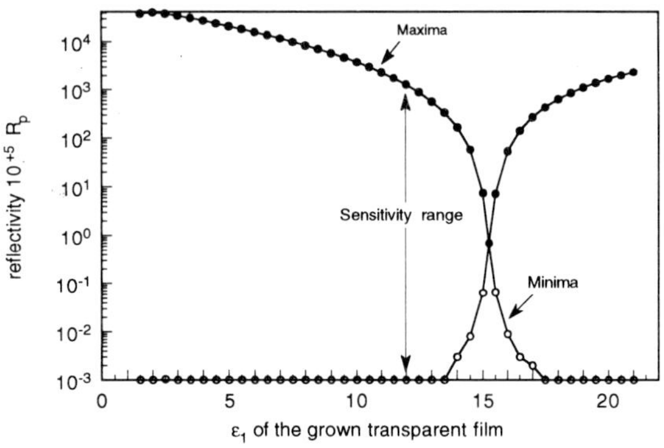

Fig. 3. Changes in reflectivity for the minima and maxima versus optical constant ϵ_{f1} of the grown film at $\varphi_B = 75.637°$.

the optical constants of the film are equal to the optical constant of the substrate.

Fig. 4 shows a contour plot, where changes in the calculated reflectivity minima and maxima positions are drawn as a function of the optical constants of the layer versus thickness of the film. Using both the value of reflectivity R_p (fig. 3) and the positions of the reflectivity maxima and minima (fig. 4), the optical constant ϵ_{f1} of the film can be extrapolated.

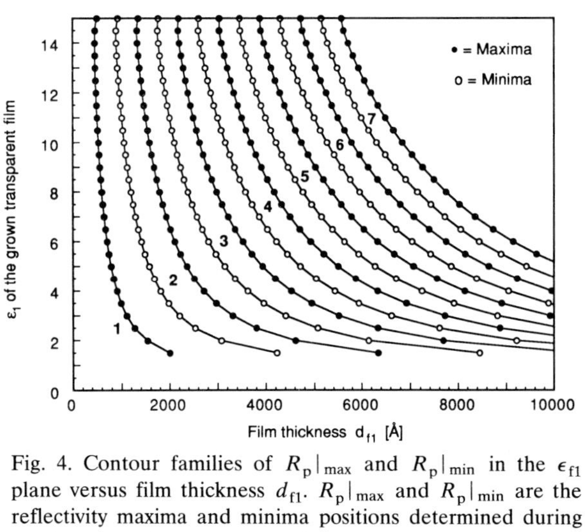

Fig. 4. Contour families of $R_p|_{max}$ and $R_p|_{min}$ in the ϵ_{f1} plane versus film thickness d_{f1}. $R_p|_{max}$ and $R_p|_{min}$ are the reflectivity maxima and minima positions determined during the film growth, respectively (experimental conditions: $\varphi = 75.637°$, $\lambda = 632.8$ nm, silicon substrate: $\epsilon_{s1} = 15.25$ and $\epsilon_{s2} = 0.17$). Curves 1 to 7 indicate the order of the occurring reflectivity maxima (minima).

4. Experimental limitations in the determination of layer growth

The sensitivity of the method during layer growth is dependent on the resolution of the occurring reflection minima which are in the order of 10^{-8} R_p. The depolarization ratio (R_s/R_p) of a Glan–Thompsen prism is better than 10^{-6}. Even for high quality stress-free window material, the polarization ratio is reduced by approximately one order of magnitude. The total depolarization can be estimated to be 10^{-5} multiplied by the reflectivity R_s of the ambient/film/substrate system.

Estimation of the error induced by the angle divergence of the light beam can be analyzed as follows: around the Brewster angle the reflectivity can be expressed as

$$R(\varphi) = R|_{\varphi_B} + \frac{\Delta R}{\theta^2}(\varphi - \varphi_B)^2. \tag{4}$$

Here φ is the angle of incidence, φ_B the Brewster angle of the substrate, $R|_{\varphi_B}$ the reflectivity at this angle and ΔR the difference between $R|_{\varphi_B-\theta}$ and $R|_{\varphi_B}$. The interval of the angle, defined by 2θ, lies in the order of 0.2° up to over 2°, depending on the absorption of the material [8]. The experimentally resolved reflectivity $\bar{R}$ at the Brewster angle can therefore be calculated as

$$\begin{aligned}\bar{R}|_{\varphi_B} &= \frac{1}{2\alpha}\int_{\varphi_B-\alpha}^{\varphi_B+\alpha} R(\varphi)\,\mathrm{d}\varphi = R|_{\varphi_B} + \frac{\Delta R\alpha^2}{3\theta^2} \\ &= R|_{\varphi_B} + \frac{|R|_{\varphi_B-\alpha} - R|_{\varphi_B}|}{3}.\end{aligned} \tag{5}$$

Here, 2α is the interval of angle divergence with $\alpha \le \theta$. For any other angle of incidence φ inside the interval $[\varphi_B \pm \theta]$ the experimentally resolved reflectivity can be estimated by

$$\begin{aligned}\bar{R}|_{\varphi_x} &= R|_{\varphi_B} + \frac{\Delta R}{6\theta^2\alpha}\left[(\varphi_{x+\alpha} - \varphi_B)^3 + (\varphi_{x-\alpha} - \varphi_B)^3\right] < R|_{\varphi_x} \\ &\quad + \frac{|R|_{\varphi_{x+\alpha}} + R|_{\varphi_{x-\alpha}} - 2R|_{\varphi_x}|}{6},\end{aligned} \tag{6}$$

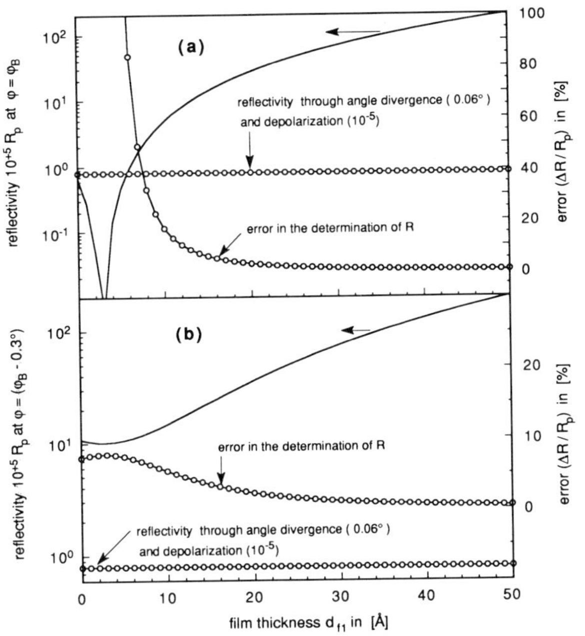

Fig. 5. (a) Changes of reflectivity during the initial stages of film growth on a Si-substrate at the Brewster angle $\varphi_B = 75.637°$ ($\lambda = 632.8$ nm). Due to depolarization and angular divergence, a large error in reflectivity occurs. (b) Changes in reflectivity by monitoring the changes of the reflected intensity at the angle $\varphi = (\varphi_B - 0.3°)$. The calculated error due to experimental limitation is reduced to less than 10% during the initial growth.

where $\varphi_{x \pm \alpha}$ lies inside the angle interval $[\varphi_B \pm \theta]$.

For any angle of incidence φ outside the interval $[\varphi_B \pm \theta]$ the reflectivity can be estimated through a linear function for angle divergences in the order of 0.2° or less. A linear function approximation, however, means that no significant error occurs during a distribution of symmetrical angle divergence. For a more analytical approximation an additional, nearly Gaussian angle divergence distribution can be assumed.

In a first approach, the worst possible conditions (eqs. 5 and 6) were assumed with a distribution of linear angle divergence of 0.06° for the He/Ne-laser. Fig. 5a shows the changes in reflectivity during the layer growth up to 50 Å. The angle of incidence was chosen at the Brewster angle φ_B. Under the assumed experimental limitations, a large error in the determination of the reflectivity occurs below 10 Å thickness. Fig. 5b shows the reflectivity behavior during layer growth for an angle of incidence 0.3° smaller than the Brewster angle.

Here, the reflectivity for the uncoated substrate starts at $R_p \approx 10^{-4}$. Therefore, the error in the determination of the reflectivity is 5–10% below 10 Å thickness. The limitations are strongly related to the depolarization ratio R_s/R_p. In case of depolarization ratios higher than 10^{-5}, the influence of beam divergence can be neglected.

While growing a layer up to several micrometers, a change of the angle of incidence occurs. The change in the angle, β, can be estimated by $\beta = \arctan(d_{f1}/L_{SD} \cos \varphi)$, where d_{f1} is the thickness of the layer; L_{SD} being the distance from the sample to the detector and φ the angle of incidence. For a film thickness of 2 μm and $L_{SD} = 20$ cm, changes in the angle of 0.002° can be observed. Compared to the 0.06° angle divergences of a He/Ne-laser, the occurring error due to the film growth can therefore be neglected.

5. Imperfect film growth

In a simplified approach, the changes in reflectivity due to imperfect layer growth can be modelled by replacing the complex dielectric function of the film, ϵ_f, with an effective complex dielectric function ϵ_e [9]:

$$\epsilon_e = \epsilon_a \frac{\epsilon_f(1-2q) + 2\epsilon_a(1+q)}{\epsilon_f(1-q) + \epsilon_a(2+q)}. \qquad (7)$$

Here q describes an inhomogeneous film in which q varies from 0 to 1; ϵ_a, ϵ_f, ϵ_e are the complex dielectric functions of the ambient, the film and the effective medium, respectively. In the case of non-epitaxially growing layers, it should be possible to fit the observed behavior with the data of an effective medium. In principle, growth modes such as Frank–van der Merwe, Stranski–Krastanov and Vollmer–Weber [6] should be distinguishable.

6. Conclusion

The new method allows the correlation of film growth with the slope of the reflectivity R_p as a function of d_{fl}. For larger thicknesses the optical constants of the film can be determined (figs. 2a and 2b). Further information on the growth process can be obtained from changes in the R_p (d_{fl}) function.

References

[1] M.A. Herman and H. Sitter, Molecular Beam Epitaxy, Fundamentals and Current Status, Vol. 7 (Springer, Berlin, 1989).

[2] C. Agrile and G.E. Rhead, Surf. Sci. Rep. 10 (6/7) (1989) 277–356.

[3] U. Bardi, Appl. Surf. Sci. 51 (1991) 89.

[4] H.J. Lewerenz and N. Dietz, Appl. Phys. Lett. 59 (1991) 1470.

[5] N. Dietz and H.J. Lewerenz, Appl. Phys. Lett. 60 (1992) 2403.

[6] J.H. van der Merwe, Vol. 35 of Springer Series in Chemical Physics, Eds. V.R. Vanselow and R. Howe (Springer, Berlin, 1980).

[7] R.M.A. Azzam and N.M. Bashara, Ellipsometry and Polarized Light (North-Holland, Amsterdam, 1987).

[8] N. Dietz and H.J. Lewerenz, in preparation.

[9] F.L. McCrakin and Colson, in: Ellipsometry in the Measurement of Surfaces and Thin Films, Eds. R.R.S.E. Passaglia and J. Kruger (U.S. Government Printing Office, Washington, DC, 1964) p. 61.

Applied Surface Science 69 (1993) 355–358
North-Holland

applied
surface science

Determination of the refractive index of $In_{0.53}Al_{0.11}Ga_{0.36}As$ on InP in the wavelength range from 280 to 1900 nm by spectroscopic ellipsometry

H.W. Dinges, H. Burkhard, R. Lösch, H. Nickel and W. Schlapp
Deutsche Bundespost Telekom Forschungszentrum beim FTZ, Am Kavalleriesand 3, D-6100 Darmstadt, Germany

Received 2 June 1992; accepted for publication 30 October 1992

The refractive index of the quarternary material $In_{0.53}Al_{0.11}Ga_{0.36}As/InP$ is measured for the first time by spectroscopic ellipsometry in the wavelength range from 280 to 1900 nm. In previous papers we found out that between InP and MBE-grown $In_{0.52}Al_{0.48}As$ or $In_{0.53}Ga_{0.47}As$ an interface layer exists, due to the interaction of arsenic with InP in the preheat phase of the MBE growth. Therefore we grew three layers of $In_{0.53}Al_{0.11}Ga_{0.36}As$ on InP with different thicknesses to determine the refractive indices and exact values of the thicknesses.

1. Introduction

With a bandgap of 0.889 eV at 300 K $In_{0.53}Al_{0.11}Ga_{0.36}As$ is a promising quarternary material for optoelectronic devices. To our best knowledge its refractive index is not yet determined with ellipsometry. In a previous paper [1] the wavelength dependent refractive index of MOVPE grown $In_{0.52}Al_{0.48}As$ layers on InP was given for the first time with single wavelength ellipsometry in the range from 400 to 800 nm. Spectroscopic ellipsometry is a non-destructive tool to determine optical properties of heterostructures. Since the two measured angles ψ and Δ at a single wavelength are insufficient to determine more than two unknown parameters like the thickness d of the natural oxide layer and the refractive index $n-ik$, we apply spectroscopic ellipsometry on thick single layer materials. Therefore, we measured [2] the ψ, Δ values of 0.5 and 2 μm thick layers of MBE-grown $In_{0.53}Ga_{0.47}As$ and $In_{0.52}Al_{0.48}As$ layers on InP in the wavelength range from 245 to 845 nm, and calculated n, k dispersion curves. We observed that between InP substrate and the layers of $In_{0.53}Ga_{0.47}As$ or $In_{0.52}Al_{0.48}As$ an about 1 nm thick interface layer exists. To determine the refractive index in the wavelength range from 900 to 1900 nm exactly, we have grown 10, 20 and 200 nm thick layers of $In_{0.52}Al_{0.48}As$ [3]. We found out, that from a wavelength of 400 nm on, the interface layer can be fitted very well by the refractive index of InAs and a thickness of 0.5 to 1.2 nm.

2. Experimental

2.1. MBE-growth

The layers were grown in a Varian MOD GENII MBE system on SI InP(100) substrates at growth rates between 0.2 to 0.4 nm/s. The growth temperatures varied from 500 to 530°C and the As/(In + Al) beam equivalent pressure (BEP) ratio from 50 to 70; but during one growing procedure all parameters were constant. The thicknesses of the layers were 25, 80 and 2000 nm. The interaction of arsenic with InP in the preheat phase of MBE-growth was also investigated.

3. Spectroscopic ellipsometry

3.1. Instrumentation and measurements

The samples were characterized by two Rudolph Research ellipsometers, the manual 436-type for the whole wavelength range from 250 to 1900 nm with Babinet Soleil Compensator, interference filters, photomultiplier R 928 and infrared detector, and the Spectro 2000 with rotating polarizer, monochromator from 250 to 840 nm and computer assisted measurement and evaluation performance. To determine ψ and Δ with the manual ellipsometer we used the classical four-quadrant method [2]. With the same method and the Babinet Soleil compensator instead of the rotating polarizer we discovered [2] that on Spectro 2000 the true ψ, Δ angles at each wavelength are the average of the ψ, Δ values measured at the two positions A and B of fixed polarizer and fixed analyzer.

A: fixed polarizer angle 0°,
fixed analyzer angle +45°;

B: fixed polarizer angle 0°,
fixed analyzer angle −45°.

3.2. Evaluation of ellipsometric measurements

ψ, Δ wavelength scans of samples can be done rapidly with the Spectro 2000, but data interpretation is complex and time-consuming [2]. For the determination of n, k graphs and thickness of layered structures with Fresnel formulas multiple layer models with abrupt interfaces are necessary. In our first measurements [2,4] we observed, that InP has a lower refractive index than both ternary compounds $In_{0.53}Ga_{0.47}As$ and $In_{0.52}Al_{0.48}As$. In the wavelength range from 250 to 900 nm a 2 μm layer of $In_{0.53}Ga_{0.47}As$ is thick enough to absorb the light completely. For this reason the interface of the layer and the bulk material has no influence. For wavelengths larger than 1000 nm light penetrates the interface and even reaches the bulk rear side. Therefore, the rear side of the substrate must be rough enough or inclined, so that the reflected light from this side cannot influence the measurements. In our first measurements of the refractive index of $In_{0.52}Al_{0.48}As$ it became evident that even a 2 μm thick $In_{0.52}Al_{0.48}As$ layer is only thick enough for wavelength up to 550 nm. Therefore, we had to determine the three thicknesses of native oxide, $In_{0.52}Al_{0.48}As$, and interface (InAs) layer in the wavelength range from 300 to 550 nm. With the three thicknesses and the refractive indices of the native oxide layer (1.9) [4], of InP, and of InAs [3–7] we could determine the wavelength dependent refractive index of $In_{0.52}Al_{0.48}As$ in the whole wavelength range from 550 to 1900 nm by minimizing the quantity [3]

$$\sum_{\lambda}\left[(\psi_m-\psi_c)^2+(\Delta_m-\Delta_c)^2\right]$$

(m = measured, c = calculated).

According to its bandgap of ~1.4 eV a 2000 nm thick layer of the quarternary material $In_{0.53}Al_{0.11}Ga_{0.36}As$ prevents light from penetrating the interface with the substrate in the wavelength range from 300 to 850 nm. To determine the thickness of the oxide layer and the refractive index of $In_{0.53}Al_{0.11}Ga_{0.36}As$ only a three layer model consisting of air–oxide–$In_{0.53}Al_{0.11}Ga_{0.36}As$ is needed. Our results are given in fig. 1. The refractive index of $In_{0.53}Al_{0.11}Ga_{0.36}As$ is higher than that of InP by 0.31 and smaller than that of $In_{0.53}Ga_{0.47}As$ by 0.095 at a wavelength of 1.5 μm. Our dispersion curves $n(\lambda)$ and $k(\lambda)$ for the investigated samples differed by ± 0.010. These

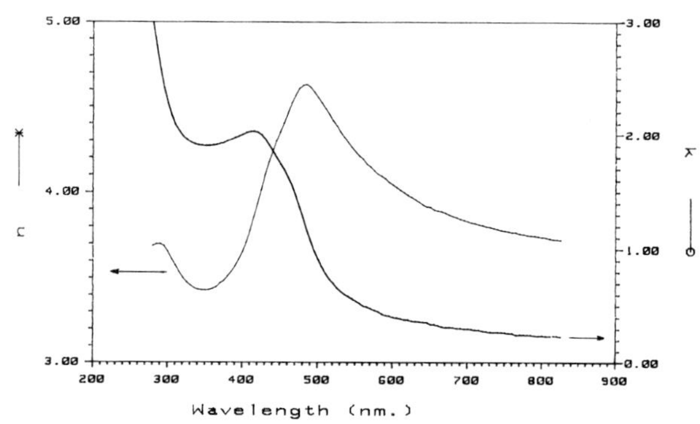

Fig. 1. Dispersion curve of a 2 μm thick sample of $In_{0.53}Al_{0.11}Ga_{0.36}As$/InP. The sample was thick enough, so that in the wavelength range from 320 to 840 nm the influence of the interface and the substrate did not exist.

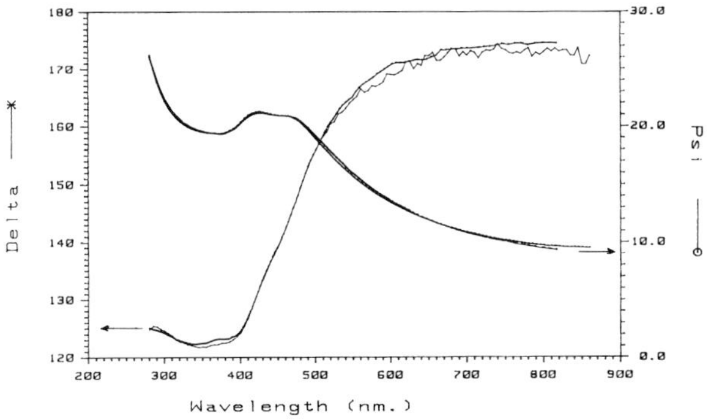

Fig. 2. Comparison of ψ, Δ curves measured on a 27 nm thick $In_{0.53}Al_{0.11}Ga_{0.36}As$/InP at an angle of incidence of 70° with a fit air–oxide layer–quarternary layer–InAs–InP. The small deviations in the ψ fit are due to the interface.

small deviations may be due to our estimation that the thickness of the natural oxide layer, 7 min after exposure to air, is in the same range (0.8 nm) as those of InP and GaAs.

With the refractive index of the thick layer we fitted the thickness of both thin layers 25 and 80 nm and also that of the interface layers. The fits were now done with a five-layer model air–oxide–$In_{0.53}Al_{0.11}Ga_{0.36}As$–InAs–InP. Fig. 2 shows the ψ, Δ fit for the smaller layer compared to the measured ψ, Δ curves between $\lambda = 320$ nm and $\lambda = 860$ nm. From $\lambda = 450$ nm on, the agreement is very good.

To get more information about the interface layer we have simulated the preheat phase of the MBE growth. We observed that in the wavelength range from 400 to 1900 nm the interface layer can be fitted very well by the refractive index of InAs and a thickness of 0.5 to 1.2 nm. An interface layer of InAs of the same thickness is also necessary to fit high-resolution X-ray diffraction (HRXRD) measurement [6] on the 25 nm thick $In_{0.53}Al_{0.11}Ga_{0.36}As$ layer.

Our investigations on semi-insulating InP and GaAs bulk samples in the infrared wavelength regime show very good agreement [2] with the results of Pettit and Turner [7] for InP and Palik [8] for GaAs. Table 1 gives the refractive indices of $In_{0.53}Al_{0.11}Ga_{0.36}As$ and $In_{0.53}Ga_{0.47}As$ in the wavelength range from 546.1 to 1900 nm. The refractive index of $In_{0.53}Al_{0.11}Ga_{0.36}As$ is higher than that of InP throughout the wavelength range, but smaller than that of $In_{0.53}Ga_{0.47}As$. To obtain better fits in the wavelength range from 1000 to 1900 nm, it is necessary to measure the refractive index of InAs in this regime since there exist only a few values given by Palik [8].

Table 1
Refractive indices of $In_{0.53}Al_{0.11}Ga_{0.36}As$ and $In_{0.53}Ga_{0.47}As$ on InP from 546.1 to 1900 nm; k values smaller than 0.01 are not shown

Wavelength (nm)	$In_{0.53}Al_{0.11}Ga_{0.36}As$		$In_{0.53}Ga_{0.47}As$	
	n	k	n	k
546.1	4.190	0.494	4.216	0.718
632.8	3.876	0.332	3.930	0.466
765.0	3.686	0.241	3.731	0.330
850.0	3.640	0.220	3.665	0.307
940.0	3.610	0.180	3.637	0.272
1020.0	3.586	0.140	3.618	0.246
1090.0	3.570	0.115	3.606	0.224
1200.0	3.550	0.090	3.590	0.194
1300.0	3.520	0.075	3.580	0.168
1400.0	3.500	0.017	3.577	0.145
1500.0	3.480		3.575	0.123
1600.0	3.454		3.575	0.098
1700.0	3.425		3.565	
1800.0	3.385		3.460	
1900.0	3.330		3.380	

4. Conclusion

The refractive indices of $In_{0.53}Al_{0.11}Ga_{0.36}As$/InP are measured for the first time in the wavelength range from 280 to 1900 nm with spectroscopic ellipsometry. E.g. at 1550 nm the refractive index is 3.46 ± 0.005. Our results on InP and GaAs are in good agreement with the measurements of Aspnes [4], Pettit and Turner [6], and Palik [8], but for InAs further measurements must be done. The interface layer due to the interaction of arsenic with InP in the preheat phase of the MBE growth exists and was also found by high resolution X-ray diffraction.

References

[1] J. Krauser, C. Bornholdt, H. Künzel, H.W. Dinges, M. Druminski, R. Gessner and K.-H. Zschauer, Proc. 4th Eur. Conf. Int. Optics (SETG, Glasgow, 1987) pp. 75–77.

[2] H.W. Dinges, H. Burkhard, R. Lösch, H. Nickel and W. Schlapp, Appl. Surf. Sci. 54 (1992) 477.
[3] H.W. Dinges, H. Burkhard, R. Lösch, H. Nickel and W. Schlapp, Exmatec Lyon 20/22.05.92, to be published.
[4] H. Burkhard, H.W. Dinges and E. Kuphal, J. Appl. Phys. 53 (1982) 655.
[5] S.M. Kelso, D.E. Aspnes, M.A. Pollack and R.E. Nahory, Phys. Rev. B 26 (1982) 6669.
[6] A. Pöcker and H.W. Dinges, to be published.
[7] G.D. Pettit and W.J. Turner, J. Appl. Phys. 36 (1965) 2081.
[8] E.D. Palik, Handbook of Optical Constants of Solids (Academic Press, London, 1985).

Applied Surface Science 69 (1993) 359–364
North-Holland

applied
surface science

Laser ablation of graphite targets

C. Germain, C. Girault, J. Aubreton and A. Catherinot
URA 320 CNRS, Faculté des Sciences, 123 Avenue Albert Thomas, 87060 Limoges Cedex, France

Received 2 June 1992; accepted for publication 30 October 1992

The plasma plume created during the interaction between a graphite target and a KrF laser beam (248 nm) is studied mainly by time and spatially-resolved spectroscopy. High resolution spectra are obtained and expansion velocities are deduced as a function of laser fluence and ambient pressure. An estimation of the vibrational and rotational temperatures as a function of ambient pressure is given. Preliminary results on the deposition of "hard" carbon films on ⟨100⟩Si substrates are presented.

1. Introduction

In the past few years there has been a growing interest in elaboration and application of diamond and diamond-like carbon (DLC) thin films. Indeed these films present particularly interesting characteristics such as extreme hardness, electrical insulation, high thermal conductivity, chemical inertness and optical transparency in the infrared and partly in the visible spectrum. Various techniques such as plasma-enhanced chemical vapor deposition, sputtering and ion beam deposition have been employed to deposit thin films. Among these techniques, pulsed laser evaporation of carbon targets [1] presents some advantages and particularly hydrogen-free films with good adherence may be grown. While most of the work dealing with this problem is devoted to the optimization of the deposition parameters to obtain films with "good" qualities [2,3], only a few studies have been undertaken to understand the transport phenomena of the carbon species from the target to the substrate [4,5].

This paper deals both with the investigation of the plasma plume created by KrF excimer laser ablation of a graphite target and with the deposition of thin films. Analysis of the temporal evolution of the spectral line and molecular band intensities emitted in the plasma plume has allowed the study of the emitting ejected species, and the influence of the ambient pressure. Finally, first results concerning the deposition of "hard" carbon thin films are presented.

2. Experimental setup

The spectroscopic investigations of the plasma plume are done using the experimental setup schematically shown in fig. 1. The laser source is an excimer laser (Lambda Physik EMG 101) operating at 248 nm (KrF), at a repetition rate of 10 Hz, with a pulse duration of 15 ns and an energy per pulse in the range 20–300 mJ. After spatial filtering, the laser beam is focused onto the target, using a Suprasil quartz lens, located within a high-vacuum stainless-steel vessel. The high-purity graphite target is translated using stepping motors between each laser shot, thereby avoiding deep crater formation. The laser-induced plasma plume is imaged to produce a 1 : 1 image onto the entrance slit of a high-resolution spectrometer (THR 1000 Jobin Yvon, resolving power = 100 000), allowing a spatial resolution better than 0.1 mm. The spectrometer is equipped with an RCA 7265 photomultiplier tube (rise time < 2 ns) connected to a boxcar averager (EG & G PAR 4400, temporal resolution = 2 ns) for spectra recording, or to a Tektronix 2440 digitizer (500

0169-4332/93/$06.00

MHz) for temporal evolution studies. The electronic devices are synchronized with the laser pulse using a delay generator (Stanford DG 535).

Thin film deposition and ion detection are realized in an ultra-high vacuum vessel, evacuated down to 10^{-6} Pa by a turbomolecular pump. The cell is equipped with an introduction chamber which allows target and substrate manipulations. Depositions are performed on monocrystalline silicon substrates located above and in front of the target surface at a distance of 34 mm. The substrate temperature is in the range 25 to 800°C and controlled by an electric heater. The high-purity graphite target is rotated between each laser shot. The laser fluence is varied between 10 and 100 J/cm^2. Deposition time is always 15 min, with a laser repetition rate of 10 Hz. The pressure within the experimental vessel is 10^{-6} Pa during deposition. For the ion detection the substrate is replaced by a collector located at a distance of 400 nm from the target and the electric signal collected by this ion probe is analyzed using a Tektronix digitizer 2440.

3. Results

3.1. Spectroscopic investigations

Temporally and spatially resolved spectroscopy has been used to provide information on the nature and velocity of the emitting ablated species and velocities of ejected particles have been particularly investigated as a function of the argon ambient pressure.

Typical emission spectra obtained with various excimer laser fluences are shown in fig. 2. Assignments of the spectral lines [6] and emission bands [7] are made according to standard tabulations. The spectra contain a strong continuum background emission next to the target ($d < 2$ mm) over the wavelength range 400–650 nm. It is certainly due to the emission of large carbon clusters. The intensity of this continuum emission increases when the argon ambient pressure is higher than 50 Pa. For low laser fluence (typically < 7 J/cm^2) the emission spectrum is dominated by the emission band of C_2 molecules emanating

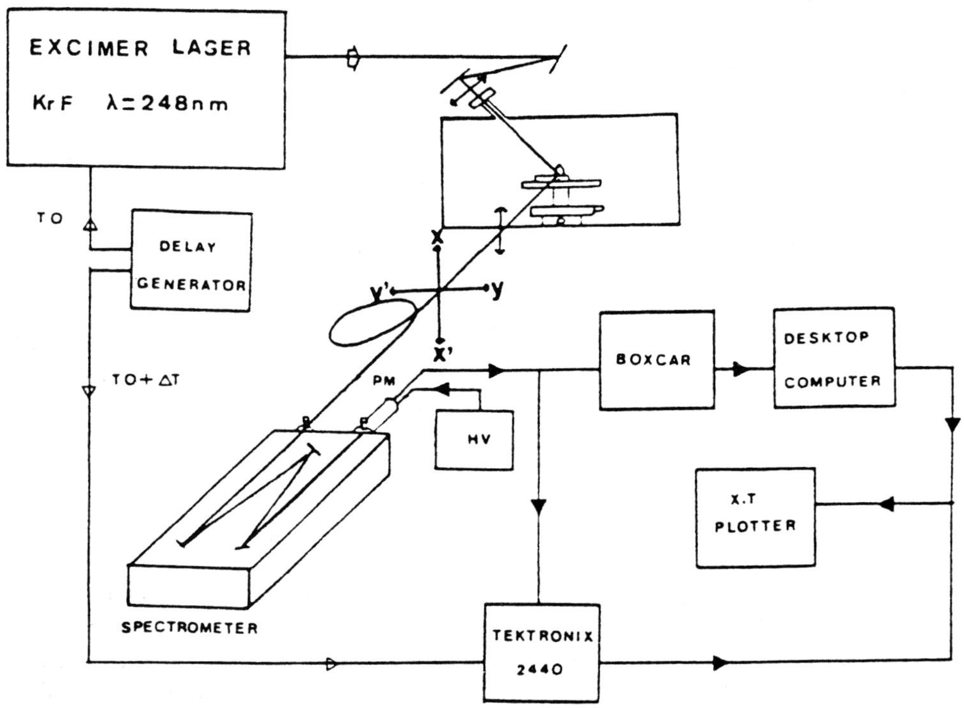

Fig. 1. Experimental setup for spectroscopic investigations.

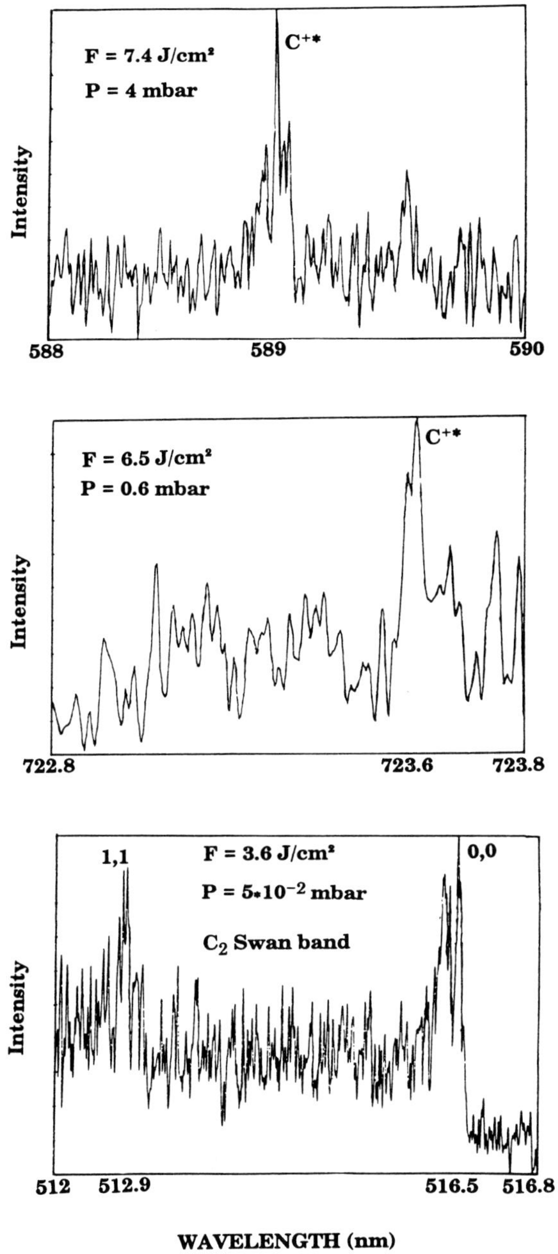

Fig. 2. Spectrum of the KrF laser-induced plasma plume on a graphite target. Observations are performed for a distance between the observed plasma slice and the target surface of about 1 mm, and for varying argon ambient pressure (P) and laser fluence (F).

from the Swan system $A^3\Pi_g \rightarrow X\,^3\Pi_u$ ($v'=0$, $v''=0$ at 516.52 nm and $v'=1$, $v''=1$ at 512.93 nm). This indicates a large concentration of C_2 within the plume, even at low laser fluence. The C_2 molecules may be either directly ejected from the graphite target or created by elementary reactive processes within the plasma. For higher fluence values emission originating from C^{+*} (at 588.9 and 723.6 nm) and C^{2+*} (at 589.4 nm) ions becomes more easily detected. Emission originating from ionic species appears when the laser fluence is sufficient (>7 J/cm^2) to create a largely ionized plasma medium; indeed, 24.376 and 47.864 eV, respectively, are necessary to create C^{+*} and C^{2+*} [8]. Surprisingly, emission originating from excited atomic carbon (C^*) is relatively difficult to detect in the wavelength range under study. The origin of this problem comes from the continuum background which is limiting our apparatus resolution, and the weak concentration of C^* within the plasma plume in our experimental conditions. These species have been observed whatever the distance of the observed plasma slice from the target ($d<2$ mm), and we have not noted a significant change of the emitted intensities with the argon ambient pressure. Moreover, the nature of the ablated products seems to be independent of the residual atmosphere. Indeed, similar investigations were performed under H_2 and $He+O_2$ atmospheres giving very similar results. Analysis of the temporal evolution of the emission intensity has been performed for C^{+*} and C_2^* as a function of the distance d of the observed slice from the graphite target surface. It is clear from these kinetics that the temporal location of the maximum emission intensity is proportional to the distance d, leading to an estimation of the particle velocity. Analysis of the measured kinetics indicates no significant change of the expansion velocity in the 5–10^{-3}–500 Pa pressure range. A similar experiment has been performed for C_2 molecules (Swan band $A^3\Pi_g$, $v'=0 \rightarrow X\,^3\Pi_u$, $v''=0$ at 516.52 nm) and results point out that as for C^{+*} no significant evolution of the expansion velocity is observed in the pressure range under study. The velocity values for C_2 molecules and C^+ ions are presented in table 1 for different argon ambient pressures.

The detection of band emission originating from C_2 is of particular interest since it provides an estimation of the "plume temperature". The C_2 molecular spectra (Swan band) were analyzed in order to determine the vibrational and rotational temperatures. The vibrational temperature (deduced from the Boltzmann diagram) is about 5000 ± 500 K and the rotational temperature (obtained by a comparison between simulated spectra and experimental ones) is higher than 6000 K (we cannot give an accurate value because of the bad signal-to-noise ratio leading to a lack of rotational resolution). From these spectroscopic investigations, it appears that in the range of pressure 4×10^{-3}–100 Pa, laser fluence 3–10 J/cm^2 and a distance from the target surface lower than 2 mm, T_{rot} and T_{vib} do not vary significantly. Moreover, even taking into account the relatively large experimental error bars, the measured values of T_{rot} are systematically higher than the T_{vib} values. This indicates that the plasma medium for $d < 2$ mm probably presents some deviations from local thermodynamic equilibrium.

3.2. *Time resolved ion detection*

Time resolved ion detection has been undertaken in order to determine time-of-flight ionized species velocities to compare with spectroscopic results. This experiment has been performed in the ultra-high vacuum vessel, using in this case an EMG 150 MSC KrF laser. Indeed, the low beam divergence of this laser allows very high fluences on the graphite target (at least up to 120 J/cm^2, the time pulse duration is 17 ns). Our ion detector does not allow any discrimination between the various ionized species (C^+, C^{2+}, C^{3+}, etc.) but just permits the observation of the total ion current. Fig. 3 displays the fluence dependence of the time-of-flight ion velocity deduced from the peak ionic signal. The expansion velocity increases with increasing fluence and is of the same order of magnitude as the value found by emission spectroscopy. Moreover, the integrated ion signal increases with the laser fluence indicating that the number of collected charges grows. This phenomenon is probably due to the increase of both:

- the number of ionized species according to that of the material ejected,
- the ionization degree within the plasma,
- the "mean charge value" of ionized species.

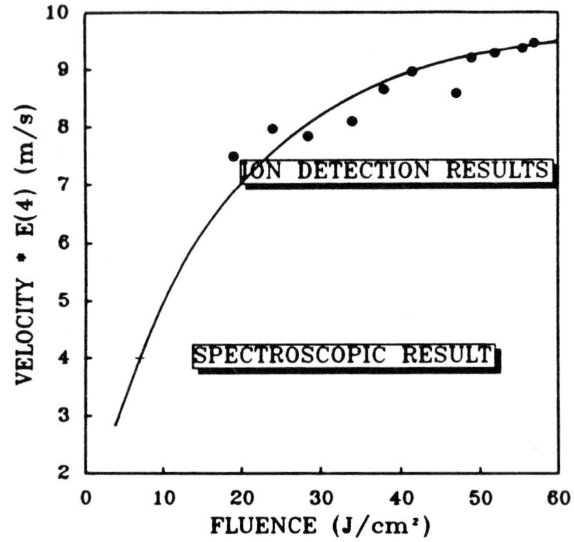

Fig. 3. Evolution of the expansion velocity of the ions with the laser fluence. Comparison of the results obtained by ion detection and those obtained by spectroscopic measurements.

Table 1
C_2 molecules and C^+ ions expansion velocities for different Ar ambient pressures, for a laser fluence value of 9, respectively 7 J/cm^2 and near the target surface ($d < 2$ mm)

C_2		C^+	
Pressure (Ar; mbar)	Velocity (m/s)	Pressure (Ar; mbar)	Velocity (m/s)
5×10^{-5}	9.4×10^3	5×10^{-5}	2×10^4
5×10^{-4}	9×10^3	5×10^{-2}	2×10^4
2×10^{-3}	2×10^4	0.1	3.9×10^4
0.05	1.5×10^4	0.6	3.1×10^4
0.5	1.2×10^4	1	3×10^4
1	9.4×10^3	5	2.4×10^4
5	2.3×10^4		

4. Deposition of carbon thin films

The films are brown and "mirror-like" with some "droplets" about 1 μm in size. The number of these "droplets" increases with laser fluence. The thickness of our films decreases when the laser fluence increases (fig. 4). Since the density

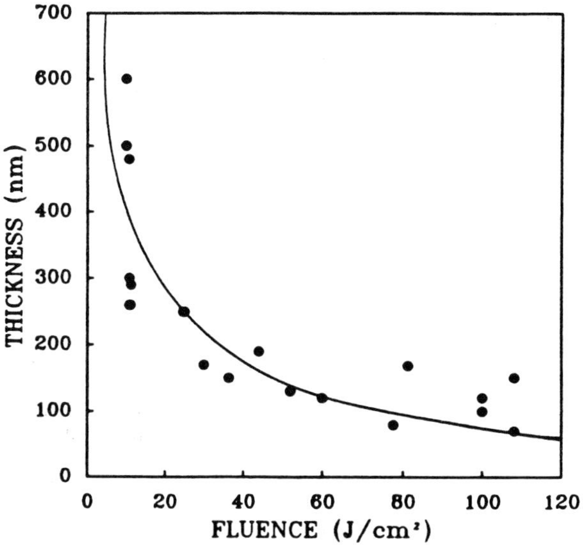

Fig. 4. Evolution of the film thickness with the laser fluence for a deposition time of 15 min.

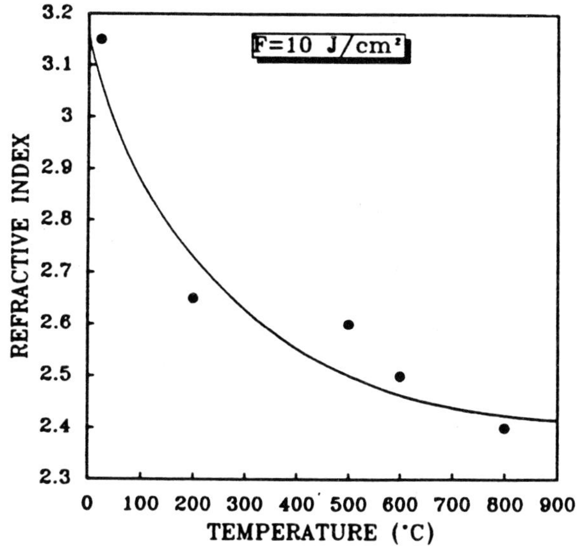

Fig. 5. Evolution of the refractive index with the substrate deposition temperature for a laser fluence of about 10 J/cm^2.

and the velocity of ions increase with laser fluence, we have ascribed these phenomena to sputtering effects of the growing film by incident particles with high energy.

The film adherence on the Si substrate is strongly depending on the laser fluence and the substrate temperature; these parameters are not independent. We have observed that the higher the laser fluence, the lower is the substrate temperature necessary to obtain a "good" adherence.

The refractive index has been measured for each film using "ex situ" monochromatic ellipsometry (6328 Å). Fig. 5 shows the evolution of the refractive index with the substrate temperature for a low laser fluence. The refractive index value decreases from 3 (graphite) down to about 2.4 (diamond). The evolution of the refractive index as a function of the laser fluence for different substrate temperatures is given in fig. 6. Despite the relatively large experimental errors bars, we observe that the refractive index tends to a value of about 2.4 (diamond) for the higher laser fluence values and this more quickly when the substrate temperature is increased. These values are larger than those currently proposed in literature [2,3,9] and are closer to diamond values.

As expected, analysis using Raman microprobes shows a spectrum similar to that of amorphous carbon. This result agrees with the measurements of Davanloo et al. [10] and is ascribed by these authors to the presence of diamond

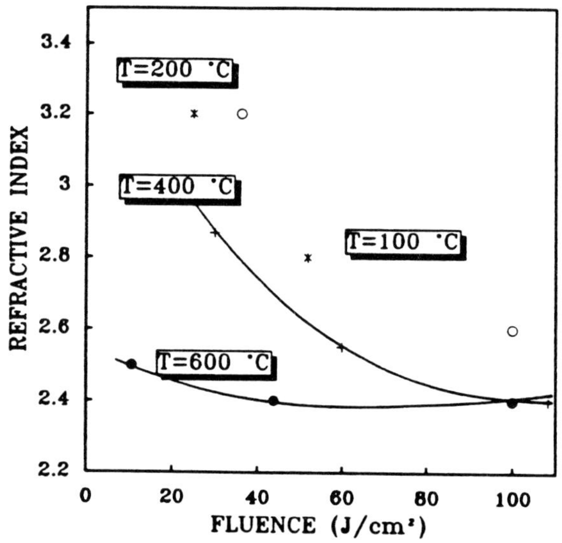

Fig. 6. Evolution of the refractive index with the laser fluence for various substrate deposition temperatures.

particles too small to be detected using this technique.

5. Conclusion

This work has been devoted to the investigation of the photoablation of graphite by a KrF excimer laser and the deposition of carbon thin films. Time and spatially resolved spectroscopic measurements and detection of the ionic signal in the gaseous phase lead to an identification of the radiative emitted species and expansion velocities. We have evaluated the rotational and vibrational temperatures of C_2 within the plasma plume. The obtained values indicate departure from local thermodynamic equilibrium. Preliminary results concerning the deposition of "hard carbon thin films" are presented.

Acknowledgement

This work is supported by DRET under contract No. 9034.

References

[1] T. Sato, S. Furuno, S. Iguchi and M. Hanabusa, Jpn. J. Appl. Phys. 26 (1987) L1487.

[2] A.A. Gorbunov and V.I. Konov, Surf. Coat. Technol. 47 (1991) 503.

[3] T. Sato, S. Furuno, S. Iguchi and M. Hanabusa, Appl. Phys. A 45 (1988) 355.

[4] X. Chen, J. Mazumder and A. Purohit, Appl. Phys. A 52 (1991) 328.

[5] A. Rohlfing, J. Chem. Phys. 89 (1988) 6103.

[6] A.R. Striganov and N.S. Sventitskii, Tables of Spectral Lines of Neutral and Ionized Atoms (IFI/Plenum, New York, 1968).

[7] R.W.B. Pearse and A.G. Gaydon, The Identification of Molecular Spectra, 4th ed. (Chapmann and Hall, London, 1976).

[8] W.L. Weise, M.W. Smith and B.M. Miles, Atomic Transition Probabilities, Vol. I, NSRDS.NBS4 (1966); Vol. II, NSRDS.NBS22 (1969).

[9] S.S. Wagal, E.M. Juengerman and C.B. Collins, Appl. Phys. Lett. 53 (1988) 187.

[10] F. Davanloo, E.M. Juengerman, D.R. Jander, T.J. Lee and C.B. Collins, J. Appl. Phys. 67 (1990) 2081.

Applied Surface Science 69 (1993) 365–369
North-Holland

applied
surface science

Laser deposition of thin films of high T_c superconductors

In situ analysis of the transient species formed in the plume and surface diagnostics of the deposited material

A. Giardini Guidoni, I. Pettiti

Dipartimento di Chimica, Università La Sapienza, P.le A. Moro 5, 00185 Rome, Italy

A. Morone, V. Marotta, M. Snels, G.P. Parisi and G. Bentivenga

Istituto Materiali Speciali, Area della Ricerca di Potenza, CNR, Via S. Loja loc. Zona Industriale, 85050 Tito (PZ), Italy

Received 2 June 1992; accepted for publication 30 October 1992

The present work reports some results on the laser deposition technique of high T_c superconducting films. In situ analysis of transient species formed in the plume and surface diagnostic of the deposited Bi–Sr–Ca–Cu–O (BSCCO) thin films has been performed. It has been ascertained that mixed oxide cluster ions are produced in the plume together with neutral and ionized atoms of constituent material. The effect of annealing parameters on the structure and quality of deposited thin films of BSCCO is reported and discussed.

1. Introduction

In the last few years many research groups applied different techniques (molecular beam epitaxy, sputtering, etc.) for the production of high T_c superconducting thin films with the aim to optimize the deposition processes and the thin film quality [1]. In our laboratory a systematic study of thin film production by laser ablation is in progress [2]. The interaction of intense laser radiation with solid matter causes several processes to occur: energy transfer, evaporation of solid material, generation of a dense plasma, formation of charged and neutral particles. The study of all these phenomena is rather complicated and theoretical models based on local thermodynamic equilibrium and hydrodynamical equations of motion can give only a qualitative explanation of the processes taking place during laser deposition. Measurements of space and time dependent emission spectra of the species present in the plume provide a key to understand the dynamics and reactivity of the ablated materials in gas phase. Evidence of chemical reactions in the plume has been previously ascertained. In particular it has been shown that the addition of copper oxide to second group elements, Ca, Ba, Sr, oxides leads to the formation of mixed ions at the expenses of pure oxide clusters [3]. The aim of this work is to throw some light on the role played by the transient species in the aggregation process and by the annealing parameters on the structure and quality of films.

2. Experimental

The laser ablation of BSCCO superconducting and nonsuperconducting targets has been performed using a frequency doubled ($\lambda = 532$ nm, pulse duration 10 ns) or quadrupled ($\lambda = 266$ nm) Nd–YAG laser (Quantel 581). The vacuum chamber for ablation and deposition, already described [2], is equipped with a quartz laser entrance window and with rotatable support for target and substrate holder. The pressure in the

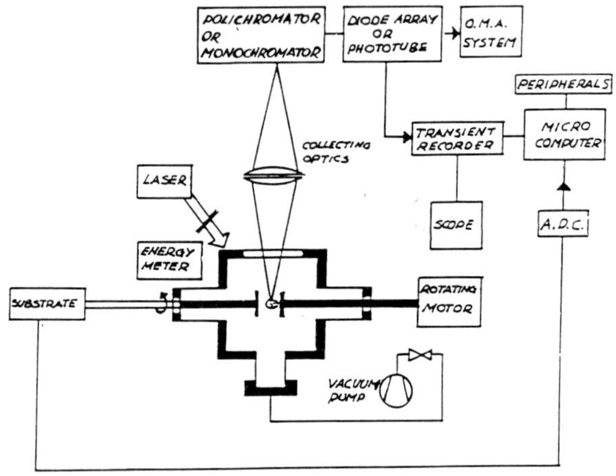

Fig. 1. Scheme of the experimental apparatus for laser ablation and deposition.

chamber is kept at 1.5×10^{-6} mbar during the ablation process. The laser energy is about 2.5–3 J/cm^2, the incidence angle with respect to the target surface is 45° (fig. 1). The ablated material originates a luminous plume which extends about 2 cm from the surface. Time-of-flight mass spectroscopy [3] and luminescence measurements [4] were utilized to obtain information on the intermediate species, neutral and ionized, present in the plume. Emission spectra of the species present have been measured through an optical multichannel analyzer (EG & G OMA 1460) mounted on a small (30 cm) monocromator (Jarrel Ash). The spectral resolution of this detection system is about 0.3 nm, the temporal resolution about 100 ns.

An MgO or $SrTiO_3$ substrate, glued by silver paint to a mechanical support, was positioned in front of the target at a distance of about 2.5 cm and kept at room temperature during the ablation experiment. Thin films of $Bi_2Sr_2CaCu_2O_x$ material were deposited on $SrTiO_3$ (100) or MgO (200) substrates by laser ablation of superconducting target of BSCCO synthesized by calcining mixtures of Bi_2O_3, $CaCO_3$ and CuO at 840°C. Other deposits were produced by laser ablation of stoichiometric 2:2:1:2 nonsuperconducting pellets. A deposition rate of 0.19 nm/s has been measured. The thickness of thin films, measured with an alfa-step profilometer was typically 1 μm after 2 hours of irradiation.

The deposited films were annealed in air in the temperature range between 840 and 880°C, for different duration time. X-ray analysis of the target and deposited films was performed with a Philips diffractometer model PW 1710. Analysis

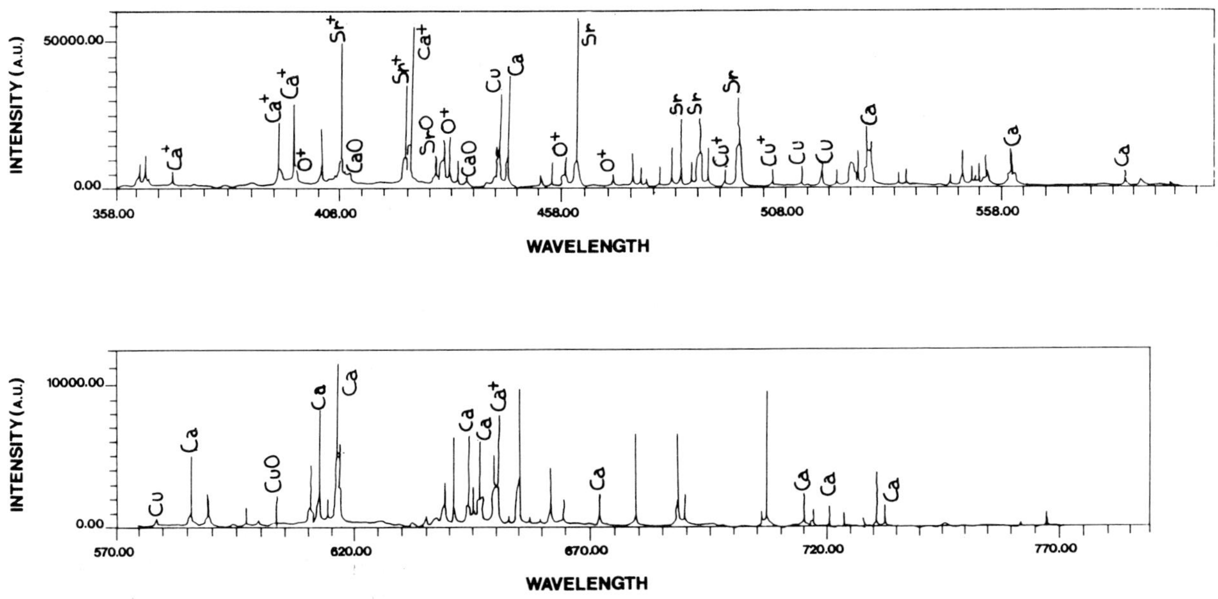

Fig. 2. Optical spectra of laser produced plume from BSCCO target at pressure of 10^{-6} Torr.

of standard pellets and of the produced thin film either "as deposited" and after annealing was performed by SEM and XPS. XPS spectra were collected by an LH XI Leybold spectrometer using unmonochromatized Mg Kα radiation (1253.6 eV). A standard four probe technique was used to measure the resistance of the annealed samples.

3. Results and discussion

Time-of-flight mass analysis of the ionic species produced by laser near the surface (1 mm) in superconducting BSCCO targets have been already reported [4] and the results will be here only summarized. Sr, Bi, Ca, Cu and simple oxide ions predominate in the mass spectrum and masses of simple and mixed positive cluster oxide ions are present. The mass spectra of nonsuperconducting specimens are similar in type of ions and relative intensity. This feature is characteristic also of the other metal oxide mixtures and of oxide mixtures of superconductor precursors [3]. The mass spectrometric plume analysis was performed in the region near to the laser spot on the surface but it cannot be excluded that heavier aggregates are formed at a larger distance from the target. They could eventually reach the substrate and initiate the film growth. The emission spectrum from the laser produced plume in a superconducting target has been already reported [4] and does not essentially differ from that of a superconducting specimen.

The total emission of the plume at the distance of 5 mm from the target is shown in fig. 2. Atom and ion emissions are strongly predominant. Very weak emission features attributable to CuO, CaO and SrO molecules could be present. Time resolved spectra measurements for most of the emission lines show a prompt and delayed emission which have already been attributed to laser and collisional excitation of ablated species, respectively [5]. For example the optical emission as a function of time, of the Sr^{3+} ion ($\lambda = 346.6$ nm) at an observation distance of 5 mm from the target surface, is reported in fig. 3. The delayed emission disappears at 1 μs, is peaked at around

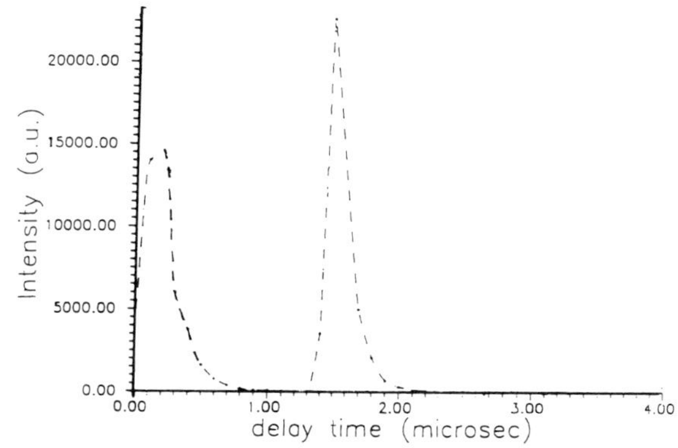

Fig. 3. Time resolved optical emission of Sr^{2+} ions ($\lambda = 346.4$ nm) at 0.5 cm distance from a BSCCO target.

0.2 μs. The low time resolution allowed only an estimation of time-of-flight velocity which is of the order of 10^6 cm/s in qualitative agreement with previous measurements [5]. XPS surface analysis of films obtained from a superconducting target confirmed the presence of all the constituent elements. XPS analysis of the Cu 2p region of the target and deposited film before and after annealing showed that the copper valence is restored after the annealing treatment in the presence of oxygen [6]. Analogous results are obtained also in the case of a nonsuperconducting target. Annealing duration and temperature affect the quality of the deposited thin films. In table 1 the results obtained on some samples originated from a superconducting and nonsuperconducting target in different annealing conditions are reported. The resistivity versus temperature characteristics of the samples A to F of the

Table 1
Results obtained on some examples originated from a superconducting and nonsuperconducting target in different annealing conditions

Sample	Target type	Dep. time (h)	Dep. temperature (K)	Annealing time (h)	T_c (K)	Major phase
A	Superc.	2	855	2	73	2212
B	Superc.	1	855	12	45	2212
C	Superc.	3	840	1	65	2212
D	Superc.	3	855	6	65	2212
E	Superc.	3	880	12	65	2201
F	Nonsuperc.	2	850	4	–	2212

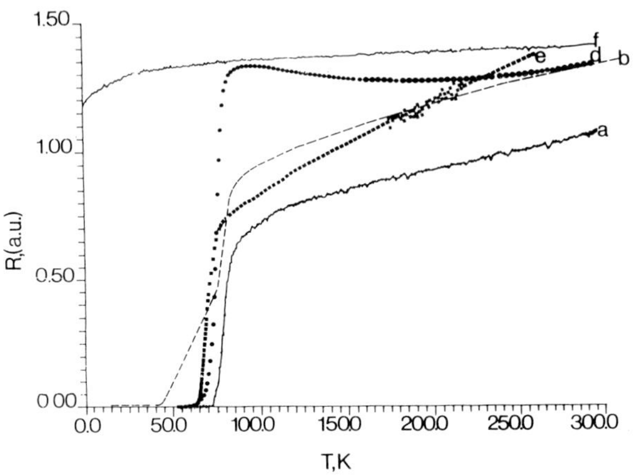

Fig. 4. Thin time $R-T$ curves at different annealing conditions: (a) sample A; (b) sample B; (c) sample D; (e) sample E; (f) sample F.

table 1 are plotted in fig. 4. It can be seen that a long annealing time (sample B) appears to deteriorate the quality of the film.

A higher annealing temperature produces a similar behaviour (sample E) as can be observed in the resistance measurements and in the X-ray diffraction patterns reported in fig. 5. The diffraction patterns of sample A and D show a series of strong peaks at $2\theta = 5.47°$, 29.3° and 35.3°. These peaks are considered to correspond to (001) series of a unit cell with its long axis perpendicular to the substrate. The observed peaks are related to the 2:2:1:2 phase (T_c 85 K). However, the spectra show that the film is polycrystalline with only a slight preferential c-axis orientation of the grains. When the BSCCO thin film was annealed near the melting point at 880°C the presence of the $2\theta = 7.2°$ and 21.9° peaks indicates a phase change from 2:2:1:2 to 2:2:0:1. A similar behaviour was previously discussed [7]. The attempt to obtain high T_c superconducting film from a nonsuperconducting target of 2:2:1:2 phase gave poor results probably due to the not-optimized heating conditions.

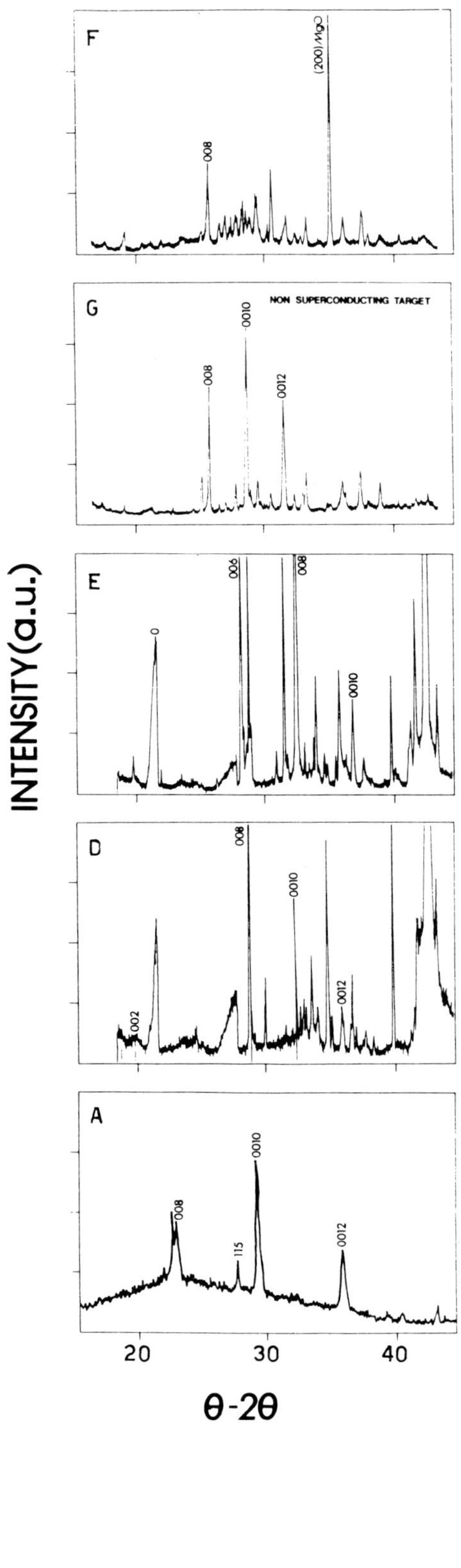

Fig. 5. X-ray diffraction patterns for BSCCO thin film and target at different annealing conditions: (A) sample A; (D) sample D; (E) sample E; (G) nonsuperconducting target; (F) sample F.

In conclusion the present investigation clearly demonstrates that the plume of evaporated material is mainly formed by atoms, ions and few oxide clusters. Its composition appears to be not always representative of the composition of the target, however, the net result of deposition is stoichiometric. To obtain high quality thin films not only a suitable target but also a good choice of post annealing treatment is needed.

References

[1] D. Bäuerle, Laser processing and diagnostic (Springer, Berlin, 1984) p. 166.

[2] L. Correra, Ed., High temperature superconductor thin film (Elsevier, Amsterdam, 1991) p. 451.

[3] A. Mele, A. Giardini Guidoni, D. Consalvo, D. Stranges and R. Teghil, Int. J. Mass Spectrom. Ion Processes 91 (1989) 319.

[4] A. Giardini Guidoni, E. Desimoni, A.M. Salvi, R. Teghil, M. Ambrico, S. Piccirillo and M. Snels, Proc. Int. Conf. Laser 90, San Diego (1990) p. 331.

[5] K.L. Saenger, J. Appl. Phys. 66 (1989) 4435;
C. Girault, D. Damiani, J. Aubreton and A. Catherinot, Appl. Phys. Lett. 55 (1989) 2;
H. Fucumera, H. Nakaminami, S. Eura and H. Masuhara, Appl. Phys. Lett. 58 (1991) 22;
J.P. Zheng, Z.Q. Huang, D.T. Shaw and H.S. Kwak, Appl. Phys. Lett. 54 (1989) 3.

[6] A. Giardini Guidoni, A. Morone, M. Snels, E. Desimoni, A.M. Salvi, R. Fantoni, W.C.M. Berden and M. Giorgi, Appl. Surf. Sci. 46 (1990) 321.

[7] M. Knai, T. Kawai, M. Kawai and C. Kawai, Jpn. Appl. Phys. 27 (1988) L1293.

Applied Surface Science 69 (1993) 370–375
North-Holland

applied surface science

Acoustic in situ monitoring of excimer laser ablation of different ceramics

Ladislav Grad and Janez Možina
Faculty of Mechanical Engineering, University of Ljubljana, P.O. Box 394, 61000 Ljubljana, Slovenia

Received 2 June 1992; accepted for publication 30 October 1992

Measurements of excimer laser generated acoustic transients were performed both on the back surface of various ceramic samples and in the surrounding gas. Several techniques of acoustic signal analysis were used to study the correlation between acoustic signal parameters and the ablation rate. The technique based on the defined ablation parameter E_s is proved to be suitable for monitoring the excimer laser ablation processes.

1. Introduction

Excimer laser processing of materials became a promising tool in microelectronics. To obtain good results and to avoid damages in situ monitoring of the process should be used. In this paper acoustic response of the ablated material has been used to monitor XeCl excimer laser ablation of alumina based ceramics with different amounts of ZrO_2 inclusions.

Pulsed laser manufacturing processes can be regarded as rich sources of ultrasonic waves which propagate from the interaction spot both into the sample and into the surrounding gas [1–3]. Thus, to obtain advanced information on the removal process both acoustic transients were measured and discussed in this paper.

In recent years some optoacoustic studies of laser-induced ablation have been done. It was suggested that optoacoustic signals can be used for time-resolved studies of the ablation process [4,5], for determination of the ablation threshold [6], and for optoacoustic depth profiling [7]. Our work, in some way, extends the work done previously by Ventzek and Gilgenbach [7] who used laser beam deflection measurements to determine the pulsed laser ablation rate. However, this method is based on a constant sound velocity. Because ablation is often accompanied by plasma and shock wave formation, the velocity of detected waves changes and makes the method inaccurate. We propose some other optoacoustic signal parameter should be used for in situ monitoring of the ablation rate. Experiments with different ceramic samples were performed to analyze and verify relations of optoacoustic signal parameters to the ablation rate, and to demonstrate their use in monitoring the manufacturing processes.

2. Experimental procedure

Experiments were performed with the experimental setup shown in fig. 1. We used a Lambda LPX 100 excimer laser with the xenon, hydrogen chloride and neon filling, delivering pulses of 20 ns FWHM duration, 308 nm wavelength and pulse energies up to 90 mJ. The repetition rate was kept below 5 Hz. The energy of each pulse was measured by a built-in energymeter. The laser beam was shaped and attenuated by a circular iris with diameter $D = 6$ mm. An additional quartz focusing lens ($f = 110$ mm) was used to focus the

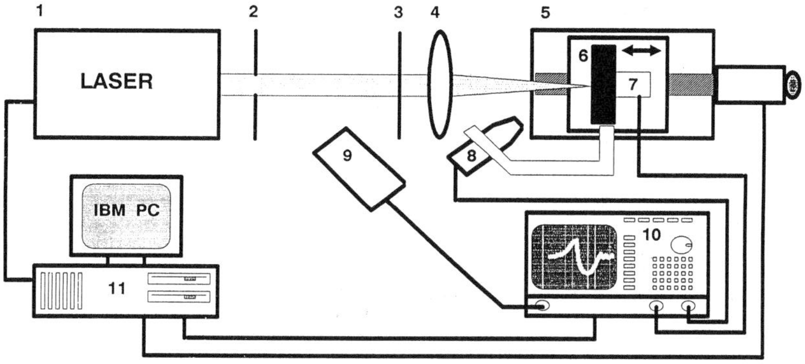

1 -LASER
2 -IRIS
3 -ATTENUATOR
4 -QUARTZ LENS
5 -NC TABLE
6 -SAMPLE
7-PIEZOELECTRIC SENSOR
8-MICROPHONE
9-PHOTODIODE
10-OSCILLOSCOPE
11-PERSONAL COMPUTER

Fig. 1. Experimental setup.

laser beam. Fluence was was kept just above the plasma formation threshold on the ceramic samples.

The experiments were performed on Al_2O_3 ceramics with 0, 2, 5 and 15% of ZrO_2 inclusions. Zirconium is often used to improve properties of alumina ceramics. Surfaces were smooth, additionally cleaned with ethanol, and visually examined to avoid possible irregularities which might influence the results.

Acoustic signals in ceramic samples were detected by a piezoelectric (PZT) sensor with a resonance frequency at 2.56 MHz, mounted on the back of the samples. A negatively polarized Brüel & Kjær 4138 microphone with a bandwidth up to 200 kHz and a preamplifier Brüel & Kjær 2619 were used to detect optoacoustic waves in the air.

Samples with the PZT sensor and the microphone were mounted on a computer-controlled translation stage. The distance between the microphone tip and the laser-irradiated area of the sample was 90 mm in all experiments.

To study the correlation between laser-induced optoacoustic signals and the ablation rate, craters were made with a different number of

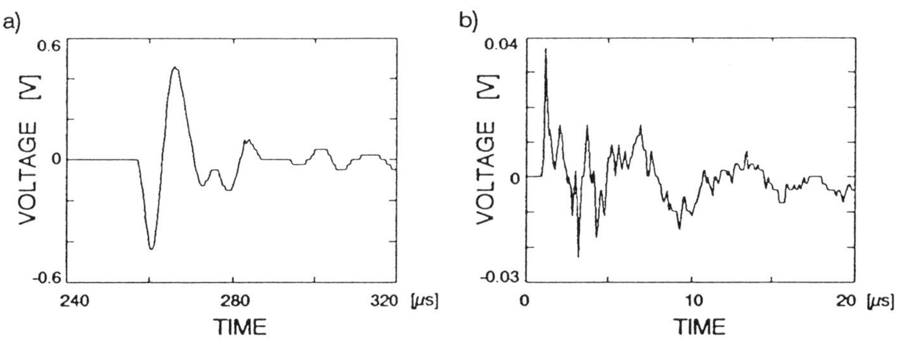

Fig. 2. Typical optoacoustic signal of the 60th laser pulse on a pure Al_2O_3 ceramic sample detected by (a) microphone and (b) PZT.

laser pulses: 60, 100, 600, 1000, 6000 and 10^4, respectively. PZT and microphone optoacoustic signals were captured with a HP-54502A oscilloscope. Detected acoustic signals, together with energy data of each laser pulse, were transferred to a PC for further processing.

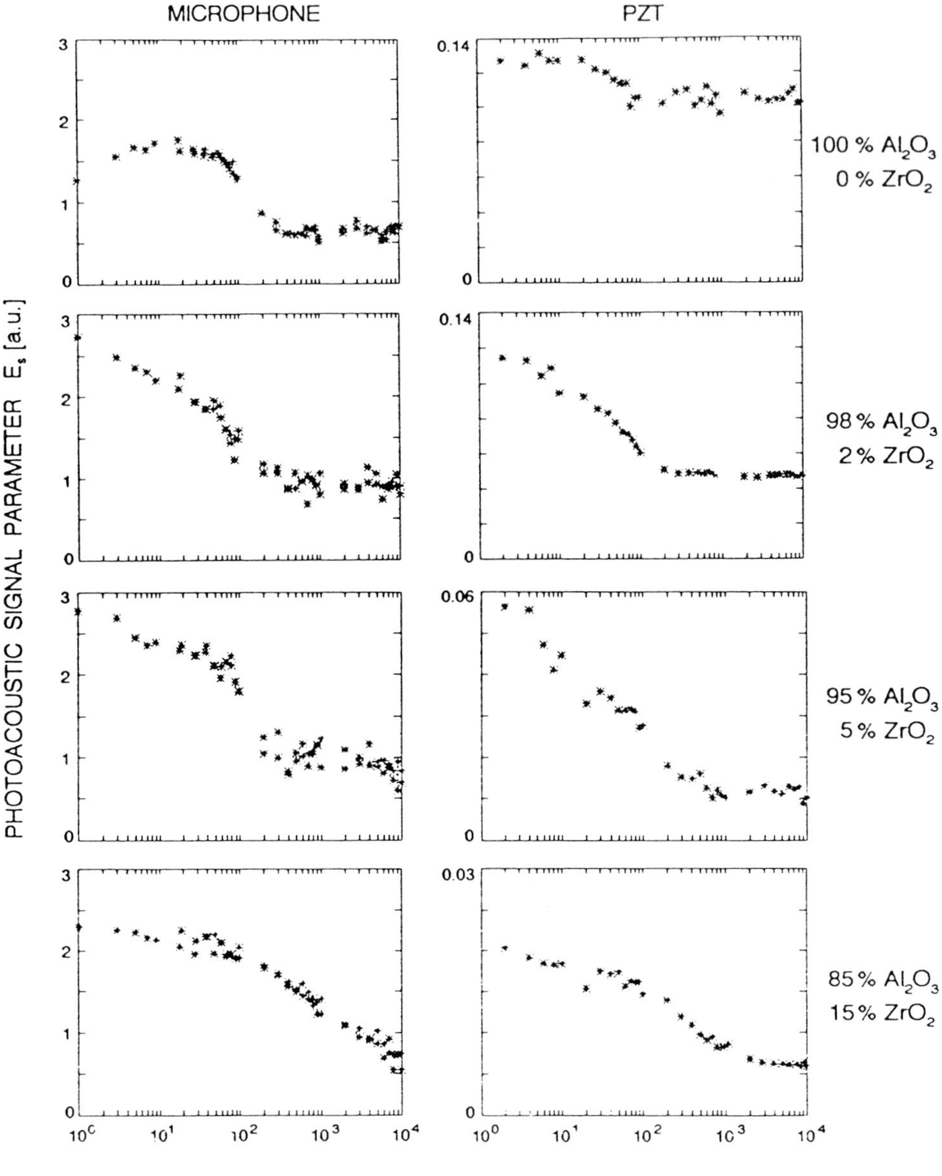

Fig. 3. Evolution of the microphone and the PZT optoacoustic signal parameter E_s during ablation of different ceramic samples with 10^4 pulses.

Depth profiles were measured after ablation with a stylus apparatus and then compared to the parameters of the detected optoacoustic waves.

3. Discussion

Typical microphone and PZT optoacoustic signals obtained during ablation of alumina ceramics are shown in fig. 2. First peak amplitude and onset times are usually chosen as characteristic parameters of each signal [8]. They are a direct measure of the wave pressure and energy deposited in the acoustic wave, so they include information about the energy transformation processes such as ablation.

In general, the microphone signal depends on the processes that occur in the air and in the solid sample. After plasma and shock wave formation the thermoelastic contribution to the signal becomes negligible. Since our experiments were performed above the plasma formation threshold we assume that the microphone signals involve just the information about the plasma and the material ejection from the sample.

Piezoelectric (PZT) signals, on the contrary, describe the effects of the processes in the solid sample. Ejection of the material during the ablation process results in a recoil momentum transfer to the sample. Its contribution dominates the signal. Ejection caused by a nanosecond laser pulse can last many microseconds [9], thus the first amplitude of the signal does not necessarily involve the whole information of the material ejection.

Several one-point parameters of the optoacoustic signals can be used to monitor the ablation process, but as mentioned above some information is always lost. We propose more complex parameters to be used. They should include information about the signal power and the entire waveform. In this paper we choose the integral of the square value of the signal $x(t)$. The integral, which is a measure of the signal energy, is expressed as

$$E_s = \int_{t_1}^{t_2} x^2(t) \, dt, \tag{1}$$

where t_1 and t_2 define the time window of integration. In our analysis of the relation between

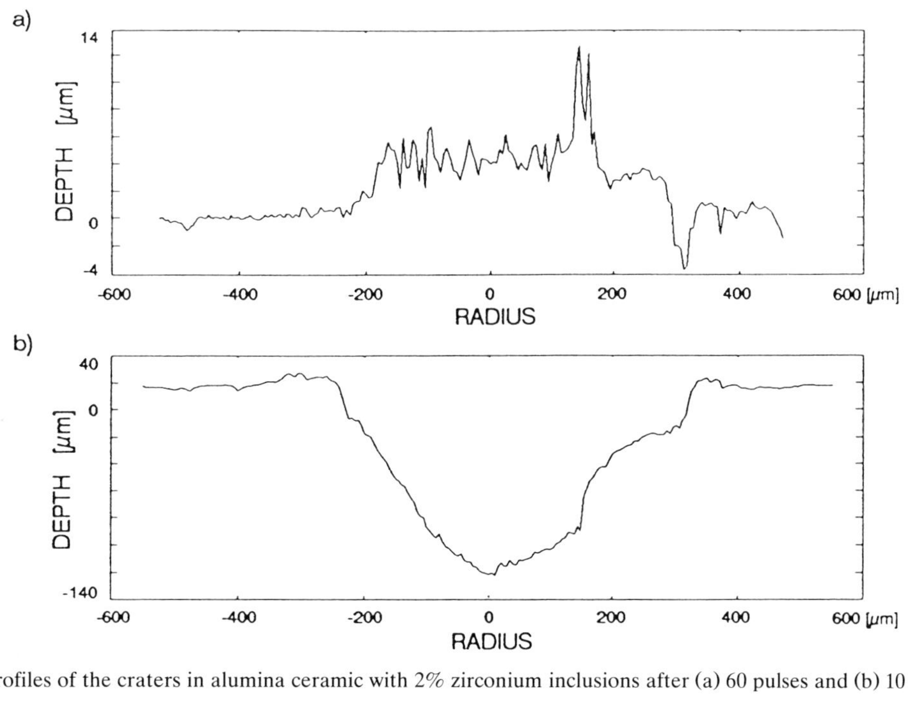

Fig. 4. Profiles of the craters in alumina ceramic with 2% zirconium inclusions after (a) 60 pulses and (b) 10000 pulses.

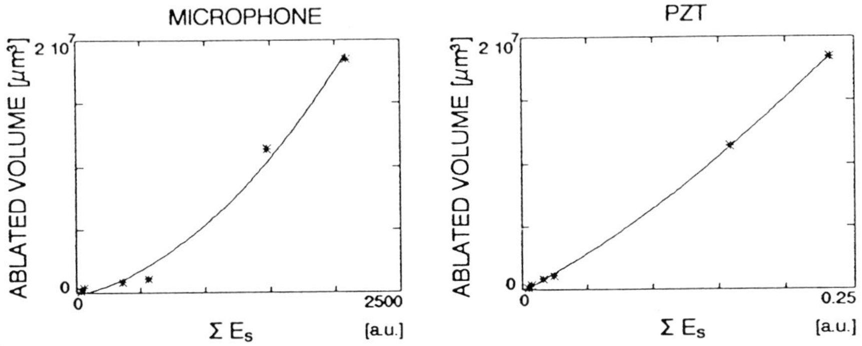

Fig. 5. Measured relation between the ablated volume of alumina ceramic with 5% of zirconium and the cumulative optoacoustic signal parameter ΣE_s for microphone and PZT signals.

the ablation rate and E_s the whole time window (as seen in fig. 2) is analyzed in detail.

The evolution of the successive microphone and PZT optoacoustic signal parameter E_s during the ablation of different ceramic samples with 10^4 pulses is shown in fig. 3. It can be explained by the laser-induced changes of the surface layer at the interaction spot. After the initial decrease of the integral E_s, a stationary regime is reached. The transition to the stationary regime is strongly dependent on the concentration of ZrO_2 in the sample.

The profiles of the craters were measured by a stylus apparatus. Measured profiles show the slight surface elevation obtained by the irradiation with the first several decades of laser shots. After the initial period, the ejection of the material contributes to the crater growth as evident

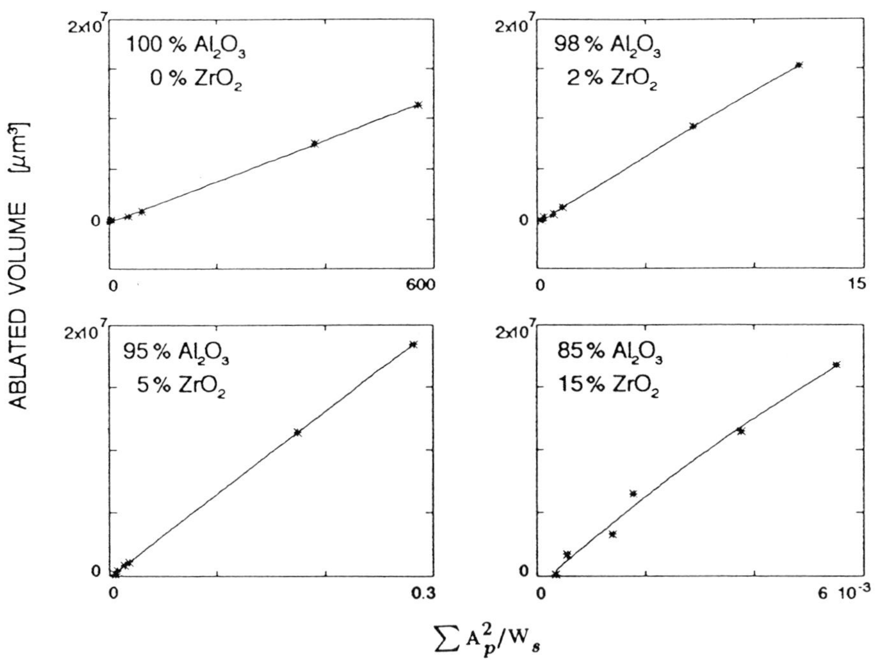

Fig. 6. Measured relation between the ablated volume of different alumina ceramics and the cumulative sum of optoacoustic ablation parameter $\Sigma A_p^2/W_s$.

from fig. 4. Typical dependence of the ablated volume on the sum of integrals E_s of all incoming laser pulses is shown in fig. 5. It is evident that both microphone and PZT signals could be used for monitoring the crater growth.

Another correlation between optoacoustic signals and ablation rate can be made using the results of Zyung et al. [4] and Grad et al. [10]. Zyung et al. showed that the shock wave energy is equal to the kinetic energy W_k:

$$W_s = W_k = mv^2/2, \quad (2)$$

where m and v stand for the mass and velocity of the ablated material. In our work we use another procedure by which it is possible to determine the shock wave energy directly from optoacoustic signals. It is based on a point explosion model using the equations of hydrodynamics for the description of blast wave propagation [7,10]. A linear relation between the amplitude A_p of the PZT signal and recoil momentum mv is assumed:

$$A_p = C_1 mv, \quad (3)$$

where C_1 is determined by the sensitivity of the experimental system. Combining eqs. (2) and (3) we get a linear relationship between the optoacoustic parameters and the mass of the ablated material:

$$m = \frac{A_p^2}{2C_1^2 W_s}, \quad (4)$$

where A_p^2/W_s can be defined as ablative parameter.

In our analysis we used the successively measured signals and the measured depth profiles. The amount of ablated material was calculated assuming the craters had angular symmetry. The results are presented in fig. 6, where the ablated volume is compared with the cumulative sum of the ablation parameters. A reasonably good linear relation is obtained for all samples. At this point we have to emphasize that optoacoustic monitoring is based on a relative comparison and that exact quantitative values are obtained by calibration.

4. Conclusions

In this paper optoacoustic monitoring of the excimer laser ablation of various ceramics was studied. To obtain advanced information about the process we captured acoustic transients in the ceramic samples and in the air. Different numbers of laser pulses up to 10^4 were used to produce the craters, and their depth profiles were measured by a stylus apparatus. Initial elevation of the surface was observed for pure alumina ceramic and for that with 2% zirconium. The stationary regime of the ablation process was reached after a certain number of laser pulses.

We proposed the integral of the square value of the optoacoustic signal as a convenient monitoring parameter E_s of the laser ablation. The corresponding sum of E_s was found to be monotonous but not a linear function of the ablated volume. In addition, microphone and piezoelectric signals were used to form the ablation parameter of the optoacoustic signals. The cumulative sum of the ablation parameters calculated for the successive signals was found to depend linearly on the ablated volume. The results confirm our presumption that by analyzing the optoacoustic signals it is possible to monitor the process of excimer laser ablation of different ceramics.

References

[1] A.P. Ghosh and J.E. Hurst, J. Appl. Phys. 64 (1988) 287.
[2] A.C. Tam, Rev. Mod. Phys. 58 (1986) 381.
[3] G. Gorodetsky, T.G. Kazyaka, R.L. Melcher and R. Srinivasan, Appl. Phys. Lett. 46 (1985) 828.
[4] T. Zyung, H. Kim, J.C. Postlewaite and D.D. Dlott, J. Appl. Phys. 65 (1989) 4548.
[5] F.W. Cross, Appl. Phys. Lett. 50 (1987) 1019.
[6] J.A. Sell and D.M. Heffelfinger, Appl. Phys. Lett. 55 (1989) 2435.
[7] P.L.G. Ventzek and R.M. Gilgenbach, J. Appl. Phys. 70 (1991) 587.
[8] J. Diaci and J. Možina, Opt. Commun., to be published.
[9] R. Srinivasan, B. Braren, K.G. Casey and M. Yeh, Appl. Phys. Lett. 55 (1989) 2790.
[10] L. Grad, J. Diaci and J. Možina, Proc. SPIE 1525 (1991) 206.

Applied Surface Science 69 (1993) 376–379
North-Holland

applied
surface science

Time resolved spectroscopy of the emission of fluorescent light upon UV laser assisted CVD of W from WF_6

P. Heszler *,[1], P. Mogyorósi [2] and J.O. Carlsson

Thin Film and Surface Chemistry Group, Department of Chemistry, Uppsala University, P.O. Box 531, S-751 21 Uppsala, Sweden

Received 2 June 1992; accepted for publication 30 October 1992

Time resolved spectroscopy of ArF excimer laser excited $WF_6/H_2/Ar$ and WF_6/Ar gas mixtures during LCVD of W has been performed with a time resolution of 20 and 5 ns, respectively. Using a high laser energy density ($\sim 1\ J/cm^2$) and a WF_6/Ar gas mixture, the recorded spectra have both line and band structures which originate from W atoms and W ions. The spectral intensity distribution followed the shape of the laser pulse (FWHM $\approx$ 20 ns). This means that the lifetime of the excited states is much shorter than that of the laser pulse duration. A broad band emission spectrum was obtained from a $WF_6/H_2/Ar$ gas mixture at a laser energy density of $\sim 100\ mJ/cm^2$. This originates from excited W clusters produced in the gas phase. Time resolved measurements of this radiation showed that the spectral intensity distribution did not change with time. This indicates that the emission process cannot be black body radiation. The time dependence of the intensity of the fluorescent light could be separated into two parts: a fast decay (life time ~ 400 ns) and a slow decay ($\sim 5.7\ \mu s$). The fast decay was explained in terms of vibrational quenching. The slow decay was assumed to be due to the spontaneous lifetime of excited states of the W clusters.

1. Introduction

Lowering of the deposition temperature and increasing deposition rates are general demands in the thin film field. For CVD processes this can be achieved, e.g., by laser induced photolysis. ArF excimer laser assisted CVD of W from WF_6 is an example of a process where the deposition temperature is lowered and the deposition rate is increased in comparison with the thermal process [1,2]. Since the laser beam is parallel to and just above the substrate surface, the gas phase processes have to be important steps in the photolytic part of the W deposition. Recently, emission of fluorescent light was observed from a $WF_6/H_2/Ar$ gas mixture [3]. This might give useful information about gas phase processes. At a lower laser fluence ($\sim 100\ mJ/cm^2$) a broad band radiation (~ 300 nm wide), centered at ~ 530 nm, was obtained. This emission primarily originates from W clusters generated in the gas phase [4]. However, it is not known if the emission comes from the W clusters as a black body radiation or originates from the fluorescence radiation of the excited states of the clusters. Moreover, it is not known if this radiation includes chemiluminescence, since, e.g., the three body reaction

$$F\cdot + H\cdot + M \rightarrow HF + M \qquad (1)$$

can yield visible luminescent light [5]. M can be Ar or H_2, the atomic F is produced by the photolysis of WF_6 while the atomic H can be formed by the reaction

$$F\cdot + H_2 \rightarrow HF + H\cdot . \qquad (2)$$

Using a high ($\sim 1\ J/cm^2$) laser fluence, a bluish spectrum, consisting of mostly separate spectral lines, is superimposed on the broad band spectrum. This spectrum, originating from W

* To whom correspondence should be addressed.

[1] On leave from the Research Group of Laser Physics of the Hungarian Academy of Sciences.

[2] Present address: Laser & Vacuum Sci., P.O. Box 1191, H-6701 Szeged, Hungary.

atoms and ions can be induced separately from a WF_6/Ar gas mixture [6].

Time resolved spectroscopy may yield useful information on the origin of the light emission and, hence, on both physical and chemical processes taking place during as well as after the laser pulse. In this paper, time resolved spectroscopic data for both WF_6/Ar and WF_6/H_2/Ar gas mixtures will be presented and discussed.

2. Experimental

The experimental arrangement is shown in fig. 1. The reactor was made in stainless steel with quartz top, front and rear windows. High purity gases were used (WF_6 99.998%; Ar 99.9997%; H_2 99.9999%). Flow rates were controlled by calibrated mass flow meters and the total pressure in the chamber was set by a gate valve at the outlet of the chamber. The noble gas was used to purge the windows to avoid deposition. The air leakage rate into the deposition system corresponded to an air impurity level of about 2 ppm in the reaction gas. Before starting the experiments the reactor was purged with the reaction gas mixture. The laser induced fluorescent light was observed perpendicular to the laser beam. The light was coupled into a quartz optical fibre which was connected to a Czerny–Turner type of grating spectrograph. A 150 mm grating was used yielding 1.8 Å spectral resolution and ±0.2 nm wavelength accuracy. The time resolution of the spectra was achieved by a M1455 G type of gateable intensified detector (EG&G PARC). The gate pulses were generated by a M1302 type of fast pulser (EG&G PARC). The timing was set in the following way. The ArF excimer laser ($\lambda = 193$ nm, FWHM 17 ns) provided electrical triggers before the UV light pulse. These triggers were delayed by a M1303 gate pulse interface (EG&G PARC) which output triggered the M1302 fast pulser. A fraction of the laser beam was coupled out onto a fast photodiode, providing a reference signal. The delay between the laser pulse and the gate pulse was measured by an oscilloscope. The spectra were analyzed by a computer through a M1461 detector interface (EG&G PARC). The spectra were not corrected for the sensitivity curve of the detector. The measurements were performed at room temperature.

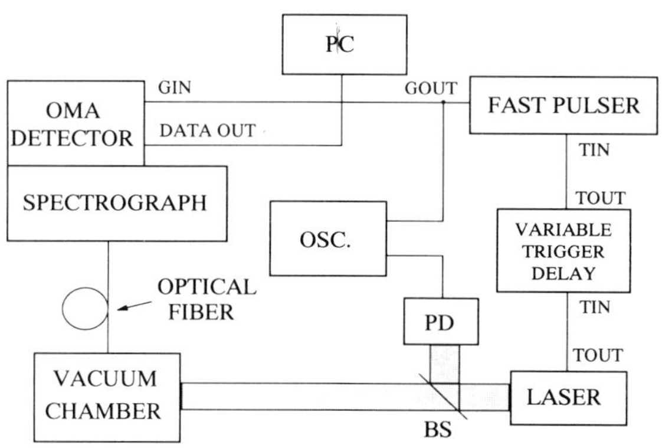

Fig. 1. Schematic diagram of the experimental setup. T: trigger pulse; G: gate pulse; BS: beam splitter; PD: photodiode; PC: computer.

The experimental parameters for the broad band measurements were as follows: WF_6/H_2/Ar gas composition 2.5/10/200 in molar ratio, total pressure 4 mbar, total flow 212.5 sccm, laser fluence ~ 100 mJ/cm^2, laser repetition rate 50 Hz, overall exposure time 5 s, detector gate pulse 20 ns. The overall system jitter was ±10 ns. A glass plate was used to filter out the strong scattered ArF line for detector reasons.

Examining the bluish light emission the following parameters were used: WF_6/Ar gas composition in 2/200 molar ratio, total pressure 6 mbar, total flow 202 sccm, laser fluence ~ 1 J/cm^2 (by using a 380 mm focal length lens), laser repetition rate 30 Hz, overall exposure time 2 s, detector gate pulse 5 ns. The overall system jitter was ±5 ns.

3. Results and discussion

3.1. Light emission at high laser fluence

The time resolved spectra of the emission of the fluorescent light from the WF_6/Ar gas mixture (blue light) are shown in fig. 2. At a delay of 0 ns only the scattered ArF laser lines (1st, 2nd and 3rd order) can be seen. The spectra for 10 and 20 ns delay, respectively, are very similar, and

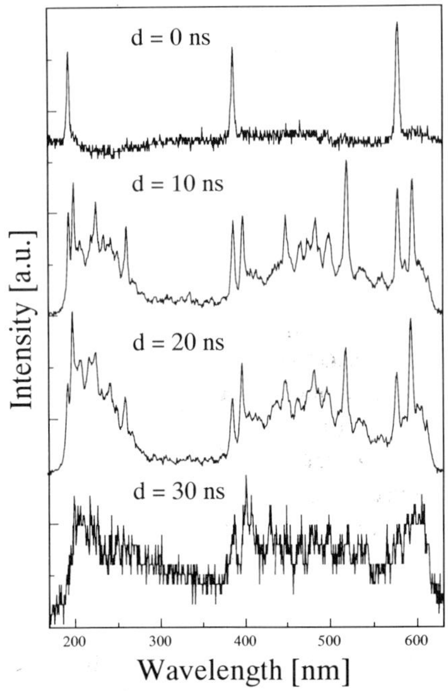

Fig. 2. Time resolved spectra using WF_6/Ar gas mixture, 0.5% WF_6 in Ar, total pressure 4 mbar, laser fluence 1 J/cm^2.

the intensity of the scattered ArF lines is lower in the spectrum for 20 ns delay. As mentioned before, the recorded spectra can be attributed to W atoms and W ions. For a delay of 30 ns the spectrum has only a few lines in common with the previous two spectra and the intensity (not indicated exactly in the figure) is much lower than for the former ones. Detailed measurements showed that the spectral intensity distributions did not change with time significantly (except for the scattered ArF laser lines). Furthermore, the time dependence of the intensity of the fluorescent light follows the time characteristics of the laser pulse within the accuracy of the measurement. This means that the life times of the excited states of the W atoms and W ions are considerably shorter than the duration of the laser pulse. This also means that the excited W atoms in the gas phase loose their activation processes before any chemical reactions can occur (see next section). However, long lifetimes for radiationless transitions cannot be excluded.

3.2. Broad band light emission

Spectra of the broad band emission from the $WF_6/H_2/Ar$ gas mixture are presented in fig. 3. In contradiction to the emission of the bluish light the broad band emission was recorded several μs after the excitation. The rise time of the fluorescence was very short ($\sim$ 25 ns). The average number of collisions per molecule at the pressure used is $\sim 10^8/s$ that is practically no chemical reactions occur during the rise time. Furthermore, the spectral intensity distribution did not change in time. These two latter statements also confirm that the broad band radiation originates from the W clusters and there is no indication of the chemiluminescence. In addition, since the spectral intensity distribution did not change with time the emission process cannot be black body radiation. In that case, the intensity maximum should shift to longer wavelengths (Wien's displacement law) as the temperature of the excited clusters decreases with time.

The time dependence of the emitted light intensity is shown in fig. 4. The emission process

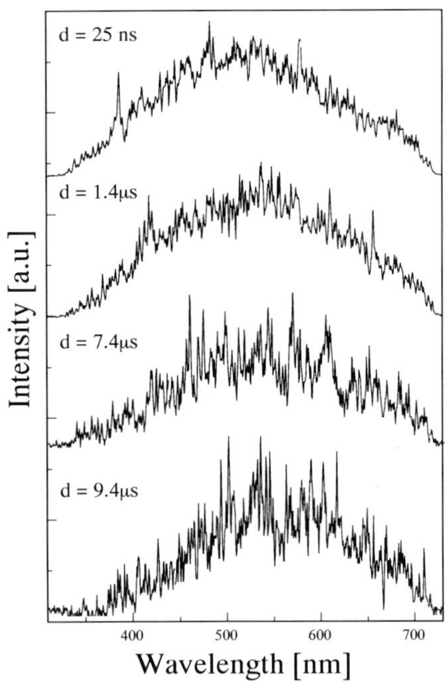

Fig. 3. Time resolved spectra recorded from $WF_6/H_2/Ar$ gas mixture, 0.5% WF_6 and 5% H_2 in Ar, 4 mbar total pressure, laser fluence 100 mJ/cm^2.

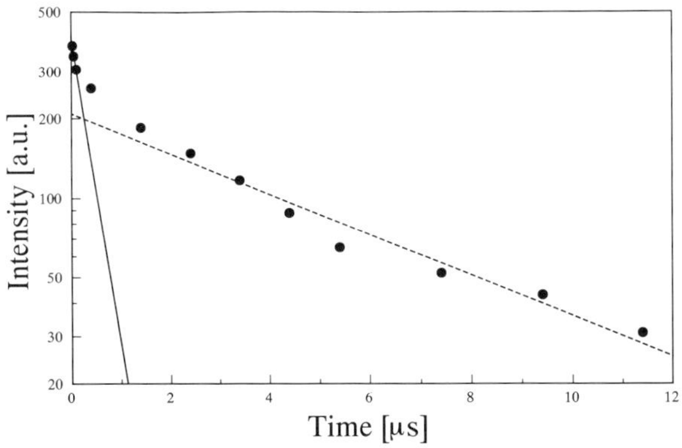

Fig. 4. Intensity dependence of the broad band radiation (see fig. 3) with time.

consists of two parts; a fast one with ~ 0.4 μs decay and a slow one with ~ 5.7 μs decay. The spontaneous lifetime of the excited states of the W clusters corresponds probably to the slow decay, while the fast decay probably is due to vibrational quenching of excited states within the W clusters. This means also that the slow decay does not start until conditions close to equilibrium, which is near room temperature, have been reached.

This investigation shows that the W clusters can be in excited states for several μs. During this period of time the clusters can be very reactive and their surfaces can catalyze further W deposition through dissociative adsorption of H_2 [7]. However, since the photolytic contribution to the W deposition rate in W LCVD is independent of the H_2 partial pressure [8], which is not the case in the thermal process [9], we believe that the catalytic role of the surface of the W clusters is not significant.

4. Conclusions

Both for the bluish and for the broad band light emission the spectral intensity distribution did not change with time. For the broad band light this means that the emission process of the excited tungsten clusters cannot be black body radiation. Furthermore, for the bluish radiation the lifetime of the excited states of W atoms and ions was much shorter than the laser pulse duration (which was 20 ns). However, the lifetime of the broad band radiation was much longer (several μs) than the duration of the laser pulse. The time dependence of the intensity of this light could be separated into two parts; a fast decay (lifetime ~ 400 ns) and a slow decay (lifetime ~ 5.7 μs). The fast decay was explained in terms of vibrational quenching. The slow decay was assumed to be due to the spontaneous lifetime of excited states in the W clusters. No sign of chemiluminescence was observed.

Acknowledgement

Financial support by the Swedish Natural Science Research Council is gratefully acknowledged.

References

[1] R.L. Krans, A. Bernstein and W.C. Sinke, Mater. Res. Soc. Symp. Proc. 181 (1990) 597.

[2] A. Shintani, S. Tsuzuku, E. Nishitani and M. Nakatani, J. Appl. Phys. 61 (1987) 2365.

[3] P. Mogyorósi, J.O. Carlsson and M. Moradi, Appl. Surf. Sci. 54 (1992) 46.

[4] P. Heszler, P. Mogyorósi and J.O. Carlsson, Mater. Res. Soc. Proc. 236 (1992) 123.

[5] P.S. Ganguli, B.L. Hertzler and M. Kaufman, Chem. Phys. Lett. 37 (1976) 319.

[6] P. Heszler, P. Mogyórosi and J.O. Carlsson, Appl. Surf. Sci., to be published.

[7] M.L. Hitchman and A.D. Jobson, Appl. Surf. Sci. 38 (1989) 312.

[8] H. Matsuhashi, S. Nishikawa and S. Ohno, Mater. Res. Soc. Symp. Proc. 129 (1989) 63.

[9] C.M. McConica and K. Krishnamani, J. Electrochem. Soc. Solid State Sci. Technol. 133 (1986) 2542.

Applied Surface Science 69 (1993) 380–387
North-Holland

applied
surface science

Laser CVD on carbon fibres: structure of layers and tensile strength of fibres

V. Hopfe [a], S. Böhm [a], G. Wieghardt [b] and A. Schulze [b]

[a] Fachbereich Chemie and [b] Fachbereich Materialwissenschaften, Technische Universität Chemnitz, P.O. Box 964, D-9010 Chemnitz, Germany

Received 2 June 1992; accepted for publication 30 October 1992

A method for continuous coating of high-tensile carbon fibre rovings as components of reinforced composites by CO_2 laser CVD has been developed. The deposition of TiC, TiB_2, SiC and pyrolytical carbon has been investigated. These materials have been selected to act as a diffusion barrier and to increase oxidation resistance. TEM/electron-diffraction measurements show well crystallized layers of crystallite size, sometimes in the order of layer thickness (30–100 nm). After the deposition of hardcoatings, the tensile strength of the fibres (3.5 GPa) decreases typically to 80%–30%, but with pyro-C strength increases to 115%. The influence of the deposition parameters on the tensile strength is rather complex, but power density of the radiation and fibre velocity rank uppermost.

1. Introduction

Carbon fibre reinforced composites with ceramic matrices are hopeful new materials suitable for high temperature applications in the 2000 K region. They have a good chance to bridge the gap between on one hand metallic superalloys with their superior mechanical properties but very restricted thermal stability and, on the other hand, ceramics with their extremely high thermal stability but intrinsic brittleness. At higher temperatures, carbon fibres must be protected by a dense coating against chemical matrix-attack or atmospheric corrosion. Obtaining good-quality and low-price coatings seems to be one of the key problems in producing fibre reinforced ceramics. Coatings for carbon fibre reinforcement are typically formed by the carbides, borides, nitrides and oxides of Ti, Si, Zr, Al or B. In addition to their ability to act as a diffusion and reaction barrier at high temperatures, they must meet a complex set of chemical and mechanical requirements. Above all, they should not diminish the intrinsically high tensile strength of the fibres and their failure probability (Weibull modulus).

Recently, we have introduced laser-driven chemical vapour deposition (LCVD) as a method for coating carbon fibres [1]. In comparison with hot-wall thermal CVD, which is now well-established in coating of fibres, the LCVD method is characterized by several significant advantages: high deposition rates (roughly 1 $\mu m/s$), small-volume cold-wall reactors, thermal reaction paths supported by photochemical channels, almost no restrictions on deposition temperatures.

In the present paper we intend to apply LCVD to the continuous coating of carbon fibre rovings. The principle of the apparatus for this process is given in section 2, together with more experimental details of the deposition and of the analytical investigations. The deposition of TiC, TiB_2, SiC and pyrolytical carbon (pyro-C) will be discussed in section 3 with special emphasis on the structure of the layers and the tensile strength of the fibres.

2. Experimental

2.1. Laser-CVD

The scheme of the LCVD reactor for continuous deposition at normal pressure is sketched in fig. 1. A 6 K roving, which has been flattened out into a tape, is transported continuously through the reaction zone via two Ar-flushed "gas windows". The beam of a technological CO_2 CW-laser (400 W, TEM00) has been shaped into a line focus perpendicular to the fibre axis which irradiates the moving fibre tape in full width. The power density may be controlled by beam-focussing from 0.1 to 10 W/mm^2. Parameters to be optimized in the LCVD experiments are: power density, fibre velocity, precursors selection, flow rate and reactants concentration. The temperature of the fibres has not yet been measured in the experiments reported here. Very recently, using a two-wavelength pyrometer, the temperature has been measured on a reactor configuration similar to the one described above, yielding temperatures in the region 1000–2400°C. Commercially available high tensile carbon fibres (T800, TORAY) have been pretreated by thermal outgassing at 660°C in flowing N_2. Pure-grade reactants have been distilled prior to being filled into their temperature-stabilized saturators.

2.2. Analytical investigations

A SEM/EDX microprobe JEOL 840A has been used for detecting the morphology, layer thickness and elemental content of the layers. To observe the end faces, the coated fibres have been cut by ordinary scissors. The structure of the layers has been investigated by a 100 keV TEM/electron diffraction device JEOL 100 CX. Bright-field images have been displayed showing the characteristic microstructure of the layers. Selected area diffraction (SAD) was used to determine the phase content of the layers. To get transparent samples for TEM, part of the layer material has been scraped off mechanically. After transferring the splinters onto an NaCl monocrystal, a carbon film has been deposited and dissolved in the usual manner.

Tensile strength measurements on single fibres have been carried out on a computerized machine Zwick 1435. A stretch velocity of 0.5 mm/min and a clamping length of 10 mm are selected. A test set of 20 individual fibres has been chosen from every deposition experiment. To estimate the tensile strength, the failure strain data have been divided by the area of the fibres derived from 120 single fibres, resulting in a mean diameter of 6.7 μm, range 5.2–8.1 μm. This great deviation from the mean diameter seems to be the most serious source of error of the resulting tensile strength data. Carbon fibres are brittle materials in character, requiring the use of Weibull statistics for data treatment. The Weibull modulus m is a quantity reflecting the uniformity of the material.

3. Results and discussion

The parameters of optimized LCVD experiments and some typical results of the extensive deposition tests are summarized in table 1. The titanium systems are characterized by a pro-

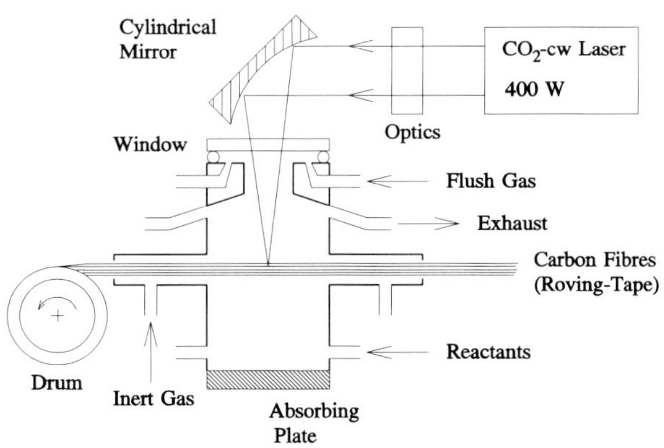

Fig. 1. Scheme of the apparatus for continuous laser CVD on fibre rovings.

nounced decrease in the deposition rate, caused by the increasing reflectivity of the deposit ("self-stop mechanism" [1]), limiting the layer thicknesses to typically 50–150 nm, depending on the power density used.

3.1. TiC_x

TiC_x layers of roughly 50 nm thickness have been deposited from $TiCl_4/H_2$ mixtures. Electron diffraction clearly shows the cubic TiC lattice (fig. 4a). By XPS, a low content of bonded oxygen is detectable, but only on the outer surface. The mean grain size of the crystallites is comparable to the layer thickness (fig. 4a), suggesting a dense "cobblestone pattern". The layers exactly copy the underlying filaments of the carbon fibre and have quite an unstructured surface, see fig. 2b.

Using conventional CVD, Mandell et al. [2] have found a dramatic decrease in the tensile strength of carbon fibres, even with TiC layers of some nm thickness. Our results with laser CVD give a less pronounced decrease in tensile strength, up to a layer thickness of roughly 100 nm. Above 150 nm, the fibres show very brittle behaviour and cannot be handled for testing. This brittleness may be due to the high degree of adhesion of the layer, which cannot prevent a crack "crossover" from the layer into the fibre. Another source of fibre failure might be the formation of fibre defects by surface etching which is caused by carbon diffusion into the layer. To prevent carbon diffusion, we have enriched the reaction atmosphere with benzene (table 1). Previous investigations have not given any evidence of the fact that the admixture of benzene had a noticeable effect on the deposition rate or morphology of the layers [1]. But now it is obvious that the admixture increases the tensile strength (table 1). An even more pronounced effect on adhesion and tensile strength is achieved by depositing a thin pyro-C interlayer between the hard coating and the fibre (section 3.4). The adhesion is obviously diminished and debonding may occur in preparing the fibres for SEM investigations (fig. 2a). The pyro-C layer acts as a sliding surface and, therefore, thermally induced tensions may relax over long distances, as seen in fig. 2b.

3.2. TiB_2

In contrast to CVD-grown layers, which normally show a rugged nodular surface, LCVD layers are smoother and denser and their nodular structure is less pronounced. Despite the fairly high deposition rate of more than 2 μm/s, which seems to be stimulated by photolytical enhancement [1], the layers only consist of a dense well-crystallized TiB_2 phase with a grain size of 40 nm, see fig. 4b. As expected, the tensile strength

Table 1
Parameters and typical results of laser CVD experiments

Layer	Reactants	Partial pressure of precursor [a] (kPa)	Fibre velocity (mm/s)	Power density [b] (W/mm^2)	Layer thickness (nm)	Tensile strength (MPa)	Weibull modulus
(Uncoated fibre)	(H_2)	–	1	10	–	3500	3.9
TiC_x	$TiCl_4/H_2$	2.3	5–1	5–10	50–100	1800–1300	2.3–1.4
TiC_x	$TiCl_4/C_6H_6/H_2$	2.0/1.0	1	5–10	50–200	2900–2300	–
TiB_2	$TiCl_4/BCl_3/H_2$	2.0/8.0	20–5	5–10	80–120	2800– 900	4.0–1.4
SiC	CH_3SiCl_3/H_2	10–5	20–5	1–5	50–100	2500– 900	4.3–1.3
Pyro-c	C_2H_2/Ar	30	30–5	5–10	–	4000–3400	–

[a] Total pressure 0.1 MPa.
[b] Power inside reactor 330 W.

is a complicated function of all deposition parameters. The most pronounced influence is exerted by those parameters which control the structure and thickness of the layers, i.e. the temperature and residence times inside the reaction zone, which are adjusted by the power density (defocusing) and the fibre velocity. Generally, the tensile strength increases with increasing beam-defocusing and velocity (fig. 3a). This effect is caused by a decrease in grain size of from 100 to about 30 nm, rather than by changes in layer thickness. In contrast to the TiC_x system, it seems to be possi-

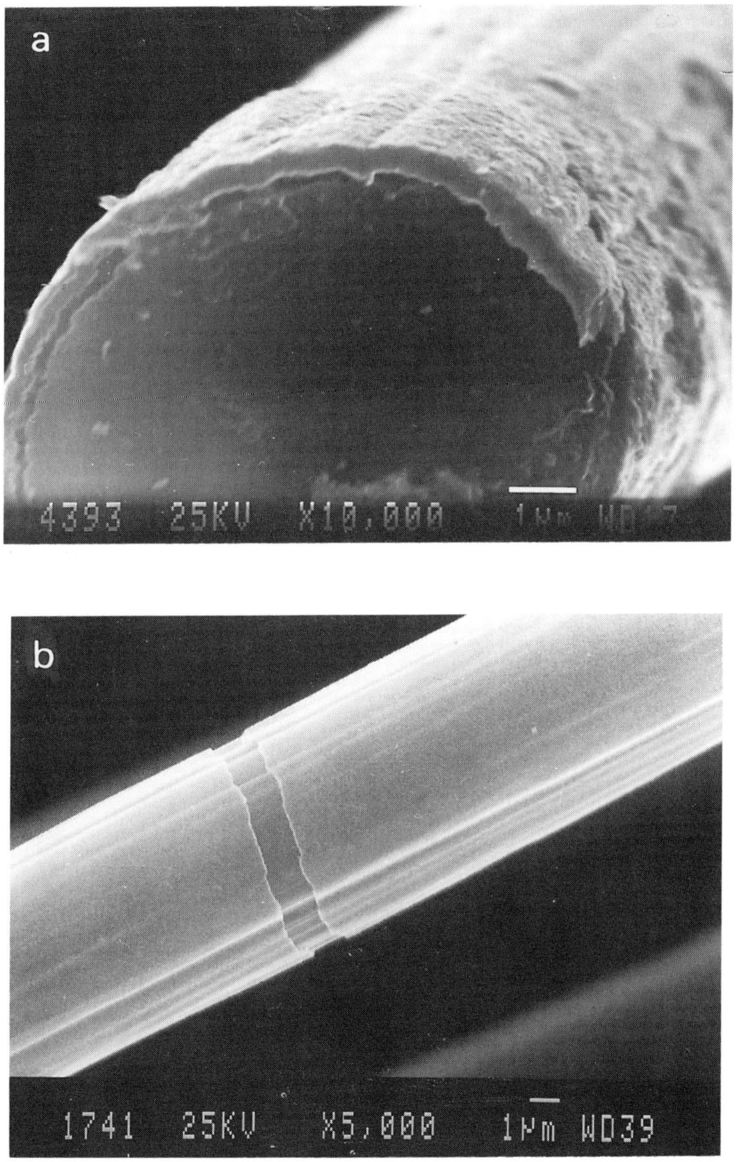

Fig. 2. SEM image of TiC onto carbon fibres with a LCVD-pyro-C interlayer: (a) cut end-face, (b) view on a relaxed layer.

ble to obtain a LCVD TiB_2 layer with roughly 100 nm thickness, which diminishes the tensile strength by less than 20%.

3.3. SiC

CVD-grown SiC layers are well established commercially available protective coatings, on graphitic materials as well as on carbon fibres [3]. Despite the fact that laser deposition from CH_3SiCl_3 is a pure photothermal process, deposition rates are typically situated in the 1 μm/s region. As this precursor yields high rates in thermal CVD too, a special gas-phase reaction mechanism seems to be responsible for the behaviour of this molecule [4]. As a consequence of roughly the same optical absorptivity of layer and fibre, there is no "self-stop mechanism" and, therefore, the layer thickness may be increased by increasing the residence times (lower fibre velocity). Deposition in the lower power-density region and at higher partial pressures (table 1) results in fine-grained β-SiC layers with a mean crystallite size of 20 nm (fig. 4c). The morphology is very smooth and the layers exactly copy the underlying fibre surface. At higher power densities the typical columnar structure appears and the surface is slightly roughened. Sometimes a periodic ripple structure with spacings in the μm region is visible, which seems to be an effect of laser radiation. The tensile strength and its dependence on LCVD parameters is comparable to that of the TiB_2 system (table 1 and fig. 3b).

3.4. Pyro-C

Pyrolytical carbon layers have been deposited using C_2H_2 or C_6H_6 (in Ar) as precursors. At higher partial pressures, the two systems tend to deposit porous layers caused by homogeneous deposition. Additionally, in the C_6H_6-system, this tendency is strengthened by laser heating of the absorbing gas phase. The prevention of gas nucleation results in the deposition of thin, smooth pyro-C layers, hardly detectable by SEM. On the other hand, TEM/SAD investigations have given evidence of very fine-grained graphitic layers with a structure comparable to that of the fibres (fig. 4d). In accordance with literature [5], the tensile strength increases even at very thin layers (fibre velocity 30 mm/s, see table 1).

4. Conclusions

The laser-CVD method developed for continuous coating of rovings holds great promise of extending the application area of conventional deposition processes, particularly with regard to high deposition rates and fine-grained but well-crystallized layer structures. A greater variety of precursors as well as coatings can be deposited, including the high temperature region and pho-

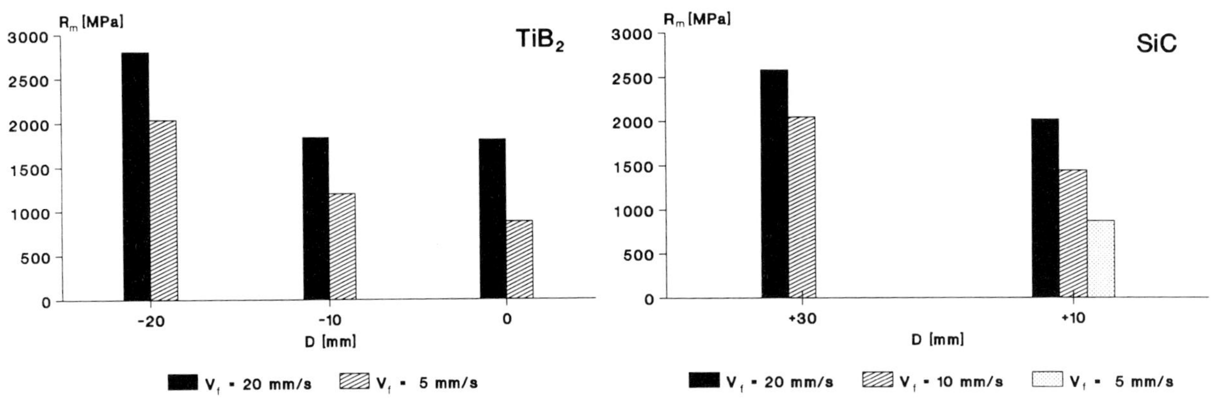

Fig. 3. Tensile strength R_m as a function of fibre velocity v_f and beam-defocussing D (a) TiB_2, (b) SiC.

tolytical reaction channels. All coatings, which may be used as a diffusion barrier, diminish the tensile strength of the fibres more or less clearly.

The reason for this behaviour may be rather complex:

(i) Surface defects from fibre pretreatment prior to the coating procedure.
(ii) Chemical reactions in between layer and fibre yields a brittle interlayer which cannot stop or reflect cracks on the interface.

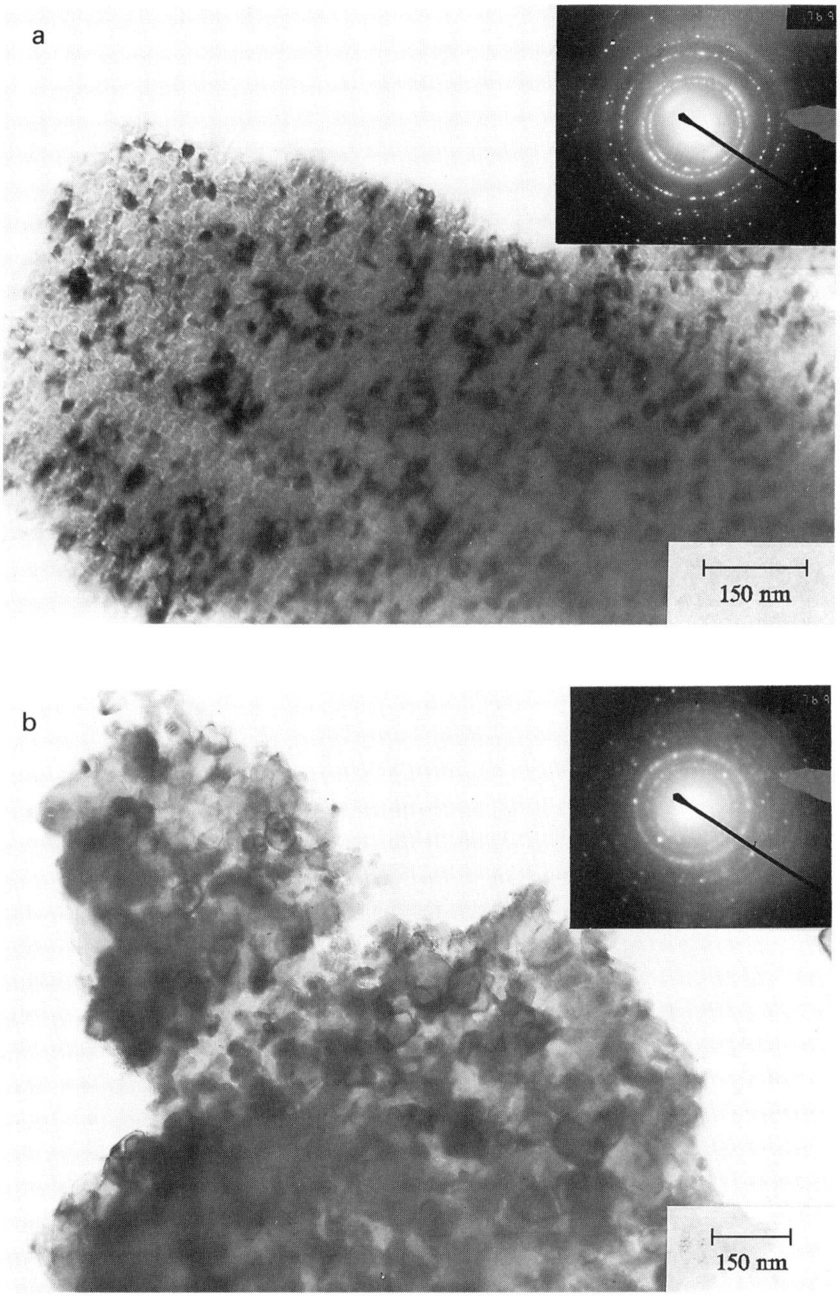

Fig. 4. TEM image and electron diffraction of LCVD layers: (a) TiC, (b) TiB_2, (c) SiC, (d) pyro-C. Figure continued on next page.

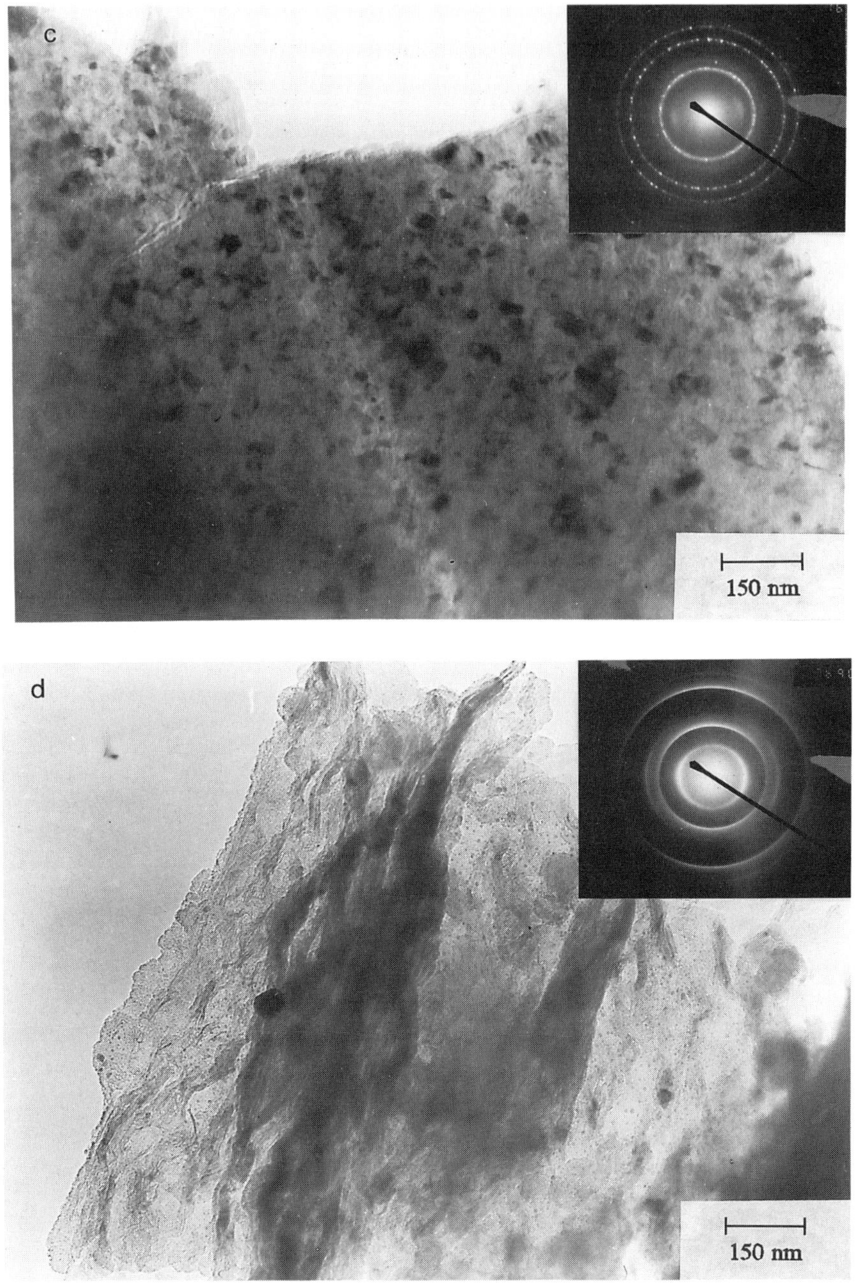

Fig. 4. Continued.

(iii) Thermally induced tension in the layer–fiber composite.

Considering the variety of parameters, the LCVD process is more difficult to optimize with respect to certain properties and the high tensile strength of the coated fibres. The most pronounced influence on the tensile strength is exerted by the thickness, structure and adhesion of the layer. Although the results presented show that, even with mono-phase single layer systems, a reasonably good tensile strength may be achieved if deposition parameters are carefully optimized, the greater potential is certainly to be found in the deposition of composite and multilayer coatings having a controlled amount of adhesion to the substrate. Experiments of this kind are currently underway.

Acknowledgements

The authors would like to thank Dr. Eva Kieselstein for scanning the electron micrographs as well as Mrs. Koestner, Mrs. Goldmann and Mr. Froehnert for their technical assistance throughout this research project.

References

[1] V. Hopfe, A. Tehel, A. Baier and J. Scharsig, Appl. Surf. Sci. 54 (1992) 78.

[2] J.F. Mandell, K.C. Hong and D.H. Grande, Ceram. Eng. Sci. Proc. 8 (1987) 937.

[3] K. Brennfleck, E. Fitzer and G. Schoch, Extended Abstr. Proc. 19th Carbon Conf., Bordeaux, 1984, p. 162.

[4] V. Hopfe, D. Wagner, P. Klobes, S. Herzig and G. Marx, J. Mol. Struct. 217 (1990) 115.

[5] E. Fitzer, Carbon Fibres and Their Composites (Springer, Berlin, 1986).

Applied Surface Science 69 (1993) 388–392
North-Holland

applied surface science

In situ second harmonic generation measurements during the electrodeposition of Ni on n-Si(111)

J. Krüger [a], N. Sorg [a], J. Reif [b] and W. Kautek [a,*]

[a] *Laboratory for Laser and Chemical Thin Film Technology, Federal Institute for Materials Research and Testing (BAM), Unter den Eichen 87, D-1000 Berlin 45, Germany*

[b] *Fachbereich Physik, Freie Universität Berlin, Arnimallee 14, D-1000 Berlin 33, Germany*

Received 2 June 1992; accepted for publication 30 October 1992

Reflection measurements of optical in situ second harmonic generation (SHG) from n-Si(111) electrodes in contact with various aqueous electrolytes under potentiostatic control in an electrochemical cell are presented. Special attention was payd to both oxide-free and oxide-covered surfaces, and, particularly, to the electroplating of Ni on Si/SiO_2. A SiO_2-covered surface yields a SH signal by an order of magnitude higher than the bare Si surface. The SH azimuthal anisotropy reflects the C_{3v} symmetry of the (111) silicon plane superimposed by an additional mirror plane which may be explained by surface step arrays due to slight misorientation of the (111) surface. With electrodeposited Ni coverage on the oxidized surface, the SH signal decreases, probably due to shadowing without substantially changing the symmetry. These facts point to the SHG originating from the Si/SiO_2 interface.

1. Introduction

Optical in situ second harmonic generation (SHG) has proved to be a versatile and extremely surface sensitive tool for the investigation of metal electrodes in electrochemistry [1]. In contrast to metal electrodes, illuminated (Si) semiconductor electrodes can show a multitude of processes like electron–hole generation and recombination in the bulk and electron–hole transfer and film formation at the interface [2,3]. A study of a Si electrode [4] and a Si wafer under wet etching conditions [5], however, did not consider these complications. They have hampered the broad application of SHG to the in situ study of the semiconductor–electrolyte interface up to now. On the other hand, SHG measurements on Si single crystals in vacuo or in air have been published [6–9]. We report in situ SH measurements of Si(111) electrodes under various electrochemical and electroplating conditions for the first time.

* To whom correspondence should be addressed.

2. Experimental

n-Si electrodes (P doping $\sim 10^{18}$ cm^{-3}, Werk für Fernsehelektronik GmbH, Berlin) were investigated in a potentiostatic standard three electrode configuration [2] in aqueous sulfate solutions. Potentials refer to the saturated calomel electrode (SCE). The electrochemical cell with a quartz window was mounted on a rotational table to allow azimuthal angle (Φ) variations. Prior to the SH measurements the open circuit potential of the cell was recorded, and the electrode was then potentiostatically kept at this potential. The native oxide was removed by etching in 2M NH_4F, adjusted to pH = 4.5 by addition of H_2SO_4. For measurements at oxide-covered electrodes, a new oxide film was grown by air contact [3]. For oxide-free surfaces, we used a 0.2M NH_4F + 0.5M K_2SO_4 electrolyte and potentiostatic control. Ni electrodeposition on oxide-covered Si was studied in 0.5M $NiSO_4$.

The SH was generated by p-polarized 532 nm radiation of a frequency-doubled Nd:YAG laser (pulse width 7 ns, repetition rate 10 Hz, average

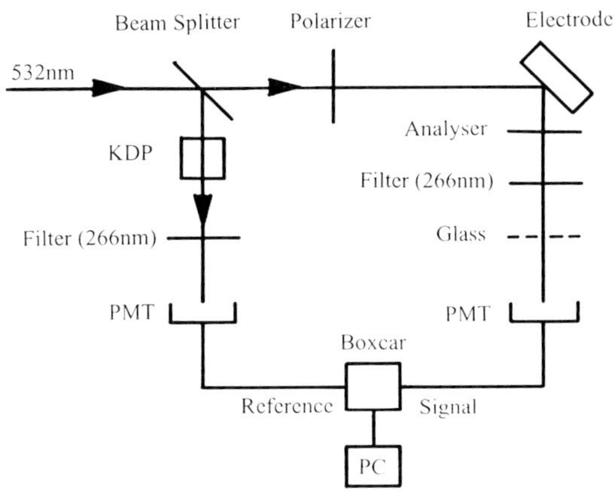

Fig. 1. Principle of the SHG setup.

power 40 mW, incident at an angle of 45° with respect to the electrode). Reflected p-polarized SH signals at 266 nm were recorded with single pulse normalization to SHG from a KDP crystal as a reference (fig. 1). The azimuthal angle dependence of the SH signals was obtained by rotating the electrode around its normal.

3. Results

In fig. 2 the azimuthal dependence of the SHG from an oxide-covered electrode in 0.5M K_2SO_4 at -0.25 V is shown. In all figures, triangles and circles symbolize measurement points obtained after variation of the azimuthal angle in opposite directions. Two theoretical curves are fitted according to different symmetries. The dotted curve was calculated using eq. (1) (see below) which corresponds to the expected threefold C_{3v} symmetry of the Si(111) surface as observed in vacuo or in air [6]. The dashed curve, calculated using eq. (2), takes into account the effect of a superimposed mirror plane along a $\langle\bar{1}\bar{1}2\rangle$ direction. In fig. 3, the SH signals from an oxide-free electrode at -0.38 V in 0.2M NH_4F + 0.5M K_2SO_4, pH = 4.5, are shown. The fitted curves were calculated using eq. (1) (dotted curve) and eq. (2) (dashed curve). The azimuthal SH behaviour of an oxide-covered Si electrode with and without electrodeposited Ni is depicted in fig. 4. Here, only eq. (2)

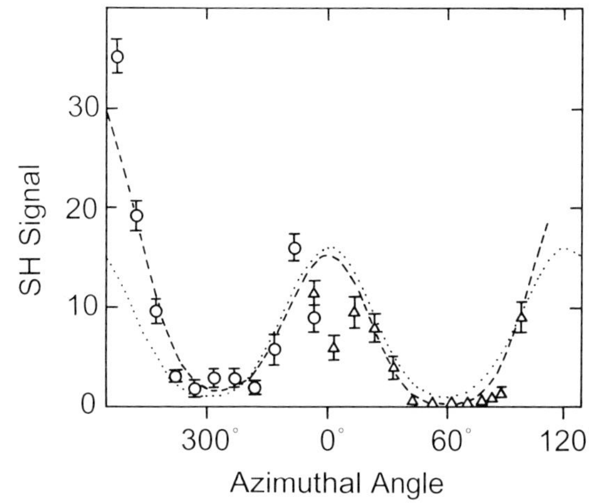

Fig. 2. SH signal from $Si/SiO_2/0.5M\ K_2SO_4$ at -0.25 V versus SCE as a function of angle Φ; dotted curve: fitted by eq. (1); dashed curve: fitted by eq. (2).

was used to calculate the curves fitted to the two sets of measurements at -0.28 V. SH signals are on the same arbitrary unit scale. This shows that the SH signal from the oxide-covered electrode is by an order of magnitude higher than that from the oxide-free electrode. The parameters a, b and c from the fits are listed in table 1. The

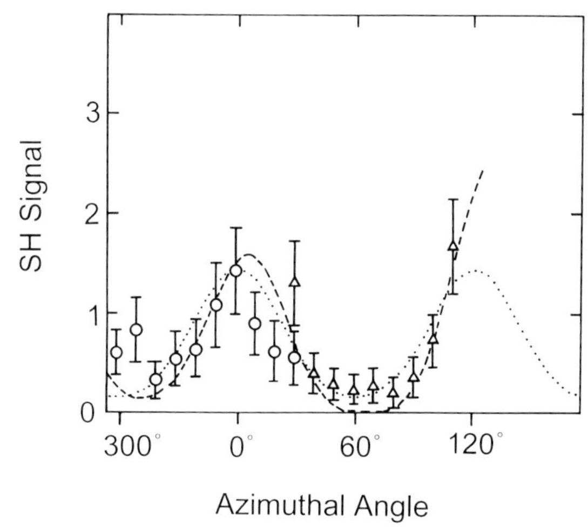

Fig. 3. SH signal from $Si/0.2M\ NH_4F+0.5M\ K_2SO_4$ at -0.38 V versus SCE as a function of angle ϕ; dotted curve: fitted by eq. (1); dashed curve: fitted by eq. (2).

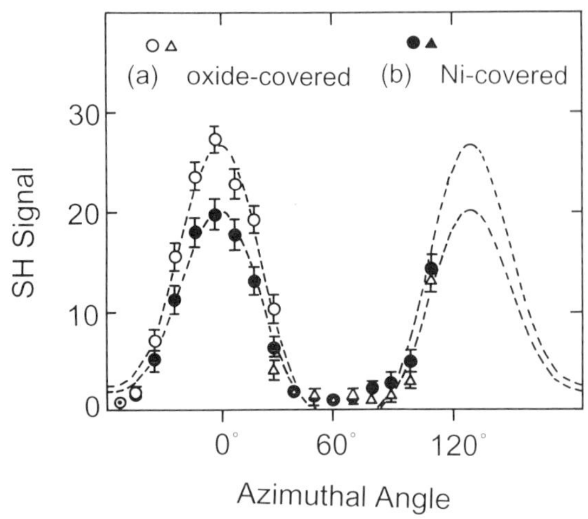

Fig. 4. SH signal from (a) $Si/SiO_2/0.5M\ NiSO_4$ (hollow symbols) and (b) $Si/SiO_2/Ni/0.5M\ NiSO_4$ (filled symbols) at −0.28 V versus SCE as a function of the angle ϕ; dashed curves: fitted by eq. (2).

Table 1
Azimuthal SH anisotropy: fit parameters (according to eqs. (1) and (2))

	a	*b*	*c*
SiO_2 (fig. 2; eq. (1))	50	–	30
SiO_2 (fig. 2)	60	24	36
Si (fig. 3; eq. (1))	16	–	8
Si (fig. 3)	20	8	12
SiO_2 (fig. 4a)	68	26	45
SiO_2/Ni (fig. 4b)	58	22	42

micrograph in fig. 5 shows a partial Ni coverage in the form of metal hemispheres.

4. Discussion

The observed azimuthal dependence of the SHG does not correspond to the C_{3v} symmetry

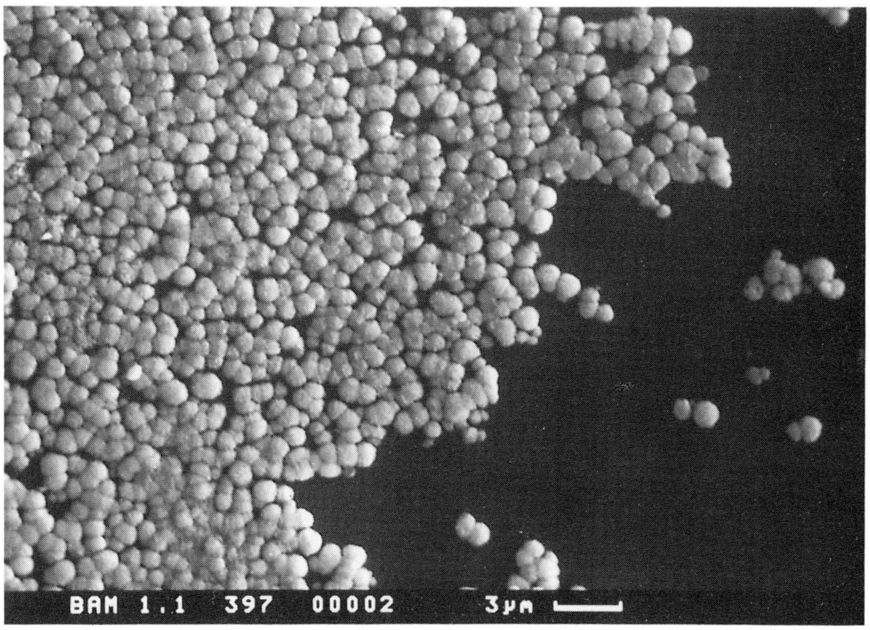

Fig. 5. SEM micrograph of the electrode surface investigated in fig. 4.

expected for Si(111)1×1, or 7×7 reconstruction prepared by heat annealing in vacuo [6–9], for which the SH intensity for p-polarization, $I(2\omega)$, is given by

$$I(2\omega) \sim (a + c \cos 3\Phi)^2 I^2(\omega), \qquad (1)$$

with Φ being the angle between the plane of incidence and the crystallographic $\langle \bar{1}10 \rangle$ direction, and a and c standing for the nonlinear polarizabilities normal and parallel to the surface, respectively. Instead, the data may be fitted by superimposing an additional mirror plane along the $\langle \bar{1}\bar{1}2 \rangle$ direction which modifies eq. (1) to

$$I(2\omega) \sim (a + b \sin \phi + c \cos 3\Phi)^2 I^2(\omega). \qquad (2)$$

This mirror plane may be readily explained by assuming the presence of parallel steps representing, e.g., $(\bar{1}\bar{1}2)$ planes [9–13]. Such steps originate from minor misorientations of the (111) surface (miscuts of the crystal). If the step fronts represent, e.g., $(\bar{1}\bar{1}2)$ faces, they should indeed result in an additional nonlinear polarizability along the $\langle \bar{1}\bar{1}2 \rangle$ direction perpendicular to the intrinsic (111) polarizability represented by an additional $\sin(\Phi)$ term in eq. (2).

As the symmetry is preserved under the oxide coverage, it is most likely that the signal originates at the Si/SiO_2 interface (fig. 6). This assumption is supported by several facts. The oxide

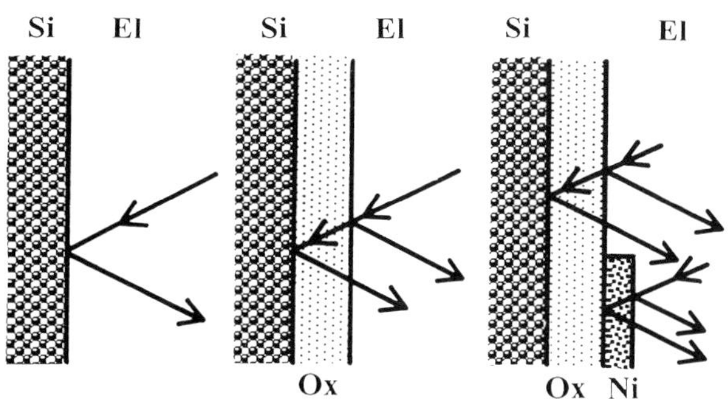

Ox ... SiO_2, SiO_x, $Si(OH)_x$
El .... Electrolyte

Fig. 6. Possible contributions to the SH signal in the systems studied: (a) oxide-free Si, (b) oxide-covered Si, (c) additional Ni deposit on (b).

is transparent to both fundamental 532 nm and second harmonic 266 nm wavelengths, thus allowing to observe interaction at the interface. SiO_2 grown on Si(111) is known to be amorphous [14]. It therefore excludes bulk SHG as well as any signature of order at the SiO_2/electrolyte interface.

The stronger SH signal from the Si/SiO_2 interface, as compared to the Si/F^--electrolyte interface, is due to either an enhancement of the polarizability at the Si/SiO_2 interface or to a reduced polarizability at the Si/F^--electrolyte interface which most probably shows Si–H termination [3,15].

5. Summary

(1) We clearly observe a signal originating at the crystal surface, because the influence of the steps and the SH magnification by the presence of the oxide exclude major bulk contributions.

(2) At the oxide-free electrode, the SH signals originate at the Si(111)/electrolyte interface. At the oxide-covered electrode they originate at the Si/SiO_2 interface.

(3) The influence of additional Ni coverage on the oxide appears to be a shadowing effect. This is concluded from the decrease in signal without changing the symmetries. The oxide separates the Si from the metal so that neither electronic effects of the Ni on the Si/SiO_2 interface, nor structural information from the Si substrate to the Ni should occur. Therefore, it is most important to study Ni deposits on the oxide-free Si surface. These experiments will be published soon [16].

Acknowledgements

This work was partially supported by the DFG (German Science Foundation). We thank E. Matthias and his coworkers (Free University Berlin) for their steady support, and R. Blumenau, J. Stange and O. Ball (Federal Institute for Materials Research and Testing) for valuable technical assistance.

References

[1] G.L. Richmond, Adv. Electrochem. Sci. Eng. 2 (1992) 142.

[2] W. Kautek, N. Sorg and W. Paatsch, Electrochim. Acta 36 (1991) 1803.

[3] N. Sorg, W. Kautek and W. Paatsch, Ber. Bunsenges. Phys. Chem. 95 (1991) 1501.

[4] O.A. Aktsipetrov and E.D. Mishina, Sov. Phys. Dokl. 29 (1984) 37.

[5] L.L. Kulyuk, D.A. Shutov, E.E. Strumban and O.A. Aktsipetrov, J. Opt. Soc. Am. B 8 (1991) 1766.

[6] H.W.K. Tom, T.F. Heinz and Y.R. Shen, Phys. Rev. Lett. 51 (1983) 1983.

[7] D. Guidotti, T.A. Driscoll and H.J. Gerritsen, Solid State Commun. 46 (1983) 337.

[8] T.F. Heinz, M.M.T. Loy and W.A. Thompson, Phys. Rev. Lett. 54 (1985) 63.

[9] J.F. McGilp, J. Phys. (Condens. Matter) 2 (1990) 7985.

[10] G. Tarrach, R. Wiesendanger, D. Bürgler, L. Scandella and H.-J. Güntherodt, J. Vac. Sci. Technol. B 9 (1991) 677.

[11] X.-S. Wang, J.L. Goldberg, N.C. Bartelt, T.L. Einstein and E.D. Williams, Phys. Rev. Lett. 65 (1990) 2430.

[12] J.L. Goldberg, X.-S. Wang, N.C. Bartelt and E.D. Williams, Surf. Sci. Lett. 249 (1991) L285.

[13] W. Ranke and M. Henzler, in: The Physics of Solid Surfaces, Landolt-Börnstein, New Series, Vol. III/24, Ed. G. Chiarotti (Springer, Berlin, 1992).

[14] Ph. Avouris, I.-W. Lyo and F. Bozso, J. Vac. Sci. Technol. B 9 (1990) 424.

[15] J.J. Boland, J. Vac. Sci. Technol. B 9 (1991) 764.

[16] N. Sorg, J. Krüger, J. Reif and W. Kautek, Ber. Bunsenges. Phys. Chem., in press.

Applied Surface Science 69 (1993) 393–397
North-Holland

applied
surface science

Direct photo-deposition of silicon dioxide films using a xenon excimer lamp

P. Bergonzo [a], U. Kogelschatz [b] and I.W. Boyd [a]

[a] Electronic and Electrical Engineering, University College London, Torrington Place, London WC1E 7JE, UK
[b] Asea Brown Boveri, Corporate Research, CH-5405 Baden, Switzerland

Received 2 June 1992; accepted for publication 30 October 1992

Recently excimer lamps have opened up the field of intense vacuum ultra-violet (VUV) light generation. With theoretical efficiencies reaching 40%, the power available from such lamps based on dielectric barrier discharge generation can be superior to those of typical low pressure mercury lamps with shorter UV wavelengths generated. Here we present, for the first time, the use of these lamps for the direct photo-deposition of silicon dioxide from silane and nitrous oxide mixtures. Deposition rates achieved on our unoptimised system are comparable with those obtained with low pressure mercury lamps. The results indicate promising further applications of such lamps towards semiconductor and optoelectronic materials processing.

1. Introduction

Over the last decade, a general requirement for low temperature processing in the semiconductor industry has become apparent. This is due, in part, to the continued reduction in device geometries and, amongst other reasons, the emergence of temperature sensitive materials such as amorphous silicon, gallium arsenide and indium phosphide (a-Si:H, GaAs, InP) [1]. Of the low temperature processing techniques now being investigated, photo-enhanced processing of materials is very promising since the processed surfaces and growing films are not subjected to damaging ionic bombardment such as can be present in plasma-assisted processing systems [2,3].

The recent development of novel excimer lamps [4] has opened up the field for relatively inexpensive, large area, direct photo-CVD to wafer scale technologies. As new powerful sources of low wavelength photons, these lamps have received considerable attention during the past few years [5,10], producing wavelengths extending to the vacuum ultra-violet domain (VUV) at higher intensities than can otherwise be currently routinely achieved. In fact, such lamps have already been used successfully to enhance various photochemical reactions such as ozone generation and metallorganic processing [7–9].

The principle of the excimer lamp relies on the radiative decomposition of excimer states created by a dielectric barrier discharge in a rare gas such as Xe, Kr, or Ar [25]. Such a principle has of course already been used in excimer lasers with rare gas halogen mixtures [11], but simplicity and reliability as well as scalability give these lamps a definite advantage over lasers when large scale industrial processes are envisaged. In photo-CVD there is no need for purely monochromatic coherent light, since the UV absorption mechanisms associated with the gas phase exhibit extended band-like properties rather than discrete transitions. Furthermore, the intense excitation levels required in such laser cavities limit their emission to a pulsed mode, hence requiring long exposure times for most reactions. Another advantage of lamps is that the excitation densities required are much lower than for lasers and, therefore, pure rare gas excitation can be successfully obtained, enabling shorter wavelengths to be produced.

At present, intensities of 100 mW/cm^2 are achievable at the excimer lamp surface at reasonable cost, compared with excimer lasers [4], and this bodes for photo-CVD applications. Till now, no such source has been used for direct deposition of dielectric films relevant to VLSI devices, such as silicon dioxide, oxynitride, or nitride. Direct photo deposition has previously been obtained with the 185 nm radiation of the low pressure mercury lamps [12,13] and with various discharge lamps (nitrogen, argon, krypton, xenon). However, the very low efficiencies involved with these systems have somewhat limited the deposition rate of the films [14,15].

Silicon dioxide has been extensively used and studied in semiconductor manufacturing applications. Highly compatible with silicon-based technologies because of its highly desirable chemical, electrical, and mechanical properties, it is used for gate and field oxides, and masks and passivating layers, as well as for tunnel oxide applications. For such a material, conventional pyrolytic deposition occurs above 600°C [24] while traditional oxidation occurs at temperatures around and above 900°C. The aim of this paper is to demonstrate the possibility and inherent advantages of depositing silicon dioxide at low temperatures with excimer lamps. With this "proof of principle" shown, the case for other VLSI materials to be deposited with deeper wavelengths with these lamps can readily be made.

2. Experimental

Excimer lamps, particularly those using the Xe_2^* generated radiation and the reaction kinetics of excimer formation, have been extensively described elsewhere [6,10,16]. The second continuum of xenon emitting at 172 nm, when excited by non-equilibrium high pressure discharges, enables high UV power levels to be generated. In this study we have used 172 nm power densities around 10 mW/cm^2. The lamp system used comprised two sections. In the first, a cell containing the pure gas in a suprasil envelope generated the radiation which was directed through a MgF_2 window. The transmitted light entered a second chamber, a reaction cell. Here, gas mixture of pure silane and nitrous oxide were introduced via mass flow controllers through a toric shaped shower onto the samples. In these experiments, sections of p-type (100) silicon wafers were used and their temperature was maintained between 200 and 300°C. Thermocouple control ensured that, although essentially negligible, the low intensity of the lamp did not increase the surface temperature of the samples during film growth. The shower–substrate and window–substrate distances were kept constant at 1 and 4 cm, respectively. Prior to deposition the samples were cleaned in a propanol solution in ultrasonic bath and the chamber was evacuated to 10^{-6} mbar. The 172 nm emission from the lamp was monitored with a MacPherson VUV monochromator.

The VUV induced reaction scheme between SiH_4 and N_2O has been recently studied by other groups. The absorption coefficient of N_2O at the wavelength used is typically around 2 cm^{-1} atm^{-1} [17,18]. At the window–substrate distance used, and with the pressure around 100 mb, the absorption by N_2O of the radiation entering the processing chamber is about 50%. In the work reported here, mixtures containing between 0.5 to 10% silane in nitrous oxide were introduced into the chamber at a constant flow of 50 sccm. Silane is already known not to be dissociated at these wavelengths and intensities. A simplified reaction scheme can be written as follows:

$$N_2O + h\nu \rightarrow N_2 + O, \quad (1)$$

$$SiH_4 + O \rightarrow SiH_3 + OH, \quad (2)$$

$$SiH_4 + OH \rightarrow SiH_3 + H_2O, \quad (3)$$

$$SiH_4 + O \rightarrow SiH_2 + H_2O, \quad (4)$$

$$SiH_2 + O \rightarrow SiH_2O, \quad (5)$$

$$SiH_3 + O \rightarrow SiH_2O + H, \quad (6)$$

$$SiH_2O + O_2 \rightarrow SiH_2O_2 + O, \quad (7)$$

$$SiH_2O_2 + O_2 \rightarrow SiO_2 + H_2O + O. \quad (8)$$

Basically, nitrous oxide is photo-dissociated (1), leading to the generation of oxygen radicals which then react with silane, (2)–(4). Branching reactions (5)–(7) subsequently lead to SiO_2 formation (8). A more complete model for this scheme has been recently detailed by Petitjean [13,19]. Film growth proceeds in a continuous manner, like with the laser-initiated layer situation, which by necessity is controlled in a pulsed growth mode.

After continuous irradiation of the silicon for times up to 30 min, the layers grown were characterised using FTIR spectrometry while thickness measurements were carried out with a Rudolf Autoell Ellipsometer. The electrical properties of the films will form the basis of a later paper, but we report here that their characteristics are quite similar to those of conventionally deposited films.

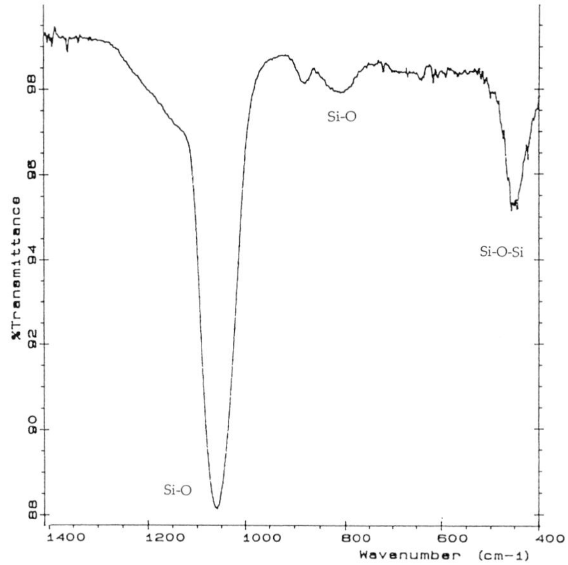

Fig. 2. FTIR absorption spectrum of a 400 Å silicon dioxide layer deposited at 300°C.

3. Results

For exposure times up to 30 min, film thicknesses of 50 nm were grown at temperatures below 300°C, producing a growth rate of around 0.3 Å/s. Fig. 1a shows the variation of the deposition rate with intensity for the 172 nm radiation used. As expected, it is proportional to the incident flux, and also clearly shows that any thermally controlled deposition of the films was insignificant under these operating conditions. At total gas pressures exceeding 150 mbar, powder was formed as a consequence of gas phase reactions being stimulated. In contrast, the deposition rate was found to be very low below pressures of about 100 mbar. At intermediate pressures and temperatures between 200 and 300°C, the films formed were scratch resistant and adherent. Typical deposition rates varied from 10 to 15 Å/min. Fig. 1b shows the variation of the deposition rate with R, the SiH_4/N_2O ratio. For $R < 2\%$, the rate is limited by the silane concentration. For $R > 2\%$ gas phase reactions leading to powder formation also limit film deposition. Refractive indices of 1.458 were obtained on optimised samples, values always ranging between 1.4 and 1.5 according to experimental conditions (higher values of 1.6 were obtained for higher ratio R). The structural properties of the oxide films deposited were characterised by their vibrational spectra. Fig. 2 shows a FTIR spectrum of an oxide film photo-deposited at a substrate temperature of 300°C. The spectrum consists of three characteristic absorption peaks at 1065 cm^{-1} (Si–O–Si stretching), 800 cm^{-1} (O–Si–O bending), and 480 cm^{-1} (Si–O–Si rocking) [20,21]. The full width at half maximum (FWHM) for the dominant IR

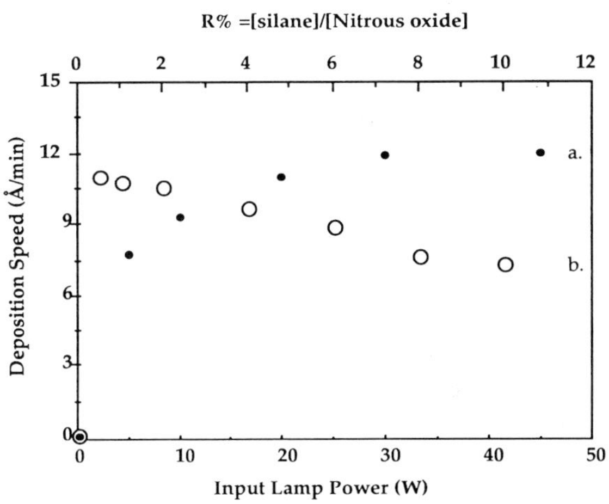

Fig. 1. Deposition speed for silicon dioxide samples deposited at 300°C versus (a) (●): VUV light exposure, and (b) (○) precursor mixture ratio.

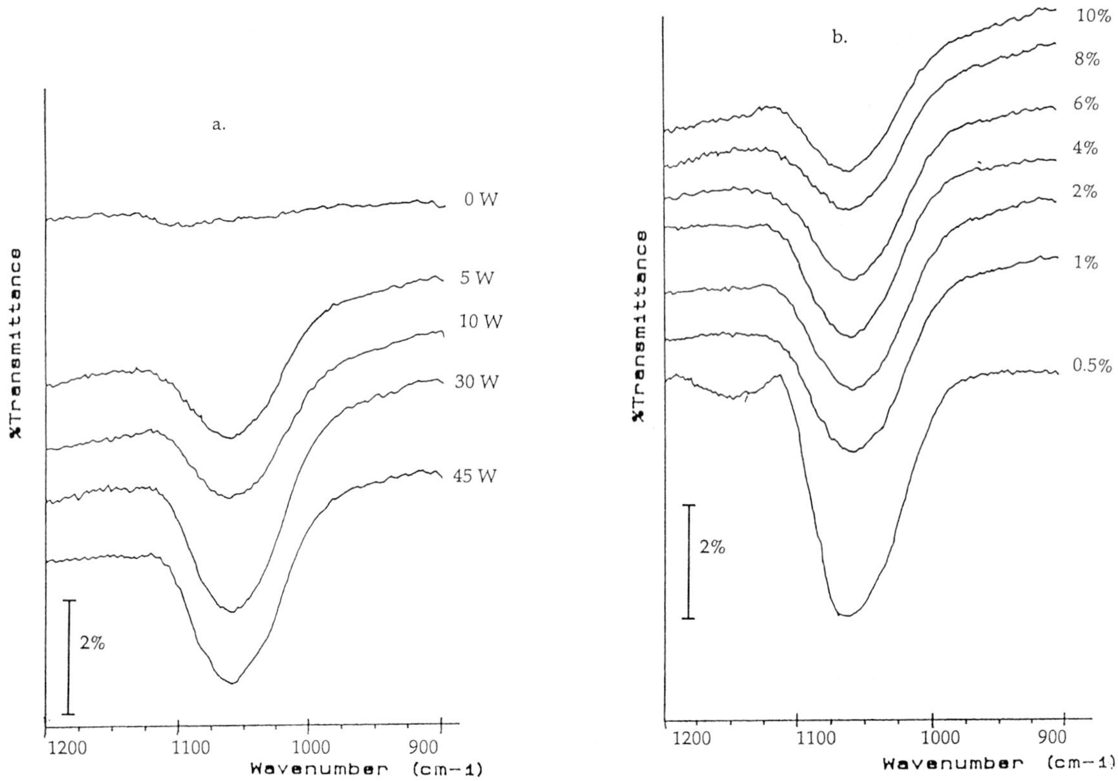

Fig. 3. Sets of FTIR spectra of the 1065 cm^{-1} peak for various (a) input lamp power, (b) precursor mixture ratio R (average thickness: 100 Å).

mode around 1065 cm^{-1} has a value of 70 cm^{-1} and remains essentially the same for all the oxides studied. The small peak centered around 885 cm^{-1} can be attributed either to Si–H, Si–OH, Si_2O_3 or to non-bridging oxygen [26]. However, the presence of Si–H and Si–OH bonds would have led to the presence of other peaks centered around 2270 and 3620 cm^{-1}, which we did not observe. This peak can, therefore, be most likely attributed to a defect state of SiO_2 [27] although further investigations are underway to confirm this point.

The absorption spectra of figs. 3a and 3b show that the quality of the film is not significantly affected by either lamp power or the precursor mixture ratio used. These parameters, however, affect the deposition rate (e.g. 6 Å/min for $R = 10\%$) much more appeciably. A plot of absorbance of the band near 1065 cm^{-1} against film thickness (measured ellipsometrically) is shown in fig. 4 for different samples (Lambert–Bouguer plot). The apparent absorption coefficient α_{app} is calculated from the slope of best fit through the

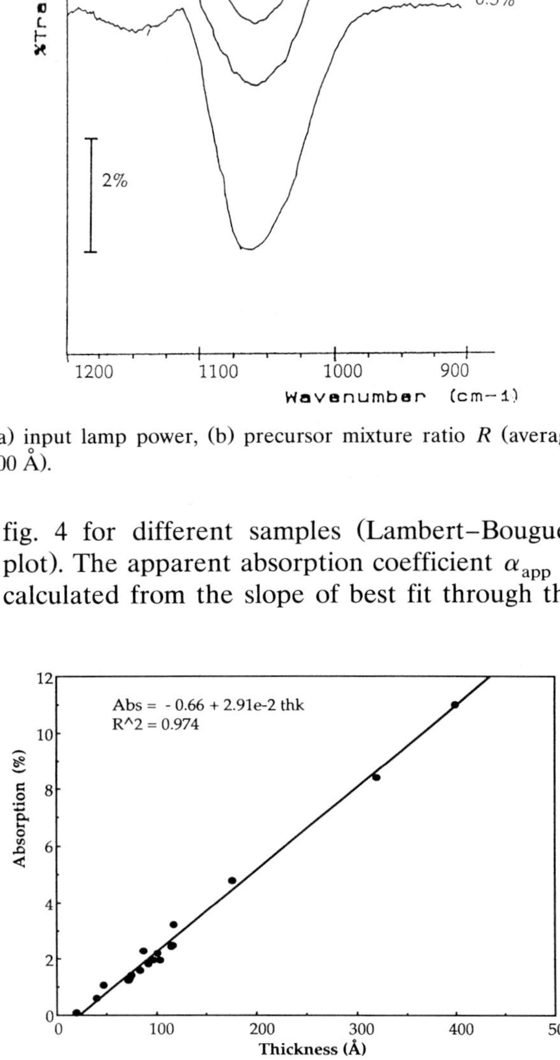

Fig. 4. Lambert–Bouguer plot of absorbance of the Si–O stretching band near 1065 cm^{-1} against film thickness for oxide layers on various silicon substrates.

data points. The intersection with the abscissa indicates the thickness of native oxide on the bare Si sample. This curve gives a native film thickness of 20 Å and and $\alpha_{app} = 2.91 \times 10^4$ cm^{-1} in good agreement with published data for much thicker films obtained from pyrolytic processes [22,23].

4. Conclusion

We have demonstrated the possibility of depositing good quality silicon dioxide films with a photochemical process based on the 172 nm wavelength from the excimer (or second excimer continuum) of xenon in an excimer silent discharge lamp. The deposition rates obtained are comparable to those from similar processes using low pressure mercury lamps [15]. This is very promising considering the fact that our lamp system has not been optimised in terms of the VUV output. This work is thus a proof of principle which can extend the use of excimer lamps to the field of material deposition for semiconductor applications. We are currently studying the application of this technique for the deposition of other material such as silicon nitride and silicon oxynitride and multi-layer dielectrics. We have also recently developed lamps of deeper wavelengths radiating at the Ar_2^* and Kr_2^* continua (126 and 146 nm). This will allow *direct* photodissociation of silane and, hence, amorphous silicon films to be processed enabling in-situ multilayer transistor fabrication.

Acknowledgements

We would like to acknowledge here funding, in part, from the EEC SCIENCE programme, SERC grant No. GR/E 16090, and ABB Corporate Research, Baden, Switzerland. Special thanks are also given to Professor M. Green and Dr. S. Best for the respective use of the ellipsometer and the FTIR spectrometer.

References

[1] R.R. Chang, R. Iyer and D.L. Lile, J. Appl. Phys. 61 (1987) 1995.
[2] C. Vinkier and S. De Jaegere, J. Electrochem. Soc. 137 (1990) 628.
[3] D.A. Buchanan, Appl. Phys. Lett. 56 (1990) 1037.
[4] U. Kogelschatz, Appl. Surf. Sci. 54 (1992) 410.
[5] B. Eliasson and B. Gellert, J. Appl. Phys. 68 (1990) 2026.
[6] B. Eliasson and U. Kogelschatz, Appl. Phys. B 46 (1988) 299.
[7] B. Eliasson and U. Kogelschatz, Ozone 13 (1991) 365.
[8] H. Esrom, J. Demmy and U. Kogelschatz, Chemtronics 4 (1989) 202.
[9] H. Esrom and U. Kogelschatz, Appl. Surf. Sci. 46 (1990) 158.
[10] B. Gellert and U. Kogelschatz, Appl. Phys. B 53 (1991) 14.
[11] Ch.K. Rhodes, Excimer Lasers, Vol. 30 of Topics in Applied Physics (Springer, Berlin, 1984).
[12] Y.I. Nissim, J.M. Moison, F. Houzay, F. Leblanc, C. Licoppe and M. Bensoussan, Appl. Surf. Sci. 46 (1990) 175.
[13] M. Petitjean, N. Proust and J.F. Chapeaublanc, Appl. Surf. Sci. 46 (1990) 189.
[14] P. Patel and I.W. Boyd, Appl. Surf. Sci. 46 (1990) 352.
[15] P.A. Robertson, The Photo-Enhanced Deposition of Amorphous Silicon and Silicon Oxide Thin Films, PhD Thesis, University of Cambridge, 1987.
[16] D.J. Eckstrom, H.N. Nakano, D.C. Lorents, T. Rothem, J.A. Betts, M.E. Lainhart, D.A. Dakin and J.E. Maenchen, J. Appl. Phys. 64 (1988) 1679.
[17] J.G. Calvert and J.N. Pitts, Photochemistry (Wiley, New York, 1966).
[18] D.L. Baulch, R.A. Cox, P.J. Crutzen, R.F. Hampson, J.A. Kerr, J. Troe and R.T. Watson, J. Phys. Chem. Ref. Data 11 (1982) 359.
[19] M. Petitjean, Optimisation des Structures MIS sur InP, Rapport de Thèse, Université Paris 6, 1991.
[20] I.W. Boyd and J.I.B. Wilson, J. Appl. Phys. 53 (1982) 4166.
[21] G. Lucovsky, M.J. Manitini, J.K. Srivastava and E.A. Irene, J. Vac. Sci. Technol. B 5 (1987) 530.
[22] I.W. Boyd and J.I.B. Wilson, J. Appl. Phys. 62 (1987) 3195.
[23] J.E. Dial, R.E. Gong, J.N. Fordemalt, J. Electrochem Soc. 115 (1968) 327.
[24] C.J. Giunta, J.D. Chapple Sokol and R.G. Gordon, J. Electrochem Soc. 137 (1990) 3237.
[25] Y. Tanaka, J. Opt. Soc. Am. 45 (1955) 710.
[26] I.W. Boyd, Laser Processing of Si and the Growth and Structure of Si Oxides, PhD Thesis, Herriot Watt University, 1982.
[27] R.J. Bell, N.F. Bird and P. Dean, J. Phys. C 2 (1968) 229.

Applied Surface Science 69 (1993) 398–402
North-Holland

applied
surface science

Rapid thermal and large area processing of thin films with a line electron beam

M. Pauli, G. Dähn, J. Müller and M. Döscher

Technical University Hamburg-Harburg, Eissendorferstrasse 42, D-2100 Hamburg 90, Germany

Received 2 June 1992; accepted for publication 30 October 1992

The rapid thermal and large area processing with a line electron beam has been used for the recrystallization of silicon films on graphite substrates and for the hetero-epitaxial growth of β-$FeSi_2$ films on (111) silicon wafers. A method has been developed to calculate the differential power of the line electron beam along the line direction from the measured beam profile. This differential power in conjunction with the FWHM of the beam profile determines the characteristics of the line heat source in the thin film. Using these parameters the temperature distribution in the thin film was simulated numerically.

1. Introduction

Thin films often require a heat treatment after deposition to improve their crystalline quality or to change their chemical composition. For this purpose the rapid thermal processing is a promising technique as it significantly decreases the process time compared to furnace annealing. Such a technique using the radiation of tungsten lamps or graphite strip heaters has been established especially for the recrystallization of silicon films or the annealing of ion implantation induced defects [1–3]. The disadvantages are the limitation of the maximum radiation power and therefore of the maximum temperature in the thin film and the dependence of the energy deposition on the optical properties of the thin film which may change during the process. In our study a heat treatment system was developed using the radiation from a line electron beam. The beam energy is easy to adjust by the acceleration voltage and the energy deposition depends mainly on the mass of the thin film so that reliable and versatile rapid thermal and large area processing of thin films is possible.

2. The line beam electron gun and the set-up of measurement

The line beam electron gun is shown schematically in fig. 1: electrons emitted from the heated line cathode are focussed by two electrodes to form a line in y-direction and are accelerated in z-direction towards a flat anode [4]. During the process the substrate lies directly on the anode and both are moved at a constant velocity in x-direction.

The beam profile of the line electron beam is measured as follows: a small cylindrical hole in the anode allows a well defined circular shaped part of the line electron beam to pass through the anode towards a metallic cylinder which is electrically insulated from the anode and connected to a measuring set-up (fig. 2a). The metallic cylinder is moved with the anode at a constant velocity beneath the line electron beam in x-direction. The current I_M passing through the circular hole in the anode is measured during the movement of the anode and recorded as a function of the x-coordinate. Dividing $I_M(x)$ by πr^2, the circular area of the cylindrical hole, yields approximately

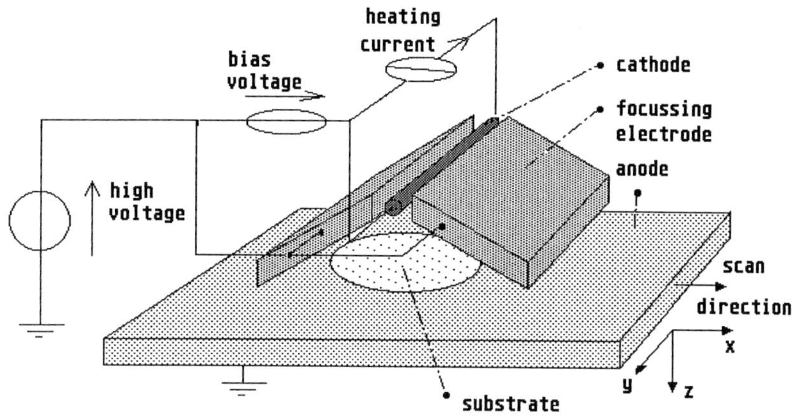

Fig. 1. Scheme of the line electron beam system.

the current density of the electron beam $i(x)$. The x-dependent curve of the current density distribution $i(x)$ represents the beam profile which is shown in fig. 2b at four different points of the y-direction. Fig. 2b shows that the emission of the line electron gun is constant along a length of 7 cm in the y-direction [4].

The FWHM of the beam profile and the power of the line electron beam determine the line heat source characteristics in the thin film. The power is taken into account differentially as power per unit length in the y-direction $\bar{P}$ and calculated as the product of the current per unit length in the y-direction $\bar{I} = \int_{-\infty}^{+\infty} i(x)\,dx$ and the accelerating voltage U_H (see also fig. 2b) [4].

3. Process requirements

The process parameters must be chosen such that the maximum temperature in the thin film fits into the temperature interval which is re-

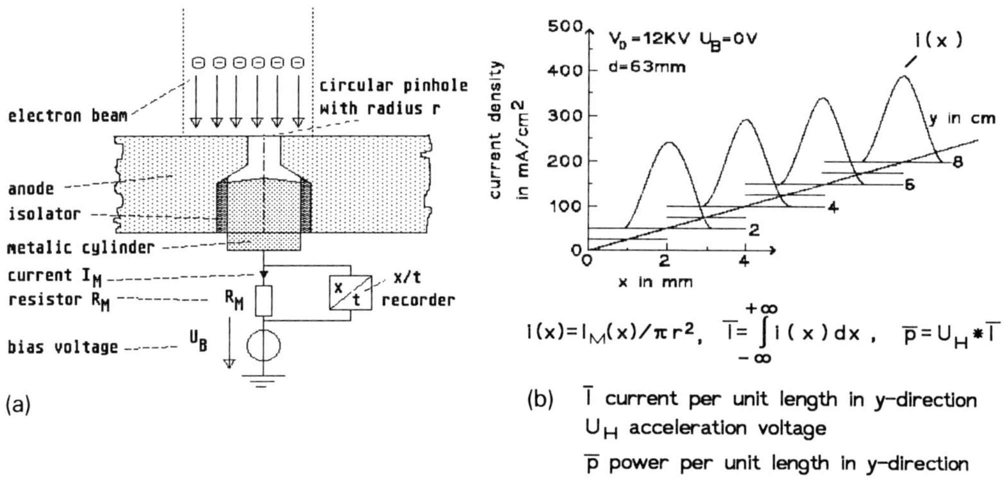

Fig. 2. (a) Set-up of measurement for the current density distribution of the line electron beam and (b) typical measured beam profiles as a function of the y-direction and the necessary equations to calculate the differential power per unit length in the y-direction.

quired by the specific thermal process. For the recrystallization of silicon films from the melt the lower temperature limit is the melting point of silicon (1410°C) [5]. The upper limit is given by the destruction of the thin film or the substrate. For the formation of β-$FeSi_2$ the temperature is determined by the temperature range in which β-$FeSi_2$ is formed, i.e., between 450°C and 950°C [6–8]. The temperature can be achieved by varying the scan velocity and selecting the appropriate differential power of the line electron beam.

4. Numerical simulation of the temperature distribution

The temperature distribution in the thin film and its development in time are determined by

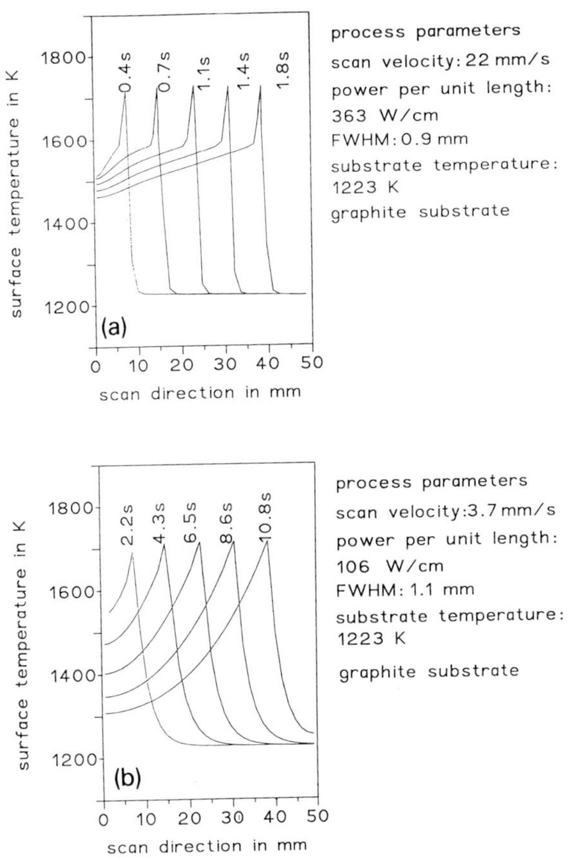

Fig. 3. Simulated temperature distribution on the substrate surface for several times. For $t = 0$ the line electron beam is positioned at $x = 0$ (x-direction = scan direction).

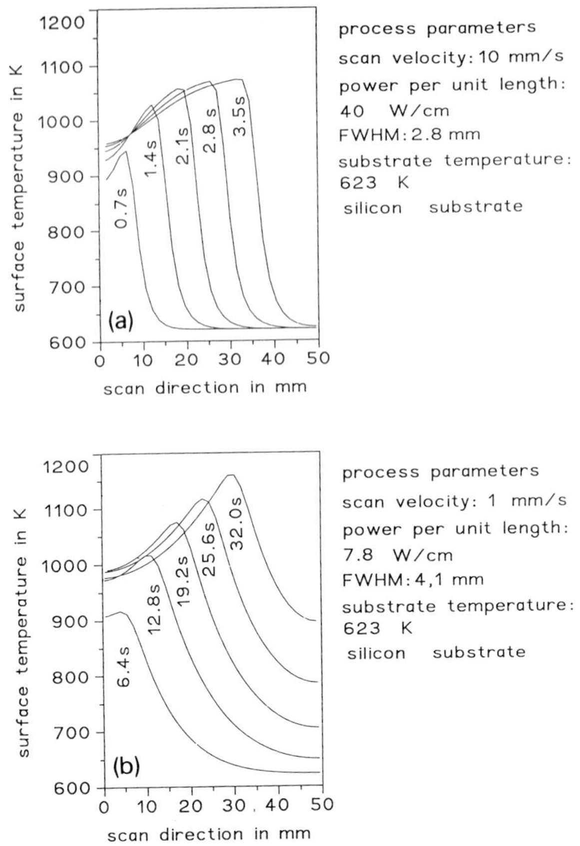

Fig. 4. Simulated temperature distribution on the substrate surface for several times. For $t = 0$ the line electron beam is positioned at $x = 0$ (x-direction = scan direction).

the parameters of the line heat source, the scan velocity and the thermal properties of the substrate and the film. A two-dimensional numerical simulation taking into account the x- and z-direction was made to calculate this temperature distribution. The fundamentals and assumptions of the simulation are given in ref. [9]. It was found that the thin film does not influence the temperature distribution. Therefore, only the process parameters and the substrate properties must be taken into account.

The numerically simulated temperature distribution on the substrate surface and its development in process time are shown in figs. 3a and 3b for a 1 mm graphite substrate and in figs. 4a and 4b for a 400 μm silicon substrate each for different process parameters. On the graphite substrate (figs. 3a and 3b) a constant temperature maximum is reached when the line electron beam has passed about 6 mm of the substrate. This temperature distribution remains nearly constant when moving across the substrate surface and a zone refining process is possible.

The temperature distributions in fig. 4 represent more a rapid but homogeneous heating than a zone refining process. Nevertheless, the temperature in scan direction stays within the temperature interval which is required for the formation of β-$FeSi_2$.

5. Recrystallization of silicon on graphite substrates

Graphite was chosen as a substrate because it is resistant against the high temperature gradients which will occur during recrystallization (fig. 3). Amorphous silicon films of 3 μm thickness were sputtered onto the graphite substrates. The silicon films were recrystallized with the process parameters given in figs. 3a and 3b. As predicted by the simulation the melting point of silicon is reached after the line electron beam has passed about 6 mm of the substrate. Beyond this point the silicon film is recrystallized across the whole wafer. Figs. 5a and 5b show the X-ray diffraction patterns of the silicon films recrystallized with the parameters given in figs. 3a and 3b, respectively. Using a low scan velocity, a silicon carbide reflection appears (fig. 5a) whereas for the recrystallization with the higher scan velocity no such reflection occurs. The formation of SiC is a parasitic process during the recrystallization of the silicon film and can be avoided by decreasing the

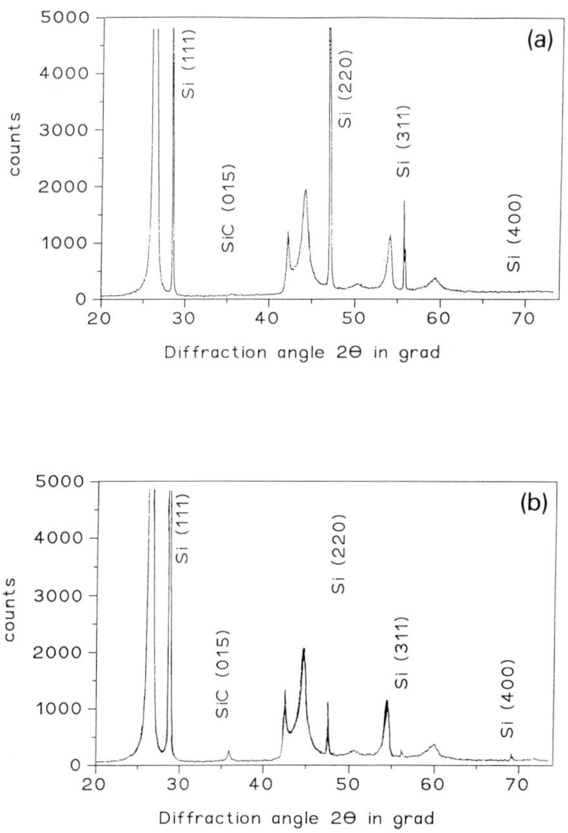

Fig. 5. X-ray diffraction pattern of a silicon film on a graphite substrate, recrystallized with the process parameters given in fig. 3.

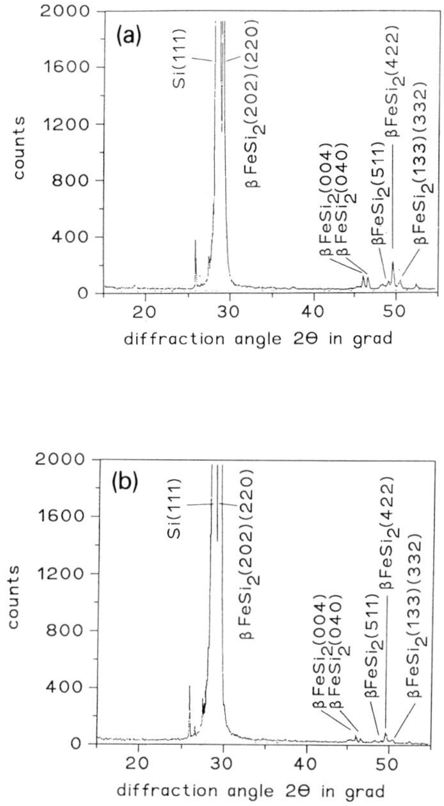

Fig. 6. X-ray diffraction pattern of a line electron beam heatened Fe + 2Si film deposited on a (111) oriented silicon substrate. Process parameters as depicted in fig. 4.

process time significantly (from 20 to 3.5 s for a 3″ graphite substrate). The recrystallization with $v = 3.7$ mm/s leads to a strong (111) preferential orientation of the recrystallized film. At the higher scan velocity ($v = 20$ mm/s) also the (110) orientation is represented strongly in the recrystallized film.

6. Hetero-epitaxial growth of β-$FeSi_2$

A 1 μm thick amorphous but stoichiometric film of the composition Fe + 2Si was sputtered onto a (111) oriented silicon substrate from a high purity $FeSi_2$ target. Before the deposition the silicon substrate was cleaned for two minutes in an argon plasma to remove the native oxide. It is known that hetero-epitaxial growth of $FeSi_2$ in the β-phase is enabled on (111) planes of silicon [10]. Fig. 6 shows the X-ray diffraction pattern of the processed Fe + 2Si films for the process parameters given in figs. 4a and 4b. Only the β-$FeSi_2$ phase is present and both films show a pronounced (202) and (220) orientation resulting from the hetero-epitaxial growth of the β-$FeSi_2$ on the (111) surface of the silicon substrate. The intensities of the other β-$FeSi_2$ reflexions are larger for the higher scan velocity (fig. 6b) compared to the lower one in fig. 6a.

7. Conclusions

Thermally activated thin film processing requires a thin film temperature which is situated in a temperature interval determined by the intended process. This criterion, however, does not take into account the process duration and the thermal gradients occurring in the films. Although the temperature criterion may be fulfilled the crystal growth conditions vary or parasitic processes occur. The dependence of the crystal growth conditions on the process parameters are the subject of further studies.

References

[1] L. Pfeiffer, K.W. West, S. Paine and D.C. Joy, Mater. Res. Soc. Symp. Proc. 35 (1985) 583.

[2] M. Haond, D.P. Vu and D. Bensahel, J. Appl. Phys. 54 (1983) 3892.

[3] V.E. Borissenko, V.V. Gribkovskii, V.A. Labunov, V.A. Samuilov and K.D. Yashin, Phys. Status Solidi (a)75 (1983) 117.

[4] M. Pauli, J. Müller, K. Hartkopf and T. Barth, Rev. Sci. Instrum. 63 (1992) 2288.

[5] Metals Handbook, 8th ed., Vol. 8, Ed. T. Lyman (American Society for Metals, Metals Park, OH, 1973).

[6] N. Cherief, C. D'Anterroches, R.C. Cinti, T.A. Nguyen Tan and J. Derrien, Appl. Phys. Lett. 55 (1989) 1671.

[7] S.S. Lau, J.S.-Y. Feng, J.O. Olowolcfe and M.-A. Nicolet, Thin Solid Films 25 (1975) 415.

[8] I. Nishida, Phys. Rev. B 7 (1973) 2710.

[9] M. Pauli, G. Dähn and J. Müller, Appl. Surf. Sci. 54 (1992) 386.

[10] H.C. Cheng, L.J. Chen and T.R. Your, Mater. Res. Soc. Symp. Proc. 25 (1984) 441.

Applied Surface Science 69 (1993) 403–406
North-Holland

applied
surface science

Real time emission spectroscopy during deposition of germanium oxide films by laser ablation

F. Vega, C.N. Afonso and J. Solis
Instituto de Optica, CSIC, Serrano 121, 28006 Madrid, Spain

Received 2 June 1992; accepted for publication 30 October 1992

The dynamics of the plume produced during excimer laser deposition of germanium oxide films by excimer laser ablation of a Ge target in the presence of an oxygen atmosphere is studied by real time emission spectroscopy. It is found that the velocity of the species in the plume (both neutral and ionized) remains unaffected under the presence of an oxygen pressure ranging from 10^{-4} up to 1 mbar, whereas the oxygen content of the films increases with the oxygen pressure up to the stoichiometric value (GeO_2) for pressures higher than 5×10^{-3} mbar. The spectral emission features show no evidence of oxidized species in the gas phase even at high oxygen pressures. It is then concluded that the incorporation of oxygen to the growing film occurs mainly by reactions at the substrate site.

1. Introduction

Laser ablation is nowadays a technique widely applied for the preparation of new materials in thin film form [1–3]. It has been found that good quality superconducting films with a composition very close to that of the target can be grown without post-annealing [4]. Many efforts have been done to correlate the final film properties to the dynamics of the species which are present in the plume [5]. The presence of oxidized species either directly ejected from the target surface or formed by reactions in the gas phase when using oxidized targets like YBaCuO, has been reported [6–8]. Their velocity was strongly dependent on the oxygen pressure applied during deposition [9] because of collisional processes with the foreign atoms.

Ge dioxide (GeO_2) is used for integrated optics and infrared optical applications [10], whereas substoichiometric oxide films (GeO_x, $x < 2$) are promising as phase change optical storage media [11]. In a previous work [12], we have shown that GeO_x films can be grown by ablation of a pure Ge target in an oxygen atmosphere. The dynamics of the plume in vacuum and in an oxygen atmosphere at pressures up to 10^{-3} mbar was analyzed by means of real time measurements of the light emitted by the plume without spectral resolution. The dynamics of the plume was found to be independent of the ambient. We have recently obtained stoichiometric GeO_2 films by ablation of Ge in an oxygen atmosphere at a pressure of 5×10^{-3} mbar [13].

The aim of this work is to correlate the properties of the laser deposited Ge oxide films to the dynamics of the species present in the plume. The light emitted by the plume is now analyzed by real time measurements with space and spectral resolution as a function of the oxygen pressure. The results show no evidence of oxidized species in the plume, which follows a collisionless expansion with the foreign atoms, even at high oxygen pressures. It is then concluded that the incorporation of oxygen to the films occurs mainly by reactions at the substrate site.

2. Experimental

The experimental set-up used for laser deposition of Ge oxide films has been described in detail elsewhere [12,13]. It includes an ArF excimer laser (193 nm, 12 ns) focused on a rotating Ge target (99,999%). The laser repetition rate is 10 Hz and the energy measured at the target surface is 0.03 J which yields to a fluence of 10 J/cm^2. The films are grown onto (100) Si substrates held at room temperature. The deposition process is carried out in an oxygen pressure ranging from 10^{-4} to 1 mbar. The stoichiometry of the as-deposited films was measured by simultaneous RBS and NRA measurements as reported in ref. [13].

The plume formed during the ablation process is imaged onto the entrance slit of a SPEX spectrometer (wavelength resolution of 0.05 nm) with a 4 × magnification. The wavelength interval between 200–340 nm where several emission bands of GeO molecules have been reported [14] was scanned, but no bands or lines which could be straightforwardly related to such molecules were found. We focused the present study on the 400–650 nm spectral range where several emission lines from Ge neutrals and ions are present [15]. The plume is scanned along the normal to the target surface with resolution of 160 μm. A photomultiplier (15 ns rise time) is connected to a boxcar averager for spectrum recording or to a 500 MHz transient digitizer for transient emission measurements. The time at which the laser pulse reaches its maximum is chosen as the temporal origin of the real time measurements.

3. Results

The emission spectrum of the plume formed during ablation of Ge consists of several lines superimposed over an intense continuum. The continuum emission intensity is very strong for distances to the target surface shorter than 0.5 mm and decreases abruptly as the spectrum is recorded further away from the target surface. Its contribution to the emission spectrum recorded in vacuum has been ruled out in fig. 1 by placing the boxcar gate around 75 ns after the laser pulse. Several emission lines which arise mainly from neutral (Ge I) and ionized atomic species (Ge II) emission are observed. There are some less intense lines which could not be attributed to any known Ge transition and may be related to the presence of excited clusters [16]. No extra emission lines are observed when the spectrum is recorded in the presence of an oxygen pressure.

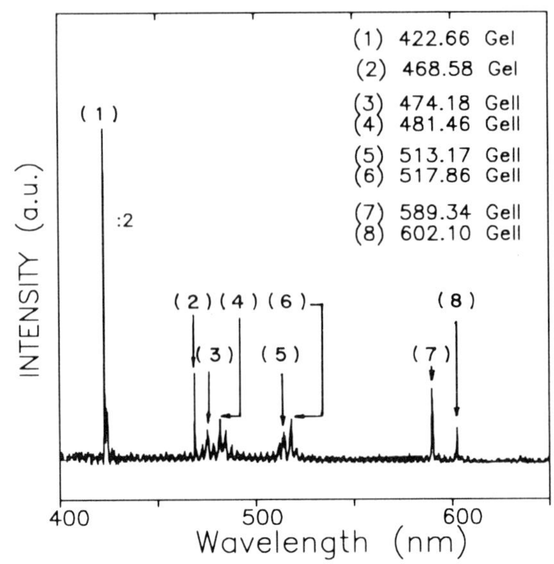

Fig. 1. Emission spectrum of the plume induced by laser ablation of a Ge target in vacuum. The identified lines come mainly from neutral (Ge I) and ionized (Ge II) species.

Fig. 2a is a plot of the transient emission from Ge I at $\lambda = 422.66$ nm for two differents distances to the target surface. It shows a sharp peak which broadens and occurs later as the emission is recorded away from the target surface, followed by a long decay tail. Fig. 2b shows the transient emission from Ge II at $\lambda = 589.34$ nm. Its emission consists of a sharp peak shortly delayed with respect to the laser pulse and a less intense, broader, and longer delayed peak. The intensity maximum also occurs later and broadens, as the emission is recorded along the plume axis and away from the target surface.

The most probable velocity of the species moving along the normal to the target surface is determined from the delay of the emission maximum and the distance along the plume axis at which the emission is recorded. All the Ge II lines show the same behaviour leading to a velocity of 1.7×10^6 cm/s which is in the order of the velocity typically reported for ionized species in laser ablation experiments [17,18]. The velocity obtained from the emission at $\lambda = 422.66$ nm (Ge I) shows a "fast" component with a value of 2.9×10^6 cm/s at distances to the target surface shorter than 0.75 mm, and a "slow" one, with a value of 6.2×10^5 cm/s at larger distances. The velocity obtained from the emission at $\lambda = 468.58$ nm (also Ge I) exhibits only the slow component.

The velocity of the species in the plume as a function of the oxygen pressure for Ge I ($\lambda =$ 422.66 nm) and Ge II ($\lambda =$ 589.34 nm) is plotted in fig. 3a. It is seen that the velocity of the emitting species both neutrals (Ge I) and ionized (Ge II) remains unaffected for oxygen pressures up to 1 mbar. In fig. 3b, the oxygen content of the as-deposited films as a function of the oxygen pressure used during the deposition process is shown. It can be seen that stoichiometric GeO_2 films are obtained at pressures higher than 5×10^{-3} mbar. The deposition rate remains constant up to 1 mbar of oxygen pressure above which an abrupt decrease is observed related to the reduction of the species velocity (fig. 3a) due to hydrodynamic effects [19].

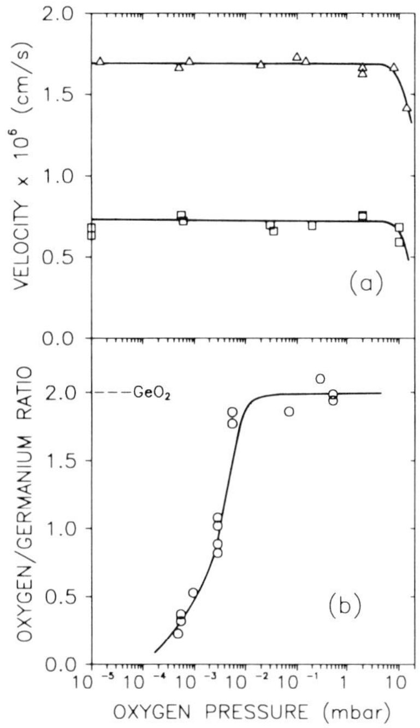

Fig. 3. Velocity of the Ge I (–□–□–□–) and Ge II (–△–△–△–) in the Ge plume (a), and oxygen to Ge ratio in the grown films (b), both as a function of the oxygen pressure applied during the deposition process.

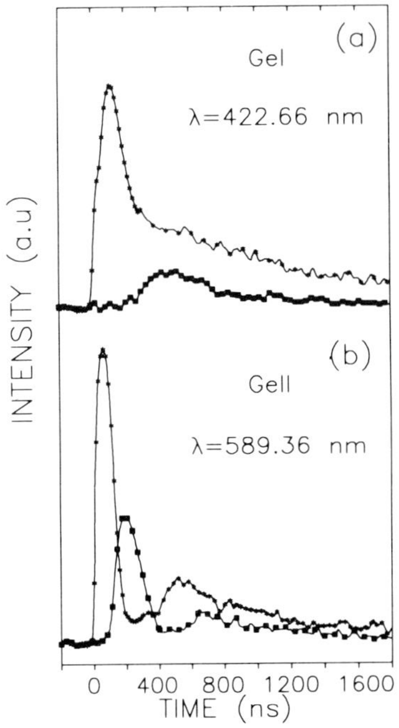

Fig. 2. Real time emission of Ge I ($\lambda = 422.66$ nm) and Ge II ($\lambda = 589.36$ nm) at two different distances to the target surface: (–*–*–*–) 0.5 mm and (–■–■–■–) 3.0 mm.

4. Discussion

It has been shown that the Ge plume emission consists of several lines coming from electronic excitational emission of neutral (Ge I) and ionized (Ge II) species, superimposed to a continuum due to free–free and free–bound electronic recombination [20]. No special spectral features or species velocity changes are found when the experiments are performed in an oxygen background pressure up to 1 mbar. In contrast, the oxygen content of the films increases with the oxygen pressure, reaching the stoichiometric GeO_2 value for pressures higher than 5×10^{-3} mbar.

Previous results on time resolved emission spectroscopy studies of the plume created above an YBaCuO target have detected the presence of oxide compounds in the plume [7,8]. It was also found that the velocity of the plume species both atomic and diatomic, was very sensitive to the oxygen pressure in the chamber [9]. Our results show that no new emitting species are present in the plume when the deposition process is performed in an oxygen atmosphere. It can be argued that oxide species may be present in the plume but they are either in the ground state or they decay through a non-radiative process. They cannot then be detected by optical emission spectroscopy. However the lack of influence of the oxygen pressure on the velocity of any of the emitting species in the plume indicates that collisions of the plume species with the foreign atoms or molecules should be practically absent. So, it is concluded that the oxidation reaction takes place mainly at the substrate.

In a separate work [13], we have studied the optical properties of laser deposited GeO_x films by means of spectroscopy ellipsometry. The use of the effective medium modelling to simulate the ellipsometric parameters of the films have shown that for films with a poor oxygen content ($x < 1$), the oxygen content was not homogeneous in depth. The presence of depth compositional changes was associated to in-diffusion of oxygen atoms that reach the surface of the growing film [13]. This result further supports our conclusion since the existence of preferential oxide formation in the plume should originate a deposit with no depth compositional inhomogeneity. Thus, the oxygen content of the films is mainly determined by the flux ratio of oxygen to Ge atoms impinging the substrate surface.

5. Conclusions

Stoichiometric GeO_2 films have been grown by laser ablation in an oxygen atmosphere at pressures higher than 5×10^{-3} mbar. The velocity of the species in the laser induced plume remains unaffected for oxygen pressures up to 1 mbar. The lack of influence of the oxygen pressure on the plume species dynamics allow us to conclude that the oxidation reaction should take place mainly at the substrate. The oxygen content of the films is therefore determined by the flux ratio of oxygen to Ge atoms impinging the substrate surface.

Acknowledgment

This work was partially supported by CYCYT (Spain) under TIC-90 program.

References

[1] E. Jacobsohn, J. Zahavi, A. Rosen and S. Nadiv, J. Mater. Sci. 26 (1991) 1861.
[2] A. Kumar and J. Narayan, Appl. Phys. Lett. 59 (1991) 785.
[3] A. Slaoui, E. Fogarassy, C. Fuchs and P. Siffert, J. Appl. Phys. 71 (1992) 590.
[4] A. Inam, M.S. Hedge, X.D. Wu, T. Venkatesan, P. England, P. Miceli, E.W. Chase, C.C. Chang, J.M. Tarascon and J.B. Wachtman, Appl. Phys. Lett. 53 (1988) 908.
[5] H. Sankur and J.T. Cheung, Appl. Phys. A 47 (1988) 271.
[6] D. Fried, G.P. Reck, T. Kushida and E.W. Rothe, J. Appl. Phys. 70 (1991) 2337.
[7] C. Girault, D. Damiani, C. Champeaux, P. Marchet, C. Germain, J.P. Mercurio, J. Aubreton and A. Catherinot, Appl. Surf. Sci. 46 (1990) 78.
[8] T. Venkatesan, X.D. Wu, A. Inam, Y. Jeon, M. Croft, E.W Chase, C.C. Chang, J.B. Wachtman, R.W. Odom, F. Radicati and C.A. Magee, Appl. Phys. Lett. 53 (1988) 1431.
[9] C. Girault, D. Damiani, J. Aubreton and A. Catherinot, Appl. Phys. Lett. 54 (1989) 2035.
[10] P. Klocek, Mater. Res. Soc. Bull. 11 (1986) 41.
[11] T. Ohta, M. Takenaga, N. Akahira and T. Yamashita, J. Appl. Phys. 53 (1982) 8497.
[12] C.N. Afonso, F. Vega, J. Solis, F. Catalina, C. Ortega and J. Siejka, Appl. Surf. Sci. 54 (1992) 175.
[13] F. Vega, J.C. Sande, C.N. Afonso, C. Ortega and J. Siejka, Appl. Opt., in press.
[14] R.W.B. Pearse and A.G. Gaydon, The Identification of Molecular Spectra (Chapman and Hall, London, 1976).
[15] J. Reader, C. Corliss, W.L. Wiese and G.A. Martin, Wavelengths and Transitions Probabilities for Atoms and Atomic Ions, NSRDS December 1980.
[16] A. Kasuya and Y. Nishina, Mater. Res. Soc. Symp. Proc. 191 (1990) 73.
[17] K.L. Saenger, J. Appl. Phys. 66 (1989) 4435.
[18] R.M. Gilgenbach and P.L.G. Ventzek, Appl. Phys. Lett. 58 (1991) 1597.
[19] W. Marine, M. Gerry, J.M. Scotto d'Aniello, M. Sentis, P. Delaporte, B. Forestier and B. Fontaine, Appl. Surf. Sci. 54 (1992) 264.
[20] H.R. Griem, Mater. Sci. Eng. A 139 (1991) 1.

Applied Surface Science 69 (1993) 407–411
North-Holland

applied
surface science

A new encapsulation method of InP during post implantation annealing

A. Kadoun [a], J. Tardy [b], I. Thomas [b], M. Gendry [b], V. Drouot [b], G. Bremond [a], D. Barbier [a] and A. Laugier [a]

[a] *Laboratoire de Physique de la Matière, URA CNRS 358, INSA-Lyon, 69621 Villeurbanne Cedex, France*
[b] *Laboratoire d'Electronique, URA CNRS 848, Ecole Centrale de Lyon, 69131 Ecully Cedex, France*

Received 2 June 1992; accepted for publication 30 October 1992

This paper describes a new encapsulation method of the surface of InP during post implantation annealing with a 1 μm thick InGaAs layer grown by molecular beam epitaxy. Secondary ion mass spectrometry microanalysis shows that the InGaAs/InP interface remains very stable upon rapid thermal annealing. However, capacitance–voltage measurements on MIS Schottky diodes reveal a wide depletion region in the underlying InP due to donor compensation. The proposed process has been successfully integrated in our technology of InP MISFET.

1. Introduction

The protection of InP during thermal processes is of a major concern for the technology, especially when high temperatures are required. This problem has been adressed for years and a great variety of methods were tested more or less successfully. One of them, widely used, consists in covering the sample by an insulating film such as AlN, Al_2O_3, Si_3N_4 or SiO_2. However, the heating treatment induces stresses at the interface which are detrimental for electrical characteristics of the substrate [1,2]. Moreover, substrate contamination by the cap atoms [3] and phosphorus outdiffusion [4] have been observed.

In this paper, we report a new method of encapsulation of InP during the high temperature post implantation annealing. This method consists in using a one micrometer thick lattice matched InGaAs epitaxial layer grown by molecular beam epitaxy (MBE) and takes advantage of the absence of stress at the interface, due to InGaAs matching to InP substrate surface, and of the fact that the InGaAs cap can, at the same time, act as an implantation mask for the transistor channel area.

The following study is focused on physical effects induced by rapid thermal annealing (RTA), in the context of this particular capping technique.

Processed annealed and unannealed samples have been characterized by means of optical (Nomarski type) microscopy, secondary ion mass spectrometry (SIMS) and capacitance–voltage measurements (C–V).

2. Experimental

The annealing treatments were carried out in a commercial Addax R1000 System wherein heating is performed using a 30 kW bank of tungsten–halogen lamps (maximum power density on a 4 in. diameter silicon wafer = 25 W/cm^2) [5]. The typical annealing cycle used is shown in fig. 1. During annealing, the sample was sandwiched front side down between two silicon wafers. The temperature was monitored by a thermocouple embedded in a 4 in. diameter silicon wafer which was used as holder for the sandwich as shown in the fig. 1 insert. RTA was performed at 750°C for 15 s under a high purity Ar gas flow. However,

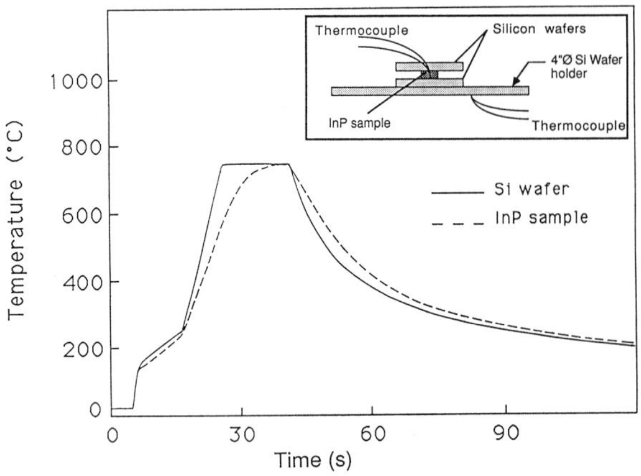

Fig. 1. Typical annealing cycle of the InP specimen compared to that of the 4 in. Si wafer holder using the experimental sandwich arrangement shown in the insert.

owing to limited heat transfer by thermal conduction, it can be seen in fig. 1 that the temperature annealing cycle measured by a thermocouple embedded in a witness InP sample, shows an effective plateau duration of only about 5 s for a programmed 15 s duration. With similar time and temperature conditions, the literature-reported electrical activations of implanted Si in InP, with doses close to that employed in this work, vary over a wide range, and can reach 60% [6,7], depending essentially on the annealing procedure.

The starting material was Fe-doped semi-insulating (SI) (100) oriented InP substrate for SIMS profiling and a non-intentionally doped (NID) (n-type) InP substrate, for electrical measurements. The InP surfaces were initially cleaned in sequential boiling baths of trichloroethylene, acetone, and methanol, followed by a native oxide etching in dilute HF, rinsed in deionized H_2O, and finally dried in a N_2 flow. A 1 μm thick lattice matched InGaAs layer ($\Delta a/a < 5 \times 10^{-4}$) was grown on the substrate by MBE at 545°C.

Si^+ ions were then implanted at room temperature (RT) into the InGaAs/(SI) InP structure placed in a non-channeling orientation, at an energy of 150 keV and a dose of 3×10^{14} cm^{-2}. Afterwards, the sample has been cleaved into two pieces; one of them has been annealed in the conditions described above, the other one being kept as a reference unannealed sample. SIMS atomic profiles were obtained by means of a Cameca IMS 4f system using an O_2^+ primary beam.

For electrical measurements, InGaAs-covered and etched specimens, both taken from the NID InP substrate, were annealed in the lamp furnace using the above described sandwich configuration (see fig. 1). These specimens will be referred to in the following as InP/IGA/RTA and InP/RTA, respectively. Unannealed specimens, either capped with InGaAs or not, were used as references and will be referred to in the following as InP/IGA and InP/Virgin, respectively. Afterwards, the InGaAs layers were chemically removed from the InP/IGA/RTA and InP/IGA samples in order to allow diode realization onto the InP substrate. To perform (MIS) tunnel Schottky devices, ohmic contacts were achieved on the back side of the substrate by evaporating Au–Ge/Au in vacuum, followed by annealing at 380°C for 10 min. The sample surface was then slightly etched in a phosphoric acid solution. An interfacial oxide 30 Å thick was grown on the etched and oxide free surface by using UV assisted oxidation. Diodes gates were then delineated by Au evaporation and lift off.

3. Results

Fig. 2 shows the depth-distibution of the different constituants of an InGaAs/(SI) InP im-

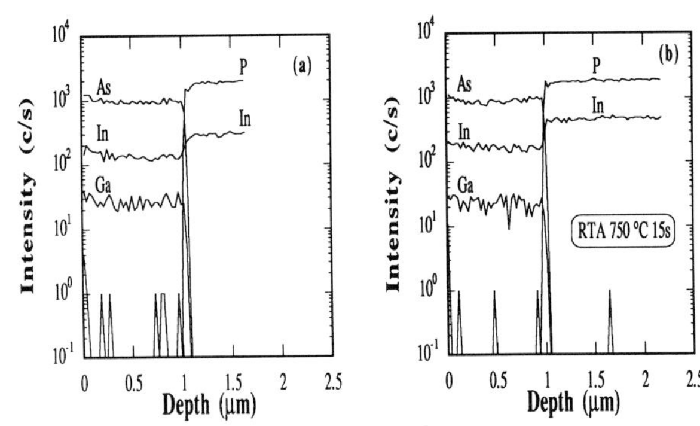

Fig. 2. SIMS profiles of the chemical constituants of the InGaAs/(SI) InP structure RT-implanted with Si^+ ions: (a) unannealed; (b) annealed (750°C/15 s).

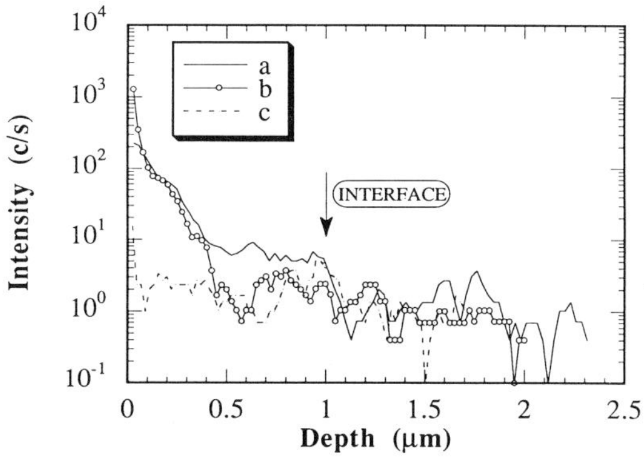

Fig. 3. SIMS profiles of RT-implanted Si in an InGaAs/(SI) InP structure (150 keV, 3×10^{14} atoms cm^{-2}): (a) as implanted; (b) after implantation and RTA at 750°C for 15 s. The dashed line (c) corresponds to the Si signal of a non-implanted and annealed stucture.

planted structure before (a) and after annealing (b). All species are shown except Si and Fe which profiles will be shown separately. After annealing, the interface obviously remains stable in depth, indicating that no interdiffusion occurred at the InGaAs/InP boundary.

The Si profile in an implanted InGaAs/InP structure is shown in fig. 3. In the InP substrate close to the interface, the weak Si SIMS signal level is buried into the Si background noise of a reference non-implanted and annealed InGaAs/InP structure also shown in the same figure. This points out the expected InGaAs cap efficiency against Si implantation and RTA, since no significant Si concentration is found inside the InP substrate beneath the cap layer before and after annealing. In fig. 4, Fe is shown accumulating in the surface region after penetrating through the InGaAs/InP interface and depleting in the first 500 nm from the interface, the Fe pile-up being more marked in the case of the annealed sample. Although less pronounced Fe pile-up is observed near the surface in the unannealed sample, it must be noted that thermal processing such as MBE, which is usually achieved for a long duration (1 h in the present case) at a temperature in the 500–570°C range, might be sufficient to cause the observed Fe migration and the surface accumulation into the InGaAs layer, which was several times reported in bare (SI) InP [8–10].

In fig. 4, the Fe distribution in a non-implanted and annealed InGaAs/InP structure is also shown. Compared to the Fe profiles in the implanted annealed or unannealed sample, a small surface accumulation seems to appear within about 10 nm and no Fe depletion in the vicinity of the interface is observed. Anyhow, the existence of the Fe surface pile-up in a non-implanted and annealed specimen may be doubtful, as far as it is confined precisely within a 10 nm surface layer, wherein the reliability of SIMS data is questionable, as previously noticed [8].

After RTA and InGaAs removal, the InP surface inspected by optical microscopy appeared perfectly smooth with no phosphorus loss-related degradation. However, free carrier profiling reported in fig. 5, and deduced from capacitance–voltage measurements showed a compensating effect extending a few tenths of a micrometer beneath the surface. This phenomenum is slightly deeper in the case of the InP/IGA/RTA sample (curve d) than in the InP/RTA specimen (curve c), and not observed in the InP/Virgin or in the InP/IGA samples (curves a and b, respectively). It is important to note that the depth of the compensated layer shown in fig. 5 is overestimated, due to errors introduced in the capacitance measurements by the presence of a compensated-insulating surface region. Moreover, a slight decrease of the carrier concentration level

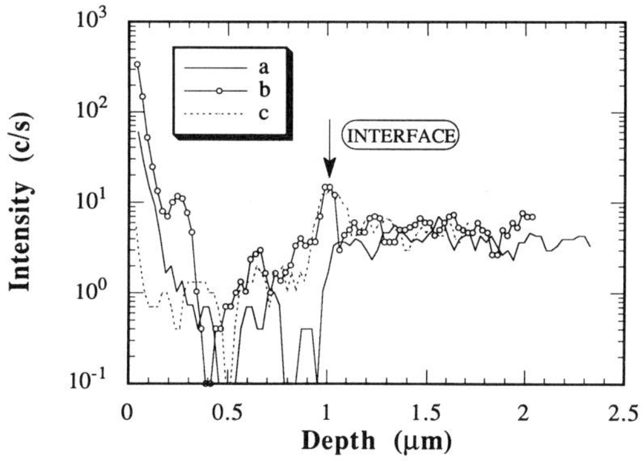

Fig. 4. SIMS Fe profiles in an InGaAs/(SI) InP structure, RT-implanted with Si^{+} ions (150 keV, 3×10^{14} atoms cm^{-2}): (a) as implanted; (b) after implantation and RTA at 750°C for 15 s; (c) non-implanted and annealed structure.

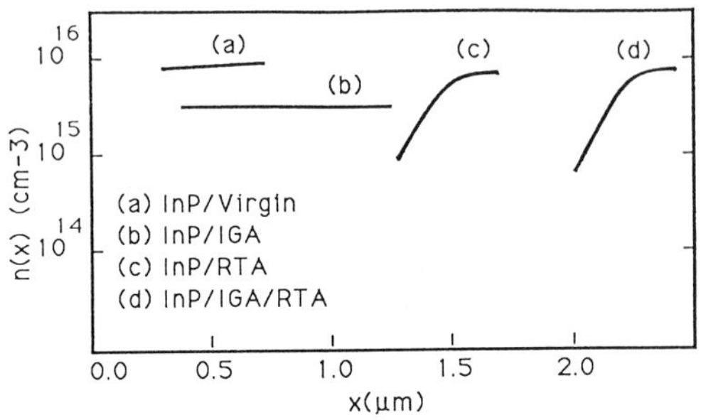

Fig. 5. Free carrier profiles in the NID InP substrate determined by $C-V$ measurements on (MIS) tunnel Schottky diodes: (a) bare InP, unannealed; (b) InGaAs layer removed, unannealed; (c) InGaAs layer removed + RTA (750°C/15 s); (d) InGaAs layer not removed + RTA (750°C/15 s).

can be noticed after InGaAs capping (curve b), compared to that of the virgin substrate (curve a).

4. Concluding remarks

In this paper a new method for the encapsulation of InP has been investigated, which provides a very effective and reproducible protection against phosphorous loss during post implantation RTA.

From the depth distribution of In, P, Ga, and As atoms (fig. 2), it was observed that neither cap species indiffusion into the InP crystal, nor substrate host atom outdiffusion into the cap occurred after annealing. It has also been verified that a thickness of 1 μm of the InGaAs material is quite sufficient to retain the Si implant during annealing, and therefore constitute an excellent dopant mask for the channel area of the transistors. However, it is clear that Fe depth-redistribution occurred when the sample was submitted to a thermal process such as MBE and/or annealing. Since the surface pile-up of Fe is much more weaker in non-implanted samples, one may conclude that Fe redistribution and surface accumulation is likely enhanced by residual post-implantation annealing damage. This latter assumption is consistent with previous experiments, using hot implantation of Si in (SI) InP [8], to avoid the formation of an amorphous layer. However, the Fe redistribution is confined within the InGaAs layer which is anyhow removed after annealing for the device realisation. Furthermore, the initial Fe concentration in the InP substrate seems not to be affected by this Fe outdiffusion.

Moreover, despite the mirror-like aspect exhibited by the InP surface, defect formation occurs during annealing as already discussed [11]. Indeed, the cap layer obviously did not prevent the formation of a wide depletion region, due to donor compensation, which appears during annealing, in the vicinity of the NID InP surface. Work is in progress to investigate the effects of RTA on the electrical properties of NID or (SI) InP substrates.

Finally, it has been shown that the Fe-redistribution and carrier compensation effects observed in this work are certainly not strong enough to affect MISFET device performances. Indeed, the described annealing procedure has been successfully integrated in the process of fabrication of an InP MISFET, which showed excellent static characteristics [12].

Acknowledgements

Thanks are due to J. Dupin and M. Monchanin for ion implantations and to C. Dubois for SIMS profiles.

References

[1] L.S. Vanasupa, M.D. Deal and J.D. Plumner, Appl. Phys. Lett. 55 (1989) 274.
[2] G.J. Valco and V.J. Kapoor, J. Electrochem. Soc. 134 (1987) 569.
[3] C.W. Farley, T.S. Kim, B.G. Streetman, T. Lareau and P. Williams, Thin Solid Films 146 (1986) 221.
[4] C.W. Farley and B.G. Streetman, J. Electrochem. Soc. 134 (1987) 498.
[5] D. Barbier, M. Remram, J.F. Joly and A. Laugier, J. Appl. Phys. 61 (1987) 156.
[6] Jesus A. del Alamo and Takashi Mizutani, J. Appl. Phys. 62 (1987) 3456.
[7] J.P. Lorenzo, D.E. Davies, K.J. Soda, T.G. Ryan and P.J. McNally, Mater. Res. Soc. Symp. Proc. 13 (1983) 683.

[8] G. Bahir and J.L. Merz, J. Appl. Phys. 65 (1989) 1009.
[9] N. Duhamel, P. Krauz and Y. Gao, J. Appl. Phys. 66 (1989) 1855.
[10] H. Kamada, S. Shinoyama and A. Katsui, J. Appl. Phys. 55 (1984) 2881.
[11] B. Sartorius, M. Schlak, M. Rosenzweig and K. Pärschke, J. Appl. Phys. 63 (1988) 4677.
[12] J. Tardy, I. Thomas, P. Viktorovitch, V. Drouot, B. Agius, F. Plais, J.P. Dupin and D. Barbier, Conf. Proc. 4th Int. Conf. Indium Phosphide and Related Materials, Newport, USA, 1992, p. 306.

Applied Surface Science 69 (1993) 412–417
North-Holland

applied
surface science

Laser direct writing of titanium silicide thin films

G. Reisse, F. Gänsicke, A. Fischer
Ingenieurhochschule Mittweida, Technikumplatz 17, P.O. Box 91, D-9250 Mittweida, Germany

and

H. Johansen
Max-Planck-Institut für Mikrostrukturphysik, Weinberg 2, D-4050 Halle, Germany

Received 2 June 1992; accepted for publication 30 October 1992

Investigations concerning the laser-induced chemical vapour deposition of $TiSi_2$ thin films from SiH_4 and $TiCl_4$ using a direct writing method are presented. A CO_2-laser at $\lambda = 10.6$ μm and an argon ion laser in the VIS (488–527 nm, multiline mode) were used for the preferably thermal deposition on SiO_2 and c-Si substrates. The influence of the deposition parameters gas composition, laser power, scan velocity and laser wavelength on structural, geometric, and electrical properties is reported.

1. Introduction

Titanium silicide $TiSi_2$ is because of its low film resistivity of about 25 $\mu\Omega \cdot$cm a promising alternative to the doped p-Si in gate electrodes and metallization schemes in integrated circuits as well as in micromechanical and micro-optical devices [1]. The direct writing method as variant of laser-induced chemical vapour deposition offers the advantage of depositing thin film structures with small lateral dimensions without photomasks [2]. The availability of different laser wavelengths allows the use of a variety of source gases on different substrates depending on their optical properties.

We investigated the laser-induced chemical vapour deposition of $TiSi_2$ thin films on silicon and silicon dioxide substrates using a CO_2-laser and an argon ion laser. Typical CVD gases like silane SiH_4 as silicon source and H_2, N_2, or Ar as carrier gases were used in combination with titanium tetrachloride $TiCl_4$ as titanium source. Differences between the experimental results in the visible and in the far infrared wavelength range are analyzed and reported here.

2. Experimental

The direct writing experiments were carried out using a stainless steel reaction chamber mounted on a computer-controlled translation stage. The deposition arrangement can be operated in a flow regime. The working pressure was maintained by an oil-sealed fore pump in combination with an oil diffusion pump. The silane flow was measured and adjusted by a flow meter with needle valve. The titanium tetrachloride which is a liquid with a vapour pressure of about 1.33 kPa at room temperature (20°C) was stored in a gas bubbler near the reaction chamber. The $TiCl_4 \cdot H_2/SiH_4$ ratio in the reaction chamber can be varied by controlling the SiH_4 gas flow and the H_2 gas flow through the bubbler. Thereby, the $TiCl_4$ content has a non-linear dependence on the H_2 flow.

The laser beam of a 200 W CW-CO_2-laser was focussed onto the surface of a SiO_2 substrate resulting in a spot with a Gaussian radius of $r_G = 500$ μm. A 3 W argon ion laser ILA 120-1 was used for the deposition on c-Si and oxidized c-Si substrates. Three exchangeable objectives producing different spots of $r_G = 4$, 7 and 15 μm were applied. More details concerning the experimental apparatus can be found in ref. [3].

3. CO_2-laser-induced deposition of $TiSi_2$

The deposition of $TiSi_2$ on SiO_2 substrates was carried out using the CO_2-laser on SiO_2 substrates because of the high absorptivity of the substrate at this wavelength. The total pressure of the reactant gas mixture of $TiCl_4$, SiH_4 and H_2 as carrier gas for the $TiCl_4$ molecules was held constant at a value of 1.2 kPa. The gas flow ratio between the $TiCl_4 \cdot H_2$ flow and the SiH_4 flow has been varied between 0.2 and 0.6 in order to study the influence on the film stoichiometry. An increase in the $TiCl_4 \cdot H_2/SiH_4$ ratio enhances the Ti content in the films (fig. 1a). This tendency is related to a negative side effect which was observed as an increasing substrate etching caused by the rise of the concentration of chlorine and the lower chlorides of the titanium within and near the deposition zone. The films grow into the substrate material. A RBS depth profile (fig. 1b) shows that the Ti content increases with film thickness. We believe that this effect is due to the inhibited nucleation of $TiSi_2$ on clean SiO_2 surfaces [4]. Therefore, a first layer of p-Si from SiH_4 is created, followed by an increasing Ti and $TiSi_2$ incorporation into the growing film.

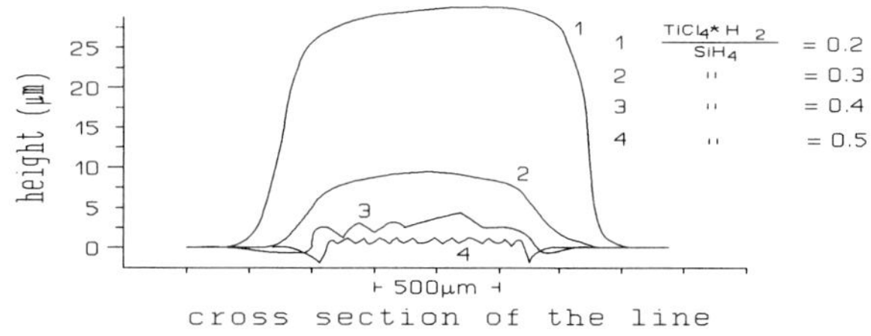

Fig. 2. Geometric line profile of $TiSi_x$ lines at different $TiCl_4 \cdot H_2/SiH_4$ gas flow ratios ($p_{tot} = 1.2$ kPa, $v_s = 0.05$ mm/s, $I = 955$ W/cm^2, $T_s = 300$ K).

The dependence of the geometric line profile on the $TiCl_4 \cdot H_2/SiH_4$ flow ratio exhibits the strengthening of the etching behaviour with the increasing $TiCl_4 \cdot H_2$ content within the gas phase (fig. 2) mentioned above.

The film deposition process is self-limiting in thickness in the centre of the lines if the $TiCl_4 \cdot H_2/SiH_4$ ratio of 0.6–0.7 is reached. At these values of the $TiCl_4 \cdot H_2/SiH_4$ ratio the lowest specific film resistance of 160 μΩ cm was measured. Lowering the flow ratio leads to an increase of the film resistivity because of the rising silicon excess in the films. At flow ratios greater than 0.7 the films become more inhomogeneous

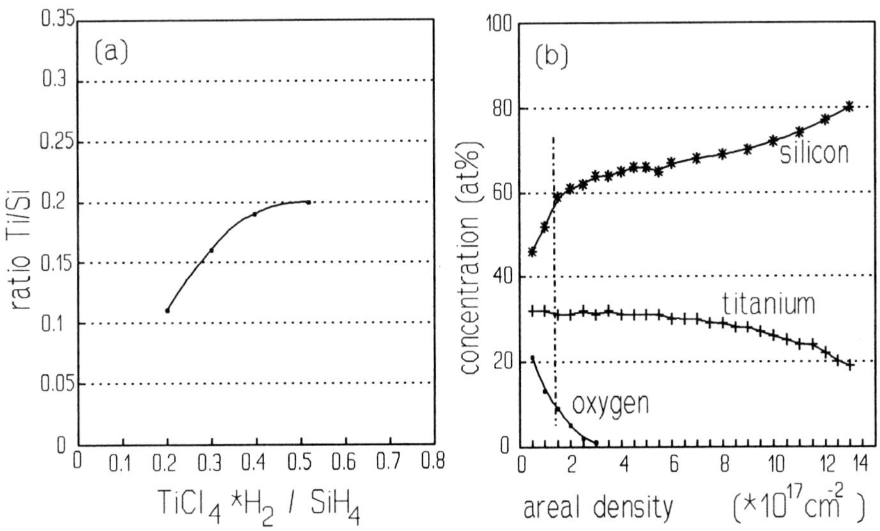

Fig. 1. (a) Dependence of the Ti/Si ratio within the film (EDX analysis) on the $TiCl_4 \cdot H_2/SiH_4$ gas flow ratio during the experiment ($p_{tot} = 1.2$ kPa, $v_s = 0.05$ mm/s, $I = 955$ W/cm^2, $T_s = 300$ K). (b) RBS analysis of a $TiSi_x$ line ($p_{tot} = 1.2$ kPa, $v_s = 0.05$ mm/s, $I = 955$ W/cm^2, $T_s = 300$ K, flow ratio $TiCl_4 \cdot H_2/SiH_4 = 0.5$).

due to local etching and the incorporation of lower chlorides of titanium and silane. Therefore, the film resistivity increases again.

A variation of the scan speed within the range of 0.02–0.2 mm/s showed no influence on the electrical and structural film properties because of the large illumination time of 50–5 s at a spot diameter of about 1 mm. The deposition rate decreased with increasing illumination time, like shown in fig. 3. We suppose that this behaviour is related to the varying temperature within the reaction zone during the film growth. At the beginning of the deposition a high absorption and low thermal conductivity of the SiO_2 substrate lead to a rapid heating of the surface within the illuminated region. After the deposition of the first Si-rich layers the reflectivity increases and the heat conductivity of the growing film, which is

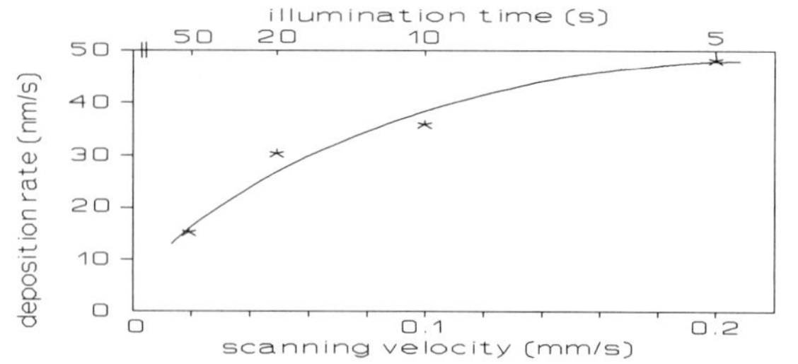

Fig. 3. Dependence of the deposition rate on the local illumination time and the scan velocity (p_{tot} = 1.2 kPa, I = 480 W/cm^2, T_s = 300 K).

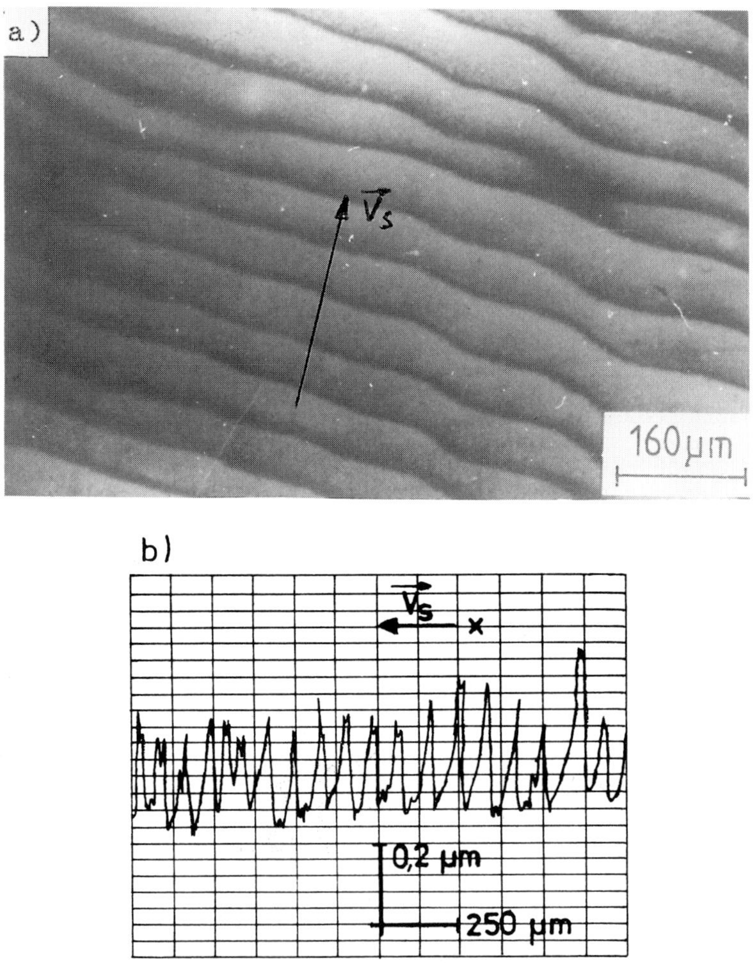

Fig. 4. (a) Optical micrograph of $TiSi_x$ films on SiO_2 substrates (p_{tot} = 1.2 kPa, v_s = 0.1 mm/s, T_s = 300 K, I = 955 W/cm^2). (b) Talystep analysis of periodic structures (parameters see fig. 4a).

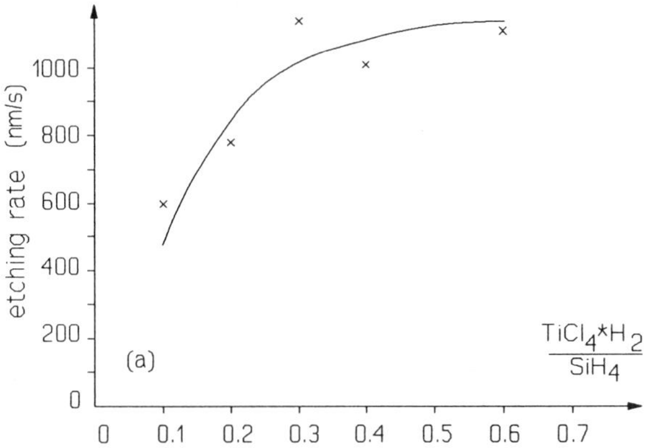

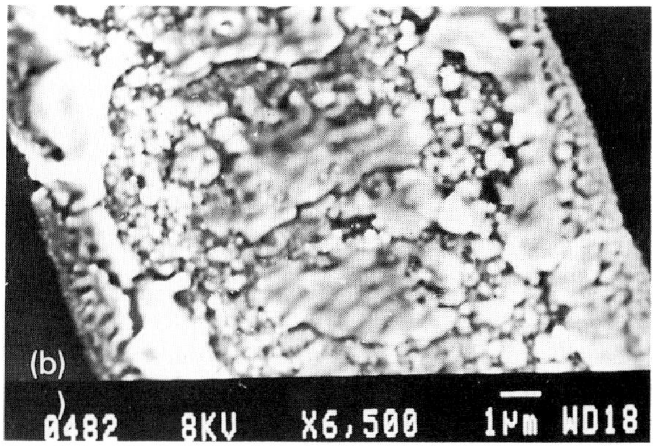

Fig. 5. (a) Etching rate as a function of the $TiCl_4 \cdot H_2/SiH_4$ flow ratio ($p_{tot} = 6.6$ kPa, $P_L = 2.25$ W, $r_G = 15$ μm, $v_s = 0.05$ mm/s, $T_s = 300$ K). (b) SEM micrograph of a $TiSi_x$ line (parameters see fig. 5a).

greater than that of the substrate, results in a lower surface temperature, therefore, the local deposition rate decreases until an equilibrium is reached.

The immediate vicinity of the $TiSi_x$ lines is contaminated with particles. These particles are probably produced by a homogeneous gas-phase reaction initiated by the absorption of the CO_2-laser radiation by the silane molecule. The films exhibit characteristic periodic structures along the scan direction. Figs. 4a and 4b show an optical micrograph and a Talystep analysis. The periodic structures depend neither on the scan step nor on a possible laser instability. We do not know the reason of this effect. We suppose that a permanent change of surface reflectivity and heat conductivity, in combination with the growth process, causes this phenomenon.

4. Argon ion laser-induced deposition of $TiSi_2$

First deposition runs were carried out using H_2 as carrier gas for the $TiCl_4$ transport into the

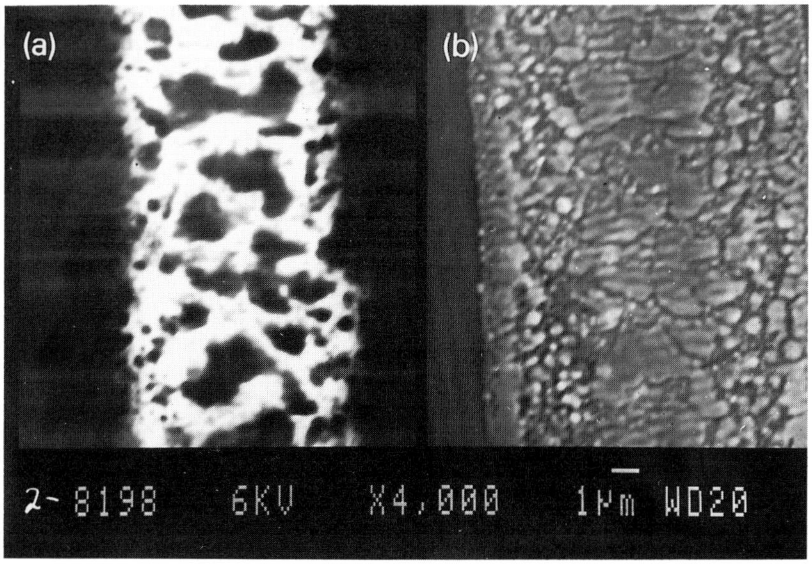

Fig. 6. (a) EBIC micrograph of $TiSi_x$ ($p_{tot} = 6.6$ kPa, $v_s = 0.05$ mm/s, $P_L = 2.25$ W, $r_G = 15$ μm, $T_s = 300$ K). (b) SEM micrograph (parameters see fig. 6a).

reaction chamber. In combination with the small spot size of about 30 μm and smaller, within the range of scan velocities of 0.02–0.2 mm/s, this carrier gas generally led to a film growth into the substrate material which was n-doped c-Si(111) in these cases, because of its absorptivity of about 63% at the argon ion laser wavelengths (VIS). The direct writing process is dominated by an etch reaction through the oxide layer into the c-Si. A film growth was observed only in the centre of the etching trench. The etching rate which was defined as the distance between the surface and the deepest point in the centre of the line depends on the $TiCl_4 \cdot H_2/SiH_4$ flow ratio (fig. 5a).

Films deposited in this manner show high film resistivities. The minimum value measured was 30 m$\Omega \cdot$cm. This poor electrical conductivity is caused by the inhomogeneous and island-like film growth. The short illumination time of $t_I \leq 0.6$ s does not allow an uniform nucleation and film growth. The EBIC micrograph (fig. 6a) shows the distribution of conducting regions and points where the passivation layer (SiO_2) was penetrated. This electrically active zone (fig. 6a) is small compared with the total line width (fig. 6b). In order to produce uniform films the local illumination time was extended by scanning the same line two times or more in the reaction gas environment.

Fig. 7 shows a Talystep analysis perpendicular to the scan direction for $TiSi_x$ films on oxidized c-Si substrates. The importance of the illumination time for the nucleation and the film growth could be demonstrated by the growth of spots. At illumination times shorter than 0.5 s in the laser power range of about 0.4–1.5 W at a Gaussian radius of $r_G = 4$ μm no film deposition has been observed. Fig. 8 shows the dependence of the film height on the illumination time and the pressure. Films deposited at the higher pressure of 2.6 kPa have a greater Si excess (88 at%, EDS) than films deposited at 1.3 kPa (75 at%, EDS). Increasing the illumination time for the growth of spots led to the formation of rods with lengths of some millimeters. A horizontal thickness modulation can be reached by varying the focal point in the z-direction (along the laser beam axis). Further experiments were done using the $TiCl_4$ vapour at the equilibrium vapour pressure (1.33 Pa). The total pressure in the reaction chamber was varied by changing the silane flow. The film resistivity could be lowered to a minimum value of 130 $\mu\Omega \cdot$cm ($p_{tot} = 2.6$ kPa, $v_s = 0.05$ mm/s, $T_s = 300$ K, $P_L = 0.9$ W, $r_G = 4$ μm, the line was scanned four time in the gas atmosphere).

All films are silicon rich (73–90 at%). Deposited $TiSi_x$ lines exhibit periodic structures within the low power range ($P_L < 1.7$ W at $r_G = 4$ μm) (fig. 9).

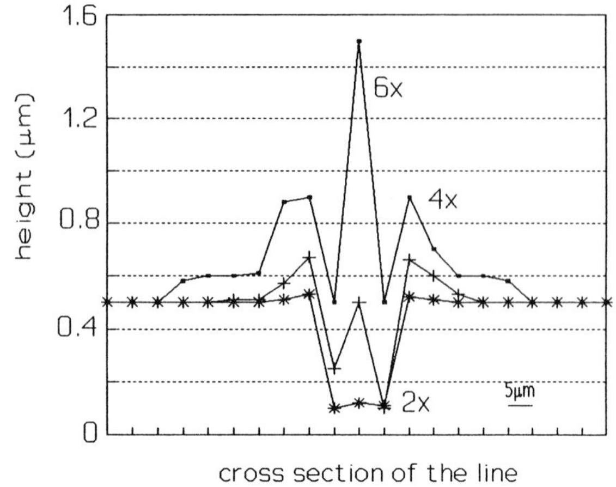

Fig. 7. Talystep analysis of $TiSi_x$. The laser was scanned two, four and six times over the same line ($p_{tot} = 3.3$ kPa, $v_s = 0.02$ mm/s, $P_L = 1.1$ W, $r_G = 4$ μm).

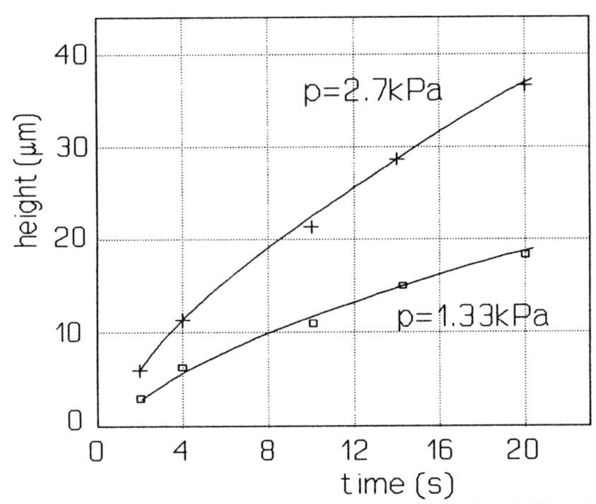

Fig. 8. Dependence of the film height on the local illumination time for the growth of spots. ($TiCl_4 \cdot H_2/SiH_4 = 0.5$, $P_L = 0.52$ W, $r_G = 4$ μm, $T_s = 300$ K).

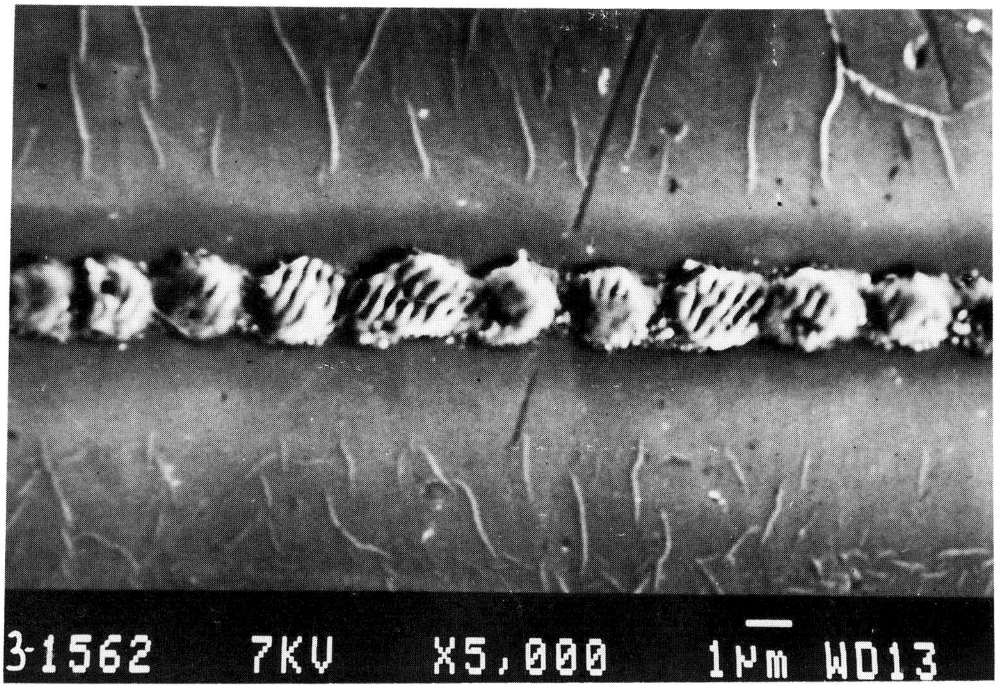

Fig. 9. SEM micrograph of $TiSi_x$ films deposited on silicon, exhibiting periodic structures (p_{tot} = 10.6 kPa, P_L = 0.85 W, r_G = 4 μm, v_s = 0.2 mm/s, T_s = 300 K).

We do not understand the reason of their formation at this time. We believe that a periodic change in reflectivity and the related temperature-dependent deposition chemistry causes this phenomenon. Periodic structures also were observed within the high power range where melting of the surface and the growing film occurs. If the melting point is reached the reflectivity changes rapidly and less energy is deposited in the film. Therefore, the molten region solidifies and the deposition of the film material begins again.

5. Conclusions

Titanium silicide thin films have been deposited using both the CO_2-laser and the argon ion laser, corresponding to the substrates to be covered (SiO_2 or c-Si). The CO_2-laser allows the production of extended thin silicide films with deposition rates of about 60 nm/s and resistivities of 160 $\mu\Omega \cdot$ cm. Using the argon ion laser the line width could be reduced to a minimum value of 1.7 μm. The deposition process strongly depends on the local illumination time. Film geometry, film composition and film morphology could be controlled by varying the deposition parameters gas composition, scan velocity and laser power. The laser deposition of $TiSi_2$ films can be developed as an alternative for the production of special devices in microsystems technology.

References

[1] R. Madar and C. Bernard, Appl. Surf. Sci. 53 (1991) 1.
[2] D. Bäuerle, Chemical Processing with Lasers, Vol. 1 of Springer Series in Materials Science (Springer, Berlin, 1986).
[3] K. Zimmer, Beiträge zur Laserinduzierten Chemischen Schichtabscheidung von Silicium, Dissertation, IH Mittweida, 1989.
[4] V. Ilderem and R. Reif, J. Electrochem. Soc. 135 (1988) 2590.

Applied Surface Science 69 (1993) 418–423
North-Holland

applied
surface science

Determination of UV-laser induced surface structures by atomic force microscopy

Lothar Wefers [a], Dirk Bosbach [b], Werner Rammensee [b] and Eckhard Schollmeyer [a]

[a] *Deutsches Textilforschungszentrum Nord-West e.V., Frankenring 2, D-4150 Krefeld, Germany*
[b] *Mineralogisch-Petrographisches Institut der Universitüt zu Köln, Zülpicher Strasse 49, D-5000 Köln 1, Germany*

Received 2 June 1992; accepted for publication 30 October 1992

The changes in surface topography due to UV-laser treatment are of high significance for the determination of material destruction thresholds and for surface reconstruction. A combined arrangement of an UV-laser and an atomic force microscope (AFM) allows for the first time a highly sensitive surface characterisation of the very same area of the irradiated sample between different laser pulses. Irradiation of poly(ethylene terephthalate) (PET) with pulsed UV-laser light of 248 nm, modifies the smooth surface of the polymer into a well-oriented structured surface. The evolution of such structures was studied. A change in surface topography was found in relation to fluence and number of pulses applied. A statistical interpretation of AFM images gives some indications to the mechanism of material processing.

1. Introduction

Excimer laser irradiation has become a proven valuable tool for material treatment like cutting and surface modification [1], i.e. excimer laser irradiation of polymers yields the removal of material in the form of ablation [2,3]. The progress of this technique also requires the development of high fluence resistant optical materials.

Direct measurements of surface changes inside an AFM induced by UV-laser treatment were tested on PET films. Several authors describe a structuring effect on polymers which is related to a cooperative material transport driven by tension fields and laser induced temperature fields [4]. The evolution of these structures on polymeric fibres, once formed, due to further laser pulses, has been studied by Kesting and Schollmeyer [5] and can be explained by the thermoplastic behaviour of the polymer fibre. The starting stage of the evolution of the surface structure and data of material change still remains uncertain.

The AFM investigation of surface structures between laser pulses in the very same area and the relating data to statistical description indicates the complex changes of the surface. Coupling an excimer laser into an AFM offers the chance to image the characteristic changes within the very same surface area from pulse to pulse. Inhomogeneities in the surface topography of untreated samples influencing the development of the structures and the damage threshold of optical materials can be neglected.

2. Experimental

An AFM (Nanoscope II from Digital Instruments) was combined with a Lambda Physik excimer laser (LPX 200) via an optical fibre. The laser operates with a KrF gas mixture and produces pulses of 23 ns duration at 248 nm. Fig. 1 schematically illustrates the general experimental configuration. The amount of energy could be controlled by the use of a variable beam splitter BS1. After the first stop, the UV-laser light was reflected by the beam splitter BS2 into the fibre optic connector unit with a 105/160 μm optical fibre (UV 105/125A, Ceram Optec). Laser fluences were measured after the beam splitter BS2. The loss of fluence on the fibre connector was estimated. The optical arrangement remained constant for all experiments allowing comparison between the measurements. The fibre optics were coupled into the modified AFM using a micropositioner for three-dimensional adjustment. A 10 mW HeNe laser, which was connected to the same optical fibre, was used to monitor the area of treatment and to adjust the end of the optical fibre above the cantilever. The pulse repetition rate was kept constant at 1 Hz to avoid heat accumulation. Before irradiation the samples were cleaned with pure ethyl-alcohol. The AFM used in this study operates in the contact mode by scanning a sample under tip in a physical contact with the surface. A triangular shaped 200 μm Si_3N_4 cantilever from Park Scientific Instruments was used having a spring constant of about 0.12 N/m. It has been shown that a non-destructive imaging of PET surfaces with AFM is possible using loading forces up to 50 nN [6]. All images were taken at a constant force mode. During laser treatment the cantilever had to be slightly removed from the surface because of destructive interactions between the laser light and cantilever. This gave rise to some inaccuracy in repositioning the cantilever after each laser pulse.

3. Results and discussion

Results of AFM imaging are shown in fig. 2. The scanned area of the sample in fig. 2 is approximately 7×7 μm^2. This high lateral resolution is used to follow the evolution of one point marked on the pictures. The results obtained therefore not only demonstrate the structure formation as a function of fluence and number of pulses applied, but the method also allows one to investigate changes within the same area and gives a statistical description of structure formation. The untreated sample has no irregularities and no decisive damages such as scratches. The surface is very smooth and a similar topography is measurable on a number of different samples. After the first laser pulse of approximately 5 mJ/cm^2 the surface topography had changed completely. The treated surface is characterized by summits and valleys. Additional laser pulses increase the summit and valley structure but a transition into another topography has not been detected. Examination of the marks in the pictures of fig. 2 shows an inaccuracy in repositioning the cantilever after a laser pulse. Particularly in the fourth (3 pulses) and the fifth picture (4 pulses) the scan area is slightly displaced to the bottom and to the left. The evolution of the

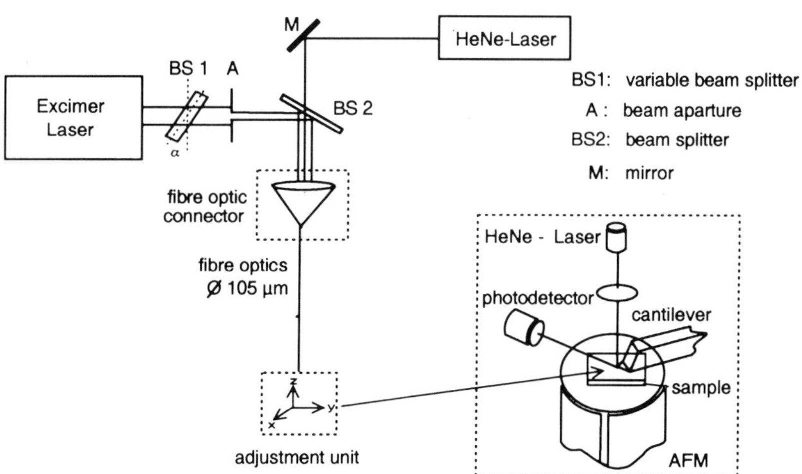

Fig. 1. Schematic of the experimental configuration.

structure is determined by the structure after the first pulse. The shifting of material within the irradiated area, or ablation of different amounts of polymeric material in summits and valleys, still remains uncertain.

The use of another piezoscanner with a maximum scansize of $63 \times 63\ \mu m^2$ allows a significant statistical approach to a quantitative description of the sample topography. PET films were irradiated with 35 and 91 mJ/cm^2. Fig. 3 gives an overview of the 35 mJ/cm^2 series in two different magnifications of 63×63 and $25 \times 25\ \mu m^2$. The untreated sample (fig. 3) shows inhomogeneities in its surface topography. These defects are still recognizable after irradiation with up to three laser pulses. After the fourth laser pulse the initial surface topography is completely "over-printed". Furthermore, fig. 3 indicates the deterministic development of laser induced surface structures. After the initial generation by the first pulse the structure spacing was constant but a change in surface roughness was detected. The R_a roughness calculated from AFM images of samples treated with up to five pulses with fluences of 35 and 91 mJ/cm^2 is shown in fig. 4. R_a clearly increases with the number of pulses and/or increasing fluence. In a statistical approach the profile heights are considered as random variables and their statistical representation in terms of the probability density function is known as the height distribution [7]. The root-mean-square (RMS) roughness is equal to the standard deviation σ for a symmetrical Gaussian distribution and acts as a link between the statis-

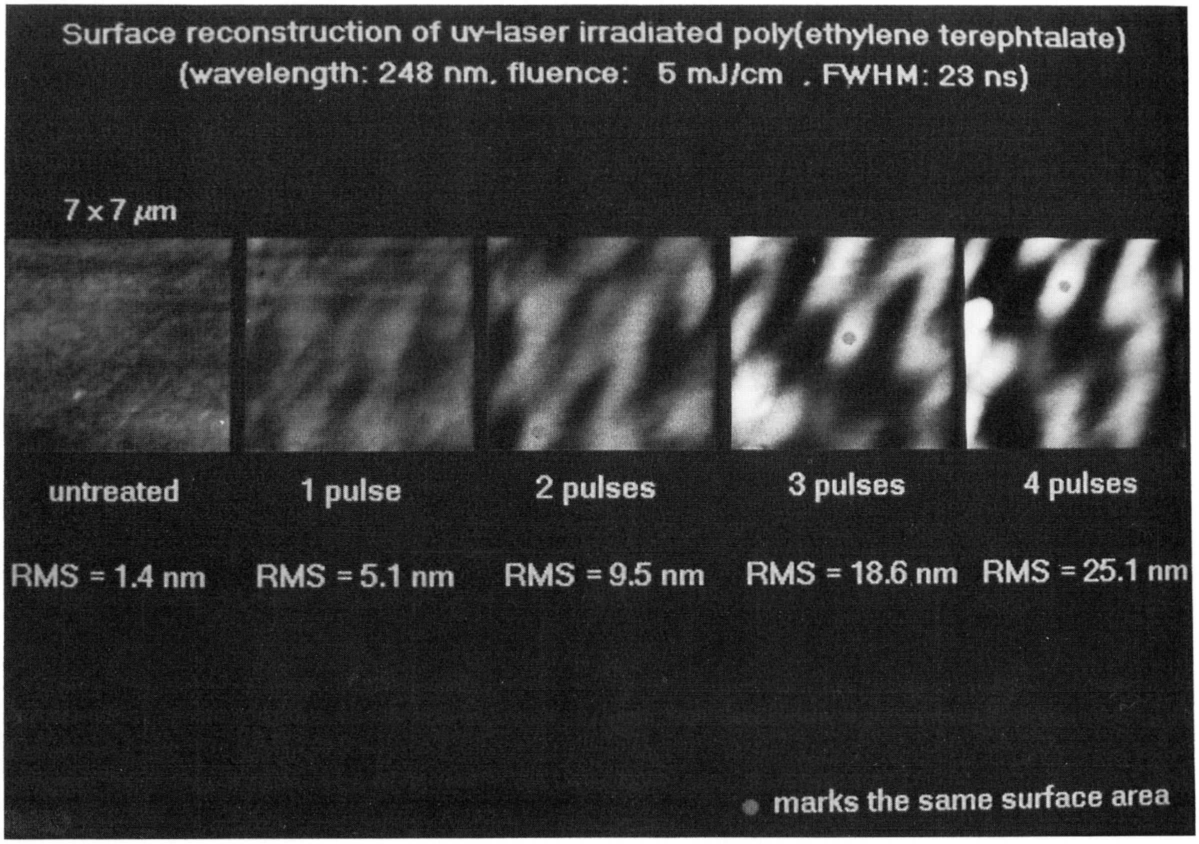

Fig. 2. Topview of UV-laser (248 nm, 5 mJ/cm^2) treated PET films. The scan size is $7 \times 7\ \mu m^2$ and the resolution is 400 points per line. From the untreated surface up to four laser pulses the structure spacing is about 1.5 μm. For better orientation the marked point indicates always the same surface area. All images were taken in the constant force mode (loading force: 50 nN).

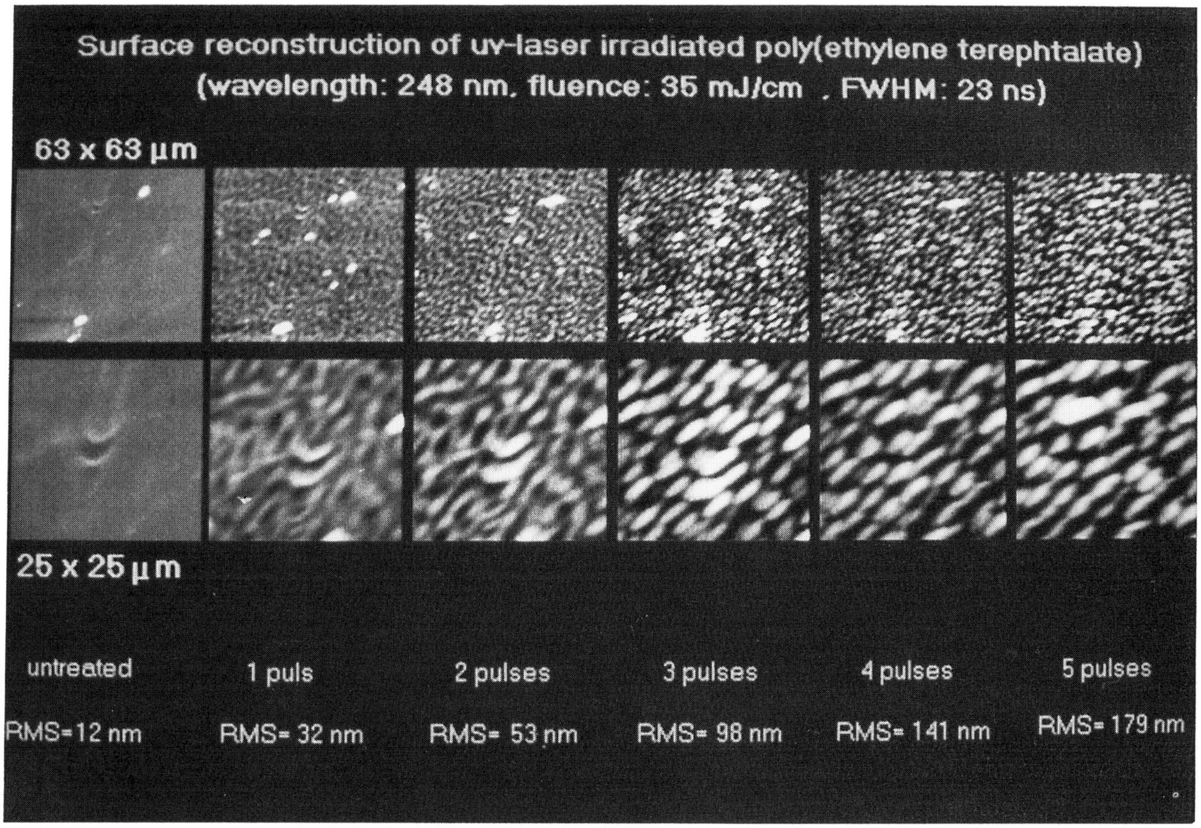

Fig. 3. Topview of UV-laser (248 nm, 35 mJ/cm^2) treated PET films. The scan size is 63×63 μm^2 and the resolution is 400 points per line. Structure spacing is about 1.5 μm. All images were taken in the constant force mode (loading force: 50 nN).

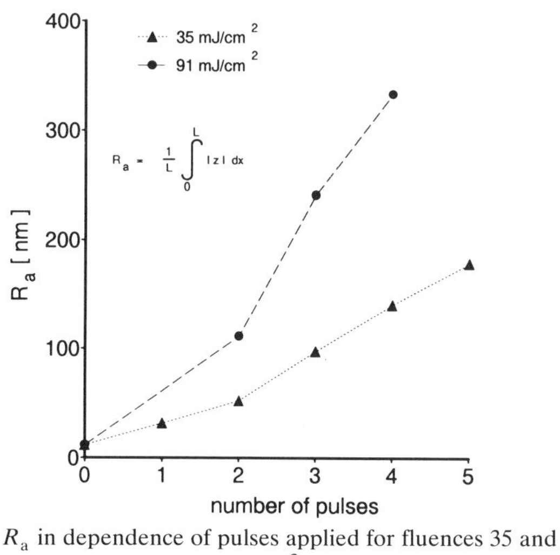

Fig. 4. R_a in dependence of pulses applied for fluences 35 and 91 mJ/cm^2.

tical approach of the roughness description similar to the centre average line R_a with $R_a = \sqrt{2/\pi} \times \text{RMS}$. Calculating higher order moments as the variance, which is the second moment and also the square of the standard deviation σ^2, gives a measure of the spread of the distribution. A useful parameter in defining the asymmetric spread of the data is the third moment, the "skewness". The fourth moment is called the "kurtosis" and represents the difference between the actual distribution and a symmetric Gaussian distribution.

Although skewness and kurtosis represent statistical properties, which may vary locally, significant changes in these moments after irradiation indicate changes in the surface structure. Most surfaces wear under mechanical claim into a neg-

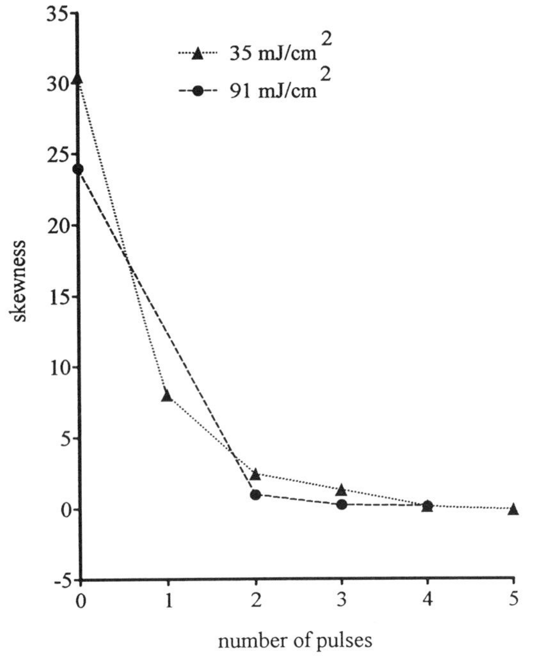

Fig. 5. Skewness in dependence of pulses applied for fluences 35 and 91 mJ/cm^2.

atively skewed form as the summits are worn down and the valleys still remain unchanged. This could be extremely important for the surface life of the technical component. In the same way, change in skewness and kurtosis measured from the very same area after laser treatment indicates material transport. The change in skewness after each laser pulse is shown in fig. 5. The samples are irradiated with 35 and 91 mJ/cm^2. In each series, the skewness reduces from high positive values to values near zero. At the same time the pulse dependent change in kurtosis indicates that the height distribution changes to a Gaussian distribution. Therefore, the calculated changes of the skewness indicate that the structure is formed by a material transport from valleys to summits.

Fig. 6 shows the marked profile of the picture in the lower left. The boundary between the treated and untreated surface area after five laser pulses (248 nm, 35 mJ/cm^2) is reconstructed. Untreated material can be seen on the right side of the profile and on the lower right of the

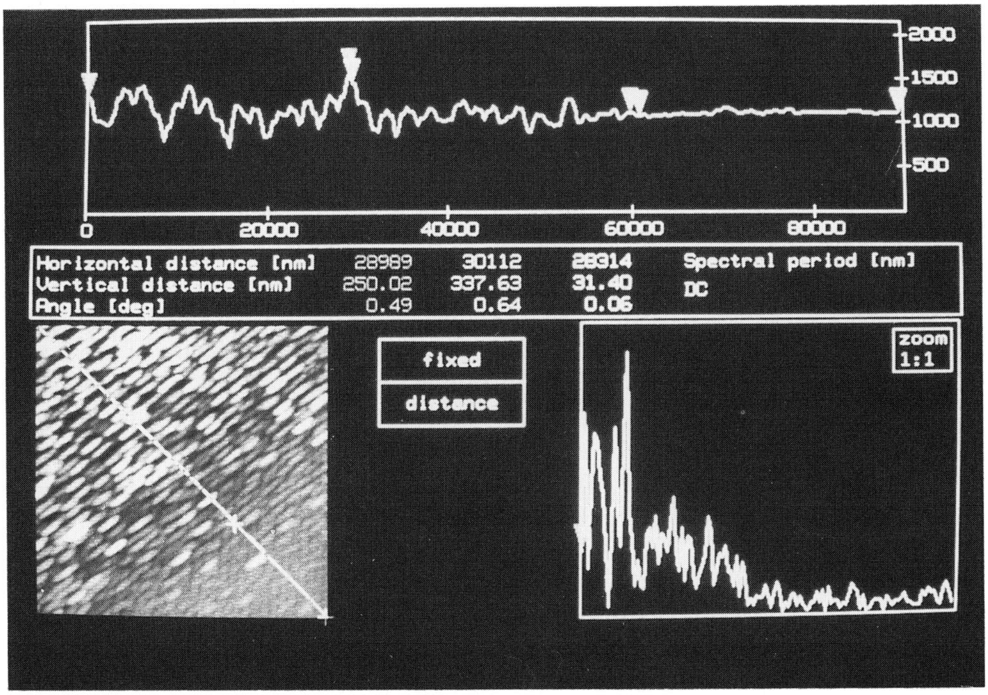

Fig. 6. Topview and height profile of the boundary area between the untreated and the treated surface after five laser pulses (248 nm, 35 mJ/cm^2).

picture. The increasing roughness mirrors the energy distribution of the laser beam. The lateral structure spacing remains the same in the whole illuminated area but the structure height decreases rapidly from the centre.

4. Conclusion

It has been shown that the combination of AFM and excimer laser is a valuable tool for imaging surface changes due to UV-laser irradiation. This experimental tool enables the non-destructive test of laser damage thresholds of optical materials, as well as the determination of laser induced structures after irradiation.

Investigation of the laser treatment of a polymeric film showed that the surface roughness results from the formation of valleys and summits and increases with increasing fluence and number of laser pulses. The structure generated by the first laser pulse determines the structure formation and the spacing after further laser pulses. The structure spacing, i.e. the summit to summit or valley to valley distance, is about 1–3 μm and seems to differ slightly with the fluence. The statistical behaviour of surface roughness, taken from the change in skewness and kurtosis, indicates a material transport from valleys to summits.

Acknowledgement

The authors acknowledge financial support by the Bundesministerium für Forschung und Technologie (BMFT) (BMFT-No. 13 N 5634).

References

[1] I.W. Boyd, Laser Processing of Thin Films and Microstructures (Springer, Berlin, 1987).
[2] Y. Kawamura, K. Toyoda and S. Namba, Appl. Phys. Lett. 40 (1982) 374.
[3] R. Srinivasan and V. Mayne-Banton, Appl. Phys. Lett. 43 (1983) 717.
[4] T. Bahners, W. Kesting and E. Schollmeyer, Proc. SPIE 1503 (1991) 206.
[5] W. Kesting and E. Schollmeyer, Makrom. Chem. 193 (1991) 169.
[6] L. Wefers, D. Knittel, D. Bosbach, W. Rammensee and E. Schollmeyer, Appl. Surf. Sci. 59 (1992) 267.
[7] T.R. Thomas, Rough Surfaces (Longman, New York, 1982).

Applied Surface Science 69 (1993) 424–428
North-Holland

applied
surface science

Internal second-harmonic generation in CW AlGaAs SQW lasers: the facet degradation monitoring

I.V. Kravetsky [a], L.L. Kulyuk [a], A.T. Lupu [b], D.A. Shutov [a], G.I. Suruceanu [b], A.V. Syrbu [b] and V.P. Yakovlev [b]

[a] *Institute of Applied Physics, Moldova Academy of Sciences, Academic Street 5, Kishinev 277028, Moldova*
[b] *Kishinev Polytechnical Institute, 168 Stefan cel Mare Street, Kishinev 277012, Moldova*

Received 2 June 1992; accepted for publication 30 October 1992

The spectra of the second-harmonic (SH) radiation, generated in the separate-confinement single quantum well (SQW) semiconductor AlGaAs/GaAs laser have been measured at various levels of the injection current. The observed correlation between laser and SH mode intensities is explained by taking into consideration the dispersion of the second-order nonlinear susceptibility of the GaAs active region. It is shown, that due to the high sensitivity of the SH signal to the semiconductor near-surface layer condition, the internal SH generation can be used for the in situ diagnostics of the laser facet degradation processes.

The internal second-harmonic generation (SHG) in semiconductor lasers was observed for the first time almost thirty years ago [1–5]. The efficiency of this process is very low and contrary to the other nonlinear-optical phenomena attending the laser generation in the semiconductor heterojunctions [6] (nonlinear refraction for example) the SHG does not influence the fundamental beam characteristics [7]. This is a possible reason why the interest to the internal SHG in semiconductor lasers decreased rapidly after the first publications.

The low efficiency of SHG in heterojunction lasers is connected with the high absorbtion at the SH frequency, which exceeds the band gap in GaAs two times. In this case the weak SH radiation, which can be observed in the output beam spectrum along with the fundamental frequency, originates in a thin near-surface layer, adjacent to the mirror facet of a heterojunction laser. The thickness of this layer is determined by the adsorption constant of the active media for the SH frequency and which is for GaAs, very small: $l_{SH} \simeq \alpha^{-1}$ $(2\hbar\omega \simeq 3$ eV$) \simeq 17$ nm [8].

The internal SH generation was studied in the AlGaAs lasers of the buried heterojunction type formed by a melt-etching and liquid phase epitaxial regrowth on the initial laser structure grown by MOCVD [9]. The initial laser structure with a single quantum well consisted of the following layers: 2 μm thick n-$Al_{0.6}Ga_{0.4}As$ cladding layer, n-$Al_xGa_{1-x}As$ $(x = 0.47)$ waveguide layer with the thickness $W/2 = 0.1 + 0.15$ μm, n-GaAs active region with the thickness $d = 6$–20 nm, symmetrical n-waveguide and p-cladding layers, respectively. The threshold current and the output power of the CW operating laser were about 10 mA and 50 mW, respectively.

The fundamental and SH spectra were studied with the help of a double monochromator with a cooled photomultiplier tube and a standard photon-counting system. The resolution of the SH spectral measurements was 0.1 meV. Part of the fundamental beam was directed to the input split of the second monochromator SPM-2 in order to control its spectrum and the integral intensity simultaneously with the SH characteristics.

The spectra of the fundamental and SH radiation of the laser diode with 6 nm thick active region at four levels of the injection current J are shown on figs. 1a and 1b. The polarization plane of the SH radiation was perpendicular to the polarization of the fundamental beam, which corresponded to the fundamental transverse TE-mode of the waveguide structure. When the laser generation was not the single-frequency, along with the double-frequency lines $\omega_{kk} = 2\omega_k$, the spectrum contains the sum frequency (SF) components $\omega_{ij} = \omega_i + \omega_j$, $(i \neq j)$.

In the case of a laser diode with generation representing a sequence of N quasi-equidistant longitudinal modes (fig. 2a), the spectral structure of the SH radiation includes at least $2N - 1$ components. Every fundamental mode is accompanied by several weak satellites, which causes the broadening of the SH and SF lines and gives rise to the broad-band pedestal (fig. 2b).

It is seen from figs. 1a and 1b that the long-wave lines 2 and 3, which have equal amplitudes in the fundamental spectrum $I(\omega_2) = I(\omega_3)$, correspond to the SH lines with different intensities $I_{SH}(\omega_{22}) > I_{SH}(\omega_{33})$. With the increasing pump current up to 80 mA lines 2 and 3 in fundamental spectrum are no longer equal and $I(\omega_3) > I(\omega_2)$, while in the SH spectrum the reverse tendency can be observed: $I_{SH}(2\omega_3) < I_{SH}(2\omega_2)$.

A phenomenological description of the behavior of the observed SH spectra can be provided under the assumption that the light interacting near the facet mirror of the waveguide structure can be considered as a superposition of collinear plane waves with equal phase.

When N non-equidistant laser modes with frequencies ω_k $(k = 1, 2, \ldots, N)$ are interacting in the thin layer adjacent to the surface, it can be shown, that the SF spectrum should contain $N(N+1)/2$ lines. For the condition of the strong SH adsorption [10] the intensity of the SH modes is

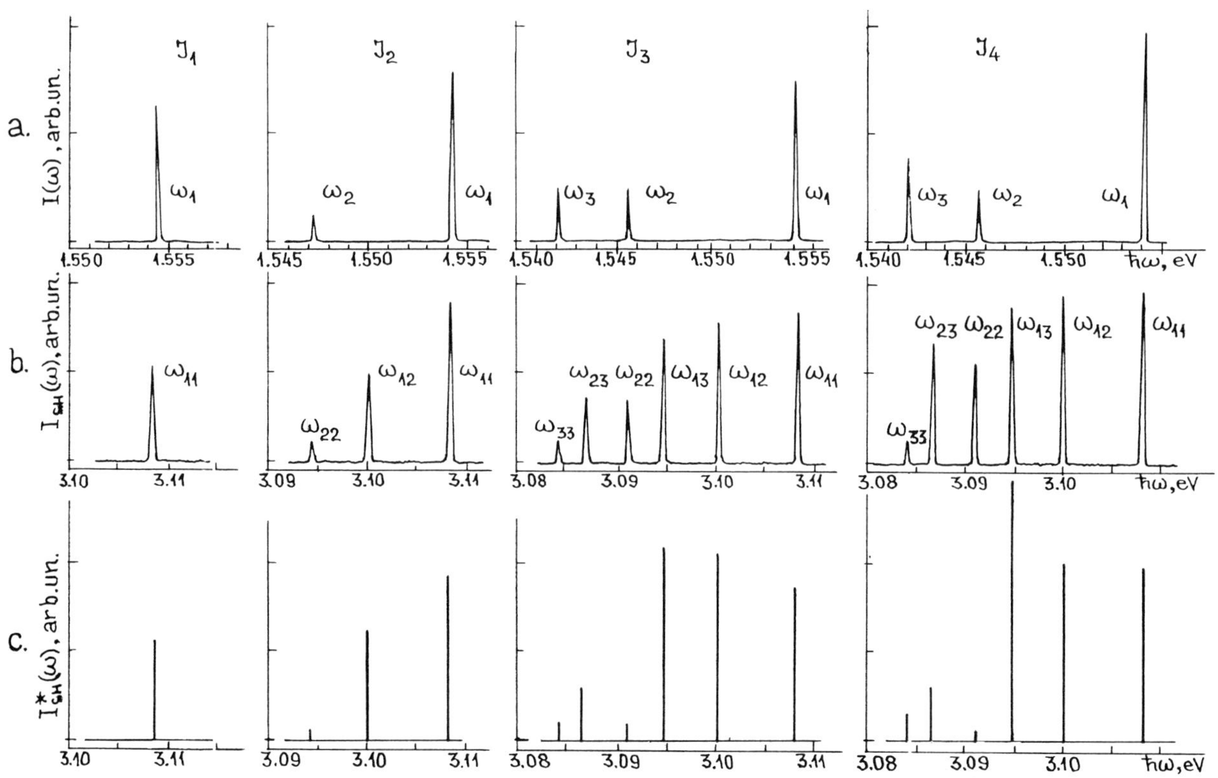

Fig. 1. Fundamental (a) and second-harmonic (b) emission spectra for a CW operating SQW laser with a 6 nm well thickness at four values of the injection current: $J_1 = 40$, $J_2 = 60$, $J_3 = 70$, $J_4 = 80$ mA. (c) Second-harmonic spectra calculated using expression (1).

connected with the intensity of pairwise interacting fundamental modes following the expression:

$$I_{sh}(\omega) = \frac{32\pi^3}{c^3}\left[\sum_k\left[(2\omega_k)^2|\chi^{(2)}(\omega;\,\omega_k,\omega_k)|^2 \times I^2(\omega_k)\delta(\omega-2\omega_k)\right]\times\left[n^2(\omega_k)n(2\omega_k)\alpha^2(2\omega_k)\right]^{-1} + 4\sum_{l,m}{}'\left[(\omega_l+\omega_k)^2|\chi^{(2)}(\omega;\,\omega_l,\omega_m)|^2 \times I(\omega_l)I(\omega_m)\delta(\omega-\omega_l-\omega_m)\right]\times\left[n(\omega_l)n(\omega_m)n(\omega_l+\omega_m)\times\alpha^2(\omega_l+\omega_m)\right]^{-1}\right] \quad (1)$$

where $n(\omega)$, $\alpha(\omega)$ are $\chi^{(2)}$ $(\omega;\,\omega_i,\omega_j)$ are the refraction index, the adsorption constant and the second-order nonlinear susceptibility for the corresponding frequencies; the δ-function describes the shape of the spectral lines. The first term corresponds to the SH contribution and the second one to SF contribution modes. Generally, the difference between SH and SF lines is not substantial because $(\omega_N-\omega_1)/\omega_k \leqslant 10^{-2}$.

In the semiconductor heterojunction laser with a waveguide structure the mode interaction takes place inside the waveguide which includes the active GaAs layer and wide band gap $Al_xGa_{1-x}As$ framing layers ($E_g = 1.9$ eV at $x = 0.47$ [11]). For a proper understanding of the observed behavior of the SH signal, it is necessary to determine the main source of nonlinearity in the waveguide

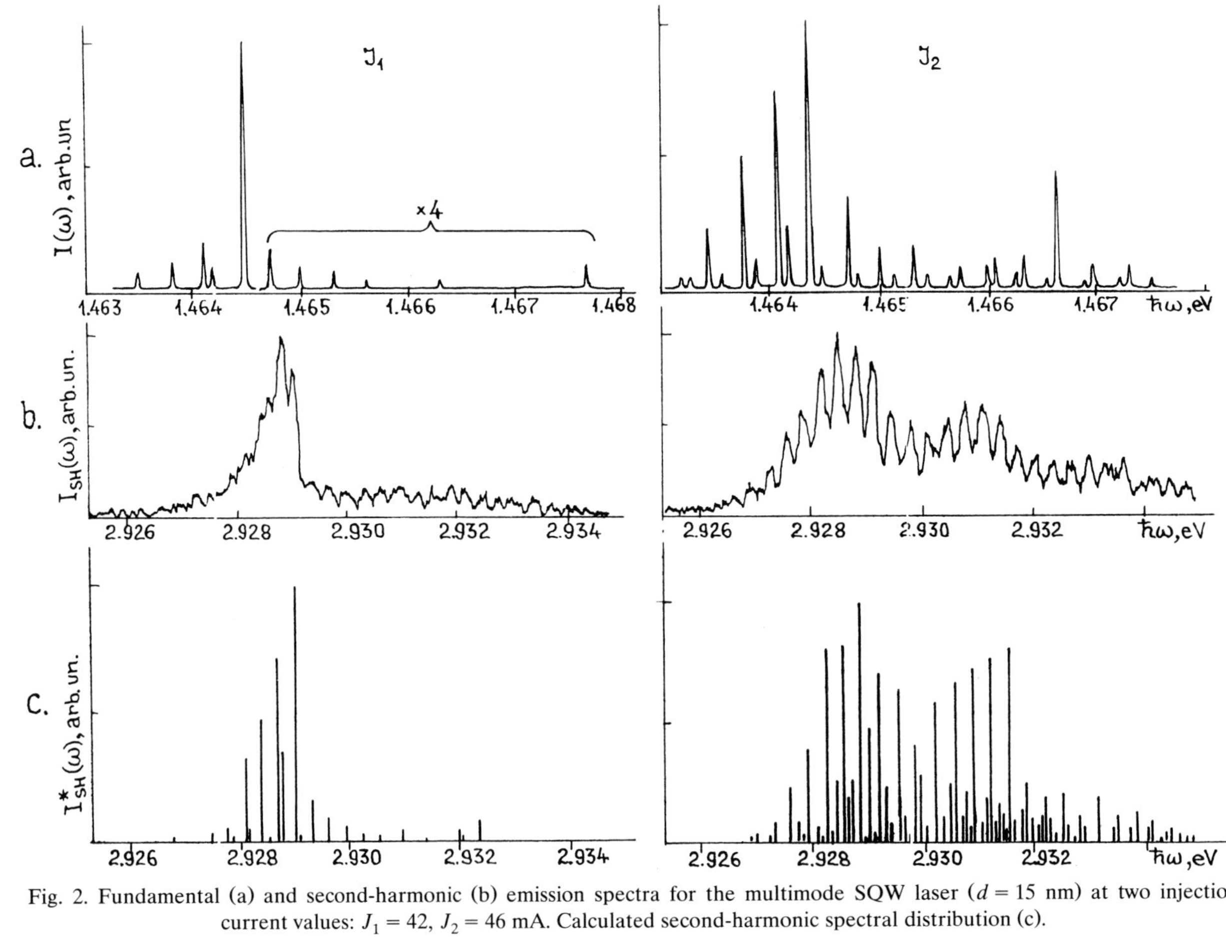

Fig. 2. Fundamental (a) and second-harmonic (b) emission spectra for the multimode SQW laser ($d = 15$ nm) at two injection current values: $J_1 = 42$, $J_2 = 46$ mA. Calculated second-harmonic spectral distribution (c).

structure. At first glance, the contribution of the waveguide layers with a thickness exceeding the thickness of the active region more than one order dominates. If so, in the absence of resonance in AlGaAs the linear and nonlinear susceptibilities both depend weakly on the frequency.

Figs. 1c and 2c represent the SH spectra calculated with expression (1) using the experimental intensities values of fundamental modes (figs. 1a and 2a) under the assumption that the nonlinear contribution of the waveguide layers dominates. The comparison of the experimental spectra in fig. 1b and calculated spectra in fig. 1c shows the discrepancy. Thus, it is not possible to describe the anomalous correlation of the fundamental and SH mode intensities without taking into account the dispersion of the linear and nonlinear optical constants and their dependence on the injection current.

A different situation exists in the active GaAs region, where the nonlinear interaction exhibits a resonance character. In spite of the small thickness of this layer, its contribution to the SH signal can be dominant, because the SH generation occurs under the double resonance. Besides the resonance at the fundamental frequency, the SH frequency corresponds to the energy of the transition $\hbar\omega \simeq E_1 + \Delta \simeq 3.1$ eV between the top of the valence band, split by the spin–orbital interaction, and the bottom of the conductivity band in the Λ symmetry direction of the GaAs Brillouin zone [12–14]. The nonlinear susceptibility $\chi^{(2)}(2\omega;\ \omega,\omega)$ is much more sensitive to the peculiarities of the band structure and to the electrons transitions matrix elements, than the linear optical constant $\epsilon(\omega)$ [13,14]. Therefore, the step-like shape of the density of states associated with the quasi-two-dimensional structure of the SQW leads to a significant sharpening of the resonance peaks of the $\chi^{(2)}(2\omega;\ \omega,\omega)$ spectrum in comparison with the bulk GaAs. Besides this, according to the estimation in ref. [15], the second-order susceptibility in GaAs exceeds several times the corresponding value in AlGaAs solid solutions even out of resonance ($\hbar\omega < E_g$). The maximum of the

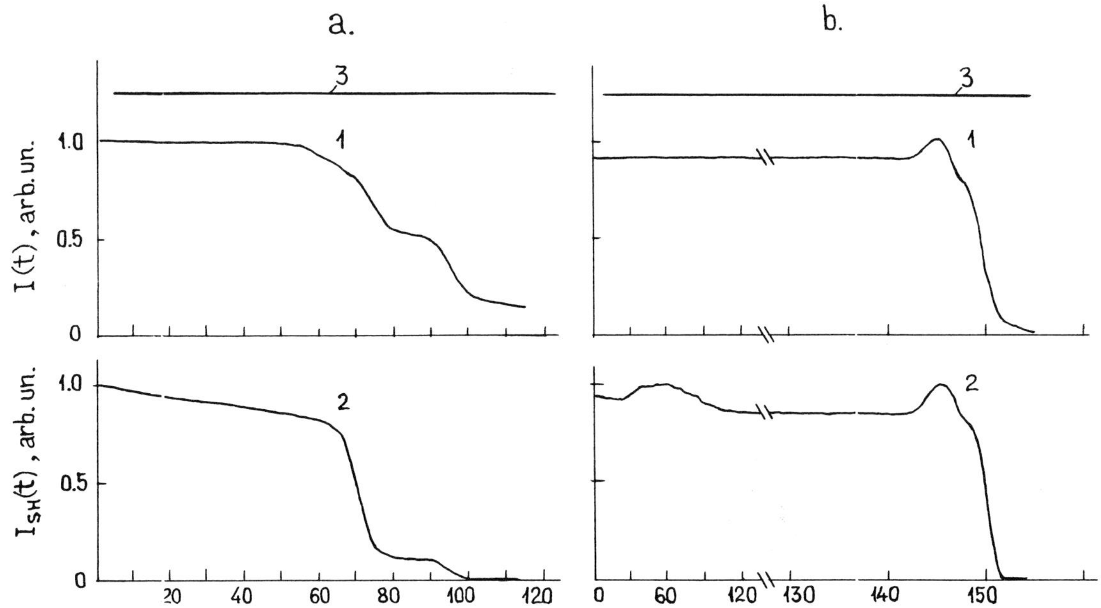

Fig. 3. Output laser intensity (1), second harmonic intensity (2) and driving current (3) as a function of time for two laser diodes with uncoated facets operating in stress regime ($J = 4J_{\text{threshold}}$).

fundamental wave intensity spatial distribution takes place in the active region, which increases the corresponding contribution to the SH signal.

The dependence of the second-order susceptibility on the injection current is complicated by the influence of the odd-order optical constants. The $\chi^{(2)}(2\omega;\ \omega,\omega)$ spectrum in a rough approximation is determined by the imaginary part of the dielectric function $\epsilon_2(\omega)$, which is unambiguously associated with the gain spectrum and therefore depends on the injection current. At the same time, the influence of the injected non-equilibrium carriers can be accomplished through the resonance of the third-order susceptibility (nonlinear refraction) in GaAs [16].

Finally, it is necessary to note, that due to the surface recombination the temperature increases much faster in the region adjacent to the facet mirror with the injection current than in the bulk of the active region, and can achieve a value of several hundred centigrade [17].

The latter circumstance along with the high sensitivity of SH signal to the semiconductor surface properties [18] allows one to use the intrinsic SHG for the effective control of the facet mirror degradation processes occurring during heterojunction laser operation, i.e., for the real-time diagnosis of the catastrophic failure of the semiconductor lasers.

The time dependence of the SH and fundamental radiation integral intensity for a heterojunction laser operating at the pump current exceeding the threshold level 4 times is illustrated by fig. 3a. On curve 1, which corresponds to the fundamental radiation, the signs of the catastrophic failure do not appear for about one hour. Contrary, the SH amplitude (curve 2) starts to decrease from the beginning of lasering in the stress regime. For another laser diode (fig. 3b) the SH intensity indicates the approach of the failure one hour before the final catastrophic degradation. It is important to note that the driving current was constant during all measurements, proving that the degradation process took place on the facets [19].

Hence, the high sensitivity of the SH signal to the facet damage is conditioned by the fact that it originates in the thin near-surface layer of the active region, where the initial facet degradation process takes place. The proposed method, based on the internal SH generation, can be used for in situ selective control of the semiconductor laser degradation.

References

[1] J.A. Armstrong, M.I. Nathan and A.W. Smith, Appl. Phys. Lett. 3 (1963) 68.

[2] L.D. Malmstrom, J.J. Schlickman and R.H. Kingston, J. Appl. Phys. 35 (1964) 248.

[3] A.W. Smith, M.I. Nathan, J.A. Armstrong, A.E. Michel and K. Weiser, J. Appl. Phys. 35 (1964) 733.

[4] W.E. Engeler and M. Garfinkel, J. Appl. Phys. 36 (1963) 1734.

[5] Yu.I. Krujilin and Yu.I. Koloskov, Zh. Prikl. Spektrosk. 12 (1970) 334.

[6] P.G. Eliseev, A.P. Bogatov, Tr. Fiz. Inst. Akad. Nauk SSSR 166 (1966) 15.

[7] P.G. Eliseev, Introduction in the Physics of the Injection Lasers (Nauka, Moscow, 1983).

[8] D.E. Aspnes and A.A. Studna, Phys. Rev. B 27 (1983) 985.

[9] A. Syrbu, SPIE Symp. on Compound Semicond. Phys. and Dev., 22–26 March 1992, Somerset, NJ (1992) p. 36.

[10] Y.R. Shen, The Principles of the Nonlinear Optics (Wiley, New York, 1984).

[11] H.C. Casey, Jr. and M.B. Panich, Heterostructure Lasers, Part B (Academic Press, New York, 1978).

[12] H. Lotem, G. Koren and Y. Yacoby, Phys. Rev. B 9 (1974) 3532.

[13] C.Y. Fong and Y.R. Shen, Phys. Rev. B 12 (1975) 2325.

[14] D.J. Moss, J.E. Sipe and H.M. van Driel, Phys. Rev. B 36 (1987) 9708.

[15] D. Vakhshoori, R.J. Fisher, M. Hang, D.L. Sivco, G.J. Zidzik, G.N.S. Chu and A.Y. Cho, Appl. Phys. Lett. 59 (1991) 896.

[16] Y.H. Lee et al., Phys. Rev. Lett. 57 (1986) 2446.

[17] F.R. Gfeller and D.J. Webb, Jpn. J. Appl. Phys. 29 (1990) L788.

[18] Y.R. Shen, J. Vac. Sci. Technol. B 3 (1985) 1464.

[19] O. Ueda, J. Electrochem. Soc. 135 (1988) C11.

Applied Surface Science 69 (1993) 429–434
North-Holland

applied surface science

Author index

Applied Surface Science 69 (1993) 435–450
North-Holland

applied surface science

Subject index

Auger electron spectroscopy

Borides

Boron

Carbon

Ceramics

Chemical vapour deposition

Cobalt

Copper

Diamond

Doping effects

Electrical properties

Electron bombardment

Electron diffraction

Electron microscopy

Ellipsometry

Epitaxy

Etching

Gallium

Gallium arsenide

Germanium

Glass

Gold

Halogenides

Halogens

Hydrogen

Indium phosphide

Infrared spectroscopy

Ion implantation

Ion scattering

Iron

Lanthanides

Laser processing

Magnesium oxide

Manganese

Mass spectroscopy

Metals

Molybdenum

Mössbauer spectroscopy

Multilayers

Nickel

Photon emission

Plasma processing

Platinum

Polymers

Pulsed laser ablation

Pulsed laser deposition

Quantum effects

Raman scattering

Rhodium

Schottky barrier

Selenium

Semiconductors

Silane

Silicides

Silicon

Silicon carbide

Silicon nitride

Silicon oxide

Silver

Strontium

Superconductivity

Surface morphology

Surface structure

Tellurium

Thin films

Titanium